Frontiers in Plant–Soil Interaction

Molecular Insights into Plant Adaptation

Frontiers in Plant–Soil Interaction

Molecular Insights into Plant Adaptation

Edited by

Tariq Aftab
Department of Botany, Aligarh Muslim University, Aligarh, India

Khalid Rehman Hakeem
Department of Biological Sciences, Faculty of Science,
King Abdulaziz University, Jeddah, Saudi Arabia

ELSEVIER

ACADEMIC PRESS
An imprint of Elsevier

Academic Press is an imprint of Elsevier
125 London Wall, London EC2Y 5AS, United Kingdom
525 B Street, Suite 1650, San Diego, CA 92101, United States
50 Hampshire Street, 5th Floor, Cambridge, MA 02139, United States
The Boulevard, Langford Lane, Kidlington, Oxford OX5 1GB, United Kingdom

British Library Cataloguing-in-Publication Data
A catalogue record for this book is available from the British Library

Library of Congress Cataloging-in-Publication Data
A catalog record for this book is available from the Library of Congress

ISBN: 978-0-323-90943-3

For Information on all Academic Press publications
visit our website at https://www.elsevier.com/books-and-journals

Publisher: Mica Haley
Acquisitions Editor: Charlotte Cockle
Editorial Project Manager: Andrea Dulberger
Production Project Manager: Omer Mukthar
Cover Designer: Miles Hitchen

Typeset by MPS Limited, Chennai, India

Contents

List of contributors

Muhammad Adnan College of Agriculture, University of Sargodha, Sargodha, Pakistan

Tariq Aftab Department of Botany, Aligarh Muslim University, Aligarh, India

Adeel Ahmad Institute of Soil and Environmental Science, University of Agriculture Faisalabad, Faisalabad, Pakistan

Hamaad Raza Ahmad Institute of Soil and Environmental Sciences, University of Agriculture Faisalabad, Faisalabad, Pakistan

Shakeel Ahmad State Key Laboratory of Rice Biology, China National Rice Research Institute, Hangzhou, P.R. China

Zahoor Ahmad Department of Botany, University of Central Punjab, Punjab Group of College Bahawalpur, Bahawalpur, Pakistan

Khan Bilal Mukhtar Ahmed Plant Physiology Section, Department of Botany, Faculty of Life Sciences, Aligarh Muslim University, Aligarh, India

Sharif Ahmed International Rice Research Institute, Dhaka, Bangladesh

Md. Jahangir Alam On-Farm Research Division, Bangladesh Agricultural Research Institute, Gaibanda, Bangladesh

Ahmad Ali Department of Life Sciences, University of Mumbai, Mumbai, India

Zulfiqar Ali Institute of Plant Breeding and Biotechnology (IPBB), MNS University of Agriculture, Multan, Pakistan

Ameena A. AL-surhanee Biology Department, College of Science, Jouf University, Sakaka, Saudi Arabia

Ajin S. Anil Division of Soil Science and Agricultural Chemistry, ICAR-Indian Agricultural Research Institute, New Delhi, India

Mohsina Anjum Plant Nutrition Laboratory, Department of Soil Science and Agriculture Chemistry, University of Agricultural Sciences, GKVK, Bengaluru, India

Mohammad Ashfaq Multidisciplinary Research Institute for Science and Technology, IIMCT, University of La Serena, La Serena, Chile; School of Life Science, BS Abdur Rahman Institute of Science and Technology, Chennai, India

Muhammad Zohaib Aslam Institute of Soil and Environmental Sciences, University of Agriculture Faisalabad, Faisalabad, Pakistan

Rana Muhammad Atif Department of Plant Breeding and Genetics, University of Agriculture, Faisalabad, Pakistan; Center for Advanced Studies in Agriculture and Food Security (CAS-AFS), University of Agriculture, Faisalabad, Pakistan

Muhammad Ashar Ayub Institute of Soil and Environmental Sciences, University of Agriculture Faisalabad, Faisalabad, Pakistan; Indian River Research and Education Centre, Institute of Food and Agricultural Sciences, University of Florida (UF-IFAS), Fort Pierce, FL, United States

Diksha Bagal Department of Botany, University of Delhi, New Delhi, India

Aditya Banerjee Department of Biotechnology, St. Xavier's College (Autonomous), Kolkata, India

Preetha Bhadra Department of Biotechnology, Centurion University of Technology and Management, Paralakhemundi, India

Karma L. Bhutia Department of Agricultural Biotechnology & Molecular Breeding, College of Basic Science and Humanities, Dr. Rajendra Prasad Central Agricultural University, Pusa, India

Muhammad Adnan Bukhari Department of Agronomy, University College of Agriculture and Environmental Sciences (UCAES), The Islamia University of Bahawalpur, Bahawalpur, Pakistan

Subhash Chand ICAR-Indian Grassland and Fodder Research Institute, Jhansi, India

Arghya Chattopadhyay Department of Soil Science and Agricultural Chemistry, Institute of Agricultural Sciences, Banaras Hindu University, Varanasi, India

Divya Chauhan Department of Chemical and Biomedical Engineering, University of South Florida, Tampa, FL, United States

Jyoti Chauhan Institute of Agriculture Sciences, Banaras Hindu University, Varanasi, India

Basant Kumar Dadrwal Institute of Agriculture Sciences, Banaras Hindu University, Varanasi, India

Prajjal Dey Faculty of Agriculture, Sri Sri University, Cuttack, India

Asik Dutta Crop Production Division, ICAR-Indian Institute of Pulse Research, Kanpur, India

Zia Ur Rahman Farooqi Institute of Soil and Environmental Sciences, University of Agriculture Faisalabad, Faisalabad, Pakistan

Abdelhalim I. Ghazy Plant Production Department, Food Science and Agricultural College, King Saud University, Riyadh, Saudi Arabia

Akash Hidangmayum Institute of Agriculture Sciences, Banaras Hindu University, Varanasi, India

Akbar Hossain Bangladesh Wheat and Maize Research Institute, Dinajpur, Bangladesh

Muhammad Mahroz Hussain Institute of Soil and Environmental Sciences, University of Agriculture Faisalabad, Faisalabad, Pakistan

Indu ICAR-Indian Grassland and Fodder Research Institute, Jhansi, India

Muhmmad Zaffar Iqbal Agricultural Biotechnology Research Institute, Ayub Agricultural Research Institute, Faisalabad, Pakistan

Deepti Jain Interdisciplinary Centre for Plant Genomics, Department of Plant Molecular Biology, Delhi University South Campus, New Delhi, India

Shakra Jamil Agricultural Biotechnology Research Institute, Ayub Agricultural Research Institute, Faisalabad, Pakistan

Hanuman Singh Jatav Department of Soil Science and Agricultural Chemistry, Institute of Agricultural Sciences, Banaras Hindu University, Varanasi, India; Sri Karan Narendra Agriculture University, Jobner, Jaipur, India; Department of Soil Science and Agricultural Chemistry, Sri Karan Narendra Agriculture University, Jaipur, India

Shamsa Kanwal Agricultural Biotechnology Research Institute, Ayub Agricultural Research Institute, Faisalabad, Pakistan

Radhika Keshan Department of Botany, University of Delhi, New Delhi, India

Asif Ali Khan Institute of Plant Breeding and Biotechnology (IPBB), MNS University of Agriculture, Multan, Pakistan

M. Masroor A. Khan Plant Physiology Section, Department of Botany, Faculty of Life Sciences, Aligarh Muslim University, Aligarh, India

Nazish Huma Khan Department of Environmental Sciences, University of Swabi, Swabi, Pakistan

Vipin Kumar Department of Agricultural Chemistry, Chaudhary Shivnath Singh Shandilya (P.G.) College, Meerut, India

Vivek Kumar Institute of Agriculture Sciences, Banaras Hindu University, Varanasi, India

Arnab Kundu Department of Agricultural Chemistry and Soil Science, Bidhan Chandra Krishi Viswavidyalaya, Mohanpur, India

Dalpat Lal Jagannath University, Jaipur, India

Saima Liaqat Department of Botany, School of Chemical and Life Sciences, Jamia Hamdard, New Delhi, India

Sagar Maitra Department of Agronomy, Centurion University of Technology and Management, Paralakhemundi, India

Udit Nandan Mishra M.S. Swaminathan School of Agriculture, Centurion University of Technology and Management, Paralakhemundi, India

Kiran Kumar Mohapatra Department of Soil Science and Agricultural Chemistry, College of Agriculture, Odisha University of Agriculture and Technology, Bhubaneswar, India

Debojyoti Moulick Plant Stress Biology and Metabolomics Laboratory, Central Instrumentation Laboratory, Assam University, Silchar, India

Mohammad Mukarram Plant Physiology Section, Department of Botany, Faculty of Life Sciences, Aligarh Muslim University, Aligarh, India

Muntazir Mushtaq School of Biotechnology, Sher-e-Kashmir University of Agricultural Sciences and Technology of Jammu, Jammu, India

Aarifa Nabi Plant Physiology Section, Department of Botany, Faculty of Life Sciences, Aligarh Muslim University, Aligarh, India

M. Naeem Plant Physiology Section, Department of Botany, Faculty of Life Sciences, Aligarh Muslim University, Aligarh, India

Mohammad Nafees Department of Environmental Sciences, University of Peshawar, Peshawar, Pakistan

Maira Naveed Institute of Soil and Environmental Sciences, University of Agriculture Faisalabad, Faisalabad, Pakistan

Amina Nisar Department of Plant Breeding and Genetics, University of Agriculture, Faisalabad, Pakistan

Salama A. Ouf Botany and Microbiology Department, Faculty of Science, Cairo University, Giza, Egypt

Deepu Pandita Government Department of School Education, Jammu, India

Rukhsar Parwez Plant Physiology Section, Department of Botany, Faculty of Life Sciences, Aligarh Muslim University, Aligarh, India

Abhik Patra Department of Soil Science and Agricultural Chemistry, Institute of Agricultural Sciences, Banaras Hindu University, Varanasi, India

Biswajit Pramanick Department of Agronomy, Dr. Rajendra Prasad Central Agricultural University, Pusa, India

Jos T. Puthur Plant Physiology and Biochemistry Division, Department of Botany, University of Calicut, Malappuram, India

Abdul Qadeer Institute of Soil and Environmental Sciences, University of Agriculture Faisalabad, Faisalabad, Pakistan

Ragini Department of Botany, Hansraj College, University of Delhi, New Delhi, India

Muhammad Khashi u Rahman College of Horticulture and Landscape Architecture, Northeast Agricultural University, Harbin, P.R. China

Shabir A. Rather College of Life Sciences, Northwest A&F University, Xianyang, P.R. China

Ravindra Kumar Rekwar Division of Soil Science and Agricultural Chemistry, ICAR-Indian Agricultural Research Institute, New Delhi, India

Aryadeep Roychoudhury Department of Biotechnology, St. Xavier's College (Autonomous), Kolkata, India

Yawar Sadiq Plant Physiology Section, Department of Botany, Faculty of Life Sciences, Aligarh Muslim University, Aligarh, India

Tooba Saeed National Centre of Excellence in Physical Chemistry, University of Peshawar, Peshawar, Pakistan

Peer Saffeullah Department of Botany, School of Chemical and Life Sciences, Jamia Hamdard, New Delhi, India

Bedabrata Saha Department of Weed Research, Newe Ya'ar Research Centre, Agricultural Research Organization, Ramat Yishai, Israel

Debanjana Saha M.S. Swaminathan School of Agriculture, Centurion University of Technology and Management, Paralakhemundi, India

Sahil Department of Botany, University of Delhi, New Delhi, India

Suman Samanta Division of Agricultural Physics, Indian Agricultural Research Institute, New Delhi, India

Nair G. Sarath Plant Physiology and Biochemistry Division, Department of Botany, University of Calicut, Malappuram, India

A.M. Shackira Department of Botany, Sir Syed College, Taliparamba, Kannur, India

Rahil Shahzad Agricultural Biotechnology Research Institute, Ayub Agricultural Research Institute, Faisalabad, Pakistan

Zoya Shaikh Department of Life Sciences, University of Mumbai, Mumbai, India

Laimayum Devarishi Sharma Department of Soil Science and Agricultural chemistry, Multi-Technology Testing Centre and Vocational Training Centre, Central Agricultural University, Selesih, India

Vinod Kumar Sharma Division of Soil Science and Agricultural Chemistry, ICAR-Indian Agricultural Research Institute, New Delhi, India

Tariq Omar Siddiqi Department of Botany, School of Chemical and Life Sciences, Jamia Hamdard, New Delhi, India

Faisal Siddique Department of Plant Pathology, Northeast Agricultural University, Harbin, P.R. China

Ankita Singh Institute of Agriculture Sciences, Banaras Hindu University, Varanasi, India

Archana Singh Department of Botany, Hansraj College, University of Delhi, New Delhi, India

Indrakant K. Singh Molecular Biology Research Lab, Department of Zoology, Deshbandhu College, University of Delhi, New Delhi, India

Ragini Singh Department of Botany, Hansraj College, University of Delhi, New Delhi, India

Sarika Singh Plant Physiology Section, Department of Botany, Faculty of Life Sciences, Aligarh Muslim University, Aligarh, India

Satish Kumar Singh Department of Soil Science and Agricultural Chemistry, Institute of Agricultural Sciences, Banaras Hindu University, Varanasi, India

Rajesh Kumar Singhal ICAR-Indian Grassland and Fodder Research Institute, Jhansi, India

Mona H. Soliman Botany and Microbiology Department, Faculty of Science, Cairo University, Giza, Egypt; Biology Department, Faculty of Science, Taibah University, Yanbu, Saudi Arabia

Palliyath Sruthi Plant Physiology and Biochemistry Division, Department of Botany, University of Calicut, Malappuram, India

Alviya Sultana School of Sciences, Monash University, Subang Jaya, Malaysia

Saurav Suman Department of Botany, Hansraj College, University of Delhi, New Delhi, India

Neetu Talreja Multidisciplinary Research Institute for Science and Technology, IIMCT, University of La Serena, La Serena, Chile

Ankita Trivedi Division of Soil Science and Agricultural Chemistry, ICAR-Indian Agricultural Research Institute, New Delhi, India

Moin Uddin Botany Section, Women's College, Aligarh Muslim University, Aligarh, India

Shahid Umar Department of Botany, School of Chemical and Life Sciences, Jamia Hamdard, New Delhi, India

Wajid Umar Doctoral School of Environmental Sciences, Szent Istvan University, Godollo, Hungary

Amjad ur Rahman Department of Botany, University of Swabi, Swabi, Pakistan

Shabir Hussain Wani Mountain Research Centre for Field Crops, Khudwani Sher-e-Kashmir University of Agricultural Sciences and Technology of Kashmir, Srinagar, India

Abbu Zaid Plant Physiology and Biochemistry Section, Department of Botany, Aligarh Muslim University, Aligarh, India

Muhammad Zia ur Rehman Institute of Soil and Environmental Sciences, University of Agriculture Faisalabad, Faisalabad, Pakistan

Chapter 1

Plant–soil interactions in a changing world: a climate change perspective

Abhik Patra[1], Hanuman Singh Jatav[1,2], Kiran Kumar Mohapatra[3], Arnab Kundu[4], Satish Kumar Singh[1], Vipin Kumar[5], Laimayum Devarishi Sharma[6] and Mohsina Anjum[7]

[1]*Department of Soil Science and Agricultural Chemistry, Institute of Agricultural Sciences, Banaras Hindu University, Varanasi, India,* [2]*Department of Soil Science and Agricultural Chemistry, Sri Karan Narendra Agriculture University, Jaipur, India,* [3]*Department of Soil Science and Agricultural Chemistry, College of Agriculture, Odisha University of Agriculture and Technology, Bhubaneswar, India,* [4]*Department of Agricultural Chemistry and Soil Science, Bidhan Chandra Krishi Viswavidyalaya, Mohanpur, India,* [5]*Department of Agricultural Chemistry, Chaudhary Shivnath Singh Shandilya (P.G.) College, Meerut, India,* [6]*Department of Soil Science and Agricultural chemistry, Multi-Technology Testing Centre and Vocational Training Centre, Central Agricultural University, Selesih, India,* [7]*Plant Nutrition Laboratory, Department of Soil Science and Agriculture Chemistry, University of Agricultural Sciences, GKVK, Bengaluru, India*

1.1 Introduction

Our planet Earth is habitable because of its relative position to the sun and because of the natural greenhouse effect of its atmosphere. Nature has resistance as well as resilience against the dynamic changes that occur within it. But, in the postindustrialization epoch with the ever-increasing population, we are just exhausting the natural resources (coal, petroleum, and other raw materials for the industry) and emitting a bulk of greenhouse gases into the atmosphere (Yang & Teng, 2018). Without any doubt, it is evident that anthropogenic activities are chiefly responsible for the changes in the composition of the atmosphere, and today greenhouse gases are the largest human influence on the global climate (Prakash & Srivastava, 2019). Carbon dioxide has increased 31% since preindustrial times and the emissions of carbon dioxide and methane are the most significant anthropogenic sources of greenhouse gases (72% and 19%, respectively) (Mariotte, Canarini, & Dijkstra, 2017).

Frontiers in Plant–Soil Interaction. DOI: https://doi.org/10.1016/B978-0-323-90943-3.00004-3

The higher CO_2 and other greenhouse gases have led to a 0.8°C rise in mean annual global temperatures (Dusenge, Duarte, & Way, 2019).

Anthropogenic climate change affects a variety of natural processes and various elements of the ecosystem. Then increase of Earth's mean temperature is a point of major concern. The goal of the 2015 Paris Agreement was to "keep a global temperature rise this century well below 2°C above preindustrial levels and to pursue efforts to limit the temperature increase even further to 1.5°C." For many regions of the world, achieving the global 2°C targets would still imply substantially higher average temperatures (Seneviratne, Donat, Pitman, Knutti, & Wilby, 2016). Warming of the Earth results in higher evaporation of surface water and water vapor influx to the atmosphere, which contribute to the erratic distribution of precipitation, increase in sea-level water due to the melting of glaciers, and the depletion in the quantity of freshwater. Salinization of aquifers, depletion of groundwater table, shifts in the timing of river flow, increased runoff, accumulation of pollutants, frequent flood and droughts, etc. hamper the water availability for crops (Matesanz & Ramírez-Valiente, 2019).

Climate modifications are likely to affect the composition and structure of communities by abiotic changes as well as by modifying the direction and intensity of species interactions (Eller et al., 2017). Moderate warming may frequently benefit or occasionally limit crop and pasture yields in temperate regions; on the contrary negative impacts on yields are very common for semiarid and tropical regions (Challinor et al., 2014). Elevated CO_2 concentrations stimulate photosynthesis, which results in increased plant productivity and modified water and nutrient cycles. However, an increase in crop yield is less than the photosynthetic response (Kirschbaum, 2011). FACE experiments confirmed the positive response of plants for elevated CO_2 concentrations in the presence of higher N content in the soil. Biological N-fixation can be hampered when less available N is present in soil under elevated CO_2 concentration. There is little evidence that atmospheric CO_2 enrichment will increase total soil organic matter content as a major portion of C flow into the soil and subsequently stimulate the soil food web, but often leading to an equivalent rise in soil CO_2 efflux (Jastrow, Miller, & Owensby, 2000). Furthermore, the vulnerability of organic carbon pools to climate change has important repercussions for land sustainability and climate mitigation actions.

Global warming significantly contributes to the shrink in the terrestrial C sink and enrichment of atmospheric CO_2. Furthermore, depletion in organic carbon stocks from climate change has vital consequences for land sustainability and climate mitigation actions. Changes in elements of climate along with environmental pollutions can affect plants by changes in soil processes (Classen et al., 2015), plant responses (Pérez-Ramos et al., 2020), species composition, distribution, and interaction (Schleuning et al., 2020), increased plant susceptibility to stresses (Pareek, Dhankher, & Foyer, 2020), increased fuel buildup and forest fire danger (Lacroix, Gifford, & Rush, 2020), water resources (Eckstein, Künzel, Schäfer, & Winges, 2019), recreation value, etc. (Dwire, Mellmann-Brown, & Gurrieri, 2018). Various scientists have shown changes in ecosystem productivity and nutrient cycling, as well as in the ecophysiology and growth of species, particularly in herbaceous communities. Soil organic matter drives the majority of soil functions; any diminution can lead to a reduction in fertility and biodiversity, harm soil structure, reduce water retention capacity, and enhance the risk of erosion and compaction. Erratic precipitation will affect the soil formation and other functions within it.

The way out from this uprising crisis in the near future will be to provide a scientific basis for the sustainable use and development of Earth's biosphere. For doing so, the understanding of ecology has an important role to play by developing predictive models that couple the global terrestrial carbon (C) cycle with atmospheric circulation. Plant–soil interactions are the soul of global biogeochemical and hydrological cycles and are affected by changing climate as a result of atmospheric warming and alterations in precipitation patterns. That is why knowledge regarding the expression of plants and soil, and their interaction are very much essential for managing and formulating the crop and soil, constructing suitable models, and ultimately to combat change. Changes of plant functional traits to climate change are very obvious as they respond to environmental changes, and the functional characteristics along with soil are attracting more and more attention from ecologists. Therefore we will understand the impact of climate change on plant functional traits by studying the indicative role of plants in climate change and their interaction with the soil.

1.2 Scenario of global climate change

Climate is considered as a long-term weather trend over a particular location, whereas "climate change" leads to a long-term alteration in the status of the climate which might be detected by fluctuations in the mean and variation of its properties (IPCC, 2014). Naturally, climate change happens due to the alteration in the incident angle of the Sun's energy or through various anthropogenic activities such as emission of greenhouse gases, sulfate aerosol, carbon particles, deforestation, and land-use change (Parikh & Parikh, 2002). Global climate change has been going on for a long time and is affecting the activity of peoples as well as natural and managed ecosystems. The global climate change mainly hampers the food production capability of agricultural practices, more particularly soil pollution, shortfall of water resources, heat and water stress, ecological imbalances, low production of agricultural commodities, high cost of cultivation, etc. (McMichael, Woodruff, & Hales, 2006).

Climate change is a global challenge to the modern world and there are so many explanations for this issue. Nevertheless, there are three vital statements on climate change that can be made with some certainty: (1) the greenhouse gas emissions levels have steadily risen since the onset of the industrial revolution; (2) the natural effect of greenhouse gases are believed to affect the Earth's energy cycle, making the earth warmer than it might previously have been; and (3) the Earth has certainly been warming over the past century, particularly over the last two and a half decades (IPCC, 2001). The man-made emission of greenhouse gases, particularly carbon dioxide (CO_2), methane (CH_4), nitrous oxide (N_2O), and chlorofluorocarbon (CFC) take a major role (Table 1.1) in climate change over the last century (IPCC, 2001). Therefore anthropogenic releases of greenhouse gases tend to change our terrestrial energy equilibrium and manifest themselves in large-scale warming of the Earth. As the release of greenhouse gases continues the intensity of the greenhouse effect and surface warming increases; the crucial effects of climate change are observed on a global scale (Smerdon, 2018).

In the last few decades, global warming has been increasing at an alarming rate, raising the atmospheric temperature and sea level in an unprecedented manner (IPCC, 2014). In comparison to the previous 1000 years, the levels of greenhouse gases have expanded enormously over the latest two centuries, particularly in the phase of the industrial revolution. Impacts have become most noticeable in the natural environment over the past 30 years and can impact all levels of life, from the individual, population species community and ecosystem to the ecoregion level (Amedie, 2013). This has resulted in an unprecedented increase in the Earth's temperature compared to past days, and human activities are the main culprit for changes in the global climate (IPCC, 2018).

The work on global climate change was first evidenced in the early 1980s by evaluating the interaction of rising CO_2 and other greenhouse gas levels with atmospheric components. Forecasting the present and future atmospheric changes are related to various human activities likely agriculture, fishery, forestry, etc., which pose a serious risk to agrobiodiversity and ecological functions (Lepetz, Massot, Schmeller, & Clobert, 2009). Agricultural activity has a prime connection with climate and it is the only means to supply food and nutrition to the human being, thus agriculture has certainly been a core focus of early climate change impact studies. Soon thereafter, research started to predict the impact of climate change on various agricultural activities. In most of the cases, the climate change predictions are fraught and inaccurately forecast the intensity and extent of temperature and precipitation changes over the upcoming years. However, this leads to an erroneous assessment of how atmospheric functions change, how population growth, economic, and technological advancement happen, and which land-use strategies will continue in the upcoming years (IPCC, 2014; Venkatramanan, Shah, & Prasad, 2020).

1.3 Plant responses toward changing climate

Ecosystems as well as biodiversity are directly and indirectly affected by climate change and global warming. In the circumstances of the changing

TABLE 1.1 Major greenhouse gases, current level, global warming potential and their life time in the atmosphere.

Greenhouse gases	Chemical formula	Atmospheric life-time	Recent level	Global warming potential	Main source	Harmful effects on human being
Carbon dioxide	CO_2	100 years	390 ppm	1	Fossil fuel combustion, deforestation, cement production	Suffocation by displacement of air, incapacitation and unconsciousness, headaches, vertigo and double vision, inability to concentrate, tinnitus, seizures
Methane	CH_4	12 years	1803 ppb	25	Wet rice paddies field, wastes, livestock, fossil fuel production, agriculture, landfills	Skin and eye burns, suffocation, loss of consciousness, headache and dizziness, nausea and vomiting, weakness, loss of coordination, increased breathing rate
Nitrous oxide	N_2O	114 years	324 ppb	298	Fertilizer application, fossil fuel and biomass combustion, industrial processes	Irritation of the respiratory system, eyes, and skin, aggravation of respiratory diseases, coughing and choking, nausea, headache, abdominal pain, difficulty breathing
Ozone	O_3	Hours to days	34 ppb	n. a.	Fuel combustion, organic compound evaporation	Chest pain, coughing and wheezing, difficulty breathing, irritation of the lungs and throat, congestion

(*Continued*)

TABLE 1.1 (Continued)

Greenhouse gases	Chemical formula	Atmospheric life-time	Recent level	Global warming potential	Main source	Harmful effects on human being
Chlorofluorocarbon-12	$CC_{l2}F_2$	100 years	527ppt	10,900	Industries, old air conditioner, refrigerators	Symptoms of intoxication, reduced coordination, light-headedness and headache, tremors and convulsion, irregular heartbeat
Hydrofluorocarbon-23	CHF_3	270 years	24.0ppt	14,800	Refrigerants	
Sulfur hexafluoride	SF_6	3200 years	7.30ppt	22,800	Electricity transmission	Skin and eye burns, irritation in nose and throats, shortness of breathing, headache, confusion, dizziness, suffocation
Nitrogen trifluoride	NF_3	740 years	0.86ppt	17,200	Semiconductor manufacturing	Skin and eye irritation, headache, dizziness, shortness of breathing

climate, plants are constantly exposed to many biotic and abiotic stresses. Plants demonstrate different morphological, physiological, biochemical, and molecular responses to extreme environments and climate change (Lengyel, Lázár, Trájer, & Stenger-Kovács, 2020). Plants are generally autotrophic; by sequestering carbon dioxide (CO_2) from the atmosphere using photosynthesis, they synthesize their food. Based on their photosynthetic metabolic processes, plants are classified into C_3, C_4, and CAM plants, and it is usually assumed that increasing CO_2 levels in the atmosphere will increase crop dry matter production (Lemon, 2019). Wand, Midgley, Jones, and Curtis (1999) reported that levels of CO_2 fixation improved by 33% and 25% simultaneously in both C_3 and C_4 plants, while stomatal activity decreased by 24% and 29%, respectively, for C_3 and C_4 plants. An extensively researched plant growth and reproductive production model responds to sustainable plant yield by increasing the number and size of leaves, increasing the root volume, nodulation ability, and root distribution depth for successful nutrient and water absorption, and finally increasing the yield potential by increasing the number of pods. In contrast, the elevated rates of ambient CO_2 also have significant impacts on the growth and development of plants as well as on agricultural productivity (FAO, 2018). Gupta (2020) estimated that, at the end of the 21st century, elevated concentrations of ambient CO_2 and temperature will decrease food grain production by up to 30% depending on the plant community (C_3 and C_4 crops). Researchers have observed that leaf region stimulation by high CO_2 declines when plants are exposed to drought in combination with high CO_2 levels in numerous plant species (*Platanus occidentalis*, *Liquidambar styraciflua*, and *Acer saccharum* Marsh.) (Tschaplinski, Stewart, & Norby, 1995). Changes in crop plant flowering periods arise as a consequence of significant climate change (Fitter & Fitter, 2002). Castro, Dohleman, Bernacchi, and Long (2009) concluded that elevated CO_2 disrupted the overall plant reproductive growth cycle. The duration of the growing season of pulse crops grown with elevated CO_2 has been increased by introducing new leaf nodes, instead of increasing the life of individual leaves. Although the overall plant growth and development cycle are vulnerable to climate change, growing climatic conditions, particularly with rising temperatures, have more significantly affected the reproductive stage of plants (Lohani, Singh, & Bhalla, 2020). For example, in rice, the maximum vegetative proliferation is reported at 33°C, whereas with exposure of temperature above 25°C the pollination, grain formation, and grain yield are adversely affected (Lohani et al., 2020). A massive reduction in floral buds and abortion of flowers or pollen sterility may result in a minor change in temperature during the reproductive phase, resulting in less grain yield or sometimes plants becoming sterile (Borghi, Perez de Souza, Yoshida, & Fernie, 2019). More climate change may cause discrepancies between the flowering period and pollinator behavior (Forrest, 2015). Leaf morphology, as well as root distribution, is responsive to elevated temperature in a specific context and has major effects on root respiration and absorption of nutrients and water (Awal & Ikeda, 2003). Reproductive growth is the most vulnerable stage to pressure from drought (Saini & Westgate, 1999). Male floral organs and gametophytes grow particularly susceptible to the consequences of drought. Via changes in biotic interactions, climatic extremities can have an indirect yet heavy effect on plant characteristics, fitness, and survival. The significance of the ecoevolutionary implications of altered species interactions should therefore not be underestimated, since they can be comparable or perhaps greater in magnitude relative to the direct impact on physiological perspectives to climate change (Alexander, Diez, & Levine, 2015). Photosynthesis, the backbone of plant physiological cycles, is significantly damaged by climate change. Changes in the rate of photosynthesis and respiration of crops as a function of temperature (Porter & Semenov, 2005) and can affect stomatal conductance by altering the activity of enzymes, electron transport, and leaf temperature (leaf-to-air vapor pressure difference). Photosynthesis rates decrease and breathing rates increase with heating relative to ambient conditions (Fig. 1.1) resulting in a reduced carbon gain. In most plants, as the temperature rises with optimum range, the respiration rate, as well as the rate of metabolism, increases due to improved respiration effects with higher available energy, which means that as long as nutrients are available the metabolism processes within the plant will also increase. While increasing atmospheric CO_2 levels and temperatures decrease the ratio of photorespiratory carbon losses to photosynthetic gains, especially in C_3 plants, above limited temperature rises delay photosynthetic rates and plant growth to fatal levels (Tkemaladze & Makhashvili, 2016). Climate change significantly affects the photosynthetic potential of

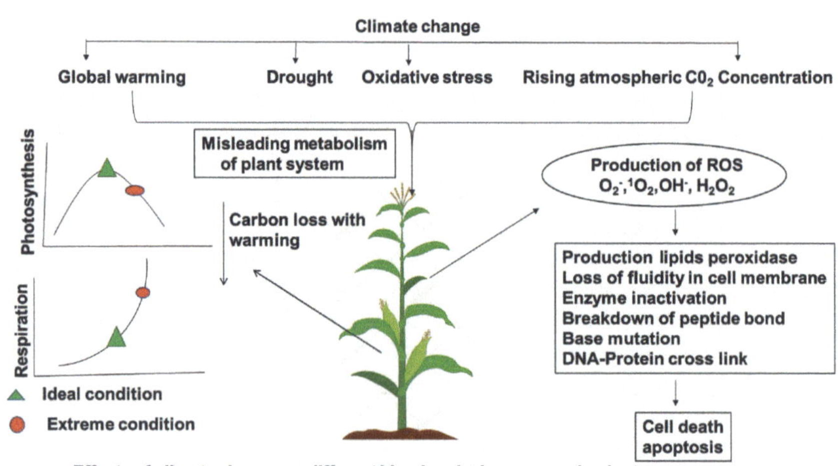

FIGURE 1.1 Effects of climate change on different biochemical processes in plant.

plants because both light-dependent reactions in grana and carbon assimilation (light-independent reactions) in the stroma are particularly vulnerable to elevated temperatures (Dutta, Chakraborti, Chaudhuri, & Mondal, 2020). Therefore minor temperature elevations result in the deactivation of the ribulose-1,5-bisphosphate carboxylase/oxygenase (Rubisco enzyme), which is primarily associated with CO_2 fixation and CO_2 processing into complex energy-rich molecules (Nagarajan & Gill, 2018). Rubisco does not function properly at an elevated temperature leading to a degradation of the Rubisco itself, which eventually contributes to the production of xylulose-1,5 bisphosphate, a photosynthetic inhibitory compound (Sage, Way, & Kubien, 2008; Ahmad, Ahanger, Alyemeni, & Alam, 2020). Grossiord et al. (2020) demonstrated that stomatal activity decreases under high vapor pressure deficiency (due to high temperature) and raises transpiration in most organisms up to a defined VPD threshold, resulting in a cascade of subsequent impacts including decreased photosynthesis and development, and higher rates of carbon deprivation and hydraulic failure. Likewise, drought stress suppresses photosystem PS II's, especially photochemical output, by reducing electron transfer, eliminating external proteins, and freeing calcium and magnesium ions from binding (Zlatev & Lidon, 2012). Very extreme circumstances of drought reduce photosynthesis due to reduced activity in Rubisco (Bota, Medrano, & Flexas, 2004). Drought stress has resulted in changes in the chlorophyll a and b ratios and carotenoids (Farooq, Wahid, Kobayashi, Fujita, & Basra, 2009). Most significantly, changes in metabolic pathways through uncoupling enzymes may lead to the development of harmful reactive oxygen species (ROS) and free radicals to climatic stress that trigger damage to the membrane, degradation of proteins, inactivation of enzymes, and thus induce oxidative stress (Dutta et al., 2020). Environmental extremes thus stimulate the formation of reactive oxygen species such as superoxide anions (O_2^-), singlet oxygen ($^1O_2$), hydrogen peroxide (H_2O_2), and hydroxyl radicals (OH). ROS is formed by a range of cellular reactions including β-oxidation of fatty acids, enhanced photorespiration, deceptive mitochondrial or chloroplast electron transport chain, and numerous enzymes such as NADPH oxidase (NOX), xanthine oxidase, lipoxygenases, and peroxidase (AbdElgawad, Zinta, Beemster, Janssens, & Asard, 2016). ROS may potentially inflict harm on the cell membrane by inducing lipid peroxidation or by interacting with biomolecules such as proteins, amino acid, fatty acid, lipids, nucleic acids, etc., and as such cellular or metabolic processes are likely to be significantly disrupted by ROS production. Consequently, under a broad variety of environmental conditions, oxidative disruption is directly related to excessive ROS generation (Fig. 1.1). In addition, climate change or more specifically temperature amplification can cause native misfolding and protein aggregation resulting in the loss of protein biochemical processes, and eventually leading to cell death (Somero, 2020). In addition, exposure of high-temperature stress can disrupt the membrane fluidity, resulting in changes in membrane-related processes and eventually complete disruption of membrane operation. At the latter stage of embryogenesis, a group of hydrophilic proteins such as late-embryogenesis abundant (LEA) proteins typically accumulate with various abiotic stresses (Andleeb et al., 2020). This shows the cellular dehydration reaction of LEA proteins and their defensive role as chaperones against tissue damage (Umezawa, Fujita, Fujita, Yamaguchi-Shinozaki, & Shinozaki, 2006). The changing climatic condition can also stimulate the production of secondary plant metabolites and decrease the nutritional quality of plants due to the increased ratio of carbon leaves to nitrogen, particularly with elevated carbon footprint (Dutta et al., 2020).

1.4 Climate change and soil interrelation

Climate is one of the five soil-forming factors and has a substantial effect on the properties of soil. The latest report of the Intergovernmental Panel on Climate Change (IPCC) states that the average temperature of the Earth will possibly increase between 1.1°C and 6.4°C by 2090–99 compared to 1980–99, with the most expected rise between 1.8°C and 4.0°C (IPCC 2007). Climate change might also affect the patterns of precipitation by modifying the amount and distribution of precipitation (IPCC 2007). With this climate change, there might be effects on the environment, inclusive of soil (Brevik, 2012). Climate change has the potential to jeopardize food security via its effect on soil properties and processes (Brevik, 2013a). To consider these effects and what should be done to adjust to them we require an understanding of in what way climate and soils interact and how climate changes may lead to resultant changes in the soil. Thus we focus on what we know about the interactions of soil and climate, in what manner soil properties and processes might be altered by climate change, and what that might purport for soil erosion and food security in the future (Brevik. 2013b).

1.4.1 Soil formation and development

Formation of soil is regulated by several aspects comprising climatic factors like temperature as well as precipitation. Soil formation is directly influenced by climatic parameters because they provide biomass and circumstances for weathering. The key parameters of climate that have a direct impact on soil formation are the summation of dynamic temperatures as well as a precipitation–evaporation ratio. These factors regulate the value of energy consumption for soil formation and water balances in soil, mechanism of organic–mineral exchanges, the transformation of organic and mineral materials, and courses of soil solutions. Stable and progressive climate warming leads to permanent modifications in the mineral matrix of soil. Alterations in external factors (temperatures and precipitation) of soil formation will result in the transformation of internal factors (energy, hydrological, biological). Climate change increases the energy of the destruction of minerals which results in the accumulation of minerals tolerant to weathering. Consequently, soil function for fertility maintenance will be lost and dependency on mineral fertilizers will be high (Pareek, 2017).

Soil development is largely controlled by three factors, that is, climate, parent material, and vegetation. The effect of climate change on soil development is projected primarily through modification in soil moisture, an increase in the soil temperature, and carbon dioxide. Climate change affects soil moisture level through direct climatic effect (temperature effect on evaporation, precipitation), rates of soil water extraction by plants, aberrations in vegetation due to climate, plant growth rates, and the consequence of raised carbon dioxide level on transpiration. Alterations in the flux of soil water could also be responsible for climate and may contribute to drought conditions by declining moisture availability, changing transmission patterns, and a rise in air temperature. Among the different factors that control the soil developmental process, climate shows a key character in the weathering of rocks and minerals. Different climate change aspects, especially rainfall, as well as temperature direct numerous stages of weathering of parent materials resulting in several changes in the mineralogy and chemistry of soil-forming rocks. Water plays a vital role in chemical weathering and therefore an increase in rainfall hastens weathering. Similar kinds of primary minerals may produce different types of secondary minerals in different weathering atmospheres. Accordingly similar rocks experiencing a different climate of weathering may give rise to distinctive soil profiles (Pareek, 2017).

1.4.2 Soil fertility and productivity

The causes of climate change like temperature, moisture, and CO_2 exert variable impacts on different soil properties and processes with significance to soil fertility and productivity. Though the effects of these factors of climate change cannot be regarded independently, the existence of one factor affects the other and subsequent effects might be multifarious. Further, these all effects may be extremely region-specific, which may depend on soil properties, degree of climate change, and climatic conditions. India is conferred with nine of 12 soil orders that occur in the world and 15 agroclimatic zones, with varied crops, farming systems, and seasons. Climate change may have indirect and direct effects on soil developmental processes and properties associated with crop production affecting the livelihoods of lots of peoples in the country (Pareek, 2017).

1.4.3 Nutrient transformation, availability, and acquisition

Plants accrue nutrients from the pool of soil solution and nutrients need to be present in solution to be considered as mobile. The biological transformation between organic and inorganic pools is intensely affected by temperature and moisture, and hence, climate change could impact N and S concentrations in solution. Some researchers suggest that an increase in carbon dioxide may not exercise a significant influence on N mineralization but concomitant warming may be responsible for enhanced N mineralization, resulting in increased solution-phase N. Extents of adsorption/desorption reactions will go faster with higher temperature, and alterations in soil moisture may further change reactions through modifying the ionic strength of soil solution (Pareek, 2017).

Nutrients' availability in soil is determined by soil chemical properties as well as the positions of ions in relation to the surface of the root and length of the pathway that nutrients need to travel in the soil to access the root surface. The rise in temperature of air and variations in precipitation have substantial impacts on the moisture regime and temperature of the root zone. It is acknowledged that soil moisture and temperature are a major factor for determining root growth and development, nutrient availability, and carbon allocation to roots directs nutrient acquisition. It is rational to believe that process consequences may be reflective of the change in the climate. The nature and degree of variation in these two parameters might be site and soil specific. It has been recommended that climate change influences on nutrient use efficiency are chiefly affected through direct impact on the surface area of the root and the rate of influx (Pareek, 2017).

1.4.4 Soil carbon content

Early outlooks were that increased atmospheric CO_2 would lead to increased plant productivity coupled with increased C sequestration by soil, meaning improved plant growth and soil–plant system may help to compensate for increasing atmospheric CO_2 (Hattenschwiler et al., 2002). This rise in the growth of the plant is termed as the CO_2 fertilization effect. However, modern studies have pointed out that the CO_2 fertilization effect may not be as great as initially believed. An increase in ozone because of climate change may neutralize the CO_2 fertilization effect resulting in moderated plant growth under raised CO_2. Also, the adverse effect of increased temperature may terminate any CO_2 fertilization effect on plant growth that does come about (Jarvis, Ramirez, Anderson, Leibing, & Aggarwal, 2010). Nitrogen restrictions might also adversely affect plant growth (Hungate, Dukes, Shaw, Luo, & Field, 2003) and the modeling of C dynamics point to less C sequestration by soil that is affected by N compared to that originally expected due to given carbon dioxide fertilization (Zaehle, Friedlingstein, & Friend, 2010). The long-term experiment in the grasslands ecosystem on elevated carbon dioxide suggested that N and P became preventive factors within 2 years, and also limiting of plant biomass as a reaction to elevated carbon dioxide (Niklaus & Korner, 2004). They concluded that an increase in productivity was due to soil moisture status as opposed to the carbon dioxide fertilization effect. Research on the decomposition of plant materials under elevated carbon dioxide in the atmosphere also specifies that increased carbon dioxide is emitted during that decomposition (Kirkham, 2011). Carney, Hungate, Drake, and Megonigal (2007) also observed that soil organic C levels were declining under increased atmospheric carbon dioxide because of improved microbial activity. Hence, elevated carbon dioxide levels may not essentially lead to increased soil C sequestration, but may instead result in more C turnover (Eglin et al., 2011).

Increased temperature is expected to have a negative effect on C distribution to the soil, leading to declines in soil organic C and creating a positive-feedback in the global C cycle (increase in temperature leads to an increase in CO_2 release from soil to the atmosphere, which results in a further increase in temperature) as global temperature rises (Wan, Lin, Xiong, Li, & Guo, 2011). Link, Smith, Halverson, and Bolton (2003) observed that soil warming and drying led to a 32% decline in soil C over a period of 5 years, a much more rapid decline in soil C compared with the reductions that have been noticed due to increased tillage in soils of a semiarid steppe. A modeling study of C responses to climate change predicted a decrease in soil organic C from 2.0%–11.5% in the North Central United States by 2100 compared to the 1990s value of C (Grace, Garcia, Gage, Robertson, & Safir, 2006).

1.4.5 Soil nitrogen content

With an increase in the C:N ratio of soil due to CO_2 enrichment, decomposing organisms in soil require more N, which can reduce N mineralization (Reich et al., 2006). If N mineralization is reduced, it may be predicted that the plant-available N in the soil will also be decreased and plant productivity will be adversely affected. Holland (2011) reported on the N limitation of CO_2 fertilized plants in their study, but an increase in temperature stimulates N availability in the soil leading to more uptake of native C. However, the provoked C uptake is not adequate to compensate for N inadequacy and the net result is the quiet upsurge in atmospheric CO_2 and an overall reduction in soil C level. Some researchers have reported that N mineralization increases with increasing temperature (Reich et al., 2006), which might have a positive impact on plant growth. However, it was revealed that N mineralization was stimulated in the first year but was depressed afterwards.

1.4.6 Soil erosion

Through climate change and anthropogenic actions, many soils of the world have supposedly become vulnerable to erosion by wind and/or water (Sivakumar, 2011). Chiew, Whetton, McMahon, and Pittock (1995) showed that an increase in rainfall owing to climate change possibly will lead to a substantial increase in runoff, with intensification larger in arid areas (five times more runoff than percent rise in rainfall) than in wet and temperate areas (twice as much runoff as a percent change in rainfall). The greater runoff will be assumed to cause an increase in erosion. Models of water erosion projected that 10% enhancement in winter rainfall may raise annual soil erosion by 150%. Li, Lui, Zhang, and Zheng (2011) predicted alteration in water erosion of −5%–195% for conventional tillage and 26%–77% for conservation tillage in the Loess Plateau region of China due to climate change.

Assessment of soil erosion pointed out that, despite less rainfall under future climate scenarios, the erosive power of surface runoff may increase and produce higher soil losses from agricultural used fields. Under extreme conditions, average soil erosion may also surpass the soil loss tolerance level. The decline in rainfall, particularly in summer and autumn, can decrease water retention in the landscape and diminish the water available for plant production (Klik & Eitzinger, 2010).

1.4.7 Soil biological activities

Feedback of soil microorganisms to changes in plant production due to raised carbon dioxide is significantly flexible because of diverse plant carbon distribution patterns in different plant–soil systems. Gross N mineralization, microbial biomass, microbial immobilization, and net N mineralization under higher carbon dioxide illustrate a high range of variability. Soil and microbial respiration rates are usually rapid under the high level of carbon dioxide, specifying that improved plant growth in elevated carbon dioxide proliferates the volume of carbon entering the soil, thus stimulating the microbial activity of soil. Soil microorganisms are often C-limited and therefore increased carbon availability stimulates microbial activity and growth. It is commonly expected that a carbon dioxide-induced increase in soil carbon availability will raise fungal biomass compared to the biomass of bacteria because of an increase in the concentration of dissolved organic C in the rhizosphere as well as the dissolved organic N in soil water. Fungal groups might have a strong influence on soil functioning due to the important roles of fungi in the degradation of organic matter, plant nutrition, nutrient cycling, and soil aggregate formation. Moreover, an increase in fungi under lower N availability at elevated carbon dioxide might be explained up to some extent, because fungi have a higher C/N ratio in comparison to bacteria and therefore have lower nitrogen demand compared to bacteria. Fungi and bacteria, the early consumers of soil organic matter, themselves are substrates for a multitude of small predators and grazers, together with nematodes, protozoa, and arthropods, which comprise the soil food web (Pareek, 2017).

The effect of elevated atmospheric carbon dioxide on the soil microbial community is usually characterized by an increase in mycorrhizal colonization because it increases the demand for plants for nutrients, coupled with increased carbon assimilation rates. Carbon dioxide enrichment ought to raise the biomass of mycorrhiza because plant demand for N and P rises simultaneously with carbon assimilation rates, and plants will distribute excess photosynthates for roots and mycorrhizal fungi to aid in satisfying this enhanced nutrient demand. Larger fine root mass and mycorrhizal infection encourage P uptake in mycorrhizal plants grown under a higher level of carbon dioxide. It appears sensible to suppose that at raised carbon dioxide levels, the biomass of mycorrhizal fungi will increase as carbon becomes relatively less limiting and soil nutrients become more limiting to plant growth. Though, information available in the literature is not always consistent on this point (Pareek, 2017).

1.5 Plant–soil feedbacks in changing climate

Interaction between plant and biotic and abiotic properties of soil is called plant–soil feedbacks (PSFs) which regulate the plant performance. The chemical, physical, and biological properties of soil determine the growth and productivity of plant and relative performance of the existing species. Plant impacts the soil properties by addition/depletion of nutrients and organic matter, in turn influencing the soil temperature and water-holding capacity as well as serving as the reservoir of resources and/or providing the habitat for the soil fauna (Bardgett & Wardle, 2010). The PSFs have an evolutionary implication and play a major role in plant communities' succession, plant fitness, invasiveness, and coexistence (Van der Putten, Bradford, Brinkman, van de Voorde, & Veen, 2016) and the response of terrestrial ecosystems to global climate change. The inextricable link between plant–soil generates a complex reaction that regulates the plant community dynamics and ecosystem. A rapidly changing climate threatens the terrestrial ecosystem and affects the growth and species distribution, hence there is a need to understand the PSFs under changing climate. The marked changes in the climate have a long-lasting impact on the abundance and distribution of the plant species. Factors that affect plant photosynthesis, such as rising CO_2 concentration in the atmosphere, soil moisture, and temperature, will affect the soil biota and hence potentially alter the PSFs. Climate change alters the plant's physiological traits, which leads to a change in the soil microbial population, hence a change in PSFs (Chomel et al., 2019). According to Wardle (2002) the three main components of the soil biota are (1) enemies (pathogenic soil microbes, nematodes, insect larvae and other invertebrates); (2) symbionts (mycorrhizal and nonmycorrhizal fungi, endophytic fungi and bacteria, nitrogen-fixing microbes, and plant growth-promoting microbes); and (3) decomposers (detritivores). Van der Putten et al. (2016) proposed a PSFs triangle which includes all three components of the soil biota and their impact on the PSFs. The bottom corner represents enemies, which is in agreement with their contribution toward negative PSFs and the top two opposing corners represent symbionts and decomposers because they generate positive PSFs (Ehrenfeld, Ravit, & Elgersma, 2005). The plant species located on the downward side of the triangle are strongly controlled by the soil-borne pathogens, which negatively affect the activity of symbionts and decomposers, thus signifying negative PSFs. One of the peculiar signs of climate change is the changes in rainfall pattern which modify the direction and intensity of the PSFs as soil organisms are very sensitive to moisture. Positive PSFs increase under drought or low moisture, in the general efficiency of the mycorrhiza fungi and other symbiotic association (Mariotte et al., 2017) such as rhizobacteria. This positively influences the water uptake (Mohan et al., 2014), water use efficiency, and the water retention capacity of the crops (Rubin, van Groenigen, & Hungate, 2017), thus increasing the overall fitness of the plant (Lau & Lennon, 2012). Whereas under high moisture condition plants become increasingly susceptible to the pest and diseases (Van der Putten et al., 2016) due to the increased activity of soil pathogens (Tedersoo et al., 2014). Change in soil moisture to either extreme also affects the litter decomposition and nutrient recycling, thus decreasing the quality of the litter (Reich, 2014). Based on the previous study Van der Putten et al. (2016) showed the changes in PSFs in altered moisture regimes due to climate change (Fig. 1.2A). Increasing temperature in general increases the positive PSFs for plant litter decomposition and the pathogenic microorganism load in soil. This will accelerate the decomposition of organic matter and release of CO_2 (Denman et al., 2007) in turn labile

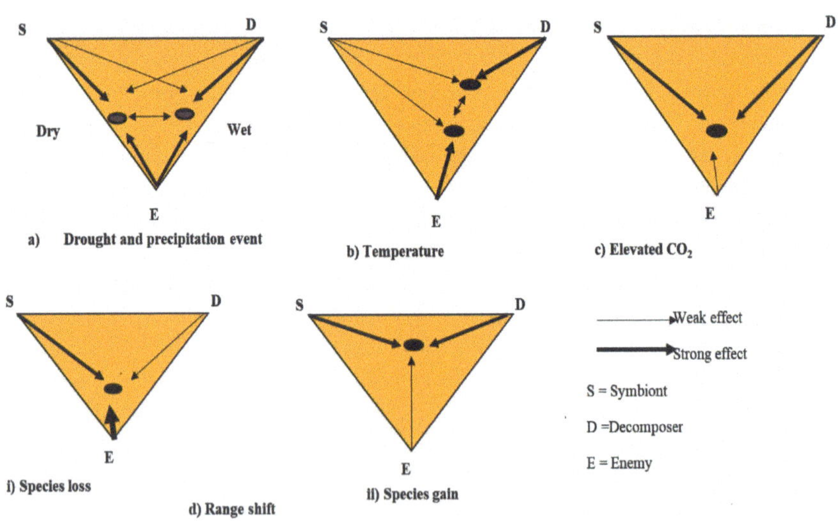

FIGURE 1.2 Plant-soil feedback triangle under global climate change. (A) Extreme drought and precipitation. (B) In case of elevated CO_2, (C) Elevated temperature. (D) Range shift. (i) Species loss (ii) Species gain.

carbon compounds will be decomposed faster than the more recalcitrant compounds. An increase in temperature directly reduces the life cycle of the soil pathogens (Chakraborty, 2013), which leads to an increment in the soil pathogens' population and results in negative PSFs. Also, it indirectly affects the plant root activity and decomposition (Bragazza, Parisod, Buttler, & Bardgett, 2013) which will increase nutrient recycling such as nitrogen (N) and phosphorus (P) (Pugnaire et al., 2019).

In general, bacterial activity increases at higher temperature, whereas it decreases the mycorrhiza activity and weakens the positive PSFs, corresponding to a higher soil bacterial: fungal ratio (Cregger, Sanders, Dunn, & Classen, 2014). However, a few of the researchers have reported that it might act otherwise, having either a neutral or negative effect due to a depleted soil carbon pool and the mutualistic relationship in arbuscular mycorrhiza and the decomposers in soil (Clemmensen et al., 2015) in the long term will be affected. Reich (2014) in a 20-year-long free-air CO_2 enrichment (FACE) study reported a shift in the N cycle from positive to negative over time in C3 grass communities and vice versa in C4 grass communities. In a nutshell, an increase in temperature increases the positive PSFs toward nutrient recycling and the pathogenic microorganism loads, whereas conversely it decreases the mutualistic and symbiotic soil biota (Fig. 1.2C). In response to climate change a shift in the distribution of plant species and microbes has been observed locally as well as globally. The range shift of plant species, that is, gain and loss of the plant species as a repercussion of global climate change has been noticed, but in general plant species loss has been extensively studied by researchers. Species loss enhances a negative PSF owing to a decrease in symbionts such as the arbuscular mycorrhizal fungi which have lower host specificity (Klironomos, 2002), the buildup of pathogens, and decreased decomposers in soil (Fig. 1.2D). On the other hand, in species gain, the exotic species have to adapt to the new situation to shift their distribution; the capacity of the species to adjust in the new habitat depends on the biotic and abiotic characteristics of the new habitat (Fig. 1.2D). Chen, Hill, Ohlemuller, Roy, and Thomas (2011) did a metaanalysis and found the range shift varied extensively among different species and noticed a shift toward higher altitude. In the new habitat, the range-shifting population can be invasive which can be a potential threat to the local species (Bardgett & van der Putten, 2014). Range shift has a lesser negative impact on plant, symbiote, and decomposer association because of lower host specificity. The timing and the kind of climate decide the range shift and expansion of the species. Bardgett and van der Putten (2014) during their study concluded that the effect of range shift under climate change on PSFs is context-dependent. The PSFs in the home range were found to be negative, whereas in the new range PSFs became less negative to positive, but over a longer period it can be more negative because of the evolution in the new pathogens. Therefore range expanders respond according to the weather and invasive species have the potential to alter the nutrient cycling and the above and below ground biota (Sardans et al., 2017). The management of the ecosystem and soil biodiversity will ensure soil functioning under a variety of novel climatic conditions (Wall, Nielsen, & Six, 2015). Due to climate change, PSFs between soil biota gets altered by modifying the nutrient availability to the plant (Pugnaire et al., 2019). The response of native species and exotic species varies, while the native plant promotes the accumulation of the soil pathogen (Lozano, Hortal, Armas, & Pugnaire, 2019) which negatively affects the invasive plant species. Native species also reported to symbiotic association with N-fixing bacteria (Gei et al., 2018) in drought conditions in tropical soil, which benefits the later species. The late-successional species form a close association with mycorrhizal fungi in increased temperature in temperate soils which are reported to slow down the succession. The significant effects of climate change are in driving how soil biota and PSFs will propel the selection and succession of species which will or will not be favored by future climate. Drivers that operate at larger spatial scales have the potential to change the components of PSF at the ecosystem level, but how these drivers interact with other drivers across scales might vary. Thus the spatial and temporal effect will decide the relative importance of single and interactive effects of climate change on PSFs.

1.6 Use of crop models to monitor plant–soil responses in climate change scenario

In the present day, study on the models of crop development based on the climate change scenario is essential. Agriculture is suffering a lot owing to an unpredictable and changing climate. An increase in global warming exposes significant crop temperature stress, therefore resulting in a decrease in the production of crops for the region. Farmers from developing countries are socioeconomically backward, often with a small piece of land holdings, using rain-fed farming, and with a limited capacity, which are some of the reasons that impel the community of farmers to opt for practices of sustainable agriculture. The climate modeling and crop simulation modeling may be expressed together with the issue of sustainability of agriculture, the security of food, methods of adaptation, and approaches for the enhancement of input production, earnings of farmers, conservations of water/environment, policy decision, researchers, economics, environmental studies, and so on (Mulla, Singh, Singh, & Bushra, 2020).

Chemical use in agriculture is responsible for the emission of greenhouse gases (GHGs) worldwide. The discharge of greenhouse gases may cause a severe hazard to the layer of ozone of the stratosphere. The release of GHGs ranging from 32.84 Gg (1980–81) to 93.82 Gg (2000–01) per year is estimated in the Indian continent likely because of the extension of area for cultivation, utilization of nitrogen as fertilizer, and animal population (Bhatia, Pathak, &

Aggarwal, 2004). The productivity of agriculture is affected by hurricanes, heat waves, floods, cold waves, and droughts. Global warming may restrict frost as well as cold waves in the future, resulting in yield loss with frost damage.

An accurate knowledge or a reasonable understanding of the impact of climatic change will undoubtedly help the scientist, agronomist, and meteorologist, to enable them to suggest to farmers how to decide on crop management and to make crop choices and use of the appropriate months and schedule for sowing seeds and irrigation. Farmers need to make changes to the conventional methods of cultivation and shift to advanced technologies to minimize the adverse effects of the changing climate. A simulation model of the crop may be used in studies of climate change to get to know the impact of climate change on the system of food. DSSAT-CROPGRO modifies a simulation of weather generators. GCMS is useful, but it has its limitations. The farmer community of the world will be benefited if the DSSAT model gave precise data using existing weather parameters. This will bring about a solution for changing climate in order to develop a high-quality production of crops both in the developed as well as underdeveloped nations (Mulla et al., 2020).

There should be an appropriate knowledge of the relationship between the growth of crop and the behavior of yields under climate change. Crop modeling is growing rapidly. It is applied in the research-oriented work, teaching to impart knowledge to the students, scientific farms, resource management, analysis of policymaking, forecasting of crop production, and so on. The modeling approach will be helpful to the farmers as an efficient tool. The simulation model allows the integration of knowledge in respect of crops as well as of a particular plant, for example, analysis of productivity, the status of change in the fertility of the soil with time. Nowadays, models of De Wit school, IBSNAT, and DSSAT are often used in the study of agrometeorology. A short description of their utility is given below.

1.6.1 The erosion–productivity impact calculator model

The United States developed the erosion–productivity impact calculator (EPIC) in 1981. This model is used in the United States to identify the correlations of productivity of plants and erosion of soil. The EPIC model can evaluate soil attrition risk. The model is used to estimate the yield of agricultural products per unit area and the whole area of the cultivable land. It is also used to study the plan of actions to manage the variation of climate and the changing climate on crop production. Further, it also assesses the harmful exposure of changing climate on agricultural products and the remedies thereof for adaptation.

1.6.2 The DSSAT model

DSSAT is a crop simulation model. It is an advanced model compared to existing models. DSSAT was introduced to reduce the shortcoming of the old models and to facilitate the maximum input details of all crops. Agriculture is the spine of the economy and the livelihood of the people of the significant parts of the world. To cope with the basic needs of the people, IBSNSAT (International Benchmark Sites Network for Agro-Technology Transfer) started using the DSSAT simulation model. The results of simulation, experiments, and execution of the product can solve the problems of universal agricultural problems for specific areas. DSSAT works as a driver proposing the management of crops, analysis of similar ecology, analysis of inherent problems of crops, and analysis of the physical properties of the soil. It also guides alternative management for better land use and with regard to preservation and sustainability. Traditional testing takes a more extended time and is costly. DSSAT minimizes the expenditure, and it also saves the time to study the cultivars that are newly developed. The DSSAT model evaluates the crop simulation model as a sensibility analysis and validation. Calibration is considered to be one of the critical aspects of the crop modeling project.

Prediction is obtained from the model with the help of calibration models. DSSAT computes the potential heat unit (PHU). Any crop variety needs an appropriate quantity of PHU at different growing stages and also at its ripening stages. The uses may opt for PHU when generating the new operation schedule file (OPS). DSSAT also analyzes sensitivity. The weather data archive is used during the experiment of sensitivity. DSSAT also analyzes seasonal rainfall data. The documentation of data that are maintained in the station with respect to climate and daily weather is required to analyze the sensitivity of the rainfall data recorded around the year.

Moreover, during the rainy season, the Earth's surface gets less heat from the sun, and hence it lessens the effect of the growth of vegetation and reproduction, resulting in a comparatively lesser yield. Hence, it is the farmer who has to opt for varieties of simulated crops that mature very soon and are popular at local levels. The DSSAT model is applied as an immediate decision aid to manage crops and to assess the potential effect of changing climate on food security for the whole world. The advancement of knowledge in recent field phonemics and genomics has opened avenues for crop models to experiment with the basic science of plants.

1.6.3 The Canegro model

The SASRI (South African Sugarcane Research Institute) has developed the Canegro model. This crop model is developed for sugarcane crop, and it is also utilized in the study of agronomy and agronomic management.

1.6.4 The NTKENAK

The NTKENAK simulates the phenology development of leaf area, accumulation and positioning of biomass, soil water balance, and yields of dry matter of Kenaf plants based on the climate and input management. The two-stage evaporation model predicts soil evaporation, and daily biomass accumulation predicts plant transpiration and daily vapor pressure deficit.

1.6.5 The SALTMED model

Model are tools for the use of scheduling of irrigation, to estimate the requirement of water for crops, prediction of yield, and salinization of soil for distinct irrigation of system, categories of soil, crops, and trees. The SALTMED model of 2015 allows more or less 20 fields at a time.

1.6.6 The CERES MAIZE model

The model of CERES MAIZE improves the simulation of crop development and yield. This model is designed for simulation of MAIZE development. Six cultivar specifications and their parameters for calibration are the main requirements for the model of CERES MAIZE in DSSAT.

1.6.7 The CANEGRO model

Nassif, Marin, Filho, Resende, and Pellegrino (2012) undertook an experiment which forecast an excellent value for sugarcane in the southern part of Brazil. Apart from photosynthesis, the model simulates CO_2 effect, concerning the resistance of stomata and transpiration methodology (Long, Ainsworth, Rogers, & Ort, 2004).

Changes and variability in climate are the biggest challenges for food, the water system, and soil. Crop simulation with high temporal–spatial resolution is used to see the impact of changing climate on water availability and crop production.

1.7 Conclusions

The human population is expected to double in the future so there will be a need to double the food production by optimizing natural resources with improved management practices. But the anthropogenic change in climate causes hindrance to achieve this goal. The decline in soil carbon-stock and subsequent enrichment of CO_2 into the atmosphere is a prime fact of concern for us as by-products of climate change. The exact consequences of changing climate are very challenging to predict, chiefly due to the complexity in the understanding of atmospheric processes, due to the interactive relationships between various environmental variables, including temperature, availability of water, solar radiation, salinity, soil nutrition, etc. Under these circumstances, understanding of plant–soil interaction is essential in order to achieve sustainability in crop production. The available evidence on crop physiology and productivity in response to climate change is very much correlated with the dynamics of the soil system itself. To a further extent, it is necessary to check or avoid unwanted impacts of climate change, formulation, and implementation of reworking strategies for the best possible outcome. Greater collaboration between scientists, experimentalists, and modelers, across respective disciplines, is crucial for better understanding the current urgency and to bridge some of the existing knowledge gaps.

References

AbdElgawad, H., Zinta, G., Beemster, G. T. S., Janssens, I. A., & Asard, H. (2016). Future climate CO_2 level mitigate stress impact on plants: increased defense or decreased challenge? *Frontiers in Plant Science*, *7*, 556.

Ahmad, P., Ahanger, M. A., Alyemeni, M. N., & Alam, P. (Eds.), (2020). *Photosynthesis, productivity and environmental stress*. Wiley.

Alexander, J. M., Diez, J. M., & Levine, J. M. (2015). Novel competitors shape species' responses to climate change. *Nature*, *525*, 515–518.

Amedie, F.A. (2013). Impacts of climate change on plant growth, ecosystem services, biodiversity, and potential adaptation measures. (Master thesis). Program Study of Biological and Enviromental Science, University of Gothenberg, Sweden.

Andleeb, T., Shah, T., Nawaz, R., Munir, I., Munsif, F., & Jalal, A. (2020). *QTL mapping for drought stress tolerance in plants. Salt and drought stress tolerance in plants* (pp. 383–403). Cham: Springer.

Awal, M. A., & Ikeda, T. (2003). Effect of elevated soil temperature on radiation-use efficiency in peanut stands. *Agricultural and Forest Meteorology*, *118*, 63–74.

Bardgett, R. D., & van der Putten, W. H. (2014). Belowground biodiversity and ecosystemfunctioning. *Nature*, *515*, 505–511.

Bardgett, R. D., & Wardle, D. A. (2010). *Aboveground-belowground linkages: Biotic interactions, ecosystem processes and global change*. Oxford: OxfordUniversity Press.

Bhatia, A., Pathak, H., & Aggarwal, P. K. (2004). Inventory of methane and nitrous oxide emissions from agricultural soils of India and their global warming potential. *Current Science*, *87*, 317–324.

Borghi, M., Perez de Souza, L., Yoshida, T., & Fernie, A. R. (2019). Flowers and climate change: A metabolic perspective. *The New Phytologist*, *224*, 1425–1441.

Bota, J., Medrano, H., & Flexas, J. (2004). Is photosynthesis limited by decreased Rubisco activity and RuBP content under progressive water stress? *The New Phytologist*, *162*, 671–681.

Bragazza, L., Parisod, J., Buttler, A., & Bardgett, R. D. (2013). Biogeochemical plant-soil microbe feedback in response to climate warming in peatlands. *Nature Climate Change*, *3*, 273–277.

Brevik, E. C. (2012). Soils and climate change: Gas fluxes and soil processes. *Soil Horizons*, *53*, 12–23.

Brevik, E. C. (2013a). Climate change, soils, and human health. In E. C. Brevik, & L. C. Burgess (Eds.), *Soils and human health* (pp. 345–383). Boca Raton, FL: CRC Press.

Brevik, E. C. (2013b). The potential impact of climate change on soil properties and processes and corresponding influence on food security. *Agriculture*, *3*, 398–417.

Carney, K. M., Hungate, B. A., Drake, B. G., & Megonigal, J. P. (2007). Altered soil microbial community at elevated CO_2 leads to loss of soil carbon. *Proceedings of the National Academy of Sciences of the United States of America*, *104*, 4990–4995.

Castro, J. C., Dohleman, F. G., Bernacchi, C. J., & Long, S. P. (2009). Elevated CO_2 significantly delays reproductive development of soybean under free-air concentration enrichment (FACE). *Journal of Experimental Botany, 60*, 2945–2951.

Chakraborty, S. (2013). Migrate or evolve: Options for plant pathogens under climate change. *Global Change Biology, 19*, 1985–2000.

Challinor, A. J., Watson, J., Lobell, D. B., Howden, S. M., Smith, D. R., & Chhetri, N. (2014). A meta-analysis of crop yield under climate change and adaptation. *Nature Climate Change, 4*, 287–291.

Chen, I. C., Hill, J. K., Ohlemuller, R., Roy, D. B., & Thomas, C. D. (2011). Rapid range shifts of species associated with high levels of climate warming. *Science (New York, N.Y.), 333*, 1024–1026.

Chiew, F. H. S., Whetton, P. H., McMahon, T. A., & Pittock, A. B. (1995). Simulation of the impacts of climate change on runoff and soil moisture in Australian catchments. *Journal of Hydrology, 167*, 121–147.

Chomel, M., Lavallee, J. M., Alvarez-Segura, N., de Castro, F., Rhymes, J. M., Caruso, T., et al. (2019). Drought decreased incorporation of recent plant photosynthate into soil food webs regardless of their trophic complexity. *Global Change Biology, 25*, 3549–3561.

Classen, A. T., Sundqvist, M. K., Henning, J. A., Newman, G. S., Moore, J. A., Cregger, M. A., ... Patterson, C. M. (2015). Direct and indirect effects of climate change on soil microbial and soil microbial-plant interactions: What lies ahead? *Ecosphere, 6*, 1–21.

Clemmensen, K. E., Finlay, R. D., Dahlberg, A., Stenlid, J., Wardle, D. A., & Lindahl, B. D. (2015). Carbon sequestration is related to mycorrhizal fungal community shifts during long-term succession in boreal forests. *The New Phytologist, 205*, 1525–1536.

Cregger, M. A., Sanders, N. J., Dunn, R. R., & Classen, A. T. (2014). Microbial communities respond to experimental warming, but site matters. *Peer J, 2*, 1–13.

Denman, K. L., Brasseur, G., Chidthaisong, A., Ciais, P., Cox, P. M., Dickinson, R. E., et al. (2007). *Couplings between changes in the climate system and biogeochemistry*. Climate change 2007: The physical science basis. Contribution of working group I to the 4th assessment report of the intergovernmental panel on climate change (pp. 499–587). Cambridge: Cambridge University Press.

Dusenge, M. E., Duarte, A. G., & Way, D. A. (2019). Plant carbon metabolism and climate change: Elevated CO_2 and temperature impacts on photosynthesis, photorespiration and respiration. *The New Phytologist, 221*, 32–49.

Dutta, P., Chakraborti, S., Chaudhuri, K. M., & Mondal, S. (2020). *Physiological responses and resilience of plants to climate change. New frontiers in stress management for durable agriculture* (pp. 3–20). Singapore: Springer.

Dwire, K.A., Mellmann-Brown, S., Gurrieri, J.T., 2018. Potential effects of climate change on riparian areas, wetlands, and groundwater-dependent ecosystems in the Blue Mountains, Oregon, USA. Climate Services. 10, 44–52.

Eckstein, D., Künzel, V., Schäfer, L., Winges, M., 2019. Global climate risk index 2020.

Eglin, T., Ciasis, P., Piao, S. L., Barre, P., Belassen, V., Cadule, P., ... Smith, P. (2011). Overview on response of global soil carbon pools to climate and land-use changes. In T. J. Sauer, J. M. Norman, & M. V. K. Sivakumar (Eds.), *Sustaining soil productivity in response to global climate change: Science, policy, and ethics* (pp. 183–199). Oxford: John Wiley & Sons, Inc.

Ehrenfeld, J. G., Ravit, B., & Elgersma, K. (2005). Feedback in the plant-soil system. *Annual review of environment and resources, 30*, 75–115.

Eller, F., Skálová, H., Caplan, J. S., Bhattarai, G. P., Burger, M. K., Cronin, J. T., et al. (2017). Cosmopolitan species as models for ecophysiological responses to global change: The common reed *Phragmitesaustralis*. *Frontiers Plant Science, 8*, 1833.

FAO. (2018). *UNICEF, WFP, WHO the state of food security and nutrition in the world 2017: Building resilience for peace and food security*. Rome: Food and Agriculture Organization of the United Nations (FAO).

Farooq, M., Wahid, A., Kobayashi, N., Fujita, D. B. S. M. A., & Basra, S. M. A. (2009). *Plant drought stress: Effects, mechanisms and management. Sustainable agriculture* (pp. 153–188). Dordrecht: Springer.

Fitter, A. H., & Fitter, R. S. R. (2002). Rapid changes in flowering time in British plants. *Science, 296*, 1689–1691.

Forrest, J. R. K. (2015). Plant-pollinator interactions and phenological change: What can we learn about climate impacts from experiments and observations? *Oikos, 124*, 4–13.

Gei, M., Rozendaal, D. M. A., Poorter, L., Bongers, F., Sprent, J. I., Garner, M. D., et al. (2018). Legume abundance along successional and rainfallgradients in Neotropical forests. *Nature Ecology & Evolution, 2*, 1104–1111.

Grace, P. R., Garcia, M. C., Gage, S. H., Robertson, G. P., & Safir, G. R. (2006). The potential impact of agricultural management and climate change on soil organic carbon of the north central region of the United States. *Ecosystems, 9*, 816–827.

Grossiord, C., Buckley, T. N., Cernusak, L. A., Novick, K. A., Poulter, B., Siegwolf, R. T., ... McDowell, N. G. (2020). Plant responses to rising vapor pressure deficit. *The New Phytologist, 226*, 1550–1566.

Gupta, G. P. (2020). *Role of global climate change in crop yield reductions. Air pollution and environmental health* (pp. 87–113). Singapore: Springer.

Hattenschwiler, S., Handa, I. T., Egli, L., Asshoff, R., Ammann, W., & Korner, C. (2002). Atmospheric CO_2 enrichment of alpine tree line conifers. *The New Phytologist, 156*, 363–375.

Holland, E. A. (2011). The role of soils and biogeochemistry in the climate and earth system. In T. J. Sauer, J. M. Norman, & M. V. K. Sivakumar (Eds.), *Sustaining soil productivity in response to global climate change: Science, policy, and ethics* (pp. 155–168). Oxford: John Wiley & Sons, Inc.

Hungate, B. A., Dukes, J. S., Shaw, M. R., Luo, Y., & Field, C. B. (2003). Nitrogen and climate change. *Science (New York, N.Y.), 302*, 1512–1513.

IPCC, 2001. Third Assessment Report. Working Group 3. England: Cambridge University Press.

IPCC. (2007). Summary for policymakers. In S. Solomon, D. Qin, M. Manning, Z. Chen, M. Marquis, K. B. Averyt, & ... H. L. Miller (Eds.), *Climate change 2007: The physical science basis; contribution of working group I to the fourth assessment report of the intergovernmental panel on climate change* (pp. 1–18). Cambridge: Cambridge University Press.

IPCC, 2014. Synthesis report. Contribution of working groups I. II and III to the fifth assessment report. 151.

IPCC, 2018. Global warming of 1.5 C. An IPCC Special Report on the impacts of global warming.

Jarvis, A., Ramirez, J., Anderson, B., Leibing, C., & Aggarwal, P. (2010). Scenarios of climate change within the context of agriculture. In M. P. Reynolds (Ed.), *Climate change and crop production* (pp. 9–37). Chippenham: CPI Antony Rowe.

Jastrow, J. D., Miller, R. M., & Owensby, C. E. (2000). Long-term effects of elevated atmospheric CO_2 on below-ground biomass and transformations to soil organic matter in grassland. *Plant and Soil, 224*, 85–97.

Kirkham, M. B. (2011). *Elevated carbon dioxide*. Boca Raton, FL: CRC Press.

Kirschbaum, M. U. (2011). Does enhanced photosynthesis enhance growth? Lessons learned from CO_2 enrichment studies. *Plant Physiology, 155*, 117–124.

Klik, A., & Eitzinger, J. (2010). Impact of climate change on soil erosion and the efficiency of soil conservation practices in Austria. *The Journal of Agricultural Science*, *148*, 529–541.

Klironomos, J. N. (2002). Feedback with soil biota contributes to plant rarity and invasiveness in communities. *Nature*, *417*, 67–70.

Lacroix, K., Gifford, R., & Rush, J. (2020). Climate change beliefs shape the interpretation of forest fire events. *Climatic Change*, *159*, 103–120.

Lau, J. A., & Lennon, J. T. (2012). Rapid responses of soil microorganisms improve plant fitnessin novel environments. *Proceedings of the National Academy of Sciences of the United States of America*, *109*, 14058–14062.

Lemon, E. R. (Ed.), (2019). *CO_2 and plants: The response of plants to rising levels of atmospheric carbon dioxide*. CRC Press.

Lengyel, E., Lázár, D., Trájer, A. J., & Stenger-Kovács, C. (2020). Climate change projections for Carpathian soda pans on the basis of photosynthesis evidence from typical diatom species. *The Science of the Total Environment*, *710*136241.

Lepetz, V., Massot, M., Schmeller, D. S., & Clobert, J. (2009). Biodiversity monitoring: Some proposals to adequately study species' responses to climate change. *Biodiversity and Conservation*, *18*, 3185.

Li, Z., Lui, W. Z., Zhang, X. C., & Zheng, F. L. (2011). Assessing the site-specific impacts of climate change on hydrology, soil erosion, and crop yields in the Loess Plateau of China. *Climate Change*, *105*, 223–242.

Link, S. O., Smith, J. L., Halverson, J. J., & Bolton, H. (2003). A reciprocal transplant experiment within a climatic gradient in a semiarid shrub-steppe ecosystem: Effects on bunchgrass growth and reproduction, soil carbon, and soil nitrogen. *Global Change Biology*, *9*, 1097–1105.

Lohani, N., Singh, M. B., & Bhalla, P. L. (2020). High temperature susceptibility of sexual reproduction in crop plants. *Journal of Experimental Botany*, *71*, 555–568.

Long, S. P., Ainsworth, E. A., Rogers, A., & Ort, D. R. (2004). Rising atmospheric carbon dioxide: Plants FACE the future. *Annual Review of Plant Biology*, *55*, 591–628.

Lozano, Y. M., Hortal, S., Armas, C., & Pugnaire, F. I. (2019). Soil micro-organisms and competitive ability of a tussock grass species in a dry ecosystem. *Journal of Ecology*, *107*, 1215–1225.

Mariotte, P., Canarini, A., & Dijkstra, F. A. (2017). Stoichiometric N:P flexibility and mycorrhizal symbiosis favour plant resistance against drought. *Journal of Ecology*, *105*, 958–967.

Matesanz, S., & Ramírez-Valiente, J. A. (2019). A review and meta-analysis of intraspecific differences in phenotypic plasticity: Implications to forecast plant responses to climate change. *Global Ecology and Biogeography*, *28*, 1682–1694.

McMichael, A. J., Woodruff, R. E., & Hales, S. (2006). Climate change and human health: Present and future risks. *The Lancet*, *367*, 859–869.

Mohan, J. E., Cowden, C. C., Baas, P., Dawadi, A., Frankson, P. T., Helmick, K., et al. (2014). Mycorrhizal fungi mediation of terrestrial ecosystem responses to global change: Mini review. *Fungal Ecology*, *10*, 3–19.

Mulla, S., Singh, S. K., Singh, K. K., & Bushra, P. (2020). Climate change and agriculture: A review of crop models. In V. Venkatramanan, et al. (Eds.), *Global climate change and environmental policy* (pp. 423–435). Springer Nature Singapore Pte Ltd.

Nagarajan, R., & Gill, K. S. (2018). Evolution of Rubiscoactivase gene in plants. *Plant Molecular Biology*, *96*, 69–87.

Nassif, D. S. P., Marin, F. R., Filho, W. J. P., Resende, R. S., & Pellegrino, G. Q. (2012). Parameterization and evaluation of the DSSAT/Canegro model for Brazilian sugarcane varieties. *Pesquisa Agropecuária Brasileira*, *47*, 311–318, Brasília.

Niklaus, P. A., & Korner, C. (2004). Synthesis of a six-year study of calcareous grassland responses to in situ CO_2 enrichment. *Ecological Monographs*, *74*, 491–511.

Pareek, A., Dhankher, O. P., & Foyer, C. H. (2020). Mitigating the impact of climate change on plant productivity and ecosystem sustainability. *Journal of Experimental Botany*, *71*, 451–456.

Pareek, N. (2017). Climate change impact on soils: Adaptation and mitigation. *MOJ Ecology & Environmental Sciences*, *2*, 136–139.

Parikh, J. K., & Parikh, K. (2002). *Climate change: India's perceptions, positions, policies and possibilities*. Mumbai: Indira Gandhi Institute of Development Research.

Pérez-Ramos, I. M., Cambrollé, J., Hidalgo-Galvez, M. D., Matías, L., Montero-Ramírez, A., Santolaya, S., & Godoy, Ó. (2020). Phenological responses to climate change in communities of plants species with contrasting functional strategies. *Environmental and Experimental Botany*, *170*103852.

Porter, J. R., & Semenov, M. A. (2005). Crop responses to climatic variation. *Philosophical Transactions of the Royal Society B: Biological Sciences*, *360*, 2021–2035.

Prakash, S., & Srivastava, S. (2019). Impact of climate change on biodiversity: An overview. *International Journal Of Scientific Innovations*, *1*, 60–65.

Pugnaire, F. I., Morillo, J. A., Peñuelas, J., Reich, P. B., Bardgett, R. D., Gaxiola, A., et al. (2019). Climate change effects on plant-soil feedbacks and consequences for biodiversity and the functioning of terrestrial ecosystems. *Science Advances*, *5*, 1834.

Reich, P. B. (2014). The world-wide 'fast-slow' plant economics spectrum: A traits manifesto. *Journal of Ecology*, *102*, 275–301.

Reich, P. B., Hobbie, S. E., Lee, T., Ellsworth, D. S., West, J. B., Tilman, D., ... Trost, J. (2006). Nitrogen limitation constrains sustainability of ecosystem response to CO_2. *Nature*, *440*, 922–925.

Rubin, R. L., van Groenigen, K. J., & Hungate, B. A. (2017). Plant growth promoting rhizobacteria are more effective under drought: A meta-analysis. *Plant and Soil*, *416*, 309–323.

Sage, R. F., Way, D. A., & Kubien, D. S. (2008). Rubisco, Rubiscoactivase and global climate change. *Journal of Experimental Botany*, *59*, 1581–1595.

Saini, H. S., & Westgate, M. E. (1999). Reproductive development in grain crops during drought. *Advances in Agronomy*, *68*, 59–96.

Sardans, J., Bartrons, M., Margalef, O., Gargallo-Garriga, A., Janssens, I. A., Ciais, P., ... Peñuelas, J. (2017). Plant invasion is associated with higher plant–soil nutrient concentrations in nutrient-poor environments. *Global Change Biology*, *23*, 1282–1291.

Schleuning, M., Neuschulz, E. L., Albrecht, J., Bender, I. M., Bowler, D. E., Dehling, D. M., et al. (2020). Trait-based assessments of climate-change impacts on interacting species. *Trends in Ecology and Evolution (Personal Edition)*, *35*, 319–328.

Seneviratne, S. I., Donat, M. G., Pitman, A. J., Knutti, R., & Wilby, R. L. (2016). Allowable CO_2 emissions based on regional and impact-related climate targets. *Nature*, *529*, 477–483.

Sivakumar, M. V. K. (2011). Climate and land degradation. In T. J. Sauer, J. M. Norman, & M. V. K. Sivakumar (Eds.), *Sustaining soil productivity in response to global climate change: Science, policy, and ethics* (pp. 141–154). Oxford: John Wiley & Sons, Inc.

Smerdon, J. (2018). *Climate change: The science of global warming and our energy future*. Columbia University Press.

Somero, G. N. (2020). The cellular stress response and temperature: Function, regulation, and evolution. *Journal of Experimental Zoology Part A: Ecological and Integrative Physiology*, *333*, 379–397.

Tedersoo, L., Bahram, M., Polme, S., Koljalg, U., Yorou, N. S., Wijesundera, R., et al. (2014). Global diversity and geography of soil fungi. *Science (New York, N.Y.)*, *346*, 1078–1088.

Tkemaladze, G. S., & Makhashvili, K. (2016). Climate changes and photosynthesis. *Annals of Agricultural Sciences*, *14*, 119–126.

Tschaplinski, T. J., Stewart, D. B., & Norby, R. J. (1995). Interactions between drought and elevated CO_2 on osmotic adjustment and solute concentrations of tree seedlings. *The New Phytologist*, *131*, 169–177.

Umezawa, T., Fujita, M., Fujita, Y., Yamaguchi-Shinozaki, K., & Shinozaki, K. (2006). Engineering drought tolerance in plants: Discovering and tailoring genes to unlock the future. *Current Opinion in Biotechnology*, *17*, 113–122.

Van der Putten, W. H., Bradford, M. A., Brinkman, E. P., van de Voorde, T. F. J., & Veen, G. F. (2016). Where, when and how plant-soil feedback matters in a changing world. *Functional Ecology*, *30*, 1109–1121.

Venkatramanan, V., Shah, S., & Prasad, R. (2020). *Global climate change and environmental policy*. Singapore: Springer.

Wall, D. H., Nielsen, U. N., & Six, J. (2015). Soil biodiversity and human health. *Nature*, *528*, 69–76.

Wan, Y., Lin, E., Xiong, W., Li, Y., & Guo, L. (2011). Modeling the impact of climate change on soil organic carbon stock in upland soils in the 21st century in China. *Agriculture, Ecosystems & Environment*, *141*, 23–31.

Wand, S. J., Midgley, G. F., Jones, M. H., & Curtis, P. S. (1999). Responses of wild C_4 and C_3 grass (*Poaceae*) species to elevated atmospheric CO_2 concentration: A meta-analytic test of current theories and perceptions. *Global Change Biology*, *5*, 723–741.

Wardle, D. A. (2002). *Communities and ecosystems: Linking the aboveground and belowground components*. Princeton, NJ: Princeton University Press.

Yang, X., & Teng, F. (2018). The air quality co-benefit of coal control strategy in China. *Resources, Conservation and Recycling*, *129*, 373–382.

Zaehle, S., Friedlingstein, P., & Friend, A. D. (2010). Terrestrial nitrogen feedbacks may accelerate future climate change. *Geophysical Research Letters*, *37*.

Zlatev, Z., & Lidon, F. C. (2012). An overview on drought-induced changes in plant growth, water relations and photosynthesis. *Emirates Journal of Food and Agriculture*, *24*, 57–72.

Chapter 2

Role of carbon cycle in soil productivity and carbon fluxes under changing climate

Zia Ur Rahman Farooqi, Muhammad Mahroz Hussain, Abdul Qadeer and Muhammad Ashar Ayub
Institute of Soil and Environmental Sciences, University of Agriculture Faisalabad, Faisalabad, Pakistan

2.1 Introduction

Ecosystems play an important part in the global cycle of carbon as they sequester 45% of terrestrial carbon and the carbon store in the soil is 50% (Bonan, 2008). According to the latest global long-term inventory data, the latest reports shows that since 2000, total carbon sinks in forests have accounted for 22% of global carbon sinks and carbon sinks offset 33% of present greenhouse gas (GHGs) emissions of fossil fuels per year (Ahlström et al., 2015; Pan et al., 2011). Although, in the forests the flow and sequestration rate of carbon alters extensively with the time (e.g., every year, every decade) and space (regional, global). Both are directly related to natural events (such as climate change, drought, forest fires, pests, and diseases or epidemics), and activities of human (e.g., cutting of trees, process of growing more trees, urban expansion, practices for the management processes). The deforestation process in tropical regions can release around 1.5 Gt of C each year, and the anthropogenic emissions of carbon accounts for about 15% (Hoekstra & Wiedmann, 2014; Peters et al., 2012). As the international community began to develop adaptation plans to deal with the impacts of global climate change (IPCC, 2001), the mechanism for improving the various natural and human factors and understanding for the cycle of carbon became common (Baccini et al., 2012).

The concentration of carbon dioxide (CO_2) in the atmosphere rose from 280 ppm before the industrial revolution to over 384 ppm in 2008 (Dlugokencky & Tans, 2016; Tans & Keeling, 2009; Tans, 2016) and 414 ppm in 2020. This increase reflects only about half of the CO_2 emissions

Frontiers in Plant–Soil Interaction. DOI: https://doi.org/10.1016/B978-0-323-90943-3.00017-1

caused by human activities, and the other half is separated into sea and land (Solomon, Manning, Marquis, & Qin, 2007). Although the location and size of the continental carbon sinks are still uncertain (Heimann, 2009), it is believed that they are located in the terrestrial biosphere. We believe that inland water plays an important part in the transportation, storage, and organic carbon mineralization process. These rivers need to be integrated into the traditional carbon cycle to ensure proper CO_2 management and mitigation of climate change (Canadell et al., 2007; Raupach et al., 2007). Inland waters (such as ponds, lakes, wetlands, streams, rivers, and reservoirs) penetrate terrestrial ecosystems and often affect the Earth's landscape. Although it is believed that inland waters only cover about 1% of the Earth's surface, their collective contribution to global carbon flow is enormous, and this ratio is between 5- and 10-fold compared to marine and terrestrial ecosystems (Battin et al., 2009; Chen et al., 2014).

In the early 1990s, when ecosystem management became a new paradigm for natural resource management, researchers used rapid technological advances (such as remote sensing, vortex-covariance flow towers, analysis of stable isotopes) and a new generation of ecosystem models to address them to find new ecosystem models. Questions asked were: what regulates the carbon storage intensity of ecosystems of forests under changing methods of management? Can sequestration of carbon be increased over various comprehensive methods of management? What is the climate importance and disturbances for influencing average carbon flows and their changes? How do various methods of fragmentation change carbon flow on a landscape scale? By facilitating the exchange of data between the research laboratories around the world, the science community has made progress in a positive way to consider the difference between carbon flow and forests (Seidl et al., 2017; Turner, 2010; Xiao et al., 2011).

The size of carbon storage from the terrestrial ecosystem with the emissions from the activities of human is 2.8 Pg C per year. So, the carbon flow rate that enters and leaves inland waters has been of little consideration in the worldwide analysis. However, quantification is important to limit the estimates of terrestrial ecosystem flows, fully integrate all vertical and horizontal carbon flows at regional and global levels, and predict feedback about climate change (Ciais et al., 2008). For example, if the inland water flow is horizontal, the large differences that often exist between continental carbon balance estimates at different scales, such as balance in the carbon budgets of Europe (Ciais et al., 2008) and Amazon (Grace & Malhi, 2002). The difference is obvious. Since the water cycle is extremely sensitive to changes in climate, the carbon that is water-based, inevitably reacts to the change in climate. Larger storms lead to more erosion and sediment flows, which means that a larger proportion of the terrestrial carbon is transported into inland waters (Battin et al., 2009).

2.2 Carbon cycling in forests

Through the process of photosynthesis in the forests the carbon enters from atmosphere. The process of storage in forests is often referred to as "primary production of gross" (GPP) or "assimilation of carbon." The weathering of the bedrock (Mc) and the lateral spread of animals (Ac) and wind (Wc) also contribute very little to this. The biomass is generated by the use of GPP and the mechanism of the plant is maintained through the automatic respiration (RA) of biological tissues such as stems, roots, and leaves. Automatic respiration can roughly be divided in above-ground respiration and underground respiration (i.e., RAa and RAb) (Coleman, Callaham, & Crossley, 2017; Hanson, Edwards, Garten, & Andrews, 2000). The net primary production (NPP) corresponds to the difference between RA and GPP and can be divided into above-ground (ANPP) and underground (BNPP) components. The rest of the GPP (i.e., NPP) can be divided into above-ground carbon distribution (AG CA) and underground carbon distribution (BG CA). They serve as a food source (Ac) for animals and as a substrate for the decomposition of biodegradation products (D) before they return to the atmosphere. They were previously converted into various trace gases (such as CO_2 and CH_4). The excretion of Ac and D is called "heterotrophic respiration" (RH). The forest contains living and dead organic matter (such as obstacles, dead branches, leaves), which indicates the presence of a small amount of above-ground heterotrophic respiration (RHa). This applies to tropical and subtropical rainforests, as high temperatures lead to many epiphytes and an increased decomposition of dead organic substances on the ground (Clark, Litz & Grant, 2008).

The increase in RH and RA is the loss decrease in forest respiration, which is called respiration of ecosystem (Re). The total volume of loss of carbon in soil (the addition of underground autotrophic respiration system (RAb) and heterotrophic respiration (RHb) is referred to as "respiration of soil" (Curtis et al., 2005; Hanson et al., 2000). Mostly the forests are situated on the slopes, so in the wind the carbon side flow (Tc, like fines, leaves) and the side flow of organic matter in animals (Ac) can be large. Finally, runoff of surface (Sc) and leaching of vertical water (Gc) lead to a small amount of carbon moving into or out of the forest.

Yuan et al. (2011) reported spatial variations in GPP in coniferous forests that are part of NEP variations. In contrast, Re determines the NEP of many deciduous forests (Luyssaert et al., 2007). GPP of forests worldwide is around 880 g C m^{-2} per year, and ranges from lower than 500 g C m^{-2} per year to almost 3000 g C m^{-2} per year. Including humid tropical regions (such as the Amazon, Central Africa, Southeast), photosynthesis covers demands for both water and temperature (Sun et al., 2011). Pine (*Pinus* > 2300 g C m^{-2} per year) (Gough, Seiler, Johnsen, & Sampson, 2002) and in Brazil the eucalyptus (*Eucalyptus*; 6640 g C m^{-2} per year) have also been reported to have high GPP (Stape, Binkley, & Ryan, 2008).

The maximum occurs in tropical wet forests, and the minimum occurs in tundra and dry desert areas. Lai et al. (2012) also reported that the tropical wet forest Re was significantly higher than that of other ecosystems, resulting in average values of NEP of 275, 400, and 120 g C m^{-2} per year in the temperate, tropics and tropical regions, and the boreal forest biome (Bonan, 2008). In plantations in forests, NEP values can increase by 1000 g C m^{-2} per year, making it a better candidate for systems of bioenergy from which ethanol is produced (like poplar and eucalyptus). Because the NEP of the forest is determined by its balance, different practices for management are usually needed to increase the GPP or reduce resource consumption. Forests that were grown recently release carbon into the atmosphere, Re is higher than the production of GPP. From the different forests, the concentration of carbon released by the layers of forests RHb and RAb makes up about 60%–80% of Re. Rs strongly depends on the temperature of the soil, total organic matter in the soil, and moisture of the soil and is an important regulator in the metabolism of underground RAb and RHb (Martin, Bolstad, Ryu, & Chen, 2009).

Other terms for carbon flow are usually small, and although they are important in certain forests, they get little attention. Some research has investigated how the loss of carbon occurs through the rivers and streams via the groundwater and runoff water leaving the forests (Roulet & Moore, 2006). Two major inland US river and stream efficiency studies have shown that their carbon content is oversaturated with emissions of 97 ± 32 Tg Cper year (Butman & Raymond, 2011). However, the carbon loss is not known in many catchments of the world. Moreover, the flow of carbon combined with wind and movement of wild animals has not been examined in the context of the complete cycle of carbon (direct transport of carbon to or from forests). The size of the entire carbon cycle factors is not same but changes significantly with the period of time. However, apparent changes in the climate which are happening within years are proof that climatic variations do not occur within few years but these apparent changes have an important role in it (Richardson, Hollinger, Aber, Ollinger, & Braswell, 2007).

The average carbon pool of living trees at this location is approximately 21% and 17% above the average and national average in United States (Turner, Koerper, Harmon, & Lee, 1995), but the Pacific Northwest was below the average level at 16% (Smithwick et al., 2009). Results of a study indicate that in temperate forests 50% of carbon from the biomass is aboveground (AGB) and 50% of the biomass is underground (BGB). However, this estimate is not accurate due to the perturbation mechanism at a particular location which has different effects on the carbon inventory estimate and there are significant differences between the meanings of a few main stocks of carbon (mainly organic matter that was dead) (Bradford, Birdsey, Joyce, & Ryan, 2008; Ryan, Bradford, Birdsey, & Joyce, 2007). The size and distribution of storage of carbon in forests worldwide vary widely.

2.3 Carbon cycle of the coastal ocean

The coastal ocean consists of various but closely interconnected ecosystems, including rivers, wetlands of tides, estuaries, and continental shelves. The carbon cycle for waters from coastal areas connecting marine and terrestrial systems is the main component of the global carbon cycle (Regnier et al., 2013). The carbon flows within and between coastal subsystems and the carbon flows caused by climate and human change vary widely. It is therefore necessary to understand and explain in detail the factors that regulate these rivers and their effects on the ocean and the global carbon budget. Although assessments previously expressed by the Panel on Climate Change (IPCC) and other agencies have ignored the coast's contribution to the anthropogenic carbon budget, they have recently been identified (Ahlström et al., 2015; Aufdenkampe et al., 2011). Since it is difficult to increase the difficulty of relatively small observational studies, it has always been a challenge to limit the exchange and fate of various forms of carbon in coastal areas, and it has not been perfect so far.

In the past decade, due to the dynamic changes in coastal systems, the rapid expansion of the collection of organic and inorganic carbon data in coastal waters (particularly under the partial pressure of carbon dioxide, pCO_2) and the new biogeochemical environment have made the region exciting. At this moment, a new generation of biogeochemical models mechanically connect with the factors that regulate the dynamics of carbon, such as stoichiometry of the element, the biological renewal of the internal and external supply of nutrients and organic matter, as well as the input and retention period (Hofmann et al., 2011). Response to change in climate and human activities will be recognized by these new methods in a better way.

2.4 Carbon inputs to coastal systems

Chemical elements loaded with carbon from soil-based origin are considered very important in ocean chemistry. Although the flow estimates of organic carbon (Dai, Yin, Meng, Liu, & Cai, 2012) and inorganic carbon (Regnier et al., 2013) have been further improved by new geospatial tools, scaling, and modeling methods, these flows do not differ significantly from previous estimates. Now the per year ration of the carbon flow from all main ocean basins and oceans are usually related to river flow, unless in certain areas, like higher carbonate and peat cover and more rates of erosion in catchment areas, the input of carbon is also controlled (Raymond et al., 2007).

2.4.1 Factors affecting carbon cycles in coastal ecosystems

For a long time, people believed that the climate was an important driver for the carbon supply of rivers in coastal waters. Rivers with high rainfall rates have higher runoff rates, and due to the importance of traffic restrictions, studies have long documented the main adjustment methods for carbon emissions. Nonbiological or biological processes are the important part for the flow of water that is regulated by changes in temperature, flow path, rate of dissolution, and storage of carbon in the catchment area. Therefore the overall impact of temperature on the flow of carbon can alter between various inorganic and organic carbon forms and in-between the regions (Laudon et al., 2012). Further on the temperatures and rainfall basis, now it is obvious that hydrological "facts" like the highest rate of rainfall from the storms are particularly essential for the transportation of organic carbon in rivers. The erosive force of these storms has resulted in most of the organic carbon (POC) from the catchment area being exported to coastal waters, particularly mountain areas (Hilton et al., 2008). These events can also increase the concentration of organic carbon that is in the dissolved form (DOC) in rivers. Therefore the concentration of the DOC in rivers exported to the coastal system increases every year. For example, tropical storms can cause more than 40% of the per year DOC exports. On a 10-year scale, a major flood event can export 80%–90% of POC from mountain areas (Hilton et al., 2008).

Climate models indicate that while it is difficult to predict changes in storm frequency, the strongest storms can occur more frequently, which affects transport from the DOC and POC rivers to coastal waters (Stocker et al., 2014). We can currently estimate river traffic in coastal waters, ranging from terrestrial carbon to medium to high safety levels. However, change in climate is supposed to have a significant effect on future carbon flows and river insecurity. In the coming decades, many studies will use precipitation as the main influencing factor for these rivers (e.g., compared to temperature) (Gough et al., 2002). In addition, the rapid transportation time of rivers in connection with the main events of water will conclude in bypassing the process of carbon processing of terrestrial ecosystem in rivers or occasionally destroy the carbon budget of coastal areas and the transportation time of terrestrial carbon changes in coastal areas (Butman & Raymond, 2011; Butman, Wilson, Barnes, Xenopoulos, & Raymond, 2015). In future the ocean will be affected by changes in climate. The extent of these modifications is still difficult to analyze because the flow of river is not stimulated by the current models of Earth's system.

2.5 The role of climate: help and disturbances

Landscapes and forests do not always stay the same. On the contrary, continuous changes occur in them, in turn causes the change in flow of carbon and sequestration with time. Forces that drive these modifications normally work in a combine form (McKinley et al., 2011; Smithwick, Harmon, Remillard, Acker, & Franklin, 2002): Environmental changes in ecosystems or landscapes (such as climate, atmospheric chemistry and soil) diseases that occur naturally and practices for the management of soil.

2.5.1 Climate change and the carbon cycle

Worldwide climate changes are hazardous and have serious effects on ecosystems. The feedback mechanisms among the climate and forest are not simple; somehow the uniqueness between these mechanisms guarantees the longer life span of forest trees. Generally, trees appears to be more receptive of changes (i.e., they react relatively slowly) than shrubs and herbaceous species. Since climate factors directly control the growth of all trees, the flow and storage of carbon have a direct impact on the climate. The rapid response to changes has been widely reported with regard to the forest ecosystem (Chen et al., 2004; Chen et al., 2014). "Fertilizing" trees by increasing CO_2 in the atmosphere can increase GPP (Pan et al., 2011), but the temperature rise is caused by increased CO_2 amount and other GHGs emissions (N_2O and CH_4). High concentrations in the atmosphere can also cause respiration loss (Re), which leads to uncertain changes in the NEP (Bonan, 2008). Fig. 2.1 explains the processes occurring in the carbon cycle, the factors affected, and their management strategies.

One of the main reasons for the uncertainty is that, due to the greater modeling cost and more complex structures, with regard to future climate no study considered factors that were more than 3. Therefore these predictions must depend on models that are justified. Extreme conditions of climate appear to have the main impacts of the change in climate (Ciais et al., 2008), especially under simultaneous extreme conditions (Xiao et al., 2010; Xiao et al., 2011) (e.g., heat wave plus drought). Although many experiments on the chronic impact of warming on the ecosystem have been carried out, most of these experiments (understandably) have been carried out on short-lived

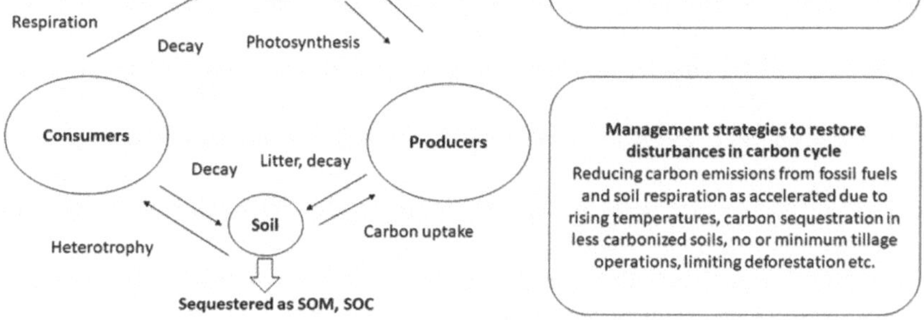

FIGURE 2.1 Carbon cycle, the factors affected, and management strategies.

plants like shrubs and vegetation (Hovenden et al., 2008). The impact of severe stress (short time or high events of temperature and drought) induced by heat is only examined by a few research works on a natural environment (Farooqi et al., 2020a; Melillo et al., 2011). Adverse impacts of stress of heat on forests and trees have been observed by the recent studies (Ahlström et al., 2015; Allen et al., 2010). Moreover, analyses have not investigated how the structures and composition of landscape respond to change in climate, which is another challenge for concluding the alterations in storage and flow rate of carbon.

Models of the ecosystem have become a significant and major part for answering different queries. Over the past 40 years, various models of ecosystems have been refined, all of which contain many components for the carbon stream and storage pools. However, comparisons between models and a review of analyzing the flow of carbon and storage of carbon from the field show that no model can positively cover all ecosystems types (Schaefer et al., 2012). The landscape consists of different ecosystems types. Therefore the community of models experiences many modifications for the development of models of new generation to do justice to this diversity. A further limitation in the response to change in climate on a landscape scale is the prediction of the specific location of the climate in future so that the model of ecosystem can be parameterized appropriately (reduction model of the region) (Spak, Holloway, Lynn, & Goldberg, 2007). The geographical resolution of the present worldwide model of circulation is too small ($>$ 100 km) to absorb the impacts of heterogeneous elements of landscape that normally only work with a resolution of 10 m. The Wisconsin Climate Change Impact Program is a well-known study that generated high-resolution regional forecasts to evaluate the impact of climate change on the ecosystem of Wisconsin. Modeling knowledge and skills of climate combined in this plan to evaluate the effects on biodiversity and the production of forests to support and develop the information on a small scale (Kucharik, Serbin, Vavrus, Hopkins, & Motew, 2010).

2.5.2 Disturbance and the carbon cycle

The reaction of the cycle of carbon in landscape forests to disturbances that occur naturally has accepted wide consideration (Anderegg, Kane, & Anderegg, 2013). This is because natural disturbances usually change the structure of the landscape immediately and cause fast modifications in the direction and size of storage and flow of carbon. The major disturbances that occur naturally are diseases and insects in the northern hemisphere, which have a serious impact on the forest cycle of carbon (Amiro et al., 2010). At the worldwide level, fires have a major role in the global vegetation pattern, particularly in the distribution of forests. With no fire, the area of forests will increase significantly from 27% surface vegetation to cover 56% (Bond, Woodward, & Midgley, 2005). For this reason, fuel has a serious impact on sequestration of carbon. The wildfire reservoir is the main cause of destruction in most of the areas. These reservoirs cause the direct loss of carbon during the various processes of combustion as well as making an environment that significantly changes the direction and size of the carbon flow. Garg, Kim, Hash, Gore, and Fisher (2008) used a model of ecosystem from the 1930s to stimulate the carbon footprints in the Canadian boreal forests. They found that the impact of temperature, precipitation, and carbon dioxide have changed each year, but large areas typically are balanced over a long period. During this time, forest fires have a direct major effect on the emissions of carbon.

Balshi et al. (2009) evaluated that emissions of CO_2 in the decade from fires will rise from 2.5–4.4 times more than the present level at the end of the current geological period. Vasileva et al. (2011) reported that wildfires in Siberia are one of the major reasons for the CO_2 concentration changes in warm seasons. Ahn et al. (2014) observed that Rs changes with intensity, but also its temporal and spatial changes largely depend on the surface pattern of the vegetation that is underground. The best example of changes in the cycle of carbon at the level of landscape is Yellowstone National Park.

However, compared to historical fire intervals, the carbon accumulation after a fire is faster. Carbon in the ecosystem is important for changes in the structure of the population (such as area of floor) and parks typically recycle about 80% of the original carbon within 50 years and 90% of the original carbon within 100 years. However, if the time interval shortens significantly with the warming of the climate, the carbon stored in the forest in the park decreases (Westerling, Turner, Smithwick, Romme, & Ryan, 2011). Deforestation through logging, reducing fuel consumption, and other types of land management are the main destructive factors that make up the carbon cycle of people in the global forest landscape. As compared to the diseases that occur naturally, there was a direct effect of activities of human on the cycle of carbon, that is serious and even permanent. Fragments of rainforest in the Amazon region have been observed to show significant reductions of conditions due to the sharp increase in tree death rate and destruction at the edge of the remaining area (Laurance, Laurancc, & Delamonica, 1998).

Management agreements at the end of the 20th century, however, aimed to significantly increase the production of timber. Erosion controls save wildlife and protect biodiversity. As awareness of people for other services of the ecosystem (like sinks of carbon) continues to grow, our present focus is to review the traditional agreements for the management at the level of landscape to attain the various goals in a sustainable manner. Our understanding of the carbon cycle in forest landscapes is not only related to the flow and storage of carbon, but also to their changes with time. Both man-made and natural diseases need to be added into the conceptual framework (Amiro et al., 2010; Anderegg et al., 2013; Melillo et al., 2011; Pan et al., 2011). Here we have a hypothetical brief discussion about NEP because most of the present study is focused on the intensity of carbon storage in forests.

Direct measurements of NEP show that there are significant differences between populations of similar ages and stages of development. Interference events are believed to move the stent forward or backward in a continuous series of time. The constant belief is that the conditional arrangement that forms the continuous sequence is same. Now we present another point of view: a conceptual framework based on three phases and depending on the changes in the NEP after the interruption. During phase 1, the type and severity of previous disturbances are the main reasons for the balance of carbon of the ecosystem. The reduction in assimilation is due to the reduction in area of leaf and the change in balance of radiation, the increase in dead organic matter, the increase in respiration caused by compaction of soil and changes in ventilation, and the change in soil energy balance (Chen et al., 2004; Chen et al., 2014).

The cycle of carbon has been a central part of various large-scale operational studies to assess the alternative options for the management. The binding capacity of forests for carbon depends roughly on the balance between photosynthesis and respiration loss. To sustain the short- or long-term rates of retention, the management of forests can be done by retaining enough plantations (like leaves) to sustain the photosynthesis, which is better for the microclimate in the subsoil and soil (e.g., reduction in respiration, reducing the temperature). The basis of this framework is that forests can be optimally managed by sustaining an enough trees (i.e., leaves) and slowing respiration (i.e., losing the ecosystem) to sustain a higher rate of photosynthesis (i.e., increasing the content of carbon, microclimate, and structure of forests and soils). Obviously, these activities for management purposes alter the overall distribution and sequestration of carbon in forests as well as the size and carbon dynamics also change with time,

Landscape management will change the heterogeneity structure of the landscape, thereby changing the carbon pool and the river. However, we have found that there are few studies regarding landscape that associate changes of structure with stocks of carbon, which prevent formulating reasonable guidelines for landscape-level management that prevents managers from adding the spatial and temporal landscape characteristics of mosaics design. Several surveys have shown that fragmentation of forest and its minor impacts will have an adverse impact on the sequestration of carbon (De Paula, Costa, & Tabarelli, 2011). Therefore management for the future should be made to lower fragmentation, and this is consistent with the conservation of biodiversity guidelines (Harper et al., 2005). However, we still lack the information into how the carbon cycle is enhanced by the alternative patterns of landscape.

2.5.3 Forests and peatlands

Characterizations of forests are done by huge storage of carbon per meter square of biomass and are susceptible to wind disasters, (ice) storms, frosts, droughts, fires, and outbreaks of pathogens or pests. It takes a long time for trees to grow and recover from extremes. This event led to a particularly long loss of forest biomass. Therefore the influence of the extreme climate on the carbon balance of the forest has both direct and backward effects and can even be maintained. Part of the effect of extreme conditions of weather is based on events that took place a few years ago. Therefore the assessment of the net impact of extreme climates on carbon storage in forests needs a clearly precise time frame within which direct or delayed reduction or addition can be analyzed. Moreover, climate protection policy based on forest biomass and longtime storage of carbon in the forest soils must prevent the possibility of accumulated carbon storage losses in the context of future events that will be at the extreme level. However, forests can be affected by various extreme events; the most common global factor is the drought that affects the carbon footprint.

In the European heat wave of 2003, inadequate rainfall (and moisture of soil) instead of inadequate temperatures was the major reason for a reduction of the flow of carbon and water in the Mediterranean and temperate forest ecosystems (Reichstein et al., 2007). Serious and repeated droughts have been known as a key factor in the recent accelerated growth in deforestation rates and forest mortality (Van Mantgem et al., 2009). In addition to the most intensively studied medium and wide belts of forest, an important operator for the physical and carbon cycle of high-yielding forests at tropics is drought. It is observed that the Amazon forest reduced by 1.6 pg (1015 g) of carbon Pg C and 2.2 Pg C (Lewis, Brando, Phillips, van der Heijden, & Nepstad, 2011) after severe droughts in 2005 and 2010, respectively. Conditions of the soil and root morphology are the major part of resilience. Cyclones and storms in the tropics often cause serious damage to the forest, which leads to the reduction of vital leaves, limbs, and widespread death (Fuhrer et al., 2006). The Superstorm called "Lothar" has reduced the forest biomass in Europe to the extent which is equal to 30% of the normal municipality's net income in Europe (Lindroth et al., 2009).

Similarly, subtropical forests are also prone to storms and cyclones (Negrón-Juárez, Baker, Zeng, Henkel, & Chambers, 2010). In January 2005, overflowing of the Amazon river caused widespread forest trees death, resulting in increased biomass (+23%) accumulation there. In the tropical forests, 16% came from fires of forest and 15% from (mainly from atmospheric lightning) supernatural fire (Van der Werf et al., 2010). The extreme climatic conditions of subtropical, boreal forests, and Mediterranean forests promote the outbreak of fire and pests, but it is not always easy to prove that a direct connection is possible.

However, facts have shown that climatic anomalies associated with the El Niño phenomenon can cause extreme fires in the Amazon rainforest and tropical forests in tropical Southeast Asia, which accounted for 66% of the abnormal growth rate of atmospheric CO_2 from 1997 to 1998. Forests have a large amount of carbon storage, but it is underground and mainly maintained by high water levels that limit decomposition. The global tropical swamp contains 90 Pg C and is particularly prone to drought and fire (Donato et al., 2011). The northern swamps contain 500 Pg C (Smith et al., 2015) and are affected by extreme factors relating to water like extreme drought and heavy rain. The combination of these factors leads to carbon decomposition, immediate release, and export, which leads to carbon loss. Organic carbon in the dissolved form loses a quarter of net production of ecosystem (Smith et al., 2015). In short, forest (and forest) ecosystems can be affected by all extreme climates and have many important processes and indirect effects.

Large carbon pools (including biomass) and carbon flows are severely affected. Forests are the most sensitive biological communities in extreme climates. However, many processes (such as death) do not result in carbon being released into the atmosphere immediately, but rather through decomposition. If the extreme situation develops quickly and carbon is generated in every cycle (e.g., through underground transportation or the formation of charcoal), the entire process can be climate-neutral and even goes toward the long-term sinks of carbon. However, this has not been proven.

2.5.4 Grasslands and savannas

Meadows are prone to drought. Unlike forests, other extreme factors (such as storms) have less impact. Fire is an internal factor of the grassland ecosystem and is usually not studied or difficult to do so. However, it helps to inhibit trees in the timber grasslands (like Miombo) and can therefore inhibit carbon buildup on the plateau. Marcolla et al. (2011) showed that grasslands can be largely adapted to annual climate change, which weakens the annual changes in the CO_2 footprint of Alpine grassland. As observed on both managed and nonmanaged meadows, meadows have high potential for the recovery of plant growth (Zavalloni et al., 2008), although the timing of drought events in the region can be a key factor (De Boeck, Dreesen, Janssens, & Nijs, 2011). However, this high flexibility differentiates forests from grasslands and explains why grasslands normally dominate in extreme drought years. As soon as the feedback of the degradation begins to take effect, drought leads to vegetation loss, high rainfalls lead to erosion, grazing in excess form is an additional pressure, and more frequent extreme events can lead to semi-arid to arid desertification of the grasslands.

2.5.5 Agricultural areas

Regarding the cycle of carbon response to severe climates, cultivated land and other ecosystems have at least three characteristics. First, the farmland system is properly managed and management measures can respond quickly (e.g., irrigation). Second, in the annual crop production, the vegetation system of soil is regularly reset by deforestation and management of agricultural (like agriculture, management of residue and irrigation) as well as being used as a sink for carbon via carbon sequestration (Farooqi, Sabir, Zeeshan, Naveed, & Hussain, 2018; Farooqi, Sabir, Zia-Ur-Rehman, & Hussain, 2020b). The duration and type of the harvest vary widely, and the soil can be exposed for a long time. The reaction to extreme events is therefore largely controlled by immediate interventions and long-term manual interventions (e.g., through a change of variety or cultivation methods) (Lobell, Sibley, & Ortiz-Monasterio, 2012). Therefore the type and extent of human intervention is one of the greatest uncertainties in assessing the impact of extreme conditions on the carbon footprint of cropland. Active interventions by farmers can also have side effects. Increasing the amount of irrigation can increase production of root biomass, activity of microbes, and the rate of erosion, leading to an increase or decrease in the storage of organic carbon in the soil. Regular harvesting and treatment of the soil make the biological legacy of long-term extreme climates less likely than forest or grassland, but the legacy effects caused by the dynamics of pathogenic bacterial populations in cultivated land cannot be removed.

Moreover, at a crucial stage in their development, species of plant can be exposed to very definite pressures, as otherwise they will no longer work the next year. Generally, the effect of severe events is a function of specific harvesting at the extreme time related to the severity of the plants to their growth phase (Reichstein et al., 2007; Reichstein et al., 2013). It should also be noted that yields of crops can significantly increase from the balance of carbon under extreme weather conditions. For example, the high carbon intake due to the unfavorable weather conditions during the ripening of the grains during the winter in Belgium did not lead to high yields on Belgian wheat farms (Matsui, Namuco, Ziska, & Horie, 1997).

2.5.6 Outlooks

Carbon research has made tremendous progress in the past two decades. This is because carbon research brings with it many pressing global social problems like climate change, energy security, depletion of natural resources, and the growth of the economy and population of the world. Forest ecosystems will play an increasingly important role in these matters, particularly because of the large flux and sink of carbon in terrestrial ecosystems. According to our review of literature, research on the carbon cycle in forest landscapes should be enhanced in the following three ways.

2.6 Humans and their interventions to carbon cycle

The relationship between the sequestration of carbon and social problems (such as global warming, management of fuel, and growth of urban areas) needs to be examined from the broader way in order to integrate people into the systems that are present naturally and to support us. The method of linking management of forest with the carbon cycle without being influenced by humans needs to be developed to build the functions related to society of human, such as the needs of society for the management of carbon (such as biological protection and bioenergy) and protective measures for other services of ecosystem.

2.6.1 Landscape-scale carbon cycles

Understanding of the flow of carbon and storage of carbon at the level of the landscape lags far behind our information at the level of ecosystem. This is partly because of the limitations of methods that existed previously and the technologies were expensive and lengthy. Since the experiments were not carried out in detail, it is not enough to test and verify the main principle and concept of ecology of landscape. Even the flow and sequestration of carbon in the vertical and horizontal directions are well coupled; this is a reasonable estimate of the horizontal carbon flow and carbon. Only a few models are used for processes on a horizontal scale. Most of the forests are managed at the level of landscape and enhance fragmentation; the lack of satisfactory prospects for the size of the landscape is particularly unfortunate. Innovative proposals are urgently needed to tackle these scientific and administrative challenges.

2.6.2 Temporal and spatial dynamics of carbon

Research into the carbon cycle for the forest ecosystems has been carried out for many years. In addition to the effects on the climate, there are still much undiscovered about the circulation of carbon, how it is modified over time, and the mechanisms for regulatory purposes. The distribution and carbon dynamics in system terrain are most of the small flows of carbon that have not yet been fully understood. There is definite data and information about the dynamics of carbon in certain components of the ecosystem (like soils at the depth, wetlands, interfaces for the urban–rural areas, land–ocean interfaces, and many key areas). However, progress in a positive way has been made in the areas of genetics, community, ecology, population, and carbon cycle research. There is no consent on the interaction in between the diversity of ecosystems of forest and functions of ecosystem. Finally, we should examine a broader understanding of the carbon cycle by including indirect drivers and feedback.

2.7 Summary

The carbon cycle has its own importance in agroecosystems due to its life-sustaining properties, and soil sustainability. The carbon cycle is disturbed through different anthropogenic activities, that is, fossil fuel burning, land-use change, and tillage practices. The balance in processes occurring in the carbon cycle is necessary. If the balance is disturbed due to some factors, serious consequences such as global warming and climate disruption may result. To avoid these consequences, different management practices are employed to rehabilitate the disturbances in the carbon cycle. These strategies are reducing carbon emissions from fossil fuels burning, zero tillage and carbon sequestration in agricultural soils, as well as reducing deforestation.

Acknowledgments

Authors want to thank Soil and Water Laboratory of Chemistry, Institute of Soil and Environmental Sciences, University of Agriculture Faisalabad, Faisalabad, Pakistan for providing access to the library for compiling the data for the chapter.

References

Ahlström, A., Raupach, M. R., Schurgers, G., Smith, B., Arneth, A., Jung, M., ... Jain, A. K. (2015). The dominant role of semi-arid ecosystems in the trend and variability of the land CO_2 sink. *Science*, *348*, 895–899.

Ahn, Y. S., Ryu, S.-R., Lim, J., Lee, C. H., Shin, J. H., Choi, W. I., ... Seo, J. I. (2014). Effects of forest fires on forest ecosystems in eastern coastal areas of Korea and an overview of restoration projects. *Landscape and Ecological Engineering*, *10*, 229–237.

Allen, C. D., Macalady, A. K., Chenchouni, H., Bachelet, D., McDowell, N., Vennetier, M., ... Hogg, E. T. (2010). A global overview of drought and heat-induced tree mortality reveals emerging climate change risks for forests. *Forest Ecology and Management*, *259*, 660–684.

Amiro, B. D., Barr, A. G., Barr, J., Black, T. A., Bracho, R., Brown, M., ... Desai, A. (2010). Ecosystem carbon dioxide fluxes after disturbance in forests of North America. *Journal of Geophysical Research: Biogeosciences*, *115*. Available from https://doi.org/10.1029/2010JG001390.

Anderegg, W. R., Kane, J. M., & Anderegg, L. D. (2013). Consequences of widespread tree mortality triggered by drought and temperature stress. *Nature Climate Change*, *3*, 30–36.

Aufdenkampe, A. K., Mayorga, E., Raymond, P. A., Melack, J. M., Doney, S. C., Alin, S. R., ... Yoo, K. (2011). Riverine coupling of biogeochemical cycles between land, oceans, and atmosphere. *Frontiers in Ecology and the Environment*, *9*, 53–60.

Baccini, A., Goetz, S., Walker, W., Laporte, N., Sun, M., Sulla-Menashe, D., ... Friedl, M. (2012). Estimated carbon dioxide emissions from tropical deforestation improved by carbon-density maps. *Nature Climate Change*, *2*, 182–185.

Balshi, M. S., McGuire, A. D., Duffy, P., Flannigan, M., Kicklighter, D. W., & Melillo, J. (2009). Vulnerability of carbon storage in North American boreal forests to wildfires during the 21st century. *Global Change Biology*, *15*, 1491–1510.

Battin, T. J., Luyssaert, S., Kaplan, L. A., Aufdenkampe, A. K., Richter, A., & Tranvik, L. J. (2009). The boundless carbon cycle. *Nature Geoscience*, *2*, 598–600.

Bonan, G. B. (2008). Forests and climate change: Forcings, feedbacks, and the climate benefits of forests. *Science.*, *320*, 1444–1449.

Bond, W. J., Woodward, F. I., & Midgley, G. F. (2005). The global distribution of ecosystems in a world without fire. *New Phytologist*, *165*, 525–538.

Bradford, J. B., Birdsey, R. A., Joyce, L. A., & Ryan, M. G. (2008). Tree age, disturbance history, and carbon stocks and fluxes in subalpine rocky mountain forests. *Global Change Biology*, *14*, 2882–2897.

Butman, D., & Raymond, P. A. (2011). Significant efflux of carbon dioxide from streams and rivers in the United States. *Nature Geoscience*, *4*, 839–842.

Butman, D. E., Wilson, H. F., Barnes, R. T., Xenopoulos, M. A., & Raymond, P. A. (2015). Increased mobilization of aged carbon to rivers by human disturbance. *Nature Geoscience*, *8*, 112–116.

Canadell, J. G., Le Quéré, C., Raupach, M. R., Field, C. B., Buitenhuis, E. T., Ciais, P., ... Marland, G. (2007). Contributions to accelerating atmospheric CO_2 growth from economic activity, carbon intensity, and efficiency of natural sinks. *Proceedings of the National Academy of Sciences*, *104*, 18866–18870.

Chen, J., Brosofske, K. D., Noormets, A., Crow, T. R., Bresee, M. K., Le Moine, J. M., ... Zheng, D. (2004). A working framework for quantifying carbon sequestration in disturbed land mosaics. *Environmental Management*, *33*, S210–S221.

Chen, J., John, R., Sun, G., McNulty, S., Noormets, A., Xiao, J., ... Franklin, J. F. (2014). *Carbon fluxes and storage in forests and landscapes. Forest landscapes and global change* (pp. 139–166). Springer.

Ciais, P., Borges, A., Abril, G., Meybeck, M., Folberth, G., Hauglustaine, D., & Janssens, I. (2008). The impact of lateral carbon fluxes on the european carbon balance. *Biogeosciences.*, *5*, 1259–1271.

Clark, C. D., Litz, L. P., & Grant, S. P. (2008). Saltmarshes as a source of chromophoric dissolved organic matter (CDOM) to Southern California coastal waters. *Limnology and Oceanography*, *53*(5), 1923–1933.

Coleman, D. C., Callaham, M. A., & Crossley, D., Jr. (2017). *Fundamentals of soil ecology*. Academic press.

Curtis, P., Vogel, C., Gough, C., Schmid, H., Su, H. B., & Bovard, B. (2005). Respiratory carbon losses and the carbon-use efficiency of a Northern Hardwood Forest, 1999–2003. *New Phytologist*, *167*, 437–456.

Dai, M., Yin, Z., Meng, F., Liu, Q., & Cai, W.-J. (2012). Spatial distribution of riverine doc inputs to the ocean: An updated global synthesis. *Current Opinion in Environmental Sustainability.*, *4*, 170–178.

De Boeck, H. J., Dreesen, F. E., Janssens, I. A., & Nijs, I. (2011). Whole-system responses of experimental plant communities to climate extremes imposed in different seasons. *New Phytologist*, *189*, 806–817.

De Paula, M. D., Costa, C. P. A., & Tabarelli, M. (2011). Carbon storage in a fragmented landscape of atlantic forest: The role played by edge-affected habitats and emergent trees. *Tropical Conservation Science*, *4*, 349–358.

Dlugokencky, E. & Tans, P. (2016). NOAA/ESRL (www.Esrl.Noaa.Gov/gmd/ccgg/trends/) 1959–1980 are based on Mauna Loa and South Pole Stations as observed by the CO_2 program at scripps institution of oceanography.

Donato, D. C., Kauffman, J. B., Murdiyarso, D., Kurnianto, S., Stidham, M., & Kanninen, M. (2011). Mangroves among the most carbon-rich forests in the tropics. *Nature Geoscience*, *4*, 293–297.

Farooqi, Z. U. R., Ayub, M. A., Zia ur Rehman, M., Sohail, M. I., Usman, M., Khalid, H., & Naz, K. (2020a). Chapter 4 - Regulation of drought stress in plants. In D. K. Tripathi, V. Pratap Singh, D. K. Chauhan, S. Sharma, S. M. Prasad, N. K. Dubey, & N. Ramawat (Eds.), *Plant life under changing environment* (pp. 77–104). Academic Press.

Farooqi, Z. U. R., Sabir, M., Zia-Ur-Rehman, M., & Hussain, M. M. (2020b). *Mitigation of climate change through carbon sequestration in agricultural soils. Climate change and agroforestry systems: Adaptation and mitigation strategies* (p. 87) CRC Press- Taylor and Francis Group.

Farooqi, Z.U.R., Sabir, M., Zeeshan, N., Naveed, K. & Hussain, M.M. (2018). Enhancing carbon sequestration using organic amendments and agricultural practices. Carbon Capture, Utilization and Sequestration.

Fuhrer, J., Beniston, M., Fischlin, A., Frei, C., Goyette, S., Jasper, K., & Pfister, C. (2006). *Climate risks and their impact on agriculture and forests in Switzerland. Climate variability, predictability and climate risks* (pp. 79–102). Springer.

Garg, R., Kim, S., Hash, D. B., Gore, J. P., & Fisher, T. (2008). Effects of feed gas composition and catalyst thickness on carbon nanotube and nanofiber synthesis by plasma enhanced chemical vapor deposition. *Journal of Nanoscience and Nanotechnology*, *8*, 3068–3076.

Gough, C., Seiler, J., Johnsen, K. & Sampson, D.A. (2002). GPP in loblolly pine: A monthly comparison of empirical and process models. General Technical Reports. SRS–48. Asheville, NC: US Department of Agriculture, Forest Service, Southern Research Station. (pp. 66–71).

Grace, J., & Malhi, Y. (2002). Carbon dioxide goes with the flow. *Nature*, *416*, 594.

Hanson, P., Edwards, N., Garten, C. T., & Andrews, J. (2000). Separating root and soil microbial contributions to soil respiration: A review of methods and observations. *Biogeochemistry*, *48*, 115–146.

Harper, K. A., Macdonald, S. E., Burton, P. J., Chen, J., Brosofske, K. D., Saunders, S. C., ... Esseen, P. A. (2005). Edge influence on forest structure and composition in fragmented landscapes. *Conservation Biology*, *19*, 768–782.

Heimann, M. (2009). Searching out the sinks. *Nature Geoscience*, *2*, 3–4.

Hilton, R. G., Galy, A., Hovius, N., Chen, M.-C., Horng, M.-J., & Chen, H. (2008). Tropical-cyclone-driven erosion of the terrestrial biosphere from mountains. *Nature Geoscience*, *1*, 759–762.

Hoekstra, A. Y., & Wiedmann, T. O. (2014). Humanity's unsustainable environmental footprint. *Science*, *344*, 1114–1117.

Hofmann, E. E., Cahill, B., Fennel, K., Friedrichs, M. A., Hyde, K., Lee, C., ... Wilkin, J. (2011). Modeling the dynamics of continental shelf carbon. *Annual Review of Marine Science*, *3*, 93–122.

Hovenden, M. J., Newton, P., Carran, R. A., Theobald, P., Wills, K. E., Vander Schoor, J. K., ... Osanai, Y. (2008). Warming prevents the elevated CO_2-induced reduction in available soil nitrogen in a temperate, Perennial Grassland. *Global Change Biology*, *14*, 1018–1024.

IPCC. (2001). *The scientific basis*. Cambridge, United Kingdom, and New York: Cambridge University Press.

Kucharik, C. J., Serbin, S. P., Vavrus, S., Hopkins, E. J., & Motew, M. M. (2010). Patterns of climate change across Wisconsin from 1950 to 2006. *Physical Geography*, *31*, 1–28.

Lai, L., Zhao, X., Jiang, L., Wang, Y., Luo, L., Zheng, Y., ... Rimmington, G. M. (2012). Soil respiration in different agricultural and natural ecosystems in an arid region. *PloS One*, *7*, e48011.

Laudon, H., Buttle, J., Carey, S. K., McDonnell, J., McGuire, K., Seibert, J., ... Tetzlaff, D. (2012). Cross-regional prediction of long-term trajectory of stream water DOC response to climate change. *Geophysical Research Letters*, *39*. Available from https://doi.org/10.1029/2012GL053033.

Laurance, W. F., Laurance, S. G., & Delamonica, P. (1998). Tropical forest fragmentation and greenhouse gas emissions. *Forest Ecology and Management*, *110*, 173–180.

Lewis, S. L., Brando, P. M., Phillips, O. L., van der Heijden, G. M., & Nepstad, D. (2011). The 2010 Amazon drought. *Science*, *331*, 554-554.

Lindroth, A., Lagergren, F., Grelle, A., Klemedtsson, L., Langvall, O., Weslien, P., & Tuulik, J. (2009). Storms can cause Europe-wide reduction in forest carbon sink. *Global Change Biology*, *15*, 346–355.

Lobell, D. B., Sibley, A., & Ortiz-Monasterio, J. I. (2012). Extreme heat effects on wheat senescence in India. *Nature Climate Change*, *2*, 186–189.

Luyssaert, S., Inglima, I., Jung, M., Richardson, A. D., Reichstein, M., Papale, D., ... Matteucci, G. (2007). CO_2 balance of boreal, temperate, and tropical forests derived from a global database. *Global Change Biology*, *13*, 2509–2537.

Marcolla, B., Cescatti, A., Manca, G., Zorer, R., Cavagna, M., Fiora, A., ... Zampedri, R. (2011). Climatic controls and ecosystem responses drive the inter-annual variability of the net ecosystem exchange of an alpine meadow. *Agricultural and Forest Meteorology*, *151*, 1233–1243.

Martin, J. G., Bolstad, P. V., Ryu, S.-R., & Chen, J. (2009). Modeling soil respiration based on carbon, nitrogen, and root mass across diverse great lake forests. *Agricultural and Forest Meteorology*, *149*, 1722–1729.

Matsui, T., Namuco, O. S., Ziska, L. H., & Horie, T. (1997). Effects of high temperature and CO_2 concentration on spikelet sterility in Indica rice. *Field Crops Research*, *51*, 213–219.

McKinley, D. C., Ryan, M. G., Birdsey, R. A., Giardina, C. P., Harmon, M. E., Heath, L. S., ... Murray, B. C. (2011). A synthesis of current knowledge on forests and carbon storage in the United States. *Ecological Applications*, *21*, 1902–1924.

Melillo, J. M., Butler, S., Johnson, J., Mohan, J., Steudler, P., Lux, H., ... Scott, L. (2011). Soil warming, carbon–nitrogen interactions, and forest carbon budgets. *Proceedings of the National Academy of Sciences*, *108*, 9508–9512.

Negrón-Juárez, R., Baker, D. B., Zeng, H., Henkel, T. K., & Chambers, J. Q. (2010). Assessing hurricane-induced tree mortality in US Gulf coast forest ecosystems. *Journal of Geophysical Research: Biogeosciences*, *115*. Available from https://doi.org/10.1029/2009JG001221.

Pan, Y., Birdsey, R. A., Fang, J., Houghton, R., Kauppi, P. E., Kurz, W. A., ... Canadell, J. G. (2011). A large and persistent carbon sink in the world's forests. *Science*, *333*, 988–993.

Peters, G., Marland, G., Le Quéré, C., Boden, T., Canadell, J., & Raupach, M. (2012). 456 correspondence: Rapid growth in CO_2 emissions after the 2008-2009 global financial 457 crisis. *Nature Climate Change*, *2*, 2–4.

Raupach, M. R., Marland, G., Ciais, P., Le Quéré, C., Canadell, J. G., Klepper, G., & Field, C. B. (2007). Global and regional drivers of accelerating CO_2 emissions. *Proceedings of the National Academy of Sciences*, *104*, 10288–10293.

Raymond, P. A., McClelland, J., Holmes, R., Zhulidov, A., Mull, K., Peterson, B., ... Gurtovaya, T. (2007). Flux and age of dissolved organic carbon exported to the Arctic Ocean: A carbon isotopic study of the five largest Arctic rivers. *Global Biogeochemical Cycles*, *21*. Available from https://doi.org/10.1029/2007GB002934.

Regnier, P., Friedlingstein, P., Ciais, P., Mackenzie, F. T., Gruber, N., Janssens, I. A., ... Andersson, A. J. (2013). Anthropogenic perturbation of the carbon fluxes from land to ocean. *Nature Geoscience*, *6*, 597–607.

Reichstein, M., Bahn, M., Ciais, P., Frank, D., Mahecha, M. D., Seneviratne, S. I., ... Frank, D. C. (2013). Climate extremes and the carbon cycle. *Nature*, *500*, 287–295.

Reichstein, M., Ciais, P., Papale, D., Valentini, R., Running, S., Viovy, N., ... Allard, V. (2007). Reduction of ecosystem productivity and respiration during the european summer 2003 climate anomaly: A joint flux tower, remote sensing and modelling analysis. *Global Change Biology*, *13*, 634–651.

Richardson, A. D., Hollinger, D. Y., Aber, J. D., Ollinger, S. V., & Braswell, B. H. (2007). Environmental variation is directly responsible for short-but not long-term variation in forest-atmosphere carbon exchange. *Global Change Biology*, *13*, 788–803.

Roulet, N., & Moore, T. R. (2006). Browning the waters. *Nature*, *444*, 283–284.

Ryan, M., Bradford, J., Birdsey, R., & Joyce, L. (2007). Tree age, disturbance history, and carbon stocks and fluxes in Subalpine Rocky Mountain forests. *Global Change Biology*, *14*, 2882–2897.

Schaefer, K., Schwalm, C. R., Williams, C., Arain, M. A., Barr, A., Chen, J. M., ... Hollinger, D. Y. (2012). A model-data comparison of gross primary productivity: Results from the North American carbon program site synthesis. *Journal of Geophysical Research: Biogeosciences*, *117*. Available from https://doi.org/10.1029/2012JG001960.

Seidl, R., Thom, D., Kautz, M., Martin-Benito, D., Peltoniemi, M., Vacchiano, G., ... Honkaniemi, J. (2017). Forest disturbances under climate change. *Nature Climate Change*, *7*, 395–402.

Smith, P., Cotrufo, M., Rumpel, C., Paustian, K., Kuikman, P., Elliott, J., ... Bustamante, M. (2015). Biogeochemical cycles and biodiversity as key drivers of ecosystem services provided by soils. *Soil Discussions*, *2*, 537–586.

Smithwick, E. A., Harmon, M. E., Remillard, S. M., Acker, S. A., & Franklin, J. F. (2002). Potential upper bounds of carbon stores in forests of the Pacific Northwest. *Ecological Applications*, *12*, 1303–1317.

Smithwick, E. A. H., Ryan, M., Kashian, D., Romme, W., Tinker, D., & Turner, M. (2009). Modeling the effects of fire and climate change on carbon and nitrogen storage in lodgepole pine (*Pinus contorta*) stands. *Global Change Biology*, *15*, 535–548.

Solomon, S., Manning, M., Marquis, M., & Qin, D. (2007). *Climate change 2007-The physical science basis: Working group I contribution to the fourth assessment report of the IPCC* (Vol. 4). Cambridge university press.

Spak, S., Holloway, T., Lynn, B., & Goldberg, R. (2007). A comparison of statistical and dynamical downscaling for surface temperature in North America. *Journal of Geophysical Research: Atmospheres*, *112*. Available from https://doi.org/10.1029/2005JD006712.

Stape, J. L., Binkley, D., & Ryan, M. G. (2008). Production and carbon allocation in a clonal eucalyptus plantation with water and nutrient manipulations. *Forest Ecology and Management*, *255*, 920–930.

Stocker, T. F., Qin, D., Plattner, G.-K., Tignor, M. M., Allen, S. K., Boschung, J., ... Midgley, P. M. (2014). *Climate change 2013: The physical science basis*. Cambridge University Press, Contribution of working group i to the fifth assessment report of IPCC the intergovernmental panel on climate change.

Sun, G., Caldwell, P., Noormets, A., McNulty, S. G., Cohen, E., Moore Myers, J., ... Xiao, J. (2011). Upscaling key ecosystem functions across the conterminous United States by a water-centric ecosystem model. *Journal of Geophysical Research: Biogeosciences*, *116*. Available from https://doi.org/10.1029/2010JG001573.

Tans, P. (2016). Annual mean atmospheric CO_2 values for Mauna Loa from Pieter Tans, NOAA/ESRL <http://www.Esrl.Noaa.Gov/gmd/ccgg/trends/> and Ralph keeling, scripps institution of oceanography (scrippsco2. Ucsd. Edu/).

Tans, P. & Keeling, R. (2009). NOAA. ESRL <www.esrl.noaa.gov/gmd/ccgg/trends/>.

Turner, D. P., Koerper, G. J., Harmon, M. E., & Lee, J. J. (1995). A carbon budget for forests of the conterminous United States. *Ecological Applications*, *5*, 421–436.

Turner, M. G. (2010). Disturbance and landscape dynamics in a changing world. *Ecology*, *91*, 2833–2849.

Van der Werf, G. R., Randerson, J. T., Giglio, L., Collatz, G., Mu, M., Kasibhatla, P. S., ... van Leeuwen, T. T. (2010). Global fire emissions and the contribution of deforestation, Savanna, forest, agricultural, and peat fires (1997-2009). *Atmospheric Chemistry and Physics*, *10*, 11707–11735.

Van Mantgem, P. J., Stephenson, N. L., Byrne, J. C., Daniels, L. D., Franklin, J. F., Fulé, P. Z., ... Taylor, A. H. (2009). Widespread increase of tree mortality rates in the Western United States. *Science*, *323*, 521–524.

Vasileva, A., Moiseenko, K., Mayer, J. C., Jürgens, N., Panov, A., Heimann, M., & Andreae, M. (2011). Assessment of the regional atmospheric impact of wildfire emissions based on CO observations at the ZOTTO tall tower station in Central Siberia. *Journal of Geophysical Research: Atmospheres*, *116*. Available from https://doi.org/10.1029/2010JD014571.

Westerling, A. L., Turner, M. G., Smithwick, E. A., Romme, W. H., & Ryan, M. G. (2011). Continued warming could transform greater Yellowstone fire regimes by mid-21st century. *Proceedings of the National Academy of Sciences*, *108*, 13165–13170.

Xiao, J., Zhuang, Q., Law, B. E., Baldocchi, D. D., Chen, J., Richardson, A. D., ... Wharton, S. (2011). Assessing net ecosystem carbon exchange of us terrestrial ecosystems by integrating eddy covariance flux measurements and satellite observations. *Agricultural and Forest Meteorology*, *151*, 60–69.

Xiao, J., Zhuang, Q., Law, B. E., Chen, J., Baldocchi, D. D., Cook, D. R., ... Ma, S. (2010). A continuous measure of gross primary production for the conterminous United States derived from modis and Ameriflux data. *Remote Sensing of Environment*, *114*, 576–591.

Yuan, W., Luo, Y., Liang, S., Yu, G., Niu, S., Stoy, P. C., ... Gough, C. M. (2011). Thermal adaptation of net ecosystem exchange. *Biogeosciences*, *8*, 1453–1463.

Zavalloni, C., Gielen, B., Lemmens, C., De Boeck, H., Blasi, S., Van den Bergh, S., ... Ceulemans, R. (2008). Does a warmer climate with frequent mild water shortages protect grassland communities against a prolonged drought? *Plant and Soil*, *308*, 119–130.

Chapter 3

Molecular advances in plant root system architecture response and redesigning for improved performance under unfavorable environments

Indu[1], Dalpat Lal[2], Basant Kumar Dadrwal[3], Debanjana Saha[4], Subhash Chand[1], Jyoti Chauhan[3], Prajjal Dey[5], Vivek Kumar[3], Udit Nandan Mishra[4], Akash Hidangmayum[3], Ankita Singh[3] and Rajesh Kumar Singhal[1]

[1]ICAR-Indian Grassland and Fodder Research Institute, Jhansi, India, [2]Jagannath University, Jaipur, India, [3]Institute of Agriculture Sciences, Banaras Hindu University, Varanasi, India, [4]M.S. Swaminathan School of Agriculture, Centurion University of Technology and Management, Paralakhemundi, India, [5]Faculty of Agriculture, Sri Sri University, Cuttack, India

3.1 Introduction

Plants, being sessile organisms, predominantly depend on their root system to adapt in contrasting environmental conditions for their morphological, anatomical, and behavioral development (Hodge, Berta, Doussan, Merchan, & Crespi, 2009; Ye et al., 2018). The plant root system architecture (RSA) comprises spatial and temporal configuration of the root system (number, biomass, volume, density, length, and fineness of lateral and primary roots) and its establishment and advancement varies with plant species, soil health and composition, availability of water and nutrients, and ubiquity and intensity of stress factors (Jin et al., 2020; Malamy, 2005). Soil is the reservoir of water and nutrients for plants on this planet and roots are the foremost channel for the conveyance of raw materials (water, vitamins, minerals, etc.) to the needy parts whereby they are converted into final products. The plants have ability to modify their RSA by instigating lateral root (LR) primordial and affect growth of their primary or secondary roots (Hodge et al., 2009). The heterogeneous nature of the surrounding soil environmental condition decides the plasticity of RSA and

Frontiers in Plant–Soil Interaction. DOI: https://doi.org/10.1016/B978-0-323-90943-3.00013-4

develops multifarious root types and patterns with specific functions (Grossman & Rice, 2012; Rogers & Benfey, 2015). An efficient RSA provides an edge to a plant for its growth and development in specific environmental conditions (such as drought, wide range of temperature fluctuations, soil salinity, and alkalinity) and is directly associated with yield attributes of the plant (Rogers & Benfey, 2015). Globally, soil fertility is deteriorating at an unprecedented rate due to improper cultural practices (high dose of chemical fertilizers, pesticides, herbicides, etc.), and extensive anthropogenic activities (Khaledian, Kiani, Ebrahimi, Brevik, & Aitkenhead-Peterson, 2017). Moreover, environmental stress, combined with poor soil fertility, reduces crop yield. Thus for global food and nutritional security, it is indispensable to identify genetic regulatory mechanism for RSA improvement to enhance food security under environmental stress (Ye et al., 2018). Environmental stimuli with plant hormones such as auxin, cytokinin (CK), jasmonic acid, and ethylene modulate root architecture and potentially affect cross regulatory phytohormonal networks which determine successful root establishment and crop stand (Qin & Huang, 2018).

The RSA is a three-dimensional structure and plays a vital role to provide strength to the plant system through the effective use of soil water and nutrients under heterogeneous conditions (Schnepf et al., 2020). LR and root hairs are key components of RSA and their development is decided by the genetic architecture of plant and their cross-talk with environmental conditions (Bray & Topp, 2018). The root metabolic activities are being regulated by oxygen availability through diffusion either from the external environment or aerial shoot portion or both (Pedersen, Sauter, Colmer, & Nakazono, 2020). Submergence stress causes a hypoxic niche which dramatically affects plant's ability to uptake water and nutrients. The hypoxic effect is further enhanced by radial oxygen loss (ROL), thus reducing the cortex to stele ratio (Yamauchi, Colmer, Pedersen, & Nakazono, 2018; Yamauchi et al., 2019). For this reason, the main emphasis is to be focused on the improvement of CSR, adventitious root primordial, and greater number of aerenchymas through flooding signals by ethylene biosynthetic pathways (Mustroph, 2018). In the model plant *Arabidopsis thaliana* ethylene regulatory transcriptional factor (ERF-VII) regulates the RSA activities through high cellular auxin signaling and maintenance of hypoxic activities (Shukla et al., 2019). The nutrient deficiency or toxicity adversely affects plant growth, and development from seed germination to plant maturity, where the roots have significant roles to play. The plants systematically change their root length, fineness, and LR density to ensure effective nutrient uptake under stress conditions (Singhal, 2015). Plant lateral growth is hampered by nitrate deficiency in soil and transporter gene *NRT1.1* performs dual functions as a nitrate sensor and hormonal signaling (Boukcim, Pagès, & Mousain, 2006; López-Bucio, Cruz-Ramırez, & Herrera-Estrella, 2003). The RSA is affected not only by soil nitrogen (Boukcim et al., 2006) and phosphorus (Péret, Clément, Nussaume, & Desnos, 2011) but also by potassium content (Sustr, Soukup, & Tylova, 2019) and their available forms.

The plant root architecture and anatomy is mostly regulated by the genes/quantitative trait loci (QTLs)/transcription factors (TFs) (Bray & Topp, 2018). Some of the important functions of RSA traits and their molecular regulatory genes and their mechanisms are highlighted in Table 3.1. The cross-talk between plant genes and environmental factors decides the programmatic changes in the plant morphology and anatomy, including root structure and ultimately affects plant growth and development (Opitz et al., 2014; Seeve et al., 2017). The observed positive changes during abiotic stresses in RSA are the increase in plant root length and width, root diameter, fineness, and root density which help the plant system to perform better by effective use of soil resources (Hu et al., 2013).

The modern "omics" techniques can be used for the identification of specific genes which regulate plant RSA and by targeting these genes a new paradigm shift can be seen against anthropogenic accelerated climate change. The TFs also play an important role in root development and tissue differentiation while interacting with external environmental stimuli and internal growth hormones (Liang et al., 2017). Thus a second "green revolution" may be targeted through a combination of effective root-trait modeling, high-throughput phenotyping, and genotyping techniques (Postma et al., 2017). The genetic manipulation can be done for RSA improvement to combat future climate changes including abiotic stresses (Khan, Gemenet, & Villordon, 2016). Below-ground trait specific program in the near future may help to divulge the role of RSA for plant growth, metabolic function, behavior, stress tolerance, development, and efficient agriculture; and will open new ways to mitigate the drastic impact of climate change.

3.2 Physiological, genetic, and molecular aspects of root system architecture regulations

3.2.1 Hormonal control of root structure architecture

Acquisition of water and nutrients adequately establish the RSA which ensures effective plant growth (Lynch, 2011). But for plant developmental biologists, a big challenge is to understand the factors that govern the patterned development of dynamic root structures. Environmental stimuli and an extensive network of plant hormones, such as Auxin (Aux), CK, jasmonates (JAs), or ethylene (ET), modulate the physiological responses of root growth. The present study, in brief, highlights the key conceptual developments in crossregulatory phytohormones networks and their role in modulating root development.

3.2.1.1 Auxin

In regulating root growth, auxin performs a pivotal role in integrating other plant hormones. There is a body of research focusing extensively on auxin synthesis, transport, and signaling in the advancement of RSA. The crucial

TABLE 3.1 Highlights the functional characterization of some important root system architecture traits, their proposed function, and regulatory mechanisms.

Root characteristics	Crop	Gene/transcription factor/transporters	Mechanism	Proposed function	Reference
Root length	*Medicago truncatula*	*SUNN*	CLV1-like leucine-rich repeat receptor kinase	Potential for absorption of soil moisture and nutrients in deeper soil layer	Schnabel, Journet, de Carvalho-Niebel, Duc, and Frugoli (2005)
Root branching	*Arabidopsis thaliana*	*AtHB23*	Controlled the expression of *LBD 16* and *LAX3*	Power of soil exploration (the major contribution to total root length)	Perotti, Ribone, Cabello, Ariel, and Chan (2019)
Root diameter	Rice	OsNAC5	Control of either the root specific (*RCc3*) or constitutive (*GOS2*) promoters	Potential for penetration ability, branching, hydraulic conductivity	Jeong et al. (2010)
Root hair density	*A. thaliana*	TGA1/TGA4 and CPC	Suppression of longitudinal elongation of trichoblast cells	Water and miner uptake	Canales, Contreras-López, Álvarez, and Gutiérrez (2017)
Root dry weight				To explore a greater soil volume	
Lateral root density	Wheat	*LRD*	By mediation of GA hormone	Rate of water and nutrient uptake	Placido et al. (2020)
Deep root to shoot ratio	Wheat	TaZFP34	*Downregulation* of *GA3- ox* and expansions	Potential for absorption of soil moisture and nutrient in deeper soil layers	Chang et al. (2016)
Hydraulic conductivity	*A. thaliana*	NRT2.1	Expression of plasma membrane aquaporin's	Rate of water uptake	Li, Tillard, Gojon, and Maurel (2016)
Lateral root formation	*A. thaliana*	*LBD18/ASL20*	Enhance EXPA17 expression during auxin response	Enhance rate of water and nutrient uptake	Lee and Kim (2013)
Root length and biomass	Wheat	*TaNAC69-1*	By repress transcription of negative growth regulators TaSHY2 and TaIAA7	To explore water in deep soil in greater volume	Chen et al. (2016)

roles of auxin in elongation and the branching of root hairs are well documented (Qin & Huang, 2018). Consequently, this has been involved in the control of drought stress tolerance via the RSA arrangement. LR initiation and promotion is induced by an important signaling molecule, auxin response factor 7 (*ARF7/ARF19*) which is induced on degradation of SLR protein (Wilmoth et al., 2005). In auxin root mutant arf7 and arf19, the LR initiation and branching are nearly absent (Vanneste et al., 2005). Further recent studies have demonstrated that the degradation of *IAA14* is crucial for root primordial initiation (Vanneste et al., 2005). Biochemical genetic studies have shown that regulation of ARF through the phosphorylation of *ARF7* and *ARF 19* potentially initiates auxin signaling to promote LR organogenesis (Cho et al., 2014). The inductive role of auxin for LR emergence by PIN3 and LAX3 signaling cascade are also reported (Perrine-Walker & Jublanc, 2014). Overall, these studies show that both auxin signaling and transport are crucial in the architecture of the root system.

3.2.1.2 Cytokinin

CK acts as an important hormone in LR formation and initiation. Several studies have been published describing the adverse effects of CK on LR initiation and development antagonistic to the inductive effect of auxin (To et al., 2004). *Arabidopsis* response regulator (ARR) type A has been shown to negatively regulate CK. An analysis of cytokinin oxidase (CKX) revealed that overexpression of CK-degrading enzyme resulted in decreased levels of CK to boost root branching (Riefler, Novak, Strnad, & Schmülling, 2006). Additionally IPT3, IPT5, and IPT7 contributes to CK biosynthesis in young primordial establishing an inhibitory field which suppresses initiation and formation of LR (Chang, Ramireddy, & Schmülling, 2015). Further, AHP6 suppresses CK signaling displaying defects in pericyclic cell division, which initiates development of LR (Moreira, Bishopp, Carvalho, & Campilho, 2013). Chang et al. (2015) stated that the mutation of *CYP735A* gene caused alteration in LR positioning implying the importance of CKs in regulating root development. Briefly, CKs acts as an important signal along with auxin in promoting and regulating LR initiation and development. However in order to unravel the cross-talk between auxin and CK, synthesis and perception of hormone in tandem with the transport is required to control LR development.

3.2.1.3 Ethylene

Ethylene works as a regulatory molecule during the process of root development. Three growth responses such as rapid downregulation of cell elongation, increased root width, and ectopic root hair induction are shown by *Arabidopsis* seedling root during treatment with ethylene or ACC (1-aminocyclopropane-1-carboxylic acid) precursor of ET. This will help plants to control root growth more dynamically (Swarup et al., 2007). Mendez-Bravo et al. (2019) found that EIN2, ethylene affects auxin level by modulating the expression of PIN2 auxin receptor. An overexpressed ethylene mutant eto1 had been identified to inhibit ethylene directed auxin biosynthesis (He et al., 2011). Kim et al. (2018) also stated that an oxidase known as MINE generating pyridoxal 5-phosphate (PNP) is induced by ethylene. This PNP acts as a cofactor of auxin biosynthesis, namely TAA1, which shows that ET-Aux cross-talk affects root initiation and growth. Further, it is suggested that ET promotes auxin conversion in turn regulating adventitious roots (ARs) thus influencing ethylene auxin cross-talk (Fattorini et al., 2018; Guan et al., 2019).

3.2.1.4 Jasmonic acid

One of the first discovered properties of JA was the inhibition of root growth. A variety of factors regulating JA signaling were exposed through screening insensitive mutants such as JAR1 (RESISTANT 1 of JASMONATE). JA stimulates MYC2, leading to PLETHORA1 (PLT1) and PLETHORA2 (PLT2) repression in the base of the stem cell (Chen et al., 2011). PLTs encode TFs such as AP2/EREBP that are essential effectors during embryonic template forming stem growth. Jasmonic acid may have detrimental effects on ARs development (Gutierrez et al., 2012). Gretchen Hagen3.3 (GH3.3), GH3.5, and GH3.6 operate upstream of ARF6, ARF8, and ARF17. By conjugating JA to amino acids Asp, Met, and Trp, these three GH3s inactivate JA and thus encourage ARs (Gutierrez et al., 2012). Jasmonic acid plays a crucial role in AR growth (Fattorini et al., 2018) along with IBA. When JA is being applied with IBA, it enhances AR formation which does not require expression of ARF6 or ARF8 (Fattorini et al., 2018). Similarly, Guan et al. (2019) recently stated that JA and the auxin signaling pathway trigger root cell stimulation, facilitating a new generation of roots. Regulation of RSA traits via phytohormones are highlighted in Fig. 3.1.

3.2.2 Environmental factors in regulation of root architecture

RSA is a three dimensional structure which includes LRs and root hairs. As we know, nonphotosynthetic organ roots are dependent on both oxygen diffusion from external environment and aerial shoot parts for their proper functioning of metabolic activities. Waterlogged soils are devoid of oxygen, which severely affects the proper development of RSA, and further it impacts the nutrient uptake, seedling establishment, and impaired plant growth, therefore RSA is a very important trait which contributes to the adaptation of plants under normal to adverse conditions (Pedersen et al., 2020). The ROL is a major barrier in waterlogged soil that interrupts the oxygen diffusion across the root and restricts the root length. Root length under flooded soil depends on the ratio of cortex to stele (CSR) and barrier of ROL. The CSR can be a major trait for root length study under flooded

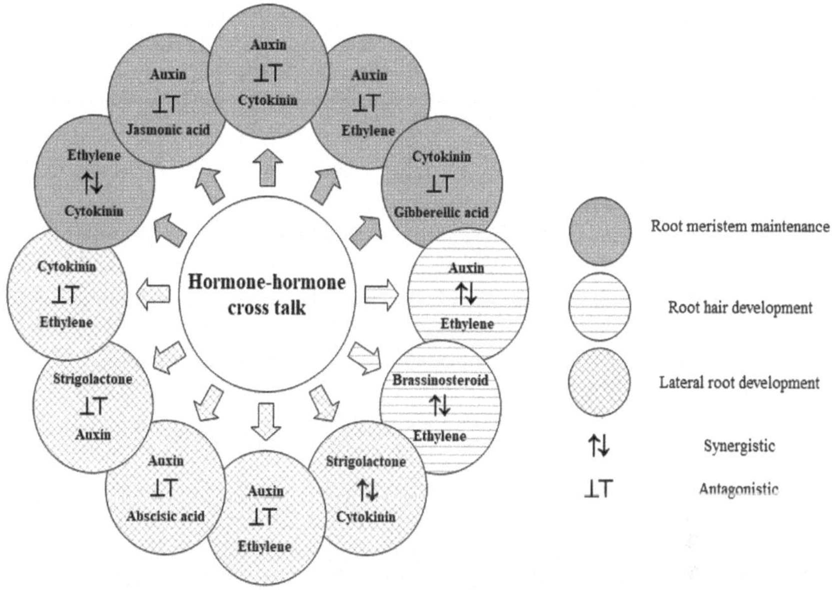

FIGURE 3.1 Simplified view of phytohormones cross-talk in root system architecture regulation. In this Aux, GA, CK, ET, ABA, SL, and BRs have synergistic or antagonistic effects on root meristem development, root hair development, and lateral root development.

conditions. Recent studies revealed that a small increase in the CSR ration of plants having low CSR like wheat and maize can be beneficial for better RSA establishment under waterlogged soils, whereas rice plants already have inherent large CSR. (Yamauchi et al., 2018, 2019). Under waterlogged conditions certain plants tend to overcome the problem by forming AR primordia and a greater number of aerenchymas to form porous roots to survive the conditions. This process is carried out by ET biosynthetic pathways that are also known as a "flooding signal" (Mustroph, 2018).

It has been stated in earlier studies that insufficient oxygen availability to the new developing cells/tissues of the plant causes the formation of low oxygenic niches in the new tissue to satisfy the oxygen demand throughout the developmental activities. In an experiment with *A. thaliana* under anaerobic conditions it has been reported that these kinds of niches occur at a specific stage of LR primordial (LRP) development. The experimental results showed that a group of ERF-VII TF control the RSA by acting in a cell under high auxin signaling and mediate the hypoxic activities. It has also found that auxin-simulative gene expression *LBD16*, *LBD18*, and *PUCHI* are repressed by ERF-VII transcriptional factors. These genes are essential for LRP development as they bind to promoter area (Shukla et al., 2019). The formation of AR in maize is promoted by *ZmEREN180* gene which is also a member of ERFVII TF that regulates the oxygen diffusion under waterlogged conditions. (Yu et al., 2019).

Constitutive expression of aerenchyma cell is noted in the case of submerged rice (Yamauchi et al., 2019). The constitutive aerenchyma are regulated through the auxin synthetic pathway, whereas the auxin pathway is regulated via indole-3-acetic acid protein and other auxin signaling factors. Auxin signaling followed by protein stimulation upregulates the TF lateral organ boundaries domain (LBD) having LBD1–8 protein, which controls constitutive aerenchyma formation and LRs formation in roots.

Roots are important organs in plants for nutrient acquisition. The deficiency of any major nutrients in soil may cause a hazardous impact on RSA which further leads to impaired growth of aerial plant parts. Plants change their growth patterns, such as stopping shoot growth and enhancing the RL and LRs density to ensure the proper nutrient uptake. Erratic distribution of nutrients particularly nitrate (NO_3^-), ammonia (NH_4^+), and phosphate in the soil affects root activities due to nutrient heterogeneity. Therefore the plant exploits the concentrated nutrient by modifying their RSA. Several studies demonstrated that LRs growth is restricted under soil nitrate deficiency. The transporter gene *NRT1.1* performs dual functions as a nitrate sensor and hormonal signaling. Under low soil nitrate concentration *NRT1.1* modifies the auxin transportation away from LRs, thus leading to inhibition of the LRs elongation and further growth. Furthermore, a higher concentration of nitrate restricts the LRs growth (López-Bucio et al., 2003; Boukcim et al., 2006). Plants grown in phosphorous (P)-deprived soil conditions exhibit restricted elongation of primary root, whereas they increase LRs density and length. In many cases root clusters are formed from a single parent root under P deficiency (Péret et al., 2011). To study the effect of nitrate flux across the RSA, Sinha et al. (2020) performed an experiment on wheat by taking two levels of nitrogen, optimum and limited externally at two stages of seedling growth. Results showed that roots in limited nitrate availability invariably increase the size of the total root system. They also observed the nitrate uptake, translocation depends on external nitrate concentration, growth stage, and root architecture system. The upregulation was observed for the gene *TaNRT2.1* under low nitrate concentration initially, then after 21 days expression was downregulated. The genes *TaNRT6.3* and *TaNRT6.4* showed low affinity for nitrate under higher external nitrate concentration. Strigolactones (SLs) are the plant growth hormones which are involved in root and shoot development. Under nutrient deficiency in the soil or water, SLs trigger the symbiotic association with certain microbes through elongation of root length (Marzec & Melzer, 2018). Recent studies demonstrated that SLs have a pleiotropic effect on root development, like exposure of roots to the SLs, and also accelerate the secondary metabolites accumulation. In an experiment with *A. thaliana* under P starvation treatment, *racGR24* enhanced the LRs primordial (Ruyter-Spira et al., 2011).

Other environmental factors like drought also affect the RSA, as under drought conditions PR elongation starts, whereas LRs growth gets inhibited. Many drought-tolerant plants like sorghum have a predominantly vertical primary root system (Singh et al., 2010). Soil salinity also has negative effect on RSA because it disrupts the osmotic potential balance across the root and surrounding soil environment. The impact of soil salinity on RSA is contrary to drought, under high salinity there is a disparity in the growth between primary and LR such that the primary root shows inhibition in the length, whereas elongation of LRs is promoted (Galvan-Ampudia & Testerink, 2011). Root phene-dependent RSA regulation by interphene and phene–environment interaction under stress are represented in Fig. 3.2.

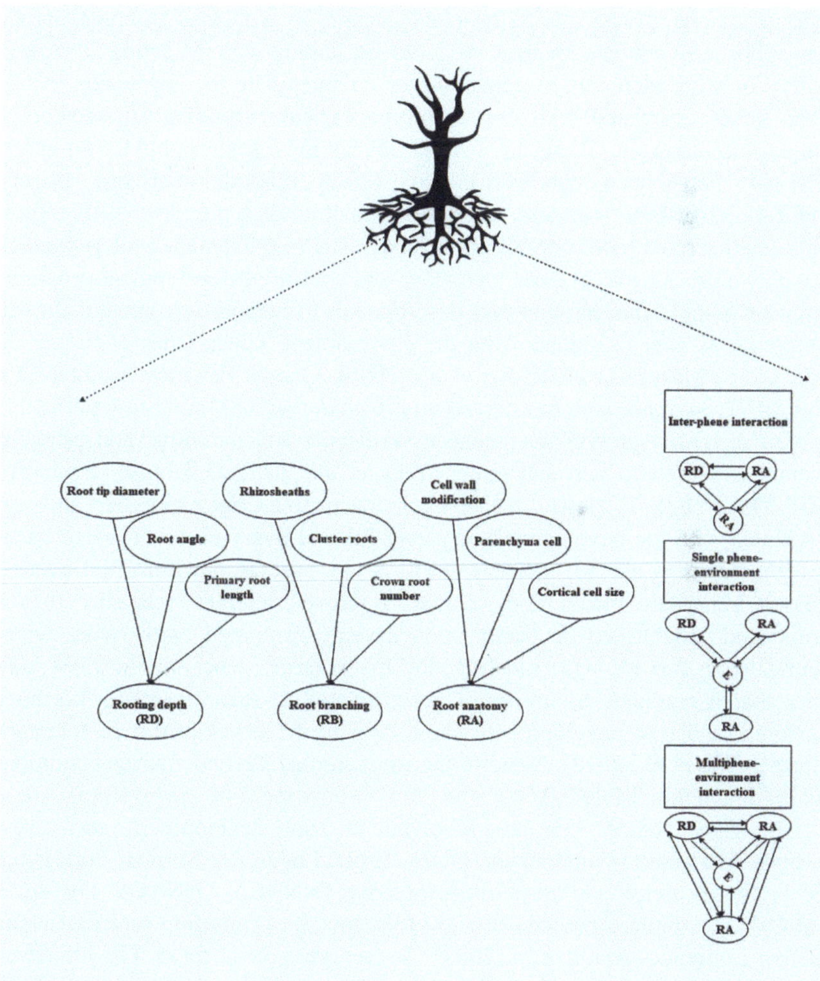

FIGURE 3.2 Simplified vie of root phene-dependent root system architecture regulation by interphene and phene–environment interaction under stress.

3.2.3 Omics and molecular aspects of root system architecture

Root anatomy and architecture in plant system is predominantly coordinates by genes and TFs. The recent "omics" techniques can be employed for the identification of RSA under unfavorable conditions. Under stress conditions RSA may regulate genes/QTLs/TFs. Integrated molecular studies enlighten complicated mechanisms, which can overcome the lacuna existing in our current knowledge about stress tolerance in plants. Abiotic stress tolerance is coordinated by a number of genes and TFs, that facilitate plants to endure unfavorable conditions, and these have potential options for their extensive application in molecular breeding. These TFs regulate the expression of diverse stress tolerance genes and in turn make the plant capable to withstand multiple abiotic stress. TFs may regulate an array of genes via linking with cis-acting DNA regulatory or other elements to upregulate or downregulate the expression of targeted genes associated with critical morphological, physiological, metabolic, and cellular changes. There are several TF families, such as NAC (no apical meristem cup-shaped cotyledon), Dof (DNA binding with one finger), APETALA2/EREBP (ethylene response element binding protein), WRKY (proteins characterized by the highly conserved WRKYGQK amino acid sequence), that play a crucial role in gene regulatory and control of development programs and root tissue differentiation and development in response to internal growth hormones, as well as signals from the environment (Chen, Chai, McIntyre, & Xue, 2018; Ma et al., 2018; Xu et al., 2006). One of the most studied TFs, ZmbZIP72, in maize was transferred into *Arabidopsis* and found to be effective in regulating root growth maintenance via deeper water mining, and provided tolerance to drought and salt stress (Ying et al., 2012). Likewise, a putative bZIP TF, ZmbZIP4, showed the upregulation of transcript expression and was associated with the development of young PR and ovary tissue in maize under drought (Opitz et al., 2014; Seeve et al., 2017). Further, the ZmbZIP4 overexpressed transgenic maize lines showed improved tolerance capacity to salt stress and established a better root system compared with wild types. Furthermore, genome-wide identification of the target genes of ZmbZIP4 indicates that it regulates a number of genes related to stress response, hormone signaling, and root development, which help in the development of tolerance capacity (Ma et al., 2018). Some of the most studied TF under stress conditions are highlighted in Table 3.2.

Similarly, NACs TFs are involved in root development and stress responses in many plant species (Olsen, Ernst, Leggio, & Skriver, 2005), and the higher expression of drought-responsive OsNAC5, OsNAC6, OsNAC9, and OsNAC10 genes provide drought tolerance by promoting root expansion (Saibo, Lourenco, & Oliveira, 2009). Overexpression of these TFs improved the root diameter and stress tolerance. However, it is expected that NAC-dependent root growth and abiotic stress tolerance are associated with increased root diameter, which affect the root penetration availability of

TABLE 3.2 Regulation of root system architecture traits by transcription factors under unfavorable situation.

Crop	Gene/ transcription factors	Effect on root	Stress tolerance	Reference
Rice	*SUB 1*	Crown root primordia development and outgrowth	Submergence	Xu et al. (2006)
Rice	*LACS2, SUB1*	Decreased LR and PR growth and lateral growth emergence, whereas ABA stimulate PR elongation and LR emergence	Drought	Wiegers, Cheer, and Silk (2009)
Wheat	*TaRNAC1*	Dehydration tolerance and enhance root growth	Drought	Chen et al. (2018)
Maize	*ZmbZIP4*	Longer root with increasing number of LR	Drought/salt	Ma et al. (2018)
Rice	*DRO1*	Enhance root growth	Drought	Uga et al. (2013)
Arabidopsis	*WRKY22*	Stand in flood	Submergence	Hsu et al. (2013)
Alfalfa	*HB1*	Lateral root primordia	Salt	Ariel et al. (2010)
Arabidopsis	*SOS1*	Cytosolic Ca^{2+} out of cell	Salt	Quintero, Ohta, Shi, Zhu, and Pardo (2002)
Arabidopsis	*NHX1*	Na^{+} sequestration in vacuoles	Salt	Gaxiola et al. (1999)
Rice	*ART1*	Modulating root elongation and cell wall elasticity	Heavy metals	Yamaji et al. (2009)

plants (Hu et al., 2013). Therefore the molecular understanding of TF/genes can be a pragmatic approach to future stress tolerance or to enhance stress potential in agricultural crops.

3.2.4 Signaling compound and mechanism in regulation of root system architecture

Plant root organogenesis includes cell division, cell differentiation, and regulation of signaling agents. Various signaling agents, such as Aux, ET, CKs, nitric oxide (NO), reactive oxygen species (ROS), and environmental constraints play crucial role in RSA. The RSA is determined by the PR, AR, LR, and root hair, helping the plant to withstand various environmental conditions and enable the plant to absorb water and nutrients (Casson & Lindsey, 2003). The primary root is originated from the basal cells of the embryo during the heart stage, while the pericycle and endodermis establish new LR meristems by triggering the cell cycle in a specific group of cells called "founder cells" (Scheres et al., 1994). LRs formation is promoted and regulated by auxin transporter AUX1 and PINs, and auxin signaling mediated by transcription regulators $SCF^{TIR1/AFB1s}$ complexes and ARF-Aux/IAA, and the *OsPIN1* gene, involved in the AR formation in rice (Xu, Zhu, Shou, & Wu, 2005). NPA (N-1-naphthylphthalamic acid), an inhibitor of auxin transporter, reduces the number of LRs. Nutrient availability and other hormones, such as CKs, ABA, and BRs, affect LR initiation. The Aux/CK balance states the root meristem size and regulates the cyclins and cyclin-dependent kinases (CDKs) with other genes that regulate the cell cycle (Malamy, 2005).

NO and ROS act as central hubs in the regulation of stresses and plant growth, and upregulate or downregulate the gene expression of auxin synthesis (Kushwaha et al., 2019). Additionally, NO is also involved in the auxin-mediated signaling pathway during LR and AR formation by modulating enzymes, and ROS regulate PR, while promoting LRs primordia in concert with NO. During LR formation, NO generation is dependent on indole-3-butyric acid (IBA)-induced nitrate reductase enzymes (NR). NO production improves nitrogen uptake capability and efficiency (Sun et al., 2015). NR regulates the synthesis of NO and its signaling is associated with phosphorylation of MAP kinase-6 in *Arabidopsis* (Wang, Du, & Song, 2011). In *Arabidopsis* NO synthase enzyme-dependent NO production interrupted under salt, while in the cold NR-dependent NO production leads to the upregulation of *NIA1* gene expression with modulation in proline accumulation (Zhao, Chen, Zhang, & Zhang, 2009; Zhao, Tian, & Zhang, 2007). Enhancement of NO accumulation promotes the AR number after *Arabidopsis* roots treatment with lipid compound alkamide N-isboutyl decanamide (Campos-Cuevas et al., 2008). Moreover, in AR development, auxin is linked to NO-dependent MAPK signaling (Pagnussat, Lanteri, Lombardo, & Lamattina, 2004). An endogenous level of NO is implicated in regulating the composition of lignin and determines the root branching in sunflowers (Monzón et al., 2014), while NO and ABA are involved in the initiation of root hair formation through auxin signaling (Lombardo & Lamattina, 2018). High NO level has reduced the PIN1-dependent auxin transport, decreased Aux concentration, and inhibited cell division in the meristematic tissue of root (Yuan & Huang, 2016). Li, Xue, Xu, Feng, and An (2009) observed that H_2O_2 application initiates ARs formation by downstream Aux signaling pathway. Ca^{2+} influx is essential for root tip development, which is stimulated by ROS (Takeda et al., 2008). There is a notable role of ROS concentration in *Arabidopsis* root tip in regulating the equilibrium between cell division and differentiation across meristemetic tissue and elongation zone (Hernández-Barrera et al., 2015; Tsukagoshi, Busch, & Benfey, 2010). H_2O_2 controls the regulation of TF UPBEAT1 (UPB1) and other gene such as *LTP5 (Lipid Transfer Protein 5*), *LTPG1* and *LTPG2 (*GPI-anchored *Lipid Transfer Protein 1* and 2), *PME44 (Pectin Methyl esterase 44*), and *GRP5 (Glycine-Rich Protein 5*) in root tip tissues, which are responsible for the regulation of peroxidase expression level (Mabuchi et al., 2018).

Under iron starving conditions, a higher NO level increased the ferric chelate reductase activity, thus improving Fe uptake in *Arabidopsis* and increased gene expression such as *Basic Helix–loop–helix transcription factor* (*FIT*) and *Ferric Reduction Oxidase 2* (*FRO2*) with downstream auxin action that helped proper development of roots (Chen et al., 2010). LR formation is altered by the action of TIR1 auxin receptor under low phosphorus (P) condition (Pérez-Torres et al., 2008). Auxin and ET are involved in root formation in tomato plants under flooded condition (Vidoz, Loreti, Mensuali, Alpi, & Perata, 2010). During cadmium stress situation, NO level is decreased and that affects crown root (CR) formation in rice seedling, which proves that NO is essential for the formation of root primordia (Xiong et al., 2009). NO donor sodium nitroprusside (SNP) brought about the formation of ARs, increased PR biomass accumulation PRBA in rice, and optimized the redox equilibrium of ascorbate and cell cycle dynamics under arsenate stress. NO synthase inhibitor NG-nitro-L-arginine methyl ester (L-NAME) and NO scavenger 2–4-carboxyphenyl-4,4,5,5-tetramethylimidazoline-1-oxyl 3 oxide (cPTIO) blocked the formation of ARs and reduced PRBA, either in the presence or absence of SNP (Kushwaha et al., 2019). Further study of the root system in plants will lead to an understanding of the action of a signaling agent and their response in the root architecture.

3.3 Molecular basis of improved root system architecture under multiple abiotic stresses

3.3.1 Improved root system architecture under salt stress

In modern agriculture, salt stress is a major challenge affecting 20% of the world's cultivated area and half of the irrigated farmland (OECD/Food & Agriculture Organization of the United Nations, 2015). In order to increase their salinity tolerance, plants may adopt multiple strategies, such as decreased growth rate, ion compartmentalization, or production of osmolytes (Munns & Gilliham, 2015). Plants suffer dehydration, nutrient shortages, membrane dysfunction, and oxidative stress under salt stress, resulting in tissue damage or early senescence (Essah, Davenport, & Tester, 2003; Katori et al., 2010). Exclusion of Na^+ from the shoot is linked with higher photosynthesis, growth, and yield retention rates (Munns & Tester, 2008). Plants must take up no more than 3% of the rhizospheric Na^+ to prevent accumulation of toxic sodium (Na^+) levels in shoots, and several protective processes are then triggered (Zhang et al., 2008). The role of HKT1 and CIPK13 has previously been identified in Na^+ exclusion, and thus they provide salinity tolerance (Roy et al., 2013). The root is the foremost plant organ sensor for salt stress; signal transduction to the shoot tissue and modulations in RSA provide salinity tolerance (Galvan-Ampudia & Testerink, 2011; Robbins, Trontin, Duan, & Dinneny, 2014). Saline conditions drastically affect the cell cycle in the root meristem and cause a reduction in growth (West, Inzé, & Beemster, 2004). Likewise, ABA signaling also interestingly is involved in the regeneration of LR development, and initiates the quiescence of LRs (Duan et al., 2013). The long-term study of salt stress on *A. thaliana* accession Col-0 (Julkowska et al., 2014) showed that salt decreases the PR growth more strongly than LR elongation, causing RSA remodeling. Similarly, other accessions of *Arabidopsis* show various RSA remodeling structures in response to salt stress, suggesting that natural fluctuations can be exploited by genome-wide association studies (GWAS) for targeted gene identification. A study on 347 HapMap population accessions of *Arabidopsis* (Horton et al., 2012; Li, Huang, Bergelson, Nordborg, & Borevitz, 2010) experimental results showed that an increase in phenotypic variation in PR and LR length is associated with candidate genetic loci (*HKT1* and *CYP79B2/B3*), which coordinates RSA remodeling in response to salt. The patterning of LR founder cell (FC) populations takes place at regular temporal intervals in *A. thaliana* and contributes to the specification of FC populations that subsequently grow into LR primordia (De Smet et al., 2007; Moreno-Risueno et al., 2010). The tissue-specific role of ABA signaling reveals a salt-mediated mechanism of root architectural regulation in the seedling. LR patterning is defined near the root tip, and later development occurs to initiate LR primordia from founder cells (Overvoorde, Fukaki, & Beeckman, 2010). In *Arabidopsis*, salt stress induces *NIT1* and *NIT2* nitrilase genes, which are associated with the biosynthesis of IAA. Furthermore, Wang, Li, and Li (2009) noted that increased salt exposure in PR tips suppressed the initiation of LR and organogenesis, which correlated with the concomitant decrease in expression of the auxin-inducible reporter DR5:GUS. Mutations in the transport or signaling of auxins (*AUX1*, *EIR1*, and *AXR1*) genes have been shown to decrease the efficiency of *Trichoderma* spp. in terms of their ability to promote root growth and development. DR5 was enabled by plant roots colonization by fungal hyphae:uidA inoculation, linked with enhanced proliferation of LR tip cells. Activation of DR5:uidA expression was demonstrated by the application of all three indole compounds using *Arabidopsis* seedlings, which showed the dose-dependent effect on biomass production. To examine RSA response under salt stress, trifoliate orange (*Poncirus trifoliata* L. Raf.) seedlings were inoculated with or without an arbuscular mycorrhizal (AM) fungus and it was hypothesized that a symbiotic relationship improved root H^+ effluxes and increased RSA of trifoliate orange seedlings, that was associated with host plant salinity tolerance (Wu & Zou, 2013).

RSA parameters and RSA gene expression was studied in cucumber seedlings grown under four treatments, and the results showed that treatment with NaCl + salicylic acid significantly increased RSA attributes such as total root length, root surface area, primary and secondary LR number, and total length. NaCl + SA treatment notably upregulated the expression of NAC1, NAC2, GL2, EXP, and downregulated RHD2, although both NaCl and SA treatments downregulated *GL2* and upregulated *RHD2* gene. Increased RSA gene expression, total length and number of primary and secondary LRs mediated by SA might be associated with the improved RSA (Miao et al., 2020). The oxidative-responsive regulatory mechanism includes autonomical expression (expressed even in the absence of osmotic stress) of a phytohormone ABA and a gene, *LRD2*, that regulates RSA (Deak & Malamy, 2005). In *Arabidopsis*, the osmotic ability of the soil changes with the depth of the root system, root biomass, root elongation rate, and the number of LRs (van der Weele, Spollen, Sharo, & Baskin, 2000). Cloning and characterization of qSOR1 (SOIL SURFACE ROOTING 1 quantitative trait locus) showed that a shallow root growth angle (RGA) in saline paddies could increase rice yields. Auxin, primarily expressed in root columella cells, and involved in gravitropism responses of roots, is negatively regulated by qSOR1. A homolog of DRO1 (DEEPER ROOTING 1), which is known to regulate RGA, was found to be qSOR1. Assays from CRISPR-Cas9 showed that other DRO1 homologs were also involved in RGA (Kitomi et al., 2020). Therefore RSA modulation has a very crucial role during salt stress adaptations.

3.3.2 Improved root system architecture under drought

Modulating RSA is an important adaptive response in plants subjected to drought stress in soil. RSA includes many traits governing the plant's productivity during normal and stressed conditions. Such traits include overall RSA such as root density, deep root weight, specific root length (SRL), and root surface area responsible for the acquisition of water and nutrients from the soil and the root hair (density and length), which are responsible for enlarging the surface of soil−root contact and improving uptake of soil water (Comas, Becker, Cruz, Byrne, & Dierig, 2013; Reddy et al., 2019).

Alteration of RSA depends on the variation of certain factors including the number, extension, and placement and growth direction of these traits. For instance, plants featuring effective root system size and deep-rooting ability could absorb more water from deep soil especially in arid and drought-prone regions. Likewise, plants known to tolerate drought features have extensive hairy roots which improve their penetration into smaller pores, thus contributing to more efficient water uptake and increased survival and tolerance to drought conditions (Barrios-Masias, Knipfer, & McElrone, 2015). These alterations of RSA contribute to developments of aboveground biomass and enhance tolerance to water stress.

The regulation of RSA under drought is controlled by a number of molecular mechanisms including various sensing and signaling events determining the soil–plant water status and changes in gene expression (Comas et al., 2013). Several specific genes regulating RSA have been identified through the application and advancements of molecular marker/QTLs (Janiak, Kwaśniewski, & Szarejko, 2016; Reddy et al., 2019). There is insufficient information regarding the phenotypes and different genetic backgrounds, which need to be quantified to better understand the molecular mechanism governing those traits (Khan et al., 2016). RSA has been the main target for research and breeding for drought-tolerant cultivars. In QTL analysis, RSA regulator DRO1 (DEEPER ROOTING 1) has been identified to affect the RGA (Uga et al., 2013). A japonica upland rice named Kinandang Patong, which contained a DRO1 copy, exhibits deeper RSA, whereas IR64, an indica lowland rice containing a truncated copy, exhibits shallower roots (Uga et al., 2013). The change in RSA on rice plants especially deep roots was found to be beneficial in improving harvest index, nutrient uptake, and CK flux from root to shoot during grain filling under drought (Arai-Sanoh et al., 2014).

In *Arabidopsis* and *Prunus* species orthologs of DRO1 influence the RSA as observed from deep-rooting phenotypes (Guseman, Webb, Srinivasan, & Dardick, 2017). The ability of DRO1 to give deeper rooting traits enables the doubling of the yield under moderate and severe drought stress (Uga et al., 2013). Thicker roots facilitate deeper penetration (Zheng et al., 2000). Proliferative rooting mainly characterizes LR number, root length density, and root surface area and volume. They have highly efficient water uptake capacity in water limited conditions. Chickpea lines with high root length density exhibit the higher yield potential and drought tolerance (Jaganathan et al., 2015). Drought stress induces plasticity responses such as increased number of fibrous roots, decreased LRs diameter, and alterations in root biomass (Meister, Rajani, Ruzicka, & Schachtman, 2014; Salazar-Henao, Vélez-Bermúdez, & Schmidt, 2016). Garay and Wilhelm (1983) observed that drought-tolerant lines exhibit better rooting depth than sensitive lines significantly under drought stress in legumes. Plants exhibiting aerenchyma formation as part of the adaptive mechanism save energy inputs for improved soil penetration and mitigate water deficit conditions (Addington et al., 2006). Metaxylems were associated with water transportation in the root and correlated with the improvement of water stress conditions in soybean (Prince et al., 2017).

Plant response to drought is manifested through different developmental processes and one important aspect is enhanced root growth while suppressing shoot growth, resulting in a greater root:shoot ratio (Xu et al., 2013). In wheat, a 20% faster root descent and more efficient RSA results in an improved yield of 0.32–0.44 t ha^{-1} due to the enhanced water extraction from deep layers (Lilley & Kirkegaard, 2011). In addition, 50% increase in root:shoot ratio was observed in drought stress on wheat (Rauf, Munir, ulHassan, Ahmad, & Afzal, 2007). Increased ABA levels corresponding to increased root growth was observed while simultaneously repressing stem growth (Sharp et al., 2004; Xu et al., 2013). Plants' ability to uptake efficient water from a large volume of soil is crucial for the reduction of yield penalties under drought conditions (Pask & Reynolds, 2013).

Efficient uptake of soil water with the increase in root length density is observed in sorghum, wheat, and barley under drought conditions (Carvalho, Azam-Ali, & Foulkes, 2014). In addition, a steeper root angle was associated with larger RL under water stress conditions (Lynch, 2013; Trachsel, Kaeppler, Brown, & Lynch, 2013). Also, the depth of root is positively correlated with root angles among different genotypes of maize (*Zea mays* L.) (Trachsel et al., 2013). Gene/QTL controlling RGA have been identified in rice (*Oryza sativa* L.) and sorghum (Lobet, Couvreur, Meunier, Javaux, & Draye, 2014; Mace et al., 2012; Uga et al., 2013). Overexpression of these genes regulating RSA cause steeper RGA for deep penetration into the soil profile (Uga et al., 2013). Furthermore, longer RL, that is, RL per unit of root biomass (specific RL, SRL) was observed to increase soil-water uptake (Comas et al., 2013). Drought tolerance in legumes correlates with the metaxylem vessels diameter and distribution, which governs the root hydraulic conductivity (Purushothaman et al., 2013). Deep root is a complex trait that is associated with RGA and RL (Araki, Morita, Tatsumi, & Iijima, 2002). The distribution of root in the soil is governed by the direction of root angle and it is recognized as one of the crucial traits for drought avoidance in sorghum (Mace et al., 2012), rice (Uga et al., 2013), and wheat (Christopher, Christopher, & Jennings, 2013). Therefore RSA are very crucial for improving root traits under drought stress and to reduce yield penalties.

3.3.3 Improved root system architecture under nutrient deficient conditions

In nature, plants have to withstand light, temperature, and nutrient variations. Plants undergo these biotic and abiotic stresses that occurs in multiple variations. A suitable example of cumulative stress would be heterogeneous abundance of different nutrients in the soil, known as nutrient deficit condition (Bouain, Krouk, Lacombe, & Rouached, 2019). In order to complete their life cycle, plants require 16 essential nutrients to produce a yield, particularly macronutrients, such as phosphorus (P) and nitrogen (N), and micronutrients such as zinc. It is most likely, therefore that a deficiency in P, N, or Zn has a serious effect on crop growth and production, posing a risk to food safety. A significant field of study has addressed the elementary function of mineral nutrient in the aboveground plant organs. However, there has been little knowledge on changes in RSA owing to nutrient deficiency (López-Bucio et al., 2003). Plant roots perform many important adaptive functions, including water and nutrient uptake from the soil. Changes in the RSA have a profound effect on a plant's ability to consume nutrients and water. Therefore it is of critical importance to consider that nutritional cross-talk modulates the growth and development of the plant root. Here we focused on how the RSA is modulated by the mineral nutritional cross-talk in plants. Liao, Ge, and Yan (2001) reported that under drought stress coupled with P deficiency, RSA performs a vital role for soil P acquisition. Expansion genes like *GmEXLB1* have been reported by Zhang et al. (2019) to play an important role in altering the RSA providing assistance for P acquisition in *Arabidopsis*. Interestingly, low P availability in soil triggers the expression of several genes associated with P uptake, further modulating the root architecture. Despite the current information on genes and their expression levels, there is still little clarity regarding the signaling pathway. There are very few studies demonstrating consensuses about the signal pathway along with the genetic expression of RSA under P deficiency (Mukatira, Liu, Varadarajan, & Raghothama, 2001; Zhang et al., 2019). A recently published research (Bouain et al., 2019; Cardoso, Pinto, & Paiva, 2019) reported a detailed study of genetic coexpression network in phosphate and aluminum providing an insight on nutrient signaling cross-talk.

In *Arabidopsis*, the variation in nitrate and phosphate supply have been shown to play opposite effects on LR development and elongation (Linkohr, Williamson, Fitter, & Leyser, 2002). *NRT1.1* and *NRT1.5* are nitrate carriers which drive root to shoot transport of NO_3^-. Phosphorus starvation induced the expression of *NRT1*. The upregulated expression resulted in the production of long PRs along with decreased LR density (Cui et al., 2019). Over recent years, the molecular studies of various nitrate transporters such as *NRT 1.1, NRT 2.1* have produced significant interest in the cross-talk of mineral nutrition and root architecture (Naz et al., 2019). But the functional recognition and classification of different genes associated with nitrate signaling pathway is the next step to understand the possible function in RSA.

Zinc is considered as the yield-limiting nutrient following nitrogen in several plant species such as rice (Quijano-Guerta, Kirk, Portugal, Bartolome,

& McLaren, 2002). Current approaches depend on excessive Zn application to manage low zinc levels in soil (Nie et al., 2019). Though the strategy is considered successful, current studies have reported nitrogen to be a significant participant in altering and accumulating zinc in plants (Erenoglu, Kutman, Ceylan, Yildiz, & Cakmak, 2011). Molecular studies, particularly transcriptomic data published by Varala et al. (2018) and Azevedo et al. (2016), provided important information on N and Zn interactions in plants. Similarly, Wang, Cheng, Chen, and Tsay (2018) provided a detailed description of upregulation of NRT 2.4 (nitrate transporter), a marker for N-deficit conditions, induced by Zn starvation condition. The study indicated that the transporter of nitrate (*NRT 2.4*) relies on the Zinc transcription factor (*bZIP19* and *bZIP23*) for its induction and that plants without it cannot react to Zn starvation. Further inquiries are needed to solve nitrogen and zinc relay signaling in relation to the transcription machinery. In order to unravel the cross-talk of various mineral nutrients, biochemical methods in tandem with the most advanced molecular mechanisms should make the full use of current phenotypic facilities for addressing crop improvement in an age of precision farming.

3.3.4 Improved root system architecture under heavy metal stress

With industrial development, heavy metal (HMs) contamination has become a global issue. HM-induced consequences include oxidative stress, genotoxicity, photosynthetic suppression, and suppression of root metabolism (Andresen & Küpper, 2013). HM toxicity affects grain yield, seed development, chlorophyll synthesis, hormone proteins, and membrane functions that affect root formation (Arao & Ae, 2003; Villiers et al., 2011). For a broad variety of physiological processes, HMs such as Fe, Cu, and Zn are required but may be excessively toxic. Other metals such as Cd, Al, As, Pb, and Hg are not required and can be extremely plant toxic. HM that collect and move in the soil can only impair plant growth and development, but may also concentrate in the human body via the food chain, presenting a major hazard to human health (Gielen, Remans, Vangronsveld, & Cuypers, 2012; Wagner, 1993). Lead has no established biological function but has numerous adverse effects on plant morphology, physiology, and biochemistry (Fahr et al., 2013). Lead soil deposition primarily impacts plants through root systems. Plant root can respond quickly to Pb stress by adjusting the growth rate and branching pattern (Fahr et al., 2013; Islam et al., 2007). Remodeling RSA plays important roles in the tolerance of tall fescue to Pb stress when grown hydroponically with Pb (0, 50, 200, and 1000 mg L^{-1}) for 2 weeks shows increased cell wall (CW) components, which lead to limiting high Pb in roots. Additionally, pectin, hemicellulose 1 (*HC1*), and hemicellulose 2 (*HC2*) uronic acid content increased under Pb tension. Pb adsorption potential also improved in vitro under H_2O_2 therapy. HO controls pectin and HC2 esterification degrees. Large concentrations of low esterified pectin and HC2 give more carboxyl groups, have more Pb binding sites, and limit more Pb in CW, which could improve tolerance of large fescue to Pb stress (Zhang et al., 2020). Poplars are promising for phytoremediation considering their strong capacity to absorb HMs into plant organs, rapid growth, deeply embedded characteristics, significant economic benefit, and low effect on food chains (Chen et al., 2014; Marmiroli et al., 2011). *Populus deltoids* is used as model to examine sexual variations in root architecture traits such as root volume, root surface area, diameter and biomass. Root Pb storage roots were lowest. In lower orders, except for root diameter, Pb also produced more important root length, surface area, and Pb concentration effects than higher root orders. Women are more prone to Pb tension than males. *P. deltoids* males can concentrate Pb in lower-order roots, suppressing high-root Pb transport (Xu et al., 2016). Aluminum tension in plants directly affects early root growth responses to Al. Aluminum exclusion is regarded by root exudates and rhizodepositions, as well as internal detoxification pathways. The essential function of Al chelation by organic acid anions and flavonoid-type phenolics outside and within plant cells is granted particular importance (Barcelo & Poschenrieder, 2002). Fractal geometry is used to describe the complexity of HMs root system anatomy, including Cd, Cu, and Zn. A calculation called the RCI is a ratio dependent on fractal dimension (FD) before and after HMs exposure. FDs is measured using a box-counting method with digitized and skeletonized photographs of wheat plant roots developed in hydroculture. RCI and relative weight were calculated at various concentrations of Cd, Cu, and Zn showing decreased Cd tension RCI at 0.01 and 0.05, all Cu concentrations, and promotion at all zinc concentrations. No improvements were observed in traditional relative weight measurements at low concentrations of Cu, Cd, and Zn. RCI was more responsive and accurate in representing HM impact than the standard scale. It means that RCI may be an important indicator of the harmful and positive effects of HM on the RSA function under HMs concentrations (Li, Rajagoplan, & Kadono, 2019). Plants use complex gene regulation pathways under HMs stress. High-throughput sequencing has contributed to a growing amount of HMs-responsive micro-RNAs in plants. Metal-regulated Mi-RNAs and their target genes are part of a dynamic regulatory network that regulates numerous biological processes, including HMs absorption and distribution, protein folding and assembly, metal chelation, ROS scavenging, hormone signaling, and microRNA biogenesis. The amount of isopentyl transferase (*IPT*) transcripts was upregulated in soybean after Cd application, leading to increased levels of CK and consequently increased resistance and inhibition of PR elongation under Cd stress (Merewitz, Gianfagna, & Huang, 2011). Vacuolar antiporters are essential elements in cation sequestration mediation (Busch & Saier, 2002). These antiporters are energized across the proton gradient through the vacuolar membrane, enabling the rapid transportation of cations through the vacuole. CAXs (for cation exchanger), among such vacuolar antiporters, was involved in several cellular reactions. *Arabidopsis* vacuolar cation exchangers (CAXs) play a vital function in mediating the infusion of cation into the vacuole. The expression is increased in *Arabidopsis CAX4* in root apex and LR primordia if the amount of Ni^{2+} or Mn^{2+} is increased. CAX4 is an antiporter that performs an essential root production role under conditions of HM stress (Mei et al., 2009).

3.3.5 Root system architecture improved under future climate change scenario

Root-traits modeling under climate change, collectively with high-throughput phenotyping, molecular techniques, and robust databases, may offer an important base for a truly inclusive "green revolution." Awareness on molecular genetics of RSA can be manipulated to improve their tolerance to multiple abiotic stresses (Khan et al., 2016). Currently in *Arabidopsis*, RSA variations have been mapped under different nutrient concentrations (Gruber, Giehl, Friedel, & von Wirén, 2013). Each nutrient concentration had a diverse response to RSA development. Auspiciously, the RSA responses to these nutrient deficiencies have been mapped extensively in crop species, providing us many insights into the understanding of RSA development (Koevoets, Venema, Elzenga, & Testerink, 2016). RSA play a crucial role in crop plants' responses to abiotic stresses and climate change conditions. Underground roots are the first to sense abiotic stresses and alter their genetics for postembryonic development (Lynch, 1995). Changes in RSA therefore affect the expansion and development of stem biomass by altering carbon partioning to shoots and activating signaling pathways involving PGRs, proteins, and RNAs (DoVale & Fritsche-Neto, 2015; Paez-Garcia et al., 2015). However, roots ultimately control leaf stomatal conductance and affect blade posture and photosynthetic rate under unfavorable situations. Ideotype breeding considers the relationships between multiple root traits along with individual traits (Rasmusson, 1987). A potato root "ideotype" is architecturally plastic, long, deep, and thin, as well as being able to self-protect against biotic stress. Ideotypes with root plasticity in terms of their architecture are preferred due to their ability to withstand the drought stress, along with transport of ABA to shoots, which facilitates the proficient closing of stomata. This is further backed by a sequence of anatomical changes like aerenchyma formation and/or increasing RL and/or increasing root surface area or AR exudates and/or symbioses in the rhizosphere (Wishart et al., 2013).

In sweet potato an association was found between RSA, biotic stress, and availability of N thereby specifying the stipulation of a more systematic approach toward managing yield constraints (Villordon & Clark, 2014). An ideotype for sweet potato, therefore in addition to being efficient in water and mineral uptake and utilization, and being pathogen resistance as indicated above for potato, should even be tolerant to flood and have evenhanded biomass partitioning between below-ground and aboveground biomass. A research program studied root phenes in terms of breeding approaches for manipulation of RSA with regard to future climate change (York, Nord, & Lynch, 2013). Model-assisted phenotyping may broaden our understanding of the complexity of the roots' stress signaling pathways (Chickarmane et al., 2010). According to a dynamic RSA model, plants with high transpiration rate had more root axes and more available water distribution. Water-saving genotypes which are tolerant to drought stress had lower root conductance than the drought susceptible genotypes (Dunbabin et al., 2013; Leitner, Felderer, Vontobel, & Schnepf, 2014). QTL analyses and gene mutants are helpful to identify different specific genes associated with RSA in crop plants. In rice, *Deeper Rooting 1* (DRO1) gene was identified to affect root angle and was introduced into elite lines of rice and sorghum through backcrossing, whereas protein kinase gene *Phosphorus Starvation 1* (PSTOL1), a *pup-1*-specific permits early root growth for P-acquisition (Gamuyao et al., 2012; Hufnagel et al., 2014).

CO_2 enrichment stimulates primary biomass in forests (Norby et al., 2005). However, the capability of forest trees to maintain high biomass in response to elevated CO_2 probably depends on the capacity of their RSA to make available plenty of belowground resources (Luo et al., 2004). RSA and root function efficacy is directly interconnected (Harper, Jones, & Sackville Hamilton, 1991; Lynch, 1995). Just as the leaves array and position on a branch control the CO_2 and sunlight capture, like the architecture of fine roots affects minerals and water absorption (Bentley et al., 2013). Addressing both RSA complexity and performing multiple tasks simultaneously can be difficult to link RSA and function (Farnsworth & Niklas, 1995; Pregitzer et al., 2002). The individual root function is determined by both the connectivity among root segments and the location of a particular portion on the larger root branching system (root order), (Pregitzer et al., 2002). From the root architectural point of view, the function (absorption or transport) and durability (days to years) of roots can be affected largely based on the location and type of root segment (distal and/or first-order root and/or an interior and/or fifth-order root) (Eissenstat & Volder, 2005; Guo et al., 2008; Pregitzer et al., 2002). Therefore RSA improvement has an important role during multiple abiotic stresses and the integrated mechanism of RSA traits improvements under unfavorable conditions are represented in Fig. 3.3.

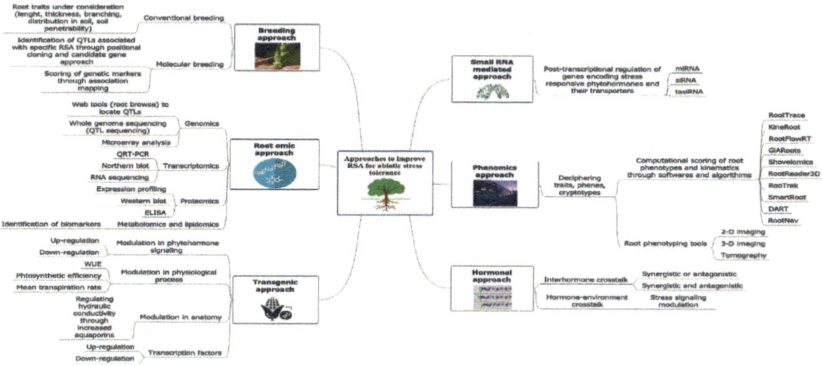

FIGURE 3.3 Simplified view of integrated approach to improve root system architecture traits under abiotic stresses. This includes the omics, transgenic, phenomics, hormones, and breeding, which improve root traits and can be helpful for future root improvement research.

3.4 Summary

The root is an important organ of a plant, which is specified for water and mineral nutrition transport, with a support function in soil. The study of RSA traits is very crucial for plant improvement program, but due to their below soil nature, the research on root traits is very complex. Through recent advancement in plant physiological and imaging technology, it is possible to estimate the actual root growth and developments. In this chapter we discussed the hormonal regulation of RSA traits at the molecular level and how hormonal cross-talk and signaling regulate root growth under various climatic conditions. Then we explored the environment regulation of RSA and discussed thoroughly how critically they regulate root growth under diverse situations. Later in this chapter we discussed the molecular mechanisms of RSA traits and we explored how root traits at molecular levels can be improved under multiple abiotic stresses and how these strategies are crucial for future crop improvement in adverse climatic situations.

References

Addington, R. N., Donovan, L. A., Mitchell, R. J., Vose, J. M., Pecot, S. D., Jack, S. B., ... Oren, R. (2006). Adjustments in hydraulic architecture of *Pinus palustris* maintain similar stomatal conductance in xeric and mesic habitats. *Plant Cell and Environment*, *29*, 535–545.

Andresen, E., & Küpper, H. (2013). Cadmium toxicity in plants. In A. Siegel, H. Siegel, & R. K. O. Siegel (Eds.), *Cadmium: From toxicity to essentiality* (pp. 395–413). Dordrecht: Springer.

Arai-Sanoh, Y., Takai, T., Yoshinaga, S., Nakano, H., Kojima, M., & Sakakibara, H. (2014). Deep rooting conferred by DEEPER ROOTING 1 enhances rice yield in paddy fields. *Scientific Reports*, *4*, 5563. Available from https://doi.org/10.1038/srep05563.

Araki, H., Morita, S., Tatsumi, J., & Iijima, M. (2002). Physiomorphological analysis on axile root growth in upland rice. *Plant Production Science*, *5*, 286–293.

Arao, T., & Ae, N. (2003). Genotypic variations in cadmium levels of rice grain. *Soil Science and Plant Nutrition*, *49*, 473–479.

Ariel, F., Diet, A., Verdenaud, M., Gruber, V., Frugier, F., & Chan, R. (2010). Environmental regulation of lateral root emergence in *Medicago truncatula* requires the HD-Zip I transcription factor HB1. *The Plant Cell*, *22*, 2171–2183.

Azevedo, H., Azinheiro, S. G., Muñoz-Mérida, A., Castro, P. H., Huettel, B., Aarts, M. G., & Assunção, A. G. (2016). Transcriptomic profiling of *Arabidopsis* gene expression in response to varying micronutrient zinc supply. *Genomics Data*, *7*, 256–258.

Barcelo, J., & Poschenrieder, C. (2002). Fast root growth responses, root exudates, and internal detoxification as clues to the mechanisms of aluminium toxicity and resistance: A review. *Environmental and Experimental Botany*, *48*, 75–92.

Barrios-Masias, F. H., Knipfer, T., & McElrone, A. J. (2015). Differential responses of grapevine rootstocks to water stress are associated with adjustments in fine root hydraulic physiology and suberization. *Journal of Experimental Botany*, *66*(19), 6069–6078.

Bentley, L. P., Stegen, J. C., Savage, V. M., Smith, D. D., von Allmen, E. I., Sperry, J. S., … Enquist, B. J. (2013). An empirical assessment of tree branching networks and implications for plant allometric scaling models. *Ecology Letters*, *16*, 1069–1078.

Bouain, N., Krouk, G., Lacombe, B., & Rouached, H. (2019). Getting to the root of plant mineral nutrition: Combinatorial nutrient stresses reveal emergent properties. *Trends in Plant Science*, *24*(6), 542–552.

Boukcim, H., Pagès, L., & Mousain, D. (2006). Local NO_3^- or NH_4^+ supply modifies the root system architecture of *Cedrus atlantica* seedlings grown in a split-root device. *Journal of Plant Physiology*, *163*, 1293–1304.

Bray, A. L., & Topp, C. N. (2018). The quantitative genetic control of root architecture in maize. *Plant and Cell Physiology*, *59*, 1919–1930.

Busch, W., & Saier, M. H. (2002). The transporter classification (TC) system. *Critical Reviews in Biochemistry and Molecular Biology*, *37*, 287–337.

Campos-Cuevas, J. C., Pelagio-Flores, R., Raya-González, J., Méndez-Bravo, A., Ortiz-Castro, R., & López-Bucio, J. (2008). Tissue culture of *Arabidopsis thaliana* explants reveals a stimulatory effect of alkamides on adventitious root formation and nitric oxide accumulation. *Plant Science (Shannon, Ireland)*, *174*, 165–173.

Canales, J., Contreras-López, O., Álvarez, J. M., & Gutiérrez, R. A. (2017). Nitrate induction of root hair density is mediated by TGA 1/TGA 4 and CPC transcription factors in *Arabidopsis thaliana*. *The Plant Journal*, *92*, 305–316.

Cardoso, T. B., Pinto, R. T., & Paiva, L. V. (2019). Analysis of gene co-expression networks of phosphate starvation and aluminium toxicity responses in Populus spp. *PLoS One*, *14*, e0223217.

Carvalho, P., Azam-Ali, S., & Foulkes, M. J. (2014). Quantifying relationships between rooting traits and water uptake under drought in Mediterranean barley and durum wheat. *Journal Integrative Plant Biology*, *56*, 455–469. Available from https://doi.org/10.1111/jipb.12109.

Casson, S. A., & Lindsey, K. (2003). Genes and signaling in root development. *New Phytologist*, *158*, 11–38.

Chang, H., Chen, D., Kam, J., Richardson, T., Drenth, J., Guo, X., … Xue, G. P. (2016). Abiotic stress upregulated TaZFP34 represses the expression of type-B response regulator and SHY2 genes and enhances root to shoot ratio in wheat. *Plant Science*, *252*, 88–102.

Chang, L., Ramireddy, E., & Schmülling, T. (2015). Cytokinin as a positional cue regulating lateral root spacing in *Arabidopsis*. *Journal of Experimental Botany*, *66*, 4759–4768.

Chen, D., Chai, S., McIntyre, C. L., & Xue, G.-P. (2018). Overexpression of a predominantly root-expressed NAC transcription factor in wheat roots enhances root length, biomass and drought tolerance. *Plant Cell Reports*, *37*, 225–237.

Chen, D., Richardson, T., Chai, S., Lynne McIntyre, C., Rae, A. L., & Xue, G. P. (2016). Drought-up-regulated TaNAC69-1 is a transcriptional repressor of TaSHY2 and TaIAA7, and enhances root length and biomass in wheat. *Plant and Cell Physiology*, *57*, 2076–2090.

Chen, L., Gao, S., Zhu, P., Liu, Y., Hu, T., & Zhang, J. (2014). Comparative study of metal resistance and accumulation of lead and zinc in two poplars. *Physiologia Plantarum*, *151*, 390–405.

Chen, Q., Sun, J., Zhai, Q., Zhou, W., Qi, L., & Xu, L. (2011). The basic helix-loop-helix transcription factor MYC2 directly represses PLETHORA expression during jasmonate-mediated modulation of the root stem cell niche in *Arabidopsis*. *The Plant Cell*, *23*, 3335–3352.

Chen, W. W., Yang, J. L., Qin, C., Jin, C. W., Mo, J. H., Ye, T., & Zheng, S. J. (2010). Nitric oxide acts downstream of auxin to trigger root ferric-chelate reductase activity in response to iron deficiency in *Arabidopsis*. *Plant Physiology*, *154*, 810–819.

Chickarmane, V., Roeder, A. H. K., Tarr, P. T., Cunha, A., Tobin, C., & Meyerowitz, E. M. (2010). Computational morphodynamics: A modeling framework to understand plant growth. *Annual Review of Plant Biology*, *61*, 65–87.

Cho, H., Ryu, H., Rho, S., Hill, K., Smith, S., Audenaert, D., ... Bennett, M. J. (2014). A secreted peptide acts on BIN2-mediated phosphorylation of ARFs to potentiate auxin response during lateral root development. *Nature Cell Biology*, *16*, 66–76.

Christopher, J., Christopher, M., & Jennings, R. (2013). QTL for root angle and number in a population developed from bread wheats (*Triticum aestivum*) with contrasting adaptation to water-limited environments. *Theoretical and Applied Genetics*, *126*, 1563–1574.

Comas, L. H., Becker, S. R., Cruz, V. M. V., Byrne, P. F., & Dierig, D. A. (2013). Root traits contributing to plant productivity under drought. *Frontiers Plant Sciences*, *4*, 442. Available from https://doi.org/10.3389/fpls.2013.00442.

Cui, Y. N., Li, X. T., Yuan, J. Z., Wang, F. Z., Wang, S. M., & Ma, Q. (2019). Nitrate transporter NPF7. 3/NRT1. 5 plays an essential role in regulating phosphate deficiency responses in *Arabidopsis*. *Biochemical and Biophysical Research Communications*, *508*, 314–319.

Deak, K. I., & Malamy, J. (2005). Osmotic regulation of root system architecture. *The Plant Journal*, *43*, 17–28.

De Smet, I., Tetsumura, T., De Rybel, B., dit Frey, N. F., Laplaze, L., Casimiro, I., & Inzé, D. (2007). Auxin-dependent regulation of lateral root positioning in the basal meristem of *Arabidopsis*. *Development (Cambridge, England)*, *134*, 681–690.

DoVale, J., & Fritsche-Neto, R. (2015). Root phenomics. In R. Fritsche-Neto, & A. Borém (Eds.), *Phenomics* (pp. 49–66). Cham: Springer International Publishing.

Duan, L., Dietrich, D., Ng, C. H., Chan, P. M. Y., Bhalerao, R., Bennett, M. J., & Dinneny, J. R. (2013). Endodermal ABA signaling promotes lateral root quiescence during salt stress in *Arabidopsis* seedlings. *The Plant Cell*, *25*, 324–341.

Dunbabin, V. M., Postma, J. A., Schnepf, A., Pagès, L., Javaux, M., & Wu, L. (2013). Modelling root–soil interactions using three-dimensional models of root growth, architecture and function. *Plant and Soil*, *372*, 93–124.

Eissenstat, D. M., & Volder, A. (2005). The efficiency of nutrient acquisition over the life of a root. *Ecological Studies*, *181*, 185–220.

Erenoglu, E. B., Kutman, U. B., Ceylan, Y., Yildiz, B., & Cakmak, I. (2011). Improved nitrogen nutrition enhances root uptake, root-to-shoot translocation and remobilization of zinc (65Zn) in wheat. *New Phytologist*, *189*, 438–448.

Essah, P. A., Davenport, R., & Tester, M. (2003). Sodium influx and accumulation in Arabidopsis. *Plant Physiology, 133*, 307–318.

Fahr, M., Laplaze, L., Bendaou, N., Hocher, V., El Mzibri, M., Bogusz, D., & Smouni, A. (2013). Effect of lead on root growth. *Frontiers in Plant Science, 4*, 1–7.

Farnsworth, K. D., & Niklas, K. J. (1995). Theories of optimization, form and function in branching architecture in plants. *Functional Ecology, 9*, 355–363.

Fattorini, L., Hause, B., Gutierrez, L., Veloccia, A., Della Rovere, F., & Piacentini, D. (2018). Jasmonate promotes auxin-induced adventitious rooting indark-grown *Arabidopsis thaliana* seedlings and stem thin cell layers by a crosstalk with ethylene signalling and a modulation of xylogenesis. *BMC Plant Biology, 18*, 182.

Galvan-Ampudia, C. S., & Testerink, C. (2011). Salt stress signals shape the plant root. *Current Opinion in Plant Biology, 14*, 296–302.

Gamuyao, R., Chin, J. H., Pariasca-Tanaka, J., Pesaresi, P., Catausan, S., & Dalid, C. (2012). The protein kinase Pstol1 from traditional rice confers tolerance of phosphorus deficiency. *Nature, 488*, 535–539.

Garay, A. F., & Wilhelm, W. W. (1983). Root system characteristics of two soybean isolines undergoing water stress conditions. *Agronomy Journal, 75*, 973–977.

Gaxiola, R. A., Rao, R., Sherman, A., Grisafi, P., Alper, S. L., & Fink, G. R. (1999). The *Arabidopsis thaliana* proton transporters, AtNhx1 and Avp1, can function in cation detoxification in yeast. *Proceedings of the National Academy of Sciences of the United States of America, 96*, 1480–1485.

Gielen, H., Remans, T., Vangronsveld, J., & Cuypers, A. (2012). MicroRNAs in metal stress: Specific roles or secondary responses? *International Journal of Molecular Sciences, 13*, 15826–15847.

Grossman, J. D., & Rice, K. J. (2012). Evolution of root plasticity responses to variation in soil nutrient distribution and concentration. *Evolutionary Applications, 5*, 850–857.

Gruber, B. D., Giehl, R. F., Friedel, S., & von Wirén, N. (2013). Plasticity of the *Arabidopsis* root system under nutrient deficiencies. *Plant Physiology, 163*, 161–179.

Guan, L., Tayengwa, R., Cheng, Z. M., Peer, W. A., Murphy, A. S., & Zhao, M. (2019). Auxin regulates adventitious root formation in tomato cuttings. *BMC Plant Biology, 19*, 435.

Guo, D., Xia, M., Wei, X., Chang, W., Liu, Y., & Wang, Z. (2008). Anatomical traits associated with absorption and mycorrhizal colonization are linked to root branch order in twenty-three Chinese temperate tree species. *New Phytologist, 180*, 673–683.

Guseman, J. M., Webb, K., Srinivasan, C., & Dardick, C. (2017). DRO 1 influences root system architecture in *Arabidopsis* and *Prunus* species. *The Plant Journal, 89*(6), 1093–1105.

Gutierrez, L., Mongelard, G., Flokova, K., Pacurar, D. I., Novak, O., & Staswick, P. (2012). Auxin controls *Arabidopsis adventitious* root initiation by regulating jasmonic acid homeostasis. *The Plant Cell, 24*, 2515–2527.

Harper, J., Jones, M., & Sackville Hamilton, N. (1991). The evolution of roots and the problems of analysing their behaviour. In D. Atkinson (Ed.), *Plant root growth, an ecological perspective* (pp. 3–22). Oxford: Blackwell Scientific Publications.

He, W., Brumos, J., Li, H., Ji, Y., Ke, M., & Gong, X. (2011). A small-molecule screen identifies L-kynurenine as a competitive inhibitor of TAA1/TAR activity in ethylene-directed auxin biosynthesis and root growth in Arabidopsis. *The Plant Cell, 23*, 3944–3960.

Hernández-Barrera, A., Velarde-Buendía, A., Zepeda, I., Sanchez, F., Quinto, C., Sánchez-Lopez, R., & Cardenas, L. (2015). Hyper, a hydrogen peroxide sensor, indicates the sensitivity of the *Arabidopsis* root elongation zone to aluminum treatment. *Sensors, 15*, 855–867.

Hodge, A., Berta, G., Doussan, C., Merchan, F., & Crespi, M. (2009). Plant root growth, architecture and function. *Plant and Soil, 321*, 153–187.

Horton, M. W., Hancock, A. M., Huang, Y. S., Toomajian, C., Atwell, S., Auton, A., & Nordborg, M. (2012). Genome-wide patterns of genetic variation in worldwide *Arabidopsis thaliana* accessions from the RegMap panel. *Nature Genetics, 44*, 212–216.

Hsu, F. C., Chou, M. Y., Chou, S. J., Li, Y. R., Peng, H. P., & Shih, M. C. (2013). Submergence confers immunity mediated by the WRKY22 transcription factor in *Arabidopsis. The Plant Cell, 25*, 2699–2713.

Hu, W., Huang, C., Deng, X., Zhou, S., Chen, L., Li, Y., ... Wang, Y. (2013). TaASR1, a transcription factor gene in wheat, confers drought stress tolerance in transgenic tobacco. *Plant, Cell & Environment, 36*, 1449–1464.

Hufnagel, B., de Sousa, S. M., Assis, L., Guimaraes, C. T., Leiser, W., & Azevedo, G. C. (2014). Duplicate and conquer: Multiple homologs of PHOSPHORUS-STARVATION TOLERANCE1 enhance phosphorus acquisition and sorghum performance on low-phosphorus soils. *Plant Physiology, 166*, 659–677.

Islam, E., Yang, X., Li, T., Liu, D., Jin, X., & Meng, F. (2007). Effect of Pb toxicity on root morphology, physiology and ultrastructure in the two ecotypes of *Elsholtzia argyi. Journal of Hazardous Materials, 147*, 806–816.

Jaganathan, D., Thudi, M., Kale, S., Azam, S., Roorkiwal, M., Gaur, P. M., ... Varshney, R. K. (2015). Genotyping-by-sequencing based intra-specific genetic map refines a 'QTL-hotspot' region for drought tolerance in chickpea. *Molecular Genetics and Genomics, 290*, 559–571.

Janiak, A., Kwaśniewski, A. M., & Szarejko, I. (2016). Gene expression regulation in roots under drought. *Journal of Experimental Botany, 67*(4), 1003–1014.

Jeong, J. S., Kim, Y. S., Baek, K. H., Jung, H., Ha, S. H., Do Choi, Y., ... Kim, J. K. (2010). Root-specific expression of OsNAC10 improves drought tolerance and grain yield in rice under field drought conditions. *Plant Physiology, 153*, 185–197.

Jin, W., Aufrecht, J., Patino-Ramirez, F., Cabral, H., Arson, C., & Retterer, S. T. (2020). Modeling root system growth around obstacles. *Scientific Reports, 10*, 1–16.

Julkowska, M. M., Hoefsloot, H. C., Mol, S., Feron, R., de Boer, G. J., Haring, M. A., & Testerink, C. (2014). Capturing *Arabidopsis* root architecture dynamics with ROOT-FIT reveals diversity in responses to salinity. *Plant Physiology, 166*, 1387–1402.

Katori, T., Ikeda, A., Iuchi, S., Kobayashi, M., Shinozaki, K., Maehashi, K., & Taji, T. (2010). Dissecting the genetic control of natural variation in salt tolerance of *Arabidopsis thaliana* accessions. *Journal of Experimental Botany, 61*, 1125–1138.

Khaledian, Y., Kiani, F., Ebrahimi, S., Brevik, E. C., & Aitkenhead-Peterson, J. (2017). Assessment and monitoring of soil degradation during land use change using multivariate analysis. *Land Degradation & Development, 28*, 128–141.

Khan, M. A., Gemenet, D. C., & Villordon, A. (2016). Root system architecture and abiotic stress tolerance: Current knowledge in root and tuber crops. *Frontiers in Plant Science, 7*, 1584.

Kim, G., Jang, S., Yoon, E. K., Lee, S. A., Dhar, S., & Kim, J. (2018). Involvement of Pyridoxine/Pyridoxamine 5'-Phosphate Oxidase (PDX3) inethylene-induced auxin biosynthesis in the *Arabidopsis* root. *Molecules and Cells, 41*, 1033–1044.

Kitomi, Y., Hanzawa, E., Kuya, N., Inoue, H., Hara, N., Kawai, S., & Sakamoto, S. (2020). Root angle modifications by the DRO1 homolog improve rice yields in saline paddy fields. *Proceedings of the National Academy of Sciences, 117*, 21242–21250.

Koevoets, I. T., Venema, J. H., Elzenga, J. T., & Testerink, C. (2016). Roots withstanding their environment: Exploiting root system architecture responses to abiotic stress to improve crop tolerance. *Frontiers in Plant Science, 7*, 1335.

Kushwaha, B. K., Singh, S., Tripathi, D. K., Sharma, S., Prasad, S. M., Chauhan, D. K., & Singh, V. P. (2019). New adventitious root formation and primary root biomass accumulation are regulated by nitric oxide and reactive oxygen species in rice seedlings under arsenate stress. *Journal of Hazardous Materials*, *361*, 134–140.

Lee, H. W., & Kim, J. (2013). EXPANSINA17 up-regulated by LBD18/ASL20 promotes lateral root formation during the auxin response. *Plant and Cell Physiology*, *54*, 1600–1611.

Leitner, D., Felderer, B., Vontobel, P., & Schnepf, A. (2014). Recovering root system traits using image analysis exemplified by two-dimensional neutron radiography images of lupine. *Plant Physiology*, *164*, 24–35.

Li, G., Tillard, P., Gojon, A., & Maurel, C. (2016). Dual regulation of root hydraulic conductivity and plasma membrane aquaporins by plant nitrate accumulation and high-affinity nitrate transporter NRT2. 1. *Plant and Cell Physiology*, *57*, 733–742.

Li, S. W., Xue, L., Xu, S., Feng, H., & An, L. (2009). Hydrogen peroxide acts as a signal molecule in the adventitious root formation of mung bean seedlings. *Environmental and Experimental Botany*, *65*, 63–71.

Li, T., Rajagoplan, U. M., & Kadono, H. (2019). Fractal based complexity analysis of wheat root system under different heavy metals. *Plant Biotechnology*, 19–0301.

Li, Y., Huang, Y., Bergelson, J., Nordborg, M., & Borevitz, J. O. (2010). Association mapping of local climate-sensitive quantitative trait loci in *Arabidopsis thaliana*. *Proceedings of the National Academy of Sciences*, *107*, 21199–21204.

Liang, C., Li, A., Yu, H., Li, W., Liang, C., Guo, S., ... Chu, C. (2017). Melatonin regulates root architecture by modulating auxin response in rice. *Frontiers in Plant Science*, *8*, 134.

Liao, H., Ge, Z., & Yan, X. (2001). Ideal root architecture for phosphorus acquisition of plants under water and phosphorus coupled stresses: From simulation to application. *Chinese Science Bulletin*, *46*, 1346–1351.

Lilley, J., & Kirkegaard, J. (2011). Benefits of increased soil exploration by wheat roots. *Field Crops Research*, *122*, 118–130. Available from https://doi.org/10.1016/j.fcr.2011.03.010.

Linkohr, B. I., Williamson, L. C., Fitter, A. H., & Leyser, H. O. (2002). Nitrate and phosphate availability and distribution have different effects on root system architecture of *Arabidopsis*. *The Plant Journal*, *29*, 751–760.

Lobet, G., Couvreur, V., Meunier, F., Javaux, M., & Draye, X. (2014). Plant water uptake in drying soils. *Plant Physiology*, *164*, 1619–1627. Available from https://doi.org/10.1104/pp.113.233486.

Lombardo, M. C., & Lamattina, L. (2018). Abscisic acid and nitric oxide modulate cytoskeleton organization, root hair growth and ectopic hair formation in Arabidopsis. *Nitric Oxide: Biology and Chemistry / Official Journal of the Nitric Oxide Society*, *80*, 89–97.

López-Bucio, J., Cruz-Ramırez, A., & Herrera-Estrella, L. (2003). The role of nutrient availability in regulating root architecture. *Current Opinion in Plant Biology*, *6*, 280–287.

Luo, Y., Su, B., Currie, W. S., Dukes, J. S., Finzi, A., Hartwig, U., ... Parton, W. J. (2004). Progressive nitrogen limitation of ecosystem responses to rising atmospheric carbon dioxide. *Bioscience*, *54*, 731–739.

Lynch, J. (1995). Root architecture and plant productivity. *Plant Physiology*, *109*, 7–13.

Lynch, J. P. (2011). Root phenes for enhanced soil exploration and phosphorus acquisition: Tools for future crops. *Plant Physiology*, *156*, 1041–1049.

Lynch, J. P. (2013). Steep, cheap and deep: An ideotype to optimize water and N acquisition by maize root systems. *Annals of Botany*, *112*, 347–357. Available from https://doi.org/10.1093/aob/mcs293.

Ma, H., Liu, C., Li, Z., Ran, Q., Xie, G., Wang, B., ... Zhang, J. (2018). ZmbZIP4 contributes to stress resistance in maize by regulating ABA synthesis and root development. *Plant Physiology*, *178*, 753–770.

Mabuchi, K., Maki, H., Itaya, T., Suzuki, T., Nomoto, M., Sakaoka, S., & Tsukagoshi, H. (2018). MYB30 links ROS signaling, root cell elongation, and plant immune responses. *Proceedings of the National Academy of Sciences*, *115*, 4710–4719.

Mace, E. S., Singh, V., Van Oosterom, E. J., Hammer, G. L., Hunt, C. H., & Jordan, D. R. (2012). QTL for nodal root angle in sorghum (*Sorghum bicolor* L. Moench) co-locate with QTL for traits associated with drought adaptation. *Theoretical and Applied Genetics*, *124*, 97–109.

Malamy, J. E. (2005). Intrinsic and environmental response pathways that regulate root system architecture. *Plant, Cell and Environment*, *28*, 67–77.

Marmiroli, M., Pietrini, F., Maestri, E., Zacchini, M., Marmiroli, N., & Massacci, A. (2011). Growth, physiological and molecular traits in Salicaceae trees investigated for phytoremediation of heavy metals and organics. *Tree Physiology*, *31*, 1319–1334.

Marzec, M., & Melzer, M. (2018). Regulation of root development and architecture by strigolactones under optimal and nutrient deficiency conditions. *International Journal of Molecular Sciences*, *19*, 1887.

Mei, H., Cheng, N. H., Zhao, J., Park, S., Escareno, R. A., Pittman, J. K., & Hirschi, K. D. (2009). Root development under metal stress in *Arabidopsis thaliana* requires the H + /cation antiporter CAX_4. *New Phytologist*, *183*, 95–105.

Meister, R., Rajani, M. S., Ruzicka, D., & Schachtman, D. P. (2014). Challenges of modifying root traits in crops for agriculture. *Trends in Plant Science*, *19*, 779–788.

Mendez-Bravo, A., Ruiz-Herrera, L. F., Cruz-Ramirez, A., Guzman, P., Martinez-Trujillo, M., & Ortiz-Castro, R. (2019). CONSTITUTIVE TRIPLERESPONSE1 and PIN2 act in a coordinate manner to support the indeterminate root growth and meristem cell proliferating activity in *Arabidopsis* seedlings. *Plant Science (Shannon, Ireland)*, *280*, 175–186.

Merewitz, E. B., Gianfagna, T., & Huang, B. (2011). Photosynthesis, water use, and root viability under water stress as affected by expression of SAG12-ipt controlling cytokinin synthesis in *Agrostis stolonifera*. *Journal of Experimental Botany*, *62*, 383–395.

Miao, Y., Luo, X., Gao, X., Wang, W., Li, B., & Hou, L. (2020). Exogenous salicylic acid alleviates salt stress by improving leaf photosynthesis and root system architecture in cucumber seedlings. *Scientia Horticulturae*, *272*, 109577.

Monzón, G. C., Pinedo, M., Di Rienzo, J., Novo-Uzal, E., Pomar, F., Lamattina, L., & de la Canal, L. (2014). Nitric oxide is required for determining root architecture and lignin composition in sunflower. Supporting evidence from microarray analyses. *Nitric Oxide: Biology and Chemistry / Official Journal of the Nitric Oxide Society*, *39*, 20–28.

Moreira, S., Bishopp, A., Carvalho, H., & Campilho, A. (2013). AHP6 inhibits cytokinin signaling to regulate the orientation of pericycle cell division during lateral root initiation. *PLoS One*, *8*, e56370.

Moreno-Risueno, M. A., Van Norman, J. M., Moreno, A., Zhang, J., Ahnert, S. E., & Benfey, P. N. (2010). Oscillating gene expression determines competence for periodic *Arabidopsis* root branching. *Science (New York, N.Y.)*, 1306–1311.

Mukatira, U. T., Liu, C., Varadarajan, D. K., & Raghothama, K. G. (2001). Negative regulation of phosphate starvation-induced genes. *Plant Physiology*, *127*, 1854–1862.

Munns, R., & Gilliham, M. (2015). Salinity tolerance of crops–what is the cost? *New Phytologist*, *208*, 668–673.

Munns, R., & Tester, M. (2008). Mechanisms of salinity tolerance. *Annual Review of Plant Biology*, *59*, 651–681.

Mustroph, A. (2018). Improving flooding tolerance of crop plants. *Agronomy*, *8*, e160.

Naz, M., Luo, B., Guo, X., Li, B., Chen, J., & Fan, X. (2019). Overexpression of nitrate transporter OsNRT2. 1 enhances nitrate-dependent root elongation. *Genes*, *10*, 290.

Nie, Z., Zhao, P., Shi, H., Wang, Y., Qin, S., & Liu, H. (2019). Nitrogen supply enhances zinc uptake and root-to-shoot translocation via up-regulating the expression of TaZIP3 and TaZIP7 in winter wheat (*Triticum aestivum*). *Plant and Soil*, *444*, 501–517.

Norby, R. J., DeLucia, E. H., Gielen, B., Calfapietra, C., Giardina, C. P., King, J. S., ... Ceulemans, R. (2005). Forest response to elevated CO_2 is conserved across a broad range of productivity. *Proceedings of the National Academy of Sciences*, *12*, 18052–18056.

OECD/Food and Agriculture Organization of the United Nations. (2015). *OECD-FAO agricultural outlook*. Paris: OECD Publishing.

Olsen, A. N., Ernst, H. A., Leggio, L. L., & Skriver, K. (2005). NAC transcription factors: Structurally distinct, functionally diverse. *Trends in Plant Science*, *10*, 79–87.

Opitz, N., Paschold, A., Marcon, C., Malik, W. A., Lanz, C., Piepho, H. P., & Hochholdinger, F. (2014). Transcriptomic complexity in young maize primary roots in response to low water potentials. *BMC Genomics*, *15*, 741.

Overvoorde, P., Fukaki, H., & Beeckman, T. (2010). Auxin control of root development. *Cold Spring Harbor Perspectives in Biology*, *2*, a001537.

Paez-Garcia, A., Motes, C. M., Scheible, W., Chen, R., Blancaflor, E. B., & Monteros, M. J. (2015). Root traits and phenotyping strategies for plant improvement. *Plants*, *4*, 334–355. Available from https://doi.org/10.3390/plants4020334.

Pagnussat, G. C., Lanteri, M. L., Lombardo, M. C., & Lamattina, L. (2004). Nitric oxide mediates the indole acetic acid induction activation of a mitogen-activated protein kinase cascade involved in adventitious root development. *Plant Physiology*, *135*, 279–286.

Pask, A., & Reynolds, M. (2013). Breeding for yield potential has increased deep soil water extraction capacity in irrigated wheat. *Crop Science*, *53*, 2090–2104. Available from https://doi.org/10.2135/cropsci2013.01.0011.

Pedersen, O., Sauter, M., Colmer, T. D., & Nakazono, M. (2020). Regulation of root adaptive anatomical and morphological traits during low soil oxygen. *New Phytologist*.

Péret, B., Clément, M., Nussaume, L., & Desnos, T. (2011). Root developmental adaptation to phosphate starvation: Better safe than sorry. *Trends in Plant Science*, *16*, 442–450.

Pérez-Torres, C. A., López-Bucio, J., Cruz-Ramírez, A., Ibarra-Laclette, E., Dharmasiri, S., Estelle, M., & Herrera-Estrella, L. (2008). Phosphate availability alters lateral root development in *Arabidopsis* by modulating auxin sensitivity via a mechanism involving the TIR1 auxin receptor. *The Plant Cell*, *20*, 3258–3272.

Perotti, M. F., Ribone, P. A., Cabello, J. V., Ariel, F. D., & Chan, R. L. (2019). AtHB23 participates in the gene regulatory network controlling root branching, and reveals differences between secondary and tertiary roots. *The Plant Journal*, *100*, 1224–1236.

Perrine-Walker, F. M., & Jublanc, E. (2014). The localization of auxin transporters PIN3 and LAX3 during lateral root development in *Arabidopsis thaliana*. *Biologia Plantarum*, *58*, 778–782.

Placido, D.F., Sandhu, J., Sato, S., Nersesian, N., Quach, T., Clemente, T.E., ... Walia, H., (2020). The LATERAL ROOT DENSITY gene regulates root growth during water stress in wheat.

Postma, J. A., Kuppe, C., Owen, M. R., Mellor, N., Griffiths, M., Bennett, M. J., ... Watt, M. (2017). OpenSimRoot: Widening the scope and application of root architectural models. *New Phytologist*, *215*, 1274–1286.

Pregitzer, K. S., DeForest, J. L., Burton, A. J., Allen, M. F., Ruess, R. W., & Hendrick, R. L. (2002). Fine root architecture of nine North American trees. *Ecological Monographs*, *72*, 293–309.

Prince, S. J., Murphy, M., Mutava, R. N., Durnell, L. A., Valliyodan, B., Shannon, J. G., & Nguyen, H. T. (2017). Root xylem plasticity to improve water use and yield in water-stressed soybean. *Journal of Experimental Botany*, *68*, 2027–2036.

Purushothaman, R., Mainassara, Z. A., Nalini, M., Rajaram, P., Krishnamurthy, L., & Cholenahalli, L. G. (2013). Root anatomical traits and their possible contribution to drought tolerance in grain legumes. *Plant Production Science*, *16*, 1–8.

Qin, H., & Huang, R. (2018). Auxin controlled by ethylene steers root development. *International Journal of Molecular Sciences*, *19*, 3656.

Quijano-Guerta, C., Kirk, G. J. D., Portugal, A. M., Bartolome, V. I., & McLaren, G. C. (2002). Tolerance of rice germplasm to zinc deficiency. *Field Crops Research*, *76*, 123–130.

Quintero, F. J., Ohta, M., Shi, H., Zhu, J. K., & Pardo, J. M. (2002). Reconstitution in yeast of the *Arabidopsis* SOS signalling pathway for Na + homeostasis. *Proceedings of the National Academy of Sciences of the United States of America*, *99*, 9061–9066.

Rasmusson, D. C. (1987). An evaluation of ideotype breeding. *Crop Science*, *27*, 1140–1146.

Rauf, M., Munir, M., ulHassan, M., Ahmad, M., & Afzal, M. (2007). Performance of wheat genotypes under osmotic stress at germination and early seedling growth stage. *African Journal of Biotechnology*, *6*, 971–975. Available from https://doi.org/10.5897/AJB2007.000-2119.

Reddy, S. H., Kambalimath, S. K., Singhal, R. K., Chikkakariyappa, M. K., Muthurajan, R., Rajanna, M. P., ... Chinnusamy, V. (2019). Allele-specific analysis of single parent backcross population identifies HOX10 transcription factor as a candidate gene regulating rice root growth. *Physiologia Plantarum*, *166*, 596–611.

Riefler, M., Novak, O., Strnad, M., & Schmülling, T. (2006). *Arabidopsis* cytokinin receptor mutants reveal functions in shoot growth, leaf senescence, seed size, germination, root development, and cytokinin metabolism. *The Plant Cell*, *18*, 40–54.

Robbins, N. E., Trontin, C., Duan, L., & Dinneny, J. R. (2014). Beyond the barrier: Communication in the root through the endodermis. *Plant Physiology*, *166*, 551–559.

Rogers, E. D., & Benfey, P. N. (2015). Regulation of plant root system architecture: Implications for crop advancement. *Current Opinion in Biotechnology*, *32*, 93–98.

Roy, S., Huang, W., Evrard, A., Schmoeckel, S., Zafar, Z., Tester, M., (2013). A novel protein kinase involved in Na^+ exclusion revealed from positional cloning.

Ruyter-Spira, C., Kohlen, W., Charnikhova, T., van Zeijl, A., van Bezouwen, L., de Ruijter, N., ... Bouwmeester, H. (2011). Physiological effects of the synthetic strigolactones analog GR24 on root System architecture in arabidopsis: Another belowground role for strigolactones. *Plant Physiology*, *155*, 721–734.

Saibo, N. J., Lourenco, T., & Oliveira, M. M. (2009). Transcription factors and regulation of photosynthetic and related metabolism under environmental stresses. *Annals of Botany*, *103*, 609–623.

Salazar-Henao, J. E., Vélez-Bermúdez, I. C., & Schmidt, W. (2016). The regulation and plasticity of root hair patterning and morphogenesis. *Development (Cambridge, England)*, *143*, 1848–1858.

Scheres, B., Wolkenfelt, H., Willemsen, V., Terlouw, M., Lawson, E., Dean, C., & Weisbeek, P. (1994). Embryonic origin of the *Arabidopsis* primary root and root meristem initials. *Development (Cambridge, England)*, *120*, 2475–2487.

Schnabel, E., Journet, E. P., de Carvalho-Niebel, F., Duc, G., & Frugoli, J. (2005). The *Medicago truncatula* SUNN gene encodes a CLV1-like leucine-rich repeat receptor kinase that regulates nodule number and root length. *Plant Molecular Biology*, *58*, 809–822.

Schnepf, A., Black, C. K., Couvreur, V., Delory, B. M., Doussan, C., Koch, A., ... Lobet, G. (2020). Call for participation: Collaborative benchmarking of functional-structural root architecture models. The case of root water uptake. *Frontiers in Plant Science*, *11*. Available from https://doi.org/10.1101/808972.

Seeve, C. M., Cho, I. J., Hearne, L. B., Srivastava, G. P., Joshi, T., Smith, D. O., ... Oliver, M. J. (2017). Water-deficit-induced changes in transcription factor expression in maize seedlings. *Plant, Cell & Environment*, *40*, 686–701.

Sharp, R. E., Poroyko, V., Hejlek, L. G., Spollen, W. G., Springer, G. K., & Bohnert, H. J. (2004). Root growth maintenance during water deficits: Physiology to functional genomics. *Journal of Experimental Botany*, *55*, 2343–2351. Available from https://doi.org/10.1093/jxb/erh276.

Shukla, V., Lombardi, L., Iacopino, S., Pencik, A., Novak, O., Perata, P., ... Licausi, F. (2019). Endogenous hypoxia in lateral root primordia controls root architecture by antagonizing auxin signaling in *Arabidopsis*. *Molecular Plant*, *12*, 538–551.

Singh, V., van Oosterom, E. J., Jordan, D. R., Messina, C. D., Cooper, M., & Hammer, G. L. (2010). Morphological and architectural development of root systems in sorghum and maize. *Plant and Soil*, *333*, 287–299.

Singhal, R.K., (2015) Characterization of HOX 10 mutant allele by selective genotyping of F2 MUTMAP population in rice (*Oryza sativa* L.). UAS, GKVK, Bengaluru <http://krishikosh.egranth.ac.in/handle/1/5810027410>.

Sinha, S. K., Kumar, A., Tyagi, A., Venkatesh, K., Paul, D., Singh, N. K., & Mandal, P. K. (2020). Root architecture traits variation and nitrate-influx responses in diverse wheat genotypes under different external nitrogen concentrations. *Plant Physiology and Biochemistry*, *148*, 246–259.

Sun, H., Li, J., Song, W., Tao, J., Huang, S., Chen, S., & Zhang, Y. (2015). Nitric oxide generated by nitrate reductase increases nitrogen uptake capacity by inducing lateral root formation and inorganic nitrogen uptake under partial nitrate nutrition in rice. *Journal of Experimental Botany*, *66*, 2449–2459.

Sustr, M., Soukup, A., & Tylova, E. (2019). Potassium in root growth and development. *Plants*, *8*, 435.

Swarup, R., Perry, P., Hagenbeek, D., Van Der Straeten, D., Beemster, G. T., & Sandberg, G. (2007). Ethylene upregulates auxin biosynthesis in *Arabidopsis* seedlings to enhance inhibition of root cell elongation. *The Plant Cell*, *19*, 2186–2196.

Takeda, S., Gapper, C., Kaya, H., Bell, E., Kuchitsu, K., & Dolan, L. (2008). Local positive feedback regulation determines cell shape in root hair cells. *Science (New York, N.Y.)*, *319*, 1241–1244.

To, J. P., Haberer, G., Ferreira, F. J., Deruere, J., Mason, M. G., Schaller, G. E., ... Kieber, J. J. (2004). Type-A *Arabidopsis* response regulators are partially redundant negative regulators of cytokinin signaling. *The Plant Cell*, *16*, 658–671.

Trachsel, S., Kaeppler, S. M., Brown, K. M., & Lynch, J. P. (2013). Maize root growth angles become steeper under low N conditions. *Field Crops Research*, *140*, 18–31. Available from https://doi.org/10.1016/j.fcr.2012.09.010.

Tsukagoshi, H., Busch, W., & Benfey, P. N. (2010). Transcriptional regulation of ROS controls transition from proliferation to differentiation in the root. *Cell*, *143*, 606–616.

Uga, Y., Sugimoto, K., Ogawa, S., Rane, J., Ishitani, M., & Hara, N. (2013). Control of root system architecture by DEEPER ROOTING 1 increases rice yield under drought conditions. *Nature Genetics*, *45*, 1097–1102. Available from https://doi.org/10.1038/ng.2725.

van der Weele, C. M., Spollen, W. G., Sharo, R. E., & Baskin, T. I. (2000). Growth of *Arabidopsis thaliana* seedlings under water deficit studied by control of water potential in nutrient-agar media. *Journal of Experimental Botany*, *51*, 1555–1562.

Vanneste, S., De Rybel, B., Beemster, G. T., Ljung, K., De Smet, I., Van Isterdael, G., ... Inze, D. (2005). Cell cycle progression in the pericycle is not sufficient for SOLITARY ROOT/IAA14-mediated lateral root initiation in *Arabidopsis thaliana*. *The Plant Cell*, *17*, 3035–3050.

Varala, K., Marshall-Colón, A., Cirrone, J., Brooks, M. D., Pasquino, A. V., Léran, S., ... Ruffel, S. (2018). Temporal transcriptional logic of dynamic regulatory networks underlying nitrogen signaling and use in plants. *Proceedings of the National Academy of Sciences*, *115*, 6494–6499.

Vidoz, M. L., Loreti, E., Mensuali, A., Alpi, A., & Perata, P. (2010). Hormonal interplay during adventitious root formation in flooded tomato plants. *The Plant Journal*, *63*, 551–562.

Villiers, F., Ducruix, C., Hugouvieux, V., Jarno, N., Ezan, E., Garin, J., ... Bourguignon, J. (2011). Investigating the plant response to cadmium exposure by proteomic and metabolomic approaches. *Proteomics*, *11*, 1650–1663.

Villordon, A. Q., & Clark, C. A. (2014). Variation in virus symptom development and root architecture attributes at the onset of storage root initiation in 'Beauregard' sweet potato plants grown with or without nitrogen. *PLoS One*, *9*, e107384.

Wagner, G. J. (1993). Accumulation of cadmium in crop plants and its consequences to human health. *Advances in Agronomy*, *51*, 173–212.

Wang, P., Du, Y., & Song, C. P. (2011). Phosphorylation by MPK6: A conserved transcriptional modification mediates nitrate reductase activation and NO production? *Plant Signaling & Behavior*, *6*, 889–891.

Wang, Y., Li, K., & Li, X. (2009). Auxin redistribution modulates plastic development of root system architecture under salt stress in *Arabidopsis thaliana*. *Journal of Plant Physiology*, *166*, 1637–1645.

Wang, Y. Y., Cheng, Y. H., Chen, K. E., & Tsay, Y. F. (2018). Nitrate transport, signaling, and use efficiency. *Annual Review of Plant Biology*, *69*, 85–122.

West, G., Inzé, D., & Beemster, G. T. (2004). Cell cycle modulation in the response of the primary root of *Arabidopsis* to salt stress. *Plant Physiology*, *135*, 1050–1058.

Wiegers, B. S., Cheer, A. Y., & Silk, W. K. (2009). Modelling the hydraulics of root growth in three dimensions with phloem water sources. *Plant Physiology*, *150*, 2092–2103.

Wilmoth, J. C., Wang, S., Tiwari, S. B., Joshi, A. D., Hagen, G., Guilfoyle, T. J., ... Reed, J. W. (2005). NPH4/ARF7 and ARF19 promote leaf expansion and auxin-induced lateral root formation. *The Plant Journal*, *43*, 118–130.

Wishart, J., George, T. S., Brown, L. K., Ramsay, G., Bradshaw, J. E., & White, P. J. (2013). Measuring variation in potato roots in both field and glasshouse: The search for useful yield predictors and a simple screen for root traits. *Plant and Soil*, *368*, 231–249.

Wu, Q. S., & Zou, Y. N. (2013). Mycorrhizal symbiosis alters root H + effluxes and root system architecture of trifoliate orange seedlings under salt stress. *Journal of Animal and Plant Sciences*, *23*, 143–148.

Xiong, J., Lu, H., Lu, K., Duan, Y., An, L., & Zhu, C. (2009). Cadmium decreases crown root number by decreasing endogenous nitric oxide, which is indispensable for crown root primordia initiation in rice seedlings. *Planta*, *230*, 599–610.

Xu, K., Xu, X., Fukao, T., Can- las, P., Maghirang-Rodriguez, R., & Heuer, S. (2006). Sub1Ais an ethylene-response-factor-likegene that confers submergence tolerance to rice. *Nature*, *442*, 705–708.

Xu, M., Zhu, L., Shou, H., & Wu, P. (2005). A PIN1 family gene, OsPIN1, involved in auxin-dependent adventitious root emergence and tillering in rice. *Plant and Cell Physiology*, *46*, 1674–1681.

Xu, W., Jia, L., Shi, W., Liang, J., Zhou, F., & Li, Q. (2013). Abscisic acid accumulation modulates auxin transport in the root tip to enhance proton secretion for maintaining root growth under moderate water stress. *The New Phytologist*, *197*, 139–150. Available from https://doi.org/10.1111/nph.12004.

Xu, Z., Chen, L., Tang, S., Zhuang, L., Yang, W., Tu, L., ... Zhang, L. (2016). Sex-specific responses to Pb stress in *Populus deltoides*: Root architecture and Pb translocation. *Trees, 30*, 2019–2027.

Yamaji, N., Huang, C. F., Nagao, S., Yano, M., Sato, Y., & Nagamura, Y. (2009). A zinc finger transcription factor ART1 regulates multiple genes implicated in aluminum tolerance in rice. *The Plant Cell, 21*, 3339–3349.

Yamauchi, T., Colmer, T. D., Pedersen, O., & Nakazono, M. (2018). Regulation of root traits for internal aeration and tolerance to soil waterlogging-flooding stress. *Plant Physiology, 176*, 1118–1130.

Yamauchi, T., Tanaka, A., Inahashi, H., Nishizawa, N. K., Tsutsumi, N., Inukai, Y., & Nakazono, M. (2019). Fine control of aerenchyma and lateral root development through AUX/IAA- and ARF-dependent auxin signaling. *Proceedings of the National Academy of Sciences of the United States of America, 116*, 20770–20775.

Ye, H., Roorkiwal, M., Valliyodan, B., Zhou, L., Chen, P., Varshney, R. K., & Nguyen, H. T. (2018). Genetic diversity of root system architecture in response to drought stress in grain legumes. *Journal of Experimental Botany, 69*, 3267–3277.

Ying, S., Zhang, D. F., Fu, J., Shi, Y. S., Song, Y. C., Wang, T. Y., & Li, Y. (2012). Cloning and characterization of a maize bZIP transcription factor, ZmbZIP72, confers drought and salt tolerance in transgenic *Arabidopsis*. *Planta, 235*, 253–266.

York, L. M., Nord, E. A., & Lynch, J. P. (2013). Integration of root phenes for soil resource acquisition. *Frontiers in Plant Science, 4*.

Yu, F., Liang, K., Fang, T., Zhao, H., Han, X., Cai, M., & Qiu, F. (2019). A group VII ethylene response factor gene, ZmEREB180, coordinates waterlogging tolerance in maize seedlings. *Plant Biotechnology Journal, 17*, 2286–2298.

Yuan, H. M., & Huang, X. (2016). Inhibition of root meristem growth by cadmium involves nitric oxide-mediated repression of auxin accumulation and signalling in *Arabidopsis*. *Plant, Cell & Environment, 39*, 120–135.

Zhang, H., Kim, M. S., Sun, Y., Dowd, S. E., Shi, H., & Paré, P. W. (2008). Soil bacteria confer plant salt tolerance by tissue-specific regulation of the sodium transporter HKT1. *Molecular Plant-Microbe Interactions, 21*, 737–744.

Zhang, J., Qian, Y., Chen, Z., Amee, M., Niu, H., Du, D., ... Sun, J. (2020). Lead-induced oxidative stress triggers root cell wall remodeling and increases lead absorption through esterification of cell wall polysaccharide. *Journal of Hazardous Materials, 385*, 121524.

Zhang, X., Cui, Y., Yu, M., Su, B., Gong, W., Baluška, F., ... Lin, J. (2019). Phosphorylation-mediated dynamics of nitrate transceptor NRT1. 1 regulate auxin flux and nitrate signaling in lateral root growth. *Plant Physiology, 181*, 480–498.

Zhang, Y., Zhou, Y., Chen, S., Liu, J., Fan, K., Li, Z., ... Lin, W. (2019). Gibberellins play dual roles in response to phosphate starvation of tomato seedlings, negatively in shoots but positively in roots. *Journal of Plant Physiology, 234*, 145–153.

Zhao, M. G., Chen, L., Zhang, L. L., & Zhang, W. H. (2009). Nitric reductase-dependent nitric oxide production is involved in cold acclimation and freezing tolerance in *Arabidopsis*. *Plant Physiology, 151*, 755–767.

Zhao, M. G., Tian, Q. Y., & Zhang, W. H. (2007). Nitric oxide synthase-dependent nitric oxide production is associated with salt tolerance in *Arabidopsis*. *Plant Physiology, 144*, 206–217.

Zheng, H. G., Babu, R. C., Pathan, M. S., Ali, L., Huang, N., Courtois, B., & Nguyen, T. (2000). Quantitative trait loci for root-penetration ability and root thickness in rice: Comparison of genetic backgrounds. *Genome / National Research Council Canada = Genome / Conseil National de Recherches Canada, 43*, 53–62.

Chapter 4

Nanotechnology-based biofortification: a plant–soil interaction modulator/enhancer

Alviya Sultana[1], Neetu Talreja[2], Divya Chauhan[3] and Mohammad Ashfaq[2,4]

[1]School of Sciences, Monash University, Subang Jaya, Malaysia, [2]Multidisciplinary Research Institute for Science and Technology, IIMCT, University of La Serena, La Serena, Chile, [3]Department of Chemical and Biomedical Engineering, University of South Florida, Tampa, FL, USA, [4]School of Life Science, BS Abdur Rahaman Institute of Science and Technology, Chennai, India

4.1 Introduction

Nanotechnology in recent years has emerged as a promising field, finding its applications through science and technology and making it a powerful tool of interdisciplinary research. Apart from revolutionizing various sectors including health care, materials, medical diagnosis, energy, and textiles, nanoscale approaches have proven their potential in providing improvement and sustenance to the agriculture sector (Afreen et al., 2018; Afreen, Talreja, Chauhan, Ashfaq, & Abd-Elsalam, 2020; Irsad et al., 2020; Mondal, 2020; Patra et al., 2018). Sustainable agriculture proposes a solution to hunger and poverty in developing countries, in response to the probable increase in the world's population, reaching 10 billion by the year 2050, as predicted by the Food and Agriculture Organization (Diaz-Ambrona & Maletta, 2014). Moreover, for developed countries the agriculture sector aims at maximizing the yield by attaining pest and drought-resistant crops (Sundström et al., 2014). Nanotechnology produces and employs particles which are of a nano-scale size and are referred to as nanomaterials and engineered nanomaterials. Nanomaterials' production could be incidental or engineered, incidental materials are produced unintentionally through environmental pollution or pollutants, whereas engineered particles are intentionally designed and produced by humans. Owing to their small size, high surface ratio, hydrophilicity, targeted delivery, slow-release strategy, these nanoparticles (NPs) attain

Frontiers in Plant–Soil Interaction. DOI: https://doi.org/10.1016/B978-0-323-90943-3.00022-5

unique physical as well as chemical properties and are reaching heights of popularity in various fields. Their value was estimated in the global nanomaterial market to be US$ 8.5 billion in 2019 and is expected to grow at a 13.1% CAGR rate (compound annual growth rate) from 2020 to 2027. Besides accelerating raw materials' supply for the food and feed industry, the present agriculture system also needs a reduction in the number of utilized resources. A recent report indicates annual consumption of fertilizer and pesticides to be 187 and 4 million tons, respectively, for crop production globally. Recent advances of nanotechnology in the agriculture sector have shown promising potential for improved agricultural production (Boverhof et al., 2015; Jeevanandam, Barhoum, Chan, Dufresne, & Danquah, 2018; Ray, Yu, & Fu, 2009).

The introduction of NPs to agriculture not only benefits plants but also interacts with the soil system and helps in its enrichment. Alongside plants, maintaining and enhancing soil fertility is crucial for developing countries (Chauhan, Afreen, Talreja, & Ashfaq, 2020; Mohammad Ashfaq, Chuahan, & Srituravanich, 2020; Omar, Afreen, Talreja, Chauhan, & Ashfaq, 2019a). Similar to gene delivery vectors, NPs also act as carriers/vehicles and transport nutrients to the plant and soil environment. Nanotechnology-based products have been developed which include, nanoparticles derived from zinc and titanium oxides and acting as growth promoters as well as antioxidants in tomato plants (Raliya, Biswas, & Tarafdar, 2015). NPs are also been reported to boost up the absorption as well as assist in the site-specific release of macronutrient fertilizers upon coating them with ZnO NPs (Milani et al., 2015). Besides acting as carriers or fertilizers, NPs also participate in physiological processes such as photosynthesis, which in turn help to increase the biomass. An essential micronutrient, iron, when introduced with NPs, that is, nano-Fe_2O_3, has been reported to increase photosynthesis in peanut and soybean plants (Alidoust & Isoda, 2013; Rui et al., 2016). The latest study has shown the interaction of iron-based NPs with plants, at the molecular level. They were reported to aid significantly in biomass increase of maize plants, by tentatively altering certain pathways, such as amino acid biosynthesis, photorespiration, and alanine metabolism. Some of the other NPs known for acting at the cellular level and enhancing photosynthetic activities are carbon dots, carbon nanotubes, and nanoceria (Giraldo et al., 2014; Li et al., 2018b; Wu, Tito, & Giraldo, 2017). These materials target and improve light-harvesting, scavenging reactive oxygen species, and electron transport rate, which in turn give a higher photosynthetic rate and yield increased plant biomass.

In soil, carbon nanotubes (CNTs) have shown promising results in improving compressive strength by approximately 120% as compared to original soil. Engineered NPs are well-known to alter bacterial communities inhabiting soil (Colman et al., 2013; Tong et al., 2012). CNTs reportedly altered plant–microbe interactions and helped in nitrogenase activity and nodulation in *Lotus japonicas*, a legume (Yuan et al., 2017). Multiwalled CNTs (MWCNTs) when applied to red clover also showed a slight increment in nitrogen fixation capabilities (Moll, Gogos, Bucheli, Widmer, & van der Heijden, 2016). NPs, specifically iron-derived nanoparticles, have been intensively employed for soil decontamination as they are excellent absorbants and possess reductive abilities (Guerra, Attia, Whitehead, & Alexis, 2018). Products derived from nanotechnology have also entered the field of microbial sensing, for example, carbon nanotubes, quantum dots, metal, and silica nanoparticles (Anjum et al., 2015; Prasad, Bhattacharyya, & Nguyen, 2017). Fig. 4.1 shows the graphical illustration of the relationships between NPs synthesis, their application, transfer into the "soil–plant system," and consumers.

Nevertheless, there are concerns of negative impacts from these materials, environmentally. The technology has an impact on plant physiological processes, fruit biomass enhancement, and microbiome enhancement, yet the long-term toxicity needs to be evaluated. This book chapter will review and discuss various nanoparticles used in agronomy. Their interaction with soil, microbiome interaction, and nutrient enrichment in the soil system will be discussed. There will be a discussion about nanomaterials interacting with

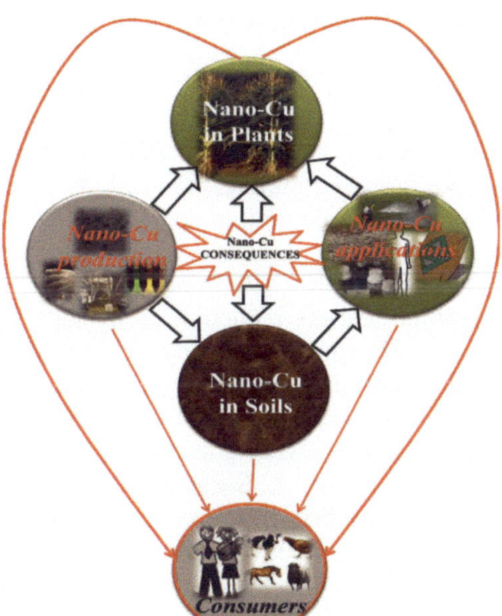

FIGURE 4.1 Graphical illustration of the relationships between nanoparticles (NPs) synthesis, their application, transfer into "soil–plant system," and consumers. *The image was taken with permission from Anjum et al., 2015, Nanoscale copper in the soil–plant system—Toxicity and underlying potential mechanisms. Environmental Research, 138, 306–325. Copyright © 2015 Elsevier Inc. All rights reserved.*

plants, their translocation to plant parts, and the molecular involvement of nanoparticles. Further, their applications in the agriculture system and various products developed will be described. Finally, the impact of these developed particles and technologies will be assessed.

4.2 Nanotechnology agents at work

4.2.1 Nanomaterials

Nano is simply a prefix used to refer to "billionth part of" and therefore the technology which develops and deals with nanostructured materials, is termed and referred to as nanotechnology. A characteristic feature of these materials is to at least have one dimension in the nanometer range (10^{-9} m). Their distinctive feature lies in the fact that materials at the nanometer scale, exhibit physical properties different from their bulk material, whereas at the microscale, physical properties remain similar to that of the bulk form (Iqbal, Singh, Singh, & Ansari, 2020; Nafees et al., 2020).

Nanoscience/nanotechnology is broadly based on two approaches, one of which is bottom-up and the second one is the top-down approach. The top-down approach is based on the miniaturization of components, the theoretical foundation of which was laid by Feynman in 1959, where he made the renowned statement, "there is plenty of room at the bottom" (Bayda, Adeel, Tuccinardi, Cordani, & Rizzolio, 2019). The other method is the converse, where molecular components self-assemble and form a suprastructure. Methods involved in the top-down approach are various decomposition techniques like grinding/milling, sonication, and homogenization under high pressure, while the bottom-up approach utilizes techniques such as sedimentation, spinning, biochemical, and green synthesis (Chauhan et al., 2020; Omar et al., 2019b). NPs from coconut shells were prepared using the milling method and the reduction in crystallite size of NPs was confirmed by scanning electron microscopy (SEM). Moreover, spherical particles of colloidal carbon and magnetite have also been reported to be prepared using the destructive approach of the top-down method (Bello, Agunsoye, & Hassan, 2015; Garrigue et al., 2004; Priyadarshana, Kottegoda, Senaratne, de Alwis, & Karunaratne, 2015). NPs of titanium dioxide anatase, on the other hand, were prepared using a bottom-up synthesis process. NPs synthesized from alizarin and titanium (IV) isopropoxide precursors, were confirmed to be growing in size upon temperature elevation, through SEM (Mogilevsky et al., 2014). Low-density lipoprotein (LDL) NPs' preparation, used for drug delivery in cancer treatment, is another example of utilizing a bottom-up synthesis approach. LDL NPs were synthesized by the solvent-exchange method, without involving the use of phospholipid and yet the NPs possessed the hydrophobicity required for drug delivery (Alimardani, Abolmaali, Tamaddon, & Ashfaq, 2020;

Needham et al., 2016). Besides chemical synthesis, the bottom-up approach also utilizes green and biogenic methods of NPs synthesis. Biological synthesis is an environment-friendly and cost-effective method and it employs bacteria, yeast, algae, fungi, plants, etc. NPs have been reported to be synthesized from wheat and oat biomass, and also by using plant extracts, whereas both bottom-up and top-down techniques could be used for preparing spherical monodispersed Bi NPs, of size ranging from 100 to 500 nm. In the bottom-up methodology, the precursor was boiled while bismuth was firstly melted down and then emulsified to produce nanoparticles through top-down technology (Ahmed, Annu, Ikram, & Yudha, 2016; Parveen, Banse, & Ledwani, 2016).

NPs can be classified based on various categories including size, chemical properties, morphology, and dimensions. Based on dimensions, nanoparticles could be zero-, one-, two-, or three-dimensional. One-dimensional include the formation of thin films, while 2D formation offers carbon nanotubes and 3D particles are quantum dots (QDs), dendrimers, and fullerenes. Based on physical and chemical properties, NPs are classified into these major categories: Carbon-based, lipid-based, metal NPs, ceramic NPs, polymeric, and semiconductor NPs. Lipid-based NPs are primarily used in the biomedical field as they contain lipid moieties. These NPs mainly comprise two parts, that is, solid core and a matrix, the solid core comprises lipid molecules whereas molecules of soluble lipophilic components contribute to matrix formation (MK, Jain A Fau-Singh, & S, 2011; Tiwari, Tiwari, & Kim, 2012). These NPs are popularly involved in various applications such as in cancer therapy and are used for carrying RNA as well as performing the role of drug delivery/carriers (Gujrati et al., 2014). As the name suggests, metal NPs are made out of metals, particularly alkali metals, and from copper, zinc, silver, and gold (noble metals). NPs from semiconductor materials possess characteristics that fall between metals and nonmetals. Similar to lipid-based, polymeric NPs are mostly spherical and composed of a core and an outer layer and are either nanosphere or nanocapsules shaped, whereas particles derived from metal and nonmetal materials form solids which are inorganic and nonmetallic, and are termed as ceramic NPs (Bayda et al., 2019; Chauhan et al., 2020; Khot, Sankaran, Maja, Ehsani, & Schuster, 2012; Omar et al., 2019a). Carbon-based NPs are primarily described as fullerenes, carbon nanofibers (CNFs), and CNTs. Structurally, fullerenes contain globular and hollow cages, whereas CNTs are elongated and tubular. The tubular structure of CNTs could be single-, double-, or multilayered and hence the names derived, that is, single-layered, double-layered, and multilayered carbon nanotubes. Finally, based on size/morphology nanoparticles can also be classified into nanoparticles, nanotubes, nanofibers, nanofilm, nanoblock, nanocomposites, and nanocrystalline solids (Ashfaq et al., 2019; Zaytseva & Neumann, 2016). Usually, NPs' characterization is based on surface charge, particle size, and morphology. Several advanced techniques offer characterization such as SEM, X-ray diffraction, X-ray photoelectron spectroscopy, and particle size analysis.

4.3 Nanomaterials and soil system

Soil stands as a major requirement for almost all crops. Not only does it hold plants but it also acts as a reservoir for plants' major necessities, that is, essential nutrients. However, due to various environmental conditions and extensive farming as a result of increasing population, soil fertility and its condition have degraded enormously over the past few decades. As reported, intensive farming practice has led to serious degradation of approximately 40% of total agricultural land globally, affecting soil fertility.

Agrochemicals have been a prominent choice to restore nutrient imbalance but are usually lost due to leaching, volatilization, microbial degradation, and photolysis (Dimkpa & Bindraban, 2016; Neina, 2019). Therefore an advanced approach to sustainable agriculture development is required.

4.3.1 Soil nutrient enrichment

Element fertilization, that is, supplementing essential nutrients to the soil, is considered to be inevitable to improve soil fertility and also crop yield. Engineered nanoparticles have shown a new path for the agricultural sector in producing food with limited sources available. Initially, nanotechnology-based nutrient enrichment was done by producing nanoparticles through top-down (physical) as well as bottom-up (chemical) techniques. However, later, these are produced by incorporating nutrients with desired nanomaterials. In general, targeted nutrients are either coated, encapsulated, or delivered as particles. The coating is done using thin polymer film, and nanoporous materials are used for encapsulation purposes, while emulsions of cationic nutrients produced at nanoscale dimension are used and surface modification are done for anionic nutrients (Ca^{2+}, Mg^{2+}, NO^{3-}, PO^{4-}) (Godfray et al., 2010; Li, Li, Li, & Fu, 2018a; Panpatte, Jhala, Shelat, & Vyas, 2016).

In a recent study, engineered nanomaterials, namely TiO_2, SiO_2, and Fe_3O_4, when introduced to soil, have been reported to alter the metabolite profile of the rhizosphere zone. Several metabolites, both hydrophilic as well as hydrophobic analytes, which are categorized as nucleotides, amino acids, sugar, lipids, organic acid, have been monitored to increase significantly. Among them, levoglucosan, which is associated with increased root length of plants and also acts as a signaling molecule in response to engineered NMs exposure, was found to be increased significantly. Apart from that, linolenic acid, methyl-beta-D-galactopyranoside, beta mannosylglycerate, gluconic acid, 4-hydroxycinnamic acid, methyl phosphate, and allo-inositol were also found to be in elevated quantities. It has been assessed that the increase in soil metabolite profile is probably attributed to the enrichment of the soil organic carbon pool. Moreover, it has been documented that all the compounds, except levoglucosan, are involved in promoting plant and microbial growth (Zhao et al., 2019). Another study reported obtaining an increase in soil organic matter content upon the introduction of zinc oxide nanomaterials. Organic matter enrichment was observed to rise from 2.8% to 18% and this enrichment is known for improving nutrient availability. Besides, functional groups present on organic matters act as binding sites for Zn compounds and hence help to increase their bioavailability (Aiken, Hsu-Kim, & Ryan, 2011; Wiatrowska & Komisarek, 2019). Another metabolomic study performed with unplanted and cucumber-planted soils, amended with silver nanoparticles (AgNPs) reported showing an altered profile of metabolites. The metabolites involved mainly in two metabolic pathways, namely, amino acid-related and sugar metabolism, were found to be significantly altered in both the soils, that is, planted as well as unplanted. It has been anticipated that altered metabolite composition led to increased soil pH which in turn, lowered water-soluble elements in both the soils. In planted soil, the presence of AgNPs significantly increased both the micro- and macroelements (Cd, Cu, Al and Si, Mg, Na, K, respectively) (H et al., 2020).

4.3.2 Interaction with soil microorganisms

It is quite evident from the literature and present situation how important it is to improve soil health and nutrition for a sustainable agriculture system. For any purpose, green technologies are always preferred over chemical alternatives. Similarly, soil microorganisms play a key role in the process of fertilization since they act on dead animal/plant materials and convert them into an absorbable form, to be readily absorbed/uptake by growing crops. Hence, any disturbance in soil microbial consortia could in turn negatively impact soil fertility, plant growth, and productivity (Davidson & Janssens, 2006; Y, Wang J Fau - Xiu, Xiu Z Fau-Alvarez, & PJ, 2013).

Nanomaterials in soil are localized in microaggregates and also approximately about 70% of total soil bacteria are also present in these aggregates, therefore NPs are in direct contact with most microbial communities inhabiting these soil microaggregates (Vittori Antisari, Antonietta, Gilmo, Paolo, & Serena, 2013). A study revealed that Fe_3O_4 exposure to maize cultivated soil abundantly reduced nitrogen-fixation-related bacteria while increased the bacteria associated with plant growth promotion and carbon cycling (H et al., 2020). The community structure of Gram-positive bacteria, *Actinobacteria*, was found to be stimulated and altered by iron oxide NPs. These metal oxide nanoparticles were involved in stimulating enzymatic activities of urease and invertase enzymes (He et al., 2011). Sensitivity of an intercellular enzyme, that is, dehydrogenase, toward heavy metals, acts as an indicator of soil conditions. Nano-CuO has been reported in a recent study to increase dehydrogenase activity by twofold at low concentration, in comparison to control soil. Moreover, nano-CuO addition to soil also resulted in a significant increase in the numbers of fungi as compared to untreated soil (Jośko et al., 2019). Nanocopper oxide's stimulating effects on dehydrogenase and the fungal population is indicative of copper as a microelement and its role as a cofactor of enzyme activity (Bruins, Kapil, & Oehme, 2000).

Mitigation of toxicity of widely used antimicrobial AgNPs in soils was obtained with a change in soil components toward a beneficial bacterium *Pseudomonas chlororaphis O6* (Calder et al., 2012). Ge and colleagues (2014) reported direct evidence of plant–soil interaction by studying the effects of nano-CeO_2 and nano-ZnO in planted and unplanted soils. The presence of soybean plants in soil helped and diminished the effects of nano-ZnO on bacterial diversity in soil (Ge et al., 2014).

In another way, soil microorganisms interact with surrounding NPs which support "green nanotechnology" and that is the biotransformation of these particles. *Trabusiella guamensis*, a soil bacteria, resistant to MWCNTs, performed biotransformation of carbon nanotubes through surface oxidation, as confirmed by various techniques, including transmission electron microscopy (TEM), which confirmed structural changes in the concentric walls of nanotubes (Chouhan et al., 2016b). In another study, Chouhan and colleagues showed effective degradation of graphene oxide nanosheets evidenced by structural changes and exfoliation of sheets, proven by TEM images (Chouhan, Pandey, Qureshi, Ozguz, & Niazi, 2016a). Engineered nanoparticles from Ag have shown strong antimicrobial activities for crop protection against bacteria, actinomycetes, and fungi (Judy et al., 2015).

4.4 Nanomaterial and plant system

Nanomaterials have been synthesized to exhibit novel properties like high surface area, reactivity, tunable porous texture, and structural changes. Nanomaterials have been extensively used in various end applications like environmental remediation, energy, antibiotic materials, sensor, and agriculture (Ashfaq, Verma, & Khan, 2017b; Ashfaq, Verma, & Khan, 2018; Bhadauriya et al., 2018; Kumar & Talreja, 2019; Talreja, Verma, & Kumar, 2016). Nanomaterials exhibit numerous characteristics like pesticides, antibacterial, herbicides, and fertilizers that target specific sites within the plants. Interestingly, nanosized materials easily translocate within the plants that create the water channels, and subsequently enhance water uptake and improve the development and growth of the plants.

4.4.1 Uptake and translocation mechanism

The uptake and translocation of the nanomaterials within the plants have been dependent on mainly two pathways based on plant tissue: (1) apoplastic pathways, through the interconnected cell walls or microscopic channels. Nanomaterials less than 50 nm easily enter through the plasmodesmata through the ions or water; and (2) symplasm pathways, more than 50 nm particle size nanomaterials easily enter through the symplastic pathways. On the other hand, the translocation of nanomaterials within the plants based on nanomaterials mainly depends on the two properties: (1) size of the nanomaterials is an important factor for translocation and uptake within the plants. Approximately up to 500 nm size nanomaterials easily enter within the plants through water channels; and (2) the surface charge of the nanomaterials is also a very important factor for the translocation and uptake of nanomaterials within the plants. If the surface charge of the nanomaterials is positively charged then nanomaterials accumulate on the root surface, and thereby there is less uptake and translocation ability of the nanomaterials within the plants. On the other hand, a negatively charged surface of the nanomaterials easily translocates within the plant because negatively charged nanomaterials and negatively charged plant cells have a repulsive force, thereby there is less accumulation on the root surface and higher translocation ability, subsequently providing advantages in the development and growth of the plants (Lin et al., 2009).

Numerous nanomaterials, especially carbon-based nanomaterials (CNTs, and CNFs) act as carriers for the growth and development of plants. CNFs are relatively newer nanomaterials compared with CNTs. CNFs hold metal-NPs and the release of metal ions is slow and has insignificant toxicity, therefore this is frequently used in biological applications including agriculture (Ashfaq, Singh, Sharma, & Verma, 2013). For example, Ashfaq et al. synthesized Cu-CNFs on an Activated carbon fiber surface by using a simple impregnation process (for Cu loading) followed by a catalytic vapor deposition process and tested against chickpea seeds. The data suggested that the Cu-CNFs efficiently translocate within the plants through root to shoot to leaf, which is confirmed from the SEM, energy dispersive X-rays (EDX), and elemental carbon (EC)/organic carbon (OC) analysis. Interestingly, higher uptake of Cu within the plant's shoot and leaf was also observed due to effective translocation and less accumulation ability of Cu-CNFs. The surface charge of the Cu-CNFs is strongly negatively charged, thereby easily translocated within the plants (Ashfaq, Verma, & Khan, 2017a). Another study of a similar group synthesized Cu/Zn-CNFs and encapsulated them with PVA−starch formulation. The synthesized PVA−starch incorporated Cu/Zn-CNFs formulation was tested against chickpea plants. The data suggested the effective translocation ability of CNF within the plants. The PVA−starch formulation easily biodegrades and the release of CNFs is slow (Kumar, Ashfaq, & Verma, 2018). In general, the studies suggested that there are two important pathways for translocation and uptake of nutrients within the plants—symplastic and apoplastic pathways—whereas based on nanomaterials, there are two main factors like size and surface charge of the nanomaterials that are very important for the translocation of nutrients within the plants.

4.5 Nanotechnology-based biofortification tools

The interference of nanotechnology in agricultural application opens the possibility of advancement for crop protection and production. There are several nanolevel techniques listed below which help to improve crop protection and production.

4.5.1 Nanofertilizers

Recently, organic and mineral fertilizers have been used to provide essential nutrients to the plant system resulting in severe damage to the plants and the overall environment. In this context, the application of nanofertilizers to provide plant nutrition has the potential ability to improve crop production and protection, together with environmental protection from organic fertilizers. Nanofertilizers are made of nanomaterials and their composites size below 100 nm (Kopittke, Lombi, Wang, Schjoerring, & Husted, 2019; Shang et al., 2019; Usman et al., 2020). Several nanomaterials are composed of polymers encapsulated over the surface of nanomaterials by utilizing the surface properties of the nanomaterial to enhance the controlled release of nutrients as required by the particular variety of plant. Several unique properties of nanofertilizers, including surface charge, size, and surface area, can be utilized to enhance crop protection and production. Surface charge enhances nutrients uptake, utilizing the interaction between nanomaterial and soil particles. Polymer coating over nanomaterial helps in the slow release of nanomaterial. The controlled release of fertilizers or micronutrients or agrochemicals incorporated with nanomaterials for reducing surface area might be providing an excellent source for plant nutrients (Duhan et al., 2017; Mani & Mondal, 2016; Prasad et al., 2017; Shakiba et al., 2020; Shebl et al., 2019). In summary, nanofertilizers serve as an excellent source of providing plant nutrients without deteriorating the environment and ecosystem.

4.5.2 Nanopesticides

Plant species needs protection from pests to avoid plant diseases. In this context, the use of several toxic chemicals not only impacts plants but also severely deteriorates the ecosystem and subsequently the environment. Avoidance of excessive chemical use and growth of the best quality crop with higher yield nanotechnology has been taken a step ahead by synthesizing nanopesticides. Nanopesticides are the nanorange of pest control formula applied to enhance crop protection and production. Nanopesticides are formulated by emulsions (e.g., nanoemulsions), nanocapsules, and inorganic nanoparticles including metals, metal oxides, and nanoclays. Nanopesticides in sizes ranging from 1 to 100 nm offer several advantages, such as less amount of active ingredient, enhanced crop production, and they represent an emerging technological development toward the future agricultural system (Köhl, Kolnaar, & Ravensberg, 2019; Ramezani, Ramezani, & Gerami, 2019; Younas et al., 2020).

4.5.3 Biosensors

Recently, biosensors are emerging as tools in the detection of plant disease and pesticides over the plant surface. The applicability of biosensors for quick detection of the particular biological component in plants within seconds opens their usefulness to fight against the particular components and spread of bacterial resistance. Several types of biosensors have been synthesized utilizing several materials, such as activated carbon, CNTs, CNFs, graphene, etc. Different biosensors can quickly detect bacteria within crops and soil. Specific properties of nanomaterials such as surface area and large pore size distribution can be utilized to capture a specific target and respond toward that target in terms of electronic signals. Surface charge over nanomaterials enhances interaction toward the biological component and they form a complex which subsequently produces color and helps in naked-eye detection within seconds. The applicability of biosensors for the detection of plant disease and checking soil quality will help in future agricultural advancement (Bobrinetskiy & Knezevic, 2018; Fang & Ramasamy, 2015; Hasan et al., 2014; Mehrotra, 2016; Srinivasan & Tung, 2015).

4.6 Environmental impact of nanomaterials

The excessive use of nanomaterials poses a risk to the environment. Nanoscience offer a great deal of high-quality advancement in numerous areas of research. However, uncertainties and the data gap between the nanomaterials and environment remains a concern. The novel characteristics of nanotechnology are required for the development of newer technology and innovations like larger specific surface area and electrical and optical properties.

Researchers continue to focus on the environmental impacts of nanomaterials and decide how to manage nanomaterials or their dissolved phases. Usually, nanomaterials pose a risk to the aquatic and terrestrial ecosystems through the novel behavior of nanomaterials like translocation ability, interaction with cellular compartment, and reactivity that might lead to more serious effects on the organism. For example, Levard et al. synthesized Ag-NPs for ascertaining the environmental transformation in terms of stability and toxicity. The data suggested that the Ag binds to the sulfur in the natural system and wastewater plants. The sulfidation of Ag decreases the toxicity of the Ag ions due to their low solubility (Levard, Hotze, Lowry, & Brown, 2012). Xiu et al. tested $AgNO_3$ against *E. coli* bacterial strain. The Ag shows the toxicity against *E. coli* bacteria and the accumulation of Ag ions on the extracellular polymeric membranes, suggesting protective barriers (Z et al., 2014). Keller et al. used Cu for the assessment of environmental fate and toxicity. The data suggested that the exposure to Cu shows the metabolic effects due to the accumulation of Cu and the generation of ROS (Keller et al., 2017). Ottoni et al. synthesized biogenic Ag-NPs by using the fungus *A. tubingensis* for ascertaining the environmental impact. The data suggested that the biogenic Ag-NPs was synthesized for soil microbial community in comparison with that of $AgNO_3$. Moreover, the biogenic Ag-NPs show the dose-dependent inhibitory effects on the germination rate (Ottoni et al., 2020). Wu et al. used Cu and CuO-NPs for the assessment of ecological response. The data suggested that environmental toxicity depends on various factors like aggregation, dissolution, and generation of reactive oxygen species. Moreover, both Cu and CuO-NPs show a higher uptake in *D. magna* and zebrafish. The particle-specific mechanism of toxicity to the *D. magna*, and zebrafish was observed (Wu, Harper, Crandon, & Harper, 2020). In general, environmental toxicity depends on several factors such as aggregation of the nanomaterials, reactivity, dissolution, and generation of ROS. Moreover, biogenic NPs might be less harmful compared with chemically synthesized nanomaterials.

4.7 Physiological and molecular response of plant using nanomaterials

The understanding of the interaction of plants and nanomaterials at physiological and molecular response—mainly uptake, cellular toxicity or phytotoxicity, and translocation—are necessary for the development and growth of the plants. Numerous nanomaterials (categorized based on shape, size, composition, and surface coating, etc.) such as metal-NPs (Cu, Zn, Ce, Fe, and Mg), metal-oxide-NPs (CuO, ZnO, MgO, and AuO), and carbon-based materials (CNTs, CNFs, graphene, and fullerene) have been used for the development and growth of the plants. However, the physiological and molecular response remains a concern, thereby it is difficult to highlight the consistent endpoint. The measurement of the metabolites or other physiological parameters and molecular response reflect biological dose and effects on the exposure of nanomaterials. Numerous studies have shown the physiological and molecular responses of the plants using different nanomaterials. For example, Ma et al. used La_2O_3-NPs and tested them against cucumber plants to ascertain the phytotoxicity and biotransformation of NPs in plants. The data suggested that the La_2O_3-NPs show phytotoxic effects. The dissolution of La_2O_3-NPs at the root surface induces the organic acid that plays the main role in phytotoxicity (Ma et al., 2011). Dimpka et al. used CuO and

ZnO NPs and tested them against wheat plants to ascertain phytotoxicity. The data suggested that the CuO NPs show phytotoxicity against wheat plants, whereas ZnO did not affect the plant growth. Moreover, an increased level of lipid peroxidase and glutathione indicated the increased production of reactive oxygen species (ROS) upon exposure of the NPs (Dimkpa et al., 2012). Ma et al. used CeO_2, and In_2O_3-NPs and tested them against the model plant *Arabidopsis thaliana* to ascertain both physiological and molecular response. The data suggested that both CeO_2, and In_2O_3-NPs increase the fresh biomass of the plants, whereas root length decreased with increasing the concentration. Both CeO_2, and In_2O_3-NPs increase the malondialdehyde level. Both CeO_2, and In_2O_3-NPs increase the antioxidant-related gene expression in the plant (Ma et al., 2013). Fig. 4.2. shows the *Arabidopsis*

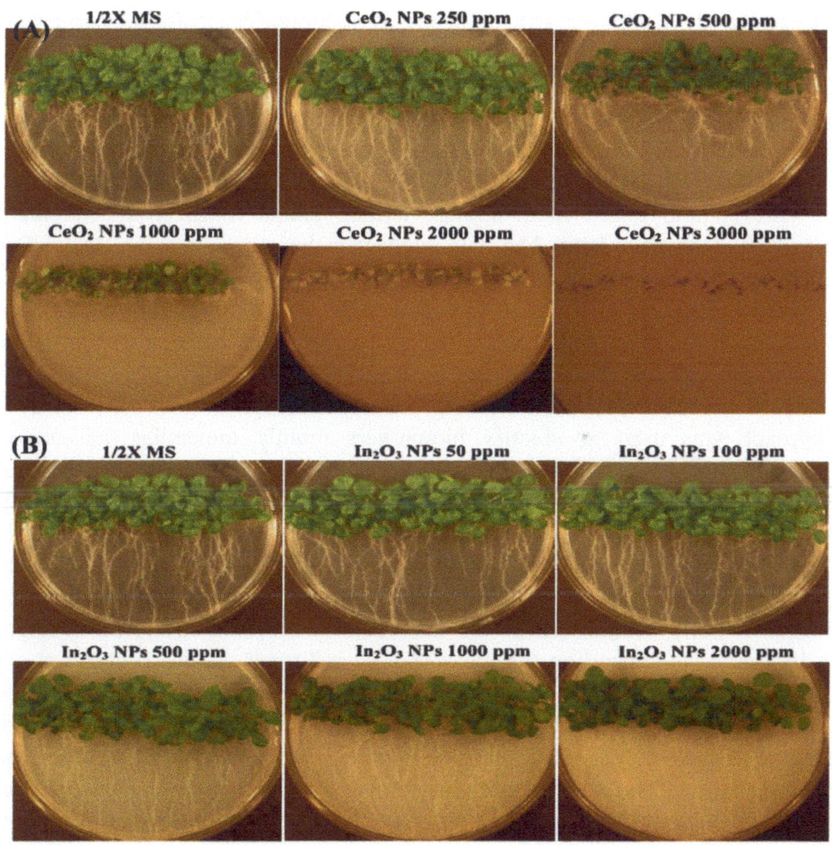

FIGURE 4.2 *Arabidopsis thaliana* exposed to different concentrations of CeO_2 and In_2O_3-NPs. *The image was taken with permission from Ma et al., 2013, Physiological and molecular response of* Arabidopsis thaliana *(L.) to nanoparticle cerium and indium oxide exposure. ACS Sustainable Chemistry & Engineering, 1(7), pp 768–778. Copyright © 2013, American Chemical Society.*

thaliana exposed with different concentrations of CeO_2, and In_2O_3-NPs. Marmiroli et al. used CdS QDs and tested them against the model plant *Arabidopsis thaliana* to ascertain the tolerance based on gene expression. The data suggested that the *atnp01* gene enhances the uptake of CdS-QDs that increases the sulfur uptake. Moreover, the Cd ions effect is less and the effect of CdS-QDs against plants cannot be due to the toxicity of Cd ions (Marmiroli, Pagano, Savo Sardaro, Villani, & Marmiroli, 2014). Hong et al. used nano-Cu, bulk Cu, nano-CuO, bulk CuO, $Cu(OH)_2$, and $CuCl_2$ to test against lettuce and alfalfa plants to ascertain the toxicity effect. The data suggested that all materials increased the Cu, P, and S content in alfalfa shoots, whereas they decreased the P and Fe content in lettuce shoots. Moreover, Cu not only decreased shoot content but also decreased the nutrient value of the plants (Hong et al., 2015). Majumdar et al. used CeO_2-NPs and tested them against kidney beans as a model plant for physiological responses with the help of proteomic analysis. The data suggested that the CeO_2-NPs might change the nutrient quality of the seeds. The proteomic analysis data indicated that the upregulation of stress-related genes at the exposure of CeO_2-NPs. Moreover, no toxic effect was observed upon exposure to CeO_2-NPs (Majumdar et al., 2015). De La Torre Roche et al. used La_2O_3-NPs and tested them against lettuce plants to ascertain the plant growth and phytotoxicity. The data suggested that upon exposure to La_2O_3-NPs plant biomass is reduced (De la Torre Roche et al., 2015). Pagano et al. use CuO, La_2O_3, and CeO_2-NPs and tested them against model plant *Arabidopsis thaliana* to ascertain the molecular response of the plant. The data suggested that the numerous genes were involved in metabolic functions, stress response, synthesis of protein, DNA repair, and transport. This approach was used to identify biomarkers mainly molecular markers in *Solanum lycopersicum* L, and *Cucurbita pepo* L plants upon the exposure of CuO, La_2O_3, and CeO_2-NPs. The data suggested that the CuO-NPs physiological response was similar to that of La_2O_3-NPs, while the transcriptomic response was different depending on the plant species. Moreover, the nonresponse depends on the NPs types and elements, which might indicate the unique toxicity mechanism of the nanomaterials. Additionally, the changes in the response depend on the types of NPs and plant species (Pagano et al., 2016). Another study of the same group shows the binary combination of CuO, La_2O_3, CeO_2, ZnO-NPs, and CdS-QDs and tested them against *Cucurbita pepo* to ascertain the physiological and molecular response. The data suggested that all treatment shows the downregulated expression that might be potential biomarkers for physiological, and molecular response. Moreover, the interaction of the nanomaterials might depend on the various factors, including surface area, surface charge, and stability of the nanomaterials (Pagano et al., 2017). Table 4.1 summarized the different nanomaterials tested against different plants and their physiological and molecular response. The data suggested that the most of the nanomaterials show

TABLE 4.1 Different nanomaterials were tested against different plants and their physiological and molecular response.

S. no.	Nanomaterials	Plant	Responses	References
1.	La_2O_3	Cucumber	La_2O_3 shows phytotoxicity.	Ma et al. (2011)
2.	CuO and ZnO	*Triticum aestivum*	CuO shows phytotoxicity while using ZnO shows no effects on the growth of the plants.	Dimkpa et al. (2012)
			NPs increases the generation of reactive oxygen species.	
3.	CeO_2, and In_2O_3-NPs	*Arabidopsis thaliana*	Increase in fresh biomass of plants.	Ma et al. (2013)
			Increase in antioxidant-related gene expression.	
4.	CdS QDs	*Arabidopsis thaliana*	The toxicity is not associated with Cd ions.	Marmiroli et al. (2014)
5.	Cu	Lettuce, and alfalfa	Increase in Cu, P, and S content in lettuce shoot.	Hong et al. (2015)
			Decrease in P and Fe content in alfalfa shoot.	
6.	CeO_2	Kidney beans	Changes in the nutrient quality of seed.	Majumdar et al. (2015)
			Upregulation of stress-related genes.	
7.	La_2O_3	Lettuce	Decrease in plant biomass upon exposure.	De la Torre Roche et al. (2015)
8.	CuO	*Solanum Lycopersicum* L, and *Cucurbita pepo* L	Significantly reduces the root and shoot growth.	Pagano et al. (2016)

(*Continued*)

TABLE 4.1 (Continued)

S. no.	Nanomaterials	Plant	Responses	References
			Downregulation gene expression at 500 mg/L.	
9.	La_2O_3	*Solanum Lycopersicum* L, and *Cucurbita pepo* L	Significantly reduces the root and shoot growth.	Pagano et al. (2016)
			Downregulation gene expression at 500 mg/L.	
10.	CeO_2	*Solanum Lycopersicum* L, and *Cucurbita pepo* L	No effect on physiological response.	Pagano et al. (2016)
			Produces changes in gene expression.	
11.	Binary combination of CuO, La_2O_3, CeO_2, ZnO-NPs, and CdS-QDs	*Cucurbita pepo*	The interaction of the nanomaterials depends on the surface area, charge, and stability.	Pagano et al. (2017)
			Downregulated gene expression was observed.	

phytotoxic effects. Moreover, upregulated or downregulated gene expression was also observed in the plants. The phytotoxic effects and gene expression might depend on the types of the nanomaterials, different plants, and plant species. The physiological and molecular response of the plants act as biomarkers to understand the exact mechanism of nanomaterials and plant interactions.

4.8 Conclusion

In the current era, nanotechnology in agriculture arises due to the increasing human population and decreasing farmland fertility. To combat nutrient deficiency, fertilizers are fortified for plants requiring nutrients. The technology has an impact on plant physiological processes, fruit biomass enhancement, and microbiome enhancement, yet the long-term toxicity needs to be evaluated. The understanding of interactions with soil, microbiome interaction, and nutrient enrichment in the soil system was necessary for the development and growth of the plants. The studies suggested that there are two important pathways for the translocation and uptake of nutrients within plants, symplastic and apoplastic pathways, whereas based on nanomaterials there are two main factors like size and surface charge of the nanomaterials that are very important for the translocation of nutrients within the plants. On the other hand, the phytotoxic effects and gene expression might depend on the types of the nanomaterials, different plants, and plant species. The physiological and molecular responses of the plants act as biomarkers to understand the exact mechanism of nanomaterials and plant interactions. Therefore focused continuous studies are required to fill the gap of knowledge to understand the exact mechanism between plants and nanomaterials.

Acknowledgment

The authors acknowledge the financial support by CONICYT through FONDECYT project No. 3190515 and 319581.

References

Afreen, S., Omar, R. A., Talreja, N., Chauhan, D., & Ashfaq, M. (2018). Carbon-based nanostructured materials for energy and environmental remediation applications. In R. Prasad, & E. Aranda (Eds.), *Approaches in Bioremediation: The New Era of Environmental Microbiology and Nanobiotechnology*. Cham: Springer International Publishing.

Afreen, S., Talreja, N., Chauhan, D., & Ashfaq, M. (2020). Chapter 15—Polymer/metal/carbon-based hybrid materials for the detection of heavy metal ions. In K. A. Abd-Elsalam (Ed.), *Multifunctional Hybrid Nanomaterials for Sustainable Agri-Food and Ecosystems*. Elsevier.

Ahmed, S., Annu., Ikram, S., & Yudha, S. S. (2016). Biosynthesis of gold nanoparticles: A green approach. *Journal of Photochemistry and Photobiology B: Biology*, *161*, 141–153.

Aiken, G. R., Hsu-Kim, H., & Ryan, J. N. (2011). Influence of dissolved organic matter on the environmental fate of metals, nanoparticles, and colloids. *Environmental Science & Technology*, *45*(8), 3196–3201.

Alidoust, D., & Isoda, A. (2013). Effect of $\gamma Fe2O3$ nanoparticles on photosynthetic characteristic of soybean (Glycine max (L.) Merr.): foliar spray versus soil amendment. *Acta Physiologiae Plantarum*, *35*(12), 3365–3375.

Alimardani, V., Abolmaali, S. S., Tamaddon, A. M., & Ashfaq, M. (2020). Recent advances on microneedle arrays-mediated technology in cancer diagnosis and therapy. *Drug Delivery and Translational Research*.

Anjum, N. A., Adam, V., Kizek, R., Duarte, A. C., Pereira, E., Iqbal, M., ... Ahmad, I. (2015). Nanoscale copper in the soil–plant system—Toxicity and underlying potential mechanisms. *Environmental Research*, *138*, 306–325.

Ashfaq, M., Singh, S., Sharma, A., & Verma, N. (2013). Cytotoxic evaluation of the hierarchical web of carbon micronanofibers. *Industrial & Engineering Chemistry Research*, *52*(12), 4672–4682.

Ashfaq, M., Talreja, N., Chuahan, D., & Srituravanich, W. (2019). Carbon nanostructure-based materials: A novel tool for detection of Alzheimer's disease. In G. M. Ashraf, & A. Alexiou (Eds.), *Biological, Diagnostic and Therapeutic Advances in Alzheimer's Disease: Non-Pharmacological Therapies for Alzheimer's Disease*. Singapore: Springer Singapore.

Ashfaq, M., Verma, N., & Khan, S. (2017a). Carbon nanofibers as a micronutrient carrier in plants: efficient translocation and controlled release of Cu nanoparticles. *Environmental Science: Nano*, *4*(1), 138–148.

Ashfaq, M., Verma, N., & Khan, S. (2017b). Highly effective Cu/Zn-carbon micro/nanofiber-polymer nanocomposite-based wound dressing biomaterial against the P. aeruginosa multi- and extensively drug-resistant strains. *Materials Science and Engineering: C*, *77*, 630–641.

Ashfaq, M., Verma, N., & Khan, S. (2018). Novel polymeric composite grafted with metal nanoparticle-dispersed CNFs as a chemiresistive non-destructive fruit sensor material. *Materials Chemistry and Physics*, *217*, 216–227.

Bayda, S., Adeel, M., Tuccinardi, T., Cordani, M., & Rizzolio, F. (2019). The history of nanoscience and nanotechnology: From chemical–physical applications to nanomedicine. *Molecules (Basel, Switzerland)*, *25*(1), 112.

Bello, S. A., Agunsoye, J. O., & Hassan, S. B. (2015). Synthesis of coconut shell nanoparticles via a top down approach: Assessment of milling duration on the particle sizes and morphologies of coconut shell nanoparticles. *Materials Letters*, *159*, 514–519.

Bhadauriya, P., Mamtani, H., Ashfaq, M., Raghav, A., Teotia, A. K., Kumar, A., & Verma, N. (2018). Synthesis of yeast-immobilized and copper nanoparticle-dispersed carbon nanofiber-based diabetic wound dressing material: Simultaneous control of glucose and bacterial infections. *ACS Applied Bio Materials*, *1*(2), 246–258.

Bobrinetskiy, I. I., & Knezevic, N. Z. (2018). Graphene-based biosensors for on-site detection of contaminants in food. *Analytical Methods*, *10*(42), 5061–5070.

Boverhof, D. R., Bramante, C. M., Butala, J. H., Clancy, S. F., Lafranconi, M., West, J., & Gordon, S. C. (2015). Comparative assessment of nanomaterial definitions and safety evaluation considerations. *Regulatory Toxicology and Pharmacology*, *73*(1), 137–150.

Bruins, M. R., Kapil, S., & Oehme, F. W. (2000). Microbial resistance to metals in the environment. *Ecotoxicology and Environmental Safety*, *45*(3), 198–207.

Calder, A. J., Dimkpa, C. O., McLean, J. E., Britt, D. W., Johnson, W., & Anderson, A. J. (2012). Soil components mitigate the antimicrobial effects of silver nanoparticles towards a beneficial soil bacterium, *Pseudomonas chlororaphis* O6. *Science of The Total Environment*, *429*, 215–222.

Chauhan, D., Afreen, S., Talreja, N., & Ashfaq, M. (2020). Chapter 8—Multifunctional copper polymer-based nanocomposite for environmental and agricultural applications. In K. A. Abd-Elsalam (Ed.), *Multifunctional Hybrid Nanomaterials for Sustainable Agri-Food and Ecosystems*. Elsevier.

Chouhan, R. S., Pandey, A., Qureshi, A., Ozguz, V., & Niazi, J. H. (2016a). Nanomaterial resistant microorganism mediated reduction of graphene oxide. *Colloids and Surfaces B: Biointerfaces*, *146*, 39–46.

Chouhan, R. S., Qureshi, A., Yagci, B., Gülgün, M. A., Ozguz, V., & Niazi, J. H. (2016b). Biotransformation of multi-walled carbon nanotubes mediated by nanomaterial resistant soil bacteria. *Chemical Engineering Journal*, *298*, 1–9.

Colman, B. P., Arnaout, C. L., Anciaux, S., Gunsch, C. K., Hochella, M. F., Jr., Kim, B., ... Bernhardt, E. S. (2013). Low concentrations of silver nanoparticles in biosolids cause adverse ecosystem responses under realistic field scenario. *PLoS One*, *8*(2), e57189.

Davidson, E. A., & Janssens, I. A. (2006). Temperature sensitivity of soil carbon decomposition and feedbacks to climate change. *Nature*, *440*(7081), 165–173.

De la Torre Roche, R., Servin, A., Hawthorne, J., Xing, B., Newman, L. A., Ma, X., ... White, J. C. (2015). Terrestrial trophic transfer of bulk and nanoparticle La2O3 does not depend on particle size. *Environmental Science & Technology*, *49*(19), 11866–11874.

Diaz-Ambrona, C. G. H., & Maletta, E. (2014). Achieving global food security through sustainable development of agriculture and food systems with regard to nutrients, soil, land, and waste management. *Current Sustainable/Renewable Energy Reports*, *1*(2), 57–65.

Dimkpa, C. O., & Bindraban, P. S. (2016). Fortification of micronutrients for efficient agronomic production: A review. *Agronomy for Sustainable Development*, *36*(1), 7.

Dimkpa, C. O., McLean, J. E., Latta, D. E., Manangón, E., Britt, D. W., Johnson, W. P., ... Anderson, A. J. (2012). CuO and ZnO nanoparticles: Phytotoxicity, metal speciation, and induction of oxidative stress in sand-grown wheat. *Journal of Nanoparticle Research*, *14*(9), 1125.

Duhan, J. S., Kumar, R., Kumar, N., Kaur, P., Nehra, K., & Duhan, S. (2017). Nanotechnology: The new perspective in precision agriculture. *Biotechnology Reports*, *15*, 11–23.

Fang, Y., & Ramasamy, R. P. (2015). Current and prospective methods for plant disease detection. *Biosensors*, *5*(3).

Garrigue, P., Delville, M.-H., Labrugère, C., Cloutet, E., Kulesza, P. J., Morand, J. P., & Kuhn, A. (2004). Top-down approach for the preparation of colloidal carbon nanoparticles. *Chemistry of Materials*, *16*(16), 2984–2986.

Ge, Y., Priester, J. H., Van De Werfhorst, L. C., Walker, S. L., Nisbet, R. M., An, Y.-J., ... Holden, P. A. (2014). Soybean plants modify metal oxide nanoparticle effects on soil bacterial communities. *Environmental Science & Technology*, *48*(22), 13489–13496.

Giraldo, J. P., Landry, M. P., Faltermeier, S. M., McNicholas, T. P., Iverson, N. M., Boghossian, A. A., ... Strano, M. S. (2014). Plant nanobionics approach to augment photosynthesis and biochemical sensing. *Nature Materials*, *13*(4), 400–408.

Godfray, H. C. J., Beddington, J. R., Crute, I. R., Haddad, L., Lawrence, D., Muir, J. F., ... Toulmin, C. (2010). Food security: The challenge of feeding 9 billion people. *Science (New York, N.Y.)*, *327*(5967), 812.

Guerra, F. D., Attia, M. F., Whitehead, D. C., & Alexis, F. (2018). Nanotechnology for environmental remediation: materials and applications. *Molecules (Basel, Switzerland)*, *23*(7), 1760.

Gujrati, M., Malamas, A., Shin, T., Jin, E., Sun, Y., & Lu, Z.-R. (2014). Multifunctional cationic lipid-based nanoparticles facilitate endosomal escape and reduction-triggered cytosolic siRNA release. *Molecular Pharmaceutics*, *11*(8), 2734–2744.

Huiling, Z., Min, H., Wenhui, Z., Jorge L, G.-T., Id, O., JC., ... Lijuan, Z. (2020). Silver nanoparticles alter soil microbial community compositions and metabolite profiles in unplanted and cucumber-planted soils. *Environmental Science & Technology*, *54*(6), 3334–3342.

Hasan, A., Nurunnabi, M., Morshed, M., Paul, A., Polini, A., Kuila, T., ... Jaffa, A. A. (2014). Recent advances in application of biosensors in tissue engineering. *BioMed Research International*, *2014*, 307519.

He, S., Feng, Y., Ren, H., Zhang, Y., Gu, N., & Lin, X. (2011). The impact of iron oxide magnetic nanoparticles on the soil bacterial community. *Journal of Soils and Sediments*, *11*(8), 1408–1417.

Hong, J., Rico, C. M., Zhao, L., Adeleye, A. S., Keller, A. A., Peralta-Videa, J. R., & Gardea-Torresdey, J. L. (2015). Toxic effects of copper-based nanoparticles or compounds to lettuce (Lactuca sativa) and alfalfa (Medicago sativa). *Environmental Science: Processes & Impacts*, *17*(1), 177–185.

Iqbal, M. S., Singh, A. K., Singh, S. P., & Ansari, M. I. (2020). Nanoparticles and plant interaction with respect to stress response. In I. Bhushan, V. K. Singh, & D. K. Tripathi (Eds.), *Nanomaterials and Environmental Biotechnology*. Cham: Springer International Publishing.

Irsad., Talreja, N., Chauhan, D., Rodríguez, C. A., Mera, A. C., & Ashfaq, M. (2020). Nanocarriers: An emerging tool for micronutrient delivery in plants. In T. Aftab, & K. R. Hakeem (Eds.), *Plant Micronutrients: Deficiency and Toxicity Management*. Cham: Springer International Publishing.

Jeevanandam, J., Barhoum, A., Chan, Y. S., Dufresne, A., & Danquah, M. K. (2018). Review on nanoparticles and nanostructured materials: History, sources, toxicity and regulations. *Beilstein Journal of Nanotechnology*, *9*, 1050–1074.

Jośko, I., Oleszczuk, P., Dobrzyńska, J., Futa, B., Joniec, J., & Dobrowolski, R. (2019). Long-term effect of ZnO and CuO nanoparticles on soil microbial community in different types of soil. *Geoderma*, *352*, 204–212.

Judy, J. D., Kirby, J. K., Creamer, C., McLaughlin, M. J., Fiebiger, C., Wright, C., ... Bertsch, P. M. (2015). Effects of silver sulfide nanomaterials on mycorrhizal colonization of tomato plants and soil microbial communities in biosolid-amended soil. *Environmental Pollution*, *206*, 256–263.

Keller, A. A., Adeleye, A. S., Conway, J. R., Garner, K. L., Zhao, L., Cherr, G. N., ... Zuverza-Mena, N. (2017). Comparative environmental fate and toxicity of copper nanomaterials. *NanoImpact*, *7*, 28–40.

Khot, L. R., Sankaran, S., Maja, J. M., Ehsani, R., & Schuster, E. W. (2012). Applications of nanomaterials in agricultural production and crop protection: A review. *Crop Protection*, *35*, 64–70.

Köhl, J., Kolnaar, R., & Ravensberg, W. J. (2019). Mode of action of microbial biological control agents against plant diseases: Relevance beyond efficacy. *Frontiers in Plant Science*, *10*, 845.

Kopittke, P. M., Lombi, E., Wang, P., Schjoerring, J. K., & Husted, S. (2019). Nanomaterials as fertilizers for improving plant mineral nutrition and environmental outcomes. *Environmental Science: Nano*, *6*(12), 3513–3524.

Kumar, D., & Talreja, N. (2019). Nickel nanoparticles-doped rhodamine grafted carbon nanofibers as colorimetric probe: Naked eye detection and highly sensitive measurement of aqueous Cr3 + and Pb2 +. *Korean Journal of Chemical Engineering*, *36*(1), 126–135.

Kumar, R., Ashfaq, M., & Verma, N. (2018). Synthesis of novel PVA–starch formulation-supported Cu–Zn nanoparticle carrying carbon nanofibers as a nanofertilizer: Controlled release of micronutrients. *Journal of Materials Science*, *53*(10), 7150–7164.

Levard, C., Hotze, E. M., Lowry, G. V., & Brown, G. E. (2012). Environmental transformations of silver nanoparticles: Impact on stability and toxicity. *Environmental Science & Technology*, *46*(13), 6900–6914.

Li, C., Li, Y., Li, Y., & Fu, G. (2018a). Cultivation techniques and nutrient management strategies to improve productivity of rain-fed maize in semi-arid regions. *Agricultural Water Management*, *210*, 149–157.

Li, W., Wu, S., Zhang, H., Zhang, X., Zhuang, J., Hu, C., ... Wang, X. (2018b). Enhanced biological photosynthetic efficiency using light-harvesting engineering with dual-emissive carbon dots. *Advanced Functional Materials*, *28*(44), 1804004.

Lin, S., Reppert, J., Hu, Q., Hudson, J. S., Reid, M. L., Ratnikova, T. A., ... Ke, P. C. (2009). Uptake, translocation, and transmission of carbon nanomaterials in rice plants. *Small (Weinheim an der Bergstrasse, Germany)*, *5*(10), 1128–1132.

Ma, C., Chhikara, S., Xing, B., Musante, C., White, J. C., & Dhankher, O. P. (2013). Physiological and molecular response of *Arabidopsis thaliana* (L.) to nanoparticle cerium and indium oxide exposure. *ACS Sustainable Chemistry & Engineering*, *1*(7), 768–778.

Ma, Y., He, X., Zhang, P., Zhang, Z., Guo, Z., Tai, R., ... Chai, Z. (2011). Phytotoxicity and biotransformation of La2O3 nanoparticles in a terrestrial plant cucumber (Cucumis sativus). *Nanotoxicology*, *5*(4), 743–753.

Majumdar, S., Almeida, I. C., Arigi, E. A., Choi, H., VerBerkmoes, N. C., Trujillo-Reyes, J., ... Gardea-Torresdey, J. L. (2015). Environmental effects of nanoceria on seed production of common bean (*Phaseolus vulgaris*): A proteomic analysis. *Environmental Science & Technology*, *49*(22), 13283–13293.

Mani, P. K., & Mondal, S. (2016). Agri-nanotechniques for plant availability of nutrients. In C. Kole, D. S. Kumar, & M. V. Khodakovskaya (Eds.), *Plant Nanotechnology: Principles and Practices*. Cham: Springer International Publishing.

Marmiroli, M., Pagano, L., Savo Sardaro, M. L., Villani, M., & Marmiroli, N. (2014). Genome-wide approach in *Arabidopsis thaliana* to assess the toxicity of cadmium sulfide quantum dots. *Environmental Science & Technology*, *48*(10), 5902–5909.

Mehrotra, P. (2016). Biosensors and their applications—A review. *Journal of Oral Biology and Craniofacial Research*, *6*(2), 153–159.

Milani, N., Hettiarachchi, G. M., Kirby, J. K., Beak, D. G., Stacey, S. P., & McLaughlin, M. J. (2015). Fate of zinc oxide nanoparticles coated onto macronutrient fertilizers in an alkaline calcareous soil. *PLoS One*, *10*(5), e0126275.

Manoj K, R., Jain, A., & Singh, S. (2011). Studies on binary lipid matrix based solid lipid nanoparticles of repaglinide: in vitro and in vivo evaluation. *Journal of Pharmaceutical Sciences*, *100*(6), 2366–2378.

Mogilevsky, G., Hartman, O., Emmons, E. D., Balboa, A., DeCoste, J. B., Schindler, B. J., ... Karwacki, C. J. (2014). Bottom-up synthesis of anatase nanoparticles with graphene domains. *ACS Applied Materials & Interfaces*, *6*(13), 10638–10648.

Mohammad Ashfaq, N.T., D. Chuahan, W. Srituravanich 2020. Polymeric Nanocomposite-Based Agriculture Delivery System: Emerging Technology for Agriculture, Farrukh Jamal: IntechOpen.

Moll, J., Gogos, A., Bucheli, T. D., Widmer, F., & van der Heijden, M. G. A. (2016). Effect of nanoparticles on red clover and its symbiotic microorganisms. *Journal of Nanobiotechnology*, *14*(1), 36.

Mondal, S. (2020). Potential of nanotechnology for rural applications. *Arabian Journal for Science and Engineering*, *45*(7), 5011–5042.

Nafees, M., Ali, S., Rizwan, M., Aziz, A., Adrees, M., Hussain, S. M., ... Junaid, M. (2020). Effect of nanoparticles on plant growth and physiology and on soil microbes. In I. Bhushan, V. K. Singh, & D. K. Tripathi (Eds.), *Nanomaterials and Environmental Biotechnology*. Cham: Springer International Publishing.

Needham, D., Arslanagic, A., Glud, K., Hervella, P., Karimi, L., Høeilund-Carlsen, P.-F., ... Walke, P. (2016). Bottom up design of nanoparticles for anti-cancer diapeutics: "put the drug in the cancer's food". *Journal of Drug Targeting*, *24*(9), 836–856.

Neina, D. (2019). The role of soil pH in plant nutrition and soil remediation. *Applied and Environmental Soil Science*, *2019*, 5794869.

Omar, R. A., Afreen, S., Talreja, N., Chauhan, D., & Ashfaq, M. (2019a). Impact of nanomaterials in plant systems. In R. Prasad (Ed.), *Plant Nanobionics: 1, Advances in the Understanding of Nanomaterials Research and Applications*. Cham: Springer International Publishing.

Omar, R. A., Afreen, S., Talreja, N., Chauhan, D., Ashfaq, M., & Srituravanich, W. (2019b). Impact of nanomaterials on the microbial system. In R. Prasad (Ed.), *Microbial Nanobionics: Volume 1, State-of-the-Art*. Cham: Springer International Publishing.

Ottoni, C. A., Lima Neto, M. C., Léo, P., Ortolan, B. D., Barbieri, E., & De Souza, A. O. (2020). Environmental impact of biogenic silver nanoparticles in soil and aquatic organisms. *Chemosphere*, *239*, 124698.

Pagano, L., Pasquali, F., Majumdar, S., De la Torre-Roche, R., Zuverza-Mena, N., Villani, M., ... Marmiroli, N. (2017). Exposure of Cucurbita pepo to binary combinations of engineered nanomaterials: Physiological and molecular response. *Environmental Science: Nano*, *4*(7), 1579–1590.

Pagano, L., Servin, A. D., De La Torre-Roche, R., Mukherjee, A., Majumdar, S., Hawthorne, J., ... Marmiroli, N. (2016). Molecular response of crop plants to engineered nanomaterials. *Environmental Science & Technology*, *50*(13), 7198–7207.

Panpatte, D. G., Jhala, Y. K., Shelat, H. N., & Vyas, R. V. (2016). Nanoparticles: The next generation technology for sustainable agriculture. In D. P. Singh, H. B. Singh, & R. Prabha (Eds.), Microbial Inoculants in Sustainable Agricultural Productivity: 2*: Functional Applications*. New Delhi: Springer India.

Parveen, K., Banse, V., & Ledwani, L. (2016). Green synthesis of nanoparticles: Their advantages and disadvantages. *AIP Conference Proceedings*, *1724*(1), 020048.

Patra, J. K., Das, G., Fraceto, L. F., Campos, E. V. R., Rodriguez-Torres, M. D. P., Acosta-Torres, L. S., ... Shin, H.-S. (2018). Nano based drug delivery systems: recent developments and future prospects. *Journal of Nanobiotechnology*, *16*(1), 71.

Prasad, R., Bhattacharyya, A., & Nguyen, Q. D. (2017). Nanotechnology in sustainable agriculture: recent developments, challenges, and perspectives. *Frontiers in Microbiology*, *8*, 1014.

Priyadarshana, G., Kottegoda, N., Senaratne, A., de Alwis, A., & Karunaratne, V. (2015). Synthesis of magnetite nanoparticles by top-down approach from a high purity ore. *Journal of Nanomaterials*, *2015*, 317312.

Raliya, R., Biswas, P., & Tarafdar, J. C. (2015). TiO2 nanoparticle biosynthesis and its physiological effect on mung bean (*Vigna radiata* L.). *Biotechnology Reports*, *5*, 22–26.

Ramezani, M., Ramezani, F., & Gerami, M. (2019). Nanoparticles in pest incidences and plant disease control. In D. G. Panpatte, & Y. K. Jhala (Eds.), *Nanotechnology for Agriculture: Crop Production & Protection*. Singapore: Springer Singapore.

Ray, P. C., Yu, H., & Fu, P. P. (2009). Toxicity and environmental risks of nanomaterials: Challenges and future needs. *Journal of Environmental Science and Health. Part C, Environmental Carcinogenesis & Ecotoxicology Reviews*, *27*(1), 1–35.

Rui, M., Ma, C., Hao, Y., Guo, J., Rui, Y., Tang, X., ... Zhu, S. (2016). Iron oxide nanoparticles as a potential iron fertilizer for peanut (*Arachis hypogaea*). *Frontiers in Plant Science*, *7*, 815.

Shakiba, S., Astete, C. E., Paudel, S., Sabliov, C. M., Rodrigues, D. F., & Louie, S. M. (2020). Emerging investigator series: Polymeric nanocarriers for agricultural applications: synthesis, characterization, and environmental and biological interactions. *Environmental Science: Nano*, *7*(1), 37–67.

Shang, Y., Hasan, M. K., Ahammed, G. J., Li, M., Yin, H., & Zhou, J. (2019). Applications of nanotechnology in plant growth and crop protection: A review. *Molecules (Basel, Switzerland)*, *24*(14), 2558.

Shebl, A., Hassan, A. A., Salama, D. M., Abd El-Aziz, M. E., & Abd Elwahed, M. S. A. (2019). Green synthesis of nanofertilizers and their application as a foliar for *Cucurbita pepo* L. *Journal of Nanomaterials*, *2019*, 3476347.

Srinivasan, B., & Tung, S. (2015). Development and applications of portable biosensors. *Journal of Laboratory Automation*, *20*(4), 365–389.

Sundström, J. F., Albihn, A., Boqvist, S., Ljungvall, K., Marstorp, H., Martiin, C., ... Magnusson, U. (2014). Future threats to agricultural food production posed by environmental degradation, climate change, and animal and plant diseases—A risk analysis in three economic and climate settings. *Food Security*, *6*(2), 201–215.

Talreja, N., Verma, N., & Kumar, D. (2016). Carbon bead-supported ethylene diamine-functionalized carbon nanofibers: An efficient adsorbent for salicylic acid. *CLEAN – Soil, Air, Water*, *44*(11), 1461–1470.

Tiwari, J. N., Tiwari, R. N., & Kim, K. S. (2012). Zero-dimensional, one-dimensional, two-dimensional and three-dimensional nanostructured materials for advanced electrochemical energy devices. *Progress in Materials Science*, *57*(4), 724–803.

Tong, Z., Bischoff, M., Nies, L. F., Myer, P., Applegate, B., & Turco, R. F. (2012). Response of soil microorganisms to as-produced and functionalized single-wall carbon nanotubes (SWNTs). *Environmental Science & Technology*, *46*(24), 13471–13479.

Usman, M., Farooq, M., Wakeel, A., Nawaz, A., Cheema, S. A., Rehman, H. U., . . . Sanaullah, M. (2020). Nanotechnology in agriculture: Current status, challenges and future opportunities. *Science of The Total Environment*, *721*, 137778.

Vittori Antisari, L., Antonietta, G., Gilmo, V., Paolo, N., & Serena, C. (2013). Toxicity of metal oxide (CeO2, Fe3O4, SnO2) engineered nanoparticles on soil microbial biomass and their distribution in soil. *Soil Biology and Biochemistry*, *60*, 87–94.

Wiatrowska, K., & Komisarek, J. (2019). Role of the light fraction of soil organic matter in trace elements binding. *PLoS One*, *14*(5), e0217077.

Wu, F., Harper, B. J., Crandon, L. E., & Harper, S. L. (2020). Assessment of Cu and CuO nanoparticle ecological responses using laboratory small-scale microcosms. *Environmental Science: Nano*, *7*(1), 105–115.

Wu, H., Tito, N., & Giraldo, J. P. (2017). Anionic cerium oxide nanoparticles protect plant photosynthesis from abiotic stress by scavenging reactive oxygen species. *ACS Nano*, *11*(11), 11283–11297.

Yang, Y., Wang, J., Xiu, Z., & Alvarez, P. J. J. (2013). Impacts of silver nanoparticles on cellular and transcriptional activity of nitrogen-cycling bacteria. *Environmental Toxicology and Chemistry*, *32*(7), 1488–1494.

Younas, A., Yousaf, Z., Rashid, M., Riaz, N., Fiaz, S., Aftab, A., & Haung, S. (2020). Nanotechnology and plant disease diagnosis and management. In S. Javad (Ed.), *Nanoagronomy*. Cham: Springer International Publishing.

Yuan, Z., Zhang, Z., Wang, X., Li, L., Cai, K., & Han, H. (2017). Novel impacts of functionalized multi-walled carbon nanotubes in plants: promotion of nodulation and nitrogenase activity in the rhizobium-legume system. *Nanoscale*, *9*(28), 9921–9937.

Z, X., Liu Y Fau-Mathieu, J., Mathieu J Fau-Wang, J., Wang J Fau-Zhu, D., Zhu D Fau-Alvarez, P. J. J., & PJ, A. (2014). Elucidating the genetic basis for *Escherichia coli* defense against silver toxicity using mutant arrays. *Environmental Toxicology and Chemistry*, *33*(5), 993–997.

Zaytseva, O., & Neumann, G. (2016). Carbon nanomaterials: Production, impact on plant development, agricultural and environmental applications. *Chemical and Biological Technologies in Agriculture*, *3*(1), 17.

Zhao, L., Zhang, H., White, J. C., Chen, X., Li, H., Qu, X., & Ji, R. (2019). Metabolomics reveals that engineered nanomaterial exposure in soil alters both soil rhizosphere metabolite profiles and maize metabolic pathways. *Environmental Science: Nano*, *6*(6), 1716–1727.

Chapter 5

Transcription factors involved in plant responses to stress adaptation

Sahil[1], Radhika Keshan[1] and Shabir A. Rather[2]
[1]*Department of Botany, University of Delhi, New Delhi, India,* [2]*College of Life Sciences, Northwest A&F University, Xianyang, P.R. China*

5.1 Introduction

The exposure of various organisms to the ever-changing external environmental conditions resulting in various stresses forces them to employ mechanisms to either survive under these stresses or escape these stresses. Among all the different organisms, because of their immobile nature plants are predominantly exposed to these stresses. Therefore, to minimize the repercussions of these stresses, plants have evolved to cope with the changing environmental conditions and the stresses caused after that. Many stress response mechanisms evolved in plants based on the timing and duration of stress to adapt and survive under these stressful conditions. The plant stress responses involve changes at different levels, either at a physiological, morphological, biochemical, or molecular level. Changes at the molecular level involve perception, signal transduction, and changes in expression of genes modulated by the regulation at different levels of central dogma like transcriptional level, posttranscriptional level, and posttranslational level. Genes induced by environmental stress are mainly categorized into two categories based on the protein products encoded by these genes: the first group includes protein products that are directly involved in providing resistance like Heat Shock Proteins (HSPs) or chaperones that play a role in protein folding, protein refolding, and removal of unfolded protein, late embryogenesis abundant (LEA), some osmoregulators, etc. The second group includes genes whose coded protein product is responsible for regulating the expression of other genes and stress signaling pathways, usually the transcriptional factors. At the transcriptional level, various transcription factors (TFs) are either positive or negative regulators of the expression of stress-responsive

Frontiers in Plant–Soil Interaction. DOI: https://doi.org/10.1016/B978-0-323-90943-3.00018-3

genes. So, for a better understanding of molecular changes in plants exposed to various stresses, regulation by the TFs needs to be extensively studied.

TFs are the proteins not only required for the initiation of transcription but also for the positive or negative regulation of gene expression. TFs with their DNA binding domain can bind to a particular DNA sequence of the promoter region and either positively or negatively regulate the expression of that particular gene with the help of their transcription activation or repression domain. Until now, various TFs are known to be involved in different abiotic stresses, for example, bHLH (basic Helix-Loop-Helix) (X. B. Xie et al., 2012; Q. Zhao et al., 2016; C. Zhao et al., 2016), bZip (basic region/leucine Zipper) (An et al., 2017), C-repeat Binding Factor/Dehydration Responsive Element-Binding Protein (CBF/DREB) (S. Yang, Tang, Ma, Wang, & Meng, 2011), Zinc Finger Protein (ZFP) (Fang et al., 2019), Abscisic acid-Response Element-Binding Protein/Abscisic acid-Response Factor (AREB/ARF) (Y. R. Ren et al., 2019), MYB (Myeloblastosis) (An et al., 2020; L. J. Zhou et al., 2017), and NAC [NAM (no apical meristem), ATAF1 and 2, and CUC2 (cup-shaped cotyledon)] families (An et al., 2018; X. D. Li et al., 2016).

5.2 Stress signaling in plants

The initial stage of stress signaling in plants involves the perception of a stress stimulus by different plant sensors followed by the induction of secondary messengers like nitric oxide (NO), cyclic adenosine monophosphate (cAMP), etc. These secondary messengers further activate different pathways, including the mitogen-activated protein kinase (MAPK) pathway and various other pathways, resulting in either activation or inactivation of TFs. These TFs further modulate the transcription of stress-responsive genes by interacting with the *cis*-regulatory elements of the promoter of target genes (Danquah, de Zelicourt, Colcombet, & Hirt, 2014). The TFs themselves are regulated by the binding of upstream elements to a specific region in their promoter and by various posttranscription modifications like sumoylation and ubiquitination etc., thus regulating the plant stress responses (C. H. Dong, Agarwal, Zhang, Xie, & Zhu, 2006; Mizoi et al., 2013).

5.2.1 Stress perception in plants

Any changes in the plant resulting from stressful environmental conditions can only occur if the plant perceives the stress stimulus. This emphasizes the presence of sensors in a plant that perceive the stimulus, but only a few of them have yet been identified. For example, a putative hyperosmotic sensor in *Arabidopsis* is OSCA1 (reduced hyperosmolality-induced calcium increase1) (Yuan et al., 2014). All the different stresses like heat stress, cold stress, salinity stress, osmotic stress, and heavy metal stress caused a rise in the concentration of cytosolic free calcium. This spike in concentration was reduced in *osca1* loss of function mutants. OSCA1 encodes hyperosmolality-gated calcium-permeable channel proteins. However, the mechanism of osmotic stress sensing by OSCA1 is still unknown. During stress, the presence of cyclic nucleotide-gated channels and glutamate receptor-like (GLR) channels in plants prove crucial in producing cytosolic Ca^{2+} signals (Swarbreck, Colaço, & Davies, 2013).

The cold stress sensing in rice was found to be moderated by another putative sensor named COLD1. COLD1 interacts with the alpha subunit of heteromeric G-protein, thereby regulating chilling tolerance in plants (Y. Ma et al., 2015). COLD1 is speculated to either regulate a calcium channel or itself be a calcium channel (Y. Ma et al., 2015). However, the downstream mechanism of cold stress tolerance is still unclear.

Temperature stress, either low or high, alters the membrane fluidity (Sangwan, Örvar, Beyerly, Hirt, & Dhindsa, 2002). Hence, this type of abiotic stress can be sensed by the integral membrane proteins, transporters, channels, etc. Heat stress can be sensed by HSPs that bind to the unfolded proteins present due to heat stress (Scharf, Berberich, Ebersberger, & Nover, 2012). Once the HSPs bind to unfolded proteins, the Heat Shock Factors (HSFs) that were earlier inactivated by the binding HSPs now get released and further activate stress-responsive genes including HSPs.

5.2.2 Signal transduction

Once the plants perceive the stress stimulus, the extracellular signal gets converted into an intracellular one by secondary messengers such as calcium ions, NO, and cyclic nucleotides (cAMP and cGMP). These secondary messengers further initiate signal transduction pathways including a series of events involving phosphorylation and dephosphorylation of proteins by enzymes, namely kinases and phosphatases. For example, the MAPK pathway and calcium-dependent protein kinase (CDPK) pathway are activated by secondary messengers in response to abiotic stress (G. T. Huang et al., 2012; Schaller, Kieber, & Shiu, 2008).

In the MAPK pathway, receptor tyrosine kinase (RTK) gets activated in response to a particular stimulus such as light, that in turn activates a monomeric G-protein present downstream to RTK and termed as RAS by exchanging the already present GDP with GTP using Guanine nucleotide Exchange Factor (GEF) and SOS. Interaction of RTK with GEF occurs through an adaptor protein, GRB2. This activation of RAS triggers a downstream kinase cascade ultimately leading to the activation of MAPK. RAF, a serine/threonine kinase, is the first kinase present just downstream to RAS (Morris, 2001). RAF is a serine/threonine kinase then phosphorylates two serine residues and thereby activates MEK (MAP-kinase-kinase). MEK is a dual-specificity enzyme that phosphorylates threonine and tyrosine and activates MAP kinase.

MAPKs are serine/threonine kinases that can phosphorylate particular protein at a specific serine/threonine residue, thereby activating or inactivating TFs that can attach to the promoter of stress-responsive genes and modulate their expression. MAPKs are thought to function in two ways: either moving inside the nucleus and phosphorylating a particular transcription factor or staying in the cytoplasm and passing the signal to cytoskeletal binding proteins or certain enzymes like protein kinases, phosphatases, etc., further activating a downstream signal transduction pathway.

Similar to the MAPK pathway, the CDPK pathway, and other signal transduction pathways activated during stress involve a series of phosphorylation and dephosphorylation events leading to activation or inactivation of target genes.

5.2.3 Expression of stress-responsive genes

The phosphorylation cascade ends up in either activation or inactivation of TFs that further associate with the specific sequences in the promoter region of stress-responsive genes, thereby regulating their expression (Danquah et al., 2014).

Genes activated by environmental stress are mainly categorized into two categories based on the protein products encoded by these genes: the first group includes genes, which includes protein products that are directly involved in providing resistance like HSPs or chaperones involved in protein folding, protein refolding, and removal of unfolded protein, LEAs, some osmoregulators, etc. The second group includes genes whose coded protein product is further responsible for regulating the expression of other genes and stress signaling pathways, usually the transcriptional factors.

In general, abiotic stresses are known to vary the amount of many transcripts and proteins by controlling gene expression and turnover of protein, implying an essential role of transcriptional and posttranscriptional regulation played in the protection of plants from various stresses at the cellular level. The generation of important metabolic proteins such as HSPs and proteins participating in the modulation of stress signaling pathways, such as kinases like MAPK or TFs are required for responses to different abiotic stresses (Wimalasekera & Scherer, 2018). To maintain a balance in stress adaptation, there is an instant production of new transcripts and after some time, a steady state is reached (Fig. 5.1).

5.3 Role of transcription factors in stress conditions

Stress genes are induced primarily at the level of transcription when spatial and temporal regulation of specific stress genes is a must for plant stress response (Rushton & Somssich, 1998). A massive region of the plant genome is devoted to transcription, as seen in *Arabidopsis*, in which the

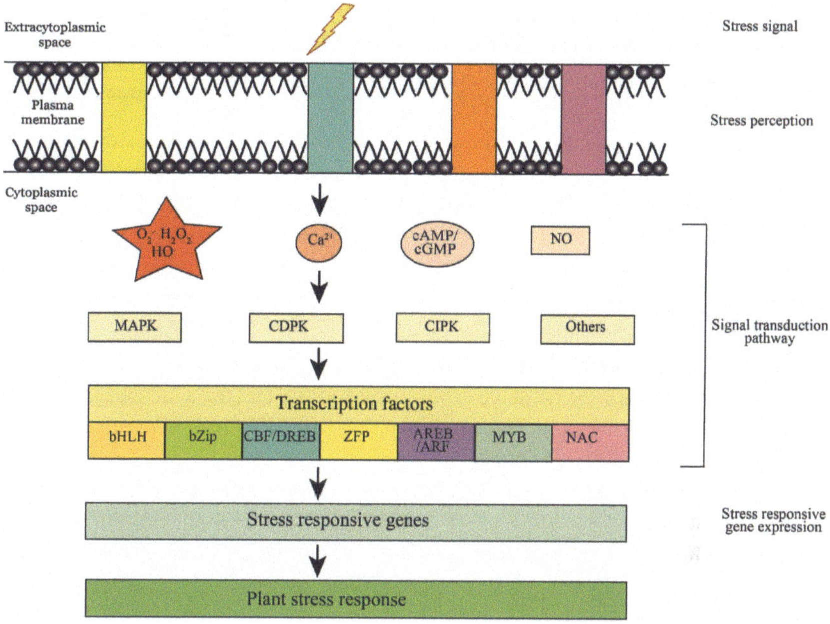

FIGURE 5.1 Plant stress signaling pathway leading to stress response in plants. *AREB/ARF*, Abscisic acid-Response Element-Binding Protein/Abscisic acid-Response Factor; *CBF*, C-repeat Binding Factor; *CDPK*, calcium-dependent protein kinase; *DREB*, Dehydration Responsive Element-Binding Protein; *MAPK*, mitogen-activated protein kinase; *MYB*, myeloblastosis oncogene; *NO*, nitric oxide; *ZFP*, Zinc Finger Protein.

number of TFs coded is more than 1500 (Riechmann et al., 2000). TFs mostly are part of large gene families that sometimes may be unique to plants. In plants, nearly 10% of genes encode TFs to regulate specific signaling-mediated function at different stages/points (Gonzalez, 2016). There are numerous TF databases available that are unique to different crops, like the grass TF database and plant transcription factor database. Major TF families, such as bZIPs (basic domain leucine zipper), WRKY, NAC (NAM, ATAF, and CUC), myeloblastosis oncogene (MYB), and AP2/ERF (ethylene-responsive-element-binding factors), are central regulators of different stress-related genes that lead to the optimal choice of genetic engineering for improving plant resistance to various stresses and their stimuli (H. Wang, Wang, Shao, & Tang, 2016).

5.3.1 bZIP transcription factors

More than seventy-five members are reported in *Arabidopsis* as belonging to the bZIPs, which in plants make up a large family of TFs (Jakoby et al., 2002). A group of stress-responsive bZIP proteins consists of TGA/octopine synthase (OCS)-element binding factor (OBF) proteins. Their binding to the activation sequence-1 (as-1)/ocs element modulates the stress-responsive gene expression like the GLUTATHIONE S-TRANSFERASE-6 and PR-1 genes (Chen & Singh, 1999; Lebel et al., 1998). In *Arabidopsis*, seven members of the TGA/OBF family have been reported, which are important in development, plant defense, and stress response. An essential element of the SA defense signaling pathway, NON-EXPRESSOR OF PR1 (NPR1), interacts with TGA/OBF family members in the nucleus. The interaction is enhanced by the treatment of SA and increases the proteins' DNA-binding activity. DNA binding specificity, properties of protein and protein interactions, and expression pattern changes of individual TGA/OBF proteins. The PR expression can also reduce by SA treatment. Therefore TGA/OBF factors play negative as well as positive roles in response to stress in plants (Fan & Dong, 2002; Subramaniam, Desveaux, Spickler, Michnick, & Brisson, 2001; J. M. Zhou et al., 2000)

ABRE binding factor ((ABF)/ABA-responsive-element-binding (AREB)) proteins, a conserved *cis*-element (PyACGTGG/TC), respond to abiotic stresses like salt and drought stress at both levels of transcription as well as post-transcription. The ABF/AREB proteins play a role in the abscisic acid (ABA)-dependent signal transduction pathway (Choi, Hong, Ha, Kang, & Kim, 2000; Uno et al., 2000). Reported in *Arabidopsis*, the overexpression of either ABF3 or ABF4 proteins may express ABA-associated phenotypes like alteration of ABA/stress-regulated genes; hypersensitivity toward ABA and increased tolerance to drought by a reduction in transpiration (Kang, Choi, Im, & Kim, 2002)

In a study, regulation at the level of transcription shows the intricacy of bZIP as pathogen attacks result in rising protection from UV via an inverse relation to an ACE promoter motif or ACGT-containing element (ACE). However, two similar ACE motifs contribute to the regulation of a UV-responsive element and also an elicitor responsive element that has an adverse action, allowing the closure of less important protection against UV when the plant is under biotic stress like pathogen attack (Logemann & Hahlbrock, 2002).

5.3.2 NAC transcription factors

NAC stands for the three genes that were first found to have the NAC domain: NAM (No Apical Meristem), ATAF1 and -2, and CUC2 (Cup-shaped Cotyledon). NAC TFs are involved in the plant's innate immune system, basal defense, and systemic acquired resistance (SAR) (Nuruzzaman, Sharoni, & Kikuchi, 2013).

Six OsNAC genes in rice seedlings are induced by the biotic introduction of Rice stripe virus and Rice tungro spherical virus (Nuruzzaman et al., 2010). Defense against *Magnaporthe grisea* is regulated by a number of regulatory proteins like OsNAC6 (Nakashima et al., 2007). Mutation of rim1-1 in rice plants makes it infection-resistant to the Dwarf virus (Satoh et al., 2011; Yoshii et al., 2010). In *Solanum tuberosum*, infection by *Phytophthora infestans* induces StNAC genes as a stress response (Collinge & Boller, 2001).

Besides biotic stress, many NAC genes initiate responses to abiotic stresses in plants, like cold, salinity, flood, and drought (Jeong et al., 2010). NAC family TFs activate the ERD1 (Early Responsive to Dehydration stress 1) gene, studied by ERD1 promoter analysis (Tran et al., 2007). Activation of PR genes induces hypersensitive response, and also cell death at infected regions occurs by the action of NAC proteins, which positively regulates the plant defenses (Jensen et al., 2008; Kaneda et al., 2009). SNAC1 functions in guard cells during dehydration (Hu et al., 2006). BnNAC gene of *Brassica* and ANAC019, ANAC055, and ANAC072 genes from *Arabidopsis* are observed to play a role in establishing a response against environmental stresses (Hegedus et al., 2003; Tran et al., 2004). AtNAP is activated during leaf senescence (Guo & Gan, 2006). In conditions of limited water supply, the transgenic *Arabidopsis* and cotton have shown increased root growth due to GhNAC2 (Gunapati et al., 2016). Resistance to chilling and drought stress is obtained by overexpressing PbeNAC1 in *Pyrus betulifolia* via interaction with PbeDREBs (C. Jin et al., 2017). Negative regulation of drought stress can be seen in *Arabidopsis* and tomato by overexpression of few NAC genes like AtNAC016 and ShNAC1 (H. Liu et al., 2018; Sakuraba, Kim, Han, Lee, & Paek, 2015)

5.3.3 AP2/ERF transcription factors

The APETALA 2/ETHYLENE-RESPONSIVE FACTOR (AP2/ERF) transcription factor family includes many plant-specific TFs. It includes four important subfamilies- DEHYDRATION-RESPONSIVE ELEMENT-BINDING PROTEIN (DREB), ERF, RELATED TO ABSCISIC ACID INSENSITIVE 3/VIVIPAROUS 1 (RAV), and AP2 (Yoon et al., 2020). In the N-terminal region of the AP2/ERF TFs is the approximately 60-amino acid-long AP2 domain helping in the binding of DNA and at the C-terminal is a regulatory domain important for transcriptional regulation (Z. Xie, Nolan, Jiang, & Yin, 2019).

In cold conditions, DREB1s/CBFs expression is highly upregulated, and by overexpressing DREB1s, the freezing stress tolerance of the transgenic *Arabidopsis*, the plant is enhanced. At the same time, the DREB1s mutant plants with knockout genes showed very little tolerance to chilling. The study emphasizes the importance of DREB1 to impart tolerance in cold stress (Q. Zhao et al., 2016; C. Zhao et al., 2016). A few similar reports display the overexpression of *DREB1B/CBF1* or *DREB1A/CBF3* enhancing abiotic stress tolerance like dryness, salt, and cold stress indicating that multiple genes are affected by DREBs/CBFs (Jaglo-Ottosen, Gilmour, Zarka, Schabenberger, & Thomashow, 1998; Q. Liu et al., 1998). A role in signal transduction during plant stress of *OsDREB1F* occurs in both rice and *Arabidopsis* when its overexpression makes the plant tolerant to low temperature, drought, and high salinity (Q. Wang et al., 2008). Other ERF TFs in rice are OsDREB1, OsEREBP1, and OsERF71, which expressions regulate stressful conditions, and their overexpression regulates the transcription of the stress-responsive genes at the time of stress (Dubouzet et al., 2003; Jisha et al., 2015; D. K. Lee, Yoon, Kim, & Kim, 2017).

Overexpression of Tobacco stress-induced gene 1 (Tsi)—an ERF gene in tobacco—increases resistance to salinity stress and pathogen attacks in the transgenic tobacco plants (Park et al., 2001). An ERF gene of tomato, PTI5, enhanced the expression of some PR genes during pathogenic infections when overexpressed in *Arabidopsis* (P. He et al., 2001). Overexpressed ERF1 of *Arabidopsis* displays increased resistance for necrotrophic fungi; however, a decrease in resistance for *Pseudomonas syringae* is observed (Berrocal-Lobo, Molina, & Solano, 2002)

5.3.4 WRKY transcription factors

WRKY TFs consist of the WRKYGQK motif that is highly conserved at the N-terminal area and a conserved zinc-finger-like motif at the C-terminal end (Rushton, Somssich, Ringler, & Shen, 2010). WRKYs participate in generating plant responses against abiotic stresses through stress-responsive expression patterns. Via functional characterization using overexpression or knockout mechanisms, the participation of WRKYs in plant stress responses has been shown. Expression of AtWRKY25, 26, 33, and 39 in *Arabidopsis* is induced by heat stress. The double mutants of these genes had similar growth and development to the control plants in normal conditions. However, their survival rate was very low in addition to having high sensitivity toward high temperatures. However, the overexpression of AtWRKY25, 26, 33, and 39 genes demonstrated an increase in heat tolerance (S. Li, Zhou, Chen, Huang, & Yu, 2010; S. Li, Fu, Chen, Huang, & Yu, 2011). WRKYs expression also plays a role in salinity tolerance (Jiang & Deyholos, 2006). Furthermore, it is also reported that similar to NACs and AP2/ERFs, WRKYs are also involved in controlling the development of the plant under environmental stress. Stress-responsive WRKY 22 modulates senescence due to low light and WRKY 46 mediates the lateral roots' growth under stress (Z. J. Ding et al., 2015; X. Zhou, Jiang, & Yu, 2011). ABA-mediated abiotic stress signaling through WRKYs was established after reports showed that expression of AtWRKY18, AtWRKY40, AtWRKY60, AtWRKY63, and GhWRKY17 affected the ABA responses (X. Ren et al., 2010; Yan et al., 2014). In *Arabidopsis* studies of gain-of-function, tolerance in drought increased by overexpression of OsWRKY11 and 45 which are stress-induced in rice and TaWRKY1 and 33 wheat (Wu, Shiroto, Kishitani, Ito, & Toriyama, 2009; Qiu and Yu, 2009; G. H. He et al., 2016). These reports demonstrate the role of the WRKYs gene in stress response and suggest its beneficial use in the production of tolerant crops.

WRKY family TFs in plants are also part of pathogen-associated molecular pattern-triggered immunity (PAMP-triggered immunity), SAR, or effector-triggered immunity (F. Chen et al., 2017). For example, in *Cucumis sativus*, a pathogenic attack by *Pseudopernospora cubensis* shows the action of CsWRKY50 in stress (Luan et al., 2019). A gene of the JA pathway, VvWRKY1, in *Vitis vinifera* responds to stress tolerance against downy mildew (Marchive et al., 2013). Similarly, in *Nicotiana*, CaWRKY27, a JA-regulated gene, enhances resistance for infection by *Ralstonia solanacearum*. Tolerance against fungal and bacterial pathogens is enhanced by GhWRKY44, a TF important in cotton–pathogen interactions studied in *Nicotiana* (Dang et al., 2014). WRKY3 in sugarcane was upregulated on the attack by *Xanthomonas albilineans* (Ntambo et al., 2019). WsWRKY1, AcWRKYs, and GmWRKY31 play a vital role in response to pathogenic stress in various crops (H. Dong et al., 2019; Jing & Liu, 2018; Singh et al., 2017). Negative regulation is also seen in some of these WRKY TFs toward fungal pathogens. Overall regulation and signal transduction responses mediated by WRKY TFs supports the enhancement of stress tolerance due to both abiotic and biotic agents.

5.3.5 MYB transcription factors

The DNA-binding domain of MYB is conserved among the members of the large family of MYB TFs (Yanhui et al., 2006). The domain of MYB is made up of approximately 50 amino acids. Depending on MYB repeats' number and position, the family is divided into four subfamilies: 1R-MYB, R2R3-MYB, R1R2R3-, and 4R-MYB (Dubos et al., 2010). Plants have about 100 R2R3-MYBs which are plant-specific MYB TFs (Wilkins, Nahal, Foong, Provart, & Campbell, 2009). Studies report the expression of MYB genes and elements regulated by stresses to make the plant tolerant or resistant.

In *Arabidopsis*, tolerance to salt stress is observed by expression of AtMYB20, whereas its' silencing causes hypersensitivity toward salinity (Cui et al., 2013). The expression of AtMYB44 is highlighted in guard cells, which is enhanced in stressful conditions like cold, drought, and salinity. AtMYB44 regulates drought tolerance by controlling the stomatal closure (Jung et al., 2008). AtMYB96 mediates the biosynthesis of cuticular wax to further regulate the water loss through leaf surfaces in dry conditions (J. S. Seo et al., 2011; P. J. Seo et al., 2011). Similarly, in *Camelina sativa* also AtMYB96 overexpression enhances the drought tolerance by the accumulation of cuticular wax indicating that the AtMYB96 gene among plants is conserved (S. B. Lee, Kim, Kim, & Suh, 2014). MYBs in other crops like soybean and rice also have shown stress-responsive tolerance. GmMYB-76,

92, and 177 in soybean and OsMYB-4, 6, 48, and 91 in rice demonstrate abiotic stress tolerance (Yoon et al., 2020).

MYB TFs induction against the infestation of insects to reduce the proliferation of aphids was seen in *Chrysanthemum* when CmMYB15 was overexpressed (C. An et al., 2019). AtMYB96 in *Arabidopsis* plays a role in SA and ABA cross-talk as a molecular link which imparts resistance to plants against pathogens (Seo & Park, 2010). In grapevines, VdMYB—a member of R2R3-MYB TF—activated defense against *Erysiphe necator* fungus (Yu et al., 2019). Resistance via wax layering in apples is regulated by the expression of MdMYB30 (Y. L. Zhang et al., 2019). In wheat, the attack by *Rhizoctonia cerealis* is controlled by the expression of TaRIM1 (Shan et al., 2016). CaPHL8, an MYB TF in pepper plants, showed resistance against *Ralstonia solanacerum* infection (Noman et al., 2019). MYB TFs are important in plants to enhance the tolerance/resistance in contradiction of all types of stresses like biotic stresses and abiotic stresses.

5.3.6 MYC transcription factors

Universally present in both plants and animals, the family of MYC TFs is a member of bHLH TFs, which consist of a basic helix–loop–helix (bHLH) domain (Sorensen et al., 2003). Studies report the stress-responsive expression of MYC in various plants enhancing the tolerance in conditions of stress. *AtMYC2* imparts tolerance to *Arabidopsis* plants in conditions of drought. Overexpression of *35S: AtMYC2* activates ABA-stress responsive genes enhancing osmotic stress tolerance (Abe et al., 2003). bHLH TFs AtbHLH17, AtbHLH68, and AtbHLH122 of *Arabidopsis* among other 160 bHLH are responsible for increasing responses to abiotic stresses like drought, cold, and oxidative stress (Babitha et al., 2013; Le Hir et al., 2017).

5.3.7 ZF transcription factors

Common in both plants and animals, the Zinc finger TFs consist of zinc finger binding domains with cysteines and/or histidine-containing motifs that attach to zinc atom/s. Regulation of stress-responsive genes is reported by Cys2/His2 (C2H2)-type ZFPs which consists of the EAR (ERF-associated amphiphilic repression) transcriptional repressor domain (Singh et al., 2009). Heat shock TFs (HS TFs) are another type of ZF TF that are responsible for building up a tolerance against heat stress in various plant species (Table 5.1).

5.4 A case study of heat shock transcription factors

Specific TFs are involved in regulating heat shock response; for example, Hsf1 is a critical regulator in eukaryotes. In prokaryotes, σ 32 plays a role in regulating heat shock response by replacing σ70 under heat stress. Under

TABLE 5.1 A summary of various genes involved abiotic stress responses.

Family	Gene	Species	Inducible by	References
bZIP	ABF1	*Arabidopsis thaliana*	Cold	Choi et al. (2000)
	ABF2	*A. thaliana*	Salt, drought	Choi et al. (2000)
	ABF3	*A. thaliana*	Salt, drought	Choi et al. (2000), Kang et al. (2002)
	ABF4	*A. thaliana*	Drought, salt, cold	Choi et al. (2000)
	GmbZIP44	*Glycine max*	Drought, salt	Liao, Zou, Wei, et al. (2008)
	GmbZIP62	*G. max*	Cold, drought, salt	Liao, Zou, Wei, et al. (2008)
	GmbZIP78	*G. max*	Drought, salt	Liao, Zou, Wei, et al. (2008)
	GmbZIP132	*G. max*	Cold, drought, salt	Liao, Zhang, Chen, and Zhang (2008)
	Wlip19	*Triticum aestivum*	Cold, drought	Kobayashi et al. (2008)
	ZmbZIP4	*Zea mays*	Salinity	H. Ma et al. (2018)
	ZmbZIP17	*Z. mays*	Drought, heat, salt	Jia et al. (2009)
	FtbZIP5	*Fagopyrum tataricum*	Drought, salinity, oxidative	Q. Li et al. (2020)
	OsABI5	*Oryza sativa*	Salt	Zou, Guan, Ren, Zhang, and Chen (2008)
	OsbZIP12	*O. sativa*	Drought	Joo, Lee, and Song (2014)
	OsbZIP23	*O. sativa*	Drought, salt, PEG	Xiang, Tang, Du, Ye, and Xiong (2008)
	OsbZIP46	*O. sativa*	Drought	N. Tang, Zhang, Li, Xiao, and Xiong (2012)
	OsbZIP71	*O. sativa*	Drought, salinity	C. Liu et al. (2014)
	OsbZIP72	*O. sativa*	Drought	G. Lu, Gao, Zheng, and Han (2009)
	OsAREB1	*O. sativa*	Drought, heat	X. F. Jin et al. (2010)
MYC (bHLH)	AtbHLH17	*A. thaliana*	Salinity, oxidative	Babitha et al. (2013)
	AtbHLH68	*A. thaliana*	Drought	Le Hir et al. (2017)
	AtbHLH122	*A. thaliana*	Salinity	Krishnamurthy et al. (2019)
	AtMYC2	*A. thaliana*	Drought, salt, cold	Abe et al. (2003)
	OsbHLH148	*O. sativa*	*O. sativa*	J. S. Seo et al. (2011), P. J. Seo et al. (2011)
	FtbHLH2	*F. tataricum*	Freezing	P. Yao et al. (2018)
	FtbHLH3	*F. tataricum*	Drought, oxidative	Y. Yao et al. (2017), P. F. Yao, Li, et al. (2017)

(Continued)

TABLE 5.1 (Continued)

Family	Gene	Species	Inducible by	References
MYB	AtMYB2	*A. thaliana*	Drought, salt	Abe et al. (2003)
	AtMYB4	*A. thaliana*	Salt, ethylene, JA, SA	Yanhui et al. (2006)
	AtMYB6	*A. thaliana*	Salt, ethylene, JA, SA	Yanhui et al. (2006)
	AtMYB7	*A. thaliana*	Salt, ethylene, JA, SA	Yanhui et al. (2006)
	MYB15	*A. thaliana*	Drought, salt, cold	M. Agarwal et al. (2006), Z. Ding et al. (2009)
	AtMYB44	*A. thaliana*	Salt, ethylene, JA, SA	Yanhui et al. (2006)
	AtMYB73	*A. thaliana*	Salt, ethylene, JA, SA	Yanhui et al. (2006)
	AtMYB96	*A. thaliana*	Drought	P. J. Seo et al. (2009)
	OsMYB3R-2	*O. sativa*	Drought, salt, cold	Dai et al. (2007)
	OsMYB4	*O. sativa*	Freezing	Vannini et al. (2004)
	OsMYB6	*O. sativa*	Drought, salinity	Y. Tang et al. (2019)
	OsMYB48-1	*O. sativa*	Drought, salinity	Xiong et al. (2014)
	OsMYB91	*O. sativa*	Drought, salinity	N. Zhu et al. (2015)
	GmMYB76	*G. max*	Salt	Liao, Zou, Wang, et al. (2008)
	GmMYB92	*G. max*	Cold, salt	Liao, Zou, Wang, et al. (2008)
	GmMYB177	*G. max*	Drought, salt	Liao, Zou, Wang, et al. (2008)
	AtMYB102	*A. thaliana*	Pathogen	L. Zhu, Guo, Ma, Wang, and Zhou (2018)
	AtMYB96	*A. thaliana*	Pathogen	Seo and Park (2010)
	VdMYB1	*Vitis vinifera*	Pathogen	Yu et al. (2019)
AP2/ERF	DREB1A	*A. thaliana*	Cold	Q. Liu et al. (1998)
	DREB2A	*A. thaliana*	Drought, salt	Q. Liu et al. (1998)
	DREB2C	*A. thaliana*	Salt, mannitol, cold	S. J. Lee et al. (2010)
	CBF1	*A. thaliana*	Cold, salinity	Gilmour et al. (1998)
	CBF2	*A. thaliana*	Cold, salinity	Gilmour et al. (1998)
	CBF3	*A. thaliana*	Cold	Gilmour et al. (1998)
	CBF4	*A. thaliana*	Drought	Haake et al. (2002)
	ERF1	*A. thaliana*	Necrotrophic fungi	Berrocal-Lobo et al. (2002)
	AtERF53	*A. thaliana*	Heat	Hsieh, Cheng, and Lin (2013)
	AtERF74	*A. thaliana*	Drought	Y. Yao et al. (2017), P. F. Yao et al. (2017)

(*Continued*)

TABLE 5.1 (Continued)

Family	Gene	Species	Inducible by	References
	OsERF71	*O. sativa*	Drought, salt, cold	D. K. Lee et al. (2017)
	GmERF3	*G. max*	Salinity, drought, heat	G. Zhang et al. (2009), M. Zhang, Liu, Bi, and Wang (2009)
	SlERF5	*Solanum lycopersicum*	Drought, salinity	Pan et al. (2012)
	OsEREBP1	*O. sativa*	Drought, salt, cold	Jisha et al. (2015)
	OsDREB1A	*O. sativa*	Cold, salt, wounding	Dubouzet et al. (2003)
	OsDREB1B	*O. sativa*	Cold	Dubouzet et al. (2003)
	OsDREB1C	*O. sativa*	Drought, salt, cold, wound	Dubouzet et al. (2003)
	OsDREB1D	*O. sativa*	None	Dubouzet et al. (2003)
	OsDREB1F	*O. sativa*	Drought, salt, cold	Q. Wang et al. (2008)
	OsDREB2A	*O. sativa*	Drought, salt, faintly to cold	Dubouzet et al. (2003)
	OsDREB2B	*O. sativa*	Heat, cold	Matsukura et al. (2010)
	OsDREB2C	*O. sativa*	None	Matsukura et al. (2010)
	OsDREB2E	*O. sativa*	None	Matsukura et al. (2010)
	TaDREB1	*T. aestivum*	Cold, drought	Shen et al. (2003)
	WDREB2	*T. aestivum*	Drought, salt, cold	Egawa et al. (2006)
	HvDRF1	*Hordeum vulgare*	Drought, salt	Xue and Loveridge (2004)
	HvDREB1	*H. vulgare*	Drought, salt, cold	Xu et al. (2009)
	ZmDREB2A	*Z. mays*	Drought, salt, cold, heat	Qin et al. (2007)
	PgDREB2A	*Pennisetum glaucum*	Drought, salt, cold	P. Agarwal, Agarwal, Nair, Sopory, and Reddy (2007)
	SbDREB2	*Sorghum bicolor*	Drought	Bihani, Char, and Bhargava (2011)
	SiDREB2	*Setaria italica*	Drought, salt	Lata, Bhutty, Bahadur, Majee, and Prasad (2011)
	CaDREB- LP1	*Capsicum annum*	Drought, salt, wounding	Hong and Kim (2005)
	AhDREB1	*Artiplex hortensis*	Salt	Shen et al. (2003)
	GmDREBa	*G. max*	Cold, drought, salt	X. P. Li et al. (2005)
	GmDREBb	*G. max*	Cold, drought, salt	X. P. Li et al. (2005)

(Continued)

TABLE 5.1 (Continued)

Family	Gene	Species	Inducible by	References
	GmDREBc	*G. max*	Drought, salt	X. P. Li et al. (2005)
	GmDREB2	*G. max*	Drought, salt	M. Chen et al. (2007)
	DmDREBa	*Dendronthema × moriforlium*	Cold	Y. Yang et al. (2009)
	DmDREBb	*D. × moriforlium*	Cold	Y. Yang et al. (2009)
	PNDREB1	*Arachis hypogea*	Drought, cold	Zhang et al. (2009)
	CAP2	*Cicer arietinum*	Drought, salt, auxin	Shukla, Raha, Tripathi, and Chattopadhyay (2006)
	DvDREB2A	*Dendrathema*	Drought, heat, cold	L. Liu et al. (2008)
	SbDREB2A	*Salicornia brachiata*	Drought, salt, heat	Gupta, Agarwal, Reddy, and Jha (2010)
NAC	ATAF1	*A. thaliana*	Drought	P. L. Lu et al. (2007)
	AtNAC2	*A. thaliana*	Salt	X. J. He et al. (2005)
	AtNAC019	*A. thaliana*	Drought, salt	Tran et al. (2004)
	ANAC042	*A. thaliana*	Heat	Shahnejat-Bushehri, Mueller-Roeber, and Balazadeh (2012)
	AtNAC055	*A. thaliana*	Drought, salt	Tran et al. (2004)
	AtNAC072	*A. thaliana*	Drought, salt	Tran et al. (2004)
	OsNAC6	*O. sativa*	Cold, drought, salt	Nakashima et al. (2007)
	SNAC1	*O. sativa*	Cold, drought, salt	Hu et al. (2006)
	SNAC2	*O. sativa*	Cold, drought, salt	Hu et al. (2008)
	CarNAC5	*C. arietinum*	Drought, heat	Peng et al. (2009)
	TaNAC4	*T. aestivum*	Cold, salt, wounding, ethylene, MeJA	Xia et al. (2010)
	GhNAC2	*Gossypium hirsutum*	Cold, drought	Meng, Cai, Zhang, and Guo (2009)
	GhNAC3	*G. hirsutum*	Cold	Meng et al. (2009)
	GhNAC4	*G. hirsutum*	Cold, drought, salt	Meng et al. (2009)
	GhNAC5	*G. hirsutum*	Cold, drought	Meng et al. (2009)
	GhNAC6	*G. hirsutum*	Cold, drought, salt	Meng et al. (2009)
	SiNAC	*S. italica*	Drought, salt, ethephone, MeJA	Puranik et al. (2011)
	OsNAC4	*O. sativa*	Pathogen	Kaneda et al. (2009)

(*Continued*)

TABLE 5.1 (Continued)

Family	Gene	Species	Inducible by	References
	ANAC042	*A. thaliana*	Pathogen	Saga et al. (2012)
	SlNAC1	*S. lycopersicum*	Pathogen	G. T. Huang et al. (2012), H. Huang et al. (2012)
WRKY	AtWRKY25	*A. thaliana*	Heat	S. Li et al. (2011)
	AtWRKY26	*A. thaliana*	Heat	S. Li et al. (2011)
	AtWRKY33	*A. thaliana*	Heat	S. Li et al. (2011)
	AtWRKY39	*A. thaliana*	Heat	S. Li et al. (2010)
	OsWRKY11	*O. sativa*	Drought	Wu et al. (2009)
	OsWRKY45	*O. sativa*	Drought	Qiu and Yu (2009)
	GmWRKY21	*G. max*		Q. Y. Zhou et al. (2008)
	GmWRKY54	*G. max*		Q. Y. Zhou et al. (2008)
	GmWRKY13	*G. max*		Q. Y. Zhou et al. (2008)
	TaWRKY1	*T. aestivum*	Drought	G. H. He et al. (2016)
	TaWRKY33	*T. aestivum*	Drought	G. H. He et al. (2016)
	NbWRKY	*Nicotiana benthamiana*		Archana, Rama, Mamrutha, and Nataraja (2009)
	VvWRKY1	*V. vinifera*	Downy mildew	Marchive et al. (2013)
	CaWRKY27	*Nicotiana tabacum*	Pathogen	Dang et al. (2014)
ZFP	ZPT2−3	*Petunia sp.*		Sugano et al. (2003)
	OSISAP1	*O. sativa*		Mukhopadhyay, Vij, and Tyagi (2004)
	CaZF	*C. arietinum*		Jain, Roy, and Chattopadhyay (2009)

normal conditions, σ32 gets degraded by forming a complex with DnaK (HSP70) and its cofactor DnaJ, leading HSPs to increase to a substantial amount (Rodriguez et al., 2008). According to the titration model, the accumulation of more and more unfolded proteins under stress leads to more chaperone requirement and the release of σ32 from these chaperone complexes. Similarly, Hsf1 of eukaryotes is kept inactive by forming a complex with Hsp90 and the Hsp70/Hsp40 chaperone system. Hsf1 activity is regulated by modifications like phosphorylation, sumoylation, and oligomerization (Åkerfelt, Morimoto, & Sistonen, 2010; Hietakangas et al., 2003; Prahlad & Morimoto, 2009). Heat shock factors like Hsf1 in eukaryotes are present in an inactive monomer form which upon stress gets oligomerized and converted

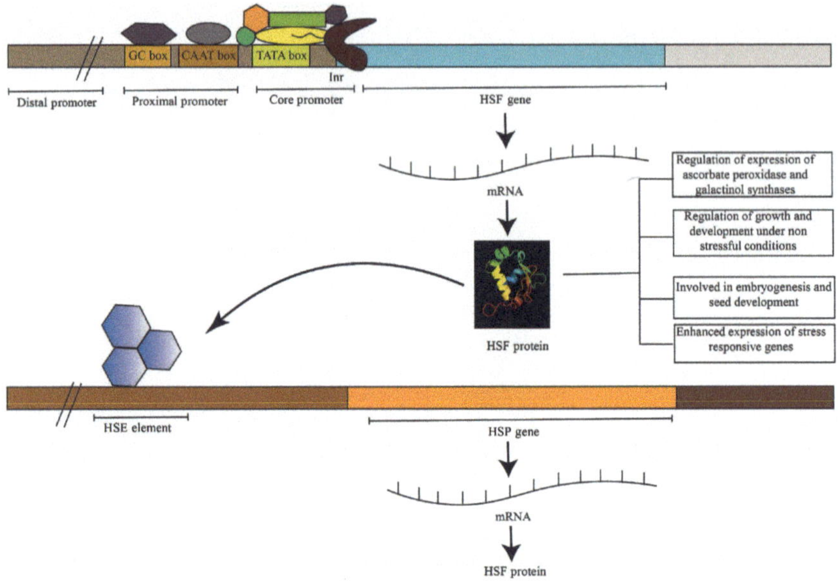

FIGURE 5.2 Pictorial representation of the role of heat shock factors (HSFs) in plants.

into active homo- or heterotrimers, transported to the nucleus. In the nucleus, they further get hyperphosphorylated and for many years, it was thought that this is required for target gene activation but comprehensive mutagenesis by Budzyński, Puustinen, Joutsen, and Sistonen (2015) revealed that there is no clear requirement of heat stress-induced phosphorylation in target gene activation. Inside the nucleus, Hsfs bind to the heat shock elements present in the promoter of stress-induced genes. The palindromic motif of nGAAn defines these HSEs. There are three different types of HSE known to date Perfect (P), Gap (G), Step (S) (Yamamoto, Mizukami, & Sakurai, 2005). Perfect (P) type HSE comprises a contiguous array of three inverted repeats of nGAAn (nTTCnnGAAnnTTCn). Gap (G) typically consists of two inverted repeats of nGAAn separated from a third unit by a gap of five base pairs (nTTCnnGAAn (5 bp)nGAAn). In Step (S) type, all three units are separated by five base pairs (nTTCn(5 bp)nTTCn(5 bp)nTTCn) (Fig. 5.2).

Heat shock factors are the TFs that play a major role in the regulation of genes responsive to heat stress and other stresses. It has been found that *Drosophila* and yeast have a single Hsf, vertebrates have four Hsfs (Nover et al., 1996). In contrast, plants can have more than 20 Hsfs; for example, 21 Hsfs have been identified in *Arabidopsis* (Scharf et al., 2012), 25 Hsfs in rice, 24 Hsfs in tomato (Scharf et al., 2012), 25 in pepper (Guo et al., 2015), and 52 in soybean (Scharf et al., 2012). Yeast Hsf is very important for its survival as it shows constitutive expression under normal conditions and is involved in many aspects like protein degradation, carbohydrate metabolism, and energy generation (Yamamoto et al., 2005).

5.5 Conclusion

The increased human population has led to an accelerated rate of change in the environment resulting in the accumulation of stressful conditions. The increase in different stresses such as high temperature stress, low-temperature stress, drought stress, and salinity stress throughout the globe has a severe effect on the agricultural crop plants. Plants first perceive the stress signal, convert it to intracellular, which leads to a signal transduction pathway that ultimately results in either activation or inactivation of genes playing a role in stress. During this process, various TFs are involved in the activation or inactivation of most of the genes.

The expression of the TFs in the plants enhances tolerance to abiotic factors and provides resistance against pathogens by regulating the expression of different stress-responsive genes. Overexpression increases tolerance, and the silencing of these genes reduces the plant's resistance. The information regarding the stress-specific TFs could be used to develop certain plants that could survive harsh environmental conditions. So, by learning about the stress-responsive genes and the TFs modulating their expression, it becomes easy to develop plants that can easily survive the onset of stressful conditions.

References

Abe, H., Urao, T., Ito, T., Seki, M., Shinozaki, K., & Yamaguchi-Shinozaki, K. (2003). *Arabidopsis* AtMYC2 (bHLH) and AtMYB2 (MYB) function as transcriptional activators in abscisic acid signaling. *The Plant Cell*, *15*(1), 63–78.

Agarwal, M., Hao, Y., Kapoor, A., Dong, C. H., Fujii, H., Zheng, X., & Zhu, J. K. (2006). A R2R3 type MYB transcription factor is involved in the cold regulation of CBF genes and in acquired freezing tolerance. *Journal of Biological Chemistry*, *281*(49), 37636–37645.

Agarwal, P., Agarwal, P. K., Nair, S., Sopory, S. K., & Reddy, M. K. (2007). Stress-inducible DREB2A transcription factor from *Pennisetum glaucum* is a phosphoprotein and its phosphorylation negatively regulate its DNA-binding activity. *Molecular Genetics and Genomics*, *277*(2), 189–198.

Åkerfelt, M., Morimoto, R. I., & Sistonen, L. (2010). Heat shock factors: Integrators of cell stress, development and lifespan. *Nature Reviews. Molecular Cell Biology*, *11*(8), 545–555.

An, C., Sheng, L., Du, X., Wang, Y., Zhang, Y., Song, A., ... Chen, S. (2019). Overexpression of CmMYB15 provides chrysanthemum resistance to aphids by regulating the biosynthesis of lignin. *Horticulture Research*, *6*(1), 1–10.

An, J. P., Li, R., Qu, F. J., You, C. X., Wang, X. F., & Hao, Y. J. (2018). An apple NAC transcription factor negatively regulates cold tolerance via CBF-dependent pathway. *Journal of Plant Physiology*, *221*, 74–80.

An, J. P., Wang, X. F., Zhang, X. W., Xu, H. F., Bi, S. Q., You, C. X., & Hao, Y. J. (2020). An apple MYB transcription factor regulates cold tolerance and anthocyanin accumulation and undergoes MIEL1-mediated degradation. *Plant Biotechnology Journal*, *18*(2), 337–353.

An, J. P., Yao, J. F., Wang, X. N., You, C. X., Wang, X. F., & Hao, Y. J. (2017). MdHY5 positively regulates cold tolerance via CBF-dependent and CBF-independent pathways in apple. *Journal of Plant Physiology*, *218*, 275–281.

Archana, K., Rama, N., Mamrutha, H.M. and Nataraja, K.N. (2009). Down-regulation of an abiotic stress related *Nicotiana benthamiana* WRKY transcription factor induces physiological abnormalities.

Babitha, K. C., Ramu, S. V., Pruthvi, V., Mahesh, P., Nataraja, K. N., & Udayakumar, M. (2013). Co-expression of AtbHLH17 and AtWRKY28 confers resistance to abiotic stress in *Arabidopsis*. *Transgenic Research*, *22*(2), 327–341.

Berrocal-Lobo, M., Molina, A., & Solano, R. (2002). Constitutive expression of ETHYLENE-RESPONSE-FACTOR1 in *Arabidopsis* confers resistance to several necrotrophic fungi. *The Plant Journal*, *29*(1), 23–32.

Bihani, P., Char, B., & Bhargava, S. (2011). Transgenic expression of sorghum DREB2 in rice improves tolerance and yield under water limitation. *The Journal of Agricultural Science*, *149*(1), 95.

Budzyński, M. A., Puustinen, M. C., Joutsen, J., & Sistonen, L. (2015). Uncoupling stress-inducible phosphorylation of heat shock factor 1 from its activation. *Molecular and Cellular Biology*, *35*(14), 2530–2540.

Chen, F., Hu, Y., Vannozzi, A., Wu, K., Cai, H., Qin, Y., ... Zhang, L. (2017). The WRKY transcription factor family in model plants and crops. *Critical Reviews in Plant Sciences*, *36*(5-6), 311–335.

Chen, M., Wang, Q. Y., Cheng, X. G., Xu, Z. S., Li, L. C., Ye, X. G., ... Ma, Y. Z. (2007). GmDREB2, a soybean DRE-binding transcription factor, conferred drought and high-salt tolerance in transgenic plants. *Biochemical and Biophysical Research Communications*, *353*(2), 299–305.

Chen, W., & Singh, K. B. (1999). The auxin, hydrogen peroxide and salicylic acid induced expression of the *Arabidopsis* GST6 promoter is mediated in part by an ocs element. *The Plant Journal*, *19*(6), 667–677.

Choi, H. I., Hong, J. H., Ha, J. O., Kang, J. Y., & Kim, S. Y. (2000). ABFs, a family of ABA-responsive element binding factors. *Journal of Biological Chemistry*, *275*(3), 1723–1730.

Collinge, M., & Boller, T. (2001). Differential induction of two potato genes, Stprx2 and StNAC, in response to infection by *Phytophthora infestans* and to wounding. *Plant Molecular Biology*, *46*(5), 521–529.

Cui, M. H., Yoo, K. S., Hyoung, S., Nguyen, H. T. K., Kim, Y. Y., Kim, H. J., ... Shin, J. S. (2013). An *Arabidopsis* R2R3-MYB transcription factor, AtMYB20, negatively regulates type 2C serine/threonine protein phosphatases to enhance salt tolerance. *FEBS Letters*, *587*(12), 1773–1778.

Dai, X., Xu, Y., Ma, Q., Xu, W., Wang, T., Xue, Y., & Chong, K. (2007). Overexpression of an R1R2R3 MYB gene, OsMYB3R-2, increases tolerance to freezing, drought, and salt stress in transgenic *Arabidopsis*. *Plant Physiology*, *143*(4), 1739–1751.

Dang, F., Wang, Y., She, J., Lei, Y., Liu, Z., Eulgem, T., ... Guan, D. (2014). Overexpression of CaWRKY27, a subgroup IIe WRKY transcription factor of Capsicum annuum, positively regulates tobacco resistance to *Ralstonia solanacearum* infection. *Physiologia Plantarum*, *150*(3), 397–411.

Danquah, A., de Zelicourt, A., Colcombet, J., & Hirt, H. (2014). The role of ABA and MAPK signaling pathways in plant abiotic stress responses. *Biotechnology Advances*, *32*(1), 40–52.

Ding, Z., Li, S., An, X., Liu, X., Qin, H., & Wang, D. (2009). Transgenic expression of MYB15 confers enhanced sensitivity to abscisic acid and improved drought tolerance in *Arabidopsis thaliana*. *Journal of Genetics and Genomics*, *36*(1), 17–29.

Ding, Z. J., Yan, J. Y., Li, C. X., Li, G. X., Wu, Y. R., & Zheng, S. J. (2015). Transcription factor WRKY 46 modulates the development of *Arabidopsis* lateral roots in osmotic/salt stress conditions via regulation of ABA signaling and auxin homeostasis. *The Plant Journal, 84*(1), 56–69.

Dong, C. H., Agarwal, M., Zhang, Y., Xie, Q., & Zhu, J. K. (2006). The negative regulator of plant cold responses, HOS1, is a RING E3 ligase that mediates the ubiquitination and degradation of ICE1. *Proceedings of the National Academy of Sciences, 103*(21), 8281–8286.

Dong, H., Tan, J., Li, M., Yu, Y., Jia, S., Zhang, C., ... Liu, Y. (2019). Transcriptome analysis of soybean WRKY TFs in response to Peronospora manshurica infection. *Genomics, 111*(6), 1412–1422.

Dubos, C., Stracke, R., Grotewold, E., Weisshaar, B., Martin, C., & Lepiniec, L. (2010). MYB transcription factors in *Arabidopsis*. *Trends in Plant Science, 15*(10), 573–581.

Dubouzet, J. G., Sakuma, Y., Ito, Y., Kasuga, M., Dubouzet, E. G., Miura, S., ... Yamaguchi-Shinozaki, K. (2003). OsDREB genes in rice, *Oryza sativa* L., encode transcription activators that function in drought-, high-salt-and cold-responsive gene expression. *The Plant Journal, 33*(4), 751–763.

Egawa, C., Kobayashi, F., Ishibashi, M., Nakamura, T., Nakamura, C., & Takumi, S. (2006). Differential regulation of transcript accumulation and alternative splicing of a DREB2 homolog under abiotic stress conditions in common wheat. *Genes & Genetic Systems, 81*(2), 77–91.

Fan, W., & Dong, X. (2002). In vivo interaction between NPR1 and transcription factor TGA2 leads to salicylic acid–mediated gene activation in *Arabidopsis*. *The Plant Cell, 14*(6), 1377–1389.

Fang, H., Dong, Y., Yue, X., Hu, J., Jiang, S., Xu, H., ... Wang, N. (2019). The B-box zinc finger protein MdBBX20 integrates anthocyanin accumulation in response to ultraviolet radiation and low temperature. *Plant, Cell & Environment, 42*(7), 2090–2104.

Gilmour, S. J., Zarka, D. G., Stockinger, E. J., Salazar, M. P., Houghton, J. M., & Thomashow, M. F. (1998). Low temperature regulation of the *Arabidopsis* CBF family of AP2 transcriptional activators as an early step in cold-induced COR gene expression. *The Plant Journal, 16*(4), 433–442.

Gonzalez, D. H. (2016). *Introduction to transcription factor structure and function. Plant transcription factors* (pp. 3–11). Academic Press.

Gunapati, S., Naresh, R., Ranjan, S., Nigam, D., Hans, A., Verma, P. C., ... Sane, V. A. (2016). Expression of GhNAC2 from G. herbaceum, improves root growth and imparts tolerance to drought in transgenic cotton and *Arabidopsis*. *Scientific Reports, 6*, 24978.

Guo, M., Lu, J. P., Zhai, Y. F., Chai, W. G., Gong, Z. H., & Lu, M. H. (2015). Genome-wide analysis, expression profile of heat shock factor gene family (CaHsfs) and characterisation of CaHsfA2 in pepper (*Capsicum annuum* L.). *BMC Plant Biology, 15*(1), 151.

Guo, Y., & Gan, S. (2006). AtNAP, a NAC family transcription factor, has an important role in leaf senescence. *The Plant Journal, 46*(4), 601–612.

Gupta, K., Agarwal, P. K., Reddy, M. K., & Jha, B. (2010). SbDREB2A, an A-2 type DREB transcription factor from extreme halophyte *Salicornia brachiata* confers abiotic stress tolerance in *Escherichia coli*. *Plant Cell Reports, 29*(10), 1131–1137.

Haake, V., Cook, D., Riechmann, J., Pineda, O., Thomashow, M. F., & Zhang, J. Z. (2002). Transcription factor CBF4 is a regulator of drought adaptation in *Arabidopsis*. *Plant Physiology, 130*(2), 639–648.

He, G. H., Xu, J. Y., Wang, Y. X., Liu, J. M., Li, P. S., Chen, M., ... Xu, Z. S. (2016). Drought-responsive WRKY transcription factor genes TaWRKY1 and TaWRKY33 from wheat confer drought and/or heat resistance in *Arabidopsis*. *BMC Plant Biology, 16*(1), 116.

He, P., Warren, R. F., Zhao, T., Shan, L., Zhu, L., Tang, X., & Zhou, J. M. (2001). Overexpression of Pti5 in tomato potentiates pathogen-induced defense gene expression and enhances disease resistance to *Pseudomonas syringae* pv. tomato. *Molecular Plant-Microbe Interactions*, *14*(12), 1453–1457.

He, X. J., Mu, R. L., Cao, W. H., Zhang, Z. G., Zhang, J. S., & Chen, S. Y. (2005). AtNAC2, a transcription factor downstream of ethylene and auxin signaling pathways, is involved in salt stress response and lateral root development. *The Plant Journal*, *44*(6), 903–916.

Hegedus, D., Yu, M., Baldwin, D., Gruber, M., Sharpe, A., Parkin, I., … Lydiate, D. (2003). Molecular characterization of *Brassica napus* NAC domain transcriptional activators induced in response to biotic and abiotic stress. *Plant Molecular Biology*, *53*(3), 383–397.

Hietakangas, V., Ahlskog, J. K., Jakobsson, A. M., Hellesuo, M., Sahlberg, N. M., Holmberg, C. I., … Sistonen, L. (2003). Phosphorylation of serine 303 is a prerequisite for the stress-inducible SUMO modification of heat shock factor 1. *Molecular and Cellular Biology*, *23*(8), 2953–2968.

Hong, J. P., & Kim, W. T. (2005). Isolation and functional characterization of the Ca-DREBLP1 gene encoding a dehydration-responsive element binding-factor-like protein 1 in hot pepper (*Capsicum annuum* L. cv. Pukang). *Planta*, *220*(6), 875–888.

Hsieh, E. J., Cheng, M. C., & Lin, T. P. (2013). Functional characterization of an abiotic stress-inducible transcription factor AtERF53 in *Arabidopsis thaliana*. *Plant Molecular Biology*, *82*(3), 223–237.

Hu, H., Dai, M., Yao, J., Xiao, B., Li, X., Zhang, Q., & Xiong, L. (2006). Overexpressing a NAM, ATAF, and CUC (NAC) transcription factor enhances drought resistance and salt tolerance in rice. *Proceedings of the National Academy of Sciences*, *103*(35), 12987–12992.

Hu, H., You, J., Fang, Y., Zhu, X., Qi, Z., & Xiong, L. (2008). Characterization of transcription factor gene SNAC2 conferring cold and salt tolerance in rice. *Plant Molecular Biology*, *67*(1-2), 169–181.

Huang, G. T., Ma, S. L., Bai, L. P., Zhang, L., Ma, H., Jia, P., … Guo, Z. F. (2012). Signal transduction during cold, salt, and drought stresses in plants. *Molecular Biology Reports*, *39*(2), 969–987.

Huang, H., Wang, Y., Wang, S., Wu, X., Yang, K., Niu, Y., & Dai, S. (2012). Transcriptome-wide survey and expression analysis of stress-responsive NAC genes in Chrysanthemum lavandulifolium. *Plant Science*, *193*, 18–27.

Jaglo-Ottosen, K. R., Gilmour, S. J., Zarka, D. G., Schabenberger, O., & Thomashow, M. F. (1998). *Arabidopsis* CBF1 overexpression induces COR genes and enhances freezing tolerance. *Science (New York, N.Y.)*, *280*(5360), 104–106.

Jain, D., Roy, N., & Chattopadhyay, D. (2009). CaZF, a plant transcription factor functions through and parallel to HOG and calcineurin pathways in Saccharomyces cerevisiae to provide osmotolerance. *PLoS One*, *4*(4), e5154.

Jakoby, M., Weisshaar, B., Dröge-Laser, W., Vicente-Carbajosa, J., Tiedemann, J., Kroj, T., & Parcy, F. (2002). bZIP transcription factors in *Arabidopsis*. *Trends in Plant Science*, *7*(3), 106–111.

Jensen, M. K., Hagedorn, P. H., De Torres-Zabala, M., Grant, M. R., Rung, J. H., Collinge, D. B., & Lyngkjaer, M. F. (2008). Transcriptional regulation by an NAC (NAM–ATAF1, 2–CUC2) transcription factor attenuates ABA signaling for efficient basal defence towards *Blumeria graminis* f. sp. hordei in *Arabidopsis*. *The Plant Journal*, *56*(6), 867–880.

Jeong, J. S., Kim, Y. S., Baek, K. H., Jung, H., Ha, S. H., Do Choi, Y., … Kim, J. K. (2010). Root-specific expression of OsNAC10 improves drought tolerance and grain yield in rice under field drought conditions. *Plant Physiology*, *153*(1), 185–197.

Jia, Z., Lian, Y., Zhu, Y., He, J., Cao, Z., & Wang, G. (2009). Cloning and characterization of a putative transcription factor induced by abiotic stress in *Zea mays*. *African Journal of Biotechnology*, *8*(24).

Jiang, Y., & Deyholos, M. K. (2006). Comprehensive transcriptional profiling of NaCl-stressed *Arabidopsis* roots reveals novel classes of responsive genes. *BMC Plant Biology*, *6*(1), 25.

Jin, C., Li, K. Q., Xu, X. Y., Zhang, H. P., Chen, H. X., Chen, Y. H., ... Zhang, S. L. (2017). A novel NAC transcription factor, PbeNAC1, of *Pyrus betulifolia* confers cold and drought tolerance via interacting with PbeDREBs and activating the expression of stress-responsive genes. *Frontiers in Plant Science*, *8*, 1049.

Jin, X. F., Xiong, A. S., Peng, R. H., Liu, J. G., Gao, F., Chen, J. M., & Yao, Q. H. (2010). OsAREB1, an ABRE-binding protein responding to ABA and glucose, has multiple functions in *Arabidopsis*. *BMB Reports*, *43*(1), 34–39.

Jing, Z., & Liu, Z. (2018). Genome-wide identification of WRKY transcription factors in kiwifruit (Actinidia spp.) and analysis of WRKY expression in responses to biotic and abiotic stresses. *Genes & Genomics*, *40*(4), 429–446.

Jisha, V., Dampanaboina, L., Vadassery, J., Mithöfer, A., Kappara, S., & Ramanan, R. (2015). Overexpression of an AP2/ERF type transcription factor OsEREBP1 confers biotic and abiotic stress tolerance in rice. *PLoS One*, *10*(6), e0127831.

Joo, J., Lee, Y. H., & Song, S. I. (2014). Overexpression of the rice basic leucine zipper transcription factor OsbZIP12 confers drought tolerance to rice and makes seedlings hypersensitive to ABA. *Plant Biotechnology Reports*, *8*(6), 431–441.

Jung, C., Seo, J. S., Han, S. W., Koo, Y. J., Kim, C. H., Song, S. I., ... Cheong, J. J. (2008). Overexpression of AtMYB44 enhances stomatal closure to confer abiotic stress tolerance in transgenic *Arabidopsis*. *Plant Physiology*, *146*(2), 623–635.

Kaneda, T., Taga, Y., Takai, R., Iwano, M., Matsui, H., Takayama, S., ... Che, F. S. (2009). The transcription factor OsNAC4 is a key positive regulator of plant hypersensitive cell death. *The EMBO Journal*, *28*(7), 926–936.

Kang, J. Y., Choi, H. I., Im, M. Y., & Kim, S. Y. (2002). *Arabidopsis* basic leucine zipper proteins that mediate stress-responsive abscisic acid signaling. *The Plant Cell*, *14*(2), 343–357.

Kobayashi, F., Maeta, E., Terashima, A., Kawaura, K., Ogihara, Y., & Takumi, S. (2008). Development of abiotic stress tolerance via bZIP-type transcription factor LIP19 in common wheat. *Journal of Experimental Botany*, *59*(4), 891–905.

Krishnamurthy, P., Vishal, B., Khoo, K., Rajappa, S., Loh, C. S., & Kumar, P. P. (2019). Expression of AoNHX1 increases salt tolerance of rice and *Arabidopsis*, and bHLH transcription factors regulate AtNHX1 and AtNHX6 in *Arabidopsis*. *Plant Cell Reports*, *38*(10), 1299–1315.

Lata, C., Bhutty, S., Bahadur, R. P., Majee, M., & Prasad, M. (2011). Association of an SNP in a novel DREB2-like gene SiDREB2 with stress tolerance in foxtail millet [*Setaria italica* (L.)]. *Journal of Experimental Botany*, *62*(10), 3387–3401.

Le Hir, R., Castelain, M., Chakraborti, D., Moritz, T., Dinant, S., & Bellini, C. (2017). At bHLH68 transcription factor contributes to the regulation of ABA homeostasis and drought stress tolerance in *Arabidopsis thaliana*. *Physiologia Plantarum*, *160*(3), 312–327.

Lebel, E., Heifetz, P., Thorne, L., Uknes, S., Ryals, J., & Ward, E. (1998). Functional analysis of regulatory sequences controllingPR-1 gene expression in *Arabidopsis*. *The Plant Journal*, *16*(2), 223–233.

Lee, D. K., Yoon, S., Kim, Y. S., & Kim, J. K. (2017). Rice OsERF71-mediated root modification affects shoot drought tolerance. *Plant Signaling & Behavior*, *12*(1), e1268311.

Lee, S. B., Kim, H., Kim, R. J., & Suh, M. C. (2014). Overexpression of *Arabidopsis* MYB96 confers drought resistance in *Camelina sativa* via cuticular wax accumulation. *Plant Cell Reports*, *33*(9), 1535–1546.

Lee, S. J., Kang, J. Y., Park, H. J., Kim, M. D., Bae, M. S., Choi, H. I., & Kim, S. Y. (2010). DREB2C interacts with ABF2, a bZIP protein regulating abscisic acid-responsive gene expression, and its overexpression affects abscisic acid sensitivity. *Plant Physiology*, *153*(2), 716–727.

Li, Q., Zhao, H., Wang, X., Kang, J., Lv, B., Dong, Q., ... Wu, Q. (2020). Tartary buckwheat transcription factor FtbZIP5, regulated by FtSnRK2. 6, can improve salt/drought resistance in transgenic *Arabidopsis*. *International Journal of Molecular Sciences*, *21*(3), 1123.

Li, S., Fu, Q., Chen, L., Huang, W., & Yu, D. (2011). *Arabidopsis thaliana* WRKY25, WRKY26, and WRKY33 coordinate induction of plant thermotolerance. *Planta*, *233*(6), 1237–1252.

Li, S., Zhou, X., Chen, L., Huang, W., & Yu, D. (2010). Functional characterization of *Arabidopsis thaliana* WRKY39 in heat stress. *Molecules and Cells*, *29*(5), 475–483.

Li, X. D., Zhuang, K. Y., Liu, Z. M., Yang, D. Y., Ma, N. N., & Meng, Q. W. (2016). Overexpression of a novel NAC-type tomato transcription factor, SlNAM1, enhances the chilling stress tolerance of transgenic tobacco. *Journal of Plant Physiology*, *204*, 54–65.

Li, X. P., Tian, A. G., Luo, G. Z., Gong, Z. Z., Zhang, J. S., & Chen, S. Y. (2005). Soybean DRE-binding transcription factors that are responsive to abiotic stresses. *Theoretical and Applied Genetics*, *110*(8), 1355–1362.

Liao, Y., Zhang, J. S., Chen, S. Y., & Zhang, W. K. (2008). Role of soybean GmbZIP132 under abscisic acid and salt stresses. *Journal of Integrative Plant Biology*, *50*(2), 221–230.

Liao, Y., Zou, H. F., Wang, H. W., Zhang, W. K., Ma, B., Zhang, J. S., & Chen, S. Y. (2008). Soybean GmMYB76, GmMYB92, and GmMYB177 genes confer stress tolerance in transgenic *Arabidopsis* plants. *Cell Research*, *18*(10), 1047–1060.

Liao, Y., Zou, H. F., Wei, W., Hao, Y. J., Tian, A. G., Huang, J., ... Chen, S. Y. (2008). Soybean GmbZIP44, GmbZIP62 and GmbZIP78 genes function as negative regulator of ABA signaling and confer salt and freezing tolerance in transgenic *Arabidopsis*. *Planta*, *228*(2), 225–240.

Liu, C., Mao, B., Ou, S., Wang, W., Liu, L., Wu, Y., ... Wang, X. (2014). OsbZIP71, a bZIP transcription factor, confers salinity and drought tolerance in rice. *Plant Molecular Biology*, *84*(1-2), 19–36.

Liu, H., Zhou, Y., Li, H., Wang, T., Zhang, J., Ouyang, B., & Ye, Z. (2018). Molecular and functional characterization of ShNAC1, an NAC transcription factor from *Solanum habrochaites*. *Plant Science*, *271*, 9–19.

Liu, L., Zhu, K., Yang, Y., Wu, J., Chen, F., & Yu, D. (2008). Molecular cloning, expression profiling and trans-activation property studies of a DREB2-like gene from chrysanthemum (*Dendranthema vestitum*). *Journal of Plant Research*, *121*(2), 215–226.

Liu, Q., Kasuga, M., Sakuma, Y., Abe, H., Miura, S., Yamaguchi-Shinozaki, K., & Shinozaki, K. (1998). Two transcription factors, DREB1 and DREB2, with an EREBP/AP2 DNA binding domain separate two cellular signal transduction pathways in drought-and low-temperature-responsive gene expression, respectively, in *Arabidopsis*. *The Plant Cell*, *10*(8), 1391–1406.

Logemann, E., & Hahlbrock, K. (2002). Crosstalk among stress responses in plants: Pathogen defense overrides UV protection through an inversely regulated ACE/ACE type of light-responsive gene promoter unit. *Proceedings of the National Academy of Sciences*, *99*(4), 2428–2432.

Lu, G., Gao, C., Zheng, X., & Han, B. (2009). Identification of OsbZIP72 as a positive regulator of ABA response and drought tolerance in rice. *Planta*, *229*(3), 605–615.

Lu, P. L., Chen, N. Z., An, R., Su, Z., Qi, B. S., Ren, F., ... Wang, X. C. (2007). A novel drought-inducible gene, ATAF1, encodes a NAC family protein that negatively regulates the expression of stress-responsive genes in *Arabidopsis*. *Plant Molecular Biology*, *63*(2), 289–305.

Luan, Q., Chen, C., Liu, M., Li, Q., Wang, L., & Ren, Z. (2019). CsWRKY50 mediates defense responses to *Pseudoperonospora cubensis* infection in *Cucumis sativus*. *Plant Science*, *279*, 59–69.

Ma, H., Liu, C., Li, Z., Ran, Q., Xie, G., Wang, B., ... Zhang, J. (2018). ZmbZIP4 contributes to stress resistance in maize by regulating ABA synthesis and root development. *Plant Physiology*, *178*(2), 753–770.

Ma, Y., Dai, X., Xu, Y., Luo, W., Zheng, X., Zeng, D., ... Xiao, J. (2015). COLD1 confers chilling tolerance in rice. *Cell*, *160*(6), 1209–1221.

Marchive, C., Léon, C., Kappel, C., Coutos-Thévenot, P., Corio-Costet, M. F., Delrot, S., & Lauvergeat, V. (2013). Over-expression of VvWRKY1 in grapevines induces expression of jasmonic acid pathway-related genes and confers higher tolerance to the downy mildew. *PLoS One*, *8*(1), e54185.

Matsukura, S., Mizoi, J., Yoshida, T., Todaka, D., Ito, Y., Maruyama, K., ... Yamaguchi-Shinozaki, K. (2010). Comprehensive analysis of rice DREB2-type genes that encode transcription factors involved in the expression of abiotic stress-responsive genes. *Molecular Genetics and Genomics*, *283*(2), 185–196.

Meng, C., Cai, C., Zhang, T., & Guo, W. (2009). Characterization of six novel NAC genes and their responses to abiotic stresses in *Gossypium hirsutum* L. *Plant Science*, *176*(3), 352–359.

Mizoi, J., Ohori, T., Moriwaki, T., Kidokoro, S., Todaka, D., Maruyama, K., ... Yamaguchi-Shinozaki, K. (2013). GmDREB2A; 2, a canonical DEHYDRATION-RESPONSIVE ELEMENT-BINDING PROTEIN2-type transcription factor in soybean, is posttranslationally regulated and mediates dehydration-responsive element-dependent gene expression. *Plant Physiology*, *161*(1), 346–361.

Morris, P. C. (2001). MAP kinase signal transduction pathways in plants. *New Phytologist*, *151*(1), 67–89.

Mukhopadhyay, A., Vij, S., & Tyagi, A. K. (2004). Overexpression of a zinc-finger protein gene from rice confers tolerance to cold, dehydration, and salt stress in transgenic tobacco. *Proceedings of the National Academy of Sciences*, *101*(16), 6309–6314.

Nakashima, K., Tran, L. S., Van Nyugen, D., Fujita, M., Maruyama, K., Todaka, D., ... Yamaguchi-Shinozaki, K. (2007). Functional analysis of a NAC-type transcription factor OsNAC6 involved in abiotic and biotic stress-responsive gene expression in rice. *The Plant Journal: For Cell and Molecular Biology*, *51*, 617–630.

Noman, A., Hussain, A., Adnan, M., Khan, M. I., Ashraf, M. F., Zainab, M., ... He, S. (2019). A novel MYB transcription factor CaPHL8 provide clues about evolution of pepper immunity againstsoil borne pathogen. *Microbial Pathogenesis*, *137*, 103758.

Nover, L., Scharf, K. D., Gagliardi, D., Vergne, P., Czarnecka-Verner, E., & Gurley, W. B. (1996). The Hsf world: Classification and properties of plant heat stress transcription factors. *Cell Stress & Chaperones*, *1*(4), 215.

Ntambo, M. S., Meng, J. Y., Rott, P. C., Henry, R. J., Zhang, H. L., & Gao, S. J. (2019). Comparative transcriptome profiling of resistant and susceptible sugarcane cultivars in response to infection by *Xanthomonas albilineans*. *International Journal of Molecular Sciences*, *20*(24), 6138.

Nuruzzaman, M., Manimekalai, R., Sharoni, A. M., Satoh, K., Kondoh, H., Ooka, H., & Kikuchi, S. (2010). Genome-wide analysis of NAC transcription factor family in rice. *Gene*, *465*(1-2), 30–44.

Nuruzzaman, M., Sharoni, A. M., & Kikuchi, S. (2013). Roles of NAC transcription factors in the regulation of biotic and abiotic stress responses in plants. *Frontiers in Microbiology*, *4*, 248.

Pan, Y., Seymour, G. B., Lu, C., Hu, Z., Chen, X., & Chen, G. (2012). An ethylene response factor (ERF5) promoting adaptation to drought and salt tolerance in tomato. *Plant Cell Reports*, *31*(2), 349–360.

Park, J. M., Park, C. J., Lee, S. B., Ham, B. K., Shin, R., & Paek, K. H. (2001). Overexpression of the tobacco Tsi1 gene encoding an EREBP/AP2–type transcription factor enhances resistance against pathogen attack and osmotic stress in tobacco. *The Plant Cell*, *13*(5), 1035–1046.

Peng, H., Cheng, H. Y., Yu, X. W., Shi, Q. H., Zhang, H., Li, J. G., & Ma, H. (2009). Characterization of a chickpea (*Cicer arietinum* L.) NAC family gene, CarNAC5, which is both developmentally-and stress-regulated. *Plant Physiology and Biochemistry*, *47*(11-12), 1037–1045.

Prahlad, V., & Morimoto, R. I. (2009). Integrating the stress response: Lessons for neurodegenerative diseases from *C. elegans*. *Trends in Cell Biology*, *19*(2), 52–61.

Puranik, S., Bahadur, R. P., Srivastava, P. S., & Prasad, M. (2011). Molecular cloning and characterization of a membrane associated NAC family gene, SiNAC from foxtail millet [*Setaria italica* (L.) P. Beauv.]. *Molecular Biotechnology*, *49*(2), 138–150.

Qin, F., Kakimoto, M., Sakuma, Y., Maruyama, K., Osakabe, Y., Tran, L. S. P., ... Yamaguchi-Shinozaki, K. (2007). Regulation and functional analysis of ZmDREB2A in response to drought and heat stresses in *Zea mays* L. *The Plant Journal*, *50*(1), 54–69.

Qiu, Y., & Yu, D. (2009). Over-expression of the stress-induced OsWRKY45 enhances disease resistance and drought tolerance in *Arabidopsis*. *Environmental and Experimental Botany*, *65*(1), 35–47.

Ren, X., Chen, Z., Liu, Y., Zhang, H., Zhang, M., Liu, Q., ... Gong, Z. (2010). ABO3, a WRKY transcription factor, mediates plant responses to abscisic acid and drought tolerance in *Arabidopsis*. *The Plant Journal*, *63*(3), 417–429.

Ren, Y. R., Yang, Y. Y., Zhang, R., You, C. X., Zhao, Q., & Hao, Y. J. (2019). MdGRF11, an apple 14-3-3 protein, acts as a positive regulator of drought and salt tolerance. *Plant Science*, *288*, 110219.

Riechmann, J. L., Heard, J., Martin, G., Reuber, L., Jiang, C. Z., Keddie, J., ... Creelman, R. (2000). *Arabidopsis* transcription factors: Genome-wide comparative analysis among eukaryotes. *Science (New York, N.Y.)*, *290*(5499), 2105–2110.

Rodriguez, F., Arsène-Ploetze, F., Rist, W., Rüdiger, S., Schneider-Mergener, J., Mayer, M. P., & Bukau, B. (2008). Molecular basis for regulation of the heat shock transcription factor σ32 by the DnaK and DnaJ chaperones. *Molecular Cell*, *32*(3), 347–358.

Rushton, P. J., & Somssich, I. E. (1998). Transcriptional control of plant genes responsive to pathogens. *Current Opinion in Plant Biology*, *1*(4), 311–315.

Rushton, P. J., Somssich, I. E., Ringler, P., & Shen, Q. J. (2010). WRKY transcription factors. *Trends in Plant Science*, *15*(5), 247–258.

Saga, H., Ogawa, T., Kai, K., Suzuki, H., Ogata, Y., Sakurai, N., ... Ohta, D. (2012). Identification and characterization of ANAC042, a transcription factor family gene involved in the regulation of camalexin biosynthesis in *Arabidopsis*. *Molecular Plant-Microbe Interactions*, *25*(5), 684–696.

Sakuraba, Y., Kim, Y. S., Han, S. H., Lee, B. D., & Paek, N. C. (2015). The *Arabidopsis* transcription factor NAC016 promotes drought stress responses by repressing AREB1 transcription through a trifurcate feed-forward regulatory loop involving NAP. *The Plant Cell*, *27*(6), 1771–1787.

Sangwan, V., Örvar, B. L., Beyerly, J., Hirt, H., & Dhindsa, R. S. (2002). Opposite changes in membrane fluidity mimic cold and heat stress activation of distinct plant MAP kinase pathways. *The Plant Journal*, *31*(5), 629–638.

Satoh, K., Shimizu, T., Kondoh, H., Hiraguri, A., Sasaya, T., Choi, I. R., ... Kikuchi, S. (2011). Relationship between symptoms and gene expression induced by the infection of three strains of Rice dwarf virus. *PLoS One*, *6*(3), e18094.

Schaller, G. E., Kieber, J. J., & Shiu, S. H. (2008). *Two-component signaling elements and histidyl-aspartyl phosphorelays*, . *The arabidopsis book* (6). American Society of Plant Biologists.

Scharf, K. D., Berberich, T., Ebersberger, I., & Nover, L. (2012). The plant heat stress transcription factor (Hsf) family: Structure, function and evolution. *Biochimica et Biophysica Acta—Gene Regulatory Mechanisms*, *1819*(2), 104–119.

Seo, J. S., Joo, J., Kim, M. J., Kim, Y. K., Nahm, B. H., Song, S. I., ... Choi, Y. D. (2011). OsbHLH148, a basic helix-loop-helix protein, interacts with OsJAZ proteins in a jasmonate signaling pathway leading to drought tolerance in rice. *The Plant Journal*, *65*(6), 907–921.

Seo, P. J., Lee, S. B., Suh, M. C., Park, M. J., Go, Y. S., & Park, C. M. (2011). The MYB96 transcription factor regulates cuticular wax biosynthesis under drought conditions in *Arabidopsis*. *The Plant Cell*, *23*(3), 1138–1152.

Seo, P. J., & Park, C. M. (2010). MYB96-mediated abscisic acid signals induce pathogen resistance response by promoting salicylic acid biosynthesis in *Arabidopsis*. *New Phytologist*, *186*(2), 471–483.

Seo, P. J., Xiang, F., Qiao, M., Park, J. Y., Lee, Y. N., Kim, S. G., ... Park, C. M. (2009). The MYB96 transcription factor mediates abscisic acid signaling during drought stress response in *Arabidopsis*. *Plant Physiology*, *151*(1), 275–289.

Shahnejat-Bushehri, S., Mueller-Roeber, B., & Balazadeh, S. (2012). *Arabidopsis* NAC transcription factor JUNGBRUNNEN1 affects thermomemory-associated genes and enhances heat stress tolerance in primed and unprimed conditions. *Plant Signaling & Behavior*, *7*(12), 1518–1521.

Shan, T., Rong, W., Xu, H., Du, L., Liu, X., & Zhang, Z. (2016). The wheat R2R3-MYB transcription factor TaRIM1 participates in resistance response against the pathogen *Rhizoctonia cerealis* infection through regulating defense genes. *Scientific Reports*, *6*(1), 1–14.

Shen, Y. G., Zhang, W. K., He, S. J., Zhang, J. S., Liu, Q., & Chen, S. Y. (2003). An EREBP/AP2-type protein in *Triticum aestivum* was a DRE-binding transcription factor induced by cold, dehydration and ABA stress. *Theoretical and Applied Genetics*, *106*(5), 923–930.

Shukla, R. K., Raha, S., Tripathi, V., & Chattopadhyay, D. (2006). Expression of CAP2, an APETALA2-family transcription factor from chickpea, enhances growth and tolerance to dehydration and salt stress in transgenic tobacco. *Plant Physiology*, *142*(1), 113–123.

Singh, A. K., Sopory, S. K., Wu, R., & Singla-Pareek, S. L. (2009). *Transgenic approaches*. *Abiotic Stress Adaptation in Plants* (pp. 417–450). Dordrecht: Springer.

Singh, A. K., Kumar, S. R., Dwivedi, V., Rai, A., Pal, S., Shasany, A. K., & Nagegowda, D. A. (2017). A WRKY transcription factor from *Withania somnifera* regulates triterpenoid withanolide accumulation and biotic stress tolerance through modulation of phytosterol and defense pathways. *New Phytologist*, *215*(3), 1115–1131.

Sorensen, A. M., Kröber, S., Unte, U. S., Huijser, P., Dekker, K., & Saedler, H. (2003). The *Arabidopsis* ABORTED MICROSPORES (AMS) gene encodes a MYC class transcription factor. *The Plant Journal*, *33*(2), 413–423.

Subramaniam, R., Desveaux, D., Spickler, C., Michnick, S. W., & Brisson, N. (2001). Direct visualization of protein interactions in plant cells. *Nature Biotechnology*, *19*(8), 769–772.

Sugano, S., Kaminaka, H., Rybka, Z., Catala, R., Salinas, J., Matsui, K., ... Takatsuji, H. (2003). Stress-responsive zinc finger gene ZPT2-3 plays a role in drought tolerance in petunia. *The Plant Journal, 36*(6), 830–841.

Swarbreck, S. M., Colaço, R., & Davies, J. M. (2013). Plant calcium-permeable channels. *Plant Physiology, 163*, 514–522.

Tang, N., Zhang, H., Li, X., Xiao, J., & Xiong, L. (2012). Constitutive activation of transcription factor OsbZIP46 improves drought tolerance in rice. *Plant Physiology, 158*(4), 1755–1768.

Tang, Y., Bao, X., Zhi, Y., Wu, Q., Guo, Y., Yin, X., ... Liu, W. (2019). Overexpression of a MYB family gene, OsMYB6, increases drought and salinity stress tolerance in transgenic rice. *Frontiers in Plant Science, 10*, 168.

Tran, L. S. P., Nakashima, K., Sakuma, Y., Simpson, S. D., Fujita, Y., Maruyama, K., ... Yamaguchi-Shinozaki, K. (2004). Isolation and functional analysis of *Arabidopsis* stress-inducible NAC transcription factors that bind to a drought-responsive cis-element in the early responsive to dehydration stress 1 promoter. *The Plant Cell, 16*(9), 2481–2498.

Tran, L. S. P., Nakashima, K., Sakuma, Y., Osakabe, Y., Qin, F., Simpson, S. D., & Yamaguchi-Shinozaki, K. (2007). Co-expression of the stress-inducible zinc finger homeodomain ZFHD1 and NAC transcription factors enhances expression of the ERD1 gene in Arabidopsis. *The Plant Journal, 49*(1), 46–63.

Uno, Y., Furihata, T., Abe, H., Yoshida, R., Shinozaki, K., & Yamaguchi-Shinozaki, K. (2000). *Arabidopsis* basic leucine zipper transcription factors involved in an abscisic acid-dependent signal transduction pathway under drought and high-salinity conditions. *Proceedings of the National Academy of Sciences, 97*(21), 11632–11637.

Vannini, C., Locatelli, F., Bracale, M., Magnani, E., Marsoni, M., Osnato, M., ... Coraggio, I. (2004). Overexpression of the rice Osmyb4 gene increases chilling and freezing tolerance of *Arabidopsis thaliana* plants. *The Plant Journal, 37*(1), 115–127.

Wang, H., Wang, H., Shao, H., & Tang, X. (2016). Recent advances in utilizing transcription factors to improve plant abiotic stress tolerance by transgenic technology. *Frontiers in Plant Science, 7*, 67.

Wang, Q., Guan, Y., Wu, Y., Chen, H., Chen, F., & Chu, C. (2008). Overexpression of a rice OsDREB1F gene increases salt, drought, and low temperature tolerance in both *Arabidopsis* and rice. *Plant Molecular Biology, 67*(6), 589–602.

Wilkins, O., Nahal, H., Foong, J., Provart, N. J., & Campbell, M. M. (2009). Expansion and diversification of the Populus R2R3-MYB family of transcription factors. *Plant Physiology, 149*(2), 981–993.

Wimalasekera, R., & Scherer, G. F. (2018). *Involvement of mitogen-activated protein kinases in abiotic stress responses in plants. Plant Metabolites and Regulation Under Environmental Stress* (pp. 389–395). Academic Press.

Wu, X., Shiroto, Y., Kishitani, S., Ito, Y., & Toriyama, K. (2009). Enhanced heat and drought tolerance in transgenic rice seedlings overexpressing OsWRKY11 under the control of HSP101 promoter. *Plant Cell Reports, 28*(1), 21–30.

Xia, N., Zhang, G., Liu, X. Y., Deng, L., Cai, G. L., Zhang, Y., ... Kang, Z. S. (2010). Characterization of a novel wheat NAC transcription factor gene involved in defense response against stripe rust pathogen infection and abiotic stresses. *Molecular Biology Reports, 37*(8), 3703–3712.

Xiang, Y., Tang, N., Du, H., Ye, H., & Xiong, L. (2008). Characterization of OsbZIP23 as a key player of the basic leucine zipper transcription factor family for conferring abscisic acid sensitivity and salinity and drought tolerance in rice. *Plant Physiology, 148*(4), 1938–1952.

Xie, X. B., Li, S., Zhang, R. F., Zhao, J., Chen, Y. C., Zhao, Q., ... Hao, Y. J. (2012). The bHLH transcription factor MdbHLH3 promotes anthocyanin accumulation and fruit colouration in response to low temperature in apples. *Plant, Cell & Environment*, *35*(11), 1884–1897.

Xie, Z., Nolan, T. M., Jiang, H., & Yin, Y. (2019). AP2/ERF transcription factor regulatory networks in hormone and abiotic stress responses in *Arabidopsis*. *Frontiers in Plant Science*, *10*, 228.

Xiong, H., Li, J., Liu, P., Duan, J., Zhao, Y., Guo, X., ... Li, Z. (2014). Overexpression of OsMYB48-1, a novel MYB-related transcription factor, enhances drought and salinity tolerance in rice. *PLoS One*, *9*(3), e92913.

Xu, Z. S., Ni, Z. Y., Li, Z. Y., Li, L. C., Chen, M., Gao, D. Y., ... Ma, Y. Z. (2009). Isolation and functional characterization of HvDREB1—a gene encoding a dehydration-responsive element binding protein in *Hordeum vulgare*. *Journal of Plant Research*, *122*(1), 121–130.

Xue, G. P., & Loveridge, C. W. (2004). HvDRF1 is involved in abscisic acid-mediated gene regulation in barley and produces two forms of AP2 transcriptional activators, interacting preferably with a CT-rich element. *The Plant Journal*, *37*(3), 326–339.

Yamamoto, A., Mizukami, Y., & Sakurai, H. (2005). Identification of a novel class of target genes and a novel type of binding sequence of heat shock transcription factor in Saccharomyces cerevisiae. *Journal of Biological Chemistry*, *280*(12), 11911–11919.

Yan, H., Jia, H., Chen, X., Hao, L., An, H., & Guo, X. (2014). The cotton WRKY transcription factor GhWRKY17 functions in drought and salt stress in transgenic *Nicotiana benthamiana* through ABA signaling and the modulation of reactive oxygen species production. *Plant and Cell Physiology*, *55*(12), 2060–2076.

Yang, S., Tang, X. F., Ma, N. N., Wang, L. Y., & Meng, Q. W. (2011). Heterology expression of the sweet pepper CBF3 gene confers elevated tolerance to chilling stress in transgenic tobacco. *Journal of Plant Physiology*, *168*(15), 1804–1812.

Yang, Y., Wu, J., Zhu, K., Liu, L., Chen, F., & Yu, D. (2009). Identification and characterization of two chrysanthemum (Dendronthema × moriforlium) DREB genes, belonging to the AP2/EREBP family. *Molecular Biology Reports*, *36*(1), 71–81.

Yanhui, C., Xiaoyuan, Y., Kun, H., Meihua, L., Jigang, L., Zhaofeng, G., ... Yunping, S. (2006). The MYB transcription factor superfamily of *Arabidopsis*: Expression analysis and phylogenetic comparison with the rice MYB family. *Plant Molecular Biology*, *60*(1), 107–124.

Yao, P., Sun, Z., Li, C., Zhao, X., Li, M., Deng, R., ... Wu, Q. (2018). Overexpression of *Fagopyrum tataricum* FtbHLH2 enhances tolerance to cold stress in transgenic *Arabidopsis*. *Plant Physiology and Biochemistry*, *125*, 85–94.

Yao, P. F., Li, C. L., Zhao, X. R., Li, M. F., Zhao, H. X., Guo, J. Y., ... Wu, Q. (2017). Overexpression of a tartary buckwheat gene, FtbHLH3, enhances drought/oxidative stress tolerance in transgenic *Arabidopsis*. *Frontiers in Plant Science*, *8*, 625.

Yao, Y., He, R. J., Xie, Q. L., Zhao, X. H., Deng, X. M., He, J. B., ... Wu, A. M. (2017). ETHYLENE RESPONSE FACTOR 74 (ERF74) plays an essential role in controlling a respiratory burst oxidase homolog D (RbohD)-dependent mechanism in response to different stresses in *Arabidopsis*. *New Phytologist*, *213*(4), 1667–1681.

Yoon, Y., Seo, D. H., Shin, H., Kim, H. J., Kim, C. M., & Jang, G. (2020). The role of stress-responsive transcription factors in modulating abiotic stress tolerance in plants. *Agronomy*, *10*(6), 788.

Yoshii, M., Yamazaki, M., Rakwal, R., Kishi-Kaboshi, M., Miyao, A., & Hirochika, H. (2010). The NAC transcription factor RIM1 of rice is a new regulator of jasmonate signaling. *The Plant Journal*, *61*(5), 804–815.

Yu, Y., Guo, D., Li, G., Yang, Y., Zhang, G., Li, S., & Liang, Z. (2019). The grapevine R2R3-type MYB transcription factor VdMYB1 positively regulates defense responses by activating the stilbene synthase gene 2 (VdSTS2). *BMC Plant Biology*, *19*(1), 478.

Yuan, F., Yang, H., Xue, Y., Kong, D., Ye, R., Li, C., ... Johnson, D. M. (2014). OSCA1 mediates osmotic-stress-evoked Ca 2 + increases vital for osmosensing in *Arabidopsis*. *Nature*, *514*(7522), 367–371.

Zhang, G., Chen, M., Li, L., Xu, Z., Chen, X., Guo, J., & Ma, Y. (2009). Overexpression of the soybean GmERF3 gene, an AP2/ERF type transcription factor for increased tolerances to salt, drought, and diseases in transgenic tobacco. *Journal of Experimental Botany*, *60*(13), 3781–3796.

Zhang, M., Liu, W., Bi, Y., & Wang, Z. (2009). Isolation and identification of PNDREB1: A new DREB transcription factor from peanut (*Arachis hypogaea* L.). *Acta Agronomica Sinica*, *35*(11), 1973–1980.

Zhang, Y. L., Zhang, C. L., Wang, G. L., Wang, Y. X., Qi, C. H., Zhao, Q., ... Hao, Y. J. (2019). The R2R3 MYB transcription factor MdMYB30 modulates plant resistance against pathogens by regulating cuticular wax biosynthesis. *BMC Plant Biology*, *19*(1), 362.

Zhao, C., Zhang, Z., Xie, S., Si, T., Li, Y., & Zhu, J. K. (2016). Mutational evidence for the critical role of CBF transcription factors in cold acclimation in *Arabidopsis*. *Plant Physiology*, *171*(4), 2744–2759.

Zhao, Q., Ren, Y. R., Wang, Q. J., Yao, Y. X., You, C. X., & Hao, Y. J. (2016). Overexpression of Mdb HLH 104 gene enhances the tolerance to iron deficiency in apple. *Plant Biotechnology Journal*, *14*(7), 1633–1645.

Zhou, J. M., Trifa, Y., Silva, H., Pontier, D., Lam, E., Shah, J., & Klessig, D. F. (2000). NPR1 differentially interacts with members of the TGA/OBF family of transcription factors that bind an element of the PR-1 gene required for induction by salicylic acid. *Molecular Plant-Microbe Interactions*, *13*(2), 191–202.

Zhou, L. J., Li, Y. Y., Zhang, R. F., Zhang, C. L., Xie, X. B., Zhao, C., & Hao, Y. J. (2017). The small ubiquitin-like modifier E3 ligase MdSIZ1 promotes anthocyanin accumulation by sumoylating MdMYB1 under low-temperature conditions in apple. *Plant, Cell & Environment*, *40*(10), 2068–2080.

Zhou, Q. Y., Tian, A. G., Zou, H. F., Xie, Z. M., Lei, G., Huang, J., ... Chen, S. Y. (2008). Soybean WRKY-type transcription factor genes, GmWRKY13, GmWRKY21, and GmWRKY54, confer differential tolerance to abiotic stresses in transgenic *Arabidopsis* plants. *Plant Biotechnology Journal*, *6*(5), 486–503.

Zhou, X., Jiang, Y., & Yu, D. (2011). WRKY22 transcription factor mediates dark-induced leaf senescence in *Arabidopsis*. *Molecules and Cells*, *31*(4), 303–313.

Zhu, L., Guo, J., Ma, Z., Wang, J., & Zhou, C. (2018). *Arabidopsis* transcription factor MYB102 increases plant susceptibility to aphids by substantial activation of ethylene biosynthesis. *Biomolecules*, *8*(2), 39.

Zhu, N., Cheng, S., Liu, X., Du, H., Dai, M., Zhou, D. X., ... Zhao, Y. (2015). The R2R3-type MYB gene OsMYB91 has a function in coordinating plant growth and salt stress tolerance in rice. *Plant Science*, *236*, 146–156.

Zou, M., Guan, Y., Ren, H., Zhang, F., & Chen, F. (2008). A bZIP transcription factor, OsABI5, is involved in rice fertility and stress tolerance. *Plant Molecular Biology*, *66*(6), 675–683.

Chapter 6

Nitrate assimilation: cross talk between plants and soil

Zoya Shaikh and Ahmad Ali
Department of Life Sciences, University of Mumbai, Mumbai, India

6.1 Introduction

Air, light, water, nutrients, and optimum temperature are vital for plant growth and sustenance. Both biotic and abiotic components of a landscape play an important role in the spreading of plant species. Absence of any one of the essential components can affect plant health. Such a component is called a limiting factor. The concept of a limiting factor is applicable to everything involved in the interaction between a plant and its habitat. In an ecosystem, any factor like water, temperature, or an essential nutrient can be a limiting factor (Jones Jr, 1997).

A total of 17 nutrients are required for optimum plant growth. While oxygen, water, and carbon dioxide are taken from the atmosphere, the majority of nutrients are obtained from the soil which is the growing medium. These essential nutrients are classified as macronutrients and micronutrients. Nitrogen, phosphorus, potassium, and magnesium are considered as macronutrients needed in large amounts (Fig. 6.1). Nitrogen is needed for healthy foliage, potassium for root growth, and phosphorus for flower development (Morgan & Connolly, 2013).

Macronutrients build up the cellular constituents of plants, such as nucleic acids and proteins; hence they are needed in a larger proportion. Micronutrients include copper, iron, zinc, and manganese; these are utilized as cofactors. These essential mineral nutrients are present under the soil as salts and are consumed by plants in the form of ions. It is important that the nutrients are present in the soil in an appropriate amount and proper ratios (Morgan & Connolly, 2013)

6.2 Nitrogen—an important element for plant growth

Nitrogen is an essential element for plants; it is a major constituent of important organic molecules like proteins and nucleic acids (Fig. 6.2). It is found

Frontiers in Plant–Soil Interaction. DOI: https://doi.org/10.1016/B978-0-323-90943-3.00002-X

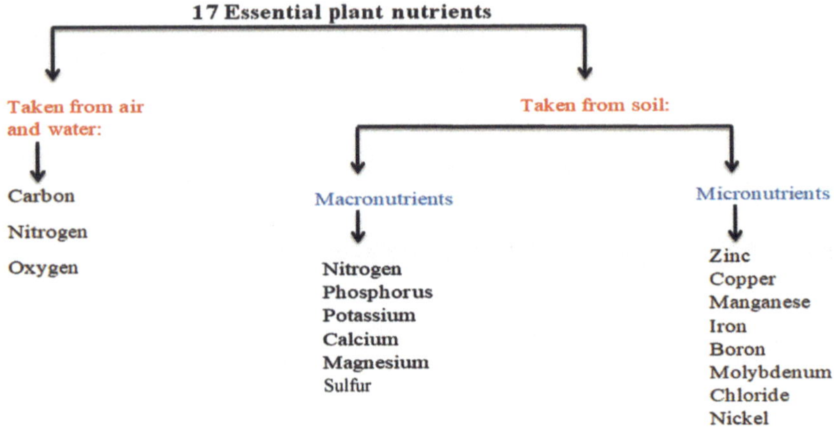

FIGURE 6.1 Seventeen essential nutrients required for plant growth.

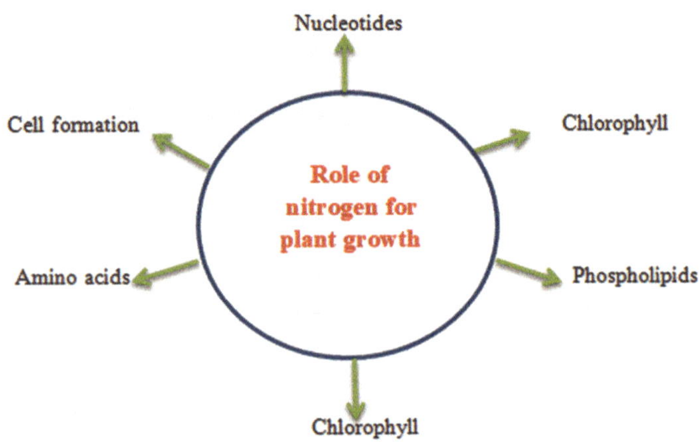

FIGURE 6.2 Nitrogen is utilized to make important plant constituents.

abundantly in the air (78% by volume) in an utilizable form; but its concentration is found to be relatively low in soil. Hence, nitrogen acts as a limiting factor for plant growth in many cases. Atmospheric nitrogen can be fixed by leguminous plants with the help of symbiotic association with nitrogen-fixing bacteria (e.g., Rhizobia). Most of the time, plants derive nitrogen from the soil because atmospheric nitrogen cannot be consumed easily due to its larger form, and a lot of energy will be required to convert it into a simpler form (Ohyama, 2010).

On the other hand, soil contains biologically useful forms of nitrogen. Soil organic matter is the major source of nitrogen, but it cannot be readily absorbed by plants. Mineralization of organic matter and decomposition of organic waste releases inorganic nitrogen compounds; these are mostly

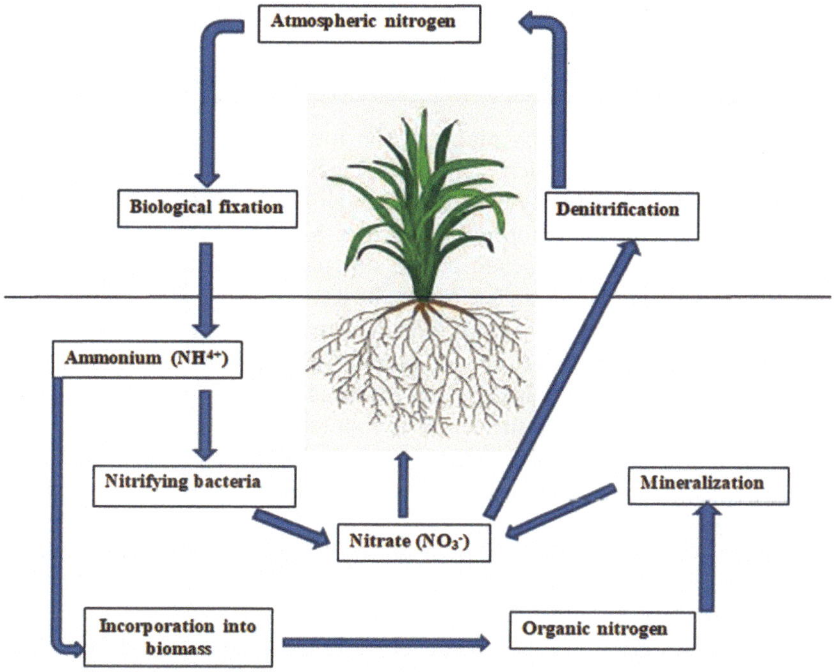

FIGURE 6.3 Plant and soil interaction for nitrogen.

nitrate and ammonium. Nitrate and ammonium are the major forms of inorganic nitrogen present in soil. These are absorbed by plants through roots (Fig. 6.3). After deriving nitrogen through nitrate reduction, ammonium absorption, or nitrogen fixation, the next step is to assimilate this nitrogen into organic compounds in which ammonium is the key component. Ammonium is added by glutamine synthetase (GS) into the amide group of glutamine. It is then incorporated into glutamate by glutamate synthase. Next, nitrogen gets assimilated into nucleic acids, proteins, and amino acids (Näsholm, Kielland, & Ganeteg, 2009).

6.3 Plant soil interaction and nitrate uptake

Plants absorb water and nutrients from the soil through roots. The physical conditions of soil like temperature, water content, and aeration influence the plant nutrient absorption. Diffusion, root interception, and mass transport are the three processes by which the plant root system acquires the nutrients from the soil. Simple diffusion is adapted for nonpolar molecules like NH_3, CO_2, and O_2. It follows the concentration gradient via the cell lipid bilayer membrane without the use of transport proteins. Facilitated diffusion is another method where there is rapid movement of ions according to the concentration gradient. This is done with the help of transport proteins. Ions are absorbed against the concentration gradient by active transport through molecular pumps which utilize ATP as the energy source (Reichardt & Timm, 2020).

Nutrients are taken up by plants in the form of small molecules (with/without charge) or ions which can be positively or negatively charged. The essential nutrients are mostly absorbed through cation exchange. In this process hydrogen ions are pumped by the root hairs into the soil via proton pumps. The cations are attached to the soil particles which are negatively charged, which are displaced by the hydrogen ions so that cations can be readily taken up by the root. Plant roots and root hairs are extremely essential for nutrient uptake. The rate of nutrient uptake can be altered by the structure of the root. As nutrients are absorbed at the roots in the form of ions, they are transported to the center of root (stele) so that the nutrients can easily reach the conducting tissues xylem and phloem. Nutrients travel from the soil to root xylem vessels (present in stele) via symplasmic and apoplastic pathways (Ohyama, 2010).

Nitrate (NO_3^-) is one of the major sources of nitrogen (N) particularly in natural ecosystems and agricultural soils which are well aerated. At the roots after the absorption, nitrate is reduced to nitrite and then into ammonia. It can also be transported to the shoots and reduced to ammonia in the leaves. Hence, ammonium is a major component in the assimilation of nitrate. A nitrate transporter present in the cell membrane absorbs nitrate ions. Incorporating an anion (NO_3^-) will require energy for absorption as the cell is negatively charged inside (low electric potential). Hence, the active transport of nitrate though the nitrate transporter is supported by cotransport of $2H^+$ with NO_3^-. Nitrate is then reduced to nitrite (NO_2^-) in the roots or leaves. This process is catalyzed by nitrate reductase (NR) enzyme in the cytosol (Ohyama, 2010; Reichardt & Timm, 2020).

Consequently, nitrite is moved to chloroplast or plastid and reduced to ammonium, which is later assimilated into glutamine by GS2 (glutamine synthetase 2) through the GS/GOGAT cycle. The reduction of nitrite to ammonia is catalyzed by nitrite reductase enzyme. After the assimilation of ammonium to Glu, the amino N of Glu is transferred to oxo-acids, leading to the production of different amino acids with the help of transaminases. If the concentration on nitrate is higher in the soil or a medium, it is stored temporarily in the vacuoles of root cortex cells (Ohyama, 2010).

Physiological studies have revealed two types of transport system by which nitrate (NO_3^-) is taken up by the roots, depending upon its concentration. Saturable high affinity transport system (HATS) and low affinity transport system (LATS). When nitrate is available at low concentrations (1 μM to 1 mM) the HATS is adapted. On the other hand, the LATS comes into play when NO_3^- concentrations are above 1 mM (Crawford & Glass, 1998). These transport systems are expressed as constitutive or inducible.

The constitutive high affinity transport system (cHATS) is present if the plants have not been exposed to NO_3^- earlier. The inducible system (iHATS) is available when NO_3^- exists in the external environment (Ali, 2020; Miller, Fan, Orsel, Smith, & Wells, 2007).

6.4 Regulation and signaling in nitrate uptake

In plants, the uptake and transport of nitrate is highly regulated. It involves maintaining a balance between nitrate influx from apoplasm to cytoplasm and efflux of nitrate in the reverse direction (Crawford & Glass, 1998; Galvan & Fernandez, 2001; Kiba & Krapp, 2016). Root nitrate uptake is upregulated by low availability of nitrogen and downregulated in the presence of high levels of nitrogen, hence feedback inhibition occurs (Imsande & Touraine, 1994; Lee, 1993). Different environmental and internal signals influence the rate of nitrate uptake, nitrate being the most important inducer in itself (Ali, 2020). Photosynthesis also plays a crucial role in regulating nitrate uptake. It follows diurnal rhythm, where nitrate uptake is at its peak during the light period and least in the dark (Delhon, Gojon, Tillard, & Passama, 1996; Ruffel, Gojon, & Lejay, 2014). A regulatory mechanism of complex nature is developed by plants to adapt to the variation in nitrogen availability, growth, and nitrogen (N)/carbon (C) metabolism. This involves short- and long-range signaling pathways. Studies based on transcriptomics have indicated that there is an interaction between hormone, C signaling pathways, and nitrogen. Therefore root nitrate uptake depends on external NO_3^- availability, N status of the plant, growth, development, and C metabolism. Since C skeletons are required to incorporate inorganic N into molecules like amino acids, proteins, and nucleic acids, a feasible C:N ratio is maintained by plants through complex regulatory networks (Alvarez, Vidal, & Gutiérrez, 2012; Forde, 2002; Imsande & Touraine, 1994; Ruffel et al., 2014).

Nitrate (NO_3^-) locally induces expression of genes which encode for nitrate transport enzymes and involved in nitrate assimilation (Crawford & Glass, 1998; Gojon, Nacry, & Davidian, 2009; R. Wang et al., 2004). This response has a duration of about a few minutes to a few hours, followed by a first supply or resupply of NO_3^- after a long period of the N-deprivation period. Some of these genes are also repressed by NO_3^- on long-term basis (Krouk, Tillard, & Gojon, 2006; Muños et al., 2004). However, an overall systemic feedback repression exists in the whole plant to regulate the root NO_3^- uptake which is linked with the high N status of the plant (Crawford & Glass, 1998; Forde, 2002; Gansel, Muños, Tillard, & Gojon, 2001). Some studies suggest the role of downstream N metabolites which are translocated from shoot to root through phloem; they mediate the downregulation of root N uptake (Ali, Raghuram, & Sivakami, 2010; Cooper & Clarkson, 1989; Muller & Touraine, 1992). In the mid-1990s, studies were initiated for the identification and characterization of genes which encode NO_3^- transporters in plants (Dreyer et al., 1999). To date, nitrate transporter1/peptide transporter family (NRT1/NPF), nitrate transporter 2 (NRT2), chloride channel, and slowly activating anion channel (SLAC/SLAH) have been discovered and characterized as the four main nitrate transporter families (Krapp, 2015).

NRT1.1/NPF6.3 was the first regulatory gene to be identified for nitrate uptake. It codes for a dual affinity NO_3^- transporter (Liu, Huang, & Tsay, 1999; Tsay, Schroeder, Feldmann, & Crawford, 1993) and acts as a NO_3^- transceptor (transporter/sensor) (Gojon, Krouk, Perrine-Walker, & Laugier, 2011). It can function in low and high nitrate affinity states (Tsay et al., 1993; R. Wang, Liu, & Crawford, 1998), is a potential nitrate sensor, and can trigger independent signaling pathways as a nitrate response. Four mechanisms of signaling are activated by *NRT1.1* (Bouguyon et al., 2015; R. Wang et al., 1998). They are the primary nitrate response (Fig. 6.4), feedback repression of *NRT2.1* due to high NO_3^- supply, the promotion of lateral root (LR) branching by *NRT1.1* in reaction to NO_3^-, and induction of genes in the presence of high NO_3^- levels (Bouguyon et al., 2015; Krouk et al., 2010).

NRT1.1 also triggers different NO_3^- signaling pathways that control the expression of the high-affinity NO_3^- transporter gene *NRT2.1*, and LRs development. Consequently, it was concluded that local induction and repression of *NRT2.1* expression by NO_3^- requires NRT1.1 (Krouk et al., 2006; Muños et al., 2004). Due to low NO_3^- concentration, NRT1.1 is phosphorylated at T101 residue by a CBL-interacting protein kinase CIPK23 (Ho, Lin, Hu, & Tsay, 2009). This phosphorylation switches NRT1.1 to the high-affinity function; hence *NRT2.1* gene expression is weakly induced. In contrast, in the case of high NO_3^- concentration, phosphorylation of T101 does not occur and NRT1.1 is switched to a low-affinity function. CIPK8 (CBL-interacting protein) (Hu, Wang, & Tsay, 2009) and NLP7 (a putative transcription factor called NIN-Like Protein7), are the recently identified components of NO_3^- signaling pathways which target *NRT2.1* expression (Ruffel et al., 2014).

Plant nutrition and growth are interconnected through a feed-forward cycle which maintains a balance at physiological and molecular levels (Krouk et al., 2011). NO_3^- and hormones interact at transporter level to integrate nutrition and development. At molecular level, the relationship between hormones and NO_3^- transport is affected by the accumulation of mRNA of genes which encode for NO_3^- transporters. For instance, *NRT1.1* gene was found to be regulated by auxin in the presence of NO_3^- (Guo, Wang, & Crawford, 2002). Cytokinin, when supplied exogenously downregulates *NRT2* genes expression. Also, abscisic acid and ethylene influence the expression of NO_3^- transporter gene expression (Ruffel et al., 2014).

NO_3^- activates *CIPK8* (calcineurin B-like protein kinase), which is downregulated in *chl1-5* mutants. In low affinity system, *CIPK8* works as a positive regulator of primary NO_3^- response. The protein kinase complex CIPK23-CBL1/CBL9 (CIPK, CBL-interacting protein kinase; CBL, calcineurin-B-like protein) is associated with dual-affinity transition changes

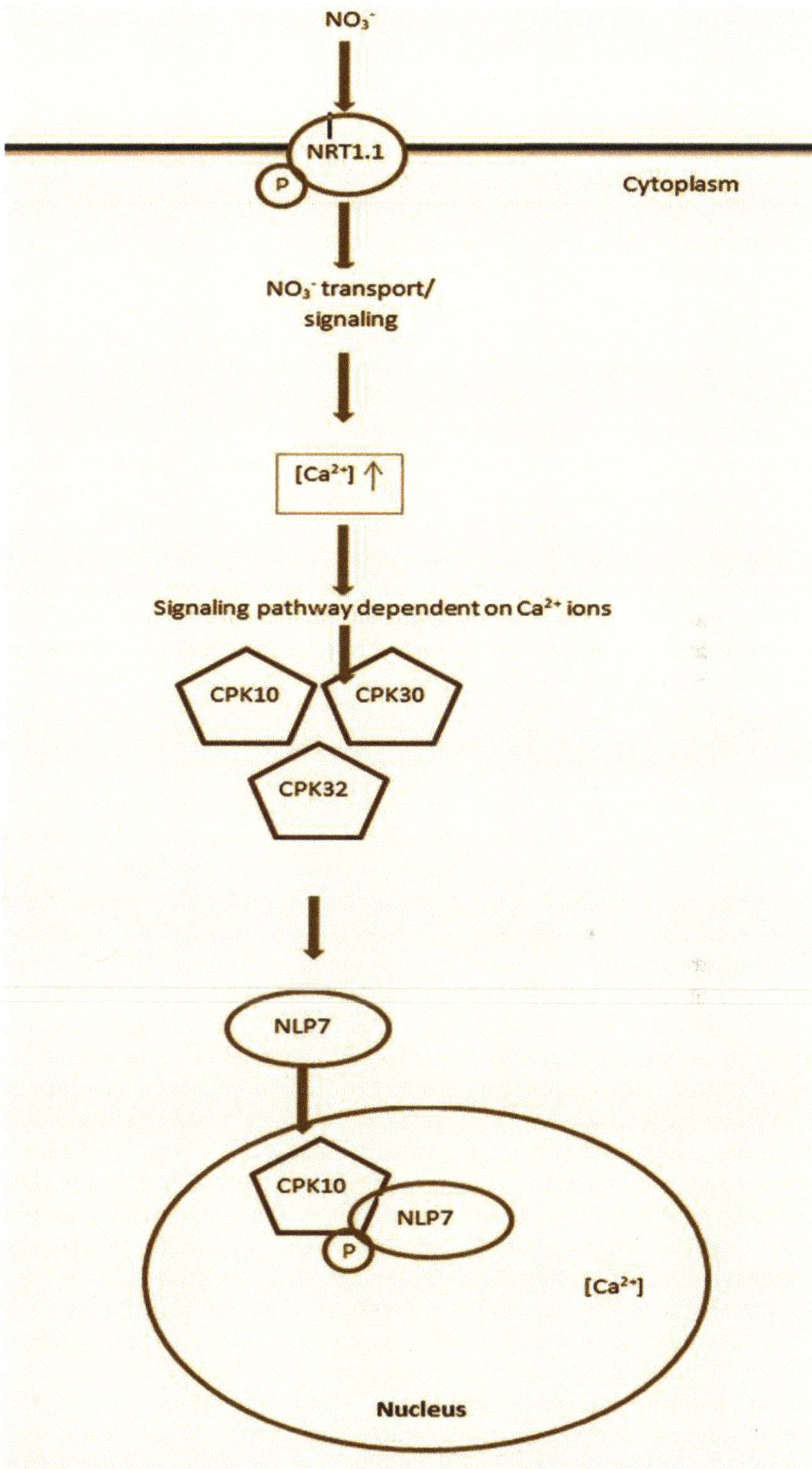

FIGURE 6.4 For primary nitrate response, CPK10, CPK30, and CPK32 are the master regulators. They connect the influx of calcium and phosphorylation of target proteins.

of *NRT1.1* through phosphorylation (Ho et al., 2009). Studies reveal that *FIP1* (factor interacting with poly (A) polymerase 1) is involved in the regulation of *CIPK8* and *CIPK23* expression related to NO_3^- signaling, Hence, nitrate mediates the signaling of CPK molecules, which leads to the phosphorylation of transcription factors. This is turn regulates the expression of downstream genes that have an impact on nitrogen assimilation (Asim et al., 2020; Liu et al., 2017).

However, increased Ca^{2+} influx affects protein phosphorylation and controls the movement of important components involved in the NO_3^- signaling pathway. *CPK10*, *30*, and *32* regulate NO_3^- response and connect the phosphorylation of the target proteins with Ca^{2+} influx (Fig. 6.3) (Armijo & Gutiérrez, 2017). Studies indicate that Ca^{2+} is a potential second messenger in NO_3^- signaling pathways (Dodd, Kudla, & Sanders, 2010). A change in cytoplasmic Ca^{2+} is perceived by Ca^{2+} sensor proteins and initiates cascades of downstream signaling which alter phosphorylation, enzymatic activity, gene expression, and cytoskeleton orientation. Phosphatidylinositol-specific phospholipase C (PI-PLC) is an important component of nitrate signaling and transport. It is modulated through the process of phosphorylation and dephosphorylating (Asim et al., 2020).

6.5 Nitrate assimilation

The biological process, by which inorganic nitrogen gets converted to ammonia, and then organic nitrogen, is known as nitrate assimilation. The conversion of nitrate to nitrite is catalyzed by enzyme NR; this could be a rate-controlling and regulated step in the nitrate assimilation process. Plants incorporate nitrate into proteins and nucleic acids through the process of nitrate assimilation (Chaffey, 2014; Solomonson & Spehar, 1977). There are two steps of nitrate reduction. First, nitrate is converted to nitrite with the help of NR and NADH/NADPH. This step occurs in the cytosol (Tischner, 2000). In the second step, nitrite is reduced to ammonia by nitrite reductase which is ferredoxin dependent, this occurs in the plastids (chloroplasts). Ammonia produced here is incorporated as the amide group of glutamine by GS, using glutamate as a substrate. After this, the amide group is transferred to 2-oxoglutarate molecule by glutamate synthase (Fd-GOGAT and NADH-GOGAT) resulting in two glutamates (Fig. 6.5). More amino acids are produced through transamination reactions using glutamine (Fariduddin, Varshney, & Ali, 2018; Lea & Miflin, 2003).

Nitrate reduction occurs in both shoots and roots. In plants, most of the nitrate assimilation occurs at roots in two cases; i.e., if the plant species belongs to temperate region or when the external availability of NO_3^- is low around the plant for them to grow (Gojon, Bussi, Grignon, & Salsac, 1991). During daytime, shoots can easily utilize sunlight for NO_3^- assimilation. Hence, less energy will be diverted away from the photosynthetic carbon

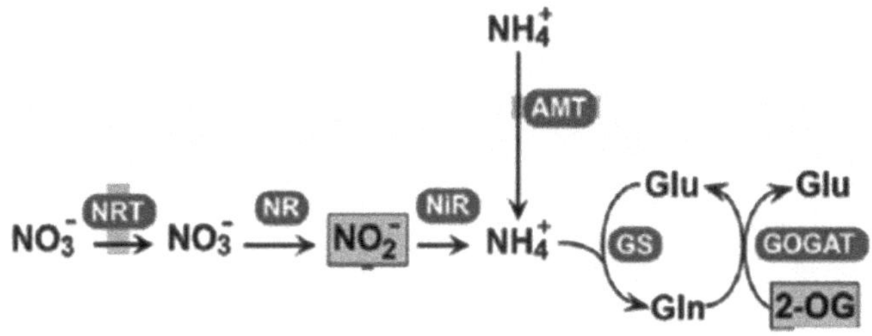

FIGURE 6.5 Schematic representation of regulation of the nitrate assimilation. *AMT*, ammonium transporter; *NRT*, nitrate transporter.

assimilation and plant growth (McDermitt & Loomis, 1981). This will result in a low energy cost for NO_3^- assimilation in the leaf compared to root, as ATP and reducing equivalents for NO_3^- assimilation can be obtained without effecting CO_2 fixation. However, during the night NO_3^- assimilation in the shoot proves disadvantageous as NO_3^- assimilation and CO_2 fixation would compete to obtain ATP and reductants produced by photosynthetic electron transport (Canvin & Atkins, 1974). This will lead to decrease in CO_2 fixation. Moreover, it is necessary to neutralize the hydroxyl ions produced during this process in the leaf, by synthesizing organic acid. This step is required for the maintenance of pH (Smirnoff & Stewart, 1985).

Previous studies reveal that there is no effect of day and night cycles on the functionality of NR in roots. Hence, the rate of nitrate assimilation is the same during both day and night. Cytosolic NO_3^- homeostasis is achieved by roots, in the case of deprivation and resupply of NO_3^-. Thus the site of nitrate assimilation along with its timing plays an important role in plant energy budget, performance, and adaptation (Garma & Bloom, 2014).

6.5.1 Enzymes of nitrate assimilation

6.5.1.1 Nitrate reductase and nitrite reductase

In higher plants, NR is a homodimeric cytosolic protein. In each monomer it has five conserved domains: a dimer interface domain, an Mo-MPT domain that contains the single molybdopterin cofactor, an FAD domain to form the cytochrome b reductase fragment, an NADH domain, and a cytochrome b domain. It catalyzes the conversion of nitrate to nitrite (Campbell, 1999). The oxygen of nitrate molecule binds to Mo in the (IV) oxidation state; this displaces a hydroxide ion. After this, a multiple bond is created between Mo (VI) and the oxygen and nitrite is ejected. NAD(P)H from intramolecular transport chain reduces the Mo(VI) double bond to oxygen (Ali, Sivakami, & Raghuram, 2007a; Fischer et al., 2005).

$$NO^{3-} + NAD(P)H + H^+ \rightarrow NO^{2-} + NAD(P)^+ + H_2O$$

The regulation of NR occurs at transcriptional and translational levels. It is induced by nitrate itself, light, and a negative feedback mechanism. Under stress conditions, the nitrate gets reduced to nitric oxide instead of getting reduced by NR for amino acids. This can cause damage to the plant. Hence, it is important to regulate reductase activity in order to minimize the production of nitric oxide. Inactivation of NR involves various steps and signals (Ali, Sivakami, & Raghuram, 2007b; Jha, Ali, & Raghuram, 2007; Y. Wang, Bouchard, & Coyne, 2018).

Nitrite reductase catalyzes the six-electron reduction of nitrite to ammonia. These enzymes are monomeric metalloproteins of size 60–64 kDa and are localized in chloroplasts of leaf and plastid of roots. Each enzyme molecule consists of a 4Fe-4S cluster at its active center, the binding site for nitrite and the siroheme prosthetic group (Ali, 2020).

$$NO^{2-} + 8H^+ + 6e^- \rightarrow NH^{4+} + 2H_2O$$

6.5.1.2 Glutamine synthetase

GS catalyzes the condensation of glutamate and ammonia to produce glutamine. This reaction is ATP dependent and utilizes the ammonia generated from nitrate reduction, photorespiration, and amino acid degradation as substrate. Co^{2+}, Mn^{2+}, or Mg^{2+} are the cofactors involved (Ali, 2020; Cren & Hirel, 1999).

$$\text{Glutamate} + NH_4 + \text{ATP} \rightarrow \text{Glutamine} + \text{ADP} + \text{Pi} + H_2O$$

In higher plants, GS consists of eight identical subunits. Its molecular mass is ~320–380 kDa and has two isoforms (GS1 and GS2). GS1 is present in leaves and nonphotosynthetic tissues' cytosol, while GS2 is found in chloroplasts, etiolated plants, and plastids of roots. The assimilation of ammonia is carried out by these isoforms. GS1 functions to assimilate ammonia (derived from soil) in the roots and then reassimilates the ammonium generated via the breakdown of nitrogenous substances at the time of germination. On the other hand, GS2 is involved in the primary assimilation of ammonia which was reduced from nitrate in chloroplast or obtained via reassimilation of photorespiratory ammonium (Ali, 2020; Cren & Hirel, 1999).

GLN2 and *GLN1* are the two classes of genes which code for GS. *GLN2* codes for chloroplastic GS2 and it is found in almost all species that have been studied to date. Chloroplastic GS2 is involved in the primary assimilation of ammonium which comes from nitrate reduction in C_3 and C_4 plants. It also plays a role in the reassimilation of ammonium which is produced through photorespiration in C_3 plants. Interestingly, the amount of ammonium flux in leaves of C_3 plants through photorespiration pathway is estimated to be five- to tenfold greater compared to the one produced from nitrate reduction (Keys et al., 1978). On the other hand, the *GLN1* gene codes for cytosolic GS1 isoforms which are found in roots or stems, and contribute to ammonium recycling at developmental stages like leaf senescence. It is also involved in glutamine synthesis for phloem sap transport (Bernard & Habash, 2009).

Three additional enzymes, namely cytosolic asparagine synthetase (AS), carbamoyl phosphate synthase (CPSase), and mitochondrial NADH-glutamate dehydrogenase, could be involved in ammonium assimilation. Cytosolic AS functions to catalyze the transfer of amide group present in glutamine to aspartate molecule (ATP-dependent), producing glutamate and asparagine (Lam et al., 2003). This suggests that AS can utilize ammonia as substrate. In addition, studies show that asparagine production does not depend on amino transferase. CPSase produces carbamoyl phosphate, which is a precursor of citrulline and arginine. This process occurs in plastids by utilizing ATP, carbonate, and ammonium or the amide group of glutamine (Potel et al., 2009). The mitochondrial NADH-glutamate dehydrogenase can incorporate ammonium into glutamate. This step occurs in response to the high levels of ammonium generated due to stress (Skopelitis et al., 2006).

Clearly, reducing powers such as NADH or ferredoxin (Fdx) are necessary for the functioning of GOGAT, NR, and NiR. Asparagine synthetase and GS require ATP. Furthermore, ketoacids and carbon skeletons are essential for the formation of organic nitrogen, such as amino acids. Thus carbon skeletons' availability for ammonium condensation and supply of Fdx, ATP, and NADH as products of photorespiration, respiration, and photosynthesis pathways are necessary for nitrogen assimilation (Ali, 2020; Masclaux-Daubresse et al., 2010).

6.5.1.3 *Glutamate synthase*

Glutamate synthase catalyzes the formation of glutamate from glutamine and ketoglutarate.

$$\text{Glutamine} + 2 - \text{OG} + 2\ \text{ferredoxinreduced} \rightarrow 2\ \text{glutamate} + 2\ \text{ferredoxinoxidised}$$

$$\text{Glutamine} + 2 - \text{OG} + \text{NAD(P)H} \rightarrow 2\ \text{glutamate} + \text{NAD(P)}^+$$

Plants possess two types of glutamate synthase: Fd-GOGAT and NADH-GOGAT. Fd-GOGAT uses ferredoxin as electron donor, and is mainly localized in leaf chloroplasts. NADH-GOGAT uses NADH as the electron donor and it is located in plastids (Vanoni, Dossena, van den Heuvel, & Curti, 2005). The two isoforms of glutamate synthase differ according to their immunological, physicochemical, and regulatory properties. Also, they are encoded by separate set of genes. The Fd-GOGAT is mainly found in chloroplasts. It takes part in assimilating ammonia derived from photorespiration and nitrate assimilation. In the majority of cases, this enzyme is found to be monomeric (145–165 kDa). The transcription of the Fd-GOGAT gene in leaves is regulated by light, increasing the enzyme activity and products of translation. NADH-GOGAT is a monomeric enzyme having molecular mass 190–240 kDa. It is present in the vascular bundles of developing leaf blades. It is less active compared to Fd-GOGAT of leaves (Ali, 2020; Ali, Jha, Sandhu, & Raghuram, 2008).

6.6 Nitrogen use efficiency and nitrate pollution

Since plants utilize nitrogen from the soil to grow, we have to replenish the used up nitrogen for new plants. If the nitrogen available is insufficient, plants become nitrogen deficient with improper growth. Generally, nitrogen is supplied to the soils through different fertilizers. Nowadays, consumption of fertilizers has increased as it improves crop yield. However, excessive usage of nitrogen fertilizers poses danger to the environment.

The fraction of applied nitrogen which is absorbed and utilized by the plant is known as nitrogen use efficiency. It is an important aspect for the improvement of crop yield and to reduce the demand for fertilizers. This will in turn help in alleviating environmental pollution which occurs due to excessive use of fertilizers. By improving the ability of plants to make use of nitrogen, we can also sustain the environment. Therefore genetic engineering along with improvement in plant breeding holds great potential for the production of plants that will efficiently utilize fertilizers (Ali, 2020). Many studies have been conducted to analyze the optimal use of fertilizers and NUE improvement of crops to increase crop yields and lessen environmental side effects. NUE of a plant is constituted by its ability for nitrogen uptake from soil and absorption. Due to population growth and higher crop demand, increased use of fertilizers containing (NO_3^-) ions is causing water pollution. Moreover, agricultural runoff can cause eutrophication of lakes and nitrate pollution of drinking water. Hence it has become imperative to decrease nitrate ions emissions into the environment (Benincasa, Guiducci, & Tei, 2011).

Overstimulation of aquatic plant and algal growth, occurring due to excess nutrients being available, especially nitrogen (including nitrates) and phosphorus, is known as eutrophication. The oxygen concentration in the water bodies can decrease due to excessive growth of aquatic plants and algae. This may deplete oxygen levels in the water and negatively affect aquatic life and diversity. Overgrowth of algal blooms and aquatic plants causes discoloration and odor in the water. Algal blooms can also release toxins into the water. Nitrate pollution of groundwater is another major issue associated with fertilizer–N use efficiency. Many studies have indicated the possibility of groundwater pollution due to nitrate in Asia and Africa (Bijay, Yadvinder, & Sekhon, 1995). Nitrate leaching occurs in the majority of regions where irrigated agriculture occurs, having greater requirement for water and nitrogen (N). This raises the possibility of the groundwater getting polluted by nitrates and arid regions. In addition to this, nitrate leaching may have drastic effects on aquatic ecosystems due to the discharge of polluted groundwater into surface water (Kraft & Stites, 2003; Shen, Lei, Yang, & Kanae, 2011).

Remobilization of nitrogen is an important aspect of NUE. NRT1.7 is a nitrate transporter which loads excess nitrate which is stored in leaves to phloem and allocates nitrate to sink leaves. When N is not available, *nrt1.7* mutant displays retarded growth. This indicates that source-to-sink remobilization of stored nitrate mediated by NRT1.7 is an important factor for sustainable plant growth. A study suggested that improving source-to-sink nitrate remobilization can help in increasing NUE and crop production (Chen, Chen, Tseng, & Tsay, 2020).

6.7 Conclusion

Since nitrate is the most important source of nitrogen for plants, nitrate assimilation is a major process through which most of the plants acquire nitrogen for growth and development. Nitrate assimilation is a multistep process involving a significant reduction pathway through which inorganic nitrogen is converted to organic molecules like amino acids, with the help of enzymes. Several studies have made it possible for us to understand the nitrate assimilation pathway and its regulation. However, comprehensive investigation is required to identify the role of components like nitrate receptors, signal transduction mechanisms, and transcription factors in the assimilatory pathway of nitrate in plants. In addition, the impact of environmental perturbations on rate of nitrate uptake in plants is another area which needs to be studied in detail. A better understanding of nitrogen use efficiency and its application can lead to the appropriate use of fertilizers and a reduction in nitrate pollution. The information obtained from previous studies regarding nitrate response in plants should be elaborated further to uncover the in-depth mechanism of nitrate signaling in plants and its integration with other plant process.

Conflict of interest

We declare that we have no conflict of interest.

References

Ali, A. (2020). Nitrate assimilation pathway in higher plants: Critical role in nitrogen signalling and utilization. *Plant Science Today*, *7*, 182–192.

Ali, A., Jha, P., Sandhu, K. S., & Raghuram, N. (2008). Spirulina nitrate-assimilating enzymes (NR, NiR, GS) have higher specific activities and are more stable than those of rice. *Physiology and Molecular Biology of Plants*, *14*, 179–182.

Ali, A., Raghuram, N., & Sivakami, S. (2010). Differential regulation of nitrate assimilatory enzymes by methionine sulfoximine in rice. *Journal of Advances in Science and Technology*, *13*, 101–104.

Ali, A., Sivakami, S., & Raghuram, N. (2007a). Effect of nitrate, nitrite, ammonium, glutamate, glutamine and 2-oxoglutarate on the RNA levels and enzyme activities of nitrate reductase and nitrite reductase in rice. *Physiology and Molecular Biology of Plants*, *13*, 17.

Ali, A., Sivakami, S., & Raghuram, N. (2007b). Regulation of activity and transcript levels of NR in rice (*Oryza sativa*): Roles of protein kinase and G-proteins. *Plant Science*, *172*, 406–413.

Alvarez, J. M., Vidal, E. A., & Gutiérrez, R. A. (2012). Integration of local and systemic signaling pathways for plant N responses. *Current Opinion in Plant Biology*, *15*, 185–191.

Armijo, G., & Gutiérrez, R. A. (2017). Emerging players in the nitrate signaling pathway. *Molecular Plant*, *10*, 1019–1022.

Asim, M., Ullah, Z., Xu, F., An, L., Aluko, O. O., Wang, Q., & Liu, H. (2020). Nitrate signaling, functions, and regulation of root system architecture: Insights from *Arabidopsis thaliana*. *Genes*, *11*, 633.

Benincasa, P., Guiducci, M., & Tei, F. (2011). The nitrogen use efficiency: Meaning and sources of variation—Case studies on three vegetable crops in Central Italy. *HortTechnology*, *21*, 266.

Bernard, S. M., & Habash, D. Z. (2009). The importance of cytosolic glutamine synthetase in nitrogen assimilation and recycling. *New Phytologist*, *182*, 608–620.

Bijay, S., Yadvinder, S., & Sekhon, G. S. (1995). Fertilizer-N use efficiency and nitrate pollution of groundwater in developing countries. *Journal of Contaminant Hydrology*, *20*, 167–184.

Bouguyon, E., Brun, F., Meynard, D., Kubeš, M., Pervent, M., Leran, S., ... Zažímalová, E. (2015). Multiple mechanisms of nitrate sensing by Arabidopsis nitrate transceptor NRT1. 1. *Nature Plants*, *1*, 1–8.

Campbell, W. H. (1999). Nitrate reductase structure, function and regulation: Bridging the gap between biochemistry and physiology. *Annual Review of Plant Biology*, *50*, 277–303.

Canvin, D., & Atkins, C. (1974). Nitrate, nitrite and ammonia assimilation by leaves: Effect of light, carbon dioxide and oxygen. *Planta*, *116*, 207–224.

Chaffey, N. (2014). Raven biology of plants, 8th edn. *Annals of Botany*, *113*(7), vii. Available from https://doi.org/10.1093/aob/mcu090, Epub 2014 May 8.

Chen, K.-E., Chen, H.-Y., Tseng, C.-S., & Tsay, Y.-F. (2020). Improving nitrogen use efficiency by manipulating nitrate remobilization in plants. *Nature Plants*, *6*, 1126–1135.

Cooper, H., & Clarkson, D. (1989). Cycling of amino-nitrogen and other nutrients between shoots and roots in cereals—A possible mechanism integrating shoot and root in the regulation of nutrient uptake. *Journal of Experimental Botany*, *40*, 753–762.

Crawford, N. M., & Glass, A. D. (1998). Molecular and physiological aspects of nitrate uptake in plants. *Trends in Plant Science*, *3*, 389–395.

Cren, M., & Hirel, B. (1999). Glutamine synthetase in higher plants regulation of gene and protein expression from the organ to the cell. *Plant and Cell Physiology*, *40*, 1187–1193.

Delhon, P., Gojon, A., Tillard, P., & Passama, L. (1996). Diurnal regulation of NO_3^- uptake in soybean plants IV. Dependence on current photosynthesis and sugar availability to the roots. *Journal of Experimental Botany*, *47*, 893–900.

Dodd, A. N., Kudla, J., & Sanders, D. (2010). The language of calcium signaling. *Annual Review of Plant Biology*, *61*, 593–620.

Dreyer, I., Horeau, C., Lemaillet, G., Zimmermann, S., Bush, D. R., Rodríguez-Navarro, A., ... Gaber, R. F. (1999). Identification and characterization of plant transporters using heterologous expression systems. *Journal of Experimental Botany*, 1073–1087.

Fariduddin, Q., Varshney, P., & Ali, A. (2018). Perspective of nitrate assimilation and bioremediation in (a non-nitrogen fixing cyanobacterium): An overview. *Journal of Environmental Biology*, *39*, 547–557.

Fischer, K., Barbier, G. G., Hecht, H.-J., Mendel, R. R., Campbell, W. H., & Schwarz, G. (2005). Structural basis of eukaryotic nitrate reduction: Crystal structures of the nitrate reductase active site. *The Plant Cell*, *17*, 1167–1179.

Forde, B. G. (2002). Local and long-range signaling pathways regulating plant responses to nitrate. *Annual Review of Plant Biology*, *53*, 203–224.

Galvan, A., & Fernandez, E. (2001). Eukaryotic nitrate and nitrite transporters. *Cellular and Molecular Life Sciences*, *58*, 225–233.

Gansel, X., Muños, S., Tillard, P., & Gojon, A. (2001). Differential regulation of the NO3−and NH4+ transporter genes AtNrt2. 1 and AtAmt1. 1 in Arabidopsis: Relation with long-distance and local controls by N status of the plant. *The Plant Journal*, *26*, 143–155.

Garma, M. B., & Bloom, A. J. (2014). Root strategies for nitrate assimilation. *Root Engineering: Basic and Applied Concepts*, *40*, 251.

Gojon, A., Bussi, C., Grignon, C., & Salsac, L. (1991). Distribution of NO_3^- reduction between roots and shoots of peachtree seedlings as affected by NO_3^- uptake rate. *Physiologia Plantarum*, *82*, 505–512.

Gojon, A., Krouk, G., Perrine-Walker, F., & Laugier, E. (2011). Nitrate transceptor (s) in plants. *Journal of Experimental Botany*, *62*, 2299–2308.

Gojon, A., Nacry, P., & Davidian, J.-C. (2009). Root uptake regulation: A central process for NPS homeostasis in plants. *Current Opinion in Plant Biology*, *12*, 328–338.

Guo, F. Q., Wang, R., & Crawford, N. M. (2002). The Arabidopsis dual-affinity nitrate transporter gene AtNRT1. 1 (CHL1) is regulated by auxin in both shoots and roots. *Journal of Experimental Botany*, *53*, 835–844.

Ho, C.-H., Lin, S.-H., Hu, H.-C., & Tsay, Y.-F. (2009). CHL1 functions as a nitrate sensor in plants. *Cell*, *138*, 1184–1194.

Hu, H. C., Wang, Y. Y., & Tsay, Y. F. (2009). AtCIPK8, a CBL-interacting protein kinase, regulates the low-affinity phase of the primary nitrate response. *The Plant Journal*, *57*, 264–278.

Imsande, J., & Touraine, B. (1994). N demand and the regulation of nitrate uptake. *Plant Physiology*, *105*, 3.

Jha, P., Ali, A., & Raghuram, N. (2007). Nitrate-induction of nitrate reductase and its inhibition by nitrite and ammonium ions in *Spirulina platensis*. *Physiology and Molecular Biology of Plants*, *13*, 163.

Jones, J. B., Jr (1997). *Plant nutrition manual*. CRC press.

Keys, A., Bird, I., Cornelius, M., Lea, P., Wallsgrove, R. M., & Miflin, B. (1978). Photorespiratory nitrogen cycle. *Nature*, *275*, 741–743.

Kiba, T., & Krapp, A. (2016). Plant nitrogen acquisition under low availability: Regulation of uptake and root architecture. *Plant and Cell Physiology*, *57*, 707–714.

Kraft, G. J., & Stites, W. (2003). Nitrate impacts on groundwater from irrigated-vegetable systems in a humid north-central US sand plain. *Agriculture, Ecosystems & Environment*, *100*, 63–74.

Krapp, A. (2015). Plant nitrogen assimilation and its regulation: A complex puzzle with missing pieces. *Current Opinion in Plant Biology*, *25*, 115–122.

Krouk, G., Lacombe, B., Bielach, A., Perrine-Walker, F., Malinska, K., Mounier, E., ... Ljung, K. (2010). Nitrate-regulated auxin transport by NRT1. 1 defines a mechanism for nutrient sensing in plants. *Developmental Cell*, *18*, 927–937.

Krouk, G., Ruffel, S., Gutiérrez, R. A., Gojon, A., Crawford, N. M., Coruzzi, G. M., & Lacombe, B. (2011). A framework integrating plant growth with hormones and nutrients. *Trends in Plant Science, 16*, 178–182.

Krouk, G., Tillard, P., & Gojon, A. (2006). Regulation of the high-affinity NO3 − uptake system by a NRT1. 1-mediated NO3 − demand signalling in Arabidopsis. *Plant Physiology*.

Lam, H.-M., Wong, P., Chan, H.-K., Yam, K.-M., Chen, L., Chow, C.-M., & Coruzzi, G. M. (2003). Overexpression of the ASN1 gene enhances nitrogen status in seeds of Arabidopsis. *Plant Physiology, 132*, 926–935.

Lea, P. J., & Miflin, B. J. (2003). Glutamate synthase and the synthesis of glutamate in plants. *Plant Physiology and Biochemistry, 41*, 555–564.

Lee, R. (1993). Control of net uptake of nutrients by regulation of influx in barley plants recovering from nutrient deficiency. *Annals of Botany, 72*, 223–230.

Liu, K.-H., Huang, C.-Y., & Tsay, Y.-F. (1999). CHL1 is a dual-affinity nitrate transporter of Arabidopsis involved in multiple phases of nitrate uptake. *The Plant Cell, 11*, 865–874.

Liu, K.-H., Niu, Y., Konishi, M., Wu, Y., Du, H., Chung, H. S., ... Maekawa, S. (2017). Discovery of nitrate–CPK–NLP signalling in central nutrient–growth networks. *Nature, 545*, 311–316.

Masclaux-Daubresse, C., Daniel-Vedele, F., Dechorgnat, J., Chardon, F., Gaufichon, L., & Suzuki, A. (2010). Nitrogen uptake, assimilation and remobilization in plants: Challenges for sustainable and productive agriculture. *Annals of Botany, 105*, 1141–1157.

McDermitt, D., & Loomis, R. (1981). Elemental composition of biomass and its relation to energy content, growth efficiency, and growth yield. *Annals of Botany, 48*, 275–290.

Miller, A. J., Fan, X., Orsel, M., Smith, S. J., & Wells, D. M. (2007). Nitrate transport and signalling. *Journal of Experimental Botany, 58*, 2297–2306.

Morgan, J., & Connolly, E. (2013). Plant-soil interactions: Nutrient uptake. *Nature Education Knowledge, 4*, 2.

Muller, B., & Touraine, B. (1992). Inhibition of NO⊟ 3 uptake by various phloem-translocated amino acids in soybean seedlings. *Journal of Experimental Botany, 43*, 617–623.

Muños, S., Cazettes, C., Fizames, C., Gaymard, F., Tillard, P., Lepetit, M., ... Gojon, A. (2004). Transcript profiling in the chl1-5 mutant of Arabidopsis reveals a role of the nitrate transporter NRT1. 1 in the regulation of another nitrate transporter, NRT2. 1. *The Plant Cell, 16*, 2433–2447.

Näsholm, T., Kielland, K., & Ganeteg, U. (2009). Uptake of organic nitrogen by plants. *New Phytologist, 182*, 31–48.

Ohyama, T. (2010). *Nitrogen as a major essential element of plants. Nitrogen assimilation in plants* (pp. 1–17), 37.

Potel, F., Valadier, M. H., Ferrario-Méry, S., Grandjean, O., Morin, H., Gaufichon, L., ... Hirose, N. (2009). Assimilation of excess ammonium into amino acids and nitrogen translocation in Arabidopsis thaliana–roles of glutamate synthases and carbamoylphosphate synthetase in leaves. *The FEBS Journal, 276*, 4061–4076.

Reichardt, K., & Timm, L. C. (2020). *How plants absorb nutrients from the soil. Soil, plant and atmosphere: Concepts, processes and applications* (pp. 313–330). Cham: Springer International Publishing.

Ruffel, S., Gojon, A., & Lejay, L. (2014). Signal interactions in the regulation of root nitrate uptake. *Journal of Experimental Botany, 65*, 5509–5517.

Shen, Y., Lei, H., Yang, D., Kanae, S. (2011). Effects of agricultural activities on nitrate contamination of groundwater in a Yellow River irrigated region.

Skopelitis, D. S., Paranychianakis, N. V., Paschalidis, K. A., Pliakonis, E. D., Delis, I. D., Yakoumakis, D. I., ... Roubelakis-Angelakis, K. A. (2006). Abiotic stress generates ROS that signal expression of anionic glutamate dehydrogenases to form glutamate for proline synthesis in tobacco and grapevine. *The Plant Cell, 18*, 2767–2781.

Smirnoff, N., & Stewart, G. (1985). Nitrate assimilation and translocation by higher plants: Comparative physiology and ecological consequences. *Physiologia Plantarum, 64*, 133–140.

Solomonson, L. P., & Spehar, A. M. (1977). Model for the regulation of nitrate assimilation. *Nature, 265*, 373–375.

Tischner, R. (2000). Nitrate uptake and reduction in higher and lower plants. *Plant, Cell & Environment, 23*, 1005–1024.

Tsay, Y.-F., Schroeder, J. I., Feldmann, K. A., & Crawford, N. M. (1993). The herbicide sensitivity gene CHL1 of Arabidopsis encodes a nitrate-inducible nitrate transporter. *Cell, 72*, 705–713.

Vanoni, M. A., Dossena, L., van den Heuvel, R. H., & Curti, B. (2005). Structure-function studies on the complex iron-sulfur flavoprotein glutamate synthase: The key enzyme of ammonia assimilation. *Photosynthesis Research, 83*, 219–238.

Wang, R., Liu, D., & Crawford, N. M. (1998). The Arabidopsis CHL1 protein plays a major role in high-affinity nitrate uptake. *Proceedings of the National Academy of Sciences, 95*, 15134–15139.

Wang, R., Tischner, R., Gutiérrez, R. A., Hoffman, M., Xing, X., Chen, M., ... Crawford, N. M. (2004). Genomic analysis of the nitrate response using a nitrate reductase-null mutant of Arabidopsis. *Plant Physiology, 136*, 2512–2522.

Wang, Y., Bouchard, J. N., & Coyne, K. J. (2018). Expression of novel nitrate reductase genes in the harmful alga, *Chattonella subsalsa*. *Scientific Reports, 8*, 1–12.

Chapter 7

Next-generation genetic engineering tools for abiotic stress tolerance in plants

Akbar Hossain[1], Sharif Ahmed[2], Zahoor Ahmad[3], Md. Jahangir Alam[4], Debojyoti Moulick[5], Bedabrata Saha[6], Suman Samanta[7], Sagar Maitra[8], Preetha Bhadra[9], Muhammad Adnan Bukhari[10] and Tariq Aftab[11]

[1]*Bangladesh Wheat and Maize Research Institute, Dinajpur, Bangladesh,* [2]*International Rice Research Institute, Dhaka, Bangladesh,* [3]*Department of Botany, University of Central Punjab, Punjab Group of College Bahawalpur, Bahawalpur, Pakistan,* [4]*On-Farm Research Division, Bangladesh Agricultural Research Institute, Gaibanda, Bangladesh,* [5]*Plant Stress Biology and Metabolomics Laboratory, Central Instrumentation Laboratory, Assam University, Silchar, India,* [6]*Department of Weed Research, Newe Ya'ar Research Centre, Agricultural Research Organization, Ramat Yishai, Israel,* [7]*Division of Agricultural Physics, Indian Agricultural Research Institute, New Delhi, India,* [8]*Department of Agronomy, Centurion University of Technology and Management, Paralakhemundi, India,* [9]*Department of Biotechnology, Centurion University of Technology and Management, Paralakhemundi, India,* [10]*Department of Agronomy, University College of Agriculture and Environmental Sciences (UCAES), The Islamia University of Bahawalpur, Bahawalpur, Pakistan,* [11]*Department of Botany, Aligarh Muslim University, Aligarh, India*

7.1 Introduction

Expanding crop creation is presently the most elevated farming need overall due to the gigantic population (FAO, 2009). Climate change and the diminished accessibility of prolific land will raise an issue for future harvest creation (Tester and Langridge, 2010). Truth be told, these burdens undermine the supportability of the horticultural industry. The test presently is to create extra foodstuffs under stress environments and in less soil. Hence, it is currently important to investigate physiological ways to adapt to this impending issue of food security. Above all else, it is critical to comprehend the idea of stress. Plants are frequently presented with a few brutal ecological

Frontiers in Plant–Soil Interaction. DOI: https://doi.org/10.1016/B978-0-323-90943-3.00008-0

conditions that conceivably create pressures. Pressures trigger a varied scope of plant reactions that adjust quality articulation and cell digestion to cause changes in development rates and harvest yields. The term, seriousness, and rate at which pressure is forced all impact on how a plant reacts. The obvious significance of abiotic stresses to horticulture can be seen and abiotic factors give the significant constraints to change yield around the world, and can decrease normal efficiency of yearly harvests by 50%–83%, contingent upon the yield (Bray, Bailey-Serres, & Weretilnyk, 2000). Subsequently, there is a critical requirement for more logical examination to build our comprehension of the physiological conduct of yields accordingly, not exclusively to a particular sort of stress, yet in addition to various communicating stressors, for example, water and temperature. It is recorded that among the abiotic stresses the following are the significant causes which antagonistically influence plant development and profitability around the world:

- Extreme temperatures (freezing, cold, heat)
- Water accessibility (dry spell, flooding)
- Ion toxicity (saltiness, substantial metals)
- Solar radiation (overabundance light, concealing, UV radiation)

Plants react to different pressure factors, for example, saltiness, heat, cold, dry spell, overabundance of water, heavy metals, injury, abundance of light, supplements, anaerobe conditions, and radiation through multidimensional physiological passageways. Plants have adjusted to react to these burdens at the subatomic, cell, physiological, and biochemical level, empowering them to endure (Nakashima et al., 2009).

In farming, the pressure features are a threat for the plants and prevent them from arriving at their full hereditary power and consequently limit the harvest efficiency around the world (Mahajan and Tuteja, 2005). This is assessed from the examination of best products, and normally produce for different harvest crop plants primarily achieve just 20% of their hereditary potential for produce because of different living and nonliving stress features. It has been evaluated that the overall abatements in potential of the most extreme yields related to nonliving stress factors changes between 54% and 82%. In this way, unmistakably there is an necessity to increase nonliving stress resistance in plants.

In addition, for the development of crop species through conventional breeding to attain the food and nutritional security of the increasing population, next-generation new biotechnological tools (NBTs), such as genetic engineering (GE), may be useful for the sustainability of crop production under the changing climate. The present chapter will discuss the concept, methods, and mechanism of NBTs/GE for improvement of crop species against abiotic stresses for meeting the food and nutritional security of the increasing population.

7.2 Consequences/adverse effects of abiotic stresses on crop productivity

Crop productivity is very sensitive to abiotic stresses. Several abiotic stresses play a crucial role on the adverse effects of crop production, mainly due to the fluctuation of environmental conditions like drought, salinity, extremely high or low temperature, metal toxicities, mineral deficiencies, etc. These abiotic stresses significantly diminish crop production. The adverse effects of some abiotic stresses have been presented in this section.

7.2.1 Drought stress effects on crop productivity

Drought is defined as the stress related to water deficit in crop production. Drought stress occurs in plants when the soil retains less moisture due to water loss from soil through evaporation and or transpiration. Drought is one of the major limiting factors that constricts plant growth and development as well as reduces the crop productivity (Araus, Slafer, Reynolds, & Royo, 2002; Karl, Melillo, & Peterson, 2009; Miao, Han, Zhang, Chen, & Ma, 2017). Drought affects both the physiological and molecular action in plants. Drought affects plants on various levels leading to the efflux of cellular water, decreased cell elongation, reduced nutrient availability in plants' root zone, damage to membranes and protein, less absorption in root hair, impaired metabolism of nutrients in plant tissues and cells, inhibited photosynthesis process, plasmolysis, and cell death, and reduced hydration with an increase of reactive oxygen species (ROS) (Artlip & Wisniewski, 2002; Farooq, Wahid, Kobayashi, Fujita, & Basra, 2009; Li et al., 2011).

Drought causes the reduction of yield in different crops species depending on their degree of severity and duration (Cattivelli et al., 2008), and different growth stages in plants with other environmental parameters (Demirevska et al., 2009). Global yield reduction is about 9%–10% as affected by drought and high temperature. In another report, Kim, Iizumi, and Nishimori (2019) reported that globally three quarters of cropped area suffer yield loss from drought. Kumaraswamy and Shetty (2016) also reported that about 30%–70% crop productivity is lost by drought stress. Drought has a significant impact on the reduction of crop yields like rice, wheat, maize, soybean, chickpea, potato, etc. There have been several studies on the effect of drought on yield reduction in different crops around the globe. Drought can trigger the reduction of maize yield ranging from 17%–60% in Southern Africa due to less germination and also less plant growth. Ray, Fares, and Risch (2018) stated that drought causes significant yield reductions of corn, cotton, sorghum, and wheat. In the case of potato, drought (water stress) imposed at vegetative and tuberization stages reduced the tuber number, tuber size, and yield (Schafleitner et al., 2007; Yuldashev & Carli, 2010). The amount of yield loss in different crops affected by drought or water deficit stress is presented in Table 7.1. The negative effect

TABLE 7.1 Yield loss (%) as affected by drought stress in different crops.

Crop species	Crop yield loss (%)	References
Rice (*Oryza sativa* L.)	53–92	Lafitte, Yongsheng, Yan, and LiZ (2007)
	48–94	Lafitte et al. (2007)
	30–55	Basnayake, Fukai, and Ouk (2006)
	60	Basnayake et al. (2006)
	12–39	Sujariya, Jongdee, Inthavong, Budhaboon, and Jongrungklang (2017)
	24–84	Venuprasad, Lafitte, and Atlin (2007)
Wheat (*Triticum aestivum* L.)	57	Balla et al. (2011)
	2.3–18.8	Potopová, Boronean, Boincean, and Soukup (2015)
	22	Akram (2011)
	21	Daryanto, Wang, and Jacinthe (2016)
Maize (*Zea mays* L.)	63–87	Kamara, Menkir, Badu-Apraku, and Ibikunle (2003)
	79–81	Monneveux, Sánchez, Beck, and Edmeades (2006)
	47–70	Chapman and Edmeades (1999)
	25–60	Atteya (2003)
	32–92	Atteya (2003)
	3.2–14	Wang et al. (2020)
	5.8–30.5	Potopová et al. (2015)
	40	Daryanto et al. (2016)
Barley (*Hordeum vulgare* L.)	49–57	Samarah (2005)
Chickpea (*Cicer arietinum* L.)	45–69	Nayyar, Kaur, Singh, and Upadhyaya (2006)
	33	Kashiwagi et al. (2015)
Soybean (*Glycine max* L.)	46–71	Samarah, Mullen, Cianzio, and Scott (2006)
	10.4–21.8	Wang et al. (2020)
Pigeon pea (*Cajanus cajan* L.)	40–55	Nam, Chauhan, and Johansen (2001)
Cowpea (*Vigna unguiculata* L.)	11–60	Ogbonnaya et al. (2003)

(*Continued*)

TABLE 7.1 (Continued)

Crop species	Crop yield loss (%)	References
Common bean (*Phaseolus vulgaris* L.)	58–87	Martínez, Silva, Ledent, and Pinto (2007)
Sunflower (*Helianthus annuus* L.)	60	Mazahery-Laghab, Nouri, and Abianeh (2003)
	Up to 61	Ghaffari, Toorchi, Valizadesh, and Shakiba (2012)
	1.5–22.3	Potopová et al. (2015)
Canola (*Brassica napus* L.)	30	Sinaki, Heravan, Rad, Noormohammadi, and Zarei (2007)
Suger beet (*Beta vulgaris*)	3.6–27.2	Potopová et al. (2015)
Sugarcane (*Saccharum officinarum* L.)	37	Vasantha, Alarmelu, Hemaprabha, and Shanthi (2005)
Potato (*Solanum teberosum* L.)	19–54	Mahmud et al. (2014)
	69	Schafleitner et al. (2007)
	13	Kawakami, Iwama, and Jitsuyama (2006)

of drought on plants is mainly due to the very poor seed germination as well as minimum crop establishment and less vigorous seedlings (Farooq et al., 2009; Kaya, Okcub, Ataka, Cikilic, & Kolsaricia, 2006). Drought impairs the growth and development of plants because of less turgor pressure (Taiz & Zeiger, 2006), cell elongation due to less water flow in the cells through xylem (Nonami, 1998), less expansion of leaves as the reason of less photosynthesis, and low turgor pressure (Rucker, Kvien, Holbrook, & Hook, 1995). Drought causes a significant yield loss in wheat due to water-stress at vegetative, pre-anthesis, and terminal or grain filling stages (Akram, 2011), and at the tasseling stage in maize (Anjum et al., 2011), and the ball production stage in cotton (Pettigrew, 2004). The reason for yield reductions in major cereals is mainly the decrease of enzymatic activity (Ahmadi & Baker, 2001).

7.2.2 Heat stress effects on crop productivity

Heat stress usually appears due to high temperatures, an enormous problem for crop productivity (Hall, 2001). Heat stress occurs when plants experience a high temperature above their adoption level, which depends on the makeup of the proteins and plant cell membranes (Sharkey & Schrader, 2006). Heat stress exhibits long-term (phonological and morphological) and short-term (changing of leaf orientation and lipid composition of plant cell membranes) effects on plants (Wahid, Gelani, Ashraf, & Foolad, 2007). It also inhibits protein synthesis, enzymatic activities, and the plant cell division process. The scorching of plant leaves and twigs, leaf senescence, less growth and development, and also the discoloration of leaves and fruits may occur due to high temperature or heat stress (Vollenweider & Gunthardt-Goerg, 2005). Heat stress reduces water and plant nutrients accessibility in the plant body and also decreases the growth and weight of roots (Huang, Rachmilevitch, & Xu, 2012; Wahid et al., 2007). Extreme heat can cause less or poor seed germination and poor crop establishment. The adverse effect of heat stress varies on the duration and severity of the heat faced by plants (Fahad, Hussain, Saud, Khan et al., 2016). High temperature adversely effects panicle number, fertilization, filled grain per panicle, and grain formation as well as yield reduction in major cereals (Fahad, Hussain, Saud, Khan et al., 2016; Prasad, Boote, Allen, Sheehy, & Thomas, 2006). The amount of yield loss due to heat stress in different crops is presented in Table 7.2. In rice, high temperature (> 30°C) shows a negative performance on the floret, elongation of the pollen tube, and stigma activity (Fahad et al., 2015; Fahad, Hussain, Saud, Khan et al., 2016). High temperature caused a significant reduction of grain weight of rice (Fahad, Hussain, Saud, Hassan et al., 2016) due to pollen sterility, poor anthesis, and less filled grain (Jagadish, Craufurd, & Wheeler, 2007; Prasad et al., 2006; Shah et al., 2011), reduced grain number and yield in wheat (Ferris, Ellis, Wheeler, & Hadley, 1998), decreased net assimilation

TABLE 7.2 Yield loss (%) as affected by heat stress or high temperature in different crops.

Crop species	Crop yield loss (%)	References
Rice (*Oryza sativa* L.)	50	Li, Peng, Xie, Han, and Tian (2010)
Wheat (*Triticum aestivum* L.)	31	Balla et al. (2011)
	40	Elbasyoni (2018)
Maize (*Zea mays* L.)	42	Badu-Apraku, Hunter, and Tollenaar (1983)
Millet (*Pennisetum glaucum* L.)	10–20	Sultan, Defrance, and Iizumi (2019)
Sorghum (*Sorghum bicolor* L.)	5–15	Sultan et al. (2019)

rate and plant growth in maize and sugarcane (Ashraf & Hafeez, 2004; Wahid & Close, 2007), and reduced protein, starch, and oil content in oilseed crops (Maestri et al., 2002; Wilhelm, Mullen, Keeling, & Singletary, 1999). In addition, heat stress also negatively impacted the yield of vegetable crops such as bean, peanut, and tomato (Rainey & Griffiths, 2005; Vara Parasad, Craufurd, & Summerfield, 1999). The reason for yield reduction in tomato was mainly due to the influences of meiosis, fertilization, and growth of embryo (Camejo et al., 2005).

7.2.3 Salinity effects on crop productivity

Salinity refers to the presence of soluble salts, especially sodium chloride, calcium, and magnesium sulfate and bicarbonates into the soil. It is also known as salinization and is expressed as electrical conductivity (EC) of the soil solution and water. Salinity stress occurs in the arid and semiarid region of the world due to higher evaporation and transpiration rates. In addition, salinity stress may be created by the use of high fertilizer application, crop irrigation by saline water, and also some soil component modifications (Carillo, Grazia, Pontecorvo, Fuggi, & Woodrow, 2011). Salt crust remains on the topsoil due to water evaporation from soil, causing a greater salinity than that of irrigation by saline water. Besides, poor irrigation water or drainage increases the trends of salinity development into the soil.

Salinity stress elicits two primary effects on plants: increasing osmotic pressure and ion toxicity due to high levels of Na^+ and Cl^- ions in the cytoplasm (Kumar, 2013). Normally, the osmotic pressure in plant cells is higher than that of the soil solution. Plants take up water and essential plant nutrients into root cells from the soil solution due to this higher osmotic pressure. But under salinity stress, the osmotic pressure in the soil solution becomes higher than the osmotic pressure in plant cells (due to the presence of high salt concentration), which restricts the ability of plants to uptake water and minerals like K^+ and Ca^{2+} and induces wilting in plants (Munns et al., 2002; Paranychianakis & Chartzoulakis, 2005). On the contrary, Na^+ and Cl^- concentration is supposed to be higher in the plant cell resulting in damage to the cell membrane and also having direct toxic effects on metabolic activities in plant cells. Salinity stress is also contributing to adverse effects on seed germination (Sosa, Llanes, Reinoso, Reginato, & Luna, 2005), poor growth and development, plant production quality, and quantity as a result of higher osmotic pressure and the specific ion toxicity (Akbari, Sanavy, & Yousefzadeh, 2007). Salinity is increasing worldwide due to both the natural and irrigation activities for crop production. Salinity has some detrimental effects on plants in physiological and metabolic processes. Salinity stress is notified by several symptoms in plants with the increase of leaf density, succulence, abscission, and root and shoot necrosis, decreased leaf area and internode growth, reduced photosynthetic activities (Flexas et al., 2007), metabolic imbalances, and deposition of ROS and free

TABLE 7.3 Yield loss (%) as affected by salinity stress in different crops.

Crop species	Crop yield loss (%)	References
Rice (*Oryza sativa* L.)	70–76	Khanam, Akhtar, Halim, and Hossain (2018)
Wheat (*Triticum aestivum* L.)	29.6	Cullu (2003)
	69	Khan, Hussain, Abid, and Imran (2004)
Maize (*Zea mays* L.)	50	Mass and Grattan (1999)
Barley (*Hordeum vulgare* L.)	65	Isla, Agragues, and Royo (1998)
Cotton (*Gossypium spp.* L)	35.4	Cullu (2003)
Carrot	32–80	Jahan, Hossain, and Karim (2019)

radicals in plant cells (Ashraf, 2009; Azevedo-Neto, Prisco, Enéas-Filho, Abreu, & Gomes-Filho, 2006). In the case of severe salinity stressed condition, there is ceased plant growth and development and ultimately plant death (Parida & Das, 2005). Salinity is an extreme threat to agricultural crop productivity around the world (Fahad et al., 2017; Gafni & Zohar, 2001; Patel, Prasher, Bonnell, & Boughton, 2002; Rogers, 2002). The amount of yield reduction in different crops due to salinity stress is presented in Table 7.3. Salt stress reduces the marketable yield of root crops, tuber crops, fruits, and vegetables and lowers the market price (Machado & Serralheiro, 2017). The amount of yield loss by salinity stress depends on the salinity level or duration or the effect of salinity effect at different stages of the crops. It also varies from crop to crop and species to species.

7.2.4 Cold effects on crop productivity

Cold stress is another important abiotic stress (chilling injury <20o̲C and freezing injury <0o̲ C), which inhibit plants growth and development by changing the physical and chemical structure of the cell membranes (Barrero-Sicilia, Silvestre, Haslam, & Michaelson, 2017), impairs seed germination and vigor of the seedlings (Kang & Saltveit, 2002; Wang et al., 2016), reduce protoplasmic streaming, causes electrolytes losses, hampers cellular and metabolic functions, and thus reduces crop productivity (Chen & Thelen, 2013; Cruz & Milach, 2004; Kasuga, Qiang, Setsuko, Shinozaki, & Kazuo, 1999; Lang, Zhang, Ebel, Dane, & Dozier, 2005; Oliver, Dennis, & Dolferus, 2007; Ruelland, Vaultier, Zachowski, & Hurry, 2009). Besides,

TABLE 7.4 Yield loss (%) as affected by cold stress in different crops.

Crop species	Crop yield loss (%)	References
Rice (*Oryza Sativa* L.)	30–40	Andaya and Mackill (2003)
	19–29	Ghadirnezhad and Fallah (2014)

cold stress also has detrimental effects on nutrient absorption, nucleic acid and protein synthesis, photosynthesis, and enzymatic activities (Takahashi, Li, Nakayama, Kawamura, & Uemura, 2013).

Cold stress also induces the abscission of flower, abortion of ovule, damage to pollen tubes, leads to reduced fertilization and poor seed set, induces metabolic changes, and reduces crop productivity, especially when affecting at the reproductive stage of plants (Thakur, Kumar, Malik, Berger, & Nayyar, 2010; Whaley, Kirby, Spink, Foulkes, & Sparkes, 2004). Cold stress accounts for a significant yield loss of crops, as presented in Table 7.4. The main cause of the yield reduction may be due to spikelets degeneration and spikelet sterility in rice (Andaya & Mackill, 2003) and significant reduction of productive tillers and young spikes in wheat (Li et al., 2015; Thakur et al., 2010).

7.2.5 Heavy metal effects on crop productivity

Heavy metals (atomic mass >20 and density >5 g cm^{-3}) cause toxicity or stress both in water or soil and are an important constraint on crop productivity throughout the world (Cai et al., 2012). There are two types of metal in the environment: one is essential (Zn, Fe, Cu, Mn, Mo, Mg, and Ni) and the other is nonessential (As, Ag, Cr, Pb, Co, Se, Hg, Cd and Sb). Plants need some of the metals in very minute quantities for their growth and development. But their excessive presence causes inhibition of growth and development as well as yield reduction and even the death of plants. Heavy metals cause physiological, morphological, and biochemical changes in plants that can play negative activities on the reduction of crop productivity. These adverse effects of heavy metals depend on the duration, intensity, crop growth stages, studied plant parts, and crop species. Heavy metals (Cu, Cd, Co, Mn, Ni, Zn, and Pb) hamper or reduce the seed germination of different crops (Chigbo & Batty, 2013; Kalai, Khasmassi, Teixeria de Silva, Gouia, & Ben-kaab, 2014; Kranner & Colville, 2011; Márquez-García et al., 2013; Wang, Wang, Li, Sun, & Xu, 2014; Munzuroglu & Geckil, 2002), change germination patterns and reduce the seedling size (Márquez-García et al., 2013), increase the abnormal embryo and decrease the seed viability (Stvolinskaya, 2000), reduce water and nutrient uptake by plant roots (Poschenrieder & Barceló, 2004), cause chlorosis and necrosis (Pandey &

Sharma, 2002; Rahman, Sabreen, Alam, & Kawai, 2005; Sabir et al., 2013; Shahid, Sabir, Ali, & Ghafoor, 2014), impair photosynthetic process (Andosch, Affenzeller, Lütz, & Lütz-Meindl, 2012; Shahid, Dumat, Pourrut, Sabir, & Pinelli, 2014; Volland, Lütz, Michalke, & Lütz-Meindl, 2012), induce lipid peroxidation (Shahid et al., 2015), reduce plant growth and nitrogen fixing capacity as well as reduce crop productivity (Poschenrieder & Barceló, 2004).

7.3 Prospects of physiological and genetic engineering tools for abiotic stress tolerance in plants

Abiotic stresses, such as salinity, drought, freezing, high temperature, and heavy metals are important challenges for sustainable crop production and cause considerable yield losses worldwide. To ensure the food security of the increasing population, it is required to explore new lands where cultivation is impractical at this stage due to different abiotic stress conditions. To utilize those unfavorable ecosystems there is the need to identify or develop cultivars of major crops which will exhibit tolerance or resistance to various stress factors and ultimately increase agricultural stable crop production (Bray et al., 2000; Mickelbart, Hasegawa, & Bailey-Serres, 2015). Genotypically advanced crop cultivars having special characters show better survival under abiotic stress conditions than other cultivars lacking this trait, therefore to sustain the growth of agricultural production, stress-tolerant or -resistant cultivars are urgent (Shinozaki & Yamaguchi-Shinozaki, 2007; Singh, Redoña, & Refuerzo, 2009). To find the stress-resistant or -tolerant crop cultivars, selection of and conventional breeding programs have been used; however, no significant success was achieved (Singla-Pareek, Reddy, & Sopory, 2001). Scientists are continuously working on developing mechanisms or new cultivars against different crop stresses, including abiotic stress, and found GE approaches are the most sustainable and eco-friendly approach to cope with this challenge (Steinwand & Ronald, 2020). Advances in GE technology using genome editing provide new opportunities for crop improvement for targeted crop traits (Lu & Zhu, 2016; Zong et al., 2017). However, for this GE technology, the selection of the candidate genes is very critical for successfully achieving the desired traits.

Abiotic stress-inducible genes are classified into two major categories, (1) structural and (2) regulatory genes. Structural genes encode proteins that provide stress tolerance directly against stress, such as antifreeze proteins, osmoprotectants, detoxification enzymes, chaperones. On the other hand, regulatory genes regulate the expression of stress-related genes and signal transduction in stress response such as transcription factors (TFs), protein kinases, etc. (Kasuga et al., 1999). The availability of the advanced genome editing tools offers excellent opportunities for trait discovery and development (Klap et al., 2017). The stress-inducible genes and their regulatory genes can be altered, transferred, and expressed in different species via an

Agrobacterium-mediated transformation system, thus inducing biochemical, molecular, and physiological alterations leading to increased tolerance against stresses including abiotic stress (Bhatnagar-Mathur, Vadez, Kiran, & Sharma, 2008).

Many physiological parameters of the crop are correlated with plant growth, development, and economic yield under abiotic stresses, and some instruments have been developed already to study the physiological state indicators which may be further used for GE programs (Waqas et al., 2019). Osmotic adjustment is an adaptive physiological mechanism involved in drought or salinity tolerance, which permits the maintenance of turgor under conditions of water deficit, as it can counteract the effects of a rapid decline in leaf water potential (Chen & Jiang, 2010). Chlorophyll (Chl) fluorescence technique is fast, and a very powerful noninvasive, nondestructive, and informative tool to study the physiological state of photosynthetic processes within thylakoid membranes, chloroplasts, plant tissues, and whole plants (Roháček, Soukupová, & Barták, 2008). It gives the ability to analyze the individual plant life processes and to develop an understanding of the plant's response to fluctuations in environmental conditions for detecting and quantifying the tolerance of plants to stress (Negrão, Schmöckel, & Tester, 2017).

For creating novel quantitative trait loci for raising novel plant types with tolerance to multiple abiotic stress tolerance, the genome editing tools can make a significant contribution. The CRISPR–Cas genome editing technology has already taken plant molecular biology beyond all expectations (Xu, Wei, & Yang, 2017). Due to the robustness and high target specificity and programmability, CRISPR–Cas allows precise genetic manipulation of crop species, which provides the opportunity to create germplasms with beneficial traits for tolerance against different stresses, including abiotic stress (Chen, Wang, Zhang, Zhang, & Gao, 2019).

7.4 Physiological mechanisms for abiotic stress tolerance in plants

7.4.1 Physiological mechanisms of plants adaptation to drought stress

From a broad perspective, the dry spell is a multidimensional wonder, incorporating water shortfalls in the dirt as well as the climate, which is principally dictated by relative humidity and air temperature. In this part, the dry season is considered as a meteorological marvel, depicted as an adequately significant stretch of unevenness among precipitation and evapotranspiration that is fit for advancing the consumption of topsoil water and initiating water pressure in plants, along these lines diminishing their development or potential yield.

Resistance is the capacity of plant species or cultivars for development and creation in dry season conditions. This relies not just upon the length of the dry spell or the existing pattern of plants and water holding limit of the soil in the root zone, it additionally relies upon plant attributes. To forestall water misfortunes, plants react by shutting the stomata. The decrease in shoot development, particularly leaves is beneficial for the plant as it reduces the upper region that is uncovered consequently limiting water misfortune. Plants may likewise forfeit or spread their more seasoned leaves, because of dry spell. Different abiotic stresses including dry spell pressure prompt the excess of toxic ROS ($O_2{\cdot}^-$, H_2O_2, O_2, and OH·) in plants which can harm proteins, lipids, starch, and DNA. ROS can be productively searched for by the plant's cancer prevention apparatus, which include enzymatic cell reinforcements, for example, SOD, CAT, APX, GR, and so on.

To adapt to dry spell pressure, plants are required to go through osmotic change whereby plants reduce their cell osmotic potential by the combination/collection of solutes comprising proline, glutamate, glycine-betaine, carnitine, mannitol, sorbitol, fructans, polyols, trehalose, sucrose, oligosaccharides, and inorganic particles like K^+. These mixes help the plant cells to keep up their water state and accordingly the capacity to provide obstruction to the dry season and cell's drying out. Additionally it has been reported that the aggregation of sinple sugars, for example, glucose and fructose, causes an expansion in the *invertase* action in leaves of uncovered vegetation in the dry season. At the cell level, the osmotic change has been measured as the most prominent acclimatization reaction of plants exposed to H_2O shortage stress, which is portrayed by a net solute increase in cells. These solutes are named viable *osmolytes* which are natural mixtures (e.g., proline, sugar alcohols, sorbitol, and mannitol) that don't affect cell digestion. The osmotic change has two primary effects: (1) retaining cell turgor with undesirable H_2O potential, and (2) providing a more prominent limit with respect to water takeup by the roots

7.4.2 Physiological mechanisms of plants adaptation to flooding

Flooding is the deadliest type of severe weather for plant production. From a broad perspective, the term flooding is frequently used to portray various circumstances in which the water abundance can vary from water immersed soil (e.g., waterlogging) to total submergence of plants. A typical reaction to flooding is the decrease in plant carbon correlation (e.g., pace of photosynthesis). Temporarily, photosynthesis can drop because of a limitation of CO_2 takeup due to stomata shutting (Mollard et al., 2010). Flooding can lead to changes in the morphology, physiology, and anatomy of the plants to shield from this unfavorable pressure (Colmer et al., 2006). The most well-known morphology reaction to flooding is the improvement of the extrinsic rootstock. Those rootstocks have a great sponginess which aids in the take-up of H_2O and supplements under the lowered condition (Zhang, Su, An, & Wu, 2008). Development of an unusual rootstock aids in the oxygenation of tissues through the parenchyma tissue and in this way builds plant biomass

(Colmer, 2003a,b). At the physiological level, overflowing adjusts the plant carbon obsession and H_2O maintenance. It additionally checks the shutting of stomata, the pace of happening, and affects photosynthesis relying upon the resilience to flooding of every plant class (Mollard et al., 2010).

7.4.3 Ion toxicity (salinity, heavy metals)

7.4.3.1 Salinity

The soils keeping great grouping of solvent salts (ECe is $\geq$ 4 dS m^{-1} that is comparable to ~ 40 mM sodium chloride and creates an osmotic weight of roughly 0.2 MPa) have alluded as saline soils and their impact brings about damage to harvest plants. Saline soils altogether limit crop creation and subsequently affect foodstuff safety. Sodic soils or basic soils contain high concentrations of carbonates and bicarbonates and an overabundance of sodium. The pH of this soil is higher than 8.5 and can reach up to 10.7. Saline soils are overwhelmed by Na^+ cations and normally solvent Cl^- and sulfate anions and the pH estimations of these soil are a lot less than in sodic soils. The amassing of salts in the soil cases a huge decrease in H_2O potential, influencing the H_2O equalization of plants and requiring the advancement of more H_2O possibilities to keep up the H_2O gradient, which permits the entry of soil H_2O into the plant (Munns et al., 2002). That osmotic impact of solutes is similar to soil H_2O shortfall, and the underlying plant reactions to execess salts are equivalent to those depicted above for water shortages.

Plants are usually isolated into two particular groups as per the level of salt resilience. Halophyte plants are local to saline soils and finish their life cycle in sodic soil conditions, and accordingly, they are more tolerant to high salt concentration in the soil. Then again, glycophytes, or nonhalophyte plants, cannot endure similar groupings of salt as halophytes.

7.4.3.2 Heavy metals

Heavy metal contamination to water and soils because of anthropogenic activities are harmful to living beings and should be handled with another developing innovation, known as phytoremediation, which offers a practical and eco-friendly path for the cleanup of sullied zones. Phytoremediation either collect or converts heavy metals to nontoxic structures (Parmar, Kanwar, & Thakur, 2013).

The components of salt resistance are associated with a few cycles: the recuperation of ionic harmony and cell homeostasis, the osmotic acclimation maintaining H_2O ingestion by the plant, and the counteraction or relief of oxidative pressure that happens similarly to that in H_2O shortages. Osmotic change can include the gathering of inorganic particles (Na^+ and Cl^-), as commonly seen in halophytes, or through the combination of viable osmolytes, a reaction generally seen in glycophytes. An extra kind of transformation to salt pressure includes the capacity to eliminate an over-abundance of salt by specific salt organs; particles are shipped to these organs, where the salt is solidified.

There is proof that salt pressure can actuate states of oxidative pressure, for example, aging, as well as the collection of ROS, including hydrogen peroxide (H_2O_2), superoxide anion ($O_2{\cdot}^-$) and hydroxyl ($OH{\cdot}$). Plants have proficient enzymatic (superoxide dismutase, SOD; catalase, CAT; ascorbate peroxidase, APX; glutathione reductase, GR; monodehydroascorbate reductase, MDHAR; dehydroascorbate reductase, DHAR; glutathione peroxidase, GPX; guaiacol peroxidase, GOPX and glutathione-S-transferase, GST) and nonenzymatic (ascorbic corrosive, ASH; glutathione, GSH; phenolic mixes, alkaloids, nonprotein amino acids and α-tocopherols) agent frameworks which work to governor the unrestrained oxidation and to shield plant cells from oxidative harm by ROS. The ROS scavengers may increase plant protection from salt pressure.

7.4.4 Extreme temperatures (freezing, cold, heat)

Among the abiotic stresses, the temperature is a key factor that plays a role in the regulation of plant physiological processes throughout the development. Any positive or negative changes in optimum temperature act as a stress stimulus that initiates the alterations in metabolic processes and promotes developmental changes reported by Kumar (2013). All plant species perform their metabolic process and complete their life cycle under an optimum temperature range. The temperature range over the ideal level is known as heat stress. Extreme temperature and high irradiances are usual phenomena under dry season and arid conditions. That is why drought stress is considered a multidimensional stress (DaMatta, 2003).

The tolerance to heat stress is generally defined as the ability of the plant to grow, flourish, and produce sustainable production under conditions of high temperature. Whereas cold/chilling stress or low-temperature stress generally results in a disturbance of cell membrane integrity leading to severe cellular dehydration and imbalance in osmotic potential. The acclimation against cold stress is achieved by the activation of different genes, which result in a restructuring of the cell membranes through alterations in the composition of lipids and synthesis of osmolytes, which protect the plant cells from dehydration and so lead to the plant stress tolerance.

Low-temperature stress moreover shows extended oxidative damage in plant species through the synthesis of ROS. In response to ROS, plants use antioxidant defense mechanisms to reduce the damaging effect of ROS when exposed to low-temperature stress (Kumar, 2013). Plants survive under low-temperature stress conditions by adopting alterations in membrane lipids, accumulation of osmolytes, synthesis of antifreeze proteins (AFP, S), and synthesis of proteins which protect membranes and macromolecules against dehydrative damage, etc. (Srivastava, 2002).

Furthermore, high temperature and irradiance are common under water deficit conditions. The increase in the capacity of the antioxidant defiance mechanism enhances the protection of plant cells, particularly by scavenging ROS. The physiology of abiotic stresses comprises the photosynthetic apparatus, and leaf area index, which subsequently leads to improved crop yield, even under drought conditions.

7.4.5 Physiological mechanisms of plants adaptation to solar radiations

Solar radiation is an important abiotic factor that plays a role in the regulation of plant physiological processes throughout the life cycle. In recent years, there has been an concern over the depletion of stratospheric ozone due to anthropogenic emissions. This has accomplished a differentiating development in solar radiation (280–320 nm). Increased levels of UV-B radiation have hazardous effects in both plants and animals. A wide range of plant responses to supplemental UV-B radiation has been seen, mostly harming the plants but sometimes with a beneficial effect. UV-B can impact plant physiological processes either through the direct effect or by strategies for numerous adjusting effects. In plants, direct effects include damage to DNA and changes in the composition of proteins, which frequently cause genetic changes influencing physiological processes, including the photosynthetic machinery. These aspects negatively influence plant progress, improvement, and morphology, particularly the production of more delicate plants (Zhang et al., 2012).

The mechanism in plants to cope with excess light stress consists of the movement of the chloroplasts. Plants are generally firm to one place, but they also show numerous well-known movements in response to stimuli, such as phototropism. At a cellular level, chloroplasts move to the sides of a plant cell with the objective that they are arranged to appear differently in the direction of incident light to avoid digestion of excessive radiation and to limit photodamage. This response is called chloroplast avoiding progress, and it is mediated by blue light and a blue light photoreceptor.

7.4.5.1 Stomatal response

Stomata open and close to regulate light, comprising blue and red light, and a distinct mechanism underlies the stomata opening via different wavelengths. The blue light is a sign and the red light is both a sign and a source of energy. High light intensity during the daytime increases the rate of transpiration to minimize the load of heat energy. However, a long time effect decreases leaf water potential and causes a reduction in the conductance of stomata and partially closing the stomata reduces transpiration, which is also known as midday depression.

7.4.5.2 Photoacclimation

Photosynthetic organisms have developed different photoprotective mechanisms in response to excess light that involve both short-term and long-term responses known as photoacclimation. Collectively, photoprotective responses attempt to avoid the synthesis of ROS that cause significant oxidative damage to the cells responsible for photosynthesis (Singhal et al., 2017)

7.4.5.3 Photo-protection

Aerobic photosynthetic living organisms have developed particular photodetected mechanisms to acclimatize to the feasibly destructive effects of light. Some green algae and plants escape from the absorption of excessive light by moving the leaves, cells, and/or chloroplasts. Inside the chloroplast, photosynthetic light attaining and electron transport changes help to use the light energy. For example, a change in the size of antenna used to harvest the light energy and capacity of photosynthesis reduce the absorption of light and enhance the utilization of light, respectively, during comparatively long-term acclimation to excessive light (Singhal et al., 2017).

7.5 Next-generation genetic engineering tools for abiotic stress tolerance in plants

The plant's growth is adversely affected by abiotic stresses like drought, salinity, heat, and chilling leading to significant damage in yield through a series of physicochemical and molecular alterations in the plants (Rai, Kalia, Singh, Gangola, & Dhawan, 2011). The pioneering technology of GE has empowered researchers to transfer genes into desirable crops from distant sources to improve plants by boosting their tolerance to abiotic stresses (Bakhsh & Hussain, 2015). Such genes have been categorized into three groups: (1) genes coding numerous osmolytes, that is, glycine betaine, proline, mannitol, and heat-shock proteins; (2) genes responsible for transporters for nutrients and water uptake, such as ion transporter, aquaporins, etc.; and (3) genes controlling transcriptional and transduction signaling mechanisms, such as DREB1, MAPK, WRKY, etc. that are employed for abiotic stress tolerance in plants (Bhatnagar-Mathur et al., 2008). Besides these GE approaches, the CRISPR/Cas9 technology, a recently proved exact gene-editing approach, has been proficiently applied for gene mutation, repression, activation, and epigenome editing for the development of transgenic plants that are tolerant to abiotic stress (Parmar et al., 2017). Also, the antisense gene and RNA meddling (RNAi) tools have extended the acceptance in plant genetics and system biology as a steady transgene method for the advancement of transgenic plants that are capable of growing under various abiotic stresses (Parmar et al., 2017). Some of the GE tools are discussed in detail in the following paragraphs.

7.5.1 CRISPR/Cas9-based genetic engineering tools

Until the last decade, zinc finger nucleases and transcription activator-like effector nucleases had been the mainstay of genome editing tools. But their knottiness made Clustered Regularly Interspaced Short Palindromic Repeats (CRISPR)/CRISPR associated protein 9 (Cas9) derived from *Streptococcus pyogenes* the powerful and ubiquitous GE tool. CRISPR/Cas9 has been part of the inherent adaptive immune system for bacteria and archaea which renders foreign invading nucleic acids to be cleaved in these organisms. This immunity is gained by integrating short fragments of invading nucleic acid fragments known as spacers in-between two contiguous repeats at the nearest end of a CRISPR locus (Mushtaq et al., 2018). During subsequent confrontations, the CRISPR array, with the integrated spacers (invasive DNA), are transcribed and processed into small interfering CRISPR RNAs (crRNAs) 40 nucleotides in length (approx.), which conflate with the trans activating CRISPR RNA (tracrRNA) to activate the Cas9 nuclease, thus cleaving the invading DNA. The presence of an protospacer adjacent motif which possesses the conserved sequence of 5′-NGG-3′ 12 bases downstream of the target DNA is a prerequisite for cleavage (Liu et al., 2008). Genome editing has been done on crop species in recent years either to engineer tolerance to abiotic stressors or for functional characterization of numerous abiotic stress responsive genes (T. Li et al., 2018; R. Li et al., 2018; Li et al., 2019; Shen et al., 2017; Shi et al., 2017; Wang et al., 2017; Zhang et al., 2019).

7.5.2 Precision genome editing in plants

Genome editing involves the introduction of a precise double-stranded break, which in turn stimulates the DNA repair mechanism. Desired modifications can be induced based on the DNA repair pathway and the DNA repair template. The two repair pathways that can be taken are nonhomologous end joining (NHEJ) and homology directed repair (HDR). NHEJ results in random insertion and deletion (indels) which causes frameshift mutations rendering the gene nonfunctional or in simple terms creating a gene knockout (Bortesi & Fischer, 2015). Alternatively, NHEJ can help mediate in targeted insertion of a DNA template into the break. When a DNA template is available with homologous region surrounding the break, the DNA break can be repaired by HDR which can be exploited to introduce precise gene modification into the genome (Zhu, Li, & Gao, 2020).

7.5.3 Multiplexed genome editing

Simultaneous mutagenesis at different loci can be accomplished by clubbing more than one target DNA template into one construct. Domesticated tomatoes today grown in the field are prone to bacterial and salinity stress, whereas a wild relative of modern tomato *Solanum pimpinellifolium* is resilient to the same but the fruit yield, shape, and size are highly compromised. Using knowledge accumulated throughout the past years' researchers have been able to utilize the multiplex CRISPR/Cas9 system to edit several genes at once to bring *S. pimpinellifolium* closure to the modern-day cultivated tomato plant (T. Li et al., 2018; R. Li et al., 2018; Zsögön et al., 2018). Likewise, the domestication of *Physalis pruinosa*, an orphan Solanaceae crop resulted in more flowers and increased fruit biomass (Lemmon et al., 2018).

7.5.4 Gene regulation

The CRISPR/Cas9 system can be repurposed for the transcriptional regulation of a gene. Fusion of a plant transcriptional activator to a deactivated Cas9 (dCas9), lacking the endonuclease activity, helps in overexpression of an endogenous gene (Zhu et al., 2020).

7.5.5 Directed evolution

Directed evolution generates user defined protein or DNA by duplicating the process of natural selection. For the modification of our gene of interest it has proved to be a powerful approach. It involves two steps: first, mutagenesis to generate diverse genotypes; and second, selection to enrich a particular phenotype under selection pressure. Use of CRISPR/Cas single guide RNA (sgRNA) library and simultaneous selection pressure enables high-throughput directed evolution. This technique has been employed in a few crop plants but its full-scale application is still lacking (Butt et al., 2019; Kuang et al., 2020)

7.5.6 Nanoparticle mediated genetic engineering

Efficiency in genetic transformation remains a challenge for GE due to the presence of a cell wall in plant cells. Either the methods of transformation are inefficient or limited to a few plant species. In this respect nanoparticles can play a crucial role because of their ability to traverse through the cell wall without any force and their ability to carry different cargoes (Cunningham, Goh, Demirer, Matos, & Landry, 2018). The first demonstration of DNA by nanoparticles in plants was shown by Torney and colleagues in 2007, though it was not a passive delivery of DNA and biolistic gun was used in the process (Torney, Trewyn, Lin, & Wang, 2007). But in due course many groups have demonstrated passive delivery of DNA via nanoparticles and further optimization of the process is still being done (Finiuk et al., 2017; Hao et al., 2013; Liu et al., 2008; Martin-Ortigosa, Valenstein, Lin, Trewyn, & Wang, 2012; Martin-Ortigosa et al., 2014; Zhao et al., 2017).

GE of plants for newer traits and improved performance under challenging environmental conditions has always been an arduous task. With the dawn of new tools for GE some of the tasks have become more convenient. The following subsections discuss the examples of transgenic plants which are tolerant to abiotic stresses, particularly, drought, heat stress, and salinity:

7.5.7 Genetic engineering for drought stress tolerance in plants

GE tools have been successfully used in several studies for drought-tolerance in plants. For example, in the banana the overexpression of MusaWRKY71 gene, which is a very potential abiotic stress-responsive WRKY TF gene, cloned from Musa species cv. Karibale Monthan, has been reported (Shekhawat & Ganapathi, 2013). Tsai-Hung, Jent-turn, Yee-yung, and Ming-Tsair (2002) transformed tomato plants with a DNA cassette containing an *Arabidopsis* C repeat/dehydration-responsive element-binding factor 1 (CBF1) cDNA and a nos terminator, driven by a cauliflower mosaic virus 35S promoter. These transgenic tomato plants were found to be more resistant to drought stress than the wild-type plants. Pasquali, Biricolti, Locatelli, Baldoni, and Mattana (2008) reported improved tolerance to cold and drought stress in transgenic apple by the overexpression of a cold-inducible Osmyb 4 gene from rice, which codes for a TF belonging to the Myb family. The overexpression of DREB1b TF gene had also been reported to induce cold tolerance and drought tolerance in a transgenic grapevine (Jin et al., 2009). Chrysanthemum is one of the leading ornamental cut flowers across the globe and its production is severely hampered by various environmental conditions (Gao, Song, Zhu, Chen, & Jiang, 2012). Drought stress harms this crop to the maximum extent by retarding its growth. WRKY TFs work as positive or sometimes negative regulators in various abiotic stress responses in plants. Fan, Song, Jiang Zhang, Sun, and Wang (2016) transformed a CmWRKY1 TF derived from *Chrysanthemum morifolium* and overexpressed it in chrysanthemum cultivar "Jinba" and found that CmWRKY1 regulates an ABA-mediated pathway by suppressing the expression levels of various genes including PP2C, ABI1, and ABI2, and activating the expression levels of genes like PYL2, SnRK2-2, ABF4, MYB2, RAB18, and DREB1A in a positive regulation. The transgenic plants displayed increased drought tolerance in polyethene glycol (PEG) stress as compared to control plants. Zhang, Li, Shu, Zhang, Ye (2011) and X. Zhang et al. (2011) transformed tomato with a miR 169 family member, Sly-miR169c, which can effectively downregulate the transcripts of the target genes—three nuclear factor Ysubunit genes (SINF-YA1/2/3) and one multidrug resistance-associated protein gene (SIMRP1), which are downregulated under drought stress.

7.5.8 Genetic engineering for heat stress tolerance in plants

Overexpression of cytosolic ascorbate peroxidase (cAPX) gene improved tolerance to heat stress in the transgenic apple, as reported by Wisniewski, Fuchigami, Wang, Srinivasan, and Norilli (2002). Wang, Wisniewski, Meilan, Cu, and Fuchigami (2006) developed transgenic tomato plants which overexpressed cAPX gene with enhanced tolerance to heat (40°C). Overexpression of Cu/Zn SOD gene (derived from *Manihot esculenta*) under an oxidative stress-inducible promoter SWPA2 in potato led to enhanced heat stress tolerance (Tang et al., 2006). Cu/Zn SOD is a ROS scavenging enzyme and thus helps in quenching free radicals released under heat stress in plants. Polyamines play an important role in imparting thermal stress tolerance in plants. S-adenosyl-I-methionine decarboxylase (SAMDC) is one of the key regulatory target enzymes in polyamines biosynthesis. Cheng et al. (2009) overexpressed SAMDC cDNA, isolated from *Saccharomyces cerevisiae*, in tomato plants for enhanced polyamines production. Song et al. (2014) overexpressed CgHSP70 gene conferring heat tolerance in chrysanthemum.

7.5.9 Genetic engineering for salinity tolerance in plants

Salinity tolerance is a complex mechanism governed by many genes (Bojorquez-Quintal et al., 2014). Husaini and Abdin (2008) overexpressed tobacco osmotin gene in strawberry (*Fragaria* × *ananasa* Duch.) and found that the transgenic strawberry plants exhibited tolerance to salt stress. Subramanyam, Sailaja, Subramanyam, Rao, and Lakshmidevi (2011) successfully improved the tolerance of chilli pepper (*Capsicum annuum* L. cv.) plants by the ectopic expression of tobacco osmotin gene via the *Agrobacterium tumefaciens*-mediated gene transfer technique. Cervera, Ortega, Navarro, Navarro, and Pena (2000) transformed *Carrizo citrange*, an excellent rootstock of citrus with a yeast-derived halo-tolerance gene, HAL2, which is involved in a salt tolerance mechanism. HAL2 gene is involved in the methionine biosynthetic pathway and confers tolerance to lithium and sodium ions. It encodes for a salt-sensitive biphosphate nucleotidase, which is required for sulfate accumulation.

The transgenic lines expressing HAL2 protein showed improved tolerance to salinity than the wild-type plants. Tomato is considered to be one of the most important vegetable crops worldwide for the commercial value it offers. Wang et al. (2005) developed transgenic tomato plants expressing tolerance to chilling and salt stress by incorporation of cytosolic ascorbate peroxidase (cAPX) gene, derived from pea (*Pisum sativum* L.). Fan et al. (2012) transformed sweet potato cv. Sushu-2 with a chloroplast betaine aldehyde dehydrogenase (SoBADH) gene from *Spinacia oleracea*, which is involved in the biosynthesis of glycine betaine. The overexpression of SoBADH gene in transgenic sweet potato improved tolerance toward salinity, oxidative stress, and low temperature by protecting against cell damage by maintaining cell membrane integrity, stronger photosynthetic activity, reduced ROS production, and activation of ROS scavenging mechanism. Bulle, Yarra, and Abbagani (2016) developed transgenic chilli pepper (*C. annuum*) plants expressing wheat Na^{+}/H^{+} antiporter gene (TaNHX2) to develop tolerance toward salinity stress. Several studies had been carried out in the past to develop suitable regeneration and genetic transformation protocols in many crop species including apple (Rustaee, Nazeri, Ghadimzadeh, & Malboobi, 2007), pomegranate (Parmar, Kamlesh, & Thakur, 2012; Parmar et al., 2013; Parmar, Kanwar, & Thakur, 2015), chilli (Khan, Siddique, Anis, & Khan, 2011; Sharma, Thakur, & Srivastava, 2006), cucumber (Vasudevan, Selvaraj, Ganapathi, & Choi, 2007), lily (Kathryn & Han, 2008), sweet orange (Singh & Rajam, 2010), broccoli (Kumar & Srivastava, 2015), date palm (Aslam, Khan, & Azad, 2015), chrysanthemum (Naing, Ai, Jeon, Lim, & Kim, 2016), etc. for enhancement of various biotic and abiotic stresses. Besides these, some of the abiotic stress-tolerant engineered crops as an enhancement of osmoprotectants have been summarized in Table 7.5. Due to the importance of gene regulation by TFs in plants, researchers have made continuous efforts to identify, characterize, clone, and transform different TFs involved in the regulation of stress-related genes (Table 7.6).

7.6 Challenges for the application of genetic engineering tools for abiotic stress tolerance in plants

Under abiotic stress conditions, various qualities are turned on in plants, bringing about different levels of proteins and metabolites, and few of them might be capable of presenting a specific level of tolerance to different stresses. Further, it is required to understand the biochemical and nuclear mechanisms that occur in response to stresses (Hrbáčková et al., 2020). Stress-actuated quality articulation in plants which can be comprehensively ordered into three gatherings. They are as follows:

- qualities encoding proteins with known enzymatic or auxiliary capacities
- proteins with so far obscure capacities
- administrative proteins

Starting endeavors to create transgenics (essential tobacco) plants for abiotic stress resistance are included "single activity qualities," that is, qualities liable for the change of a solitary metabolite that would present expanded resilience to salt or dry spell pressure stress-instigated proteins with referred to capacities, for example, water channel proteins, key compounds for osmolyte (proline, betaine, sugars, e.g., polyamines and trehalose) biosynthesis, detoxification chemicals, and transport proteins were the underlying focuses of plant change (Surabhi, Badajena, & Sahoo, 2019). Truth be told, metabolic characteristics, particularly pathways with moderately hardly any

TABLE 7.5 Enhancing osmoprotectants against abiotic stress through genetic engineering tools.

Transgenic crops/ osmoprotectants	Mechanisms	Accountable gene (s)	References
Glycine betaine (GB)			
Oryza sativa L.	Transgenic lines of rice exhibited the enhanced level of salinity tolerance	codA	Mohanty et al. (2002)
	Transgenic plants were tolerance to saline conditions	COX	Su et al. (2006)
	Enhanced tolerance to salt stress and temperature stress in the seedling stage	CMO	Shirasawa, Takabe, Takabe, and Kishitani (2006)
Zea mays L.	Transgenic maize plants showed that advanced level of GB lead to higher adaptability against drought	betA	Quan, Shang, Zhang, Zhao, and Zhang (2004)
Gossypium hirsutum L.	Enhanced tolerance to chilling temperature was recorded	betA	Zhang et al. (2012)
	The transgenic cotton plant showed 26% and 131% higher level of GB, ultimately lead to improved adaptability against salinity	AhCMO	Zhang et al. (2009)
Lycopersicon esculentum Mill	Higher accumulation of GB in the chloroplast of transgenic tomato plants enhanced salinity stress tolerance	codA	Park, Jeknic, Pino, Murata, and Chen (2007)
Brassica juncea L.	Transgenic *B. juncea* enhanced the ability to grow against to salt stress	codA	Parsad, Sharmila, Kumar, and Saradhi (2000)
Solanum tuberosum L.	Enhanced the ability of transgenic *S. tuberosum* against drought and salt stresses	COX	Ahmad et al. (2008)
Nicotiana tobaccum L.	Transgenic tobacco plant enriched the tolerance ability against drought through an excessive accumulation of GB	CMO	Zhang et al. (2008)

(Continued)

TABLE 7.5 (Continued)

Transgenic crops/ osmoprotectants	Mechanisms	Accountable gene (s)	References
Mannitol and Trehalose			
Nicotiana tobaccum L.	Transgenic tobacco plants exhibited an enhanced level of drought tolerant	TPS1	Karim et al. (2007)
Oryza sativa L.	Transgenic rice plants showed higher adaptability against drought stress	Ubi1::TPS1	Jang et al. (2003)
	Transgenic rice plants with both genes (otsA and otsB) enriched adaptability against drought stress	otsA and otsB	Garg et al. (2002)
Lycopersicon esculentum	Transgenic *L. esculentum* plants with more sugar and starch contents enhanced tolerance against drought and salinity-induced oxidative stress	Yeast TPS1	Cortina and Culiáñez-Macià (2005)
Triticum aestivum L.	Wheat transgenic plants showed improved adaptability against both drought and salt stresses	mtlD	Abebe, Guenzi, Martin, and Cushman (2003)
Proline			
Nicotiana tobaccum L.	Transgenic tobacco plants accumulated higher levels of the protein that lead to enhancing tolerance against drought	P5CR	Popisilova, Haisel, and R. Vankova (2011)
Glycine max L.	Transgenic soybean enhanced photosynthesis during heat and drought stress	P5CR	Ronde, Cress, Krugerd, Strasserd, and Van Staden (2004)
Petunia hybrida	The transgenic *P. hybrid* plants survived 14 days under drought stress	P5CS	Yamada et al. (2005)

TABLE 7.6 Proteins and transcription factors (regulatory proteins) in crop plants responsible for abiotic stress tolerance.

Transgenic crops	Mechanisms	Accountable gene	References
Transcription factors			
Oryza sativa L.	Transgenic rice with OsDREB gene showed the enhancement of adaptability against drought, salt and cold stress	OsDREB	Dubouzet et al. (2003)
	Transgenic rice with DREB1A /(CBF3) gene enhanced tolerance to drought and salt stresses	AtDREB	Oh et al. (2005)
	Transgenic rice with AP37 gene increased yield by 16% to 57% under severe drought conditions as compared to control	AP37	Kim and Kim (2009)
Zea mays L.	Transgenic maize with ZmDREB2A gene promoted tolerance against abiotic stress	ZmDREB2A	Qin et al. (2007)
Arachis hypogaea L.	Transgenic peanut with AtDREB1A increased stress tolerance through the higher accumulation of proline and antioxidants levels	AtDREB1	Bhatnagar-Mathur, Devi, Vadez, and Sharma (2009)
	Transgenic peanut with transcription factor of DREB1 increased yield up to 24% under drought stress as compared to control a variety		Bhatnagar-Mathur et al. (2014)
Nicotiana tobaccum L.	Transgenic tobacco through overexpressing of GhDREB1 exposed to enhance chilling tolerance than wild-type of plants.	GhDREB1	Shan et al. (2007)
Triticum aestivum L.	Wheat plants transmuted with atDREB1A under drought stress showed enhancement of drought tolerance	AtDREB	Pellegrineschi et al. (2004)

(*Continued*)

TABLE 7.6 (Continued)

Transgenic crops	Mechanisms	Accountable gene	References
Late embryogenesis abundant			
Brassica campestris L.	Transgenic Chinese cabbage plants exhibited the enhancement of tolerance to both drought and salt stress	ME-lea N4	Park, Liu, Kanno, and Kameya (2005b)
Triticum aestivum L.	Wheat transgenic lines embracing barley HVA1 exhibited high tolerance against drought	HVA1	Sivamani, Bahieldin, and Wraith (2000)
Oryza sativa L.	Rice transgenic lines embracing barley HVA1 coded improved tolerance against drought	HVA1	Xu et al. (1996)
	Transgenic rice enhanced the dehydration tolerance as a result of the accumulated level of wheat LEA genes, PMA80 and PMA1959	PMA1959 and PMA80 LEA	Cheng, Targolli, Huang, and Wu (2002)
	Transgenic rice showed a greater tolerance against drought as a result of the overexpression of OsLEA 3-1 gene	OsLEA 3-1	Xiao, Huang, Tang, and Xiong (2007)
Lactuca sativa L.	Drought and salt tolerance of transgenic lettuce plants was increased as a result of the overexpression of *the Brassica napus* ME-lea N4	ME-lea N4	Park, Liu, Kanno, and Kameya (2005a)

proteins, have been portrayed hereditarily and show up more manageable to controls than auxiliary and formative attributes (Basso et al., 2019). In any case, that tactic has mistreated the way that abiotic stress resilience is probably going to include numerous qualities all at once, and that solitary quality resistance is probably not going to be maintainable. Subsequently, a second "wave" of change endeavors to change plants with the third class of pressure prompted qualities; to be specific, administrative proteins has risen. Through these proteins, numerous qualities associated with pressure reaction can be all the while directed by a solitary quality encoding pressure inducible record factor, subsequently offering probability of improving resilience towards different burdens including dry spell, saltiness, and low temperature. This "second wave" has likewise harmonized with superior coordination of hereditary designing and plant physiology. Moreover, hereditary designing permits regulating the circumstance and tissue-explicitness including articulation level of the presented qualities for their ideal capacity. The essential discoveries on stress advertisers prompted a significant move in the worldview for hereditarily designing pressure open-minded yields lately (Bhui, Dinos, Galant-Miecznikowska, de Jongh, & Stansfeld, 2016). The most generally utilized promoters in producing transgenic plants are constitutively communicated, that is, they are turned on constantly and all through the vegetation cycle. In any case, in situations where the quality articulation should be custom-made to a particular organ or a particular time, such constitutive advertisers may not be a reasonable decision, particularly for the pressure incited qualities. This has the grounds that the constitutive articulation of certain pressure-actuated qualities can show genuine malicious impacts on the plant.

7.6.1 Drought

Different qualities controlling flagging and quality administrative pathways offer certain critical focuses for hereditary designing for abiotic stress resilience. Record factors (TFs) that control or shift on the statement of various qualities engaged with bestowing abiotic stress resilience in plants has been suggested as proficient focuses on hereditary change. These record factors incorporate DREB1 quality family, Myb quality family, and also very well as it has shown that changed tomato containing a DNA tape comprised of an *Arabidopsis* C rehash/parchedness reactive component restricting variable 1 (CBF1) cDNA and a nos eliminator, compelled by a cauliflower mosaic infection 35S advertiser. The transgenic tomato was extra impervious to water shortage compared to the wild plants. It has been accounted for improved resilience to cold and dry season pressure in genetically modified apple by the overarticulation of a chilly inducible Osmyb 4 quality from rice, which codes for a record factor having a place with Myb family. The overarticulation of DREB1b TF quality acquired additionally accounted for initiation of cold resistance as well as dry spell resilience in transgenic grapes.

Dry spell stress hurts the harvest of chrysanthemum to the most extreme degree by impeding its development. WRKY record factors (TFs) fill in as certain or some of the time negative controllers in different abiotic stress reactions in plants. The changed a CmWRKY1 TF got from *C. morifolium*, then overcommunicated it in chrysanthemum variety "Jinba." It was discovered that CmWRKY1 controls an ABA-interceded pathway by smothering the articulation levels of different qualities including PP2C, ABI1 and ABI2, and enacting the articulation levels of qualities like PYL2, SnRK2-2, ABF4, MYB2, RAB18 and DREB1A in a positive guideline. The genetically modified crops showed expanded dry season resistance in PEG stress when contrasted with untreated plants. Likewise, different abiotic stress resilience in banana was accounted for by the overarticulation of MusaWRKY71 quality, which is known as abiotic stress-responsive WRKY TF quality.

In the treatment for the recovery of stress, transgenic potato plants showed a more grounded cancer prevention agent action, more chlorophyll, greater effective photosynthesis and higher recuperation, nearly. Plant miniature RNA (miRNA) controls a few formative and physiological wonders inside the plants including dry spell reactions. changed tomato with a miR 169 relative, Sly-miR169c, those may successfully down-manage the records of the objective qualities—three atomic factor Ysubunit qualities (SINF-YA1/2/3) and opposition related protein quality (SIMRP1), which are down-directed under dry spell pressure. The transgenic tomato plants overcommunicating Sly-miR169c showed decreased opening of stomata, diminished leaf water misfortune and record rate with upgraded dry season resilience characteristics.

7.6.2 Salinity

Salinity is an important stress limiting crop yield. Saline soils contain more dissolvable salts with higher EC. Pungency (having an intense flavor or odor) impacted soil contains pointless dissolvable salts and compatible sodium on a shallow level, that impacts root structures of plants. As demonstrated with the salt/pungency stress caused different biochemical and physiological changes in plants, inclusive of osmatic stress, ionic off-kilter nature as well as assistant weight (Shinwari, Jan, Nakashima, & Yamaguchi-Shinozaki, 2020). High pungency interrupts the osmatic balances, thus provoking a basic decline in water and supplement take-up. Researchers recommended that pungency can agitate the overall plant advancement by influencing the unpredictable associations in supplement take-up and assembling, hormonal cumbersomeness, and oxidative weight. Generally, glycophytes are salt-delicate plants which exhibit resistance through gathering osmoprotectants; in any case, halophytes can bear and grow well under pungency stress of more than (200 mM NaCl) by using unequivocal segments to diminish the take-up and assembling of Cl^-, Na^+ particles, therefore minimizing the perilous effects. A couple of sorts of halophytes have specific organs to release Cl^-, and Na^+ particles at the leaves' surfaces, while because of plant rule of Na^+ particles out of the phone, a couple of cells have tremendous vacuoles that can go probably as a sink for the social occasion of wealth Na^+ through vehicle into vacuoles. The plants which are found to be under the stress of salinity, take-up overabundance measures of these particles, accordingly expanding their amassing in various tissues of the plants, and prompting oxidative stress. Plants react to salt pressure through a progression of systems, for example, Na^+/K^+ homeostasis and Na^+ avoidance. Also, the abundance measure of these particles in plant cells prompts the overproduction of ROS, which is profoundly harmful and causes oxidative pressure. Salt/saltiness stress caused various physiological and biochemical changes in plants, including osmatic stress, ionic lopsided characteristics, and auxiliary pressure (Bhowmik, Hassan, Molla, Rahman, & Islam, 2019). A few types of halophytes have particular organs to discharge Na^+ and Cl^- particles at the leaves surfaces, while on account of plant guideline of Na + particles out of the cell, a few cells have enormous vacuoles that can go about as a sink for the aggregation of abundance Na^+ through vehicle into vacuoles. During the most recent twenty years, analysts looked for the qualities liable for saltiness resistance. Utilizing biotechnological instruments, various qualities, record components, and miRNAs were related to consequences for saltiness stress. Various examinations revealed that HKT qualities ought to be associated with the prohibition of Na + from leaves in crops. SKC1 is a key QTL (encode OsHKT1;5) that controls salt resistance in *Oryza sativa*. The past investigation introduced that Nax1 and Nax2 loci can convey TmHKT1;4-A2 and TmHKT1;5-A (sodium carrier) and have the extraordinary capacity to confine the Na^+ aggregation in wheat leaf tissues, hence prompting upgraded salt resilience (Raza et al., 2020). Knockout plants showed more salt affectability, and the sodium particle take-up and aggregation in shoots proposed that OsHKT1;1 is conceivably associated with recuperating Na + from the leaf sharp edge. It is communicated in the vascular tissues of roots and leaves, however, its careful part in salt resistance is obscure. OsHKT1;4 is principally communicated in the leaf sheath and encodes a plasma layer confined protein. Along these lines, the HKT qualities family has incredible potential in Na + /K + homeostasis and is anticipated to show wide significance in saltiness resistance. As of late, another examination additionally investigated the part of the HKT quality in salt resistance. The scientists announced that the overexpression of zmKHT1;5 upgraded salt resilience by actuating cancer prevention agent protein exercises (SOD, CAT, POD) and diminished the collection of H_2O_2 and MDA (Amin et al., 2019).

7.6.3 Heat stress

Temperature limits may seriously harm crops plants at different growth stages, bringing about loss of efficiency (Wani et al., 2020). Vegetables, for example, chickpea, lentil, mung bean, peas, and soybean, were portrayed by various levels of affectability to high- and low-temperature stresses, and these trade-offs are exhibited at various formative stages, that is, germination, development of seedlings, growth stage, blossoming, and filling of seed or pod stage. In the case of green gram, it was noted that for every degree of fluctuation in temperature diminishes the seed yield by 35%–40%. Expanding air CO_2 focus alongside temperature that restricts plant development, especially in C3 plant species, including mung bean also. Heat stress reduces the movement of CAT, SOD, APX and expands the levels of ROS harm in the aeronautical pieces of soybean, mung bean and chickpea plants. Utilization of exogenous GSH improved heat stress resistance in mung bean seedlings by adjusting cell reinforcement frameworks, though more significant levels of cancer prevention agent metabolites have been recognized in heat-tolerant soybean, mung bean, and chickpea plants genotypes under stress.

The Heat Shock proteins (HSP) are unambiguous proteins that are prepared when cells are temporarily unprotected to unfavourable temperatures. The role of HSPs is intra-cellular chaperones for other proteins which play a significant role in protein–protein interfaces such as collapsible and supporting in the formation of proper protein shape and inhibition of undesirable protein accumulation (Pandey, Srivastava, Pandey, & Dubey, 2020). The estimated molecular weight, HSPs are clustered into five categories such as (1) Hsp100, (2) Hsp90, (3) Hsp70, (4) Hsp60 and (5) small heat-shock proteins (sHsps). The plant species with higher category have minimum 20 sHsps with 40 types. It is proved that the divergence of these HSPs redirect an alteration to adapt plants against high temperature stress. The dictation of HSPs-genes is systematized by supervisory proteins, named heat stress transcription factors (Hsfs). Higher plant species illustrate at least 21 Hsfs for regulation, but they also cooperate in all phases during heat stress for triggering, maintaining and recovery of plant cells (Visioni et al., 2019).

7.6.4 Heavy metal stress

Right now, the world contains quickly developing innovation and industrialization, as a result the harmfulness of heavy metals has created as a worldwide danger for every individual. The amassing of hefty metals makes obliteration in farming grounds. Moreover, substantial metals are essential to life with availability in the following sum. The overabundance of these is poisonous to plant body (Hasanuzzaman, Hakeem, Nahar, & Alharby, 2019). At a cell level, the enhancement amount of substantial metals limit the biological processes of the living individuals. Earlier researches proposed that weighty metals went about as exchangers of fundamental metal particles or by hindering practical gatherings and resulted in oxidative stress in plants. For example, Fe and Cu are basically redox dynamic, produced ROS straightforwardly by redox responses; conversely, different metals like Mn, Pb, Ni, Al, Cd, and Zn created ROS through the backhanded instrument. The reaction to heavy metals has been explored in vegetables, featuring the part of cancer prevention agent players. Peas (*P. sativum*) in soil Pb stress demonstrated the enhancement of APX, GR and CAT in leaves and roots of plants which are set off by jasmonic acid corrosive. Additionally, in chickpea plants exposed to expanding vanadium (V) focuses the cell reinforcement compound such as CAT, SOD, POD (Muthuramalingam et al., 2019).

7.7 Conclusion

The growth of the human population is increasing at an alarming rate and it is anticipated to reach 7.774 billion by the year 2020. To meet the food demands of an increasing population, agricultural productivity needs to be also increased. The IPCC has anticipated that due to the extreme events of abiotic stress as a consequence of changing climate, food and environmental security will be a big challenge in future. However, as sessile organisms, plants cannot escape extremely harsh environmental conditions. Under extreme stress conditions the stomata of plants remain close, thus reducing the availability of carbon dioxide, which eventually slows down the rate of biochemical reactions, that is, photosynthesis and transpiration. Genetically improved plants follow diverse physiological mechanisms to cope with abiotic stresses. On the other hand, genetic improvement of crops through conventional breeding is a slow and difficult process, causing high heterozygosity, prolonged juvenile periods, and also inconsistency. Therefore the NBTs, namely GE, may be useful for induction of desired agronomic and breeding traits through the inclusion of responsible genes into the genome. The breeding of genetically modified plants through the introduction of desirable agronomic and quality traits by utilization of NBTs/GE has been found to be apposite for many crops. For the improvement of crop species against abiotic stresses, the method of NBTs/GE should be a pioneering tool in future farming systems for meeting the food and nutritional security of the increasing population.

References

Abebe, T., Guenzi, A. C., Martin, B., & Cushman, J. C. (2003). Tolerance of mannitol accumulating transgenic wheat to water stress and salinity. *Plant Physiology*, *131*, 1748–1755.

Ahmad, R., Kim, M. D., Back, K. H., Kim, H. S., Lee, H. S., Kwon, S. Y., … Kwak, S. S. (2008). Stress-induced expression of choline oxidase in potato plant chloroplasts confers enhanced tolerance to oxidative, salt, and drought stresses. *Plant Cell Reports*, *27*, 687–698.

Ahmadi, A., & Baker, D. A. (2001). The effect of water stress on the activities of key regulatory enzymes of the sucrose to starch pathway in wheat. *Plant Growth Regulation*, *35*, 81–91.

Akbari, G., Sanavy, S. A. M. M., & Yousefzadeh, S. (2007). Effect of auxin and salt stress (NaCl) on seed germination of wheat cultivars (*Triticum aestivum* L.). *Pakistan Journal of Biological Sciences*, *10*(15), 2557–2561.

Akram, M. (2011). Growth and yield components of wheat under water stress on different growth stages. *Bangladesh Journal of Agricultural Research, 36*(3), 455–468.

Amin, A. B., Rathnayake, K. N., Yim, W. C., Garcia, T. M., Wone, B., Cushman, J. C., & Wone, B. W. (2019). Crassulacean acid metabolism abiotic stress-responsive transcription factors: A potential genetic engineering approach for improving crop tolerance to abiotic stress. *Frontiers in Plant Science, 10*, 129.

Andaya, V. C., & Mackill, D. J. (2003). Mapping of QTLs associated with cold tolerance during the vegetative stage in rice. *Journal of Experimental Botany, 54*(392), 2579–2585.

Andosch, A., Affenzeller, M. J., Lütz, C., & Lütz-Meindl, U. (2012). A freshwater green alga under cadmium stress: Ameliorating calcium effects on ultrastructure and photosynthesis in the unicellular model Micrasterias. *Journal of Plant Physiology, 169*, 1489–1500.

Anjum, S. A., Wang, L. C., Farooq, M., Hussain, M., Xue, L. L., & Zou, C. M. (2011). Brassinolide application improves the drought tolerance in maize through modulation of enzymatic antioxidants and leaf gas exchange. *Journal of Agronomy and Crop Science, 197*, 177–185.

Araus, J. L., Slafer, G. A., Reynolds, M. P., & Royo, C. (2002). Plant breeding and water relations in C3cereals: What should we breed for? *Annals of Botany, 89*, 925–940.

Artlip, T. S., & Wisniewski, M. E. (2002). Induction of proteins in response to biotic and abiotic stresses. In M. Pessarkli (Ed.), *Handbook of plant and crop physiology* (pp. 657–679). New York: Marcel Dekker.

Ashraf, M. (2009). Biotechnological approach of improving plant salt tolerance using antioxidants as markers. *Biotechnology Advances, 27*, 84–93.

Ashraf, M., & Hafeez, M. (2004). Thermo tolerance of pearl millet and maize at early growth stages: growth and nutrient relations. *Biologia Plantarum, 48*, 81–86.

Aslam, J., Khan, S. A., & Azad, M. A. K. (2015). Agrobacterium-mediated genetic transformation of datepalm (*Phoenix dactylifera* L.) cultivar ''Khalasah'' via somatic embryogenesis. *Plant Science Today, 2*, 93–101.

Atteya, A. M. (2003). Alteration of water relations and yield of corn genotypes in response to drought stress. *Bulgarian Journal of Plant Physiology., 29*, 63–76.

Azevedo-Neto, A. D., Prisco, J. T., Enéas-Filho, J., Abreu, C. E. B., & Gomes-Filho, E. (2006). Effect of salt stress on antioxidative enzymes and lipid peroxidation in leaves and roots of salt tolerant and salt-sensitive maize genotypes. *Environmental and Experimental Botany, 56*, 87–94.

Badu-Apraku, B., Hunter, R. B., & Tollenaar, M. (1983). Effect of temperature during grain filling on whole plant and grain yield in maize (Zea mays L.). *Canadian Journal of Plant Science, 63*, 357–363.

Bakhsh, A., & Hussain, T. (2015). Engineering crop plants against abiotic stress: Current achievements and prospects. *Emirates Journal of Food and Agriculture., 27*(1), 24–39. Available from https://doi.org/10.9755/ejfa.v27i1.17980.

Balla, K., Rakszegi, M., Li, Z., Bekes, F., Bencze, S., & Veisz, O. (2011). Quality of winter wheat in relation to heat and drought shock after anthesis. *Czech Journal of Food Sciences, 29*, 117–128.

Barrero-Sicilia, C., Silvestre, S., Haslam, R. P., & Michaelson, L. V. (2017). Lipid remodelling: Unravelling the response to cold stress in Arabidopsis and its extremophile relative *Eutrema salsugineum. Plant Science (Shannon, Ireland), 263*, 194–200.

Basnayake, J., Fukai, S., & Ouk, M. (2006). Contribution of potential yield, drought tolerance and escape to adaptation of 15 rice varieties in rainfed lowlands in Cambodia. In *Proceedings of the Australian agronomy conference*. Birsbane, Australia: Australian Society of Agronomy.

Basso, M. F., Ferreira, P. C. G., Kobayashi, A. K., Harmon, F. G., Nepomuceno, A. L., Molinari, H. B. C., & Grossi-de-Sa, M. F. (2019). Micro RNAs and new biotechnological tools for its modulation and improving stress tolerance in plants. *Plant Biotechnology Journal*, *17*(8), 1482–1500.

Bhatnagar-Mathur, P., Devi, M. J., Vadez, V., & Sharma, K. K. (2009). Differential antioxidative responses in transgenic peanut bear no relationship to their superior transpiration efficiency under drought stress. *Journal of Plant Physiology*, *166*, 1207–1217.

Bhatnagar-Mathur, P., Rao, J. S., Vadez, V., Dumbala, S. R., Rathore, A., Yamaguchi-Shinozaki, K., & Sharma, K. K. (2014). Transgenic peanut overexpressing the DREB1A transcription factor has higher yields under drought stress. *Molecular Breeding*, *33*, 327–340.

Bhatnagar-Mathur, P., Vadez, A. E., Kiran, A. E., & Sharma, K. (2008). Transgenic approaches for abiotic stress tolerance in plants: Retrospect and prospects. *Plant Cell Reports*, *27*, 411–424.

Bhowmik P., Hassan M. M., Molla K., Rahman M., & Islam M. T. (2019). Application of CRISPR-Cas genome editing tools for the improvement of plant abiotic stress tolerance. In *Approaches for enhancing abiotic stress tolerance in plants* (pp. 459–472). CRC Press.

Bhui, K., Dinos, S., Galant-Miecznikowska, M., de Jongh, B., & Stansfeld, S. (2016). Perceptions of work stress causes and effective interventions in employees working in public, private and non-governmental organisations: A qualitative study. *BJPsych Bulletin*, *40* (6), 318–325. Available from 10.1192/pb.bp.115.050823.

Bojorquez-Quintal, E., Velarde-Buendıa, A., Ku-Gonzalez, A., Carillo-Pech, M., Ortega-Camacho, D., Echevarrıa-Machado, I., . . . Martínez-Estévez, M. (2014). Mechanisms of salt tolerance in habanero pepper plants (Capsicum chinense Jacq.): Proline accumulation, ions dynamics and sodium root-shoot partition and compartmentation. *Frontiers in Plant Science*, *5*, 10–3389.

Bortesi, L., & Fischer, R. (2015). The CRISPR/Cas9 system for plant genome editing and beyond. *Biotechnology Advances*, *33*(1), 41–52.

Bray E. A., Bailey-Serres J., & Weretilnyk E. (2000). Responses to abiotic stresses. In *Biochemistry and molecular biology of plants* (pp. 1158–1203). American Society of Plant Physiologists.

Bulle, M., Yarra, R., & Abbagani, S. (2016). Enhanced salinity stress tolerance in transgenic chilli pepper (*Capsicum annuum* L.) plants overexpressing the wheat antiporter (TaNHX2) gene. *Molecular Breeding*, *36*, 36.

Butt, H., Eid, A., Momin, A. A., Bazin, J., Crespi, M., Arold, S. T., & Mahfouz, M. M. (2019). CRISPR directed evolution of the spliceosome for resistance to splicing inhibitors. *Genome Biology*, *20*(1), 1–9.

Cai, L., Xu, Z., Ren, M., Guo, Q., Hu, X., Hu, G., . . . Peng, P. (2012). Source identification of eight hazardous heavy metals in agricultural soils of Huizhou, Guangdong Province, China. *Ecotoxicology and Environmental Safety*, *78*, 2–8.

Camejo, D., Rodriguez, P., Morales, M. A., Dell'amico, J. M., Torrecillas, A., & Alarcon, J. J. (2005). High temperature effects on photosynthetic activity of two tomato cultivars with different heat susceptibility. *Journal of Plant Physiology*, *162*, 281–289.

Carillo, P., Grazia, M., Pontecorvo, G., Fuggi, A., & Woodrow, P. (2011). Salinity stress and salt tolerance. In A. Shanker (Ed.), *Abiotic stress in plants—mechanisms and adaptations (Internet)*. Rijeka: IntechOpen. Available from. Available from http://www.intechopen.com/books/abiotic-stress-in-plants mechanisms-and-adaptations/salinity-stress-and-salt-tolerance.

Cattivelli, L., Rizza, F., Badeck, F. W., Mazzucotelli, E., Mastrangelo, A. M., Francia, E., et al. (2008). Drought tolerance improvement in crop plants: An integrated view from breeding to genomics. *Field Crops Research, 105*(1-2), 1–14.

Cervera, M., Ortega, C., Navarro, A., Navarro, L., & Pena, L. (2000). Generation of transgenic citrus plants with the tolerance-to salinity gene HAL2 from yeast. *Journal of Horticultural Science and Biotechnology, 75*, 26–30.

Chapman, S. C., & Edmeades, G. O. (1999). Selection improves drought tolerance in tropical maize populations II. Direct and correlated responses among secondary traits. *Crop Science, 39*, 1315–1324.

Chen, H., & Jiang, J. (2010). Osmotic adjustment and plant adaptation to environmental changes related to drought and salinity. *Environmental Reviews, 18*(1), 309–319.

Chen K., Wang Y., Zhang R., Zhang H., & Gao C. (2019). *CRISPR/Cas genome editing and precision plant breeding in agriculture*. 10.1146/annurev-arplant-050718-100049.

Chen, M., & Thelen, J. J. (2013). ACYL-LIPID DESATURASE2 is required for chilling and freezing tolerance in arabidopsis. *The Plant Cell, 25*(4), 1430–1444.

Cheng, L., Zou, Y., Ding, S., Zhang, J., Yu, X., Cao, J., & Lu, G. (2009). Polyamine accumulation in transgenic tomato enhances the tolerance to high temperature stress. *Journal of Integrative Plant Biology, 51*, 489–499.

Cheng, Z., Targolli, J., Huang, X., & Wu, R. (2002). Wheat LEA genes, PMA80 and PMA1959, enhance dehydration tolerance of transgenic rice (Oryza sativa L.). *Molecular Breeding, 10*, 71–82.

Chigbo, C., & Batty, L. (2013). Effect of combined pollution of chromium and benzo (a) pyrene on seed growth of *Lolium perenne*. *Chemosphere, 90*, 164–169.

Colmer, T. D. (2003a). Long–distance transport of gases in plants: A perspective on internal aeration and radial oxygen loss from roots. *Plant, Cell & Environment, 26*, 17–36.

Colmer, T. D. (2003b). Aerenchyma and an inducible barrier to radial oxygen loss facilitate root aeration in upland, paddy and deepwater rice (*Oryza sativa* L.). *Annals of Botany, 91*, 301–309.

Colmer, T. D., Cox, M. C. H., & Voesenek, L. A. C. J. (2006). Root aeration in rice (*Oryza sativa* L.): Evaluation of oxygen, carbon dioxide, and ethylene as possible regulators of root acclimatizations. *New Phytologist, 170*, 767–778.

Cortina, C., & Culiáñez-Macià, F. A. (2005). Tomato abiotic stress enhanced tolerance by trehalose biosynthesis. *Plant Science (Shannon, Ireland), 169*, 75–82.

Cruz, R., & Milach, S. (2004). Cold tolerance at the germination stage of rice: Methods of evaluation and characterization of genotypes. *Scientia Agricola, 61*, 1–8.

Cullu, M. A. (2003). Estimation of the effect of soil salinity on crop yield using remote sensing and geographic information system. *Turkish Journal of Agriculture and Forestry, 27*(1), 23–28.

Cunningham, F. J., Goh, N. S., Demirer, G. S., Matos, J. L., & Landry, M. P. (2018). Nanoparticle-mediated delivery towards advancing plant genetic engineering. *Trends in Biotechnology, 36*(9), 882–897.

DaMatta, F. M. (2003). Drought as a multidimensional stress affecting photosynthesis in tropical tree crops. *Advances in Plant Physiology, 5*, 227–265.

Daryanto, S., Wang, L., & Jacinthe, P. A. (2016). Global synthesis of drought effects on maize and wheat production. *PLoS One, 11*, e0156362.

Demirevska, K., Zasheva, D., Dimitrov, R., Simova-Stoilova, L., Stamenova, M., & Feller, U. (2009). Drought stress effects on Rubisco in wheat: Changes in the Rubisco large subunit. *Acta Physiologiae Plantarum/Polish Academy of Sciences, Committee of Plant Physiology Genetics and Breeding, 31*, 1129–1138.

Dubouzet, J. G., Sakuma, Y., Ito, Y., Kasuga, M., Dubouzet, E. G., Miura, S., ... Yamaguchi-Shinozaki, K. (2003). OsDREB genes in rice, Oryza sativa L. encode transcription activators that function in drought- high-salt- and cold-responsive gene expression. *The Plant Journal: for Cell and Molecular Biology*, *33*, 751–763.

Elbasyoni, I. S. (2018). Performance and stability of commercial wheat cultivars under terminal heat stress. *Agron*, *8*, 37.

Fahad, S., Bajwa, A. A., Nazir, U., Anjum, S. A., Farooq, A., Zohaib, A., ... Huang, J. (2017). Crop production under drought and heat stress: Plant responses and management options. *Frontiers in Plant Science*, *8*, 1147.

Fahad, S., Hussain, S., Saud, S., Hassan, S., Tanveer, M., Ihsan, M. Z., et al. (2016). A combined application of biochar and phosphorus alleviates heat-induced adversities on physiological, agronomical and quality attributes of rice. *Plant Physiology and Biochemistry: PPB/Societe Francaise de Physiologie Vegetale*, *103*, 191–198.

Fahad, S., Hussain, S., Saud, S., Khan, F., Hassan, S., Amanullah., et al. (2016). Exogenously applied plant growth regulators affect heat-stressed rice pollens. *Journal of Agronomy and Crop Science*, *202*, 139–150.

Fahad, S., Hussain, S., Saud, S., Tanveer, M., Bajwa, A. A., Hassan, S., et al. (2015). A biochar application protects rice pollen from high-temperature stress. *Plant Physiology and Biochemistry: PPB/Societe Francaise de Physiologie Vegetale*, *96*, 281–287.

Fan, W., Zhang, M., Zhang, H., & Zhang, P. (2012). Improved tolerance to various abiotic stresses in transgenic sweet potato (*Ipomoea batatas*) expressing spinach betaine aldehyde dehydrogenase. *PLoS One*, *7*(5), e37344.

Fan, Q., Song, A., Jiang Zhang, T., Sun, H., & Wang, Y. (2016). CmWRKY1 enhances the dehydration tolerance of chrysanthemum through the regulation of ABA-associated genes. *PLoS One*, *11*, e0150572.

FAO. How to Feed the World in 2050. 2009, Available from http://www.fao.org/fileadmin/templates/wsfs/docs/expert_paper/How_to_Feed_the_World_in_2050.pdf.

Farooq, M., Wahid, A., Kobayashi, N., Fujita, D., & Basra, S. M. A. (2009). Plant drought stress: Effects, mechanisms and management. *Agronomy for Sustainable Development*, *29*(1), 185–212.

Ferris, R., Ellis, R. H., Wheeler, T. R., & Hadley, P. (1998). Effect of high temperature stress at anthesis on grain yield and biomass of field grown crops of wheat. *Annals of Botany*, *82*, 631–639.

Finiuk, N., Buziashvili, A., Burlaka, O., Zaichenko, A., Mitina, N., Miagkota, O., & Yemets, A. (2017). Investigation of novel oligoelectrolyte polymer carriers for their capacity of DNA delivery into plant cells. *Plant Cell, Tissue and Organ Culture (PCTOC)*, *131*(1), 27–39.

Flexas, J., Diaz-Espejo, A., Galmés, J., Kaldenhoff, R., Medrano, H., & Ribas-Carbo, M. (2007). Rapid variations of mesophyll conductance in response to changes in CO2 concentration around leaves. *Plant, Cell & Environment*, *30*, 1284–1298.

Gafni, A., & Zohar, Y. (2001). Sodicity, conventional drainage and bio-drainage in Israel. *Australian Journal of Soil Science*, *39*, 1269–1278.

Gao, H., Song, A., Zhu, X., Chen, F., & Jiang, J. (2012). The heterologous expression in Arabidopsis of a highly tolerant to a new CMV pathotype. *Plant Cell Reports*, *28*, 223–232.

Garg, A. K., Kim, J. K., Owens, T. G., Ranwala, A. P., Choi, Y. C., Kochian, L. V., & Wu, R. J. (2002). Trehalose accumulation in rice plants confers high tolerance levels to different abiotic stresses. *Proceedings of the National Academy of Sciences of the United States of America*, *99*, 15898–15903.

Ghadirnezhad, R., & Fallah, A. (2014). Temperature effect on yield and yield components of different rice cultivars in flowering stage. *International Journal of Agronomy., 2014*, 1–4.

Ghaffari, M., Toorchi, M., Valizadesh, M., & Shakiba, M. R. (2012). Morphological screening of sunflower inbred lines under drought stress condition. *Turkish Journal of Field Crops, 17*(2), 185–190.

Hall, A. E. (2001). *Crop responses to environment* (pp. 17–28). Boca Raton: CRC press LLC.

Hao, Y., Yang, X., Shi, Y., Song, S., Xing, J., Marowitch, J., & Chen, J. (2013). Magnetic gold nanoparticles as a vehicle for fluorescein isothiocyanate and DNA delivery into plant cells. *Botany, 91*(7), 457–466.

Hasanuzzaman, M., Hakeem, K. R., Nahar, K., & Alharby, H. F. (Eds.), (2019). *Plant abiotic stress tolerance: Agronomic, molecular and biotechnological approaches*. Switzerland AG: Springer Nature. Available from http://doi.org/10.1007/978-3-030-06118-0.

Hrbáčková, M., Dvořák, P., Takáč, T., Tichá, M., Luptovčiak, I., Šamajová, O., & Šamaj, J. (2020). Biotechnological perspectives of omics and genetic engineering methods in alfalfa. *Frontiers in Plant Science*, 11. Available from 10.3389/fpls.2020.00592.

Huang, B., Rachmilevitch, S., & Xu, J. (2012). Root carbon and protein metabolism associated with heat tolerance. *Journal of Experimental Botany, 63*(9), 3455–3465.

Husaini, A. M., & Abdin, M. Z. (2008). Overexpression of tobacco osmotin gene leads to salt stress tolerance in strawberry (*Fragaria ananassa* Duch.) plants. *Indian Journal of Biotechnology, 7*, 465–471.

Isla, R., Agragues, R., & Royo, A. (1998). Validity of various physiological traits as screening criteria for salt tolerance in barley. *Field Crops Research, 58*, 97–107.

Jagadish, S. V. K., Craufurd, P. Q., & Wheeler, T. R. (2007). High temperature stress and spikelet fertility in rice (*Oryza sativa* L.). *Journal of Experimental Botany, 58*, 1627–1635.

Jahan, I., Hossain, M. M., & Karim, M. R. (2019). Effect of salinity stress on plant growth and root yield of carrot. *Progressive Agriculture, 30*(3), 263–274.

Jang, I. C., Oh, S. J., Seo, J. S., Choi, W. B., Song, S. I., Kim, C. H., ... Kim, J. K. (2003). Expression of a bifunctional fusion of the Escherichia coli genes for trehalose-6-phosphate synthase and trehalose-6-phosphate phosphatase in transgenic rice plants increases trehalose accumulation and abiotic stress tolerance without stunting growth. *Plant Physiology, 131*, 516–524.

Jin, W. M., Dong, J., Hu, Y. L., Lin, Z. P., Xu, X. F., & Han, Z. H. (2009). Improved cold-resistant performance in transgenic grape (*Vitis vinifera* L.) overexpressing cold-inducible transcription factors AtDREB1b. *Horticultural Science, 44*, 35–39.

Kalai, T., Khasmassi, K., Teixeria de Silva, J. A., Gouia, H., & Ben-kaab, L. B. (2014). Cadmium and copper stress affect seedling growth and enzymatic activities in germinating barley seeds. *Archives of Agronomy and Soil Science, 6*, 60.

Kamara, A. Y., Menkir, A., Badu-Apraku, B., & Ibikunle, O. (2003). The influence of drought stress on growth, yield and yield components of selected maize genotypes. *Journal of Agricultural Science, 141*, 43–50.

Kang, H. M., & Saltveit, M. E. (2002). Chilling tolerance of maize cucumber and rice seedling leaves and roots are differentially affected by salicylic acid. *Physiologia Plantarum, 115*, 571–576.

Karim, S., Aronsson, H., Ericson, H., Pirhonen, M., Leyman, B., Welin, B., ... Holmström, K. (2007). Improved drought tolerance without undesired side effects in transgenic plants producing trehalose. *Plant Molecular Biology, 64*, 371–386.

Karl, T. R., Melillo, J. M., & Peterson, T. C. (Eds.), (2009). *Global climate change impacts in the United States*. Cambridge, UK: Cambridge University Press.

Kashiwagi, J., Krishnamurthy, L., Purushothaman, R., Upadhyaya, H. D., Gaur, P. M., Gowda, C. L. L., ... Varsgney, R. K. (2015). Scope for improvement of yield under drought through the root traits in chickpea (Cicer arietinum L.). *Field Crops Research*, *170*, 47–54.

Kasuga, M., Qiang, L., Setsuko, M., Shinozaki, K., & Kazuo, K. (1999). Improving plant drought, salt, and freezing tolerance by gene transfer of a single stress-inducible transcription factor. *Nature Biotechnology*, *17*, 287–291.

Kathryn, K. K., & Han, B. H. (2008). Biolistic-mediated transformation of Lilium longiflorum cv. Nellie White. *Horticultural Science*, *43*(6), 1864–1869.

Kawakami, J., Iwama, K., & Jitsuyama, Y. (2006). Soil water stress and the growth and yield of potato plants grown from microtubers and conventional seed tubers. *Field Crops Research*, *95*, 89–96.

Kaya, M. D., Okcub, G., Ataka, M., Cikilic, Y., & Kolsaricia, O. (2006). Seed treatments to overcome salt and drought stress during germination in sunflower (*Helianthus annuus* L.). *European Journal of Agronomy*, *24*, 291–295.

Khan, H., Siddique, I., Anis, M., & Khan, P. R. (2011). In vitro organogenesis from internode derived callus cultures of Capsicum annuum L. *Journal of Plant Biochemistry and Biotechnology*, *20*, 84–89.

Khan, M. A., Hussain, N., Abid, M., & Imran, T. (2004). Screening of wheat (*Triticum aestivum* L.) cultivars for saline conditions under irrigated arid environment. *Journal of Research (Science)*, *15*, 471–477.

Khanam, T., Akhtar, N., Halim, M. A., & Hossain, F. (2018). Effect of irrigation salinity on the growth and yield of two Aus rice cultivars of Bangladesh. Jahangirnagar University. *The Journal of Biological Sciences*, *7*(2), 1–12.

Kim, W., Iizumi, T., & Nishimori, M. (2019). Global patterns of crop production losses associated with drought from 1983 to 2009. *Journal of Applied Meteorology and Climatology*, *58*(6), 1233–1244.

Kim, Y. S., & Kim, J. K. (2009). Rice transcription factor transcription factor AP37 involved in grain yield increase under drought stress. *Plant Signaling & Behavior*, *4*, 735–736.

Klap, C., Yeshayahou, E., Bolger, A. M., Arazi, T., Gupta, S. K., Shabtai, S., et al. (2017). Tomato facultative parthenocarpy results from SlAGAMOUS-LIKE 6 loss of function. *Plant Biotechnology Journal*, *15*, 634–647. Available from https://doi.org/10.1111/pbi.12662.

Kranner, I., & Colville, L. (2011). Metals and seeds: biochemical and molecular implications and their significance for seed germination. *Environmental and Experimental Botany*, *72*, 93–105.

Kuang, Y., Li, S., Ren, B., Yan, F., Spetz, C., Li, X., & Zhou, H. (2020). Base-editing-mediated artificial evolution of OsALS1 in planta to develop novel herbicide-tolerant rice germplasms. *Molecular Plant*.

Kumar, M. (2013). Crop plants and abiotic stresses. *Journal of Biomolecular Research & Therapeutics.*, *3*(1), 1000e125.

Kumar, P., & Srivastava, D. K. (2015). High frequency organogenesis in hypocotyl, cotyledon, leaf and petiole explants of broccoli (Brassica oleracea L. var. italica), an important vegetable crop. *Physiology and Molecular Biology of Plants*, *21*(2), 279–285.

Kumaraswamy, S., & Shetty, P. K. (2016). Critical abiotic factors affecting implementation of technological innovations in rice and wheat production: A review. *Agricultural Reviews*, *4*, 268–278.

Lafitte, H. R., Yongsheng, G., Yan, S., & LiZ, K. (2007). Whole plant responses, key processes, and adaptation to drought stress: The case of rice. *Journal of Experimental Botany*, *58*, 169–175.

Lang, P., Zhang, C. K., Ebel, R. C., Dane, F., & Dozier, W. A. (2005). Identification of cold acclimated genes in leaves of *Citrus unshiu* by mRNA differential display. *Gene*, *359*, 111–118.

Lemmon, Z. H., Reem, N. T., Dalrymple, J., Soyk, S., Swartwood, K. E., Rodriguez-Leal, D., & Lippman, Z. B. (2018). Rapid improvement of domestication traits in an orphan crop by genome editing. *Nature Plants*, *4*(10), 766–770.

Li, C., Jiang, D., Wollenweber, B., Li, Y., Dai, T., & Cao, W. (2011). Waterlogging pretreatment during vegetative growth improves tolerance to waterlogging after anthesis in wheat. *Plant Science (Shannon, Ireland)*, *180*(5), 672–678.

Li, R., Liu, C., Zhao, R., Wang, L., Chen, L., Yu, W., & Shen, L. (2019). CRISPR/Cas9-Mediated SlNPR1 mutagenesis reduces tomato plant drought tolerance. *BMC Plant Biology*, *19*(1), 1–13.

Li, R., Zhang, L., Wang, L., Chen, L., Zhao, R., Sheng, J., & Shen, L. (2018). Reduction of tomato-plant chilling tolerance by CRISPR–Cas9-mediated SlCBF1 mutagenesis. *Journal of Agricultural and Food Chemistry*, *66*(34), 9042–9051.

Li, T., Yang, X., Yu, Y., Si, X., Zhai, X., Zhang, H., & Xu, C. (2018). Domestication of wild tomato is accelerated by genome editing. *Nature Biotechnology*, *36*(12), 1160–1163.

Li, X., Cai, J., Liu, F., Zhou, Q., Dai, T., Cao, W., & Jiang, D. (2015). Wheat plants exposed to winter warming are more susceptible to low temperature stress in the spring. *Plant Growth Regulation*, *77*, 11–19.

Li, Z., Peng, T., Xie, Q., Han, S., & Tian, J. (2010). Mapping of QTL for tiller number at different stages of growth in wheat using double haploid and immortalized F2 populations. *Journal of Genetics*, *89*, 409–415.

Liu, J., Wang, F. H., Wang, L. L., Xiao, S. Y., Tong, C. Y., Tang, D. Y., & Liu, X. M. (2008). Preparation of fluorescence starch-nanoparticle and its application as plant transgenic vehicle. *Journal of Central South University of Technology*, *15*(6), 768–773.

Lu, Y., & Zhu, J. K. (2016). Precise editing of a target base in the rice genome using a modified CRISPR/Cas9 system. *Molecular Plant*. Available from https://doi.org/10.1016/j.molp.2016.11.013.

Machado, R. M. A., & Serralheiro, R. P. (2017). Soil salinity: Effect on vegetable crop growth. management practices to prevent and mitigate soil salinization. *Horticulturae*, *3*, 30.

Maestri, E., Klueva, N., Perrotta, C., Gulli, M., Nguyen, H. T., & Marmiroli, N. (2002). Molecular genetics of heat tolerance and heat shock proteins in cereals. *Plant Molecular Biology*, *48*, 667–681.

Mahajan, S., & Tuteja, N. C. (2005). salinity and drought stresses: An overview. *Archives of Biochemistry and Biophysics*, *444*(2), 139–158.

Mahmud, A. A., Hossain, M., Bazzaz, M. M., Khan, M. S. A., Hossain, M. A., & Kadian, M. S. (2014). Tuber yield, tuber quality and plant water status of potato under drought and well watered condition. *Global Journal of Science Frontier Research*, *14*(10), 101–107.

Márquez-García, B., Márquez, C., Sanjosé, I., Nieva, F. J., Rodríguez-Rubio, P., & Muñoz-Rodríguez, A. F. (2013). The effects of heavy metals on germination and seedling characteristics in two halophyte species in Mediterranean marshes. *Marine Pollution Bulletin*, *70*, 119–124.

Martínez, J. P., Silva, H., Ledent, J. F., & Pinto, M. (2007). Effect of drought stress on the osmotic adjustment, cell wall elasticity and cell volume of six cultivars of common beans (*Phaseolus vulgaris* L.). *European Journal of Agronomy*, *26*, 30–38.

Martin-Ortigosa, S., Peterson, D. J., Valenstein, J. S., Lin, V. S. Y., Trewyn, B. G., Lyznik, L. A., & Wang, K. (2014). Mesoporous silica nanoparticle-mediated intracellular Cre protein delivery for maize genome editing via loxP site excision. *Plant Physiology*, *164*(2), 537–547.

Martin-Ortigosa, S., Valenstein, J. S., Lin, V. S. Y., Trewyn, B. G., & Wang, K. (2012). Gold functionalized mesoporous silica nanoparticle mediated protein and DNA codelivery to plant cells via the biolistic method. *Advanced Functional Materials*, *22*(17), 3576–3582.

Mass, E. V. & Grattan, S. R. 1999. *Crop yields as affected by salinity*. American Society of Agronomy, Crop Science Society of America, Soil Science Society of America, 677 S. Segoe Rd., Madison, WI 53711, USA. Agricultural Drainage, Agronomy Monograph no. 38. 55-93.

Mazahery-Laghab, H., Nouri, F., & Abianeh, H. Z. (2003). Effects of the reduction of drought stress using supplementary irrigation for sunflower (Helianthus annuus) in dry farming conditions. *Pajouhesh-Va-Sazandegi Summer*, *16*(2), 81–86 (**59** in Agronomy and Horticulture).

Miao, Z., Han, Z., Zhang, T., Chen, S., & Ma, C. (2017). A systems approach to a spatiotemporal understanding of the drought stress response in maize. *Science Reports.*, *7*, 6590.

Mickelbart, M., Hasegawa, P., & Bailey-Serres, J. (2015). Genetic mechanisms of abiotic stress tolerance that translate to crop yield stability. *Nature Reviews. Genetics*, *16*, 237–251.

Mohanty, A., Kathuria, H., Ferjani, A., Sakamoto, A., Mohanty, P., & Murata, N. (2002). Transgenics of an elite indica rice variety Pusa Basmati 1harbouring the codA gene are highly tolerant to salt stress. *TAG. Theoretical and Applied Genetics. Theoretische und Angewandte Genetik*, *106*, 51–57.

Mollard, F. P., Striker, G. G., Ploschuk, E. L., & Insausti, P. (2010). Subtle topographical differences along a floodplain promote different plant strategies among *Paspalum dilatatum* subspecies and populations. *Austral Ecology*, *35*(2), 189–196.

Monneveux, P., Sánchez, C., Beck, D., & Edmeades, G. O. (2006). Drought tolerance improvement in tropical maize source populations: evidence of progress. *Crop Science*, *46*, 180–191.

Munns, R., Husain, S., Rivelli, A. R., Richard, A. J., Condon, A. G., Megan, P. L., ... Hare, R. A. (2002). Avenues for increasing salt tolerance of crops, and the role of physiologically based selection traits. *Plant and Soil*, *247*, 93–105.

Munzuroglu, O., & Geckil, H. (2002). Effects of metals on seed germination, root elongation, and coleoptile and hypocotyl growth in *Triticum aestivum* and *Cucumis sativus*. *Archives of Environmental Contamination and Toxicology*, *43*, 203–213.

Mushtaq, M., Bhat, J. A., Mir, Z. A., Sakina, A., Ali, S., Singh, A. K., & Bhat, R. (2018). CRISPR/Cas approach: A new way of looking at plant-abiotic interactions. *Journal of Plant Physiology*, *224*, 156–162.

Muthuramalingam P., Jeyasri R., Krishnan S. R., Pandian S. T. K., Sathishkumar R., & Ramesh M. (2019) Integrating the bioinformatics and omics tools for systems analysis of abiotic stress tolerance in Oryza sativa (L.). In *Advances in plant transgenics: Methods and applications* (pp. 59–77). Singapore: Springer.

Naing, A. H., Ai, T. N., Jeon, S. M., Lim, S. H., & Kim, C. K. (2016). An efficient protocol for Agrobacterium-mediated genetic transformation of recalcitrant chrysanthemum cultivar Shinma. *Acta Physiologiae Plantarum/Polish Academy of Sciences, Committee of Plant Physiology Genetics and Breeding*, *38*, 38.

Nakashima, K., Ito, Y., & Yamaguchi-Shinozaki, K. (2009). Transcriptional regulatory networks in response to abiotic stresses in Arabidopsis and grasses. *Plant Physiology*, *149*(1), 88–95.

Nam, N. H., Chauhan, Y. S., & Johansen, C. (2001). Effect of timing of drought stress on growth and grain yield of extra-short-duration pigeon pea lines. *The Journal of Agricultural Science.*, *136*, 179–189.

Nayyar, H., Kaur, S., Singh, S., & Upadhyaya, H. D. (2006). Differential sensitivity of Desi (small-seeded) and Kabuli (large-seeded) chickpea genotypes to water stress during seed filling: effects on the accumulation of seed reserves and yield. *Journal of the Science of Food and Agriculture*, *86*, 2076–2082.

Negrão, S., Schmöckel, S. M., & Tester, M. (2017). Evaluating physiological responses of plants to salinity stress. *Annals of Botany*, *119*(1), 1–11. Available from https://doi.org/10.1093/aob/mcw191, 2017.

Nonami, H. (1998). Plant water relations and control of cell elongation at low water potentials. *Journal of Plant Research*, *111*, 373–382.

Ogbonnaya, C. I., Sarr, B., Brou, C., Diouf, O., Diop, N. N., & Roy-Macauley, H. (2003). Selection of cowpea genotypes in hydroponics, pots, and field for drought tolerance. *Crop Science*, *43*, 1114–1120.

Oh, S. J., Jeong, J. S., Kim, E. H., Yi, N. R., Yi, S. I., Jang, I. C., ... Kim, J. K. (2005). Matrix attachment region from the chicken lysozyme locus reduces variability in transgene expression and confers copy number dependence in transgenic rice plants. *Plant Cell Reports*, *24*(3), 145–154.

Oliver, S. N., Dennis, E. S., & Dolferus, R. (2007). ABA regulates apoplastic sugar transport and is a potential signal for cold-induced pollen sterility in rice. *Plant & Cell Physiology*, *48*, 1319–1330.

Pandey, N., & Sharma, C. P. (2002). Effect of heavy metals Co^{2+}, Ni^{2+}, and Cd $^{2+}$ on growth and metabolism of cabbage. *Plant Science (Shannon, Ireland)*, *163*, 753–758.

Pandey P., Srivastava S., Pandey A. K., & Dubey R. S. (2020) Abiotic-stress tolerance in plants-system biology approach. In *Plant life under changing environment* (pp. 577–609). Academic Press.

Paranychianakis, N. V., & Chartzoulakis, K. S. (2005). Irrigation of Mediterranean crops with saline water: From physiology to management practices. *Agriculture, Ecosystems & Environment*, *106*, 171–187.

Parida, A. K., & Das, A. B. (2005). Salt tolerance and salinity effects on plants: A review. *Ecotoxicology and Environmental Safety*, *60*(3), 324–349.

Park, B. J., Liu, Z., Kanno, A., & Kameya, T. (2005a). Increased tolerance to salt and water deficit stress in transgenic lettuce (*Lactuca sativa* L.) by constitutive expression of LEA. *Plant Growth Regulation*, *45*, 165–171.

Park, B. J., Liu, Z., Kanno, A., & Kameya, T. (2005b). Genetic improvement of Chinese cabbage for salt and drought tolerance by constitutive expression of a *Brassica napus* LEA gene. *Plant Science (Shannon, Ireland)*, *169*, 553–558.

Park, E. J., Jeknic, Z., Pino, M. T., Murata, N., & Chen, T. H. (2007). Glycinebetaine accumulation in chloroplasts is more effective than that in cytosol in protecting transgenic tomato plants against abiotic stress. *Plant, Cell & Environment*, *30*, 994–1005.

Parmar, N., Kamlesh, K., & Thakur, A. K. (2012). In vitro organogenesis from cotyledon derived callus cultures of Punica granatum L. cv. Kandhari Kabuli. *National Academy Science Letters*, *35*, 215–220.

Parmar, N., Kanwar, K., & Thakur, A. K. (2013). High efficiency plant regeneration via direct organogenesis in Punica granatum L. cv. Kandhari Kabuli from hypocotyl explants. *The Proceedings of the National Academy of Sciences, India, Section B: Biological Sciences*, *83*, 569–574.

Parmar, N., Kanwar, K., & Thakur, A. K. (2015). High efficiency plant regeneration from cotyledon explants of pomegranate (Punica granatum L.) cv. Kandhari Kabuli. *Vegetos*, *28*, 160–165.

Parmar, N., Singh, K. H., Sharma, D., Singh, L., Kumar, P., Nanjundan, J., ... Thakur, A. K. (2017). Genetic engineering strategies for biotic and abiotic stress tolerance and quality enhancement in horticultural crops: A comprehensive review. *3 Biotech*, *7*(4), 239. Available from 10.1007/s13205-017-0870-y.

Parsad, K. V. S. K., Sharmila, P., Kumar, P. A., & Saradhi, P. P. (2000). Transformation of Brassica juncea (L.) Czern with a bacterial codA gene enhances its tolerance to salt stress. *Molecular Breeding*, *6*, 489–499.

Pasquali, G., Biricolti, S., Locatelli, F., Baldoni, E., & Mattana, M. (2008). Osmyb4 expression improves adaptive responses to drought and cold stress in transgenic apples. *Plant Cell Reports*, *27*, 1677–1686.

Patel, R., Prasher, S., Bonnell, R., & Boughton, R. (2002). Development of comprehensive soil salinity index. *Journal of Irrigation and Drainage Engineering-ASCE*, *128*, 185–188.

Pellegrineschi, A., Reynolds, M., Pacheco, M., Brito, R. M., Almeraya, R., Yamaguchi-Shinozaki, K., & Hoisington, D. (2004). Stress-induced expression in wheat of the Arabidopsis thaliana DREB1A gene delays water stress symptoms under greenhouse conditions. *Genome/National Research Council Canada = Genome/Conseil National de Recherches Canada*, *47*, 493–500.

Pettigrew, W. T. (2004). Physiological consequences of moisture deficit stress in cotton. *Crop Science*, *44*, 1265–1272.

Popisilova, J., Haisel, D., & R. Vankova, R. (2011). Responses of transgenic tobacco plants with increased proline content to drought and/or heat stress. *American Journal of Plant Sciences*, *2*, 318–324.

Poschenrieder, C., & Barceló, J. (2004). Water relations in heavy metal stressed plants. In C. Poschenrieder, & J. Barceló (Eds.), *Heavy metal stress in plants* (pp. 207–299). Berlin: Springer.

Potopová, V., Boronean, C., Boincean, B., & Soukup, J. (2015). Impact of agricultural drought on main crop yieldsin the Republic of Moldova. *International Journal of Climatology*, 1–20.

Prasad, P. V. V., Boote, K. J., Allen, L. H., Jr, Sheehy, J. E., & Thomas, J. M. G. (2006). Species, ecotype and cultivar differences in spikelet fertility and harvest index of rice in response to high temperature stress. *Field Crops Research*, *95*, 398–411.

Qin, F., Kakimoto, M., Sakuma, Y., Maruyama, K., Osakabe, Y., Tran, L. S. P., ... Yamaguchi-Shinozaki, K. (2007). Regulation and functional analysis of ZmDREB2A in response to drought and heat stress in *Zea mays* L. *The Plant Journal: for Cell and Molecular Biology*, *50*, 54–69.

Quan, R., Shang, M., Zhang, H., Zhao, Y., & Zhang, J. (2004). Engineering of enhanced glycine-betaine synthesis improves drought tolerance in maize. *Plant Biotechnology Journal*, *2*, 477–486.

Rahman, H., Sabreen, S., Alam, S., & Kawai, S. (2005). Effects of nickel on growth and composition of metal micronutrients in barley plants grown in nutrient solution. *Journal of Plant Nutrition*, *28*, 393–404.

Rai, M. K., Kalia, R. K., Singh, R., Gangola, M. P., & Dhawan, A. K. (2011). Developing stress tolerant plants through in vitro selection—An overview of the recent progress. *Environmental and Experimental Botany*, *71*, 89–98.

Rainey, K., & Griffiths, P. (2005). Evaluation of *Phaseolus acutifolius* A. Gray plant introductions under high temperatures in a controlled environment. *Genetic Resources and Crop Evolution*, *52*, 117–120.

Ray, R. L., Fares, A., & Risch, E. (2018). Effects of drought on crop production and cropping areas in Texas. *Agricultural & Environmental Letters*, *3*, 170037.

Raza, A., Charagh, S., Zahid, Z., Mubarik, M. S., Javed, R., Siddiqui, M. H., & Hasanuzzaman, M. (2020). Jasmonic acid: A key frontier in conferring abiotic stress tolerance in plants. *Plant Cell Reports*, *1*, 29.

Rogers, M. E. (2002). Irrigating perennial pasture with saline water: Effects on soil chemistry, pasture production and composition. *Australian Journal of Experimental Agriculture*, *42*(3), 265–272.

Roháček, K., Soukupová, J., & Barták, M. (2008). Chlorophyll fluorescence: A wonderful tool to study plant physiology and plant stress. *Plant Cell Compartments Selected Topics*, *2008*, 41–104.

Ronde, J. A. D., Cress, W. A., Krugerd, G. H. J., Strasserd, R. J., & Van Staden, J. (2004). Photosynthetic response of transgenic soybean plants, containing an Arabidopsis P5CR gene, during heat and drought stress. *Journal of Plant Physiology*, *161*, 1211–1224.

Rucker, K. S., Kvien, C. K., Holbrook, C. C., & Hook, J. E. (1995). Identification of peanut genotypes with improved drought avoidance traits. *Peanut Science*, *24*, 14–18.

Ruelland, E., Vaultier, M. N., Zachowski, A., & Hurry, V. (2009). Cold signaling and cold acclimation in plants. *Advances in Botanical Research*, *49*, 35–150.

Rustaee, M., Nazeri, S., Ghadimzadeh, M., & Malboobi, M. A. (2007). Optimizing in vitro regeneration from Iranian native dwarf rootstock of apple (*Malus domestica* Borkh). *International Journal of Agriculture And Biology*, *9*, 775–778.

Sabir, M., Hakeem, K. R., Aziz, T., Zia-ur Rehman, M., Rashid, I., & Ozturk, M. (2013). Higher Ni levels could modify growth response and mineral composition of wheat (*Triticum aestivum*) cultivars grown in contaminated soil. *Clean Soil Air Water*, *42*(9), 1263–1271.

Samarah, N. H. (2005). Effects of drought stress on growth and yield of barley. *Agronomy for Sustainable Development*, *25*, 145–149.

Samarah, N. H., Mullen, R. E., Cianzio, S. R., & Scott, P. (2006). Dehydrin-like proteins in soybean seeds in response to drought stress during seed filling. *Crop Science*, *46*, 2141–2150.

Schafleitner, R., Guitierrez, R., Espino, R., Guadin, A., Perez, J., Martinez, M., ... Bonierbale, M. (2007). Field screening for variation of water stress tolerance in *Solanum tuberosum* L. by agronomical, physiological and genetic analysis. *Potato Research*, *50*, 71–85.

Shah, F., Huang, J., Cui, K., Nie, L., Shah, T., Chen, C., & Wang, K. (2011). Impact of high temperature stress on rice plant and its traits related to tolerance. *Journal of Agricultural Science*, *149*, 545–556.

Shahid, M., Dumat, C., Pourrut, B., Sabir, M., & Pinelli, E. (2014). Assessing the effect of metal speciation on lead toxicity to *Vicia faba* pigment contents. *Journal of Geochemical Exploration*, *144*, 290–297.

Shahid, M., Khalid, S., Abbas, G., Shahid, N., Nadeem, M., Sabir, M., ... Dumat, C. (2015). Heavy metal stress and crop productivity. In K. R. Hakeem (Ed.), *Crop production and global environmental issues* (pp. 1–25). Springer International Publishing.

Shahid, M., Sabir, M., Ali, M. A., & Ghafoor, A. (2014). Effect of organic amendments on phytoavailability of nickel and growth of berseem (*Trifolium alexandrinum*) under nickel contaminated soil conditions. *Chemical Speciation & Bioavailability*, *26*, 37–42.

Shan, D. P., Huang, J. G., Yang, Y. T., Guo, Y. H., Wu, C. A., Yang, G. D., Gao, Z., & Zheng, C. C. (2007). Cotton GhDREB1 increases plant tolerance to low temperature and is negatively regulated by gibberellic acid. *New Phytologist*, *176*(1), 70–81.

Sharkey, T. D., & Schrader, S. M. (2006). High temperature stress. In K. V. Madhava Rao, A. S. Raghavendra, & R. K. Janardhan (Eds.), *Physiology and molecular biology of stress tolerance in plants* (pp. 101–130). Dordrecht: Springer.

Sharma, S., Thakur, A. K., & Srivastava, D. K. (2006). Plant regeneration and Agrobacterium-mediated gene transfer in Bell pepper (*Capsicum annuum* L.). *Crop Improvement*, *33*, 1–8.

Shekhawat, U. K. S., & Ganapathi, T. R. (2013). MusaWRKY71 overexpression in banana plants leads to altered abiotic and biotic stress responses. *PLoS One*, *8*, e75506.

Shen, C., Que, Z., Xia, Y., Tang, N., Li, D., He, R., & Cao, M. (2017). Knock out of the annexin gene OsAnn3 via CRISPR/Cas9-mediated genome editing decreased cold tolerance in rice. *Journal of Plant Biology*, *60*(6), 539–547.

Shi, J., Gao, H., Wang, H., Lafitte, H. R., Archibald, R. L., Yang, M., & Habben, J. E. (2017). ARGOS 8 variants generated by CRISPR-Cas9 improve maize grain yield under field drought stress conditions. *Plant Biotechnology Journal*, *15*(2), 207–216.

Shinozaki, K., & Yamaguchi-Shinozaki, K. (2007). Gene networks involved in drought stress response and tolerance. *Journal of Experimental Botany*, *58*, 221–227.

Shinwari, Z. K., Jan, S. A., Nakashima, K., & Yamaguchi-Shinozaki, K. (2020). Genetic engineering approaches to understanding drought tolerance in plants. *Plant Biotechnology Reports*, 1–12.

Shirasawa, K., Takabe, T., Takabe, T., & Kishitani, S. (2006). Accumulation of glycine betaine in rice plants that overexpress choline monooxygenase from spinach and evaluation of their tolerance to abiotic stress. *Annals of Botany*, *98*, 565–571.

Sinaki, J. M., Heravan, E. M., Rad, A. H. S., Noormohammadi, G., & Zarei, G. (2007). The effects of water deficit during growth stages of canola (*Brassica napus* L.). *American-Eurasian Journal of Agricultural & Environmental Science*, *2*, 417–422.

Singh, R. K., Redoña, E., & Refuerzo, L. (2009). Varietal improvement for abiotic stress tolerance in crop plants: Special reference to salinity in rice. In A. Pareek, S. Sopory, & H. Bohnert (Eds.), *Abiotic stress adaptation in plants*. Dordrecht: Springer. Available from 10.1007/978-90-481-3112-9_18.

Singh, S., & Rajam, M. V. (2010). Highly efficient and rapid plant regeneration in Citrus sinensis. *Journal of Plant Biochemistry and Biotechnology*, *19*, 195–202.

Singhal, R. K., Sodani, R., Chauhan, J., Sharma, M. K., & Yashu, B. R. (2017). Physiological adaptation and tolerance mechanism of rice (*Oryza sativa* L.) in multiple abiotic stresses. *International Journal of Pure & Applied Bioscience*, *5*(3), 459–466.

Singla-Pareek, S. L., Reddy, M. K., & Sopory, S. K. (2001). Transgenic approach towards developing abiotic stress tolerance in plants. *Proceedings of the Indian National Science Academy*, *67*, 265–284.

Sivamani, E., Bahieldin, A., & Wraith, J. M. (2000). Improved biomass productivity and water use efficiency under water deficit conditions in transgenic wheat constitutively expressing the barley HVA1 gene. *Plant Science (Shannon, Ireland)*, *155*, 1–9.

Song, A., Zhu, X., Chen, F., Gao, H., Jiang, J., & Chen, S. (2014). A chrysanthemum heat shock protein confers tolerance to abiotic stress. *International Journal of Molecular Science*, *15*, 5063–5078.

Sosa, L., Llanes, A., Reinoso, H., Reginato, M., & Luna, V. (2005). Osmotic and specific ion effects on the germination of *Prosopis strombulifera*. *Annals of Botany*, *96*, 261–267.

Srivastava, L. M. (2002). *Plant growth and development: hormones and environment*. Elsevier.

Steinwand, M. A., & Ronald, P. C. (2020). Crop biotechnology and the future of food. *Nature Food*, 273–283.

Stvolinskaya, N. S. (2000). Viability of *Taraxacum officinale* Wigg. in populations of the city of Moscow in relation to motor transport pollution. *Russian Journal of Ecology*, *31*, 129–131.

Su, J., Hirji, R., Zhang, L., He, C., Selvaraj, G., & Wu, R. (2006). Evaluation of the stress-inducible production of choline oxidase in transgenic rice as a strategy for producing the stress protectant glycine betaine. *Journal of Experimental Botany*, *57*, 1129–1135.

Subramanyam, K., Sailaja, K. V., Subramanyam, K., Rao, D. M., & Lakshmidevi, K. (2011). Ectopic expression of an osmotin gene leads to enhanced salt tolerance in transgenic chilli pepper (*Capsicum annum* L.). *Plant Cell, Tissue and Organ Culture*, *105*, 181–192.

Sujariya, S., Jongdee, B., Inthavong, T., Budhaboon, C., & Jongrungklang, N. (2017). Use of water balance model for estimating variability of growing season and drought occurrence for rainfed lowland rice. In *Poster session present at the International conference on sustainable agriculture: system approach for sustainable livelihood and environment*. Chiang Mai, Thailand.

Sultan, B., Defrance, D., & Iizumi, T. (2019). Evidence of crop production losses in West Africa due to historical global warming in two crop models. *Science Reports*, *9*, 12834.

Surabhi G. K., Badajena B., & Sahoo S. K. (2019) Genome editing and abiotic stress tolerance in crop plants. In *Recent approaches in omics for plant resilience to climate change* (pp. 35–56). Springer, Cham.

Taiz, L., & Zeiger, E. (2006). *Plant physiology* (4th Edn). Sunderland, MA: Sinauer Associates Inc Publishers.

Takahashi, D., Li, B., Nakayama, T., Kawamura, Y., & Uemura, M. (2013). Plant plasma membrane proteomics for improving cold tolerance. *Frontiers in Plant Science*, *4*, 90.

Tang, L., Kwon, S. Y., Kim, S. H., Kim, J. S., Choi, J. S., Cho, K. Y., ... Lee, H. S. (2006). Enhanced tolerance of transgenic potato plants expressing both superoxide dismutase and ascorbate peroxidase in chloroplasts against oxidative stress and high temperature. *Plant Cell Reports*, *25*, 1380–1386.

Tester, M., & Langridge, P. (2010). Breeding technologies to increase crop production in a changing world. *Science*, *327*(5967), 818–822.

Thakur, P., Kumar, S., Malik, J. A., Berger, J. D., & Nayyar, H. (2010). An overview: Cold stress effects on reproductive development in grain crops. *Environmental and Experimental Botany*, *67*, 429–443.

Torney, F., Trewyn, B. G., Lin, V. S. Y., & Wang, K. (2007). Mesoporous silica nanoparticles deliver DNA and chemicals into plants. *Nature Nanotechnology*, *2*(5), 295–300.

Tsai-Hung, H., Jent-turn, L., Yee-yung, C., & Ming-Tsair, C. (2002). Heterology expression of the Arabidopsis C-Repeat/Dehydration Response Element Binding Factor 1 gene confers elevated tolerance to chilling and oxidative stresses in transgenic tomato. *Plant Physiology*, *130*, 618–626.

Vara Parasad, P. V., Craufurd, P. Q., & Summerfield, R. J. (1999). Fruit number in relation to pollen production and viability in groundnut exposed to short episodes of heat stress. *Annals of Botany*, *84*, 381–386.

Vasantha, S., Alarmelu, S., Hemaprabha, G., & Shanthi, R. M. (2005). Evaluation of promising sugarcane genotypes for drought. *Sugar Technol*, *7*, 82–83.

Vasudevan, A., Selvaraj, N., Ganapathi, A., & Choi, C. W. (2007). Agrobacterium-mediated genetic transformation in cucumber (Cucumis sativus L.). *American Journal of Biochemistry and Biotechnology*, *3*, 24–32.

Venuprasad, R., Lafitte, H. R., & Atlin, G. N. (2007). Response to direct selection for grain yield under drought stress in rice. *Crop Science*, *47*, 285–293.

Visioni A., Al-Abdallat A., Elenien J. A., Verma R. P. S., Gyawali S., & Baum M. (2019). Genomics and molecular breeding for improving tolerance to abiotic stress in barley (Hordeum vulgare L.). In *Genomics assisted breeding of crops for abiotic stress tolerance* (pp. 49–68), vol. II. Springer, Cham.

Volland, S., Lütz, C., Michalke, B., & Lütz-Meindl, U. (2012). Intracellular chromium localization and cell physiological response in the unicellular alga Micrasterias. *Aquatic Toxicology (Amsterdam, Netherlands)*, *109*, 59–69.

Vollenweider, P., & Gunthardt-Goerg, M. S. (2005). Diagnosis of abiotic and biotic stress factors using the visible symptoms in foliage. *Environmental Pollution (Barking, Essex: 1987)*, *137*, 455–465.

Wahid, A., & Close, T. J. (2007). Expression of dehydrins under heat stress and their relationship with water relations of sugarcane leaves. *Biologia Plantarum*, *51*, 104–109.

Wahid, A., Gelani, S., Ashraf, M., & Foolad, M. R. (2007). Heat tolerance in plants: An overview. *Environmental and Experimental Botany*, *61*, 199–223.

Wang, B., Wang, C., Li, J., Sun, H., & Xu, Z. (2014). Remediation of alkaline soil with heavy metal contamination using tourmaline as a novel amendment. *Journal of Environmental Chemical Engineering*, *2*, 1281–1286.

Wang, C., Linderholm, H. W., Song, Y., Wang, F., Liu, Y., Tian, J., ... Ren, G. (2020). Impacts of drought on maize and soybean production in Northeast China during the past five decades. *International Journal of Environmental Research and Public Health*, *17*, 2459.

Wang, L., Chen, L., Li, R., Zhao, R., Yang, M., Sheng, J., & Shen, L. (2017). Reduced drought tolerance by CRISPR/Cas9-mediated SlMAPK3 mutagenesis in tomato plants. *Journal of Agricultural and Food Chemistry*, *65*(39), 8674–8682.

Wang, W., Chen, Q., Hussain, S., Mei, J., Dong, H., Peng, S., et al. (2016). Pre-sowing seed treatments in direct-seeded early rice: consequences for emergence, seedling growth and associated metabolic events under chilling stress. *Sci. Rep.*, *6*, 19637.

Wang, Y., Wisniewski, M., Meilan, R., Cu, M., & Fuchigami, L. (2006). Transgenic tomato (*Lycopersicon esculentum*) overexpressing cAPX exhibits enhanced tolerance to UV-B and heat stress. *Journal of Applied Horticulture*, *8*, 87–90.

Wang, Y., Wisniewsky, M., Meilan, R., Cui, M., Webb, R., & Fuchigamy, L. (2005). Overexpression of cytosolic ascorbate peroxidase in tomato confers tolerance to chilling and salt stress. *Journal of the American Society for Horticultural Science. American Society for Horticultural Science*, *130*, 167–173.

Wani, S. H., Kumar, V., Khare, T., Tripathi, P., Shah, T., Ramakrishna, C., & Mangrauthia, S. K. (2020). miRNA applications for engineering abiotic stress tolerance in plants. *Biologia (Lahore, Pakistan)*, 1–19.

Waqas, M. A., Kaya, C., Riaz, A., Farooq, M., Nawaz, I., Wilkes, A., & Li, Y. (2019). Potential mechanisms of abiotic stress tolerance in crop plants induced by thiourea. *Frontiers in Plant Science*, *10*, 1336. Available from https://doi.org/10.3389/fpls.2019.01336.

Whaley, J. M., Kirby, E. J. M., Spink, J. H., Foulkes, M. J., & Sparkes, D. L. (2004). Frost damage to winter wheat in the UK: The effect of plant population density. *European Journal of Agronomy*, *21*, 105–115.

Wilhelm, E. P., Mullen, R. E., Keeling, P. L., & Singletary, G. W. (1999). Heat stress during grain filling in maize: Effects of kernel growth and metabolism. *Crop Science*, *39*, 1733–1741.

Wisniewski M., Fuchigami L., Wang Y., Srinivasan C., & Norilli J. (2002). Overexpression of a cytosolic ascorbate peroxidase gene in apple improves resistance to heat stress. In *XXXVI International horticultural congress and exhibition* (p 147) (Abstr.)

Xiao, B., Huang, Y., Tang, N., & Xiong, L. (2007). Overexpression of LEA gene in rice improves drought resistance under field conditions. *TAG. Theoretical and Applied Genetics. Theoretische und Angewandte Genetik*, *115*, 35–46.

Xu, D., Duan, X., Wang, B., Hong, B., Ho, T. H. D., & Wu, R. (1996). Expression of a late embryogenesis abundant protein gene, HVA1, from barley confers tolerance to water deficit and salt stress in transgenic rice. *Plant Physiology*, *110*, 249–257.

Xu, R., Wei, P., & Yang, J. (2017). Use of CRISPR/Cas genome editing technology for targeted mutagenesis in rice. In A. Reeves (Ed.), *In vitro mutagenesis. Methods in molecular biology* (vol 1498). New York, NY: Humana Press. Available from 10.1007/978-1-4939-6472-7_3.

Yamada, M., Morishita, H., Urano, K., Shiozaki, N., Yamaguchi-Shinozaki, K., Shinozaki, K., & Yoshiba, Y. (2005). Effects of free proline accumulation in petunias under drought stress. *Journal of Experimental Botany*, *56*, 1975–1981.

Yuldashev, F. & Carli, C. (2010). Enhanced food and income security in SWCA through potato varieties with improved tolerance to abiotic stress. *Evaluation of CIP-Bred clones for tolerance to abiotic stress in Central Asia. 2010. Report*. Deutsche Gesellschaft für Technische Zusammenarbeit (GTZ). Bonn, Germany.

Zhang, A., Liu, Y., Wang, F., Li, T., Chen, Z., Kong, D., & Tang, J. (2019). Enhanced rice salinity tolerance via CRISPR/Cas9-targeted mutagenesis of the OsRR22 gene. *Molecular Breeding*, *39*(3), 47.

Zhang, G. H., Su, Q., An, L. J., & Wu, S. (2008). Characterization and expression of a vacuolar Na^+/H^+ antiporter gene from the monocot halophyte Aeluropus littoralis. *Plant Physiology and Biochemistry: PPB/Societe Francaise de Physiologie Vegetale*, *46*, 117–126.

Zhang, H., Dong, H., Li, W., Sun, Y., Chen, S., & Kong, X. (2009). Increased glycine betaine synthesis and salinity tolerance in AhCMO transgenic cotton lines. *Molecular Breeding*, *23*, 289–298.

Zhang, K., Wang, J., Lian, L., Fan, W., Guo, N., & Lv, S. (2012). Increased chilling tolerance following the transfer of a betA gene enhancing glycine betaine synthesis in cotton (*Gossypium hirsutum* L.). *Plant Molecular Biology Reporter/ISPMB*, *30*, 1158–1171.

Zhang, X., Zou, Z., Gong, P., Zhang, J., Ziaf, K., Li, H., … Ye, Z. (2011). Overexpression of microRNA169 confers enhanced drought tolerance to tomato. *Biotechnology Letters*, *33*, 403–409.

Zhang, Y., Li, H., Shu, W., Zhang, C., & Ye, Z. (2011). RNA interference of a mitochondrial APX gene improves vitamin C accumulation in tomato fruit. *Scientia Horticulturae*, *129*, 222–226.

Zhao, X., Meng, Z., Wang, Y., Chen, W., Sun, C., Cui, B., & Luo, D. (2017). Pollen magnetofection for genetic modification with magnetic nanoparticles as gene carriers. *Nature Plants*, *3*(12), 956–964.

Zhu, H., Li, C., & Gao, C. (2020). Applications of CRISPR–Cas in agriculture and plant biotechnology. *Nature Reviews. Molecular Cell Biology*, *21*(11), 661–677.

Zong, Y., Wang, Y., Li, C., Zhang, R., Chen, K., Ran, Y., … Gao, C. (2017). Precise base editing in rice, wheat and maize with a Cas9- cytidine deaminase fusion. *Nature Biotechnology*, *35*, 438–440.

Zsögön, A., Čermák, T., Naves, E. R., Notini, M. M., Edel, K. H., Weinl, S., & Peres, L. E. P. (2018). De novo domestication of wild tomato using genome editing. *Nature Biotechnology*, *36*(12), 1211–1216.

Chapter 8

Omics approaches for improving abiotic stress tolerance in rice: recent advances and future prospects

Rahil Shahzad[1,*], Shakra Jamil[1,*], Shakeel Ahmad[2,*], Amina Nisar[3], Shamsa Kanwal[1], Rana Muhammad Atif[3,4], Muhmmad Zaffar Iqbal[1], Zulfiqar Ali[5], Abdelhalim I. Ghazy[6] and Asif Ali Khan[5]

[1]Agricultural Biotechnology Research Institute, Ayub Agricultural Research Institute, Faisalabad, Pakistan, [2]State Key Laboratory of Rice Biology, China National Rice Research Institute, Hangzhou, P.R. China, [3]Department of Plant Breeding and Genetics, University of Agriculture, Faisalabad, Pakistan, [4]Center for Advanced Studies in Agriculture and Food Security (CAS-AFS), University of Agriculture, Faisalabad, Pakistan, [5]Institute of Plant Breeding and Biotechnology (IPBB), MNS University of Agriculture, Multan, Pakistan, [6]Plant Production Department, Food Science and Agricultural College, King Saud University, Riyadh, Saudi Arabia

8.1 Introduction

Rice is consumed by half of the world's population as a staple food. Due to the ever-growing population, it is expected that approximately 2000 million metric tons of rice will be required by the end of the 2030. Today, rice is cultivated on an average of 158 million hectares of land with an annual production of 700 million metric tons of paddy annually (Gadal, Shrestha, Poudel, & Pokharel, 2019).

The ever-increasing global population is putting pressure on the rice crop's grain harvest and thus decreasing per capita availability. In addition to population pressure, climate change is another big problem to tackle. Actually, it is giving rise to various abiotic stresses, that is, high salinity, high and low temperature, dehydration, flooding, heavy metals, and heavy winds, that are affecting the growth and development of rice plants and ultimately decreasing the crop's yield (Fahimirad & Ghorbanpour, 2019). After the onset of various abiotic stresses, plants respond through changes at

* These authors have contributed equally.

Frontiers in Plant–Soil Interaction. DOI: https://doi.org/10.1016/B978-0-323-90943-3.00011-0

the cellular, molecular, morphological, and biochemical levels (Van Oort, 2018). Therefore there is a need to understand the mechanisms by which rice plant respond to various abiotic stresses. It will help for the breeding of abiotic stress-tolerant rice varieties through the exploitation of various omics approaches (Fig. 8.1).

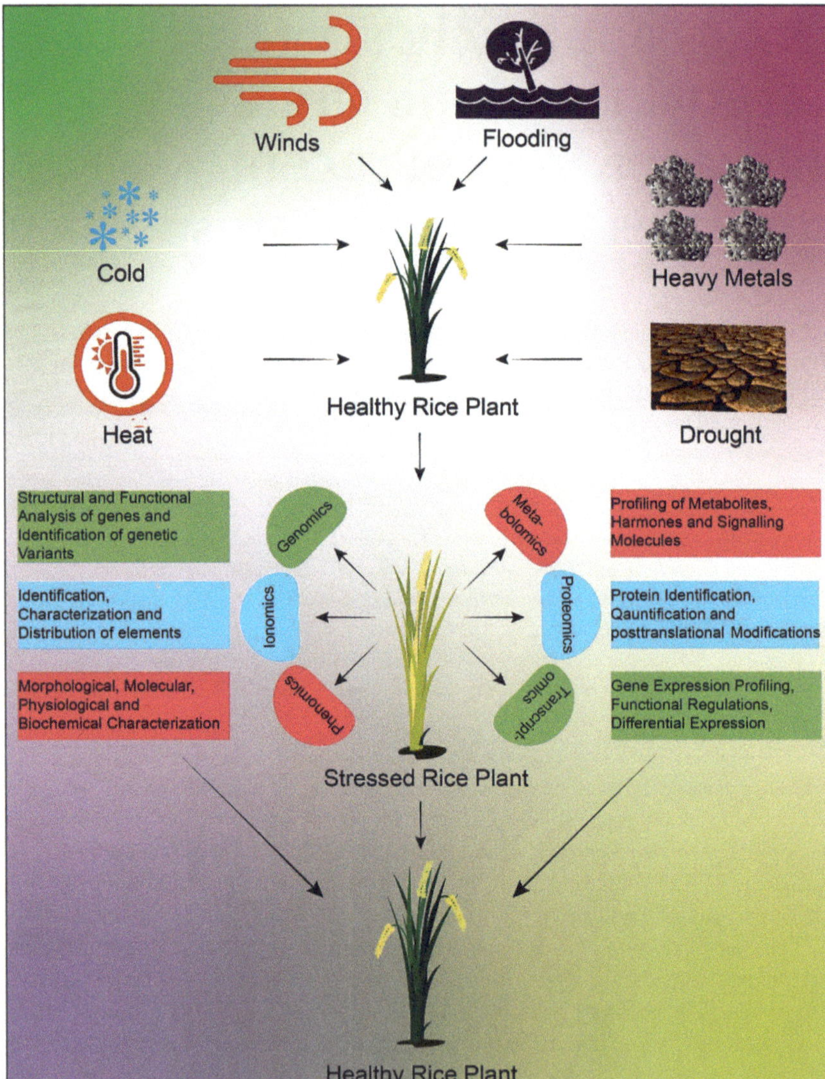

FIGURE 8.1 Effect of different abiotic stresses and role of different "omics" approaches to combat these stresses on rice plant. Different abiotic stresses, such as high and low temperature, drought, flooding, heavy metals, and heavy winds, affect the growth and productivity of rice plant. To cope with these abiotic stresses, different "omics" approaches are being utilized that resultantly give a healthy plant which can withstand abiotic stresses and produce high yield.

The "omics" deals with various discipline of science to evaluate the functioning and interaction of various biological processes and biological information in various clusters of life. It revolves around the genomics and epigenomics (study of genes), transcriptomics (the study of the transcriptome), proteomics (the study of the proteins and their role in various plant process), metabolomics (the study of various metabolites and their role in plant development), interactomics (the study of interaction of various biological molecules and their role in plant development), and phenomics (phenotypic study of plants) (Kandpal, Saviola, & Felton, 2009). Thus to answer the question "Can the omics approaches play their role in the development of climate resilient rice crop?," we have summarized the application of various omics approaches with a special focus on genomics to combat abiotic stresses for developing climate smart rice varieties for better yield and production in a changing climate.

8.2 Genomics

The science of genomics deals with the full genetic supplement of an organism. The analysis of the functions of the genomes, their interpretation, and exploiting their role under various biotic and abiotic stresses come under the scope of genomics (Tao, Zhao, Mace, Henry, & Jordan, 2019). The following sections deals with the various genomics techniques and their role in the development of climate resilient rice varieties.

8.2.1 Quantitative trait loci mapping

Quantitative Trait Loci (QTL) mapping methodology provides a basis to plant breeders for analysis of complicated stress mechanisms including salt, heat, drought, and waterlogging stresses because they have wide impacts on crop production. QTL mapping and marker assisted selection for stress-related QTLs speed up the breeding programs by the introgression of stress-related alleles and the development of stress responsive cultivars. It also help to explore the role of genotypic variation on plant traits (Taunk, Rani, Singh, Yadav, & Yadav, 2019).

QTL mapping mainly recognizes the genotypic markers, which are helpful in plant breeding approaches, and enhances our knowledge related to the genetic makeup of quantitative traits and their inheritance among species and within a species. It works by finding the linkage between genetic markers and traits of interest (quantitative traits) using diverse populations such as; backcross plants, double haploids, recombinant inbred lines (RILs), near isogenic lines (NILs) and segregating population (F2) (Semagn, Bjørnstad, & Xu, 2010). The prerequisites of QTL mapping are (1) selection of suitable mapping population, (2) phenotyping for desired traits, (3) genotyping via molecular markers, and (4) application of statistical procedures to recognize the association of molecular markers with desired traits. Multiple stress-related QTLs in various crops have been identified through QTL mapping technology, such as cold (Liu et al., 2013; Zhang et al., 2014), salinity (Fan, Shabala, Ma, Xu, & Zhou, 2015), waterlogging (Xu et al., 2006), heat (Shanmugavadivel et al., 2017), and drought (Harris et al., 2007).

A SUB1 QTL, was identified as having submergence tolerance in *Indica* variety. This QTL includes three genes named as *SUB1A*, *SUB1B*, and *SUB1C*, which code for Ethylene Responsive Factors (Xu et al., 2006). Similarly, by crossing FR13A/IR42, recombinant inbred lines of rice were developed and five submergence tolerant QTLs were detected at 1st, 4th, 8th, 9th, and 10th chromosomes (Gonzaga, Carandang, Sanchez, Mackill, & Septiningsih, 2016). Likewise, QTL mapping has been applied in numerous studies related to rice in order to know about the genetic architecture of salt stress responsive traits (Soltani et al., 2016). Identification of five QTLs for four particular salt stress traits of roots and three QTLs for shoot associated traits at different positions have been performed (Islam et al., 2011). Similarly, RIL 272F8 population was used in a study to identify the heat stress responsive QTLs in rice. Two qSSIY5.2/qSTIY5.1 and qSTIPSS9.1 were identified consisting of 54 and 65 genes, respectively, for heat stress tolerance (Shanmugavadivel et al., 2017). Several QTLs such as qCTS-2 at 2nd chromosome (Liu et al., 2013), both qCTS-1 and qSPA-1 at 1st chromosome (Ji et al., 2010), and qCTS-12 at 12th chromosome (Zhang et al., 2014) were found to be involved in chilling stress tolerance in rice. In addition, extensively grown rice cultivars have drought sensitivity. For development of drought stress resistance rice cultivars, a breeding project at International Rice Research Institute (IRRI) has recognized the significant QTLs by the application of molecular markers. The identified QTLs are qDTY12.1, qDTY3.2, qDTY3.1, qDTY2.1, qDTY6.1, and qDTY1.1 (BP et al., 2013; Venuprasad et al., 2009; Yadav et al., 2019). QTL mapping is helpful for the detection of QTL that regulate the drought stress responsive multiple traits (Barik, Pandit, Pradhan, Mohanty, & Mohapatra, 2019). Numerous QTLs locating on diverse chromosomes with tolerance for drought stress have been recognized as well as mapped via diverse mapping populations (Kam, Fofana, Zongo, & Taiwo, 2018).

8.2.2 Association mapping

Association mapping (AM) is an advanced methodology of QTL mapping depending upon linkage disequilibrium (LD). LD is a measure of association of a marker with a locus/QTL/gene or between two QTLs or genes. On the other hand, AM is the linkage among the phenotypic trait and a molecular marker. In AM, a fundamental role is played by LD. Estimation of LD, in specific species under analysis, is of particular preference because it defines the quantitative features of marker and density of experimental design under analysis. Numerous factors such as gene conversion, lower recombination frequency, increased mutation chances, and out crossing are involved in LD disturbance while a lower recombination rate, genetic drift, epistasis, population subdivisions, insignificant population size, and inbreeding play key roles in the increase of LD (Gupta, Rustgi, & Kulwal, 2005; Kim et al., 2007).

Disturbance of LD owing to genetic drift has been detected in numerous plant species, that is, *Zea mays, Arabidopsis thaliana* (Nordborg et al., 2005), *Triticum aestivum, Sorghum bicolor* (Maccaferri, Sanguineti, Noli, & Tuberosa, 2005), and *Hordeum vulgare* (Caldwell, Russell, Langridge, & Powell, 2006). Association mapping is basically classified into two groups, that is, (1) genome-wide association mapping (GWAM) and (2) candidate gene association mapping (CGAM). The GWAM is related to genotypic variation in a genome and its association with diverse traits, whereas CGAM is concerned with particular candidate genes with respect to their polymorphs (Taunk et al., 2019). Furthermore, the applications of QTL mapping and association mapping studies in rice are summarized in Table 8.1.

8.2.2.1 Genome-wide association mapping

The GWAM is an effective technique for recognizing the traits-linked alleles in the whole genome of diverse cultivars. GWAM basically signifies the association between SNPs and traits. It is a widely applied approach to develop abiotic and biotic stresses-resistant crop cultivars (Manolio, 2010). In order to study the whole genome associations, there is a need for high-density genome-wide markers that perfectly show the haplotype diversity and comprehensive genome LD structure. GWAM has been applied in diverse crops for multiple complicated traits because of a drop in genotyping as well as sequencing expenditure (Kumar et al., 2015). GWAM studies have been applied in 167 rice cultivars for genotyping the panicle micronutrient and spikelet sterility traits. A close association among these traits and 14 loci were observed. All these loci were studied for diverse tasks associated to heat shock proteins (HSP), regulating the plant responses and detection of abiotic stresses (Lafarge et al., 2017).

Rice is sensitive to low temperature stress at the seedling stage. A study consisted of 304 *Indica* rice cultivars for the detection of genomic regions that respond to low temperature stress and prevent rice plants from frost injury. Numerous QTLs associated to chilling stress tolerance in different cultivars were recognized showing genome-wide expression of linked traits. Nineteen SSR markers showed close association to low temperature stress tolerance (4°C and 8°C) at the seedling stage in rice. The following chilling stress tolerant QTLs: $qCTS_{-2}$, $qSCT_{11}$, $qCTS_{-3.1}$, $qCTS_{12.1}$, CTB_2, $qCTS_9$, $qCTS_{6.1}$, $qSCT_2$, $qSCT_{1a}$, $qCTS_{11.1}$, and $qCTS_{-1b}$ are required for development of cold stress responsive cultivars (Pandit et al., 2017). Similarly, a GWAM study explored 12 *Indica* rice traits at reproductive stage and 20 SNPs having close association with Na^+/K^+ ratios were recognized. In rice, a salt stress-tolerant QTL Saltol, was identified on chromosome number 1 through GWAM. Moreover, salinity stress-tolerant QTLs were also identified at chromosomes 4, 6, and 7 (Kumar et al., 2015).

TABLE 8.1 The summary of QTL mapping and association mapping studies in rice.

Approach	Identified genes or QTLs	Tolerance against	References
QTL Mapping	Sub1 QTL	Submergence	Xu et al. (2006)
	qSSIY5.2/qSTIY5.1, qSTIPSS9.1	Heat	Shanmugavadivel et al. (2017)
	qCTS-2, qCTS-1 and qSPA-1, qCTS-12	Cold	Ji et al. (2010); Liu et al. (2013); Zhang et al. (2014)
	qDTY12.1, qDTY3.2, qDTY3.1, qDTY2.1, qDTY6.1 and qDTY1.1	Drought	Barik, Pandit, Mohanty, Nayak, and Pradhan (2020)
	qSdw12, qSdw3b, qSZn5/qSFe5, qSdw3a	Heavy metals	Vromman et al. (2013); Zhang et al. (2017)
	qTr8, qCc3, qTr6, qCc12, qPn3-2, qLa12	Salinity	Sun et al. (2019)
	Xa21, Xa5, Xa13	Bacterial Blight	Mekonnen, Haileselassie, and Tesfaye (2017)
Association mapping	QTLs qCTS.2, qSCT11, qCTS-3.1, qCTS12.1, CTB2, qCTS9, qCTS6.1, qSCT2, qSCT1a, qCTS11.1 and qCTS-1b	Cold	Song et al. (2018)
	Saltol	Salinity	Kumar et al. (2015)
	STAR1, Nrat1, OsALS1, ASR5, ART1, STAR2, OsFRDL4, ART2 and OsMGT1.	Heavy metals (Al)	Meng et al. (2017)
	qCD8-1, OsABCB24, qCd1–3	Cd	Meng et al. (2017)

8.2.2.2 Candidate gene association mapping

The candidate gene selection is based upon already existing information about locality and function of genes involved in biochemical, physiological, and genetic pathways. Many QTLs have been discovered for aluminum (Al) tolerance in rice through diverse mapping populations (Zhang, Zhong, Tong,

Shahid, & Li, 2016). Moreover, Al stress-tolerant cloned rice genes are $STAR_1$, $Nrat_1$, $OsALS_1$, ASR_5, ART_1, $STAR_2$, $OsFRDL_4$, ART_2, and $OsMGT_1$., which were involved in diverse roles due to their varied Al concentration and mapping population. However, they also represent a complicated Al tolerance pathway in rice (Zhang, Zhong, Zhong, & Tong, 2019).

In another study, 338 *Indica* rice accessions were utilized to find out the cadmium (Cd) stress responsive genes through CGAM. The Cd-accumulation capacity of 338 diverse rice accessions was assessed by growing them in the Cd-contaminated soils with varying Cd concentrations. The results indicated that 13 rice lines including six *Indica* varieties showed relatively less Cd accumulation. The sequencing based SNP discovery approach was used for the identification of Cd stress responsive QTLs and 35 such QTLs were also identified in this study. Moreover, qCd8-1 QTL was identified in a genome-wide association study (GWAS) that was conducted in multiple years. Among them, nine QTLs were colocalized with identified genes or QTLs. A novel QTL, qCd1−3, with the lowest P value was selected for further LD decay analysis and candidate gene prediction. They found differential expression of *OsABCB24* between high-Cd-accumulative and low-Cd-accumulative accessions, suggesting it may be a candidate gene for qCd1−3 associated with low Cd accumulation. These results may be helpful for further exploiting novel functional genes related to Cd accumulation and developing a rice variety with low Cd accumulation through marker-assisted breeding (Pan et al., 2020).

8.2.3 Allele mining

Allele mining is an approach that is used to dissect naturally occurring allelic variations or suitable alleles of a candidate gene controlling key agronomic traits. Allele mining is effective for identifying the suitable alleles from diverse genotypes by using the sequence of a particular genotype (Latha, Rubia, Bennett, & Swaminathan, 2004). This plays a significant role in the detection of alleles, recognition of halophytes, and their use in marker-assisted selection (MAS) (Kumar et al., 2010). In the rice genome, salinity stress-tolerant genes were identified through allele mining via salt inducible calmodulin rice genes (Latha et al., 2004). During the allele mining of *HKT1−5* in *O. glaberrima* and *O. sativa* with respect to salt stress tolerance, three minor and seven major alleles were categorized (Platten, Egdane, & Ismail, 2013).

8.2.3.1 Targeted induced local lesion in the genome

It is one of the techniques to find out gene function, which follows a reverse genetic strategy either by the screening of a mutation in a gene with a known sequence or allows rapid mutational screening to obtain a lesion in the gene of interest. The targeted induced local lesion in the genome (TILLING) involves the creation of a mutagenic population through various mutagenic agents, that is, physical mutagens (irradiations) or chemical mutagens (ethyl methanosulfonate, EMS). The mutagenesis is followed by screening of the mutated population for the identification of the desired mutants. The DNA of the selected plants is extracted in the M_2 generation and pools are prepared for the eight plants. Pooled DNA is used for the conductance of the PCR using the fluorescent dye. This is followed by the sequence identification of the gene using the Li-COR analyzer. Finally, the gene sequence of the mutated gene is compared with the reference genome and sequences are matched through use of various bioinformatics tools and single nucleotide polymorphism (SNP) is identified and the allele is labeled (Cooper, Henikoff, Comai, & Till, 2013; Premkumar, Kumar, & Chauhan, 2020).

8.2.3.2 ECOTYPE targeting induced local lesions in genome

ECOTYPE targeting induced local lesions in genome (EcoTILLING) is a mutation identification-based reverse genetic approach to find out the superior genotypes and the functions of a gene. This methodology particularly illustrates the insertions and deletions (InDels) and SNP in naturally mutated populations (Al-Yassin & Khademian, 2015). EcoTILLING, basically, is a form of TILLING, which gives the results based on mutation-induced populations (Comai et al., 2004). EcoTILLING efficiently estimates the levels of heterozygosity in a single gene fragment. EcoTILLING was performed in the nine desired genes of the *Hordeum vulgare* for allele mining and haplotype detection. EcoTILLING recognized 46 InDels and 185 SNPs having drought stress tolerance (Cseri et al., 2011). Two salt stress responsive genes, Os*SalT* and *Oscpk17*, were analyzed in a study which used 375 accessions and identified 23 InDels and 15 SNPs, respectively, for these genes (Negrao, Almadanim, Pires, Mcnally, & Oliveira, 2011). This technique has played a prominent part in identifying the diversity of promoter sequences of 24 diverse transcriptional factor families among 95 landraces of rice. Furthermore, five genes responsible for drought tolerance level and three genes responsible for drought tolerance index (DTI) were identified in rice through this approach (Yu et al., 2012).

To find out the salt stress-tolerant alleles, the EcoTILLING approach has been used in 375 domesticated rice accessions. This study was linked with the identification of small indels and SNPs. As a result, 23 and 15 indels or SNPs were recognized. Those were involved in salt tolerant (Salt) and salt stress signal transduction (OsCPK17) pathways (Negrao et al., 2011). This approach is similar to TILLING but its main target is to detect the naturally occurring genetic variability of organisms along with the haplotype and SNP identification. Nowadays, EcoTILLING has been applied in brassica, lotus, maize, poplar, and rice. This technique is widely helpful for identifying the potent novel functional alleles in the candidate and novel genes or transcriptional factors (TFs) for the genetic improvement of crops (Khan et al., 2018).

8.2.3.3 Development of functional markers

Functional markers (FMs) have their origin in the polymorphic sequences of the allelic variants of a functional gene (Salgotra, Gupta, & Stewart, 2014). These are also known as the "perfect markers" owing to their complete association with desired traits (Akpınar, Lucas, & Budak, 2013). FMs are able to overcome the problems originated during the utilization of the random DNA markers. Desired alleles could be fixed into the genetic architecture through the FMs without further standardization (Cakir, Gillespie, & Scofield, 2010; Gupta, Kumar, Mir, & Kumar, 2010). FMs are present within the selected genes and have a direct association with the morphological traits. They lessen the wrong selection and play a significant role in the allele fixation in a population, development of FM associated haplotypes, and in combining the FM alleles linked with complicated traits (Salgotra & Stewart, 2020).

In field, *phosphate* (*P*) *uptake 1* (*OsPup1*) gene, responsible for tolerance to P-deficit stress condition, was fixed in rice backgrounds using the FMs. An InDel-based FM and *OsPup1* gene- particular SNP have been established for their transfer into the P deficiency stress susceptible rice varieties with the help of MAS (Chin et al., 2011). Numerous regions of the Asian belt face flooding in the rice growing period. However, a flooding tolerance gene (*OsSub1*) has been recognized and *OsSub1* linked FM also established. This FM is extensively utilized in Swarna rice to develop the flooding stress tolerance (Neeraja et al., 2007; Salgotra & Stewart, 2020).

8.2.4 Marker-assisted selection

Marker-assisted selection (MAS) has emerged as a powerful strategy for the development of different stress resistance crops after the identification of unique QTLs and molecular markers linked to these QTLs and different stress responsive genes. MAS is utilized for pyramiding of desirable genes into a single variety for the development of durable resistance against biotic and abiotic stresses. MAS has been very widely used in rice for the development of abiotic stress-tolerant rice varieties, as highlighted in the following examples. To combat submergence stress in rice, *OsSub1* gene located at the 9th chromosome of rice was utilized to develop resistant varieties through MAS at IRRI (Neeraja et al., 2007). Similarly for the development of submergence-tolerant rice varieties, a submergence-tolerant locus from FR13A variety was transferred into the Swarna rice variety (Taunk et al., 2019).

The first drought stress responsive aerobic variety of rice MAS 946-1 was developed through MAS and is cultivated in India (Gandhi, 2007). MAS was exploited to transfer the four QTLs of Azucena variety (specific for deep root systems) to IR64 (a rice variety having a shallow root system). The results showed that drought tolerance capacity of the MAS-derived plants was enhanced as compared to their wild types due to the acquisition of more water from the root zone (Steele, Price, Shashidhar, & Witcombe, 2006). NILs were developed in rice through the transfer of three drought tolerant QTLs from CT9993 upland variety into a lowland IR20 *Indica* cultivar (Suji et al., 2012).

Due to the huge genetic diversity in rice, MAS has become the most efficacious approach for the development of high yielding varieties in rice (Mackill et al., 2007). Salt stress responsive QTL (Saltol) was introgressed from a salt-tolerant variety FL478 to a high yielding but a salt susceptible variety ASS996 through marker-assisted backcross (MABC) (Luu Tn, Luu, Abdelbagi, & Le H, 2012). In another study, the same QTL was transferred from FL478 to another salt susceptible rice variety Bacthom7 via MABC. The MAB witnessed the recovery of 96%–100% of the blood of the parent variety Bacthom7 in the 3rd generation following crossing (Vu, Le, Ismail, & Le, 2012). Marker-associated pyramiding was used to stack salinity-tolerant qSKC-1 (responsive for K^+) and qSNC-7 (responsive for Na^+) QTLs in rice varieties. These highlighted examples depict that MAS has a wide role to play in the improvement of rice crops. The future of MAS seems bright with the development of climate resilient rice varieties.

8.2.5 Comparative genomics

Comparative genomics have made a significant comparison between the expressed genes of diverse stress-tolerant populations. This study also has provided a way for mining the particular detected targets of microbial species (Yu et al., 2010). The accessibility of plant genome diversity with genome's expression profiling data and multiple stress-linked cDNA libraries provide a strong base for the genomics-dependent detection of stress responsive genes as well as pathways (Ma, Qin, & Tran, 2012). Comparative genomics-based analysis among species or between the different genotypes of a species illustrate the particular stress responsive genes of a species. Comparative genomics have helped to identify various stress responsive TFs of soybean (*Glycine max*), sorghum (*Sorghum bicolor*), wheat (*Triticum aestivum*), barley (*Hordeum vulgare*), and maize (*Zea mays)* with the help of comparative analysis of the varied known TFs of rice (*Oryza sativa)* and Arabidopsis (*Arabidopsis thaliana*) (Mochida et al., 2011; Tran & Mochida, 2010).

Comparative genomic analysis of three Asian strains of *Ustilaginoidea virens* (Uv-Gvt, UV-8b and MAFF 236576) in rice has detected the 6.40%–19.72% repetitive DNA of the *U. virens* genome. Furthermore, this study has characterized the almost 422 host–pathogen interacting genes and diversity of pathogen population (Pramesh et al., 2020). The basic helix–loop–helix (*bHLH*) TFs have key involvement in various molecular and cellular pathways. A previous comparative genomics study has categorized 231, 183, and 571 *bHLHs* in maize, rice, and wheat, respectively. Moreover, 29, 31, and 45 *bHLHs* having differential expression in maize, rice, and wheat during the drought stress, respectively, have been discovered by the RNA-seq technique (Wei & Chen, 2018).

Through the comparative genomics approach, a study to analyze the effect on salt stress on leguminous crops was performed including the three cultivated and three model species. The results indicated that from the expressed genes approximately 60% were responsive in one or more species, however, $<1\%$ are responsive in all species (Sanchez et al., 2011).

8.3 Transcriptomics

Transcriptomics is the study of the transcriptome—the complete set of RNA transcripts that are produced by the genome, under specific circumstances or in a specific cell—using high-throughput methods, such as microarray analysis. There is a dire need to explore the functions of stress responsive genes in order to enhance the stress tolerance (Umezawa, Fujita, Fujita, Yamaguchi-Shinozaki, & Shinozaki, 2006; Valliyodan & Nguyen, 2006). The microarray-based expression profiling in different crops has led the foundation of characterization of the plant stress responsive genes. Microarrays are effective for the identification of stress responsive genes at the transcriptional level (Lee, Henderson, & Zhu, 2005; Xiong, Schumaker, & Zhu, 2002). Microarray technology not only identifies the stress responsive genes but is also effective for recognition of cis-acting elements and understanding the molecular pathways during stress responses (Simpson et al., 2003). Specific DNA microarrays store the genomic sequence data along with expression data for the identification of downstream genes of stress-associated TF and cis-acting elements (Maruyama et al., 2004; Seki et al., 2001). This forward genetic approach is extensively used to identify the several knockout mutants with stress sensitive and tolerant phenotypes (Bartels & Sunkar, 2005). Microarrays are helpful for gene identification and understanding of their responsive mechanisms during abiotic stresses. These are worthwhile for the recognition of multiple stress stimulated genes and stress responsive TFs and transcriptome analysis during abiotic stress.

Several hundred genes which respond to different abiotic stresses, that is, drought, heat, cold, heavy metals, salinity, and waterlogging, have been identified through microarray analysis in different crops (Seki et al., 2010). Similarly microarray analysis was also helpful in the identification of target genes of different transcriptional factors, cis-regulatory elements in the promotor region of different genes, coregulatory genes, and downstream or regulatory genes (Kimotho, Baillo, & Zhang, 2019). The microarray approach was used for the identification of abiotic stress responsive genes in many crops, that is, *Arabidopsis thaliana* (Ghorbani, Alemzadeh, & Razi, 2019), rice (Sirohi, Yadav, Afzal, Mani, & Singh, 2020), sorghum (Devnarain et al., 2019) cotton (Ahmad et al., 2020), potato (Kim et al., 2019), and maize (Liu et al., 2019).

8.4 Proteomics

Proteomics is the large-scale study of proteins. Proteins are vital parts of living organisms, which play key roles in several functions. The proteome is the entire set of proteins that is produced or modified by an organism or system. Proteomics has enabled the identification of ever increasing numbers of protein. Proteomics has its main involvement in the identification of molecular pathways associated with plant responses to either biotic or abiotic stresses and is also effective for enhancing the efficacy of indirect selection criteria related to inherited traits (Nouri, Toorchi, & Komatsu, 2011). Advancement of bioinformatics tools, mass spectrometry instruments, and protein profiling mechanisms lead toward high-throughput investigation. Formerly used two- dimensional gel electrophoresis technologies have been substituted by advance shotgun proteomics technology comprising of Tandem Mass Tag or label-free quantitation methodology (Cramer et al., 2013) and isobaric tag for Relative and Absolute Quantitation (Li et al., 2015).

The structural and functional abundance as well as interactions of proteins at translational and posttranslational level were studied through proteomics for the understanding of complicated biological mechanisms (Ghosh & Xu, 2014). This technique is highly efficacious for the recognition of abiotic stress responsive proteins in diverse crops (Ngara & Ndimba, 2014). A proteomic study was conducted to identify the effect of high temperature (HT) on the expression profile of storage proteins and rice grain quality. HT is responsible for decreases in amylose content, flour gel consistency, and grain weight. This also reduces the globulin, glutelins, protein disulfide isomerase, and prolamins (Lin et al., 2010).

Moreover, heat stress-related proteomic study has showed that 18 heat stress responsive proteins, known as HSPs, have been recognized. Out of these, seven HSP_{70}, three HSP_{100}, one chaperonin 60 beta precursor, and seven HSPs are identified. In the cultured rice, both cysteine proteinase inhibitor 8 and dehydrin have distinctive expression at HT (Zou, Liu, & Chen, 2011). In another proteomic study, the identification of 56 abundant proteins has been made after salinity stress treatment. Such proteins have involvement in multiple plant mechanisms like oxidative phosphorylation, protein synthesis, photosynthesis, transcription, protein degradation, signal transduction, metabolisms, actin regulation, and antioxidant activities (Sharma, Sihmar, Santal, & Singh, 2019).

A proteomic study related to chilling stress (5°C) for 48 hrs. on rice seedlings has concluded that there is a decrease in the relative abundance of defense-response–associated proteins (phenylalanine ammonia lyase and β-1–3-glucanase) while the relative abundance of energy metabolism-associated proteins (putative fructose-bisphosphate aldolase, adenylate kinase, UDP-glucose pyrophosphorylase and putative glyoxysomal malate dehydrogenease) increases. Cold stress is involved in the reduction of defense proteins and enhancement of energy proteins (Hashimoto & Komatsu, 2007).

Likewise, heavy metals, owing to their metallic features and optimum concentration in the plant metabolism, disturb the developmental stages, cellular growth, and metabolic activities. However, the first rice proteomic analysis related to heavy metals was conducted on cadmium (Cd). Cd is involved in severe disruption of the photosynthetic setup through RuBisCO degradation.

Mercury (Hg) is responsible for oxidative loss due to reactive oxygen species (ROS) development, and other heavy metals like zinc (Zn), cobalt (Co), lithium (Li), and strontium (Sr) have high association with low plant yield and inhibition of metabolic pathways and productivity (Singh & Jwa, 2013).

8.5 Metabolomics

Metabolomics is the scientific study of chemical processes involving metabolites, the small molecule substrates, intermediates, and products of metabolism. It deals with the quantitative and comprehensive analysis of the metabolome at a specific developmental stage in a particular tissue or cell for all small molecules involved in biological metabolism (Fiehn, 2001; Taunk et al., 2019). During abiotic stress, several factors disrupt the plant metabolism, such as increased demand for particular compounds, substrate shortage, inhibition of metabolism-associated enzymes, and many other factors. That is why the metabolic setups are reorganized and maintained for improving the abiotic stress tolerance in plants (Obata & Fernie, 2012). Metabolomics is a robust methodology, which illustrates metabolic regulations accompanying the gene functions like a part of system biology and functional genomics. Within all omics approaches, metabolomics is a widely applied technique in diverse organisms due to the small or no modifications in its organization (Arbona, Manzi, Ollas, & Gómez-Cadenas, 2013).

Plant metabolomics comprise the different technologies including Fourier-transform infrared spectroscopy, liquid chromatography–mass spectrometry (LC-MS), nuclear magnetic resonance spectroscopy (NMR), and gas chromatography–mass spectrometry (GC-MS) (Obata & Fernie, 2012). Owing to different abiotic stresses, disturbed plant metabolisms have been effectively utilized in the recognition of various factors as the identification of a significant compound in the plant biosynthetic pathway, severity of stress on plant metabolism and categorization of samples. Omics studies the definite factors with the help of metabolic changes as a marker involved in metabolic movements (Hirayama & Shinozaki, 2010).

The concentration of plant's primary metabolites, that is, sugar alcohol, amino acids, and sugars, is changed due to different abiotic stresses. According to the metabolic analysis, during the abiotic stress, the secondary metabolites, such as UV-irradiation, phenolics, alkaloid glycosides, glucosinolates, and carotenoids function as semiotic molecules (Munns & Tester, 2008). In *Arabidopsis thaliana*, drought stress was responsible for the accumulation of flavonoids and aliphatic glucosinolates along with the decreased accumulation of phytoalexin camalexin (Grubb & Abel, 2006). Metabolomics studies for abiotic stresses such as cold, drought, light, and salinity have recognized each abiotic stress specific metabolites (Caldana et al., 2011).

Rice is an excellent source of energy intake for about half of the world's population. Due to climate change and global warming, the traditional rice landraces are under stress that delimits their growth and productivity. To determine the number of metabolites in rice and their chemical diversity under various abiotic stresses, the metabolite composition of cultivated rice needs to be investigated with analytical techniques, that is, MS and/or NMR spectroscopy, and rice metabolite databases need to be constructed. Chemical diversity in rice based on metabolites under abiotic stress includes sugars, amino and organic acids (OAs), aromatic compounds, and phytohormones, which could be detected by GC-MS, LC-MS, or CE-MS, etc. The salt stress resulted in the accumulation of sucrose, allantoin, glutamate, and threonine in salt-stressed rice varieties. These results indicated that accumulation of metabolites under salt stress is an indication of the tolerance/response level of different varieties and may be used as criteria for selection under salt stress. In rice roots exposed to long-term mild salinity, sucrose accumulation was found to be conserved as a metabolic response. Metabolomics analysis of rice under salinity stress revealed elevation of serotonin, and gentisic acid levels in the leaves of tolerant varieties (Pradhan et al., 2019).

8.6 Phenomics

Phenomics pertains to the image analysis and integration of biological data. It combines biology and informatics to solve the problem of rapid and accurate estimation of the plant phenotype and to analyze phenotypic traits (Kumar, Mahato, Kumar, & Mishra, 2020). It helps in the identification of changes in the phenotypes that are undergone by plants under various abiotic stresses. Hence, it helps in the identification of the plant attributes/traits that are most/least affected by the abiotic stresses. High-throughput phenotyping is a bottleneck in plant genetic improvement. The leading cause of slow development of abiotic stress-tolerant crop varieties is inefficient and laborious phenotyping techniques. Our breeding programs are lacking the high-throughput and nondestructive phenotyping methods for efficient characterization of the changes undergone by plants under various abiotic stresses and various phenomena are left unnoticed. However, recent developments have paved way for mechanized phenotyping tools, and nowadays infrared thermography, spectroscopic techniques, fluorescence imaging, and integrated imaging techniques are being used by the developed breeding programs (Singh, Mishra, Bohra, Joshi, & Siddique, 2018). The use of phenomics for high-throughput phenotyping of the rice under various abiotic stresses needs be focused in each breeding program for the development of durable stress-resistant varieties.

8.7 Other omics approaches

The other omics approaches include ionomics and interactomics. The ionomics deals with the study of the changes in the ionic balance of the cells and plants owing to the distribution of the elements in the cells. Different research has been conducted in rice to study the effect of heavy metals on the ionomics of rice plants (Du, Liu, Wang, Yang, & Wang, 2020; Feng et al., 2017). The interactomics deals with the study of the interactome. In molecular biology, an interactome is the whole set of molecular interactions in a particular cell. The term specifically refers to physical interactions among molecules but can also describe sets of indirect interactions among genes. A study published in PLOS Genetics back 2011 looked at the interactome of 100 proteins of the rice through yeast2-hybrid assay. The results were based on phenotypic analyses, coexpression of transcripts, and protein–protein interaction assays. The results indicated 10 novel regulator molecules for abiotic stress tolerance, including two protein molecules not previously labeled as stress inducible proteins. Further evidence obtained through this experimentation indicated that a cross-talk exists between biomolecules under various biotic and abiotic stress responses (Seo et al., 2011).

8.8 Conclusion and future prospects

Omics approaches are overlapping and interdependent. The integration of all data obtained from omics approaches is vital to conclude and comprehend major cell response cascades, which may differ among tolerant and sensitive rice varieties during abiotic stress. Genomic studies can give scientists a general view of the most dominant responsive genes in tolerant varieties. Transcriptomics illustrates complex expression networks of RNA, which have an important role in the functioning of tolerant varieties under stress conditions. Proteomics is a key approach for discovering the major proteins participating in the defense systems of rice plants. Metabolomics exhibits crucial secondary metabolites of tolerant varieties for combating abiotic stress. Phenomics studies help breeders to link phenotypic traits and genomic data. Bioinformatics tools integrate all the data obtained from the "omics" sciences and lead to the accumulation of comprehensive knowledge about tolerance mechanisms toward abiotic stresses in rice.

A lot of focus has been given to the genomics, transcriptomics, and phenomics in rice, but despite the significance of other techniques such as metabolomics, ionomics, proteomics, and ionomics, these are relatively neglected disciplines in rice crops. For the development of durable stress-resistant varieties, which perform better under changing climate conditions, the breeding programs should be facilitated with a multiomics network. The multiomics approach would help in the identification of stress responsive genes, and their stacking in one variety for tolerance to multiple stresses. There is a need to develop multiomics networks and their integration in conventional breeding programs for the development of climate resilient rice varieties.

References

Ahmad, Z., Sadique, S., Sarwar, M. B., Rashid, B., Hassan, S., Rasheed, S., ... Husnain, T. (2020). Transcriptomic analysis of root specific drought mediated response of *G. arboreum* and *G. hirsutum*. *Biologia*, *75*, 627–636.

Akpınar, B. A., Lucas, S. J., & Budak, H. (2013). Genomics approaches for crop improvement against abiotic stress. *The Scientific World Journal, 2013.*

Al-Yassin, A., & Khademian, R. (2015). Allelic variation of salinity tolerance genes in barley ecotypes (natural populations) using EcoTILLING: A review article. *American–Eurasian Journal of Agricultural and Environmental Sciences, 15,* 563–572.

Arbona, V., Manzi, M., Ollas, C. D., & Gómez-Cadenas, A. (2013). Metabolomics as a tool to investigate abiotic stress tolerance in plants. *International Journal of Molecular Sciences, 14,* 4885–4911.

Barik, S. R., Pandit, E., Mohanty, S. P., Nayak, D. K., & Pradhan, S. K. (2020). Genetic mapping of physiological traits associated with terminal stage drought tolerance in rice. *BMC Genetics, 21,* 1–12.

Barik, S. R., Pandit, E., Pradhan, S. K., Mohanty, S. P., & Mohapatra, T. (2019). Genetic mapping of morpho-physiological traits involved during reproductive stage drought tolerance in rice. *PLoS ONE, 14,* e0214979.

Bartels, D., & Sunkar, R. (2005). Drought and salt tolerance in plants. *Critical Reviews in Plant Sciences, 24,* 23–58.

Bp, M. S., Ahmed, H. U., Henry, A., Mauleon, R., Dixit, S., Vikram, P., ... Mandal, N. P. (2013). Genetic, physiological, and gene expression analyses reveal that multiple QTL enhance yield of rice mega-variety IR64 under drought. *PLoS ONE, 8,* e62795.

Cakir, C., Gillespie, M. E., & Scofield, S. R. (2010). Rapid determination of gene function by virus-induced gene silencing in wheat and barley. *Crop Science, 50,* S-77–S-84.

Caldana, C., Degenkolbe, T., Cuadros-Inostroza, A., Klie, S., Sulpice, R., Leisse, A., ... Hannah, M. A. (2011). High-density kinetic analysis of the metabolomic and transcriptomic response of Arabidopsis to eight environmental conditions. *The Plant Journal, 67,* 869–884.

Caldwell, K. S., Russell, J., Langridge, P., & Powell, W. (2006). Extreme population-dependent linkage disequilibrium detected in an inbreeding plant species, Hordeum vulgare. *Genetics, 172,* 557–567.

Chin, J. H., Gamuyao, R., Dalid, C., Bustamam, M., Prasetiyono, J., Moeljopawiro, S., ... Heuer, S. (2011). Developing rice with high yield under phosphorus deficiency: Pup1 sequence to application. *Plant Physiology, 156,* 1202–1216.

Comai, L., Young, K., Till, B. J., Reynolds, S. H., Greene, E. A., Codomo, C. A., ... Odden, A. R. (2004). Efficient discovery of DNA polymorphisms in natural populations by Ecotilling. *The Plant Journal, 37,* 778–786.

Cooper, J. L., Henikoff, S., Comai, L., & Till, B. J. (2013). *TILLING and ecotilling for rice. Rice Protocols* (pp. 39–56). Springer.

Cramer, G. R., Van Sluyter, S. C., Hopper, D. W., Pascovici, D., Keighley, T., & Haynes, P. A. (2013). Proteomic analysis indicates massive changes in metabolism prior to the inhibition of growth and photosynthesis of grapevine (Vitis viniferaL.) in response to water deficit. *BMC Plant Biology, 13,* 49.

Cseri, A., Cserháti, M., Von Korff, M., Nagy, B., Horváth, G. V., Palágyi, A., ... Törjék, O. (2011). Allele mining and haplotype discovery in barley candidate genes for drought tolerance. *Euphytica, 181,* 341.

Devnarain, N., Crampton, B. G., Olivier, N., Van Der Westhuyzen, C., Becker, J. V., & O'kennedy, M. M. (2019). Transcriptomic analysis of a Sorghum bicolor landrace identifies a role for beta-alanine betaine biosynthesis in drought tolerance. *South African Journal of Botany, 127,* 244–255.

Du, F., Liu, P., Wang, K., Yang, Z., & Wang, L. (2020). Ionomic responses of rice plants to the stresses of different arsenic species in hydroponics. *Chemosphere, 243,* 125398.

Fahimirad, S., & Ghorbanpour, M. (2019). *Omics approaches in developing abiotic stress tolerance in rice (Oryza sativa L.). Advances in Rice Research for Abiotic Stress Tolerance* (pp. 767–779). Elsevier.

Fan, Y., Shabala, S., Ma, Y., Xu, R., & Zhou, M. (2015). Using QTL mapping to investigate the relationships between abiotic stress tolerance (drought and salinity) and agronomic and physiological traits. *BMC Genomics*, *16*, 1–11.

Feng, X., Han, L., Chao, D., Liu, Y., Zhang, Y., Wang, R., … Ding, Y. (2017). Ionomic and transcriptomic analysis provides new insight into the distribution and transport of cadmium and arsenic in rice. *Journal of Hazardous Materials*, *331*, 246–256.

Fiehn, O. (2001). Combining genomics, metabolome analysis, and biochemical modelling to understand metabolic networks. *Comparative and Functional Genomics*, *2*, 155–168.

Gadal, N., Shrestha, J., Poudel, M. N., & Pokharel, B. (2019). A review on production status and growing environments of rice in Nepal and in the world. *Archives of Agriculture and Environmental Science*, *4*, 83–87.

Gandhi, D. (2007). "UAS scientist develops first drought tolerant rice." The Hindu.

Ghorbani, R., Alemzadeh, A., & Razi, H. (2019). Microarray analysis of transcriptional responses to salt and drought stress in *Arabidopsis thaliana*. *Heliyon*, *5*, e02614.

Ghosh, D., & Xu, J. (2014). Abiotic stress responses in plant roots: A proteomics perspective. *Frontiers in Plant Science*, *5*, 6.

Gonzaga, Z. J. C., Carandang, J., Sanchez, D. L., Mackill, D. J., & Septiningsih, E. M. (2016). Mapping additional QTLs from FR13A to increase submergence tolerance in rice beyond SUB1. *Euphytica*, *209*, 627–636.

Grubb, C. D., & Abel, S. (2006). Glucosinolate metabolism and its control. *Trends in Plant Science*, *11*, 89–100.

Gupta, P., Kumar, J., Mir, R., & Kumar, A. (2010). 4 Marker-assisted selection as a component of conventional plant breeding. *Plant Breeding Reviews*, *33*, 145.

Gupta, P. K., Rustgi, S., & Kulwal, P. L. (2005). Linkage disequilibrium and association studies in higher plants: present status and future prospects. *Plant Molecular Biology*, *57*, 461–485.

Harris, K., Subudhi, P., Borrell, A., Jordan, D., Rosenow, D., Nguyen, H., … Mullet, J. (2007). Sorghum stay-green QTL individually reduce post-flowering drought-induced leaf senescence. *Journal of Experimental Botany*, *58*, 327–338.

Hashimoto, M., & Komatsu, S. (2007). Proteomic analysis of rice seedlings during cold stress. *Proteomics*, *7*, 1293–1302.

Hirayama, T., & Shinozaki, K. (2010). Research on plant abiotic stress responses in the post-genome era: Past, present and future. *The Plant Journal*, *61*, 1041–1052.

Islam, M., Hassan, L., Salam, M., Collard, B., Singh, R., & Gregorio, G. (2011). QTL mapping for salinity tolerance at seedling stage in rice. *Emirates Journal of Food and Agriculture*, 137–146.

Ji, Z.-J., Zeng, Y.-X., Zeng, D.-L., Li, X.-M., Liu, B.-X., & Yang, C.-D. (2010). Identification of QTLs for rice cold tolerance at plumule and 3-leaf-seedling stages by using QTLNetwork software. *Rice Science*, *17*, 282–287.

Kam, H., Fofana, M., Zongo, A., & Taiwo, S. (2018). Grain yield and related traits under drought and irrigated conditions. *International Journal of Current Advanced Research*, *7*, 9059–9066.

Kandpal, R. P., Saviola, B., & Felton, J. (2009). The era of omics unlimited. *BioTechniques*, *46*, 351–355.

Khan, A., Abidi, I., Bhat, M., Dar, Z., Ali, G., Shikari, A. B., & Khan, M. (2018). Tilling and eco-tilling–A reverse genetic approach for crop improvement. *Int. J. Curr. Microbiol. App. Sci*, *7*, 15–21.

Kim, C.-K., Oh, J.-H., Na, J.-K., Cho, C., Kim, K.-H., Yu, G. E., & Kim, D.-Y. (2019). The genes associated with drought tolerance by multi-layer approach in potato. *Plant Breeding and Biotechnology*, *7*, 405–414.

Kim, S., Plagnol, V., Hu, T. T., Toomajian, C., Clark, R. M., Ossowski, S., ... Nordborg, M. (2007). Recombination and linkage disequilibrium in Arabidopsis thaliana. *Nature Genetics*, *39*, 1151–1155.

Kimotho, R. N., Baillo, E. H., & Zhang, Z. (2019). Transcription factors involved in abiotic stress responses in Maize (Zea mays L.) and their roles in enhanced productivity in the post genomics era. *PeerJ*, *7*, e7211.

Kumar, G. R., Sakthivel, K., Sundaram, R. M., Neeraja, C. N., Balachandran, S. M., Rani, N. S., ... Madhav, M. S. (2010). Allele mining in crops: prospects and potentials. *Biotechnology advances*, *28*, 451–461.

Kumar, M., Mahato, A., Kumar, S., & Mishra, V. K. (2020). *Phenomics-assisted breeding: An emerging way for stress management. New Frontiers in Stress Management for Durable Agriculture* (pp. 295–310). Springer.

Kumar, V., Singh, A., Mithra, S. A., Krishnamurthy, S., Parida, S. K., Jain, S., ... Sharma, S. (2015). Genome-wide association mapping of salinity tolerance in rice (Oryza sativa). *DNA Research*, *22*, 133–145.

Lafarge, T., Bueno, C., Frouin, J., Jacquin, L., Courtois, B., & Ahmadi, N. (2017). Genome-wide association analysis for heat tolerance at flowering detected a large set of genes involved in adaptation to thermal and other stresses. *PLoS ONE*, *12*, e0171254.

Latha, R., Rubia, L., Bennett, J., & Swaminathan, M. (2004). Allele mining for stress tolerance genes in Oryza species and related germplasm. *Molecular Biotechnology*, *27*, 101–108.

Lee, B.-H., Henderson, D. A., & Zhu, J.-K. (2005). The Arabidopsis cold-responsive transcriptome and its regulation by ICE1. *The Plant Cell*, *17*, 3155–3175.

Li, W., Zhao, F. A., Fang, W., Xie, D., Hou, J., Yang, X., ... Lv, S. (2015). Identification of early salt stress responsive proteins in seedling roots of upland cotton (Gossypium hirsutum L.) employing iTRAQ-based proteomic technique. *Frontiers in Plant Science*, *6*, 732.

Lin, C.-J., Li, C.-Y., Lin, S.-K., Yang, F.-H., Huang, J.-J., Liu, Y.-H., & Lur, H.-S. (2010). Influence of high temperature during grain filling on the accumulation of storage proteins and grain quality in rice (Oryza sativa L.). *Journal of Agricultural and Food Chemistry*, *58*, 10545–10552.

Liu, F., Xu, W., Song, Q., Tan, L., Liu, J., Zhu, Z., ... Sun, C. (2013). Microarray-assisted fine-mapping of quantitative trait loci for cold tolerance in rice. *Molecular Plant*, *6*, 757–767.

Liu, X., Zhang, X., Sun, B., Hao, L., Liu, C., Zhang, D., ... Shi, Y. (2019). Genome-wide identification and comparative analysis of drought-related microRNAs in two maize inbred lines with contrasting drought tolerance by deep sequencing. *PLoS ONE*, *14*, e0219176.

Luu Tn, H., Luu, M. C., Abdelbagi, M. I., & Le H, H. (2012). Introgression the salinity tolerance QTLs Saltol into AS996, the elite rice variety of Vietnam. *American Journal of Plant Sciences*, 2012.

Ma, Y., Qin, F., & Tran, L.-S. P. (2012). Contribution of genomics to gene discovery in plant abiotic stress responses. *Molecular Plant*, *5*, 1176–1178.

Maccaferri, M., Sanguineti, M. C., Noli, E., & Tuberosa, R. (2005). Population structure and long-range linkage disequilibrium in a durum wheat elite collection. *Molecular Breeding*, *15*, 271–290.

Mackill, D., Collard, B., Neeraja, C., Rodriguez, R., Heuer, S., & Ismail, A. (2007). *QTLs in rice breeding: Examples for abiotic stresses. Rice Genetics V* (pp. 155–167). World Scientific.

Manolio, T. A. (2010). Genomewide association studies and assessment of the risk of disease. *New England journal of medicine*, *363*, 166–176.

Maruyama, K., Sakuma, Y., Kasuga, M., Ito, Y., Seki, M., Goda, H., … Yamaguchi-Shinozaki, K. (2004). Identification of cold-inducible downstream genes of the Arabidopsis DREB1A/CBF3 transcriptional factor using two microarray systems. *The Plant Journal*, *38*, 982–993.

Mekonnen, T., Haileselassie, T., & Tesfaye, K. (2017). Identification, mapping and pyramiding of genes/quantitative trait loci (qtls) for durable resistance of crops to biotic stresses. *Journal of Plant Pathology and Microbiology*, 8.

Meng, L., Wang, B., Zhao, X., Ponce, K., Qian, Q., & Ye, G. (2017). Association mapping of ferrous, zinc, and aluminum tolerance at the seedling stage in Indica Rice using MAGIC populations. *Frontiers in Plant Science*, *8*, 1822.

Mochida, K., Yoshida, T., Sakurai, T., Yamaguchi-Shinozaki, K., Shinozaki, K., & Tran, L.-S. P. (2011). In silico analysis of transcription factor repertoires and prediction of stress-responsive transcription factors from six major gramineae plants. *DNA Research*, *18*, 321–332.

Munns, R., & Tester, M. (2008). Mechanisms of salinity tolerance. *Annual review of plant biology*, *59*, 651–681.

Neeraja, C. N., Maghirang-Rodriguez, R., Pamplona, A., Heuer, S., Collard, B. C., Septiningsih, E. M., … Ismail, A. M. (2007). A marker-assisted backcross approach for developing submergence-tolerant rice cultivars. *Theoretical and Applied Genetics*, *115*, 767–776.

Negrao, S., Almadanim, C., Pires, I., Mcnally, K., & Oliveira, M. (2011). Use of EcoTILLING to identify natural allelic variants of rice candidate genes involved in salinity tolerance. *Plant Genetic Resources*, *9*, 300.

Ngara, R., & Ndimba, B. K. (2014). Understanding the complex nature of salinity and drought-stress response in cereals using proteomics technologies. *Proteomics*, *14*, 611–621.

Nordborg, M., Hu, T. T., Ishino, Y., Jhaveri, J., Toomajian, C., Zheng, H., … Goyal, R. (2005). The pattern of polymorphism in Arabidopsis thaliana. *PLoS biology*, *3*, e196.

Nouri, M.-Z., Toorchi, M., & Komatsu, S. (2011). Proteomics approach for identifying abiotic stress responsive proteins in soybean. *Soybean-molecular aspects of breeding*, 187–214.

Obata, T., & Fernie, A. R. (2012). The use of metabolomics to dissect plant responses to abiotic stresses. *Cellular and Molecular Life Sciences*, *69*, 3225–3243.

Pan, X., Li, Y., Liu, W., Liu, S., Min, J., Xiong, H., … Li, X. (2020). QTL mapping and candidate gene analysis of cadmium accumulation in polished rice by genome-wide association study. *Scientific Reports*, *10*, 1–11.

Pandit, E., Tasleem, S., Barik, S. R., Mohanty, D. P., Nayak, D. K., Mohanty, S. P., … Pradhan, S. K. (2017). Genome-wide association mapping reveals multiple QTLs governing tolerance response for seedling stage chilling stress in indica rice. *Frontiers in Plant Science*, *8*, 552.

Platten, J., Egdane, J., and Ismail, A., 2013. Salinity tolerance, Na + exclusion and allele mining of HKT1; 5. Oryza sativa.

Pradhan, A. K., Shandilya, Z. M., Lahkar, L., Hasnu, S., Kalita, J., Borgohain, D., & Tanti, B. (2019). *Comparative metabolomics approach towards understanding chemical variation in rice under abiotic stress. Advances in rice research for abiotic stress tolerance* (pp. 537–550). Elsevier.

Pramesh, D., Prasannakumar, M. K., Muniraju, K. M., Mahesh, H., Pushpa, H., Manjunatha, C., … Kumara, M. K. (2020). Comparative genomics of rice false smut fungi Ustilaginoidea virens Uv-Gvt strain from India reveals genetic diversity and phylogenetic divergence. *3 Biotech*, *10*, 1–14.

Premkumar, A., Kumar, M., & Chauhan, R. (2020). Improvement of rice variety for biotic and abiotic stress resistance using the technique of allele mining. *Journal of Pharmacognosy and Phytochemistry*, *9*, 1888–1893.

Salgotra, R., Gupta, B., & Stewart, C. (2014). From genomics to functional markers in the era of next-generation sequencing. *Biotechnology Letters*, *36*, 417–426.

Salgotra, R. K., & Stewart, C. N. (2020). Functional markers for precision plant breeding. *International Journal of Molecular Sciences*, *21*, 4792.

Sanchez, D. H., Pieckenstain, F. L., Szymanski, J., Erban, A., Bromke, M., Hannah, M. A., … Udvardi, M. K. (2011). Comparative functional genomics of salt stress in related model and cultivated plants identifies and overcomes limitations to translational genomics. *PLoS ONE*, *6*, e17094.

Seki, M., Narusaka, M., Abe, H., Kasuga, M., Yamaguchi-Shinozaki, K., Carninci, P., … Shinozaki, K. (2001). Monitoring the expression pattern of 1300 Arabidopsis genes under drought and cold stresses by using a full-length cDNA microarray. *The Plant Cell*, *13*, 61–72.

Seki, M., Okamoto, M., Matsui, A., Kim, J.-M., Kurihara, Y., Ishida, J., … Shinozaki, K. (2010). *Microarray analysis for studying the abiotic stress responses in plants. Molecular techniques in crop improvement* (pp. 333–355). Springer.

Semagn, K., Bjørnstad, Å., & Xu, Y. (2010). The genetic dissection of quantitative traits in crops. *Electronic Journal of Biotechnology*, *13*, 16–17.

Seo, Y.-S., Chern, M., Bartley, L. E., Han, M., Jung, K.-H., Lee, I., … Cao, P. (2011). Towards establishment of a rice stress response interactome. *PLoS Genetics*, *7*, e1002020.

Shanmugavadivel, P., Sv, A. M., Prakash, C., Ramkumar, M., Tiwari, R., Mohapatra, T., & Singh, N. K. (2017). High resolution mapping of QTLs for heat tolerance in rice using a 5K SNP array. *Rice*, *10*, 28.

Sharma, J. K., Sihmar, M., Santal, A. R., & Singh, N. (2019). Impact assessment of major abiotic stresses on the proteome profiling of some important crop plants: A current update. *Biotechnology and Genetic Engineering Reviews*, *35*, 126–160.

Simpson, S. D., Nakashima, K., Narusaka, Y., Seki, M., Shinozaki, K., & Yamaguchi-Shinozaki, K. (2003). Two different novel cis-acting elements of erd1, a clpA homologous Arabidopsis gene function in induction by dehydration stress and dark-induced senescence. *The Plant Journal*, *33*, 259–270.

Singh, B., Mishra, S., Bohra, A., Joshi, R., & Siddique, K. H. (2018). *Crop phenomics for abiotic stress tolerance in crop plants. Biochemical, physiological and molecular avenues for combating abiotic stress tolerance in plants* (pp. 277–296). Elsevier.

Singh, R., & Jwa, N.-S. (2013). Understanding the responses of rice to environmental stress using proteomics. *Journal of proteome research*, *12*, 4652–4669.

Sirohi, P., Yadav, B. S., Afzal, S., Mani, A., & Singh, N. K. (2020). Identification of drought stress-responsive genes in rice (*Oryza sativa*) by meta-analysis of microarray data. *Journal of Genetics*, *99*, 1–10.

Soltani, A., Kumar, A., Mergoum, M., Pirseyedi, S. M., Hegstad, J. B., Mazaheri, M., & Kianian, S. F. (2016). Novel nuclear-cytoplasmic interaction in wheat (Triticum aestivum) induces vigorous plants. *Functional & Integrative Genomics*, *16*, 171–182.

Song, J., Li, J., Sun, J., Hu, T., Wu, A., Liu, S., … Zhao, M. (2018). Genome-wide association mapping for cold tolerance in a core collection of rice (*Oryza sativa* L.) landraces by using high-density single nucleotide polymorphism markers from specific-locus amplified fragment sequencing. *Frontiers in Plant Science*, *9*, 875.

Steele, K., Price, A. H., Shashidhar, H., & Witcombe, J. (2006). Marker-assisted selection to introgress rice QTLs controlling root traits into an Indian upland rice variety. *Theoretical and Applied Genetics*, *112*, 208–221.

Suji, K., Prince, K. S. J., Mankhar, P. S., Kanagaraj, P., Poornima, R., Amutha, K., … Babu, R. C. (2012). Evaluation of rice (*Oryza sativa* L.) near iso-genic lines with root QTLs for plant production and root traits in rainfed target populations of environment. *Field Crops Research, 137*, 89–96.

Sun, J., Xie, D., Zhang, E., Zheng, H., Wang, J., Liu, H., ... Zou, D. (2019). QTL mapping of photosynthetic-related traits in rice under salt and alkali stresses. *Euphytica, 215*, 147.

Tao, Y., Zhao, X., Mace, E., Henry, R., & Jordan, D. (2019). Exploring and exploiting pan-genomics for crop improvement. *Molecular Plant, 12*, 156–169.

Taunk, J., Rani, A., Singh, R., Yadav, N. R., & Yadav, R. C. (2019). *Genomic strategies for improving abiotic stress tolerance in crop plants. Genetic Enhancement of Crops for Tolerance to Abiotic Stress: Mechanisms and Approaches* (pp. 205–230). Springer, *Vol. I.*

Tran, L.-S. P., & Mochida, K. (2010). A platform for functional prediction and comparative analyses of transcription factors of legumes and beyond. *Plant Signaling & Behavior, 5*, 550–552.

Umezawa, T., Fujita, M., Fujita, Y., Yamaguchi-Shinozaki, K., & Shinozaki, K. (2006). Engineering drought tolerance in plants: discovering and tailoring genes to unlock the future. *Current Opinion in Biotechnology, 17*, 113–122.

Valliyodan, B., & Nguyen, H. T. (2006). Understanding regulatory networks and engineering for enhanced drought tolerance in plants. *Current Opinion in Plant Biology, 9*, 189–195.

Van Oort, P. (2018). Mapping abiotic stresses for rice in Africa: Drought, cold, iron toxicity, salinity and sodicity. *Field crops research, 219*, 55–75.

Venuprasad, R., Dalid, C., Del Valle, M., Zhao, D., Espiritu, M., Cruz, M. S., ... Atlin, G. (2009). Identification and characterization of large-effect quantitative trait loci for grain yield under lowland drought stress in rice using bulk-segregant analysis. *Theoretical and Applied Genetics, 120*, 177–190.

Vromman, D., Lutts, S., Lefèvre, I., Somer, L., De Vreese, O., Šlejkovec, Z., & Quinet, M. (2013). Effects of simultaneous arsenic and iron toxicities on rice (Oryza sativa L.) development, yield-related parameters and As and Fe accumulation in relation to As speciation in the grains. *Plant and Soil, 371*, 199–217.

Vu, H. T. T., Le, D. D., Ismail, A. M., & Le, H. H. (2012). Marker-assisted backcrossing (MABC) for improved salinity tolerance in rice (*Oryza sativa* L.) to cope with climate change in Vietnam. *Australian journal of crop science, 6*, 1649.

Wei, K., & Chen, H. (2018). Comparative functional genomics analysis of bHLH gene family in rice, maize and wheat. *BMC Plant Biology, 18*, 309.

Xiong, L., Schumaker, K. S., & Zhu, J.-K. (2002). Cell signaling during cold, drought, and salt stress. *The Plant Cell, 14*, S165–S183.

Xu, K., Xu, X., Fukao, T., Canlas, P., Maghirang-Rodriguez, R., Heuer, S., ... Mackill, D. J. (2006). Sub1A is an ethylene-response-factor-like gene that confers submergence tolerance to rice. *Nature, 442*, 705–708.

Yadav, S., Sandhu, N., Majumder, R. R., Dixit, S., Kumar, S., Singh, S., ... Singh, V. K. (2019). Epistatic interactions of major effect drought QTLs with genetic background loci determine grain yield of rice under drought stress. *Scientific Reports, 9*, 1–13.

Yu, S., Chen, W., Wang, D., He, X., Zhu, X., & Shi, X. (2010). Species-specific PCR detection of the food-borne pathogen Vibrio parahaemolyticus using the irgB gene identified by comparative genomic analysis. *FEMS Microbiology Letters, 307*, 65–71.

Yu, S., Liao, F., Wang, F., Wen, W., Li, J., Mei, H., & Luo, L. (2012). Identification of rice transcription factors associated with drought tolerance using the ecotilling method. *PLoS ONE, 7*, e30765.

Zhang, J., Chen, K., Pang, Y., Naveed, S. A., Zhao, X., Wang, X., ... Li, Z. (2017). QTL mapping and candidate gene analysis of ferrous iron and zinc toxicity tolerance at seedling stage in rice by genome-wide association study. *BMC Genomics, 18*, 828.

Zhang, P., Zhong, K., Tong, H., Shahid, M. Q., & Li, J. (2016). Association mapping for aluminum tolerance in a core collection of rice landraces. *Frontiers in Plant Science*, *7*, 1415.

Zhang, P., Zhong, K., Zhong, Z., & Tong, H. (2019). Mining candidate gene for rice aluminum tolerance through genome wide association study and transcriptomic analysis. *BMC Plant Biology*, *19*, 490.

Zhang, S., Zheng, J., Liu, B., Peng, S., Leung, H., Zhao, J., ... Huang, Z. (2014). Identification of QTLs for cold tolerance at seedling stage in rice (Oryza sativa L.) using two distinct methods of cold treatment. *Euphytica*, *195*, 95–104.

Zou, J., Liu, C., & Chen, X. (2011). Proteomics of rice in response to heat stress and advances in genetic engineering for heat tolerance in rice. *Plant cell Reports*, *30*, 2155–2165.

Chapter 9

Interactions of phytohormones with abiotic stress factors under changing climate

Abbu Zaid[1], Muntazir Mushtaq[2] and Shabir Hussain Wani[3]

[1]Plant Physiology and Biochemistry Section, Department of Botany, Aligarh Muslim University, Aligarh, India, [2]School of Biotechnology, Sher-e-Kashmir University of Agricultural Sciences and Technology of Jammu, Jammu, India, [3]Mountain Research Centre for Field Crops, Khudwani Sher-e-Kashmir University of Agricultural Sciences and Technology of Kashmir, Srinagar, India

9.1 Introduction

Plants have an internal signaling network scheme by which tight regulation of the growth and developmental processes is maintained. Plants respond to various internal and external stimuli by initiating marked changes in a signaling system which directly or indirectly regulates the mechanisms. In the present era of changing climate and excess anthropogenic emissions in the environment, plants acclimatize themselves by arbitrating changes in endogenous phytohormone biosynthesis, source/sink transitions, modulation in antioxidant defense system, and nutrient and photosynthetic substrate allocations (Ahmad, Zaid, Sadiq, Bashir, & Wani, 2019; Sadiq, Zaid, & Khan, 2020; Wani, Kumar, Shriram, & Sah, 2016). Under these circumstances, plants are challenged by various abiotic and biotic stress factors. The abiotic stress conditions like salt, drought, extreme temperatures, toxic metals/metalloids (metal/s), flooding/waterlogging (WL), ultraviolet (UV) radiations, etc. are now more prevalent as a result of changing climate (Raza et al., 2019). The multitude of such abiotic stresses has posed a significant threat to sustainable productivity of crops worldwide. In plants a common induced side-effects of abiotic stress is the oxidative stress through the overproduction/enhanced biosynthesis of various reactive oxygen species (ROS) viz-superoxide radical ($O_2^{\cdot -}$), hydroxyl radical ($OH^{\cdot}$), hydroperoxyl radical ($HO_2^{\cdot}$), alkoxy radical ($RO^{\cdot}$), peroxy radical ($ROO^{\cdot}$), excited carbonyl (RO^*), hydrogen peroxide (H_2O_2), and singlet oxygen ($^1O_2$) species (Hasanuzzaman et al., 2020; Zaid & Mohammad, 2018;

Frontiers in Plant–Soil Interaction. DOI: https://doi.org/10.1016/B978-0-323-90943-3.00010-9

Zaid & Wani, 2019). Although the formation/biosynthesis of various ROS in plants is a normal cellular process, however, their overaccumulation as the result of stress conditions has been reported to severely damage vital cellular structures including carbohydrates, proteins, lipids, DNA, RNA, etc. by causing their oxidation (Gill & Tuteja, 2010; Mittler, 2017; Raja, Majeed, Kang, Andrabi, & John, 2017). As a result of oxidative stress normal metabolism is disturbed and crop productivity is decreased. However, under stress conditions, plants are equipped with several underlying strategies to cope with that. Nonetheless, phytohormone biosynthesis plays a crucial role in the stress adaptation mechanisms.

The phytohormones are considered as the most important endogenous signaling elicitors which control physiological and molecular responses under optimal and stressful conditions, which are critical requirements for the survival of plants as sessile creatures. The term "phytohormone" means a naturally transported organic signaling molecule found in plants which at low concentration regulates various vital physiological processes. Phytohormones are synthesized in one plant part and are transported to the required site (Went & Thimann, 1937).There are different kinds of phytohormones present in plants. Auxins, gibberellins (gibberellic acid; GA), cytokinins (CKs), abscisic acid (ABA), ethylene (ET), jasmonic acid (JA), nitric oxide (NO; a gaseous molecule), brassinosteroids (BRs; phytohormones of steroidal nature), polyamines (PAs; aliphatic having nitrogen structure), salicylic acid (SA), triacontanol (Tria), and strigolactones (SLs) are important classes of phytohormones which exert extensive significant roles in diverse plant developmental processes under optimal and stress conditions. The convergent sites among signal transduction pathways of phytohormones are considered as cross-talk, which constitute signaling networks in plants. The abiotic stress perception in plants initiates the activation of the machinery of signaling cross-talk of amplification and transduction pathways with phytohormones. Via these signaling mechanisms, phytohormones interact with different abiotic stress factors by which a common second messenger or a phosphorylation cascade is activated. Therefore studying the interaction of various phytohormones with a myriad of abiotic stress factors can increase our knowledge by gaining an insight into the underlying mechanisms of abiotic stress tolerance in plants. Therefore this chapter reviews the interaction/cross-talk of important phytohormones (SA, NO, and BRs) with various signaling agents during abiotic stress conditions. The collected literature could be used to construct a model of abiotic stress signaling for the potential to develop abiotic stress tolerance in crop plants.

9.2 Interaction of salicylic acid with other signaling agents under abiotic stress

Phytohormones exert a significant role in regulating plants' growth, development, and metabolism to an appreciable extent, as well as protect them against various environmental pressures. The phytohormone "salicylic acid" (SA) is an aromatic signaling molecule whose name is derived from the Latin word *Salix* (willow tree). SA in plants is found mostly in powdered crystalline form. The melting temperature of SA ranges from 157°C to 159°C and has a pH of 2.4. A German scientist namely Johann Andreas Buchner in 1828 was able to isolate a small quantity of "salicin." After then the salicin was found to be present in various other plants in addition to the willow plant (Raskin, 1992). Chemically, SA is composed of hydroxyl and carboxylic groups attached to the benzene ring. The functional groups can react with various compounds and thus the SA can either be glycosylated or methylated (Rivas-San Vicente & Plasencia, 2011). In crop plants the effective concentration of SA depends mainly on their developmental stages and shows a variation as a result of various environmental stresses. The presence of SA has been detected in diverse crop plants (reviewed by Arif, Sami, Siddiqui, Bajguz, & Hayat, 2020). SA is regarded as a multifunctional phytohormone which participates in eliciting various plant responses to salinity, metal/s, drought, cold, heat, herbicide, and UV stresses. SA modulates plant physiology via increasing enzymatic and nonenzymatic antioxidant machinery, ROS scavenging enzymes and shows a functional interplay with other phytohormones under normal and stressed conditions to maintain optimum plant growth and development. SA interacts with various signaling elicitors to control vital physiological processes in plants under abiotic stresses. In *Arabidopsis thaliana*, SA interacts with glutathione to control seed germination under salt stress (Borsani, Valpuesta, & Botella, 2001). In wheat, Alamri, Siddiqui, Al-Khaishani, and Ali (2018) applied SA which resulted in marked stimulation in germination of seeds. The ability of 25 μM SA seed priming on germination and seedling performance in *Trifolium* plants was evaluated under metal stress (Bortolin et al., 2020). SA improved the seedling performance by boosting the antioxidant enzyme activities as well as reducing the malondialdehyde (MDA) content in the tested plant. Tahjib-Ul-Arif et al. (2018) applied 1.0 mM SA to enhance the photosynthesis attributes of inhibition and antioxidant capacity for yield improvement in maize under 75 mM salinity stress. The results showed that exogenous SA ameliorated the salt-induced inhibitory responses and yield loss by stimulating photosynthetic efficiency, reducing oxidative damage, and boosting the activities of catalase (CAT) and ascorbate peroxidase (APX) enzymes. The interplay of SA with other potential phytohormones like BRs showed promising results in reducing stress-induced effects in plants. It has been found that the overexpression of BRs signaling receptor *BAK1* resulted in excess SA production (Kim, Shang, Joo, Kim, & Nam, 2017). Banerjee and Roychoudhury (2018) proposed that transcription factors (TFs) *NPR1* and *WRKY70* are common and important regulators of abiotic stress factors in SA and BRs signaling. Nonetheless, *NPR1*, which plays a significant role in SA-mediated defense mechanisms, is also a component of 24-epibrassinolide (24-EBL).

The cross-talk of *CAMTA* with SA has been reported to confer freezing stress tolerance in *Arabidopsis* (Kim, Park, Gilmour, & Thomashow, 2013). Moreover, SA also interacts with GA to improve plant defense against abiotic stress (Emamverdian, Ding, & Mokhberdoran, 2020). In a classical study representing the cross-talk of SA with GA, Alonso-Ramírez et al. (2009) found that the exogenous application of GA ameliorated the ill effects of salt, oxidative, and heat stresses in *A. thaliana*. They also observed that GA treatment increases the expression of genes related to SA biosynthesis. Furthermore, the SA impaired mutants are found to be more sensitive to salt stress than wild type even in the presence of GA. Under 50 μM Ni stress, foliar SA (10^{-5} M) supply upregulated the antioxidant defense and glyoxalase system to tackle Ni toxicity for enhanced growth and photosynthesis in Indian mustard plants (Zaid, Mohammad, Wani, & Siddique, 2019). In yet another recent study, Zaid, Mohammad, and Fariduddin (2020) observed that the application of SA (10^{-6} M) surpassed the effect of GA and Tria and consequently improved the growth, photosynthesis, mineral nutrients, and antioxidant system under Cd toxicity in *Mentha arvensis* L. plants. Mohammadi, Amirikia, Ghorbanpour, Fatehi, and Hashempour (2019) applied SA (0, 100, and 150 ppm) on different ecotypes of *Thymus vulgaris* under water stress. The results revealed that the enhanced production of secondary metabolites by SA foliar supply ameliorated the water stress effects in tested plants. Li, Huang, Liu, Li, and Tan (2020) found that the exogenous SA treatment alleviates herbicide toxicity by mediating coordination with the plant antioxidant defense system to improve photosynthesis in *Glycine max* Merr. plants. The dual application of SA and methyl jasmonate (MJ) via seed and foliar pretreatments mitigated drought-induced stress in maize plants by modulating the levels of osmolytes and endogenous ABA and antioxidant enzyme activities (Tayyab et al., 2020). The involvement of SA in light-dependent cold acclimation in maize has also been reported (Pál, Janda, Majláth, & Szalai, 2020). In drought-stressed *Silybum marianum* L. plants, foliar SA improved osmotic adjustment, water content, and yield (Estaji & Niknam, 2020). Yu et al. (2020) studied the enhancement effect of SA pretreatment in *A. thaliana* under salt stress. The SA application enhanced salt stress tolerance and the effect was significantly impaired by blocking the alternative oxidase (*AOX*) pathway. Thus the SA-mediated effect was attributed to the induction of *AOX* capacity and antioxidant defense system. Thus the above discussion clearly suggests that the interplay of SA with various signaling molecules controls stress resistance in diverse crop plants.

9.3 Interaction of nitric oxide with other signaling agents under abiotic stress

NO is a free radical gaseous growth regulator signaling molecule implicated in an array of plants' physiological functions under optimal and stressful environmental pressures. NO can induce either beneficial or detrimental effects depending on its concentration and site of production. The peculiarity of the reactive properties of NO is attributed to the presence of an unpaired electron within the π orbital of the N atom of the NO. During the last decade a large body of literature suggested NO as a key component of the signaling networks controlling a myriad of plant physiobiochemical processes and adaptation to various environmental pressures (Ahmad, Ahanger, Alyemeni, Wijaya, & Alam, 2018; Asgher et al., 2017; Dong, Jinc, Liu, Xu, & Kong, 2014; Fatma, Masood, Per, & Khan, 2016; Freschi, 2013; Hancock & Veal, 2020; Hasanuzzaman, Nahar, Alam, & Fujita, 2012; Hayat, Yadav, Wani, Irfan, & Ahmad, 2011; Rather, Mir, Masood, Anjum, & Khan, 2020; Santisree, Bhatnagar-Mathur, & Sharma, 2015; Sehar, Masood, & Khan, 2019; Shivaraj et al., 2020; Verma, Tiwari, Singh, & Prasad, 2020). In plants, NO also mediates S-nitrosylation and tyrosine nitration, which are the two important posttranslational modification processes (Nabi et al., 2019). Under NaCl (50 and 100 mM) stress, NO (50 μM) improves growth traits (shoot and root length) and biomass yield (shoot dry weight) in chickpea plants (Ahmad et al., 2016). In maize plants, it has been found that the lower dose of NO promoted root growth, whereas the higher NO concentration inhibits the growth (Gouvea, Souza, Magalhaes, & Martins, 1997). Pagnussat, Simontacchi, Puntarulo, and Lamattina (2002) unraveled the molecular events involved in adventitious root formation in cucumber. NO mediated cross-talk with the auxin was responsible for the development of adventitious root. Correa-Aragunde, Graziano, Chevalier, and Lamattina (2006) using RT-PCR technique showed that in tomato seeds NO-mediated modulated cross-talk of cell cycle regulatory genes, that is, *CYCA2;1*, *CYCA3;1*, *CYCD3;1*, *CDKA1*, and the *Kip* related protein *KRP2* induced lateral root formation. Stimulation of seed germination in *Paulonia tomentosa* was found to be associated with the involvement of NO in the phytochrome controlled manner (Giba et al., 1998). In *A. thaliana*, NO acts upstream of GA for dormancy loss (Bethke et al., 2007). Under elevated carbon dioxide concentration, Li, Li, Leng, Zeng, and Ma (2019) showed that NO acts downstream of SA to induced biosynthesis of flavonoids in *Camellia sinensis* L. Under low temperature stress, the pivotal cross-talk of NO and calcium ions (Ca^{2+}) in *Cucumis sativus* L. seedlings was explored (Zhang et al., 2020). The involvement of Ca^{2+} in exogenous NO-mediated improvement in photosynthesis under low temperature stress was further authenticated by using Ca^{2+} channel blocker and calmodulin antagonist that reversed the positive effect of NO. The interaction of NO with enzymes of sulfur assimilation in mustard was associated with the alleviation salt-induced inhibited growth and photosynthetic performance (Fatma et al., 2016). In order to study the potential mechanisms of NO cross-talk in conferring thermotolerance in *Vicia faba* L., Alamri et al. (2019) exposed the plants to three levels of heat stress regimes (32°C, 37°C, and 42°C), respectively, for 48 h. Heat stress increased the ROS, that is, H_2O_2 and $O_2^{\cdot -}$, contents of NO, proline, MDA, electrolyte leakage (EL), DNA damage and degradation of chlorophyll, but decreased water and chlorophyll content. The application of NO scavenger 2-(4-carboxyphenyl)-4,4,5,5-tetramethylimidazoline-1-oxyl-3-oxide (cPTIO) was found to decrease the content of water, proline, and NO. The increasing levels of heat stress also increased the activities of enzymes of proline metabolism (Δ^1-pyrroline-5-carboxylate synthetase, ornithine-δ-aminotransferase, and proline dehydrogenase) and also the expression of heats shock proteins (HSPs) like *Hsp17.6*, *Hsp70*, *Hsp90-1*, and *Hsp101*. The study concluded with the observation that NO showed a functional cross-talk with proline metabolism and various HSPs to promote heat stress tolerance in tested plants. In yet another experiment for deciphering the role of NO under nutrient stress-induced deficiency in tomato plants, Siddiqui et al. (2020) applied exogenous NO (100 μM: (Na_2 [Fe $(CN)_5NO].2H_2O$) in the presence or absence of cPTIO. The plants were studied to establish the role of NO in nutrient acquisition and assimilation. The nutrient (Sulfur; S) starved plants showed a concomitant decrease in growth, photosynthetic pigments, and activities of δ-aminolevulinic acid dehydratase, carbonic anhydrase (CA), and RuBisco, but an enhancement in ROS and MDA levels was observed. Application of NO to S-deficient plants reduced the contents of ROS as well as MDA but enhanced the contents of proline, NO, hydrogen sulfide (H_2S), S uptake, cysteine and reduced glutathione, ATP sulfurylase, adenosine 5-phosphosulfate reductase, sulfide reductase, and O-acetylserine (thiol) lyase and also the ascorbate content. These effects were invalidated by the application of cPTIO. Thus the cross-talk of NO with S- assimilating enzymes enhanced the tolerance of tomato seedlings to minimize the S-induced oxidative stress. On subjecting wheat plants to osmotic stress conditions, the tolerance tradeoff between NO and H_2S was studied (Khan, Mobin, Abbas, & Siddiqui, 2017). NO acts upstream of H_2S to alleviate the osmotic-induced oxidative stress by maintaining enhanced antioxidant defense enzyme activities, solute accumulation, and homeostasis of cysteine. Sun, Liu, Lu, Jin, and Lin (2018) observed that NO-mediated cross-talk of H_2O_2 enhances the metal (Al)-resistance in wheat plants. When mungbean plants were exposed to dual stress (cadmium + osmotic), Li et al. (2019) observed that NO either shows a common signaling network with auxin or acts as a downstream signaling molecule to auxin in response to dual stress. Su et al. (2018) showed the cross-talk link of hydrogen (H_2) and NO under osmotic stress tolerance in alfalfa seedlings. Plants fed with H_2 water showed an enhanced production of NO which alleviated PEG-stressed-induced alfalfa seedling growth inhibition. The role of NO was further authenticated by applying tungstate. While using NO scavenger, NO and syntheses of proline, redox balance, and osmotic stress tolerance mediated by the interplay of H_2 and NO were severely affected. Thus the current appraisal clearly suggests that NO mediates intimate cross-talk with several signaling molecules to confer tolerance against various abiotic stresses.

9.4 Interaction of brassinosteroids with other signaling agents under abiotic stress

Brassinosteroids (BRs) are a group of ubiquitous plant polyhydroxylated steroidal phytohormones that are implicated in regulating diverse and vital biological processes. 28-Homobrassinolide (28-HBR) and 24-epibrassinolide (24-EBL) are two isomers of BRs and both are known to induce abiotic stress resistance. Cross-talk between BRs signaling and abiotic stress factors have shown promising results in reducing the ill effects of various abiotic stress induced effects in crop plants. A range of literature has suggested that BRs control a broad spectrum of physiological and developmental processes under drought (Gill, Cai, Zhang, & Zeng, 2017), high-temperature (Li et al., 2018; Wu et al., 2014), cold (Fu et al., 2019; Hayat, Alyemeni, & Yadav, 2014; Singh et al., 2012), salt (Ahanger, Mir, Alyemeni, & Ahmad, 2020; De Oliveira, Lima, da Silva, Batista, & da Silva Lobato, 2019; Kaur et al., 2018; Kaya, Aydemir, Akram, & Ashraf, 2018; Kolomeichuk et al., 2020), autotoxicity (Yang, Azher Nawaz, Li, Bai, & Li, 2019), heavy metal (Da Silva Cunha et al., 2020; Jan et al., 2020; Kohli et al., 2018; Peng et al., 2020; Yadav et al., 2018), acid rain (Da Fonseca, da Silva, & da Silva Lobato, 2020), pesticide (Liu et al., 2019; Sharma, Bhardwaj, Kumar, & Thukral, 2016), as well as biotic stress (Zhang et al., 2018) in crop plants. The signal transduction and biosynthetic pathways were characterized in *Solanum lycopersicum*. The wild tomato (WT) and a BR synthesis mutant, Micro-Tom (MT), were used. The WT showed higher tolerance to drought and osmotic stresses and the converse case was found in MT tomato. However in MT various defects were completely recovered by the presence of BRs. The results suggested that the signaling pathways of BRs are chiefly involved in mediating abiotic stress responses via cross-talking with abiotic stress-related gene networks (Lee et al., 2018). Chen et al. (2019) elucidated the mechanisms of BRs-mediated drought tolerance in cotton plants, by employing a cotton BR-deficient mutant, pag1 (*pagoda1*), for analysis. It was found that the deficiency of BRs causes enhanced drought sensitivity and yield penalty in the tested plants. The mutant also showed a short development in primary roots and lateral roots. In tomato plants, 24-EBL application increased BRs but decreased the contents of ABA and ROS, which increased the stomatal aperture and drought resistance. In this case the BRs signaling receptor *BRASSINOSTEROID-INSENSITIVE 1* (*BRI1*) was found to be negatively related with the drought stress resistance (Nie et al., 2019). The effect of 0.05, 0.2, and 1.0 mg L^{-1} brassinolide (BL) application in mediating 200 mM salt stress in *Malus hupehensis* Rehd., was studied (Su, Zheng, Tian, & Wang, 2020). The salt stress decreased significantly the chlorophyll content and photosynthetic rate. The supply of 0.2 mg L^{-1} BL alleviated the ill effects of salt stress and maintained photosynthesis. The BL application boosted SOD, CAT, proline, and sugar content, thereby eliminating excess salt-induced ROS production. Nevertheless, BL-mediated decrease in the accumulation of Na^+ but increase in K^+ content was associated with regulating the expression of $Na^+(K^+)/H^+$ antiporter genes (*MhNHXs*). The key BRs signal transduction transcription factors, *MhBZR1* and *MhBZR2*, bind directly to the promoter region of *MhSOS1* and *MhNHX4-1* in inhibiting the expression of salt responses. It is known that BRs promote seed germination by inactivating ABA signaling and activating GA downstream genes. However, for elucidating the role of BR signaling pathways in the cold stratification for seed dormancy and germination, Kim, Warpeha, and Huber (2019) studied the role of BR receptor kinase (*BRI1*). The BR insensitive mutant, *bri1–5* showed delayed seed germination, however, this effect was reversed by inhibiting the ABA biosynthesis with fluridone. The mutants were less sensitive to the cold stratification release of dormancy. The cold stratification also upregulated the BR biosynthetic *DWF4* and *DET2*, GA biosynthesis, *GA3ox1*, and cold responsive genes *CBF1* and *CBF2*. In the case of *bri1*–5 seeds, this effect was ineffective. The results suggested that BR acts upstream of ABA signaling and downstream of cold stratification signaling during the processes of dormancy and germination. Fu et al. (2019) studied the BRs-mediated enhancement under cold stress tolerance in *Elymus nutans* vis-à-vis probing the role of proline. Plants were exposed to 4°C in a growth chamber. Cold stress increased the expression of BRs-biosynthetic genes, endogenous levels of BRs, MDA, and ROS accumulation but inhibited plant growth. The 24-EBL (1 μM) application increased CAT, proline levels and the activities of its biosynthesis enzymes, Δ^1-pyrroline-5-carboxylate synthetase (*P5CS*) and Δ^1-pyrroline-5-carboxylate reductase (*P5CR*), ascorbate glutathione (AsA-GSH) cycle enzymes thereby lowering ROS and plant growth inhibition and reduced the proline dehydrogenase (*PDH*) and Δ^1-pyrroline-5-carboxylate dehydrogenase (*P5CDH*) activities. The results suggested that application of BRs enhanced cold stress tolerance by inducing the expression of genes related to cold stress, regulation of antioxidant defense system, and cross-talk with proline metabolism. The involvement of ABA and antioxidant defense system in BRs (24-EBL, 0.10 μM) induced water stress tolerance in *Vitis vinifera* L was studied by investigating the role of AsA-GSH cycle (Wang, Chen, Jiang, Duan, & Xi, 2019). Water stress increased ROS but decreased the contents of ascorbic acid and reduced glutathione. The application of 24-EBL reduced ROS production and enhanced the AsA–GSH cycle, ABA content and the transcription of *VviNCED1, VviNCED2, VviZEP, VviSnRK2.6, VviPP2C4, VviABF1*, and *VviABF2* under water stress. It was found that the BRs-mediated cross-talk of ABA regulated the complex mechanisms to confer water stress tolerance in grape seedlings. In tomato plants, BRs was found to exert a positive regulatory role during photoprotection in a *PROTON GRADIENT REGULATION5* (*PGR5*)-dependent manner under chilling stress (Fang et al., 2019). The underlying mechanisms on how H_2O_2 and NO interact with each other in BRs-induced cold stress tolerance were unraveled. The cross-talk of H_2O_2 and NO with BRs signaling pathways mediates protection of the photosystem under cold stress in *Medicago truncatula* (Arfan et al., 2019). The foliar treatment of plants with 1 μM BRs induced the production of H_2O_2 and NO while the addition of an equal amount of brassinazole (*BRZ*) which is a specific inhibitor of BRs biosynthesis pathways reduced these contents appreciably. By using dimethylthiourea (DMTU, a H_2O_2 and OH^- scavenger), researchers showed the blocking of BRs-induced NO production. However, by using 2-phenyl-4,4,5,5-tetramethylimidazoline-1-oxyl-3-oxide (PTIO, a scavenger of NO) BRs-induced H_2O_2 production. In rice plants the cross-talk of BRs with melatonin-mediated skotomorphogenesis by regulating BRs biosynthetic genes including *DWARF4*, *D11*, and *RAVL1* was studied (Hwang & Back, 2018). In controlling growth of roots in *A. thaliana*, BRs-mediated cross-talk with ROS and ET synthesis was reported (Lv et al., 2018). In *Eucalyptus urophylla*, application of 50-nM 24-EBL minimizes 250 mM salt stress by enhancing ion homeostasis, modulating antioxidant defense system, and restoration of leaf anatomy (De Oliveira et al., 2019). The BRs application also reduces the ill effects of fungal pathogens.

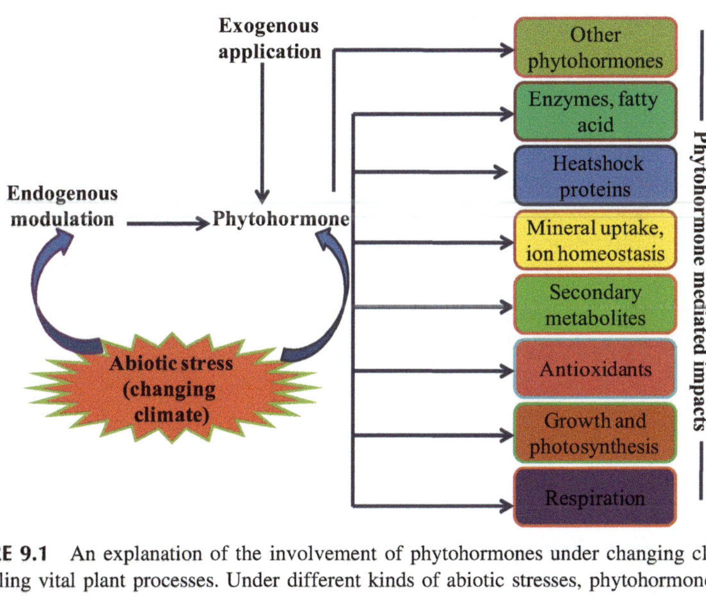

FIGURE 9.1 An explanation of the involvement of phytohormones under changing climate in controlling vital plant processes. Under different kinds of abiotic stresses, phytohormones cross-talk with other signaling agents to control plant metabolism. Phytohormone-induced impacts under changing climate among others include expression of heat shock proteins, respiration, photosynthesis, antioxidants regulation, biosynthesis of primary and secondary metabolites, mineral absorption, and enzyme and fatty acid synthesis in plants.

Zhang et al. (2018) applied 100 nM 24-EBL for enhancing plant defense against *Colletotrichum gloeosporioides* in *Camellia sinensis* L. The 24-EBL supply increased phenylalanineammonium lyase (PAL) activity by increasing the transcript levels of *CsPAL, CsC4H, Cs4CL, CsPR1*, and *CsLOX1* and caused a reduction in H_2O_2 accumulation in tea leaves. Sharma, Ching, Saini, Bhardwaj, and Pati (2013) studied the effect of exogenous application of 24-EBL in enhancing salinity stress tolerance in rice. The salt stress exposed plants showed a significant decrease in growth parameters, enhanced MDA levels and modulated activities of antioxidant enzymes. The 24-EBL supply to salt-stressed plants improved growth traits, levels of protein and proline content, antioxidant enzymes activity, and downregulates the expression levels of *OsDWF4* and *SalT* genes. These above results clearly indicate that BRs have a profound role in enhancing abiotic stress tolerance in various crop plants. A schematic below diagram indicates the major plant functions that are mediated by application of various phytohormones under abiotic stresses (Fig. 9.1).

9.5 Conclusion

The complex interregulatory interactions between various phytohormones and abiotic stress factors are presumed to be either synergistic or antagonistic in crop plants. Various abiotic stress factors under changing climate in plants cause oxidative stress by orchestrating the rate of ROS biosynthesis. The modulation of phytohormone levels either by endogenous biosynthesis or external supplementation could balance the oxidative stress state by integrating plant metabolism through controlling vital plant physiological processes including stimulation of the antioxidant defense system. Phytohormones also cross-talk with other signaling agents to mediate plant physiology by overexpression of various defense-related genes. A diagrammatic sketch showing the effect of phytohormones regulated principal plant physiological processes is represented in Fig. 9.1. The underlying mechanisms of phytohormone cross-talk need to underpin to exploit the current knowledge for advanced plant physiology under abiotic stress in the changing climate.

References

Ahanger, M. A., Mir, R. A., Alyemeni, M. N., & Ahmad, P. (2020). Combined effects of brassinosteroid and kinetin mitigates salinity stress in tomato through the modulation of antioxidant and osmolyte metabolism. *Plant Physiology and Biochemistry*, *147*, 31–42.

Ahmad, B., Zaid, A., Sadiq, Y., Bashir, S., & Wani, S. H. (2019). *Role of selective exogenous elicitors in plant responses to abiotic stress tolerance. Plant abiotic stress tolerance* (pp. 273–290). Cham: Springer.

Ahmad, P., Abdel Latef, A. A., Hashem, A., Abd-Allah, E. F., Gucel, S., & Tran, L. S. P. (2016). Nitric oxide mitigates salt stress by regulating levels of osmolytes and antioxidant enzymes in chickpea. *Frontiers in Plant Science*, *7*, 347.

Ahmad, P., Ahanger, M. A., Alyemeni, M. N., Wijaya, L., & Alam, P. (2018). Exogenous application of nitric oxide modulates osmolytes metabolism, antioxidants, enzymes of ascorbate-glutathione cycle and promotes growth under cadmium stress in tomato. *Protoplasma, 255*, 79–93.

Alamri, S. A., Siddiqui, M. H., Al-Khaishani, M. Y., & Ali, H. M. (2018). Response of salicylic acid on seed germination and physio-biochemical changes of wheat under salt stress. *Acta Scientific Agriculture, 2*(5), 36–42.

Alamri, S. A., Siddiqui, M. H., Al-Khaishany, M. Y., Khan, M. N., Ali, H. M., & Alakeel, K. A. (2019). Nitric oxide-mediated cross-talk of proline and heat shock proteins induce thermotolerance in *Vicia faba* L. *Environmental and Experimental Botany, 161*, 290–302.

Alonso-Ramírez, A., Rodríguez, D., Reyes, D., Jiménez, J. A., Nicolás, G., López-Climent, M., & Nicolás, C. (2009). Evidence for a role of gibberellins in salicylic acid- modulated early plant responses to abiotic stress in *Arabidopsis* seeds. *Plant Physiology, 150*(3), 1335–1344.

Arfan, M., Zhang, D. W., Zou, L. J., Luo, S. S., Tan, W. R., Zhu, T., & Lin, H. H. (2019). Hydrogen peroxide and nitric oxide crosstalk mediates brassinosteroids induced cold stress tolerance in *Medicago truncatula. International Journal of Molecular Sciences, 20*(1), 144.

Arif, Y., Sami, F., Siddiqui, H., Bajguz, A., & Hayat, S. (2020). Salicylic acid in relation to other phytohormones in plant: A study towards physiology and signal transduction under challenging environment. *Environmental and Experimental Botany, 175*, 104040.

Asgher, M., Per, T. S., Masood, A., Fatma, M., Freschi, L., Corpas, F. J., & Khan, N. A. (2017). Nitric oxide signaling and its crosstalk with other plant growth regulators in plant responses to abiotic stress. *Environmental Science and Pollution Research, 24*(3), 2273–2285.

Banerjee, A., & Roychoudhury, A. (2018). Interactions of brassinosteroids with major phytohormones: Antagonistic effects. *Journal of Plant Growth Regulation, 37*(4), 1025–1032.

Bethke, P. C., Libourel, I. G., Aoyama, N., Chung, Y. Y., Still, D. W., & Jones, R. L. (2007). The *Arabidopsis* aleurone layer responds to nitric oxide, gibberellin, and abscisic acid and is sufficient and necessary for seed dormancy. *Plant Physiology, 143*(3), 1173–1188.

Borsani, O., Valpuesta, V., & Botella, M. A. (2001). Evidence for a role of salicylic acid in the oxidative damage generated by NaCl and osmotic stress in *Arabidopsis* seedlings. *Plant Physiology, 126*(3), 1024–1030.

Bortolin, G. S., Teixeira, S. B., de Mesquita Pinheiro, R., Ávila, G. E., Carlos, F. S., da Silva Pedroso, C. E., & Deuner, S. (2020). Seed priming with salicylic acid minimizes oxidative effects of aluminum on *Trifolium* seedlings. *Journal of Soil Science and Plant Nutrition, 20*, 2502–2511.

Chen, E., Zhang, X., Yang, Z., Zhang, C., Wang, X., Ge, X., & Li, F. (2019). BR deficiency causes increased sensitivity to drought and yield penalty in cotton. *BMC Plant Biology, 19* (1), 220.

Correa-Aragunde, N., Graziano, M., Chevalier, C., & Lamattina, L. (2006). Nitric oxide modulates the expression of cell cycle regulatory genes during lateral root formation in tomato. *Journal of Experimental Botany, 57*(3), 581–588.

Da Fonseca, S. S., da Silva, B. R. S., & da Silva Lobato, A. K. (2020). 24-Epibrassinolide positively modulate leaf structures, antioxidant system and photosynthetic machinery in rice under simulated acid rain. *Journal of Plant Growth Regulation*. Available from https://doi.org/10.1007/s00344-020-10167-4.

Da Silva Cunha, L. F., de Oliveira, V. P., do Nascimento, A. W. S., da Silva, B. R. S., Batista, B. L., Alsahli, A. A., & da Silva Lobato, A. K. (2020). Leaf application of 24- epibrassinolide mitigates cadmium toxicity in young *Eucalyptus urophylla* plants by modulating leaf anatomy and gas exchange. *Physiologia Plantarum*. Available from https://doi.org/10.1111/ppl.13182.

De Oliveira, V. P., Lima, M. D. R., da Silva, B. R. S., Batista, B. L., & da Silva Lobato, A. K. (2019). Brassinosteroids confer tolerance to salt stress in *Eucalyptus urophylla* plants enhancing homeostasis, antioxidant metabolism and leaf anatomy. *Journal of Plant Growth Regulation*, *38*(2), 557–573.

Dong, Y. J., Jinc, S. S., Liu, S., Xu, L. L., & Kong, J. (2014). Effects of exogenous nitric oxide on growth of cotton seedlings under NaCl stress. *Journal of Soil Science and Plant Nutrition*, *14*, 1–13.

Emamverdian, A., Ding, Y., & Mokhberdoran, F. (2020). The role of salicylic acid and gibberellin signaling in plant responses to abiotic stress with an emphasis on heavy metals. *Plant Signaling & Behavior*, *15*, 1777372.

Estaji, A., & Niknam, F. (2020). Foliar salicylic acid spraying effect'on growth, seed oil content, and physiology of drought-stressed *Silybum marianum* L. plant. *Agricultural Water Management*, *234*, 106116.

Fang, P., Yan, M., Chi, C., Wang, M., Zhou, Y., Zhou, J., & Yu, J. (2019). Brassinosteroids act as a positive regulator of photoprotection in response to chilling stress. *Plant Physiology*, *180*(4), 2061–2076.

Fatma, M., Masood, A., Per, T. S., & Khan, N. A. (2016). Nitric oxide alleviates salt stress inhibited photosynthetic response by interacting with sulphur assimilation in mustard. *Frontiers in Plant Science*, *7*, 521.

Freschi, L. (2013). Nitric oxide and phytohormone interactions: Current status and perspectives. *Frontiers in Plant Science*, *4*, 398.

Fu, J., Sun, P., Luo, Y., Zhou, H., Gao, J., Zhao, D., & Hu, T. (2019). Brassinosteroids enhance cold tolerance in *Elymus nutans* via mediating redox homeostasis and proline biosynthesis. *Environmental and Experimental Botany*, *167*, 103831.

Giba, Z., Grubišić, D., Todorović, S., Sajc, L., Stojaković, Đ., & Konjević, R. (1998). Effect of nitric oxide–releasing compounds on phytochrome–controlled germination of Empress tree seeds. *Plant Growth Regulation*, *26*(3), 175–181.

Gill, M. B., Cai, K., Zhang, G., & Zeng, F. (2017). Brassinolide alleviates the drought- induced adverse effects in barley by modulation of enzymatic antioxidants and ultrastructure. *Plant Growth Regulation*, *82*(3), 447–455.

Gill, S. S., & Tuteja, N. (2010). Reactive oxygen species and antioxidant machinery in abiotic stress tolerance in crop plants. *Plant Physiology and Biochemistry*, *48*(12), 909–930.

Gouvea, C. M. C. P., Souza, J. F., Magalhaes, A. C. N., & Martins, I. S. (1997). NO·–Releasing substances that induce growth elongation in maize root segments. *Plant Growth Regulation*, *21*(3), 183–187.

Hancock, J. T., & Veal, D. (2020). Nitric oxide, other reactive signalling compounds, redox and reductive stress. *Journal of Experimental Botany*. Available from https://doi.org/10.1093/jxb/eraa331.

Hasanuzzaman, M., Bhuyan, M. H. M., Zulfiqar, F., Raza, A., Mohsin, S. M., Mahmud, J. A., & Fotopoulos, V. (2020). Reactive oxygen species and antioxidant defense in plants under abiotic stress: Revisiting the crucial role of a universal defense regulator. *Antioxidants*, *9*(8), 681.

Hasanuzzaman, M., Nahar, K., Alam, M., & Fujita, M. (2012). Exogenous nitric oxide alleviates high temperature induced oxidative stress in wheat (*Triticum aestivum* L.) seedlings by modulating the antioxidant defense and glyoxylase system. *Australian Journal of Crop Science*, *6*, 1314–1323.

Hayat, S., Alyemeni, M. N., & Yadav, S. (2014). Do brassinosteroids ameliorate freezing stress in *Cicer arietinum*. *Legume Research-An International Journal*, *37*(1), 68–71.

Hayat, S., Yadav, S., Wani, A. S., Irfan, M., & Ahmad, A. (2011). Nitric oxide effects on photosynthetic rate, growth and antioxidant activity in tomato. *International Journal of Vegetable Science*, *17*, 333–348.

Hwang, O. J., & Back, K. (2018). Melatonin is involved in skotomorphogenesis by regulating brassinosteroid biosynthesis in rice plants. *Journal of Pineal Research*, *65*(2), e12495.

Jan, S., Noman, A., Kaya, C., Ashraf, M., Alyemeni, M. N., & Ahmad, P. (2020). 24- Epibrassinolide alleviates the injurious effects of Cr (VI) toxicity in tomato plants: Insights into growth, physio-biochemical attributes, antioxidant activity and regulation of ascorbate–glutathione and glyoxalase cycles. *Journal of Plant Growth Regulation*. Available from https://doi.org/10.1007/s00344-020-10169-2.

Kaur, H., Sirhindi, G., Bhardwaj, R., Alyemeni, M. N., Siddique, K. H., & Ahmad, P. (2018). 28-homobrassinolide regulates antioxidant enzyme activities and gene expression in response to salt-and temperature-induced oxidative stress in *Brassica juncea*. *Scientific Reports*, *8*(1), 1–13.

Kaya, C., Aydemir, S., Akram, N. A., & Ashraf, M. (2018). Epibrassinolide application regulates some key physio-biochemical attributes as well as oxidative defense system in maize plants grown under saline stress. *Journal of Plant Growth Regulation*, *37*(4), 1244–1257.

Khan, M. N., Mobin, M., Abbas, Z. K., & Siddiqui, M. H. (2017). Nitric oxide-induced synthesis of hydrogen sulfide alleviates osmotic stress in wheat seedlings through sustaining antioxidant enzymes, osmolyte accumulation and cysteine homeostasis. *Nitric Oxide: Biology and Chemistry / Official Journal of the Nitric Oxide Society*, *68*, 91–102.

Kim, S. Y., Shang, Y., Joo, S.-H., Kim, S.-K., & Nam, K. H. (2017). Overexpression of *BAK1* causes salicylic acid accumulation and deregulation of cell death control genes. *Biochemical and Biophysical Research Communications*, *484*, 781–786.

Kim, S. Y., Warpeha, K. M., & Huber, S. C. (2019). The brassinosteroid receptor kinase, BRI1, plays a role in seed germination and the release of dormancy by cold stratification. *Journal of Plant Physiology*, *241*, 153031.

Kim, Y., Park, S., Gilmour, S. J., & Thomashow, M. F. (2013). Roles of *CAMTA* transcription factors and salicylic acid in configuring the low-temperature transcriptome and freezing tolerance of *Arabidopsis*. *The Plant Journal*, *75*(3), 364–376.

Kohli, S. K., Handa, N., Sharma, A., Gautam, V., Arora, S., Bhardwaj, R., & Ahmad, P. (2018). Combined effect of 24-epibrassinolide and salicylic acid mitigates lead (Pb) toxicity by modulating various metabolites in *Brassica juncea* L. seedlings. *Protoplasma*, *255*(1), 11–24.

Kolomeichuk, L. V., Efimova, M. V., Zlobin, I. E., Kreslavski, V. D., Ol'ga, K. M., Kovtun, I. S., & Allakhverdiev, S. I. (2020). 24-Epibrassinolide alleviates the toxic effects of NaCl on photosynthetic processes in potato plants. *Photosynthesis Research*, *146*, 151–163.

Lee, J., Shim, D., Moon, S., Kim, H., Bae, W., Kim, K., & Lee, Y. J. (2018). Genome-wide transcriptomic analysis of BR-deficient Micro-Tom reveals correlations between drought stress tolerance and brassinosteroid signaling in tomato. *Plant Physiology and Biochemistry*, *127*, 553–560.

Li, S. W., Li, Y., Leng, Y., Zeng, X. Y., & Ma, Y. H. (2019). Nitric oxide donor improves adventitious rooting in mung bean hypocotyl cuttings exposed to cadmium and osmotic stresses. *Environmental and Experimental Botany*, *164*, 114–123.

Li, X., Wei, J. P., Ahammed, G. J., Zhang, L., Li, Y., Yan, P., & Han, W. Y. (2018). Brassinosteroids attenuate moderate high temperature-caused decline in tea quality by enhancing theanine biosynthesis in *Camellia sinensis* L. *Frontiers in Plant Science*, *9*, 1016.

Li, Y. F., Huang, L. L., Liu, X. L., Li, X. S., & Tan, H. H. (2020). Exogenous salicylic acid alleviates halosulfuron-methyl toxicity by coordinating the antioxidant system and improving photosynthesis in soybean (*Glycine max* Merr.). *Acta Physiologiae Plantarum*, *42*(5).

Liu, S., He, Y., Tian, H., Yu, C., Tan, W., Li, Z., & Duan, L. (2019). Application of brassinosteroid mimetics improves growth and tolerance of maize to nicosulfuron toxicity. *Journal of Plant Growth Regulation*, *38*(2), 701–712.

Lv, B., Tian, H., Zhang, F., Liu, J., Lu, S., Bai, M., & Ding, Z. (2018). Brassinosteroids regulate root growth by controlling reactive oxygen species homeostasis and dual effect on ethylene synthesis in *Arabidopsis*. *PLoS Genetics*, *14*(1), e1007144.

Mittler, R. (2017). ROS are good. *Trends in Plant Science*, *22*(1), 11–19.

Mohammadi, H., Amirikia, F., Ghorbanpour, M., Fatehi, F., & Hashempour, H. (2019). Salicylic acid induced changes in physiological traits and essential oil constituents in different ecotypes of *Thymus kotschyanus* and *Thymus vulgaris* under well-watered and water stress conditions. *Industrial Crops and Products*, *129*, 561–574.

Nabi, R. B. S., Tayade, R., Hussain, A., Kulkarni, K. P., Imran, Q. M., Mun, B. G., & Yun, B. W. (2019). Nitric oxide regulates plant responses to drought, salinity, and heavy metal stress. *Environmental and Experimental Botany*, *161*, 120–133.

Nie, S., Huang, S., Wang, S., Mao, Y., Liu, J., Ma, R., & Wang, X. (2019). Enhanced brassinosteroid signaling intensity via *SlBRI1* overexpression negatively regulates drought resistance in a manner opposite of that via exogenous BR application in tomato. *Plant Physiology and Biochemistry*, *138*, 36–47.

Pagnussat, G. C., Simontacchi, M., Puntarulo, S., & Lamattina, L. (2002). Nitric oxide is required for root organogenesis. *Plant Physiology*, *129*(3), 954–956.

Pál, M., Janda, T., Majláth, I., & Szalai, G. (2020). Involvement of salicylic acid and other phenolic compounds in light-dependent cold acclimation in maize. *International Journal of Molecular Sciences*, *21*(6), 1942.

Peng, R., Sun, W., Jin, X., Yu, L., Chen, C., Yue, Z., & Dong, Y. (2020). Analysis of 2,4 epi-brassinolide created an enhancement tolerance on Cd toxicity in *Solanum nigrum* L. *Environmental Science and Pollution Research*, *27*, 16784–16797.

Raja, V., Majeed, U., Kang, H., Andrabi, K. I., & John, R. (2017). Abiotic stress: Interplay between ROS, hormones and MAPKs. *Environmental and Experimental Botany*, *137*, 142–157.

Raskin, I. (1992). Role of salicylic acid in plants. *Annual Review of Plant Physiology*, *43*, 439–463. Available from https://doi.org/10.1146/annurev.pp.43.060192.002255.

Rather, B. A., Mir, I. R., Masood, A., Anjum, N. A., & Khan, N. A. (2020). Nitric oxide pretreatment advances seed germination and alleviates copper-induced photosynthetic inhibition in indian mustard. *Plants*, *9*(6), 776.

Raza, A., Razzaq, A., Mehmood, S. S., Zou, X., Zhang, X., Lv, Y., & Xu, J. (2019). Impact of climate change on crops adaptation and strategies to tackle its outcome: A review. *Plants*, *8* (2), 34.

Rivas-San Vicente, M., & Plasencia, J. (2011). Salicylic acid beyond defence: Its role in plant growth and development. *Journal of Experimental Botany*, *62*, 3321–3338.

Sadiq, Y., Zaid, A., & Khan, M. M. A. (2020). *Adaptive physiological responses of plants under abiotic stresses: Role of phytohormones. Plant ecophysiology and adaptation under climate change: Mechanisms and perspectives I* (pp. 797–824). Singapore: Springer.

Santisree, P., Bhatnagar-Mathur, P., & Sharma, K. K. (2015). NO to drought-multifunctional role of nitric oxide in plant drought: Do we have all the answers? *Plant Science*, *239*, 44–55.

Sehar, Z., Masood, A., & Khan, N. A. (2019). Nitric oxide reverses glucose-mediated photosynthetic repression in wheat (*Triticum aestivum* L.) under salt stress. *Environmental and Experimental Botany, 161*, 277–289.

Sharma, A., Bhardwaj, R., Kumar, V., & Thukral, A. K. (2016). GC-MS studies reveal stimulated pesticide detoxification by brassinolide application in *Brassica juncea* L. plants. *Environmental Science and Pollution Research, 23*(14), 14518–14525.

Sharma, I., Ching, E., Saini, S., Bhardwaj, R., & Pati, P. K. (2013). Exogenous application of brassinosteroid offers tolerance to salinity by altering stress responses in rice variety Pusa Basmati-1. *Plant Physiology and Biochemistry, 69*, 17–26.

Shivaraj, S. M., Vats, S., Bhat, J. A., Dhakte, P., Goyal, V., Khatri, P., & Sharma, T. R. (2020). Nitric oxide and hydrogen sulfide crosstalk during heavy metal stress in plants. *Physiologia Plantarum, 168*(2), 437–455.

Siddiqui, M. H., Alamri, S., Alsubaie, Q. D., Ali, H. M., Khan, M. N., Al-Ghamdi, A., & Alsadon, A. (2020). Exogenous nitric oxide alleviates sulfur deficiency-induced oxidative damage in tomato seedlings. *Nitric Oxide: Biology and Chemistry / Official Journal of the Nitric Oxide Society, 94*, 95–107.

Singh, I., Kumar, U., Singh, S. K., Gupta, C., Singh, M., & Kushwaha, S. R. (2012). Physiological and biochemical effect of 24-epibrassinoslide on cold tolerance in maize seedlings. *Physiology and Molecular Biology of Plants, 18*, 229–236.

Su, J., Zhang, Y., Nie, Y., Cheng, D., Wang, R., Hu, H., & Shen, W. (2018). Hydrogen- induced osmotic tolerance is associated with nitric oxide-mediated proline accumulation and reestablishment of redox balance in alfalfa seedlings. *Environmental and Experimental Botany, 147*, 249–260.

Su, Q., Zheng, X., Tian, Y., & Wang, C. (2020). Exogenous brassinolide alleviates salt stress in *Malus hupehensis* rehd. by regulating the transcription of NHX-Type Na + (K +)/H + antiporters. *Frontiers in Plant Science, 11*, 38.

Sun, C., Liu, L., Lu, L., Jin, C., & Lin, X. (2018). Nitric oxide acts downstream of hydrogen peroxide in regulating aluminum-induced antioxidant defense that enhances aluminum resistance in wheat seedlings. *Environmental and Experimental Botany, 145*, 95–103.

Tahjib-Ul-Arif, M., Siddiqui, M. N., Sohag, A. A. M., Sakil, M. A., Rahman, M. M., Polash, M. A. S., & Tran, L. S. P. (2018). Salicylic acid-mediated enhancement of photosynthesis attributes and antioxidant capacity contributes to yield improvement of maize plants under salt stress. *Journal of Plant Growth Regulation, 37*(4), 1318–1330.

Tayyab, N., Naz, R., Yasmin, H., Nosheen, A., Keyani, R., Sajjad, M., & Roberts, T. H. (2020). Combined seed and foliar pre-treatments with exogenous methyl jasmonate and salicylic acid mitigate drought-induced stress in maize. *PLoS One, 15*(5), e0232269.

Verma, N., Tiwari, S., Singh, V. P., & Prasad, S. M. (2020). Nitric oxide in plants: An ancient molecule with new tasks. *Plant Growth Regulation, 90*, 1–13.

Wang, Y. T., Chen, Z. Y., Jiang, Y., Duan, B. B., & Xi, Z. M. (2019). Involvement of ABA and antioxidant system in brassinosteroid-induced water stress tolerance of grapevine (*Vitis vinifera* L.). *Scientia Horticulturae, 256*, 108596.

Wani, S. H., Kumar, V., Shriram, V., & Sah, S. K. (2016). Phytohormones and their metabolic engineering for abiotic tolerance in crop plants. *Crop Journal, 4*, 162–176.

Went, F. W., & Thimann, K. V. (1937). *Phytohormones* (p. 3) New York: Macmillan.

Wu, X., Yao, X., Chen, J., Zhu, Z., Zhang, H., & Zha, D. (2014). Brassinosteroids protect photosynthesis and antioxidant system of eggplant seedlings from high-temperature stress. *Acta Physiologiae Plantarum, 36*(2), 251–261.

Yadav, P., Kaur, R., Kanwar, M. K., Sharma, A., Verma, V., Sirhindi, G., & Bhardwaj, R. (2018). Castasterone confers copper stress tolerance by regulating antioxidant enzyme responses, antioxidants, and amino acid balance in *B. juncea* seedlings. *Ecotoxicology and Environmental Safety*, *147*, 725–734.

Yang, P., Azher Nawaz, M., Li, F., Bai, L., & Li, J. (2019). Brassinosteroids regulate antioxidant system and protect chloroplast ultrastructure of autotoxicity-stressed cucumber (*Cucumis sativus* L.) seedlings. *Agronomy*, *9*(5), 265.

Yu, L. L., Liu, Y., Zhu, F., Geng, X. X., Yang, Y., He, Z. Q., & Xu, F. (2020). The enhancement of salt stress tolerance by salicylic acid pretreatment in *Arabidopsis thaliana*. *Biologia Plantarum*, *64*, 150–158.

Zaid, A., & Mohammad, F. (2018). Methyl jasmonate and nitrogen interact to alleviate cadmium stress in *Mentha arvensis* by regulating physio-biochemical damages and ROS detoxification. *Journal of Plant Growth Regulation*, *37*(4), 1331–1348.

Zaid, A., Mohammad, F., & Fariduddin, Q. (2020). Plant growth regulators improve growth, photosynthesis, mineral nutrient and antioxidant system under cadmium stress in menthol mint (*Mentha arvensis* L.). *Physiology and Molecular Biology of Plants*, *26*(1), 25–39.

Zaid, A., Mohammad, F., Wani, S. H., & Siddique, K. M. (2019). Salicylic acid enhances nickel stress tolerance by up-regulating antioxidant defense and glyoxalase systems in mustard plants. *Ecotoxicology and Environmental Safety*, *180*, 575–587.

Zaid, A., & Wani, S. H. (2019). *Reactive oxygen species generation, scavenging and signaling in plant defense responses. Bioactive molecules in plant defense* (pp. 111–132). Cham: Springer.

Zhang, L., Ahammed, G. J., Li, X., Wei, J. P., Li, Y., Yan, P., & Han, W. Y. (2018). Exogenous brassinosteroid enhances plant defense against *Colletotrichum gloeosporioides* by activating phenylpropanoid pathway in *Camellia sinensis* L. *Journal of Plant Growth Regulation*, *37* (4), 1235–1243.

Zhang, Z., Wu, P., Zhang, W., Yang, Z., Liu, H., Ahammed, G. J., & Cui, J. (2020). Calcium is involved in exogenous NO-induced enhancement of photosynthesis in cucumber (*Cucumis sativus* L.) seedlings under low temperature. *Scientia Horticulturae*, *261*, 108953.

Chapter 10

The role of soil microbes in the plant adaptation to stresses: current scenario and future perspective

Ameena A. Al-surhanee[1], Mona H. Soliman[2,3] and Salama A. Ouf[2]
[1]Biology Department, College of Science, Jouf University, Sakaka, Saudi Arabia, [2]Botany and Microbiology Department, Faculty of Science, Cairo University, Giza, Egypt, [3]Biology Department, Faculty of Science, Taibah University, Yanbu, Saudi Arabia

10.1 Introduction

Plants are exposed to a variety of environmental factors during the developmental events of life cycle. These factors may be abiotic as well as biotic in nature. In this connection there is an immense chance of alterations in the growth pattern of plant's development. Among the key determinants of restricted growth in plants are included the existing environmental conditions which include soil health like salinity status, moisture levels, temperatures (low and high), presence of metal(loid)s, and nutrient status. Growth regulating environmental factors determine the yield productivity of the crop plants (Soliman, Alhaithloul, et al., 2020; Soliman, Elkelish, Souad, Alhaithloul, & Farooq, 2020). These environmental factors bring the modulations of the physiological and biochemical pathways of plants which are controlled at molecular levels. Such alterations result in the declined growth and yield which result in a major threat to world food security (Elkelish, Alhaithloul, Qaric, Soliman, & Hasanuzzaman, 2020; Elkelish, Soliman, Alhaithloul, & El-Esawi, 2019). The characteristics of existing soil have obvious effects on the plants growing and on the resistance mechanisms against the environmental stresses which include both abiotic or biotic thereby imparting considerable direct or indirect effects on plant growth (Ahanger, Hashem, Abd Allah, & Ahmad, 2014; Ahanger, Tyagi, Wani, & Ahmad, 2014; Begum, Qin, et al., 2019).

Frontiers in Plant–Soil Interaction. DOI: https://doi.org/10.1016/B978-0-323-90943-3.00006-7

Stress factors impede plant performance by restricting growth and such damage can be much evident in stress prone areas like arid and semiarid regions of globe and rapid changes in climate pattern has further aggravated the damage (Barnawal, Bharti, Maji, Chanotiya, & Kalra, 2014). In most areas the introduction of bad management practices has resulted in the conversion of agricultural land into unproductive lands. For example the excessive use of polluted water like salt and metal containing water for irrigating the field is one of the main contributors to this agricultural malpractice. Environmental stresses result in the toxic effects via modulations of the major physiological and biochemical processes like photosynthesis, ion homeostasis, enzyme functioning, assimilate transport, etc. which ultimately result in significant decline in growth (M. I. R. Khan, Nazir, Asgher, Per, & Khan, 2015). It has been reported that environmental stresses hamper plant photosynthesis which is largely associated with altered carbon and nitrogen assimilation. In addition stresses like salinity decline nitrogen fixation in legumes (Tejera, Campos, Sanjuan, & Lluch, 2004) and drought brings down the assimilation of nitrogen (Ahanger, Tittal, Mir, & Agarwal, 2017), considerably resulting in declined yield productivity. One of the prime reasons for declined growth and productivity due to stressful environmental conditions is the excessive generation of toxic reactive oxygen species (ROS) which pose potential damaging effects to lipids, induce protein oxidation, and destabilize nucleic acids (Ahanger, Tomar, Tittal, Argal, & Agarwal, 2017; Mittler, 2002). Among the main ROSs are included superoxide, hydrogen peroxide, hydroxyl, and peroxide radicals. They are often accumulated in sensitive tissues, including leaves, wherein they induce oxidation of macromolecular structures, therefore hampering the physiological and biochemical performance of plants. The main origins of ROS include mitochondria, chloroplast, and peroxisomes (Mittler, 2002).

10.2 Role of soil microbes in plant growth—an overview

Initially stress exposed plants tend to adapt strategies that assist in the avoidance of the deleterious effects of stresses. These adaptations include morphological, physiological, and biochemical modifications for lessening the stress damage (Fusconi and Berta, 2012). Besides having the indigenously occurring tolerance mechanisms certain beneficial microbes existing naturally in soil improve the tolerance mechanisms for better performance of plants under adverse conditions. These beneficial microbes include plant growth-promoting rhizobacteria (PGPR) and arbuscular mycorrhizal fungi (AMF). Their involvement in the regulation of growth and developmental processes under stress have been consolidated through several studies carried at physiological, biochemical, and molecular levels (Abdel Latef & Chaoxing, 2014; Begum, Ahanger, & Zhang, 2020; Begum, Qin, et al., 2019; Glick, 2012; Hameed et al., 2014; Youseif, 2018; Zahir, Ghani, Naveed, Nadeem, &

Asghar, 2009). AMF are microscopic filamentous fungi that have the capability to colonize most plants, either directly into the root tissues or the rhizosphere. AMF usually forma ramified filaments of few centimeters to benefit the host plant (Abd-Allah, Abeer, Alqarawi, & Alwathnani, 2015).

Both PGPR and AMF form symbiotic associations with a number of plant species and have the potential to improve soil structure and plant growth under different environments (Begum, Qin, et al., 2019; Youseif, 2018). They modulate the rhizospheric soil health by making significant enhancement in its properties like mineral elements, moisture content, enzyme activities, etc. Under abiotic stresses like drought, salt, and heavy metal stress, AMF act as bioameliorators and assist the existing plant species in mitigation of stress-induced damaging effects (Abd-Allah et al., 2015). They modify the root architecture to mediate the increased exploration of water and nutrients by roots for the benefit of plants (Wu, Zou, & He, 2010), and all these result in the modification of physiological performance of the host plant. It has been demonstrated that AMF inoculation results in improved leaf area and mineral content like phosphorous, nitrogen, and potassium imparting significant enhancement in the photosynthetic rate (Begum, Ahanger, et al., 2019). These beneficial soil microbes associate themselves with most of the crop plants, thereby contributing to regulation of growth through the mediation of several benefits (Ahanger, Hashem, et al., 2014; Begum et al., 2020; Begum, Ahanger, et al., 2019; Begum, Qin, et al., 2019; Hameed et al., 2014).

PGPR are beneficial plant associated bacteria aggressively colonizing the rhizosphere roots leading to beneficial effects on plants (Youseif, 2018). Both direct and indirect growth-promoting effects of PGPR have been observed (Glick, 2012). The most prominent effects of PGPR contributing to plant growth promotion and stress mitigation include (1) enhanced nutrient availability facilitation through increased nitrogen fixation, (2) phosphorous solubilization, and (3) regulation of phytohormones like auxins, abscissic acid, gibberellic acid, and cytokinin (Ahemad & Kibret, 2014; Panwar, Tewari, & Nayyar, 2014; Zaidi, Khan, Ahemad, & Oves, 2009). Interestingly PGPR reduce the damaging effects of biotic stresses including pathogens by inhibiting their growth through the production of antibiotics, siderophores, and induction of systemic acquired resistance (Glick, 2012; Labuschagne, Pretorius, & Idris, 2010; Ramamoorthy, Viswanathan, Raguchander, Prakasam, & Samiyappan, 2001; Zahir et al., 2009). Zahir et al. (2009) have demonstrated that PGPR mediate stress tolerance through the regulation of 1-aminocyclopropane-1-carboxylate (ACC) deaminase activity.

The PGPR associated with plant roots directly reflect and influence the underlying mechanisms governing the interactions between host plant and the microbe. Most of the PGPR colonize the root surfaces and develop in spaces occurring in between the root hairs and the rhizodermal layers, while some PGPR do not come in physical contact with roots (Gray & Smith, 2005; Ilangumaran & Smith, 2017). Plant root exudates act as an integral rhizosphere signal for regulating the communication for the establishment of beneficial plant–microbe association. In addition the compounds like phenolics, flavonoids, and the organic acids that are secreted by the roots are considered as chemical signals for mediating the bacterial chemotaxis, exopolysaccharides secretion, sensing of quorum, and the formation of biofilm (Badri, Weir, van der Lelie, & Vivanco, 2009). PGPR isolated from soils are screened in vitro for the plant growth-promoting characteristics and later those beneficial effects are observed in experimental trials before the commercialization of a PGPR strain. PGPR improve growth and development of plants through enhancement in the nutrient assimilation, acting as biofertilizers making enough nitrogen fixation, mediating solubilization of phosphorous, improving acquisition of iron, acting as biocontrol agents against pathogens, mediating degradation of organic pollutants, and mediating phytoremediation (Chowdhury, Hartmann, Gao, & Borriss, 2015; Kuan, Othman, Abdul Rahim, & Shamsuddin, 2016; Steenhoudt & Vanderleyden, 2006; Weyens et al., 2015; Yang, Kloepper, & Ryu, 2009; Ilangumaran & Smith, 2017). In addition PGPR enhance plant–water relations, maintain homeostasis of ions, and improve photosynthetic efficiency under adverse conditions. These stress mitigation mechanisms are intricate and are not fully understood. Mechanisms regulating the PGPR mediated growth are fine-tuned by complex signaling events leading to significant stress alleviation (Smith, Gravel, & Yergeau, 2017).

10.3 Stress tolerance mechanisms

Some indigenous tolerance mechanisms have been evolved by plants to counteract the damaging effects of stresses. These mechanisms either function individually as well as in combination leading to mitigation of damaging effects of stress. Some mechanisms are specific to stress while some are common to many stresses. In this connection, in mitigating the damage of excess ROS, the antioxidant defense system has a major role. Optimal functioning of the antioxidant system maintains the ROS levels at optimal concentrations (Elkelish et al., 2019, 2020). The antioxidant system comprises enzymatic and nonenzymatic components. Superoxide dismutase, catalase, ascorbate peroxidase, monodehydroascorbate reductase, dehydroascorbate reductase, and glutathione reductase are enzymatic, while among the nonenzymatic components are ascorbic acid, tocopherol, glutathione, and polyphenols (Mittler, 2002; Soliman, Alhaithloul, et al., 2020; Soliman, Elkelish, et al., 2020). This intriguing defense system neutralizes excess ROS for developing protection against stress-mediated oxidative damage. Besides this osmoregulation mediated by compatible osmolytes is also considered as a major contributor of stress mitigation (Hayat et al., 2012). Key osmolytes include free proline, amino acids, sugars, polyols, etc. leading to the stabilization of cellular structures and functioning (Ahanger, Tyagi, et al., 2014;

Soliman, Alhaithloul, et al., 2020). It is believed that whenever the antioxidant system fails to overcome the damaging effects of stresses osmolytes have a significant contribution to protect cells from ROS. Osmotic adjustment particularly under salinity is attained by accumulation of organic solutes and inorganic ions concomitant with ion sequestration and compartmentation (Azooz, Youssef, & Ahmad, 2011). Under salinity and metal stress the sequestration and compartmentation of toxic ions to vacuoles and other less sensitive spaces are considered major contributors to tolerance.

10.4 Environmental stresses and soil microbe-mediated plant growth regulation

10.4.1 Drought stress

Drought is the reduced availability of water and it has been reported widely to have serious damaging effects on the growth patterns of plants, resulting in a significant decline in yield productivity (Ahanger, Tyagi, et al., 2014; Jatav, Agarwal, Tomar, & Tyagi, 2014). Drought downregulates growth by influencing the key physiological and biochemical attributes like enzyme activity, uptake and assimilation of mineral elements, membrane functioning, photosynthesis, transpiration, protein synthesis, and defense systems like antioxidant and osmolyte metabolism (Ahanger & Agarwal, 2016; Hameed et al., 2014). Reduced water availability induced by drought stress is usually accompanied by a considerable rise in temperature which is more often aggravated by the changing climate. Soil desiccation accompanied by rapid transpiration rates resulting from increased temperatures accelerates the damaging effects of drought by imparting rapid oxidative damage to key organelles like chloroplasts, mitochondria etc. Contrary to this, plant species exhibiting symbiotic association with soil microbes, including PGPR and AMF, show significant growth enhancement resulting in regulation of metabolic processes through increased mineral uptake (Al-Karaki, McMichael, & Zak, 2004; Chiapero, Cappellari, Alderete, Palermo, & Banchio, 2019; Hameed et al., 2014; Kasim et al., 2013). Increased drought tolerance in *Triticum aestivum* due to inoculation of *Trichoderma harzianum* has a cumulative impact on the soil attributes inducing increased chlorophyll synthesis, uptake of nitrogen, and increased yield (Al-Ezerjawi & Kadhim, 2014). Under water deficit significant amelioration in the uptake of water and nutrients results from the deep intrusion of AMF hyphae as well as the production of glomalin (Wu & Zou, 2017). Interestingly colonization of PGPR has also been reported to significantly enhance the growth through induction of a considerable increase in water relations (Ghorbanpour et al., 2013). Beneficial soil microbes including PGPR and AMF impart enhancement in the growth of host plants by improving the soil stabilization. In addition, the increased functioning of defense systems is due to beneficial soil microbes for increased drought stress tolerance besides significant enhancement in the water use efficiency. Such upregulation of defenses systems and the water relations are reflected as reduced ROS accumulation and hence the oxidative damage (Heideri & Golpayegani, 2012). PGPR (Heideri & Golpayegani, 2012) and AMF (Gholamhoseini, Ghalavand, Dolatabadian, Jamshidi, & Khodaei-Joghan, 2013) inoculation to plants results in protection of photosynthesis by enhancement of the stomatal conductance, gas exchange, transpiration rate, and water use efficiency. Beneficial soil microbe inoculation causes an enhancement in hydraulic conductivity, accumulation of compatible osmotica accumulation, and induces morphological modifications for greater avoidance of drought-mediated damaging effects (Hameed et al., 2014; Vurukonda, Vardharajula, Shrivastava, & Ali, 2016). Yedidia, Srivastva, Kapulnik, and Chet (2001) have demonstrated greater growth of cucumber due to *T. harzianum* inoculation via increased mineral uptake, which was observed due to the direct influence on the soil health. In another study Li et al. (2013) demonstrated considerable enhancement in the expression of root aquaporin genes resulting in the enhancement of the water content in *Zea mays* and more importantly the expression was greater in hyphae intruded regions. Inoculation of PGPR imparts drought stress tolerance by greater production of exopolysaccharides, volatile compounds, phytohormones, 1-aminocyclopropane-1carboxylate deaminase, osmolyte synthesis, strengthening of antioxidant system, gene regulation, and morphological modification (Vurukonda et al., 2016). PGPR inoculation strengthens the defense systems including the antioxidant system and osmolyte metabolism, resulting in greater growth and photosynthetic protection (Heideri & Golpayegani, 2012). Increased accumulation of osmolytes including proline, glycine betaine, sugars, etc. results in maintenance of tissue water potential for the greater acquisition of water from soil solution (Ahanger, Tyagi, et al., 2014) so that drought-mediated damaging effects are lessened (Ahanger, Tittal, et al., 2017). AMF inoculation differentially regulates the synthesis of amino acids under drought stress (Ogawa & Yamauchi, 2006). N. Khan, Bano, and Zandi (2018) while working on chick pea drought stress tolerance demonstrated that PGPR and phytohormones work synergistically for improved drought stress tolerance.

In addition, PGPR and AMF-mediated growth regulation and drought stress tolerance are due to the regulation of metabolite synthesis having a key role in the maintenance of redox components like ascorbic acid, glutathione, etc. Chiapero et al. (2019) reported significant enhancement in the antioxidant system and the accumulation of nonenzymatic components like phenols, resulting in significant enhancement in growth and drought stress amelioration. In addition increased phenols and proline due to PGPR inoculation provide greater protection to membrane structural and functional stability. Inoculation of PGPB (*Bacillus amyloliquefaciens* 5113 and *Azospirillum brasilense* NO40) has been demonstrated to significantly enhance the expression levels of genes coding for the ascorbate-glutathione cycle enzymes under drought stress in wheat, which make a significant contribution to the maintenance of redox homeostasis (Kasim et al., 2013).

Drought-induced oxidative damage due to greater ROS production hinders the normal cellular functioning (Ahanger, Tomar, et al., 2017). Greater functioning of the antioxidant system reduces the drought triggered oxidative stress by maintaining redox balance and low levels of ROS (Mittler, 2002). It has been reported that both PGPR (Batool et al., 2020) and AMF (Huang, Zou, & Wu, 2017) inoculation upregulates the antioxidant system leading to a significant decline in ROS accumulation.

10.4.2 Salinity stress

Increasing soil salinity has been among the devastating environmental factors contributing significantly to yield losses. Higher soil salinity affects plant growth and development by declining the key attributes like water uptake, photosynthesis, uptake and assimilation of mineral elements, and defense mechanisms like the antioxidant system (Elkelish et al., 2019, 2020; M. I. R. Khan, Asgher, & Khan, 2014). Soil salinity impedes water accessibility to roots and also restricts its uptake and entry into plant by altering the water potential. This ultimately results in osmotic and ionic stress, resulting in oxidative damage to major cellular functioning. Increased uptake of toxic ions like Na, Cl, etc. result in considerable accumulation of ROS, thereby damaging the key macromolecules like proteins, nucleic acids, lipids, etc., hence the cellular functioning is altered (Ahanger et al., 2019; Ahanger, Mir, Alyemeni, & Ahmad, 2020; Ahmad, Hakeem, Kumar, Ashraf, & Akram, 2012; Soliman, Elkelish, et al., 2020). Several key strategies for enhancing the salinity tolerance potential of plants have been tested for their potential to protect plant growth and yield. Alteration of redox homeostasis due to salinity reduces nutrient acquisition and enzyme functioning (M. I. R. Khan et al., 2014).

The employment of novel strategies for the mitigation of damaging effects of salinity and the reclamation of saline affected soils has been the prime target of sustainable agriculture. In this connection the use of AMF and PGPR species has been extensively tested for their potential to ameliorate salinity-induced damage in plants (Abdel Latef & Chaoxing, 2014; Aroca et al., 2013; Bharti, Pandey, Barnawal, Patel, & Kalra, 2016; Elhindi, El-Din, & Elgorban, 2017; Habib, Kausar, & Saud, 2016; R. P. Singh & Jha, 2017). It is clear that the inoculation of AMF and PGPR regulates growth under salinity, ameliorating the damaging effects of salinity on the uptake of water and minerals and their use efficiency, enzyme activity, protein synthesis, and gene expression, thereby imparting obvious influence on photosynthesis and the stress tolerance mechanisms. Lin, Wang, Sun, Mu, and Yan

(2017) and Elhindi et al. (2017) have demonstrated that colonization of plants with AMF results in increased growth by regulating the morphology and hydraulic conductivity of roots, gas exchange parameters, and photosynthesis reflected in increased water use efficiency, and nitrogen and potassium uptake. The decline in uptake of mineral elements and the synthesis of chlorophyll is ameliorated by inoculation of AMF (Cekic, Unyayar, & Ortas, 2012; Ilangumaran & Smith, 2017; Scagel & Bryla, 2017; Upadhyay & Singh, 2015). Increased photosynthetic functioning due to inoculation of AMF under salinity stress is due to cumulative beneficial effects on chlorophyll synthesis, photosystem functioning, and carbonic anhydrase activity (Talaat & Shawky, 2004). On other hand PGPR-induced photosynthetic regulation has been attributed to improved gas exchange, nonstomatal regulation including photosystem functioning, osmotic balance maintenance, mineral uptake, and antioxidant metabolism (Habib et al., 2016; Ilangumaran & Smith, 2017; Marulanda, Azcon, Chaumont, Ruiz-Lozano, & Aroca, 2010; Nadeem, Zahir, Naveed, & Arshad, 2009; Numan et al., 2018). PGPR improve mineral assimilation through N fixation via the development of siderophores (Numan et al., 2018) and modulates gene expression of leaf and rhizosphere community for improved salinity tolerance in rice (Nautiyal et al., 2013). The protection of membrane lipids under salinity stress due to AMF inoculation has been observed to prevent the ROS-mediated damage to membrane structure and functioning under salinity stress (Alqarawi, Hashem, Abd_Allah, Alshahrani, & Huqail, 2014). Greater membrane stability and plasma membrane functioning due to the inoculation of beneficial soil microbes is a result of their impact on the stress tolerance mechanisms like antioxidant functioning, osmoregulation, and cellular stress signaling (Bharti et al., 2016), resulting in increased growth and developmental regulation under adverse conditions. Improved membrane functioning due to AMF (Alqarawi et al., 2014) and PGPR (R. P. Singh & Jha, 2017) is due to greater maintenance of polyunsaturated fatty acid ratio. Upregulation of ABA-signaling cascade genes like *TaABARE* and *TaOPR1* in PGPR (*Dietzia natronolimnaea* STR1) inoculated wheat prevents the oxidative effects of salinity which were attributed to upregulation of expression levels of antioxidants genes as well (Bharti et al., 2016).

PGPR assist plants to combat salinity stress by enhancing the synthesis of organic as well as inorganic compounds, including phytohormones like ABA, ethylene, volatile compounds, and antioxidants. It is established now that exploitation of halotolerant species of PGPR can better help crop plants to withstand salinity stress (Abbas et al., 2019). Inoculation of the halotolerant PGPR *Bacillus fortis* improves salinity tolerance in *Capsicum annum* by strengthening the tolerance mechanisms and the expression levels of genes like CAPIP2, CaKR1, CaOSM1, and CAChi2 to control stress tolerance (Yasin et al., 2018). The decline in the accumulation of toxic ions like Na under salinity stress due to PGPR and AMF results due to the influence on the expression patterns of their genes (Bharti et al., 2016; Evelin, Devi, Gupta, & Kapoor, 2019). Beneficial soil microbes bring ion homeostasis and maintain improved uptake of beneficial mineral ions to increase water use efficiency and osmoprotection leading to a significant increase in the photosynthetic efficiency and cellular structural stability. Overexpression of genes involved in Na extrusion result in maintenance of a favorable Na/K ratio which is mainly regulated by plasma membrane and tonoplast aquaporins (Evelin et al., 2019; reviewed in Bhat et al., 2020). Therefore the dissection and subsequent studies for further elucidation of their function in salinity tolerance as influenced by microbes colonization needs further attention. Inoculation of PGPR *B. amyloliquefaciens* SN13 imparts salinity and osmotic stress tolerance in rice by upregulating the expression of stress responsive genes and the tolerance mechanisms like antioxidant and osmolyte metabolism (Tiwari, Prasad, Chauhan, & Lata, 2017).

10.4.3 Temperature stress

Drought and salinity temperature regimes (high and low) are the prime stress factors responsible for declined growth and yield productivity across the globe. An increase in temperature results in intensifying the damaging effects of other stresses like drought, mineral stresses, etc. The damaging effects of temperature regimes on the growth and functioning of plants has been reported by several researchers. There are alterations in growth, photosynthesis, mineral uptake, osmolytes, antioxidant functioning, enzyme activity, gene expression, phytohormones metabolism, and hence yield (Bita & Gerats, 2013; Hussain et al., 2018; Pischl & Barber, 2017; Wang, Heckathorn, Mainali, & Tripathee, 2016; Zhang et al., 2018). Exposure of wheat seedlings to low temperatures has been reported to increase the activities of sucrose phosphate synthase and sucrose synthase activity significantly at 4°C and 2°C respectively while reductions in the activities of sucrose synthase and invertase were exhibited with a reduction in temperature. In addition to this, the expression of genes like *TaSuS1* was downregulated, while *TaSuS2* was upregulated and *CWINV* expression was inhibited with a significant increase in abscisic acid content and decrease in auxin and gibberellins (Zhang et al., 2018). Temperature (high and low) increased the production of ROS imparting oxidative effects. Exposure of *Hordeum spontaneum* to combination of drought and temperature stress resulted in a significant decline in biomass production and yield through inhibition of PSII functioning reflecting the altered functioning of oxygen evolving complex and PSII (Jedmowski, Ashoub, Momtaz, & Brüggemann, 2015). Temperature extremes alter the cellular osmotic potential, membrane structural, and functional integrity and restrict tolerance mechanisms, resulting in loss of cellular stability (Ritonga & Chen, 2020; Yu et al., 2020; Zhang et al., 2018). Low temperatures induce leaf area reduction, chlorosis, wilting, reduced hydraulic conductivity, impaired stomatal functioning, and photosynthesis (Liu et al., 2013; Paredes & Quiles, 2015).

New techniques like DNA technologies, mutational breeding, and genome editing have been proposed to have a promising role in the development of temperature extreme resistant crop varieties (Janni et al., 2020). However, the exploitation of beneficial microbes for reducing the damaging effects of temperature stress (low and high) have gained attention from researchers all across the globe. Liu et al. (2013) while working with oat (*Avena nuda*) have demonstrated significant reduction in the oxidative damage caused due to cold (−10°C) stress. Such effects of AMF not only depend on the type of plant but also have much relation with the temperature tolerance potential of the host plants. In this regard, Badri et al. (2009) have reported that under ambient and elevated temperatures the growth and AMF colonization pattern, like spore germination and extraradical hyphal growth, show a significant reduction in sensitive plants as compared to tolerant ones. There are reports contrary to this, wherein it has been reported that low temperatures have no effect on the AMF colonization (Heinemeyer & Fitter, 2004). The damaging effects of temperature (low and high) regimes caused due to the greater accumulation of ROS are believed to be reduced by the upregulation of the antioxidant system (X. Chen et al., 2014). Inoculation of AMF prevents the damaging effects of higher temperature and drought on mineral uptake and protein content.

Inoculation of ACC-deaminase and exopolysaccharide producing *Bacillus cereus* to *Solanum lycopersicum* significantly mitigates the damaging effects of high temperature by modulating the antioxidant system and phytohormone synthesis resulting in enhanced growth and tolerance (Mukhtar et al., 2020). In another study Ali, Sandhya, Grover, Linga, and Bandi (2011) demonstrated that inoculation of thermo-tolerant *Pseudomonas putida* strain AKMP7 to wheat results in improved growth and yield productivity through upregulation of the antioxidant system and the accumulation of osmolytes like proline, sugars, and amino acids. Zubair et al. (2019) reported significant mitigation of cold stress-mediated damaging effects of wheat by inoculation of psychrophillic *Bacillus* spp. through modulation of cold stress responsive bacterial genes and the further strengthening of tolerance mechanisms and phytohormone synthesis, reflected by quick stress signaling and subsequent amelioratory responses.

Amelioration of the damaging effects of temperature stress on the photosynthetic attributes have been reported due to AMF colonization through the accumulation of proline and sugars (Xu, Lu, & Zhu, 2016) indicating the role of AMF in mediating osmolyte signaling for improved plant growth and performance. AMF inoculated to cucumber prevents the damaging effects of excess ROS (H_2O_2) on the membrane functioning and growth by maintaining the optimal functioning of key tolerance mechanisms (S. Chen et al., 2013).

PGPR (*Burkholderia phytofirmans PsJN*) inoculation upregulates the expression pattern of genes like *RbcS, RbcL, CBF1, CBF2, CBF3, ICE1, COR15a*, and *COR78*, resulting in greater photosynthesis by improving PSII activity, Rubisco protein abundance, and pigment content (Su et al., 2015).

10.4.4 Heavy metal(loid) stress

The rapid industrialization and urbanization has posed a serious threat to sustainable agriculture through the toxic pollution caused by heavy metal and metalloids. These metal(loid)s are very toxic to plants as well as humans. Their accumulation within the edible plant parts makes them more dangerous for humans, resulting in severe diseases like renal disorders etc. (Ahmad et al., 2015; Goyal et al., 2020). Among the toxic heavy metals are cadmium, arsenic, lead, chromium, mercury, etc. Their high concentration of bioavailability results in the generation of toxic radicals resulting in uncontrolled oxidation of key macromolecules like proteins, lipids, and nucleic acids through the initiation of a chain of reactions, hence altering the normal metabolism and thereby causing the reduction in growth and biomass productivity (Goyal et al., 2020). Strategies that contribute to heavy metal tolerance include sequestration into cellular compartments like vacuoles or apoplasts and detoxification, which involves their conversion into nontoxic forms (Goyal et al., 2020). However, the strengthening of the indigenous tolerance mechanisms through biological means can be more affective. The damaging effects of heavy metals and metalloids have been widely reported (Ahmad et al., 2015; Asgher, Khan, Khan, Fatma, & Masood, 2014; Mesa-Marin et al., 2018). Heavy metals reduce the enzyme activity by showing greater affinity toward the active sites of enzymes. Reduced leaf area, declined pigment and Rubisco synthesis, altered mineral uptake and assimilation, photosynthetic arrest and enzyme activity inhibition, and reduced yield are the key effects of metal toxicity on plants (Ahanger, Aziz, Alsahli, Alyemeni, & Ahmad, 2020; Fatnassi, Chiboub, Jebara, & Jebara, 2015). Plants upregulate the tolerance mechanisms for elimination of the toxic effects resulting due to heavy metal accumulation, however there must be an effort to improve the existing tolerance mechanism for mediating quick response against any kind of ill effects resulting from a metal. In this connection the use of beneficial soil microbes can be one of the best strategies for strengthening of these tolerance mechanisms. The following section discusses some recent reports showing the PGPR and AMF-mediated growth and metal tolerance regulation in plants.

The soil-borne microbes have the capacity to alter the oxidation state of toxic metal(loid) ions, rendering them either less toxic or leading to their complete elimination by complex formation (Van Hullebusch, Lens, & Tabak, 2005). The beneficial soil microbes mediate elimination of toxic metal ions through precipitation as polyphosphate granules, chelation, or the modulations in the soil rhizospheric characters. It has been suggested that AMF-led secretion of glomalin is responsible for the restricted movement of metals like cadmium and copper (Gonzalez-Chavez, Carrillo-Gonzalez, Wright, & Nichols, 2004). The inoculation of PGPR (*Pseudomonas aeruginosa* and *Burkholderia gladioli*) restricts the entry of cadmium into *Lycopersicum esculentum* by downregulating the expression of metal transporters with significant increase in the synthesis of chelating compounds like protein and nonprotein thiols resulting in greater growth and photosynthetic protection (Khanna, Jamwal, Gandhi, Ohri, & Bhardwaj, 2019). Fatnassi et al. (2015) have demonstrated increased growth under copper stress due to inoculation of PGPR (*Enterbacter clocae* and *Pseudomonas* sp.) through upregulation of the antioxidant system and the tolerance mechanisms were further strengthened by *Rhizobium* application. Inoculation of PGPR (*Bacillus* spp.) reduces the root respiration and oxidative stress under heavy metal (cadmium, arsenic, lead, copper, nickel, and zinc) stress, and enhanced the antioxidant system (Mesa-Marin et al., 2018). The above-discussed reports do justify the potentiality of PGPR and AMF in avoiding the damaging effects of heavy metals. It can be said that the beneficial soil microbes contribute to metal ion tolerance through restricted entry of metal ions into plant tissue. On other hand, Konkolewska et al. (2020) reported that inoculation of *B. phytofirmans* $PsJN^T$ improves the phytoextraction of metal ions like Zn, Cd, and Pb by improving biomass production, thus paving the way for greater metal ion accumulation. Therefore it could be inferred that for plants having considerable efficiency to accumulate metal ions PGPR can prove beneficial in protecting soils from toxic effects, while in the case of sensitive plants they prove beneficial by restricting the entry of metals into shoot tissues. The genes encoding metallothionein from *Gigaspora margarita* have been shown to have the efficiency to bring down the metal toxicity (Lanfranco, Bolchi, Ros, Ottonello, & Bonfante, 2002). Among the AMF species the variation in production of glomalin renders them less or more efficient in the chelation of metal ions. Besides having an impact on the physiological and biochemical attributes, the beneficial soil microbes have an obvious influence at the molecular levels studied in terms of gene expression patterns. Shavalikohshori, Zalaghi, Sorkheh, and Enaytizamir (2020) have reported the upregulation of the proline biosynthesizing genes in wheat due to *Pseudomonas* inoculation.

10.4.5 Biotic stresses

Biotic stresses are other key factors affecting the growth and development of plants and posing a great threat to crop productivity. Different disease-causing fungi, viruses etc. have been isolated and observed to reduce the growth of crop plants all over the globe (Jain, Sarsaiya, Wu, Lu, & Shi, 2019; Nassar, Ibrahim, & Adss, 2016; V. K. Singh, Singh, & Upadhyay,

2017). This biotic infestation has led to serious threats on sustainable crop productivity. Many reports have demonstrated the beneficial role of PGPR and AMF in lessening the damaging effects of biotic infestation through strengthening of the tolerance mechanisms, like the inhibition of toxic microbial growth, antioxidant system, systematic acquired resistance, phytohormone and phytoalexin production, secondary metabolite synthesis, etc. (Lyu, Backer, Robuinson, & Smith, 2019; Miozzi et al., 2019; Suryadi, Susilowati, & Fauziah, 2019). Inoculation of AMF increases the production of key secondary metabolites by improving the expression of genes regulating the synthesis of flavonoids and chlorogenic acid in tomato, resulting in increased resistance against the tomato mosaic virus (Aseel, Rashad, & Hammad, 2019). Beneficial bacteria exhibit visible antagonistic effects against the charcoal rot-causing fungi *Macrophomina phaseolina* in soybean by enhancing the activity of antioxidant enzymes and phenyl alanine ammonia lyase (Choudhary, 2011).

10.5 Conclusion and future directions

Soil beneficial microbes, including PGPR and AMF, have obvious beneficial effects on the growth of host plants under a range of environmental stresses. These effects are obviously evident at the morphological levels and are also reflected in increased yield productivity. Both PGPR and AMF contribute to upregulation of the tolerance mechanisms at physiological, biochemical, and molecular levels for enhancing the tolerance to the abiotic and biotic stress factors. The modulations of phytohormones, ROS status, nutrient uptake, and assimilation, and ion sequestration contribute immensely to stress adaptation. Using the existing novel techniques for identification of key pathways and their subsequent dissection for improved tolerance can be worthwhile in improving phytoremediation potential as well as the stress-withstanding potential of crop plants. Transgenics harboring the genes that are responsible for stress tolerance and increased yield have to be investigated for further research. Microbial strains with tolerant genes for particular or combination of stresses need to be identified and exploited for improving tolerance in crops.

References

Abbas, R., Rasul, S., Aslam, K., Baber, M., Shahid, M., Mubeen, F., & Naqqash, T. (2019). Halotolerant PGPR: A hope for cultivation of saline soils. *Journal of King Saud University – Science.*, *31*(4), 1195–1201.

Abd-Allah, E. S., Abeer, H., Alqarawi, A. A., & Alwathnani, H. A. (2015). Alleviation of adverse impact of cadmium stress in sunflower (*Helianthus annuus* L.) by arbuscular mycorrhizal fungi. *Pakistan Journal of Botany*, *47*(2), 785–795.

Abdel Latef, A. A., & Chaoxing, H. (2014). Does the inoculation with *Glomus mosseae* improve salt tolerance in pepper plants? *Journal of Plant Growth Regulation*, *33*, 644–653.

Ahanger, M. A., & Agarwal, R. M. (2016). Potassium improves antioxidant metabolism and alleviates growth inhibition under water and osmotic stress in wheat (*Triticum aestivum* L). *Protoplasma*. Available from https://doi.org/10.1007/s00709-016-1037-0.

Ahanger, M. A., Aziz, U., Alsahli, A. A., Alyemeni, M. N., & Ahmad, P. (2020). Combined kinetin and spermidine treatments ameliorate growth and photosynthetic inhibition in *Vigna angularis* by upregulating antioxidant and nitrogen metabolism under cadmium stress. *Biomolecules*, *10*, 147. Available from https://doi.org/10.3390/biom10010147.

Ahanger, M. A., Mir, R. A., Alyemeni, M. N., & Ahmad, P. (2020). Combined effects of brassinosteroid and kinetin mitigates salinity stress in tomato through the modulation of antioxidant and osmolyte metabolism. *Plant Physiology and Biochemistry*, *147*, 31–42.

Ahanger, M. A., Qin, C., Maodong, Q., Dong, X. X., Ahmad, P., Abd_Allah, E. F., & Zhang, L. (2019). Spermine application alleviates salinity induced growth and photosynthetic inhibition in *Solanum lycopersicum* by modulating osmolyte and secondary metabolite accumulation and differentially regulating antioxidant metabolism. *Plant Physiology and Biochemistry*, *144*, 1–13.

Ahanger, M. A., Tittal, M., Mir, R. A., & Agarwal, R. M. (2017). Alleviation of water and osmotic stress-induced changes in nitrogen metabolizing enzymes in *Triticum aestivum* L. cultivars by potassium. *Protoplasma*, *254*(5), 1953–1963.

Ahanger, M. A., Tomar, N. S., Tittal, M., Argal, S., & Agarwal, R. M. (2017). Plant growth under water/salt stress: ROS production; antioxidants and significance of added potassium under such conditions. *Physiology and Molecular Biology of Plants*, *23*(4), 731–744.

Ahanger, M. A., Tyagi, S. R., Wani, M. R., & Ahmad, P. (2014). Drought tolerance: Roles of organic osmolytes, growth regulators and mineral nutrients. In P. Ahmad, & M. R. Wani (Eds.), *Physiological mechanisms and adaptation strategies in plants under changing environment* (Volume Ist, pp. 25–56). Springer Science + Business media, inc.

Ahanger, M. A., Hashem, A., Abd Allah, E. F., & Ahmad, P. (2014). Arbuscular mycorrhiza in crop improvement under environmental stress. In P. Ahmad, & S. Rasool (Eds.), *Emerging Technologies and management of crop stress tolerance* (Volume 2, pp. 69–95).

Ahemad, M., & Kibret, M. (2014). Mechanisms and applications of plant growth promoting rhizobacteria: Current perspective. *Journal of King Saud University – Science*, *26*, 1–20.

Ahmad, P., Hakeem, K. R., Kumar, A., Ashraf, M., & Akram, N. A. (2012). Salt induced changes in photosynthetic activity and oxidative defense system of three cultivars of mustard (*Brassica juncea* L.). *African Journal of Biotechnology*, *11*(11), 2694–2703.

Ahmad, P., Sarwat, M., Bhat, N. A., Wani, M. R., Kazi, A. G., & Tran, L. S. (2015). Alleviation of cadmium toxicity in *Brassica juncea* L. (Czern. & Coss.) by calcium application involves various physiological and biochemical strategies. *PLoS One*, *10*(1), e0114571. Available from https://doi.org/10.1371/journal.pone.0114571.

Al-Ezerjawi, N. H., & Kadhim, J. H. (2014). Effect of two isolates of *Trichoderma harzianum* on total nitrogen, chlorophyll a & b contents and yield of wheat (*Triticum aestivum* L) Class Eba'a-95. *International Journal of Science and Research*, *3*(8), 1078–1083.

Ali, S. Z., Sandhya, V., Grover, M., Linga, V. R., & Bandi, V. (2011). Effect of inoculation with a thermotolerant plant growth promoting Pseudomonasputida strain AKMP7 on growth of wheat (Triticum spp.) under heat stress. *Journal of Plant Interactions*, *6*(4), 239–246. Available from https://doi.org/10.1080/17429145.2010.545147.

Al-Karaki, G., McMichael, B., & Zak, J. (2004). Field response of wheat to arbuscular mycorrhizal fungi and drought stress. *Mycorrhiza*, *14*, 263–269.

Alqarawi, A. A., Hashem, A., Abd_Allah, E. F., Alshahrani, T. S., & Huqail, A. A. (2014). Effect of salinity on moisture content, pigment system, and lipid composition in *Ephedra alata* Decne. *Acta Biologica Hungarica, 65*(1), 61–71. Available from https://doi.org/10.1556/ABiol.65.2014.1.6.

Aroca, R., Ruiz-Lozano, J. M., Zamarreno, A. M., Paz, J. A., Garcia-Mina, J. M., Pozo, M. J., & Lopez-Raez, J. A. (2013). Arbuscular mycorrhizal symbiosis influences strigolactone production under salinity and alleviates salt stress in lettuce plants. *Journal of Plant Physiology, 170*, 47–55.

Aseel, D. G., Rashad, Y. M., & Hammad, S. M. (2019). Arbuscular mycorrhizal fungi trigger transcriptional expression of flavonoid and chlorogenic acid biosynthetic pathways genes in tomato against *Tomato Mosaic Virus. Scientific Reports, 9*, 9692. Available from https://doi.org/10.1038/s41598-019-46281-x.

Asgher, M., Khan, N. A., Khan, M. I. R., Fatma, M., & Masood, A. (2014). Ethylene production is associated with alleviation of cadmium-induced oxidative stress by sulfur in mustard types differing in ethylene sensitivity. *Ecotoxicology and Environmental Safety, 106*, 54–61.

Azooz, M. M., Youssef, A. M., & Ahmad, P. (2011). Evaluation of salicylic acid (SA) application on growth, osmotic solutes and antioxidant enzyme activities on broad bean seedlings grown under diluted seawater. *International Journal of Plant Physiology and Biochemistry, 3*(14), 253–264.

Badri, D. V., Weir, T. L., van der Lelie, D., & Vivanco, J. M. (2009). Rhizosphere chemical dialogues: Plant–microbe interactions. *Current Opinion in Biotechnology, 20*, 642–650. Available from https://doi.org/10.1016/j.copbio.2009.09.014.

Barnawal, D., Bharti, N., Maji, D., Chanotiya, C. S., & Kalra, A. (2014). ACC deaminase-containing Arthrobacter protophormiae induces NaCl stress tolerance through reduced ACC oxidase activity and ethylene production resulting in improved nodulation and mycorrhization in *Pisum sativum. Journal of Plant Physiology, 171*, 884–894.

Batool, T., Ali, S., Seleiman, M. F., Naveed, N. H., Ali, A., Ahmed, K., ... Mubushar, M. (2020). Plant growth promoting rhizobacteria alleviates drought stress in potato in response to suppressive oxidative stress and antioxidant enzymes activities. *Scientific Reports, 10*, 16975. Available from https://doi.org/10.1038/s41598-020-73489-z.

Begum, N., Ahanger, M. A., Su, Y., Lei, Y., Mustafa, N. S. A., Ahmad, P., & Zhang, L. (2019). Improved drought tolerance by AMF inoculation in maize (*Zea mays*) involves physiological and biochemical implications. *Plants, 8*, 579. Available from https://doi.org/10.3390/plants8120579.

Begum, N., Ahanger, M. A., & Zhang, L. (2020). AMF inoculation and phosphorus supplementation alleviates drought induced growth and photosynthetic decline in *Nicotiana tabacum* by upregulating antioxidant metabolism and osmolyte accumulation. *Environmental and Experimental Botany*. Available from https://doi.org/10.1016/j.envexpbot.2020.104088.

Begum, N., Qin, C., Ahanger, M. A., Raza, S., Khan, M. I., Ahmed, N., ... Zhang, L. (2019). Role of arbuscular mycorrhizal fungi in plant growth regulation: Implications in abiotic stress tolerance. *Frontiers in Plant Science*. Available from https://doi.org/10.3389/fpls.2019.01068.

Bharti, N., Pandey, S. S., Barnawal, D., Patel, V. K., & Kalra, A. (2016). Plant growth promoting rhizobacteria *Dietzia natronolimnaea* modulates the expression of stress responsive genes providing protection of wheat from salinity stress. *Scientific Reports, 6*, 34768.

Bhat, M. A., Kumar, V., Bhat, M. A., Wani, I. A., Dar, F. L., Iqra, F., ... Jan, A. T. (2020). Mechanistic insights of the interaction of plant growth-promoting rhizobacteria (PGPR) with plant roots toward enhancing plant productivity by alleviating salinity stress. *Frontiers in Microbiology*, 20 August 2020. Available from https://doi.org/10.3389/fmicb.2020.01952.

Bita, C. E., & Gerats, T. (2013). Plant tolerance to high temperature in a changing environment: Scientific fundamentals and production of heat stress-tolerant crops. *Frontiers in Plant Science*. 31 July 2013 | https://doi.org/10.3389/fpls.2013.00273.

Cekic, F. O., Unyayar, S., & Ortas, I. (2012). Effect of arbuscular mycorrhizal inoculation on biochemical parameters in *Capsicum annum* grown under long term salt stress. *Turkish Journal of Botany*, *36*, 63–72.

Chen, S., Jin, W., Liu, A., Zhang, S., Liu, D., Wang, F., ... He, C. (2013). Arbuscular mycorrhizal fungi (AMF) increase growth and secondary metabolism in cucumber subjected to low temperature stress. *Scientia Horticulturae*, *160*, 222–229.

Chen, X., Song, F., Liu, F., Tian, C., Liu, S., Xu, H., & Zhu, X. (2014). Effect of different arbuscular mycorrhizal fungi on growth and physiology of maize at ambient and low temperature regimes. *Scientific World Journal*. Available from http://doi.org/10.1155/2014/956141.

Chiapero, J., Cappellari, L. D. R., Alderete, L. G. S., Palermo, T. B., & Banchio, E. (2019). Plant growth promoting rhizobacteria improve the antioxidant status in *Mentha piperita* grown under drought stress leading to an enhancement of plant growth and total phenolic content. *Industrial Crops and Products*, *139*111553. Available from https://doi.org/10.1016/j.indcrop.2019.111553.

Choudhary, D. K. (2011). Plant growth-promotion (PGP) activities and molecular characterization of rhizobacterial strains isolated from soybean (*Glycine max* L. Merril) plants against charcoal rot pathogen, *Macrophomina phaseolina*. *Biotechnology Letters*, *33*(11), 2287–2295. Available from https://doi.org/10.1007/s10529-011-0699-0.

Chowdhury, S. P., Hartmann, A., Gao, X. W., & Borriss, R. (2015). Biocontrol mechanism by root-associated *Bacillus amyloliquefaciens* FZB42-a review. *Frontiers in Microbiology*, *6*, 780. Available from https://doi.org/10.3389/fmicb.2015.00780.

Elhindi, K. M., El-Din, A. S., & Elgorban, A. M. (2017). The impact of arbuscular mycorrhizal fungi in mitigating salt-induced adverse effects in sweet basil (*Ocimum basilicum* L.). *Saudi Journal of Biological Sciences*, *24*(1), 170–179. Available from https://doi.org/10.1016/j.sjbs.2016.02.010.

Elkelish, A. A., Alhaithloul, H. A. S., Qaric, S. H., Soliman, M. H., & Hasanuzzaman, M. (2020). Pretreatment with *Trichoderma harzianum* alleviates waterlogging-induced growth alterations in tomato seedlings by modulating physiological, biochemical, and molecular mechanisms. *Environmental and Experimental Botany*, *171*103946.

Elkelish, A. A., Soliman, M. H., Alhaithloul, H. A., & El-Esawi, M. A. (2019). Selenium protects wheat seedlings against salt stress-mediated oxidative damage by upregulating antioxidants and osmolytes metabolism. *Plant Physiology and Biochemistry*, *137*, 144–153.

Evelin, H., Devi, T. S., Gupta, S., & Kapoor, R. (2019). Mitigation of salinity stress in plants by arbuscular mycorrhizal symbiosis: Current understanding and new challenges. *Frontiers in Plant Science*, *10*, 470. Available from https://doi.org/10.3389/fpls.2019.00470.

Fatnassi, I. C., Chiboub, M., Jebara, O. S. M., & Jebara, S. H. (2015). Impact of dual inoculation with Rhizobium and PGPR on growth and antioxidant status of *Vicia faba* L. under copper stress. *Comptes Rendus Biologies*, *338*(4), 241–254.

Fusconi, A., & Berta, G. (2012). Environmental stress and role of arbuscular mycorrhizal symbiosis. In P. Ahmad, & M. Prasad (Eds.), *Abiotic Stress Responses in Plants*. New York, NY: Springer. Available from https://doi.org/10.1007/978-1-4614-0634-1_11.

Gholamhoseini, M., Ghalavand, A., Dolatabadian, A., Jamshidi, E., & Khodaei-Joghan, A. (2013). Effects of arbuscular mycorrhizal inoculation on growth, yield, nutrient uptake and irrigation water productivity of sunflowers grown under drought stress. *Agricultural Water Management*, *117*, 106–114.

Ghorbanpour, M., Hatami, M., & Khavazi, K. (2013). Role of plant growth promoting rhizobacteria on antioxidant enzyme activities and tropane alkaloid production of *Hyoscyamus niger* under water deficit stress. *Turkish Journal of Biology, 37*, 350–360. Available from https://doi.org/10.3906/biy-1209-12.

Glick, B. R. (2012). Plant growth-promoting bacteria: Mechanisms and applications. *Scientifca*, Article ID 963401. Available from http://doi.org/10.6064/2012/963401.

Gonzalez-Chavez, M. C., Carrillo-Gonzalez, R., Wright, S. F., & Nichols, K. A. (2004). The role of glomalin, a protein produced by arbuscular mycorrhizal fungi, in sequestering potentially toxic elements. *Environmental Pollution, 130*, 317–323.

Goyal, D., Yadav, A., Prasad, M., Singh, T. B., Shrivastav, P., Ali, A., ... Mishra, S. (2020). Effect of heavy metals on plant growth: An overview. In M. Naeem, A. Ansari, & S. Gill (Eds.), *Contaminants in agriculture*. Cham: Springer. Available from https://doi.org/10.1007/978-3-030-41552-5_4.

Gray, E. J., & Smith, D. L. (2005). Intracellular and extracellular PGPR: Commonalities and distinctions in the plant-bacterium signaling processes. *Soil Biology & Biochemistry, 37*, 395–412. Available from https://doi.org/10.1016/j.soilbio.2004.08.030.

Habib, S. H., Kausar, H., & Saud, H. M. (2016). Plant growth-promoting rhizobacteria enhance salinity stress tolerance in Okra through ROS-scavenging enzymes. *BioMed Research International, 2016*(2016), 6284547. Available from https://doi.org/10.1155/2016/6284547.

Hameed, A., Wu, Q. S., Abd-Allah, E. F., Hashem, A., Kumar, A., Lone, H. A., & Ahmad, P. (2014). Role of AM Fungi in alleviating drought stress in plants. In M. Miransari (Ed.), *Use of microbes for the alleviation of soil stresses*. Springer Science + Business Media New York. Available from http://doi.org/10.1007/978-1-4939-0721-2_4.

Hayat, S., Hayat, Q., Alyemeni, M. N., Wani, A. S., Pichtel, J., & Ahmad, A. (2012). Role of proline under changing environments: A review. *Plant Signaling & Behavior, 7*, 1456–1466.

Heideri, M., & Golpayegani, A. (2012). Effects of water stress and inoculation with plant growth promoting rhizobacteria (PGPR) on antioxidant status and photosynthetic pigments in basil (*Ocimum basilicum* L.). *Journal of the Saudi Society of Agricultural Sciences, 11*(1), 57–61.

Heinemeyer, A., & Fitter, A. H. (2004). Impact of temperature on the arbuscular mycorrhizal (AM) symbiosis: Growth responses of the host plant and its AM fungal partner. *Journal of Experimental Botany, 55*(396), 525–534. Available from https://doi.org/10.1093/jxb/erh049.

Huang, Y. M., Zou, Y. N., & Wu, Q. S. (2017). Alleviation of drought stress by mycorrhizas is related to increased root H2O2 efflux in trifoliate orange. *Scientific Reports, 7*, 42335. Available from https://doi.org/10.1038/srep42335.

Hussain, H. A., Hussain, S., Khaliq, A., Ashraf, U., Anjum, S. A., Men, S., & Wang, L. (2018). Chilling and drought stresses in crop plants: Implications, cross talk, and potential management opportunities. *Frontiers in Plant Science, 2018*(9), 393. Available from https://doi.org/10.3389/fpls.2018.00393.

Ilangumaran, G., & Smith, D. L. (2017). Plant growth promoting rhizobacteria in amelioration of salinity stress: A systems biology perspective. *Frontiers in Plant Science*. 23 October 2017 | https://doi.org/10.3389/fpls.2017.01768.

Jain, A., Sarsaiya, S., Wu, Q., Lu, Y., & Shi, J. (2019). A review of plant leaf fungal diseases and its environment speciation. *Bioengineered, 10*(1), 409–424.

Janni, M., Gulli, M., Maestri, E., Marmiroli, M., Valliyodan, B., Nguyen, H. T., & Marmiroli, N. (2020). Molecular and genetic bases of heat stress responses in crop plants and breeding for increased resilience and productivity. *Journal of Experimental Botany, 71*(13), 3780–3802.

Jatav, K. S., Agarwal, R. M., Tomar, N. S., & Tyagi, S. R. (2014). Nitrogen metabolism, growth and yield responses of wheat (*Triticum aestivum* L) to restricted water supply and varying potassium treatments. *Journal of the Indian Botanical Society*, *93*, 177–189.

Jedmowski, C., Ashoub, A., Momtaz, O., & Brüggemann, W. (2015). Impact of drought, heat, and their combination on chlorophyll fluorescence and yield of wild barley (*Hordeum spontaneum*). *Journal of Botany*. Available from http://doi.org/10.1155/2015/120868.

Kasim, W. A., Osman, M. E., Omar, M. N., El-Daim, I. A. A., Bejai, S., & Meijer, J. (2013). Control of drought stress in wheat using plant-growth-promoting bacteria. *Journal of Plant Growth Regulation*, *32*(2013), 122–130. Available from https://doi.org/10.1007/s00344-012-9283-7.

Khan, M. I. R., Asgher, M., & Khan, N. A. (2014). Alleviation of salt-induced photosynthesis and growth inhibition by salicylic acid involves glycine betaine and ethylene in mungbean (*Vigna radiata* L.). *Plant Physiology and Biochemistry*, *80*, 67–74.

Khan, M. I. R., Nazir, F., Asgher, M., Per, T. S., & Khan, N. A. (2015). Selenium and sulfur influence ethylene formation and alleviate cadmium-induced oxidative stress by improving proline and glutathione production in wheat. *Journal of Plant Physiology*, *173*, 9–18.

Khan, N., Bano, A., & Zandi, P. (2018). Effects of exogenously applied plant growth regulators in combination with PGPR on the physiology and root growth of chickpea (*Cicer arietinum*) and their role in drought tolerance. *Journal of Plant Interactions*, 239–247.

Khanna, K., Jamwal, V. L., Gandhi, S. G., Ohri, P., & Bhardwaj, R. (2019). Metal resistant PGPR lowered Cd uptake and expression of metal transporter genes with improved growth and photosynthetic pigments in *Lycopersicon esculentum* under metal toxicity. *Scientific Reports*, *9*, 5855. Available from https://doi.org/10.1038/s41598-019-41899-3.

Konkolewska, A., Piechalak, A., Ciszewska, L., Anto-Krzeminska, N., Skrzpcak, T., Sitko, K., ... Malecka, A. (2020). Combined use of companion planting and PGPR for the assisted phytoextraction of trace metals (Zn, Pb, Cd). *Environmental Science and Pollution Research*, *27*, 13809–13825.

Kuan, K. B., Othman, R., Abdul Rahim, K., & Shamsuddin, Z. H. (2016). Plant growth-promoting rhizobacteria inoculation to enhance vegetative growth, nitrogen fixation and nitrogen remobilisation of maize under greenhouse conditions. *PLoS One*, *11*, e0152478. Available from https://doi.org/10.1371/journal.pone.0152478.

Labuschagne, N., Pretorius, T., & Idris, A. H. (2010). Plant growth promoting rhizobacteria as biocontrol agents against soil-borne plant diseases. In D. K. Maheshwari (Ed.), *Plant growth and health promoting bacteria* (pp. 211–230). Berlin, Heidelberg: Springer Verlag.

Lanfranco, L., Bolchi, A., Ros, E. C., Ottonello, S., & Bonfante, P. (2002). Differential expression of a metallothionein gene during the presymbiotic versus the symbiotic phase of an arbuscular mycorrhizal fungus. *Plant Physiology*, *130*, 58–67.

Li, T., Hu, Y., Hao, Z., Li, H., Wang, Y., & Chen, B. (2013). First cloning and characterisation of two functional aquaporin genes from am arbuscular mychorrizal fungus Glomus intraradices. *The New Phytologist*, *197*, 617–630.

Lin, J., Wang, Y., Sun, S., Mu, C., & Yan, X. (2017). Effects of arbuscular mycorrhizal fungi on the growth, photosynthesis and photosynthetic pigments of *Leymus chinensis* seedlings under salt-alkali stress and nitrogen deposition. *Science of the Total Environment*, *576*, 234–241.

Liu, W., Yu, K., He, T., Li, F., Zhang, D., & Liu, J. (2013). The low temperature induced physiological responses of *Avena nuda* L., a cold-tolerant plant species. *The Scientific World Journal*, 2013. Available from http://doi.org/10.1155/2013/658793.

Lyu, D., Backer, R., Robuinson, W. G., & Smith, D. L. (2019). Plant growth-promoting rhizobacteria for *Cannabis* production: Yield, cannabinoid profile and disease resistance. *Frontiers in Microbiology*. Available from https://doi.org/10.3389/fmicb.2019.01761.

Marulanda, A., Azcon, R., Chaumont, F., Ruiz-Lozano, J. M., & Aroca, R. (2010). Regulation of plasma membrane aquaporins by inoculation with a *Bacillus megaterium* strain in maize (*Zea mays* L.) plants under unstressed and salt-stressed conditions. *Planta*, *232*, 533–543. Available from https://doi.org/10.1007/s00425-010-1196-8.

Mesa-Marin, J., Del-Saz, N. F., Rodriguez-Llorente, I. D., Redondo-Gomez, S., Pajuelo, E., Ribas-Carbo, M., & Mateos-Naranjo, E. (2018). PGPR reduce root respiration and oxidative stress enhancing *Spartina maritima* root growth and heavy metal rhizoaccumulation. *Frontiers in Plant Science*. Available from https://doi.org/10.3389/fpls.2018.01500.

Miozzi, L., Vaira, A. M., Catoni, M., Fiorlli, V., Accotto, G. P., & Lanfranso, L. (2019). Arbuscular mycorrhizal symbiosis: Plant friend or foe in the fight against viruses? *Frontiers in Microbiology*. Available from https://doi.org/10.3389/fmicb.2019.01238.

Mittler, R. (2002). Oxidative stress, antioxidants and stress tolerance. *Trends in Plant Science*, *7*, 405–410.

Mukhtar, T., Rehman, S., Smith, D., Sultan, T., Seleiman, M. F., Alsadon, A. A., ... Saad, M. A. O. (2020). Mitigation of heat stress in *Solanum lycopersicum* L. by ACC-deaminase and exopolysaccharide producing *Bacillus cereus*: Effects on biochemical profiling. *Sustainability*, *2020*(12), 2159. Available from https://doi.org/10.3390/su12062159.

Nadeem, S. M., Zahir, Z. A., Naveed, M., & Arshad, M. (2009). Rhizobacteria containing ACC-deaminase confer salt tolerance in maize grown on salt-affected fields. *Canadian Journal of Microbiology*, *55*, 1302–1309. Available from https://doi.org/10.1139/W09-092.

Nassar, A. M. K., Ibrahim, A. A., & Adss, I. A. A. (2016). 2,4-Dichlorophenoxy acetic acid, abscisic acid, and hydrogen peroxide induced resistance-related components against potato early blight (*Alternaria solani*, Sorauer). *Annals of Agricultural Sciences*, *61*(1), 15–23.

Nautiyal, C. S., Srivastava, S., Chauhan, P. S., Seem, K., Mishra, A., & Sopory, S. K. (2013). Plant growth-promoting bacteria *Bacillus amyloliquefaciens* NBRISN13 modulates gene expression profile of leaf and rhizosphere community in rice during salt stress. *Plant Physiology and Biochemistry*, *66*, 1–9. Available from https://doi.org/10.1016/j.plaphy.2013.01.020.

Numan, M., Bashir, S., Khan, Y., Mumtaz, R., Shinwari, Z. K., Khan, A. L., ... Al-Harrasi, A. (2018). Plant growth promoting bacteria as an alternative strategy for salt tolerance in plants: A review. *Microbial Research*, *209*, 21–32.

Ogawa, A., & Yamauchi, A. (2006). Root osmotic adjustment under stress in maize seedlings. Mode of accumulation of several solutes for osmotic adjustment in root. *Plant Production Science*, *9*, 39–46.

Panwar, M., Tewari, R., & Nayyar, H. (2014). Microbial consortium of plant growth-promoting rhizobacteria improves the performance of plants growing in stressed soils: An overview. In M. S. Khan, A. Zaidi, & J. Musarrat (Eds.), *Phosphate solubilizing microorganisms* (pp. 257–285). Springer International Publishing Switzerland.

Paredes, M., & Quiles, M. J. (2015). The effects of cold stress on photosynthesis in hibiscus plants. *PLoS One*, *10*(9), e0137472. Available from https://doi.org/10.1371/journal.pone.0137472.

Pischl, P. H., & Barber, N. A. (2017). Plant responses to arbuscular mycorrhizae under elevated temperature and drought. *Journal of Plant Ecology*, *10*(4), 692–701.

Ramamoorthy, V., Viswanathan, R., Raguchander, T., Prakasam, V., & Samiyappan, R. (2001). Induction of systemic resistance by plant growth promoting rhizobacteria in crop plants against pests and diseases. *Crop Protection*, *20*, 1–11.

Ritonga, F. N., & Chen, S. (2020). Physiological and molecular mechanism involved in cold stress tolerance in plants. *Plants, 2020*(9), 560. Available from https://doi.org/10.3390/plants9050560.

Scagel, C. F., & Bryla, D. R. (2017). Salt exclusion and mycorrhizal symbiosis increase tolerance to NaCl and $CaCl_2$ salinity in 'Siam Queen' basil. *HortScience, 52*(2), 278–287. Available from https://doi.org/10.21273/HORTSCI11256-16.

Shavalikohshori, O., Zalaghi, R., Sorkheh, K., & Enaytizamir, N. (2020). The expression of proline production/degradation genes under salinity and cadmium stresses in *Triticum aestivum* inoculated with *Pseudomonas* sp. *International Journal of Environmental Science and Technology, 17*, 2233–2242. Available from https://doi.org/10.1007/s13762-019-02551-9.

Singh, V. K., Singh, H. B., & Upadhyay, R. S. (2017). Role of fusaric acid in the development of 'Fusarium wilt' symptoms in tomato: Physiological, biochemical and proteomic perspectives. *Plant Physiology and Biochemistry, 118*, 320–332.

Singh, R. P., & Jha, P. N. (2017). Analysis of fatty acid composition of PGPR Klebsiella sp. SBP-8 and its role in ameliorating salt stress in wheat. *Symbiosis, 73*(2). Available from https://doi.org/10.1007/s13199-017-0477-4.

Smith, D. L., Gravel, V., & Yergeau, E. (2017). Editorial: Signaling in the phytomicrobiome. *Frontiers in Plant Science, 8*, 611. Available from https://doi.org/10.3389/fpls.2017.00611.

Soliman, M., Alhaithloul, H. A., Hakeem, K. R., Alharbi, B. M., El-Esawi, M., & Elkelish, A. (2020). Exogenous nitric oxide mitigates nickel-induced oxidative damage in eggplant by upregulating antioxidants, osmolyte metabolism, and glyoxalase systems. *Plants, 2019*(8), 562. Available from https://doi.org/10.3390/plants8120562.

Soliman, M., Elkelish, A., Souad, T., Alhaithloul, H., & Farooq, M. (2020). Brassinosteroid seed priming with nitrogen supplementation improves salt tolerance in soybean. *Physiology and Molecular Biology of Plants*. Available from https://doi.org/10.1007/s12298-020-00765-7.

Steenhoudt, O., & Vanderleyden, J. (2006). Azospirillum, a free-living nitrogen fixing bacterium closely associated with grasses: Genetic, biochemical and ecological aspects. *FEMS Microbiology Reviews, 24*, 487–506. Available from https://doi.org/10.1111/j.1574-6976.2000.tb00552.x.

Su, F., Jacquard, C., Villaume, S., Michel, J., Rabenoelina, F., Clement, C., ... Vaillant-Gaveau, N. (2015). *Burkholderia phytofirmans PsJN* reduces impact of freezing temperatures on photosynthesis in *Arabidopsis thaliana. Frontiers in Plant Science, 6*, 810. Available from https://doi.org/10.3389/fpls.2015.00810.

Suryadi, Y., Susilowati, D. N., & Fauziah, F. (2019). Management of plant diseases by PGPR-mediated induced resistance with special reference to tea and rice crops. In R. Sayyed (Ed.), *Plant growth promoting rhizobacteria for sustainable stress management. Microorganisms for sustainability* (vol 13). Singapore: Springer. Available from https://doi.org/10.1007/978-981-13-6986-5_4.

Talaat, N. B., & Shawky, B. T. (2004). Protective effects of arbuscular mychorrizal fungi on wheat (*Triticum aestivum* L) plants exposed to salinity. *Environmental and Experimental Botany, 98*, 20–31.

Tejera, N. A., Campos, R., Sanjuan, J., & Lluch, C. (2004). Nitrogenase and antioxidant enzyme activities in *Phaseolus vulgaris* nodules formed by *Rhizobium tropici* isogenic strains with varying tolerance to salt stress. *Journal of Plant Physiology, 161*, 329–338.

Tiwari, S., Prasad, V., Chauhan, P. S., & Lata, C. (2017). *Bacillus amyloliquefaciens* confers tolerance to various abiotic stresses and modulates plant response to phytohormones through osmoprotection and gene expression regulation in rice. *Frontiers in Plant Science*. Available from https://doi.org/10.3389/fpls.2017.01510.

Upadhyay, S. K., & Singh, D. P. (2015). Effect of salt-tolerant plant growth-promoting rhizobacteria on wheat plants and soil health in a saline environment. *Plant Biology*, *17*, 288–293. Available from https://doi.org/10.1111/plb.12173.

Van Hullebusch., Lens, P., & Tabak, H. (2005). Developments in bioremediation of soils and sediments polluted with metals and radionuclides. 3. Influence of chemical speciation and bioavailability on contaminants remobilization bio-processes. *Reviews in Environmental Science and Bio/Technology*, *4*, 185–212.

Vurukonda, S. S. K. P., Vardharajula, S., Shrivastava, M., & Ali, S. K. Z. (2016). Enhancement of drought stress tolerance in crops by plant growth promoting rhizobacteria. *Microbiological Research*, *184*, 13–24.

Wang, D., Heckathorn, S. A., Mainali, K., & Tripathee, R. (2016). Timing effects of heat-stress on plant ecophysiological characteristics and growth. *Frontiers in Plant Science*. Available from https://doi.org/10.3389/fpls.2016.01629.

Weyens, N., Beckers, B., Schellingen, K., Ceulemans, R., Van der Lelie, D., Newman, L., et al. (2015). The potential of the Ni-resistant TCE degrading *Pseudomonas putida* W619-TCE to reduce phytotoxicity and improve phytoremediation efficiency of poplar cuttings on A Ni-TCE co-contamination. *International Journal of Phytoremediation*, *17*, 40–48. Available from https://doi.org/10.1080/15226514.2013.828016.

Wu, Q. S., & Zou, Y. N. (2017). *Arbuscular mycorrhizal fungi and tolerance of drought stress in plants. Arbuscular mycorrhizas and stress tolerance of plants* (pp. 25–41). .

Wu, Q. S., Zou, Y. N., & He, X. H. (2010). Contributions of arbuscular mycorrhizal fungi to growth, photosynthesis, root morphology and ionic balance of citrus seedlings under salt stress. *Acta Physiologiae Plantarum*, *32*, 297–304.

Xu, H., Lu, Y., & Zhu, X. (2016). Effects of arbuscular mycorrhiza on osmotic adjustment and photosynthetic physiology of maize seedlings in black soils region of Northeast China. *Brazilian Archives of Biology and Technology*, *59*. Available from http://doi.org/10.1590/1678-4324-2016160392.

Yang, J., Kloepper, J. W., & Ryu, C. M. (2009). Rhizosphere bacteria help plants tolerate abiotic stress. *Trends in Plant Science*, *14*, 1–4. Available from https://doi.org/10.1016/j.tplants.2008.10.004.

Yasin, N. A., Akram, W., Khan, W. U., Ahmad, S. R., Ahmad, A., & Ali, A. (2018). Halotolerant plant-growth promoting rhizobacteria modulate gene expression and osmolyte production to improve salinity tolerance and growth in *Capsicum annum* L. *Environmental Science and Pollution Research International*, *25*(23), 23236–23250. Available from https://doi.org/10.1007/s11356-018-2381-8.

Yedidia, I., Srivastva, A. K., Kapulnik, Y., & Chet, I. (2001). Effect of *Trichoderma harzianum* on microelement concentrations and increased growth of cucumber plants. *Plant and Soil*, *235*, 235–242.

Youseif, S. H. (2018). Genetic diversity of plant growth promoting rhizobacteria and their effects on the growth of maize plants under greenhouse conditions. *Annals of Agricultural Sciences*, *63*(1), 25–35.

Yu, J., Su, D., Yang, D., Dong, T., Tang, Z., Li, H., ... Zhang, B. (2020). Chilling and heat stress-induced physiological changes and MicroRNA-related mechanism in sweetpotato (*Ipomoea batatas* L.). *Frontiers in Plant Science*. Available from https://doi.org/10.3389/fpls.2020.00687.

Zahir, Z. A., Ghani, U., Naveed, M., Nadeem, S. M., & Asghar, H. N. (2009). Comparative effectiveness of Pseudomonas and Serratia sp. containing ACC-deaminase for improving growth and yield of wheat (*Triticum aestivum* L.) under salt-stressed conditions. *Archives of Microbiology*, *191*, 415–424.

Zaidi, A., Khan, M., Ahemad, M., & Oves, M. (2009). Plant growth promotion by phosphate solubilizing bacteria. *Acta Microbiologica et Immunologica Hungarica, 56*, 263–284.

Zhang, W., Wang, J., Huang, Z., Mi, L., Xu, K., Wu, J., ... Jiang, D. (2018). Effects of low temperature at booting stage on sucrose metabolism and endogenous hormone contents in winter wheat spikelet. *Frontiers in Plant Science.* 18 April 2019 I https://doi.org/10.3389/fpls.2019.00498.

Zubair, M., Hanif, A., Farzand, A., Sheikh, T. M. M., Khan, A. R., Suleman, M., ... Gao, X. (2019). Genetic screening and expression analysis of Psychrophilic *Bacillus* spp. reveal their potential to alleviate cold stress and modulate phytohormones in wheat. *Microorganisms, 7*(9), 337. Available from https://doi.org/10.3390/microorganisms7090337.

Chapter 11

CRISPR/Cas-mediated genome editing for improved stress tolerance in plants

Deepu Pandita
Government Department of School Education, Jammu, India

11.1 Introduction

Global population is predicted to touch 9.7 billion by the year 2050, and agricultural food production will have to escalate by around 50% by 2030 and by 70%–110% by 2050, or 44 million metric tons per annum for a well-nourished world population (Alexandratos & Bruinsma, 2012; Godfray et al., 2010; Jones et al., 2014; Ray, Mueller, West, & Foley, 2013). These are real challenges and major bottlenecks because of the future expansion of arable lands on one side; secondly climate change with altered global weather patterns and increased environmental stresses which bring increased environmental stresses which affect crop production drastically (Cominelli, Conti, Tonelli, & Galbiati, 2013; Eckardt, 2009; Tester & Langridge, 2010); and thirdly the growing population may lead to future food and nutritional scarcity of the world. Abiotic stresses, for instance, heat, salinity, cold, drought, flood, heavy metals, and radiation are critical and regulating factors for global crop production (Doupis, Chartzoulakis, Beis, & Patakas, 2011; Hasanuzzaman, Nahar, & Fujita, 2013; Hasanuzzaman, Nahar, Fujita, Ahmad, Chandna, et al., 2013). Salinity and drought are the two most critical limiting factors (Munns, 2011). For sustaining the growing population in lieu of change in climatic conditions, stress-tolerant varieties need to be designed with stable or increased productivity in stress environments. The genome editing strategies allow prospects for improving crop plants through DNA modifications with potential for addressing the present challenges.

11.1.1 The journey from classical breeding to modern-day CRISPR

The process of the development of new varieties via conventional/crop breeding is a prolonged process in nature and takes 7–12 years based on the

Frontiers in Plant–Soil Interaction. DOI: https://doi.org/10.1016/B978-0-323-90943-3.00001-8

crop variety (Gilliham, Able, & Roy, 2017; Tai & Young, 1984) and further unwanted genes also get transferred along with the gene of interest (GOI) (Fig. 11.1). The conventional breeding relies on genetic diversity, which restricts crop improvement because of its low levels (Li et al., 2020). Transgenic approaches generate food crops with enriched abiotic and biotic environmental stress tolerance by overexpression of the genes (dos Reis, Lima, & de Souza, 2012; Pandey et al., 2011), which encode enzymes for mannitol (Chan, Grumet, & Loescher, 2011), glycine betaine (Fan, Zhang, Zhang, & Zhang, 2012), and polyamine biosynthesis (Cheng et al., 2009). Genetically engineered plant crops (GMOs) transfer only GOI (Fig. 11.1), but have not been extended to consumers primarily because of public nonacceptance of GMOs and political concerns (Callaway, 2018; Globus & Qimron, 2018). The different varieties of engineered endonucleases of genome-editing tools that offer targeted alterations include new breeding techniques (NBTs), for instance zinc-finger nucleases (ZFNs), transcription activator-like nucleases (TALENs) and Clustered Regularly Interspaced Short Palindromic Repeats (CRISPR)-Cas (CRISPR-associated nuclease) (CRISPR/Cas). ZFNs act as efficient and effective premeditated restriction enzymes/dimers; they were developed in the 1990s and have revolutionized the genome manipulation area (Rai et al., 2018). ZFNs have two structural

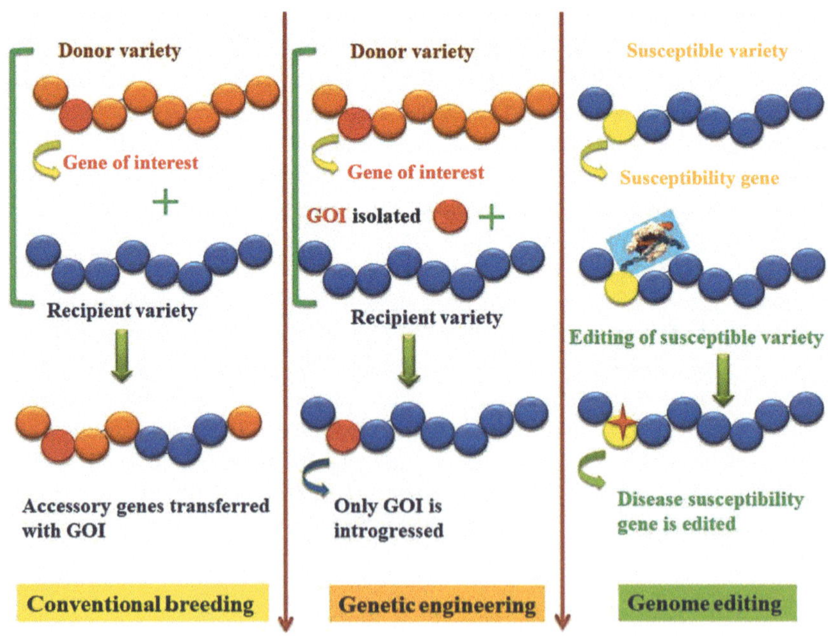

FIGURE 11.1 Strategies of gene transfer in conventional breeding, genetic engineering (GMO), and genome editing.

domains, that is, the sequence-specific DNA-binding zinc finger motif domain with 300–600 zinc-finger repeats (Carlson, Fahrenkrug, & Hackett, 2012) and a nonspecific DNA Fok1 cleavage domain of type II restriction endonuclease which creates staggered double-strand breaks (DSB) with overhangs (Carroll, Morton, Beumer, & Segal, 2006; Pabo, Peisach, & Grant, 2001). The DNA binding domain contains an array of 4–6 zinc finger repeat binding modules comprising a single zinc finger unit. Every module can monitor 9–18 DNA nucleotide sequences (Pabo et al., 2001). ZFN DNA binding domains recognize three base pairs at the target site by creating DSBs (Durai et al., 2005; Rai et al., 2018). ZFNs enabled genome modification of *Nicotiana tabacum*, *Arabidopsis thaliana*, *Zea mays*, *Oryza sativa*, soybean, etc. (Curtin et al., 2018; Osakabe, Osakabe, & Toki, 2010; Petolino et al., 2010; Baltes et al., 2014). Transcription activator-like nucleases are long fragments of transcription activator-like effector (TALE) sequences occurring in nature and fused with Fokl domain with carboxylic-terminal end of manipulated TALE repeat arrays (Christian et al., 2010). The TALENs comprise customizable DNA-binding dimeric domains (multiplex repeats of extremely conserved 33–34 amino acid sequences specific for all nucleotides) of TALE proteins joined to a Fokl nuclease domain of nonspecific nature which generate overhangs (Christian et al., 2010; Zhang, Zhang, & Yin, 2019). TALENs precisely target a single nucleotide at the target site (instead of three) and lead to nonhomologous mutations with greater efficiency in various living organisms (Boch et al., 2009; Joung & Sander, 2012). TALE targets are identified by repeat variable di-residue flanking the 12th and 13th locations of target sites (Bedell et al., 2012). Success stories with accurate genome editing by TALENs are available in angiosperms and bryophytes (Bedell et al., 2012; Kopischke, Schüßler, Althoff, & Zachgo, 2017; Li, Liu, Spalding, Weeks, & Yang, 2012). Reports of plant editing with ZFNs and TALENs are less due to the drawbacks and constraints of prolonged and costly target enzyme construction, low efficiency and validation, specificity, and higher off-target mutation rates. The novel approach of CRISPR superseded ZFNs and TALENs from "methods of the year" in year 2011 to "breakthrough of the year" in year 2015. CRISPR of prokaryotes allows genome editing in eukaryotes (Jinek et al., 2012; Nakayama et al., 2013). The CRISPR-based genome editing toolkits enable economically rational, swift, specific, multiplexing, flexible, and non-GMO modifications or editing of GOI (Fig. 11.1) at precise genome loci in an exact and expectable manner of stress-related traits in plants to the desired genotype or variety for improved resistance to stresses and crop production and remodeling of future crops (Chen, Wang, Zhang, Zhang, & Gao, 2019; Li et al., 2016). CRISPR/Cas system finds use in improvement of plant traits in nearly all major food crops for instance rice, cassava, wheat, maize, etc. (Bhowmik, Hassan, Molla, Rahman, & Islam, 2019; Chen et al., 2019) and has regulatory approval in various countries (Schmidt, Belisle, & Frommer, 2020).

11.2 Genome rewriting with the pen of genome editing CRISPR tools

CRISPR/Cas systems extend the GE toolbox with versatile applications in plant biology (Komor, Badran, & Liu, 2017) and hold the promise for a second green revolution with sustainable agriculture by modifying food crops. The fundamentals of CRISPR biology serve as accurate, simple, and efficient tools of revolutionary technology for genome editing and DNA modifications, novel allelic variant design, and exploring biological phenomena in both prokaryotic and eukaryotic living organisms (Cong et al., 2013; Jinek et al., 2012; Komor et al., 2017; Ran et al., 2013; Selle & Barrangou, 2015; Vats et al., 2019). CRISPR-based tools used for genome editing of vital traits for crop improvement include CRISPR/Cas9, CRISPR/Cas12a, CRISPR/Cas13, CRISPR interference (CRISPRi), and CRISPR-mediated activation (CRISPRa) (Adli, 2018; Chen et al., 2019; Mishra, Joshi, & Zhao, 2018; Zhang, Massel, Godwin, & Gao, 2019). Advanced techniques of CRISPR holds potential for genetic manipulations like knockout generation, precise modifications, gain-of-function, gene expression analysis, multiplex genome engineering, RNA targeting, epigenome editing, DNA imaging, and RNA editing (Chen et al., 2019). CRISPR-guided DNA transposase systems proficiently insert custom genes into the desired site without being dependent on DSB repair by the homology-directed repair (HDR) pathway of the host (Klompe, Vo, Halpin-Healy, & Sternberg, 2019; Strecker et al., 2019). CRISPRi and CRISPRa regulate gene expression at transcriptional levels (Bortesi, & Fischer, 2015; Čermák, Baltes, Čegan, Zhang, & Voytas, 2015; Gilbert et al., 2014; Lowder et al., 2015; Qi et al., 2013). CRISPR versions have the benefits of precision, efficiency, and robustness but also some drawbacks (Begemann et al., 2017a,b; Fiaz et al., 2019; Smargon et al., 2017; Zhang, Massel, Godwin, & Gao, 2018). CRISPR/Cas has applications in innumerable plants on different traits for instance, apple (Nishitani et al., 2016), *Arabidopsis* (Feng et al., 2013; Miki, Zhang, Zeng, Feng, & Zhu, 2018), rice (Begemann et al., 2017a,b; Li et al., 2018; Shan et al., 2013; Zheng et al., 2016), and maize (Svitashev et al., 2015). CRISPR-modified tomatoes have improved flavor and sugar content, and mushrooms have reduced melanin content. Drought-resistant corn and wheat resistant against powdery mildew disease are other examples (Shi et al., 2017; Tieman et al., 2017; Waltz, 2016; Wang et al., 2014). Some examples of commercialized CRISPR/Cas9-mediated edited plants are browning-resistant white button mushroom (Gill, 2016), disease and drought-resistant waxy corn (Sander, & Jeschke, 2016), gluten-free wheat (Dormehl, 2017), soybean with enhanced oleic acid content (Kim et al., 2017), and longer freshness and nonacrylamide producing ranger russet potato (Regalado, 2015).

CRISPR/Cas has two main components of Cas protein and single guide RNA (sgRNA) with a 5' sequence complementary to target and a 3' sequence that binds to the Cas protein (Vats et al., 2019). Cas induces DSB in DNA whereas 20 NT lengthy replaceable guide sequence within sgRNA directs Cas

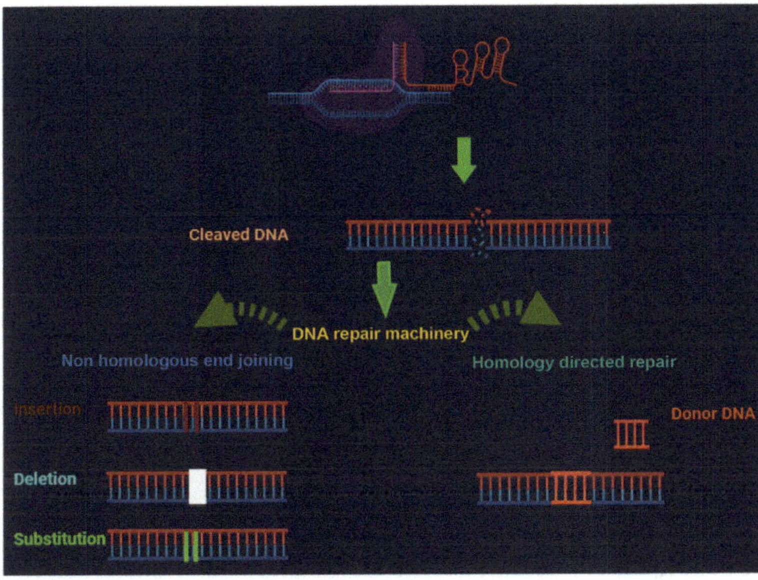

FIGURE 11.2 Mechanisms of CRISPR/Cas system.

for targeting target sites (Li et al., 2013). Cas9 from *Streptococcus pyogenes*, needs 3 nt (NGG) protospacer adjacent motif (PAM) adjacent to 20 NT long recognition site of sgRNA, which varies in different CRISPR/Cas systems (Doudna & Charpentier, 2014; Leenay & Beisel, 2017). DNA DSBs are generated at the genomic target site in CRISPR/Cas genome editing followed by two main plant innate DNA repair pathway mechanisms, that is, nonhomologous end-joining (NHEJ) and HDR for gene knockout/gene silencing or gene knock-in, respectively (Fig. 11.2) (Rong, & Golic, 2000). Cas9 produces blunt ends at DSBs and requires 5′-NGG-3′ triplet PAM (Protospacer Adjacent Motif) (Braff, Yaung, Esvelt, & Church, 2016; Chneiweiss et al., 2017; Vanegas, Jarczynska, Strucko, & Mortensen, 2019), whereas Cas12 produces staggered DSBs with overhangs via the NHEJ pathway and needs "T" rich (5′-TTTN-3′) PAM sequences (Alok et al., 2020; Gao et al., 2017; Mahfouz, 2017). Insertion and replacement of genes and insertion/deletion (indel) mutations follow NHEJ (Chen et al., 2019). With a DNA donor of an exogenous nature, gene replacements, gene insertions, and modifications can be accomplished through homology-directed repair (Mishra et al., 2018).

11.3 The CRISPR/Cas system

The CRISPR/Cas system is an acquired immunity system of bacteria and archaea that protects them from invading viruses, foreign mobile genetic

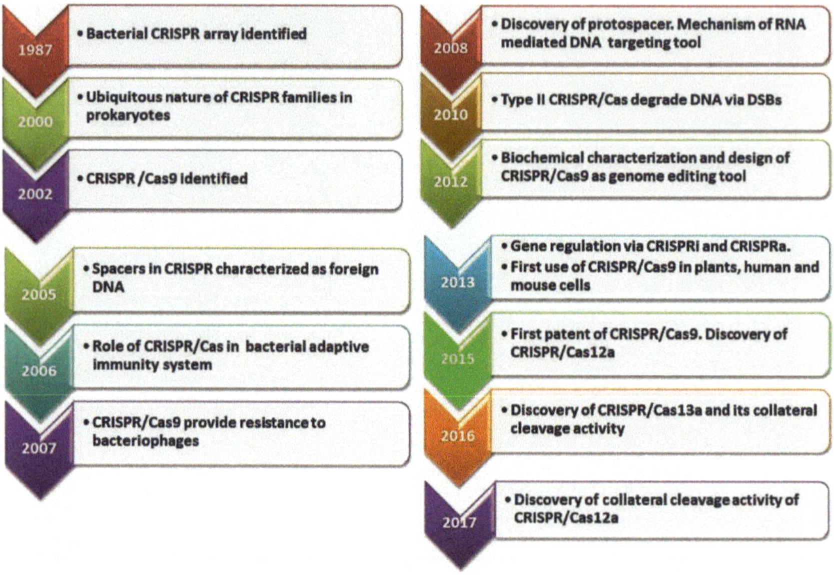

FIGURE 11.3 Timetable of CRISPR/Cas system.

elements of plasmids, and phages. Discovery of the CRISPR/Cas system, for instance, the repeat sequences type I–E CRISPR/Cas system in *Escherichia coli* was done by Ishino, Shinagawa, Makino, Amemura, and Nakata (1987). The primary functional characterization was accomplished in the 1990s and the function of CRISPR/Cas in immunity was understood in 2007 (Barrangou et al., 2007). After in vitro CRISPR/Cas-mediated gene editing, first reports of gene editing in animals (Mali et al., 2013) and *A. thaliana*, *Nicotiana benthamina* (Li et al., 2013), and *O. sativa* living systems were reported (Shan et al., 2013). The CRISPR term was coined by Jansen, Embden, Gaastra, and Schouls (2002). The detailed timeline of the CRISPR/Cas systems is given in Fig. 11.3.

11.3.1 Adaptive immunity system by CRISPR/Cas

There are three major evolutionary stages of the CRISPR/Cas system in bacteria. CRISPR/Cas-mediated acquired-immunity in prokaryotes is broadly classified into adaptation, expression, interference, and immunity phases (Chen et al., 2019; Makarova et al., 2011).

1. *CRISPR adaptation:* adaptation is the process of spacer acquisition or integration of short plasmid or virus-derived foreign invader genetic elements into a CRISPR locus as new spacer sequences into host genome

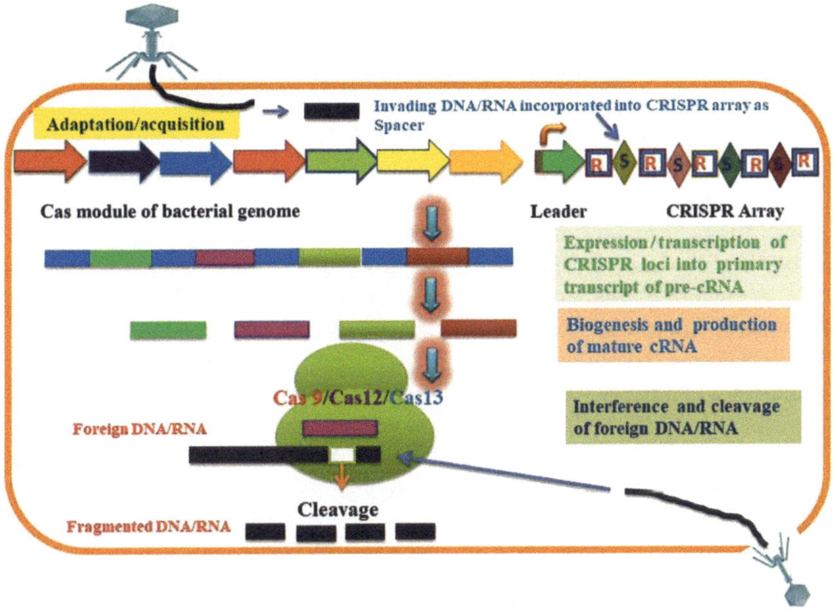

FIGURE 11.4 Outline of bacterial adaptive immunity by CRISPR/Cas.

via Cas1 and Cas2 (Chen et al., 2019; Sternberg, Richter, Charpentier, & Qimron, 2016) (Fig. 11.4).

2. *Expression/crRNA biogenesis*: expression stage consists of the transcription or expression of CRISPR array into pre-crRNA/guide RNA (gRNA) (Terns, 2018) and transcribed guide CRISPR pre-crRNA maturation into crRNAs which in turn integrate with Cas effector nucleases to form the crRNA effector surveillance complex (Chen et al., 2019; Terns, 2018) (Fig. 11.4).
3. *CRISPR interference*: in the interference and immunity step, RNA-guided CRISPR/Cas effector surveillance complex (Cas protein-gRNA complex) recognizes genomic protospacer adjacent motifs (PAM) which are target sequences complementary to crRNAs and activated nuclease activity catalyzes sequence-specific cleavage of plasmid or viral foreign invading DNA matching the spacers (Jackson et al., 2017; Jiang & Doudna 2017; Knott & Jennifer, 2018). The cleavage by recognition through PAM averts outbreaks to the CRISPR locus in bacteria (Cong et al., 2013; Gleditzsch et al., 2019; Jinek et al., 2012; Ran et al., 2013) (Fig. 11.4).

11.3.2 Classification

The CRISPR/Cas systems are functionally categorized into two major classes (class 1 and class 2), and are further divided into six types and 34 subtypes based on the structure and composition of Cas and the effector nuclease genes (Makarova, Wolf, & Koonin, 2018; Shmakov et al., 2017).

The Class 1 CRISPR/Cas system comprises three distinct types of I, III, and IV with effector composed of multisubunit multifunctional protein complexes for interference, whereas the Class 2 CRISPR/Cas system has II, V, and VI types with solitary protein subunit effector endonucleases of multidomains for interference (Makarova et al., 2015; Manghwar, Lindsey, Zhang, & Jin, 2019; Shmakov et al., 2015) and the practice of genome editing is mainly done by class 2 CRISPR/Cas. The II and V types find applications in DNA engineering whereas type VI is used exclusively for RNA engineering (Burmistrz, Krakowski, & Krawczyk-Balska, 2020; Moon, Kim, Ko, & Kim, 2019). Subtype III slices both the DNA fragments and RNA sequences (Samai et al., 2015). The outline of CRISPR/Cas system classification is drawn in Fig. 11.5.

Type II CRISPR system: the CRISPR era started by the maneuver of type II-A (SpCas9, Cas9 from *Streptococcus pyogenes*) of the Class 2 system, which also has II-B and II-C subtypes, to genome editing in eukaryotic systems (Mougiakos, Bosma, de Vos, van Kranenburg, & van der Oost, 2016). Type II-A Cas9, differs from additional Cas nucleases like Cas12a, Cas13, Cas14, CasX, and CasY (Manghwar et al., 2019), due to RNase III genes involved in the pre-crRNA maturation. The Cas9-gRNA complex allows G-rich protospacer adjacent motif recognition and targets DNA complementary to the crRNA spacer sequence. This induces unwinding of target DNA to R-loop formation in Cas between guide RNA and host DNA (Szczelkun et al., 2014). HNH and RuvC nuclease domain of Cas9 organize as bilobed architecture of recognition (REC) and nuclease (NUC) lobes which cleave the target and nontarget DNA strands, respectively (Ishino, Krupovic, & Forterre, 2018; Sternberg, LaFrance, Kaplan, & Doudna, 2015). In the past, SpCas9 was believed to create blunt-end DSBs, but now staggered ends with 5'-overhangs owing to postcleavage trimming by RuvC domain are also reported

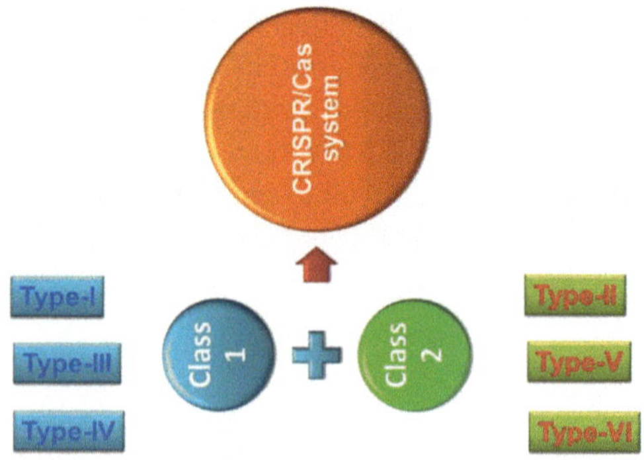

FIGURE 11.5 Classification of CRISPR/Cas system.

(Zuo & Liu, 2016). Type II-B and II-C need tracrRNA as well for recognizing the target sequence. FnCas9 of II-B uses the complex of small CRISPR/Cas-associated RNA (scaRNA) and tracrRNA to target mRNA and for mRNA degradation (Chylinski, Makarova, Charpentier, & Koonin, 2014). NmCas9 of II-C has RNase III independent RNA processing activity (Mir, Edraki, Lee, & Sontheimer, 2018). SaCas9 and CjCas9 have a reduced size compared with SpCas9 (Cebrian-Serrano, & Davies, 2017). The split system of N- and C-SpCas9 sequences suggests a genome editing substitute in viral vector-based delivery systems (Truong et al., 2015).

Type V CRISPR system: this type is represented by Cas12a (Cpf1) RNA-guided endonuclease and is further divided into 10 subtypes from A–I to U in line with the resemblance of domain organization (Yan et al., 2019). The V endonucleases have bilobed architecture of REC and NUC lobes but own only three RuvC domains in NUC of Cas12a, Cas12b, Cas12e, and Cas14 (Fernandes, Pastor, & Bochtler, 2016; Harrington et al., 2018; Liu et al., 2017; Liu et al., 2019; Swarts, van der Oost, & Jinek, 2017; Zetsche et al., 2015). Cas12a needs only crRNA and has RNA editing activity to cleave pre-crRNA into mature crRNA (Fonfara, Richter, Bratovic, Le Rhun, & Charpentier, 2016), whereas a supplementary tracrRNA is essential for Cas12b, Cas12e, and Cas14a (Yan et al., 2019). This group has thymidine T-rich (5′-TTTV-3′) PAM sequences located 3' upstream of a protospacer (Yan et al., 2019). Cas12a cleaves single-stranded DNA as well as double-stranded DNA (Chen et al., 2018). Cas14 cuts ssDNA lacking PAM specificity (Harrington et al., 2018) and dsDNA by T-rich PAM (TTTR/TTAT) sequence recognition (Karvelis et al., 2019). Structural alterations in NUC lobe after protospacer adjacent motif recognition prompt cleavage of target DNA by the RuvC domain in Cas12a and domain positions scissile phosphates without direct cleaving (Swarts et al., 2017; Yamano et al., 2016). A solitary active site in RuvC domains of the Cas12a and Cas12b cleaves target and nontarget strands (Swarts et al., 2017; Yang, Gao, Rajashankar, & Patel, 2016). Cleaved target sequences have staggered 5' 4–5 bp sticky overhangs in Cas12a, Cas12b, Cas12e, and Cas14 (Karvelis et al., 2019; Liu et al., 2017; Liu et al., 2019; Zetsche et al., 2015). Cas14 shows collateral ssDNA cleavage activity after target recognition (Harrington et al., 2018). The type V systems show CRISPR RNAs (crRNAs) maturation of pre-crRNA into mature crRNA by intrinsic Cas12a ribonuclease activity without tracrRNA and RNase III (Zetsche et al., 2015). LbCas12a variants have transmuted protospacer adjacent motif specificity (Tóth et al., 2020). GMO monitoring, gene replacements, and multiplexing have been investigated with LbCas12a in rice (Banakar et al., 2020; Zhang, Zhang, & Xie, 2020).

Type VI CRISPR system: this shows distinctive RNase activity and targets RNA exclusively. It can be used for detection and manipulation of RNA (O'Connell, 2019). Type VI has four types of different sizes and sequences, that is, VI-A (Cas13a), VI-B (Cas13b), VI-C (Cas13c), and VI-D (Cas13d)

(Shmakov et al., 2015). Cas13a grasps only crRNA with 20-nt target-binding sequence. Two conserved Higher Eukaryotes and Prokaryotes Nucleobinding (HEPN) domains instead of RuvC domain act as the RNase active center and play a role in the maturation and cleavage of target RNA (East-Seletsky, O'Connell, Burstein, Knott, & Doudna, 2017; Shmakov et al., 2015), and the targeting of ssRNA, needing a protospacer flanking sequence (PFS) as opposed to PAM. The cleavage rate of the target is greater when PFS is non-complementary to C in crRNA 5' repeat region of Cas13a (Abudayyeh et al., 2016). HEPN domain on activation by target ssRNA cleavage also shows collateral activity toward nontarget ssRNAs (O'Connell, 2019). The transacting nontarget ssRNA cleavage can be used in signal amplifications in molecular diagnostic systems (Gootenberg et al., 2017). A detailed portrayal of CRISPR/Cas in plants is discussed in various chapters by Pandita (2020 unpublished), Pandita (2020 unpublished), Pandita, Puli, Palakolanu (2020 unpublished).

11.4 Genome editing for improvement of plants

The novel CRISPR/Cas toolbox has modernized plant breeding and finds many applications in plant sciences with regard to biotic and abiotic stresses for the enhancement of plant disease resistance and designing climate smart crops and resilient crop varieties by working out the complex inheritance and manipulation of genes of abiotic and biotic stress-related traits (Fig. 11.6). Both biotic and abiotic stresses and narrow diversity at genetic

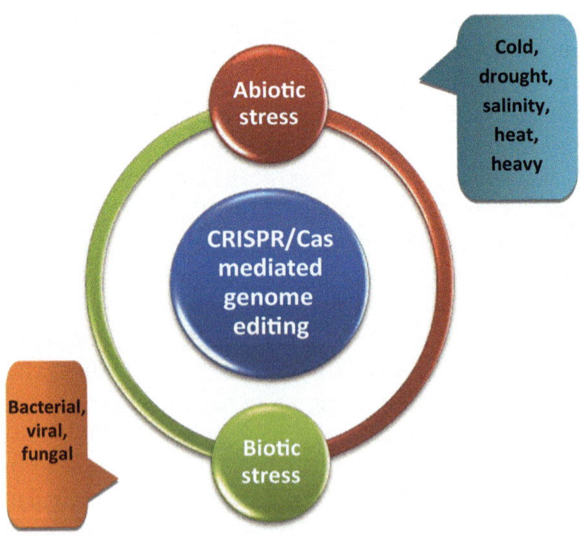

FIGURE 11.6 CRISPR/Cas-mediated genome editing of plants under different stresses.

level in cultivars are the chief limitations to extra enhancement in crop productivity. Disease resistance genes are present in plants. Similarly, genes and quantitative trait loci (QTL) have been recognized for abiotic environmental stresses, for instance, water deficiency, submergence, salinity, and low temperature (Jena & Mackill, 2008). The "rewriting of natural DNA" by CRISPR/Cas may license crafting of biotic or abiotic stress resilient and climate smart crops.

11.4.1 CRISPR-mediated genome editing strategies for abiotic stress tolerance

Global climate change is leading to increased negative impacts of abiotic stresses, for instance, drought, heat stress, salinity, heavy metals, and cold stress, to crop plants which are the most grave restraints in the production of agriculture, and these influence yield and quality of the agricultural production (Michelbart, Hasegawa, & Bailey-Serres, 2015). Abiotic stress substantially limits the yield of crops globally by decreasing the growth and development of plants (Pandey, Irulappan, Bagavathiannan, & Senthil-Kumar, 2017). Tolerance to abiotic stress is directed by countless genes which are mostly influenced by ecological factors which make it confounding for investigation (Bhat et al., 2016; Deshmukh et al., 2014). CRISPR/Cas9 editing of OsPYL abscisic acid receptor gene family and pyl1/4/6 triple knockout in *O. sativa* increased yield and tolerance to heat and exhibited reduced preharvest sprouting (Miao et al., 2018). CRISPR/Cas9-mediated OsPRX2 gene knockout revealed higher levels of antioxidant induction in parallel with the usual reactive oxygen species (ROS) accumulation (Mao et al., 2018). The knockout mutant of cell wall protein OsPRP3 (Gothandam, Nalini, Karthikeyan, & Shin, 2010) responsible for the floral organ extracellular matrix structure raises the integrity of cell walls in cold tolerant plants. OsRR22 gene translates a 696-amino acid B-type response regulator transcription factor responsible for cytokinin signal transduction and metabolism. OsRR22 and OsNAC041 knockouts show significantly increased salinity tolerance in *O. sativa* (Bo et al., 2019; Takagi et al., 2015; Zhang et al., 2019). Semirolled leaf 1, 2 (SRL1 and SRL2) gene rice mutants and the hybrids produced by mutants exhibited improved panicle number and yield per capita plant during water deficiency (Liao et al., 2019). Advanced breeding through CRISPR/Cas9 has developed a new ARGOS8 allele variant in a genome-edited corn variety with enhanced harvest under water deficiency (Shi et al., 2017). CRISPR/Cas9-mediated Drb2a and Drb2b knockouts control salt and drought stress in soybean (Curtin et al., 2018). Control of drought tolerance mechanism through the mitogen-activated protein kinase (SlMAPK3) gene which protects membrane cells against oxidation and regulates transcription genes was reported in the tomato SlMAPK3 knockout mutants under drought stress by CRISPR/Cas9 (Wang, Chen, et al., 2017). Tissue-specific AtEF1 promoter derived truncated gRNAs (tru-gRNAs) and Cas9 cause modifications in abiotic stress-responsive OST2/AHA1 causing increased stomatal responses in *Arabidopsis* (Osakabe & Osakabe, 2017). CRISPR/Cas9 has been adapted for editing the genomes of *A. thaliana*, *O. sativa*, tobacco, and citrus (Jia, Xu, Orbovic, Zhang, & Wang, 2017; Steinert, Schiml, Fauser, & Puchta, 2015). Similarly, CRISPR-Cpf1 enabled the editing of rice, *A. thaliana*, soybean, and tobacco genomes (Kim et al., 2017; Tang et al., 2017; Xu et al., 2017). The uses of the CRISPR/Cas tool for abiotic stress tolerance include herbicide tolerance in potato (Butler, Baltes, Voytas, & Douches, 2016), soybean (Li et al., 2015), rice (Butt et al., 2017; Nishizawa-Yokoi, Endo, Ohtsuki, Saika, & Toki, 2015; Shimatani et al., 2017); drought tolerance in rice (Shim et al., 2018; Tang et al., 2016); and salt tolerance in *O. sativa* (Li et al., 2014) and cotton (Chen et al., 2017). Over the years the genomes of various crop plants were edited for abiotic stress tolerance. Some of the remarkable applications of CRISPR/Cas systems in abiotic stress tolerance are summarized in Table 11.1. *Acidamino coccus* Cas12a (Cpf1) enabled multiplexed genome editing by targeting 25 multiple genes which can also be followed for targeting complex traits of plant abiotic stress (Campa, Weisbach, Santinha, Incarnato, & Platt, 2019). Aquaporins and supplementary transporter proteins are some of the distinguishable contestants for genome editing for increasing abiotic stress tolerance (Ahmad, Tripathi, Deshmukh, Pratap Singh, & Corpas, 2019; Bhat et al., 2019; Shivaraj, Deshmukh, Sonah, & Bélanger, 2019; Sonah, Deshmukh, Labbé, & Bélanger, 2017; Vishwakarma et al., 2019; Zargar et al., 2017). Thus in forthcoming era, CRISPR/Cas toolbox can be harnessed comprehensively to target minor genes of complex quantitative traits connected with pathways of various abiotic stresses (Mushtaq et al., 2018).

11.4.2 Biotic stress management through the CRISPR/Cas approach

With the novel GE technique of CRISPR/Cas, the advanced breeding of pathogen resistant transgene-free crops can be achieved in a shorter duration (Borel, 2017). CRISPR/Cas mutagenesis of OsSWEET13, which is a disease susceptibility gene encoding sucrose transporter with a central function in plant–pathogen interaction, led to enhanced resistance against bacterial blight caused by *Xanthomonas oryza* epv. *Oryzae* in rice (Zhou et al., 2015). PthXo2X increases OsSWEET13 expression in the host. CRISPR/Cas9-based genome disruption of translation initiation codons of OsERF922 gene to introduce In Dels lead to development of Blast disease resistant japonica rice plants (Han et al., 2015; Ma et al., 2017; Wang et al., 2016). OsSEC3A (with a role in exocytosis)-mutated plants increased resistance against *Magnaporthe oryzae*, enhanced salicylic acid (SA) contents, and showed stimulation of pathogenesis- and SA-linked genes (Ma et al., 2017). Rice

TABLE 11.1 Applications of CRISPR/Cas system in abiotic stress tolerance.

Genes targeted	Effect of targeted gene on plant	Type of CRISPR/Cas genome editing system	Reference
AREB1	ABA signaling-mediated drought tolerance in *Arabidopsis*	CRISPR/ dCas9HAT (epigenome editing)	Roca-Paixão et al. (2019)
ARGOS8	Drought stress tolerance in maize by homology-directed repair (HDR) based insertion of GOS2 promoter	CRISPR/Cas9	Shi et al. (2017)
Drb2a and Drb2b	Salt and drought tolerance in soybean	CRISPR/Cas9	Curtin et al. (2018)
HKT1;2 HDR	Multiple-stress tolerance in tomato	CRISPR/Cpf1	Vu et al. (2020)
MYB30 and two yield related genes	Improved rice cold tolerance and yield	CRISPR/Cas9	Zeng et al. (2020)
NPRI	Drought tolerance in tomato	CRISPR/Cas9	Erpen-Dalla Corte et al. (2019)
OPEN STOMATA 2 (OST2) (loss of function)	Drought tolerance in *Arabidopsis* through stomatal response	CRISPR/Cas9	Osakabe et al. (2016)
Oryza sativa bilateral blade senescence 1 (BBS1)	Rice sensitive to drought stress	CRISPR/Cas9	Zeng et al. (2018)
OsCOLD1	Cold tolerance in rice	CRISPR/Cas9-HDR	Ma et al. (2015)
OsDERF1, OsPMS3, OsMSH1, OsMYB5	Drought tolerance in rice	CRISPR/Cas9	Hsu, Lander, and Zhang (2014)
OsEPFL9	Regulates stomatal density in rice leaves	CRISPR-Cas12a	Yin, Anand, Quick, and Bandyopadhyay (2019)

(*Continued*)

TABLE 11.1 (Continued)

Genes targeted	Effect of targeted gene on plant	Type of CRISPR/Cas genome editing system	Reference
OsDEP1, OsPDS, and OsEPFL9	Abiotic stress tolerance in rice	FnCas12a	Zhong et al. (2018)
OsMODD	Drought tolerance in rice	CRISPR/Cas9 knock out/ CRISPRi	Tang et al. (2016)
OsMPK2, OsPDS, OsBADH2	Multiple-stress tolerance in rice	CRISPR/Cas9	Korotkova, Gerasimova, and Khlestkina (2019)
OsNAC041	Salinity tolerance in rice	CRISPR/Cas9	Bo et al. (2019)
OsRR22	Improved salt tolerance in rice	CRISPR/Cas9	Farhat et al. (2019); Zhang et al. (2019)
OsSAPK2 (loss of function)	ABA signaling-mediated drought tolerance in rice	CRISPR/Cas9	Lou, Wang, Liang, and Yu (2017)
OsSIT1	Salt tolerance in rice	CRISPR/Cas9 knock out	Li et al. (2014)
OsSRL1, OsSRL2 (loss of function)	Leaf rolling in rice for drought tolerance	CRISPR/Cas9	Liao et al. (2019)
OsTT1	Thermotolerance in African rice	CRISPR/Cas9-HDR or C>T and A>G editors	Li et al. (2015)
PYL1, PYL4, PYL6	Controls plant growth and stress responses	CRISPR/Cas9	Miao et al. (2018)
SAPK1 and SAPK2	Salinity tolerance in soybean	CRISPR/Cas9	Lou et al. (2017)
SlAGL6	Heat stress in tomato	CRISPR/Cas9	Klap et al. (2017)
SlMAPK3 (loss of function)	ABA dependent kinase signaling and drought tolerance in tomato	CRISPR/Cas9	Wang, Chen, et al. (2017); Wang, Mao, Lu, Tao, and Zhu (2017)

(*Continued*)

TABLE 11.1 (Continued)

Genes targeted	Effect of targeted gene on plant	Type of CRISPR/Cas genome editing system	Reference
SlNPR1 (loss of function)	Drought resistance in tomato	CRISPR/Cas9	Li et al. (2019)
SlHKT1;2 Salt tolerance gene 1; 2	Abiotic salt tolerance in tomato	LbCas12a	Vu et al. (2020)
TaDREB-2 and 3	Drought signaling in *Triticum*	CRISPR/Cas9	Kim, Alptekin, and Budak (2018)
TIFY1a, TIFY1b	Plant adaption to cold in rice	CRISPR/Cas9	Huang, Zeng, Li, and Zhao (2017)
ZmHKT1	Salinity tolerance in maize	CRISPR/Cas9	Zhang et al. (2018)
HSP90: Heat Shock Protein 90	Heat tolerance in *Tetraselmis suecica*	CRISPR/Cas9	Xu, Soni, Chopra, and Chan (2020)

eIF4G knockouts are resistant against rice tungro spherical virus (RTSV) (Macovei et al., 2018). OsCYP71A1 gene knockout obstructed biosynthesis of serotonin and improved salicylic acid contents in rice plants, making plants resistant against plant hoppers and stem borers (Lu et al., 2018).

The CsLOB1 genes when manipulated by CRISPR/Cas9, exhibited improved resistance to *Xanthomonas axonopodis* causing canker disease resistance in citrus (Peng et al., 2017). CRISPR/Cas9-mediated tomato yellow leaf curl virus (TYLCV)-resistant plants were designed by coat protein (CP) or replicase (Rep) targeting with single nucleotide modification (Tashkandi, Ali, Aljedaani, Shami, & Mahfouz, 2018) and tomato plants entitled "tomelo" have become powdery mildew resistant through targeting of MLO genes (Nekrasov et al., 2017). The eukaryotic translational initiation factor eIF4E gene knockout of *Cucumis sativus* L. are moderately resistant to Papaya ringspot mosaic virus-W (PRSV-W), Zucchini yellow mosaic virus (ZYMV), and Cucumber vein yellowing virus (CVYV) (Chandrasekaran et al., 2016; Meyer, Deleu, Garcia-Mas, & Havey, 2008). MLO7 susceptibility gene on targeting makes grapes powdery mildew resistant (Malnoy et al., 2016; Pessina et al., 2016). Apples resistant to fire blight phytopathogen were generated by genome editing of DIPM-1, DIPM-2, and DIPM-4 genes

TABLE 11.2 Applications of CRISPR/Cas system in biotic stress tolerance.

Genes targeted	Effect of targeted gene on plant trait	Type of CRISPR/Cas genome editing system	Reference
BeYDV	Bean Yellow Dwarf Virus (BeYDV) resistance in tomato	CRISPR/Cas9	Baltes et al. (2015)
bW2 and bE1	Corn smut	CRISPR/Cas9	Schuster, Schweizer, Reissmann, and Kahmann (2016)
CLCuD IR and Rep regions	Cotton leaf curl disease	CRISPR/Cas9	Chandrasekaran et al. (2016)
CsLOB1 (promoter)	Citrus (orange) canker resistance	CRISPR/Cas9	Iqbal et al. (2016); Peng et al. (2017)
DIPM-1, DIPM-2, and DIPM-4 genes	Resistance to fire blight disease in apple	CRISPR/Cas9	Malnoy et al. (2016)
EDR1	Improved resistance against wheat powdery mildew	CRISPR/Cas9	Zhang et al. (2017)
eIF4E /exon	Broad virus (RNA) resistance in cucumber	CRISPR/Cas9	Chandrasekaran et al. (2016)
eIF4G	Candidate rice tungro disease resistance gene	CRISPR/Cas9	Macovei et al. (2018)
eIF (iso) 4E (eIF transcription factor) eIF4E/exon	Turnip mosaic virus (TuMV) in *Arabidopsis thaliana*	CRISPR/Cas9	Pyott, Sheehan, and Molnar (2016)
Gh14-3-3	Resistance to cotton verticillium wilt	CRISPR/Cas9	Zhang et al. (2018)
HC-Pro and GFP2 genes	TuMV in *Nicotiana benthamiana*	Cas13a	Aman, Ali, and Butt (2018)
TuMV virus	TuMV resistance in *A. thaliana*	LshpCas13	Aman, Mahas, Butt, Aljedaani, and Mahfouz (2018)

(*Continued*)

TABLE 11.2 (Continued)

Genes targeted	Effect of targeted gene on plant trait	Type of CRISPR/Cas genome editing system	Reference
MLO-A1, TaMLO-B1 and TaMLO-D1	Resistance to powdery mildew in wheat	CRISPR/Cas9	Wang et al. (2014)
nCBP-1 & nCBP-2/exon	Brown streak (RNA) in cassava	CRISPR/Cas9	Gomez et al. (2019)
Nonexpressor of pathogenesis-related 3 (TcNPR3) gene	Black pod disease in cacao	CRISPR/Cas9	Fister, Landherr, Maximova, and Guiltinan (2018)
ORFs and the IR sequence sDNA of virus	Tomato yellow leaf curl virus (TYLCV) and Merremia mosaic virus (MeMV)	CRISPR/Cas9	Ali et al. (2015)
OsERF922 (ethylene responsive factor transcription factor)	Responsible for rice blast resistance	CRISPR/Cas9	Wang et al. (2016)
OsMPK5	Fungal blast and bacterial blight resistance in rice	CRISPR/Cas9	Xie and Yang (2013)
OsSWEET11, OsSWEET14 (rice bacterial blight susceptibility genes)	Bacterial blight susceptibility in rice	CRISPR/Cas9	Jiang et al. (2013); Zafar et al. (2020)
OsSWEET13	Bacterial blight tolerance in rice	CRISPR/Cas9	Zhou et al. (2015)
Pi-ta	Resistance to the rice blast disease	CRISPR/Cas9	Xu et al. (2019)
S-genes	*Phytophthora infestans* resistance in potato	CRISPR/Cas9	Andersson et al. (2017)
SlDMR6-1	Bacterial speck, blight and spot resistance in tomato	CRISPR/Cas9	Thomazella, Brail, Dahlbeck, and Staskawicz (2016)
SlJAZ2	Bacterial speck resistance in tomato	CRISPR/Cas9	Ortigosa, Gimenez-Ibanez, Leonhardt, and Solano (2019)

(*Continued*)

TABLE 11.2 (Continued)

Genes targeted	Effect of targeted gene on plant trait	Type of CRISPR/Cas genome editing system	Reference
SlMLO1	Resistance to powdery mildew in tomato	CRISPR/Cas9	Nekrasov et al. (2017)
TaMLO-A1 (Wheat mildew resistance locus1) and TaMLO-B1	Powdery Mildew resistance in *Triticum aestivum*	CRISPR/Cas9	Wang et al. (2014)
TYLCV-IR (intergenic regions), RCA regions	Leaf curl disease in *Nicotiana benthamiana*	CRISPR/Cas9	Ali et al. (2015)
VvPDS, MLO-7	Powdery mildew resistance in grapes	CRISPR/Cas9	Malnoy et al. (2016); Nakajima et al. (2017)
WRKY52	Gray mold resistance in grapes	CRISPR/Cas9	Wang et al. (2018)
WRKY70, WRKY11	JA- and SA-induced resistance to pathogens in rapeseed	CRISPR/Cas9	Sun et al. (2018)

(Malnoy et al., 2016). The CRISPR/Cas9 machinery developed resistance against cotton leaf curl disease (Iqbal, Sattar, & Shafiq, 2016), bacterial leaf blight (Jiang et al., 2013), brown streak in cassava (Gomez et al., 2019), powdery mildew in wheat (Zhang et al., 2017), beet severe curly top virus resistance (Ji, Zhang, Zhang, Wang, & Gao, 2015), and bean yellow dwarf virus in tobacco (Baltes et al., 2015). Some other examples of CRISPR/Cas-mediated biotic stress tolerance of plants are summarized in Table 11.2.

11.5 Conclusion

Global climate change is predicted to pose harsh dangers to agricultural productivity worldwide, and subsequently it poses a challenge to food and nutritional security. Conventional breeding appears incompetent to fulfill the growing demands food and feed through plant improvement for tolerance of various biotic and abiotic stresses. The revolutionary CRISPR toolbox has made genome editing an effortlessly practicable route to abiotic stress tolerance and other desirable crop trait improvement. CRISPR/Cas can enhance hybrid-breeding techniques by eradicating undesirable genes and crop traits or by accumulation of the desired traits to the best varieties in a precise manner in a single generation. CRISPR-Cas technology can improve global food security and enable the dream of sustainable agriculture to become a reality.

References

Abudayyeh, O. O., Gootenberg, J. S., Konermann, S., Joung, J., Slaymaker, I. M., Cox, D. B. T., ... Minakhin, L. (2016). C2c2 is a single-component programmable RNA-guided RNA-targeting CRISPR effector. *Science (New York, N.Y.)*, *353*, aaf5573. Available from https://doi.org/10.1126/science.aaf5573.

Adli, M. (2018). The CRISPR tool kit for genome editing and beyond. *Nature Communications*, *9*, 1–13.

Ahmad, P., Tripathi, D. K., Deshmukh, R., Pratap Singh, V., & Corpas, F. J. (2019). *Revisiting the role of ROS and RNS in plants under changing environment*. Amsterdam: Elsevier.

Alexandratos, N., & Bruinsma, J. (2012). *World agriculture towards 2030/2050: The 2012 revision*. Rome: ESA Working paper FAO.

Ali, Z., Abulfaraj, A., Idris, A., Ali, S., Tashkandi, M., & Mahfouz, M. M. (2015). CRISPR/Cas9-mediated viral interference in plants. *Genome Biology*, *16*(1), 238–249. Available from https://doi.org/10.1186/s13059-015-0799-6.

Alok, A., Sandhya, D., Jogam, P., Rodrigues, V., Bhati, K. K., Sharma, H., et al. (2020). The Rise of the CRISPR/Cpf1 system for efficient genome editing in plants. *Frontiers in Plant Science*, *11*, 264. Available from https://doi.org/10.3389/fpls.2020.00264.

Aman, R., Ali, Z., Butt, H., et al. (2018). RNA virus interference via CRISPR/Cas13a system in plants. *Genome Biology*, *19*(1), 1–9. Available from https://doi.org/10.1186/s13059-017-1381-1.

Aman, R., Mahas, A., Butt, H., Aljedaani, F., & Mahfouz, M. (2018). Engineering RNA virus interference via the CRISPR/Cas13 machinery in *Arabidopsis*. *Viruses*, *10*(12), 732. Available from https://doi.org/10.3390/v10120732.

Andersson, M., Turesson, H., Nicolia, A., Fält, A. S., Samuelsson, M., & Hofvander, P. (2017). Efficient targeted multiallelic mutagenesis in tetraploid potato (*Solanum tuberosum*) by transient CRISPR-Cas9 expression in protoplasts. *Plant Cell Reports*, *36*, 117–128.

Baltes, N. J., Gil-Humanes, J., Cermak, T., Atkins, P. A., & Voytas, D. F. (2014). DNA replicons for plant genome engineering. *Plant Cell*, *26*, 151–163. Available from https://doi.org/10.1105/tpc.113.119792.

Baltes, N. J., Hummel, A. W., Konecna, E., Cegan, R., Bruns, A. N., Bisaro, D. M., et al. (2015). Conferring resistance to geminiviruses with the CRISPR-Cas prokaryotic immune system. *Nature Plants*, *1*(10), 15145.

Banakar, R., Schubert, M., Collingwood, M., Vakulskas, C., Alan, L., Eggenberger., & Wang, K. (2020). Comparison of CRISPR-Cas9/Cas12a ribonucleoprotein complexes for genome editing efficiency in the rice phytoene desaturase (OsPDS) gene. *Rice*, *13*(1), 4. Available from https://doi.org/10.1186/s12284-019-0365-z.

Barrangou, R., Fremaux, C., Deveau, H., Richards, M., Boyaval, P., Moineau, S., ... Horvath, P. (2007). CRISPR provides acquired resistance against viruses in prokaryotes. *Science (New York, N.Y.)*, *315*, 1709–1712.

Bedell, V. M., Wang, Y., Campbell, J. M., Poshusta, T. L., Starker, C. G., Krug, R. G., ... Leung, A. Y. H. (2012). In vivo genome editing using a high-efficiency TALEN system. *Nature*, *491*, 114–118.

Begemann, M. B., Gray, B. N., January, E., Gordon, G. C., He, Y., Liu, H., ... Oufattole, M. (2017a). Precise insertion and guided editing of higher plant genomes using Cpf1 CRISPR nucleases. *Scientific Reports*, *7*, 1–6.

Begemann, M. B., Gray, B. N., January, E., Gordon, G. C., He, Y., Liu, H., ... Oufattole, M. (2017b). Precise insertion and guided editing of higher plant genomes using Cpf1 CRISPR nucleases. *Scientific Reports*, *7*, 11606.

Bhat, J. A., Ali, S., Salgotra, R. K., Mir, Z. A., Dutta, S., Jadon, V., ... Singh, P. (2016). Genomics election in the era of next generation sequencing for complex traits in plant breeding. *Frontiers in Genetics*, *7*, 221.

Bhat, J. A., Shivaraj, S., Singh, P., Navadagi, D. B., Tripathi, D. K., Dash, P. K., ... Deshmukh, R. J. P. (2019). Role of silicon in mitigation of heavy metal stresses in crop plants. *Plants*, *8*, 71.

Bhowmik, P., Hassan, M. M., Molla, K., Rahman, M., & Islam, M. T. (2019). Application of CRISPR-Cas genome editing tools for the improvement of plant abiotic stress tolerance. In M. Hasanuzzaman, K. Nahar, M. Fujita, H. Oku, & M. T. Islam (Eds.), *Approaches for enhancing abiotic stress tolerance in plants* (pp. 459–472). New York: CRC Press.

Bo, W., Zhaohui, Z., Huanhuan, Z., Xia, W., Binglin, L., Lijia, Y., ... Chunguo, W. (2019). Targeted mutagenesis of NAC transcription factor gene, OsNAC041, leading to salt sensitivity in Rice. *Rice Science*, *26*, 98–108.

Boch, J., Scholze, H., Schornack, S., Landgraf, A., Hahn, S., Kay, S., ... Bonas, U. (2009). Breaking the code of DNA binding specificity of TAL-type III effectors. *Science (New York, N.Y.)*, *326*, 1509–1512.

Borel, B. (2017). CRISPR, microbes and more are joining the war against crop killers. *Nature*, *543*(7645), 302–304. Available from https://doi.org/10.1038/543302a.

Bortesi, L., & Fischer, R. (2015). The CRISPR/Cas9 system for plant genome editing and beyond. *Biotechnology Advances*, *33*(1), 41–52.

Braff, J. L., Yaung, S. J., Esvelt, K. M., & Church, G. M. (2016). Characterization of Cas9–guide RNA orthologs. *Cold Spring Harbor Protocols*. Available from https://doi.org/10.1101/pdb.top086793.

Burmistrz, M., Krakowski, K., & Krawczyk-Balska, A. (2020). RNA-targeting CRISPR–Cas systems and their applications. *International Journal of Molecular Sciences*, *21*, 1122. Available from https://doi.org/10.3390/ijms21031122.

Butler, N. M., Baltes, N. J., Voytas, D. F., & Douches, D. S. (2016). Geminivirus-mediated genome editing in potato (*Solanum tuberosum* L.) using sequence-specific nucleases. *Frontiers in Plant Science*, *7*, 1045.

Butt, H., Eid, A., Ali, Z., Atia, M. A. M., Mokhtar, M. M., Hassan, N., et al. (2017). Efficient CRISPR/Cas9-mediated genome editing using a chimeric single-guide RNA molecule. *Frontiers in Plant Science*, *8*, 1441.

Callaway, E. (2018). CRISPR plants now subject to tough GM laws in European Union. *Nature*, *560*, 16. Available from https://doi.org/10.1038/d41586-018-05814-6.

Campa, C. C., Weisbach, N. R., Santinha, A. J., Incarnato, D., & Platt, R. (2019). Multiplexed genome engineering by Cas12a and CRISPR arrays encoded on single transcripts. *Nature Methods*, *16*, 887–893.

Carlson, D. F., Fahrenkrug, S. C., & Hackett, P. B. (2012). Targeting DNA with fingers and TALENs. *Molecular Therapy Nucleic Acids*, *1*(1), e3. Available from https://doi.org/10.1038/mtna.2011.5.

Carroll, D., Morton, J. J., Beumer, K. J., & Segal, D. J. (2006). Design, construction and in vitro testing of zinc finger nucleases. *Nature Protocols*, *1*, 1329–1341.

Cebrian-Serrano, A., & Davies, B. (2017). CRISPR-Cas orthologues and variants: Optimizing the repertoire, specificity and delivery of genome engineering tools. *Mammalian Genome: Official Journal of the International Mammalian Genome Society*, *28*, 247–261.

Čermák, T., Baltes, N. J., Čegan, R., Zhang, Y., & Voytas, D. F. (2015). High frequency, precise modification of the tomato genome. *Genome Biology*, *16*(1), 232.

Chan, Z., Grumet, R., & Loescher, W. (2011). Global gene expression analysis of transgenic, mannitol-producing, and salt-tolerant *Arabidopsis thaliana* indicates widespread changes in abiotic and biotic stress-related genes. *Journal of Experimental Botany*, *62*, 1–17.

Chandrasekaran, J., Brumin, M., Wolf, D., Leibman, D., Klap, C., Pearlsman, M., et al. (2016). Development of broad virus resistance in non-transgenic cucumber using CRISPR/Cas9 technology. *Molecular Plant Pathology*, *17*, 1140–1153. Available from https://doi.org/10.1111/mpp.12375.

Chen, J. S., Ma, E., Harrington, L. B., Da Costa, M., Tian, X., Palefsky, J. M., et al. (2018). CRISPR-Cas12a target binding unleashes indiscriminate singlestranded DNase activity. *Science (New York, N.Y.)*, *360*, 436–439. Available from https://doi.org/10.1126/science.aar6245.

Chen, K., Wang, Y., Zhang, R., Zhang, H., & Gao, C. (2019). CRISPR/Cas genome editing and precision plant breeding in agriculture. *Annual Review of Plant Biology*, *70*, 667–697.

Chen, Y., Ma, J., Zhang, X., Yang, Y., Zhou, D., Yu, Q., et al. (2017). A Novel non-specific lipid transfer protein gene from sugarcane (NsLTPs), obviously responded to abiotic stresses and signaling molecules of SA and MeJA. *Sugar Technology*, *19*, 17. Available from https://doi.org/10.1007/s12355-016-0431-4.

Cheng, L., Zou, Y., Ding, S., Zhang, J., Yu, X., Cao, J., & Lu, G. (2009). Polyamine accumulation in transgenic tomato enhances the tolerance to high temperature stress. *Journal of Integrative Plant Biology*, *51*, 489–499.

Chneiweiss, H., Hirsch, F., Montoliu, L., Muller, A. M., Fenet, S., Abecassis, M., Merchant, J., et al. (2017). Fostering responsible research with genome editing technologies: A European perspective. *Transgenic Research*, *26*, 709–713.

Christian, M., Cermak, T., Doyle, E. L., Schmidt, C., Zhang, F., Hummel, A., et al. (2010). Targeting DNA double-strand breaks with TAL effector nucleases. *Genetics*, *186*, 757–761. Available from https://doi.org/10.1534/genetics.110.120717.

Chylinski, K., Makarova, K. S., Charpentier, E., & Koonin, E. V. (2014). Classification and evolution of type II CRISPR-Cas systems. *Nucleic Acids Research*, *42*, 6091–6105.

Cominelli, E., Conti, L., Tonelli, C., & Galbiati, M. (2013). Challenges and perspectives to improve crop drought and salinity tolerance. *New Biotechnology*, *30*, 355–361.

Cong, L., Ran, F. A., Cox, D., Lin, S., Barretto, R., Habib, N., et al. (2013). Multiplex genome engineering using CRISPR/Cas systems. *Science (New York, N.Y.)*, *339*, 819–823.

Curtin, S. J., Xiong, Y., Michno, J. M., Campbell, B. W., Stec, A. O., Cermák, T., ... Stupar, R. M. (2018). CRISPR/Cas9 and TALENs generate heritable mutations for genes involved in small RNA processing of Glycine max and *Medicago truncatula*. *Plant Biotechnology Journal*, *16*, 1125–1137.

Deshmukh, R., Sonah, H., Patil, G., Chen, W., Prince, S., Mutava, R., ... Nguyen, H. (2014). Integrating omic approaches for abiotic stress tolerance in soybean. *Frontiers in Plant Science*, *5*, 244.

Dormehl, L. (2017). CRISPR's latest achievement? Taking the gluten out of wheat [Internet]. Digital Trends. <https://www.digitaltrends.com/cool-tech/crispr-gene-editing-low-gluten/> Accessed 03.03.18.

dos Reis, S. P., Lima, A. M., & de Souza, C. R. B. (2012). Recent molecular advances on downstream plant responses to abiotic stress. *International Journal of Molecular Sciences*, *13*, 8628–8647.

Doudna, J. A., & Charpentier, E. (2014). Genome editing. The new frontier of genome engineering with CRISPRCas9. *Science (New York, N.Y.)*, *346*(6213), 1258096. Available from https://doi.org/10.1126/science.1258096.

Doupis, G., Chartzoulakis, K., Beis, A., & Patakas, A. (2011). Allometric and biochemical responses of grapevines subjected to drought and enhanced ultraviolet-B radiation. *Australian Journal of Grape and Wine Research*, *17*, 36–42.

Durai, S., Mani, M., Kandavelou, K., Wu, J., Porteus, M. H., & Chandrasegaran, S. (2005). Zinc finger nucleases: Custom-designed molecular scissors for genome engineering of plant and mammalian cells. *Nucleic Acids Research*, *33*, 5978–5990.

East-Seletsky, A., O'Connell, M. R., Burstein, D., Knott, G. J., & Doudna, J. A. (2017). RNA targeting by functionally orthogonal Type VI-A CRISPR-Casenzymes. *Molecular Cell*, *66*, 373–383.

Eckardt, N. A. (2009). The future of science: Food and water for life. *Plant Cell*, *21*, 368–372.

Erpen-Dalla Corte, L., Mahmoud, L. M., Moraes, T. S., Mou, Z., Grosser, J. W., & Dutt, M. (2019). Development of improved fruit, vegetable, and ornamental crops using the CRISPR/cas9 genome editing technique. *Plants*, *8*, 601.

Fan, W., Zhang, M., Zhang, H., & Zhang, P. (2012). Improved tolerance to various abiotic stresses in transgenic sweet potato (*Ipomoea batatas*) expressing spinach betaine aldehyde dehydrogenase. *PLoS One*, *7*, e37344.

Farhat, S., Jain, N., Singh, N., Sreevathsa, R., Dash, P. K., Rai, R., ... Rai, V. (2019). CRISPR-Cas9 directed genome engineering for enhancing salt stress tolerance in rice. *Seminars in Cell and Developmental Biology*, *96*, 91–99. Available from https://doi.org/10.1016/j.semcdb.2019.05.003.

Feng, Z., Zhang, B., Ding, W., Liu, X., Yang, D. L., Wei, P., ... Mao, Y. (2013). Efficient genome editing in plants using a CRISPR/Cas system. *Cell Research*, *23*, 1229–1232. Available from https://doi.org/10.1038/cr.2013.114.

Fernandes, H., Pastor, M., & Bochtler, M. (2016). Type II and type V CRISPR effector nucleases from a structural biologist's perspective. *Postepy Biochemii*, *62*, 315–326.

Fiaz, S., Ahmad, S., Noor, M. A., Wang, X., Younas, A., Riaz, A., ... Ali, F. (2019). Applications of the CRISPR/Cas9 system for rice grain quality improvement: Perspectives and opportunities. *International Journal of Molecular Sciences*, *20*, 888.

Fister, A. S., Landherr, L., Maximova, S. N., & Guiltinan, M. J. (2018). Transient expression of CRISPR/Cas9 machinery targeting TcNPR3 enhances defense response in *Theobroma cacao*. *Frontiers in Plant Science*, *9*, 268.

Fonfara, I., Richter, H., Bratovic, M., Le Rhun, A., & Charpentier, E. (2016). The CRISPRassociated DNA-cleaving enzyme Cpf1 also processes precursor CRISPR RNA. *Nature*, *532*, 517–521.

Gao, L., Cox, D. B. T., Yan, W. X., Manteiga, J. C., Schneider, M. W., Yamano, T., et al. (2017). Engineered Cpf1 variants with altered PAM specificities increase genome targeting range. *Nature Biotechnology*, *35*, 789–792. Available from https://doi.org/10.1038/nbt.3900.

Gilbert, L. A., Horlbeck, M. A., Adamson, B., Villalta, J. E., Chen, Y., Whitehead, E. H., & Qi, L. S. (2014). Genome-scale CRISPR-mediated control of gene repression and activation. *Cell*, *159*(3), 647–661.

Gill, C. (2016). Penn state developer of gene-edited mushroom wins "Best of What''s New' award I Penn State University [Internet]. Penn State News. <http://news.psu.edu/story/432734/2016/10/19/academics/penn-statedeveloper-gene-edited-mushroom-wins-best-whats new?utm_source = newswire&utm_medium = email&utm_term = 433007_HTML&utm_con tent = 10-19-2016-21-27&utm_campaign = Penn + State + Today>. Accessed 03.03.18.

Gilliham, M., Able, J. A., & Roy, S. J. (2017). Translating knowledge about abiotic stress tolerance to breeding programmes. *The Plant Journal*, *90*, 898–917.

Gleditzsch, D., Pausch, P., Müller-Esparza, H., Özcan, A., Guo, X., Bange, G., & Randau, L. (2019). PAM identification by CRISPR-Cas effector complexes: Diversified mechanisms and structures. *RNA Biology*, *16*(4), 504–517. Available from https://doi.org/10.1080/15476286.2018.1504546.

Globus, R., & Qimron, U. (2018). A technological and regulatory outlook on CRISPR crop editing. *Journal of Cellular Biochemistry*, *119*, 1291–1298.

Godfray, H. C. J., Beddington, J. R., Crute, I. R., Haddad, L., Lawrence, D., Muir, J. F., et al. (2010). Food security: the challenge of feeding 9 billion people. *Science (New York, N.Y.)*, *327*, 812–818.

Gomez, M. A., Lin, Z. D., Moll, T., Chauhan, R. D., Hayden, L., Renninger, K., ... Staskawicz, B. J. (2019). Simultaneous CRISPR/Cas9-mediated editing of cassava eIF4E isoforms nCBP-1 and nCBP-2 reduces cassava brown streak disease symptom severity and incidence. *Plant Biotechnology Journal*, *17*(2), 421–434. Available from https://doi.org/10.1111/pbi.12987.

Gootenberg, J. S., Abudayyeh, O. O., Lee, J. W., Essletzbichler, P., Dy, A. J., Joung, J., ... Freije, C. A. (2017). Nucleic acid detection with CRISPR-Cas13a/C2c2. *Science (New York, N.Y.)*, *356*, 438–442. Available from https://doi.org/10.1126/science.aam9321.

Gothandam, K. M., Nalini, E., Karthikeyan, S., & Shin, J. S. (2010). OsPRP3, a flower specific proline-rich protein of rice, determines extracellular matrix structure of floral organs and its overexpression confers cold-tolerance. *Plant Molecular Biology*, *72*, 125.

Han, X., Liu, Z., Jo, M. C., Zhang, K., Li, Y., Zeng, Z., ... Qin, L. (2015). CRISPR-Cas9 delivery to hard-to-transfect cells via membrane deformation. *Science Advances*, *1*, e1500454.

Harrington, L. B., Burstein, D., Chen, J. S., Paez-Espino, D., Ma, E., Witte, I. P., et al. (2018). Programmed DNA destruction by miniature CRISPR-Cas14 enzymes. *Science (New York, N.Y.)*, *362*, 839–842. Available from https://doi.org/10.1126/science.aav4294.

Hasanuzzaman, M., Nahar, K., & Fujita, M. (2013). Plant response to salt stress and role of exogenous protectants to mitigate salt-induced damages. In P. Ahmad, M. M. Azooz, & M. N. V. Prasad (Eds.), *Ecophysiology and responses of plants under salt stress* (pp. 25–87). New York: Springer.

Hasanuzzaman, M., Nahar, K., Fujita, M., Ahmad, P., Chandna, R., Prasad, M. N. V., & Öztürk, M. (2013). Enhancing plant productivity under salt stress – Relevance of poly-omics. In P. Ahmad, M. M. Azooz, & M. N. V. Prasad (Eds.), *Salt stress in plants: Omics, signaling and responses* (pp. 113–156). Berlin: Springer.

Hsu, P. D., Lander, E. S., & Zhang, F. (2014). Development and applications of CRISPR-Cas9 for genome engineering. *Cell*, *157*(6), 1262–1278.

Huang, X. Z., Zeng, X. F., Li, J. R., & Zhao, D. G. (2017). Construction and analysis of tify1a and tify1b mutants in rice (*Oryza sativa*) based on CRISPR/Cas9 technology. *Journal of Agricultural Biotechnology*, *25*, 1003–1012. Available from https://doi.org/10.3969/j.issn.1674-7968.2017.06.015.

Iqbal, Z., Sattar, M. N., & Shafiq, M. (2016). CRISPR/Cas9: A tool to circumscribe cotton leaf curl disease. *Frontiers in Plant Science*, *7*, 475.

Ishino, Y., Krupovic, M., & Forterre, P. (2018). History of CRISPR-Cas from encounter with a mysterious repeated sequence to genome editing technology. *Journal of Bacteriology*, *200*, e00580–00517.

Ishino, Y., Shinagawa, H., Makino, K., Amemura, M., & Nakata, A. (1987). Nucleotide sequence of the IAP gene, responsible for alkaline phosphatase isozyme conversion in *Escherichia coli*, and identification of the gene product. *Journal of Bacteriology*, *169*, 5429–5433.

Jackson, S. A., McKenzie, R. E., Fagerlund, R. D., Kieper, S. N., Fineran, P. C., & Brouns, S. J. J. (2017). CRISPR-Cas: Adapting to change. *Science (New York, N.Y.)* (80), 356. Available from https://doi.org/10.1126/science.aal5056.

Jansen, R., Embden, J. D. V., Gaastra, W., & Schouls, L. M. (2002). Identification of genes that are associated with DNA repeats in prokaryotes. *Molecular Microbiology*, *43*, 1565–1575.

Jena, K. K., & Mackill, D. J. (2008). Molecular markers and their use in marker-assisted selection in rice. *Crop Science*, *48*, 1266–1276.

Ji, X., Zhang, H., Zhang, Y., Wang, Y., & Gao, C. (2015). Establishing a CRISPR-Cas-like immune system conferring DNA virus resistance in plants. *Nature Plants*, *1*, 15144. Available from https://doi.org/10.1038/nplants.2015.144.

Jia, H., Xu, J., Orbovic, V., Zhang, Y., & Wang, N. (2017). Editing citrus genome via SaCas9/sgRNA system. *Frontiers in Plant Science*, *8*, 2135. Available from https://doi.org/10.3389/fpls.2017.02135.

Jiang, F., & Doudna, J. A. (2017). CRISPR-Cas9 structures and mechanisms. *Annual Review of Biophysics*, *46*, 505–529.

Jiang, W., Zhou, H., Bi, H., Fromm, M., Yang, B., & Weeks, D. P. (2013). Demonstration of CRISPR/Cas9/sgRNA-mediated targeted gene modification in *Arabidopsis*, tobacco, sorghum and rice. *Nucleic Acids Research*, *41*, 188. Available from https://doi.org/10.1093/nar/gkt780.

Jinek, M., Chylinski, K., Fonfara, I., Hauer, M., Doudna, J. A., & Charpentier, E. (2012). A programmable dual-RNA-guided DNA endonuclease in adaptive bacterial immunity. *Science (New York, N.Y.)*, *337*, 816–821. Available from https://doi.org/10.1126/science.1225829.

Jones, J. D. G., Witek, K., Verweij, W., Jupe, F., Cooke, D., Dorling, S., et al. (2014). Elevating crop disease resistance with cloned genes. *Philosophical Transactions of the Royal Society B: Biological Sciences*, *369*, 20130087.

Joung, J. K., & Sander, J. D. (2012). TALENs: A widely applicable technology for targeted genome editing. *Nature Reviews. Molecular Cell Biology*, *14*, 49–55. Available from https://doi.org/10.1038/nrm3486.

Karvelis, T., Bigelyte, G., Young, J.K., Hou, Z., Zedaveinyte, R., Pociute, K., ... Siksnys, V. (2019). PAM recognition by miniature CRISPR-Cas14 triggers programmable double-stranded DNA cleavage. bioRxiv. <https://doi.org/10.1101/654897>.

Kim, D., Alptekin, B., & Budak, H. (2018). CRISPR/Cas9 genome editing in wheat. *Functional & Integrative Genomics*, *18*(1), 31–41.

Kim, H., Kim, S. T., Ryu, J., Kang, B. C., Kim, J. S., & Kim, S. G. (2017). CRISPR/Cpf1-mediated DNA-free plant genome editing. *Nature Communications*, *8*, 1–7. Available from https://doi.org/10.1038/ncomms14406.

Klap, C., Yeshayahou, E., Bolger, A. M., Arazi, T., Gupta, S. K., Shabtai, S., ... Barg, R. (2017). Tomato facultative parthenocarpy results from SlAGAMOUS-LIKE 6 loss of function. *Plant Biotechnology Journal*, *15*, 634–647.

Klompe, S. E., Vo, P. L. H., Halpin-Healy, T. S., & Sternberg, S. H. (2019). Transposon-encoded CRISPR-Cas systems direct RNA-guided DNA integration. *Nature*. Available from https://doi.org/10.1038/s41586-019-1323-z.

Knott, G. J., & Jennifer, A. D. (2018). CRISPR-Cas guides the future of genetic engineering. *Science (New York, N.Y.)*, *361*(6405), 866–869. Available from https://doi.org/10.1126/science.aat5011.

Komor, A. C., Badran, A. H., & Liu, D. R. (2017). CRISPR-based technologies for the manipulation of eukaryotic genomes. *Cell*, *168*(1-2), 20–36. Available from https://doi.org/10.1016/j.cell.2016.10.044.

Kopischke, S., Schüßler, E., Althoff, F., & Zachgo, S. (2017). TALEN-mediated genome-editing approaches in the liverwort *Marchantia polymorpha* yield high efficiencies for targeted mutagenesis. *Plant Methods*, *13*, 20.

Korotkova, A. M., Gerasimova, S. V., & Khlestkina, E. K. (2019). Current achievements in modifying crop genes using CRISPR/Cas system. *Vavilovskii Zhurnal Genet. Selektsii*, *23*, 29–37.

Leenay, R. T., & Beisel, C. L. (2017). Deciphering, communicating, and engineering the CRISPR PAM. *Journal of Molecular Biology*, *429*(2), 177–191. Available from https://doi.org/10.1016/j.jmb.2016.11.024.

Li, B., Li, Q., Mao, X., Li, A., Wang, J., Chang, X., et al. (2016). Two novel AP2/EREBP transcription factor genes TaPARG have pleiotropic functions on plant architecture and yield-related traits in common wheat. *Frontiers in Plant Science*, *7*, 1191.

Li, C., Li, W., Zhou, Z., Chen, H., Xie, C., & Lin, Y. (2020). A new rice breeding method: CRISPR/Cas9 system editing of the Xa13 promoter to cultivate transgene-free bacterial blight-resistant rice. *Plant Biotechnology Journal*, *18*, 313–315.

Li, C. H., Wang, G., Zhao, J. L., Zhang, L. Q., Ai, L. F., Han, Y. F., et al. (2014). The receptor-like kinase SIT1 mediates salt sensitivity by activating MAPK3/6 and regulating ethylene homeostasis in rice. *The Plant Cell*, *26*(6), 2538–2553. Available from https://doi.org/10.1105/tpc.114.125187.

Li, J. F., Norville, J. E., Aach, J., McCormack, M., Zhang, D., Bush, J., ... Sheen, J. (2013). Multiplex and homologous recombination-mediated genome editing in *Arabidopsis* and *Nicotiana benthamiana* using guide RNA and Cas9. *Nature Biotechnology*, *31*, 688–691. Available from https://doi.org/10.1038/nbt.2654.

Li, M., Li, X., Zhou, Z., Wu, P., Fang, M., Pan, X., et al. (2016). Reassessment of the four yield-related genes Gn1a, DEP1, GS3, and IPA1 in rice using a CRISPR/Cas9 system. *Frontiers in Plant Science*, *7*, 377. Available from https://doi.org/10.3389/fpls.2016.00377.

Li, R., Liu, C., Zhao, R., Wang, L., Chen, L., Yu, W., ... Shen, L. (2019). CRISPR/Cas9-mediated SlNPR1 mutagenesis reduces tomato plant drought tolerance. *BMC Plant Biology*, *19*, 38. Available from https://doi.org/10.1186/s12870-018-1627-4.

Li, S., Li, J., Zhang, J., Du, W., Fu, J., Sutar, S., ... Xia, L. (2018). Synthesis-dependent repair of Cpf1-induced double-strand DNA breaks enables targeted gene replacement in rice. *Journal of Experimental Botany*, *69*, 4715–4721.

Li, T., Liu, B., Spalding, M. H., Weeks, D. P., & Yang, B. (2012). High-efficiency TALEN-based gene editing produces disease-resistant rice. *Nature Biotechnology*, *30*, 390.

Li, X. M., Chao, D. Y., Wu, Y., Huang, X., Chen, K., Cui, L. G., et al. (2015). Natural alleles of a proteasome α2 subunit gene contribute to thermotolerance and adaptation of African rice. *Nature Genetics*, *47*(7), 827–833. Available from https://doi.org/10.1038/ng.3305.

Liao, S., Qin, X., Luo, L., Han, Y., Wang, X., Usman, B., ... Li, R. (2019). CRISPR/ Cas9-induced mutagenesis of semi-rolled leaf 1,2 confers curled leaf phenotype and drought tolerance by influencing protein expression patterns and ROS scavenging in rice (*Oryza sativa* L.). *Agronomy*, *9*, 728. Available from https://doi.org/10.3390/agronomy91.10728.

Liu, J., Orlova, N., Oakes, B. L., Ma, E., Spinner, H. B, Baney, K. L. M., ... Doudna, J. A. (2019). CasX enzymes comprise a distinct family of RNA-guided genome editors. *Nature, 566*, 218–223. Available from https://doi.org/10.1038/s41586-019-0908-x.

Liu, L., Chen, P., Wang, M., Li, X., Wang, J., Yin, M., & Wang, Y. (2017). C2c1-sgRNA complex structure reveals RNA-guided DNA cleavage mechanism. *Molecular Cell, 65*(2), 310–322. Available from https://doi.org/10.1016/j.molcel.2016.11.040.

Lou, D., Wang, H., Liang, G., & Yu, D. (2017). OsSAPK2 confers abscisic acid sensitivity and tolerance to drought stress in rice. *Frontiers in Plant Science, 8*, 993. Available from https://doi.org/10.3389/fpls.2017.00993.

Lowder, L. G., Zhang, D., Baltes, N. J., Paul, J. W., Tang, X., Zheng, X., ... Qi, Y. (2015). A CRISPR/Cas9 toolbox for multiplexed plant genome editing and transcriptional regulation. *Plant Physiology, 00636*.

Lu, H. P., Luo, T., Fu, H. W., Wang, L., Tan, Y. Y., Huang, J. Z., et al. (2018). Resistance of rice to insect pests mediated by suppression of serotonin biosynthesis. *Nature Plants, 4*, 338–344.

Ma, J., Chen, J., Wang, M., Ren, Y., Wang, S., Lei, C., & Cheng, Z. (2017). Disruption of OsSEC3A increases the content of salicylic acid and induces plant defense responses in rice. *Journal of Experimental Botany, 69*, 1051–1064.

Ma, Y., Dai, X., Xu, Y., Luo, W., Zheng, X., Zeng, D., ... Zhang, D. (2015). COLD1 confers chilling tolerance in rice. *Cell, 160*(6), 1209–1221. Available from https://doi.org/10.1016/j.cell.2015.01.046.

Macovei, A., Sevilla, N. R., Cantos, C., Jonson, G. B., Slamet-Loedin, I., Cermák, T., ... Chadha-Mohanty, P. (2018). Novel alleles of rice eIF4G generated by CRISPR/Cas9-targeted mutagenesis confer resistance to Rice tungro spherical virus. *Plant Biotechnology Journal, 16*, 918–1927.

Mahfouz, M. M. (2017). Genome editing: the efficient tool CRISPR–Cpf1. *Nature Plants, 3*, 17028.

Makarova, K. S., Haft, D. H., Barrangou, R., Brouns, S. J. J., Charpentier, E., Horvath, P., et al. (2011). Evolution and classification of the CRISPR-Cas systems. *Nature Reviews. Microbiology, 9*, 467–477.

Makarova, K. S., Wolf, Y. I., & Koonin, E. V. (2018). Classification and nomenclature of CRISPR-Cas systems: Where from here? *CRISPR Journal, 1*, 325–336.

Makarova, K. S., Wolf, Y. I., Alkhnbashi, O. S., Costa, F., Shah, S. A., Saunders, S. J., et al. (2015). An updated evolutionary classification of CRISPRCas systems. *Nature Reviews. Microbiology, 13*, 722–736. Available from https://doi.org/10.1038/nrmicro3569.

Mali, P., Yang, L., Esvelt, K. M., Aach, J., Guell, M., DiCarlo, J. E., ... Church, G. M. (2013). RNA-guided human genome engineering via Cas9. *Science (New York, N.Y.), 339*, 823–826.

Malnoy, M., Viola, R., Jung, M. H., Koo, O. J., Kim, S., Kim, J. S., ... Kanchiswamy, C. N. (2016). DNA-free genetically edited grapevine and apple protoplast using CRISPR/Cas9 ribonucleoproteins. *Frontiers in Plant Science, 7*, 1904.

Manghwar, H., Lindsey, K., Zhang, X., & Jin, S. (2019). CRISPR/Cas system: Recent advances and future prospects for genome editing. *Trends in Plant Science, 24*(1 2), 11, 02–25. Available from https://doi.org/10.1016/j.tplants.2019.09.006.

Mao, X., Zheng, Y., Xiao, K., Wei, Y., Zhu, Y., Cai, Q., ... Zhang, J. (2018). OsPRX2 contributes to stomatal closure and improves potassium deficiency tolerance in rice. *Biochemical and Biophysical Research Communications, 495*, 461–467.

Meyer, J. D. F., Deleu, W., Garcia-Mas, J., & Havey, M. J. (2008). Construction of a fosmid library of cucumber (*Cucumis sativus*) and comparative analyses of the eIF4E and eIF(iso)

4E regions from cucumber and melon (*Cucumis melo*). *Molecular Genetics and Genomics: MGG*, *279*, 473–480.

Miao, C., Xiao, L., Hua, K., Zou, C., Zhao, Y., Bressan, R. A., et al. (2018). Mutations in a subfamily of abscisic acid receptor genes promote rice growth and productivity. *PNAS*, *115*, 6058–6063.

Michelbart, M. V., Hasegawa, P. M., & Bailey-Serres, J. (2015). Genetic mechanisms of abiotic stress tolerance that translate to crop yield stability. *Nature Reviews. Genetics*, *16*, 237–251.

Miki, D., Zhang, W., Zeng, W., Feng, Z., & Zhu, J. K. (2018). CRISPR/Cas9-mediated genetargeting in *Arabidopsis* using sequential transformation. *Nature Communications*, *9*, 1967.

Mir, A., Edraki, A., Lee, J., & Sontheimer, E. J. (2018). Type II-C CRISPR-Cas9 biology, mechanism, and application. *ACS Chemical Biology*, *13*, 357–365.

Mishra, R., Joshi, R. K., & Zhao, K. (2018). Genome editing in rice: Recent advances, challenges, and future implications. *Frontiers in Plant Science*, *9*, 1361.

Moon, S. B., Kim, D. Y., Ko, J.-H., & Kim, Y.-S. (2019). Recent advances in the CRISPR genome editing tool set. *Experimental & Molecular Medicine*, *51*, 130. Available from https://doi.org/10.1038/s12276-019-0339-7.

Mougiakos, I., Bosma, E. F., de Vos, W. M., van Kranenburg, R., & van der Oost, J. (2016). Next generation prokaryotic engineering: The CRISPR-Cas toolkit. *Trends in Biotechnology*, *34*, 575–587.

Munns, R. (2011). Plant adaptations to salt and water stress: Differences and commonalities. In T. Ismail (Ed.), *Advances in botanical research* (57, pp. 1–32). Waltham, MA: Elseiver.

Mushtaq, M., Bhat, J. A., Mir, Z. A., Sakina, A., Ali, S., Singh, A. K., ... Bhat, R. (2018). CRISPR/Cas approach: A new way of looking at plant-abiotic interactions. *Journal of Plant Physiology*, *224*, 156–162.

Nakajima, I., Ban, Y., Azuma, A., Onoue, N., Moriguchi, T., Yamamoto, T., ... Endo, M. (2017). CRISPR/Cas9-mediated targeted mutagenesis in grape. *PLoS One*, *12*, e0177966.

Nakayama, T., Fish, M. B., Fisher, M., Oomen-Hajagos, J., Thomsen, G. H., & Grainger, R. M. (2013). Simple and efficient CRISPR-Cas9-mediated targeted mutagenesis in *Xenopus tropicalis*. *Genesis (New York, N.Y.: 2000)*, *51*, 835–843. Available from https://doi.org/10.1002/dvg.22720.

Nekrasov, V., Wang, C., Win, J., Lanz, C., Weigel, D., & Kamoun, S. (2017). Rapid generation of a transgene-free powdery mildew resistant tomato by genome deletion. *Scientific Reports*, *7*, 482.

Nishitani, C., Hirai, N., Komori, S., Wada, M., Okada, K., Osakabe, K., ... Osakabe, Y. (2016). Efficient genome editing in apple using a CRISPR/Cas9 system. *Scientific Reports*, *6*, 31481.

Nishizawa-Yokoi, A., Endo, M., Ohtsuki, N., Saika, H., & Toki, S. (2015). Precision genome editing in plants via gene targeting and piggyBac-mediated marker excision. *The Plant Journal*, *81*, 160–168.

O'Connell, M. R. (2019). Molecular mechanisms of RNA targeting by Cas13containing Type VI CRISPR–Cas systems. *Journal of Molecular Biology*, *431*, 66–87.

Ortigosa, A., Gimenez-Ibanez, S., Leonhardt, N., & Solano, R. (2019). Design of a bacterial speck resistant tomato by CRISPR/Cas9-mediated editing of SlJAZ2. *Plant Biotechnology Journal*, *17*, 665–673.

Osakabe, K., Osakabe, Y., & Toki, S. (2010). Site-directed mutagenesis in Arabidopsis using custom-designed zinc finger nucleases. *Proceedings of the National Academy of Sciences of the United States of America*, *107*, 12034–12039.

Osakabe, Y., & Osakabe, K. (2017). *Genome editing to improve abiotic stress responses in plants, . Progress in molecular biology and translational science* (149, pp. 99–109). Amsterdam: Elsevier.

Osakabe, Y., Watanabe, T., Sugano, S. S., Ueta, R., Ishihara, R., Shinozaki, K., & Osakabe, K. (2016). Optimization of CRISPR/Cas9 genome editing to modify abiotic stress responses in plants. *Scientific Reports, 6*, 26685. Available from https://doi.org/10.1038/srep2.6685.

Pabo, C. O., Peisach, E., & Grant, R. A. (2001). Design and selection of novel Cys2His2 zinc finger proteins. *Annual Review of Biochemistry, 70*, 313–340.

Pandey, P., Irulappan, V., Bagavathiannan, M. V., & Senthil-Kumar, M. (2017). Impact of combined abiotic and biotic stresses on plant growth and avenues for crop improvement by exploiting physio-morphological traits. *Frontiers in Plant Science, 8*, 537. Available from https://doi.org/10.3389/fpls.2017.00537.

Pandey, S. K., Nookaraju, A., Upadhyaya, C. P., Gururani, M. A., Venkatesh, J., Kim, D. H., & Park, S. W. (2011). An update on biotechnological approaches for improving abiotic stress tolerance in tomato. *Crop Science, 51*, 1–22.

Peng, A., Chen, S., Lei, T., Xu, L., He, Y., Wu, L., ... Zou, X. (2017). Engineering canker-resistant plants through CRISPR/Cas9-targeted editing of the susceptibility gene CsLOB1 promoter in citrus. *Plant Biotechnology Journal, 15*, 1509–1519.

Pessina, S., Lenzi, L., Perazzolli, M., Campa, M., Dalla Costa, L., Urso, S., ... Malnoy, M. (2016). Knockdown of MLO genes reduces susceptibility to powdery mildew in grapevine. *Horticulture Research, 3*, 16016.

Petolino, J. F., Worden, A., Curlee, K., Connell, J., Moynahan, T. L. S., Larsen, C., & Russell, S. (2010). Zinc finger nuclease-mediated transgene deletion. *Plant Molecular Biology, 73*, 617–628.

Pyott, D. E., Sheehan, E., & Molnar, A. (2016). Engineering of CRISPR/ Cas9mediated potyvirus resistance in transgene-free *Arabidopsis* plants. *Molecular Plant Pathology, 17*, 1276–1288. Available from https://doi.org/10.1111/mpp.12417.

Qi, L. S., Larson, M. H., Gilbert, L. A., Doudna, J. A., Weissman, J. S., Arkin, A. P., & Lim, W. A. (2013). Repurposing CRISPR as an RNA-guided platform for sequencespecific control of gene expression. *Cell, 152*(5), 1173–1183.

Rai, K.M., Ghose, K., Rai, A., Singh, H., Srivastava, R., & Mendu, V. (2018). Genome engineering tools in plant synthetic biology. In *Current developments in biotechnology and bioengineering: synthetic biology, cell engineering and bioprocessing technologies.*

Ran, F. A., Hsu, P. D., Lin, C. Y., Gootenberg, J. S., Konermann, S., Trevino, A. E., et al. (2013). Double nicking by RNA-guided CRISPR Cas9 for enhanced genome editing specificity. *Cell, 154*, 1380–1389.

Ray, D. K., Mueller, N. D., West, P. C., & Foley, J. A. (2013). Yield trends are insufficient to double global crop production by 2050. *PLoS One, 8*, e66428.

Regalado, A.A. (2015). Potato made with gene editing - MIT technology review [Internet]. <https://www.technologyreview.com/s/536756/a-potato-made-with-gene-editing/> Accessed 03.03.18.

Roca-Paixão, J. F., Gillet, F. X., Ribeiro, T. P., Bournaud, C., Lourenco-Tessutti, T., Noriega, D., ... Grossi-de-Sa, M. F. (2019). Improved drought stress tolerance in *Arabidopsis* by CRISPR/dCas9 fusion with a histone acetyl transferase. *Scientific Reports, 9*, 8080. Available from https://doi.org/10.1038/s4159.8-019-44571-y.

Rong, Y. S., & Golic, K. G. (2000). Gene targeting by homologous recombination in Drosophila. *Science (New York, N.Y.), 288*, 2013–2018.

Samai, P., Pyenson, N., Jiang, W., Goldberg, G. W., Hatoum-Aslan, A., & Marraffini, L. A. (2015). Co-transcriptional DNA and RNA cleavage during Type III CRISPR-Cas immunity. *Cell, 161*(5), 1164–1174. Available from https://doi.org/10.1016/j.cell.2015.04.027.

Sander, J., & Jeschke, M. (2016). CRISPR-Cas advanced plant breeding [Internet]. DuPont Pioneer. <https://www.pioneer.com/home/site/us/agronomy/library/crispr-cas/>. Accessed 03.03.18.

Schmidt, S. M., Belisle, M., & Frommer, W. B. (2020). The evolving landscape around genome editing in agriculture. *EMBO Reports*, *21*, 19–22. Available from https://doi.org/10.15252/embr.202050680.

Schuster, M., Schweizer, G., Reissmann, S., & Kahmann, R. (2016). Genome editing in *Ustilago maydis* using the CRISPR-Cas system. *Fungal Genetics and Biology*, *89*, 3–9.

Selle, K., & Barrangou, R. (2015). Harnessing CRISPR-Cas systems for bacterial genome editing. *Trends in Microbiology*, *23*, 225–232.

Shan, Q., Wang, Y., Li, J., Zhang, Y., Chen, K., Liang, Z., ... Qiu, J. (2013). Targeted genome modification of crop plants using a CRISPR-Cas system. *Nature Biotechnology*, *31*, 686.

Shi, J., Gao, H., Wang, H., Lafitte, H. R., Archibald, R. L., Yang, M., et al. (2017). ARGOS8 variants generated by CRISPR-Cas9 improve maize grain yield under field drought stress conditions. *Plant Biotechnology Journal*, *15*, 207–216. Available from https://doi.org/10.1111/pbi.12603.

Shim, J. S., Oh, N., Chung, P. J., Kim, Y. S., Choi, Y. D., & Kim, J. K. (2018). Overexpression of osnac14 improves drought tolerance in rice. *Frontiers in Plant Science*, *9*, 310.

Shimatani, Z., Kashojiya, S., Takayama, M., Terada, R., Arazoe, T., Ishii, H., et al. (2017). Targeted base editing in rice and tomato using a CRISPR-Cas9 cytidine deaminase fusion. *Nature Biotechnology*, *35*, 441–443. Available from https://doi.org/10.1038/nbt.3833.

Shivaraj, S., Deshmukh, R., Sonah, H., & Bélanger, R. (2019). Identification and characterization of aquaporin genes in *Arachis duranensis* and *Arachis ipaensis* genomes, the diploid progenitors of peanut. *BMC Genomics*, *20*, 222.

Shmakov, S., Abudayyeh, O. O., Makarova, K. S., Wolf, Y. I., Gootenberg, J. S., Semenova, E., ... Severinov, K. (2015). Discovery and functional characterization of diverse class 2 CRISPR-Cas systems. *Molecular Cell*, *60*(3), 385–397. Available from https://doi.org/10.1016/j.molcel.2015.10.008.

Shmakov, S., Smargon, A., Scott, D., Cox, D., Pyzocha, N., Yan, W., ... Wolf, Y. I. (2017). Diversity and evolution of class 2 CRISPR–Cas systems. *Nature Reviews. Microbiology*, *15*, 169.

Smargon, A. A., Cox, D. B. T., Pyzocha, N. K., Zheng, K., Slaymaker, I. M., Gootenberg, J. S., ... Makarova, K. S. (2017). Cas13b is a type VI-B CRISPR-associated RNA-guided RNase differentially regulated by accessory proteins Csx27 and Csx28. *Molecular Cell*, *65*, 618–630.

Sonah, H., Deshmukh, R. K., Labbé, C., & Bélanger, R. (2017). Analysis of aquaporins in Brassicaceae species reveals high-level of conservation and dynamic role against biotic and abiotic stress in canola. *Scientific Reports*, *7*, 2771.

Steinert, J., Schiml, S., Fauser, F., & Puchta, H. (2015). Highly efficient heritable plant genome engineering using Cas9 orthologues from *Streptococcus thermophilus* and *Staphylococcus aureus*. *The Plant Journal: For Cell and Molecular Biology*, *84*(6), 1295–1305. Available from https://doi.org/10.1111/tpj.13078, PMID: 26576927.

Sternberg, S. H., LaFrance, B., Kaplan, M., & Doudna, J. A. (2015). Conformational control of DNA target cleavage by CRISPR-Cas9. *Nature*, *527*, 110–113.

Sternberg, S. H., Richter, H., Charpentier, E., & Qimron, U. (2016). Adaptation in CRISPR-Cas systems. *Molecular Cell*, *61*, 797–808.

Strecker, J., Ladha, A., Gardner, Z., Schmid-Burgk, J. L., Makarova, K. S., Koonin, E. V., et al. (2019). RNA-guided DNA insertion with CRISPR-associated transposases. *Science (New York, N.Y.)*, *365*(6448), 48–53.

Sun, Q., Lin, L., Liu, D., Wu, D., Fang, Y., Wu, J., & Wang, Y. (2018). CRISPR/Cas9-mediated multiplex genome editing of the BnWRKY11 and BnWRKY70 genes in *Brassica napus* L. *International Journal of Molecular Sciences*, *19*, 2716.

Svitashev, S., Young, J., Schwartz, C., Gao, H., Falco, S. C., & Cigan, A. (2015). Targeted mutagenesis, precise gene editingandsite-specific gene insertion in maize using Cas9 and guide RNA. *Plant Physiology*, *169*, 931–945.

Swarts, D. C., van der Oost, J., & Jinek, M. (2017). Structural basis for guide RNA processing and seed-dependent DNA targeting by CRISPR-Cas12a. *Molecular Cell*, *66*, 221–233.

Szczelkun, M. D., Tikhomirova, M. S., Sinkunas, T., Gasiunas, G., Karvelis, T., Pschera, P., ... Seidel, R. (2014). Direct observation of R-loop formation by single RNA-guided Cas9 and Cascade effector complexes. *Proceedings of the National Academy of Sciences of the United States of America*, *111*(27), 9798–9803. Available from https://doi.org/10.1073/pnas.1402597111.

Tai, G. C., & Young, D. A. (1984). Early generation selection for important agronomic characteristics in a potato breeding population. *American Potato Journal*, *61*, 419–434.

Takagi, H., Tamiru, M., Abe, A., Yoshida, K., Uemura, A., Yaegashi, H., et al. (2015). Mutmap accelerates breeding of a salt-tolerant rice cultivar. *Nature Biotechnology*, *33*, 445–449.

Tang, N., Ma, S., Zong, W., Yang, N., Lv, Y., Yan, C., ... Xiang, Y. (2016). MODD mediates deactivation and degradation of OsbZIP46 to negatively regulate ABA signaling and drought resistance in rice. *The Plant Cell*, *28*(9), 2161–2177. Available from https://doi.org/10.1105/tpc.16.00171.

Tang, X., Lowder, L. G., Zhang, T., Malzahn, A. A., Zheng, X., Voytas, D. F., ... Qi, Y. (2017). A CRISPR-Cpf1 system for efficient genome editing and transcriptional repression in plants. *Nature Plants*, *17*(3), 17018. Available from https://doi.org/10.1038/nplants.2017.18.

Tashkandi, M., Ali, Z., Aljedaani, F., Shami, A., & Mahfouz, M. (2018). Engineering resistance against Tomato yellow leaf curl virus via the CRISPR/Cas9 system in tomato. *Plant Signaling & Behavior*, *13*, e1525996.

Terns, M. P. (2018). CRISPR-based technologies: Impact of RNA-targeting systems. *Molecular Cell*, *72*(3), 404–412. Available from https://doi.org/10.1016/j.molcel.2018.09.018, 01.

Tester, M., & Langridge, P. (2010). Breeding technologies to increase crop production in a changing world. *Science (New York, N.Y.)*, *327*, 818–822.

Thomazella, D.P.D.T., Brail, Q., Dahlbeck, D., & Staskawicz, B.J. (2016). CRISPR-Cas9 mediated mutagenesis of a DMR6 ortholog in tomato confers broad-spectrum disease resistance. bioRxiv.

Tieman, D., Zhu, G., Resende, M. F., Jr, Lin, T., Nguyen, C., Bies, D., et al. (2017). A chemical genetic roadmap to improved tomato flavor. *Science (New York, N.Y.)*, *355*, 391–394. Available from https://doi.org/10.1126/science.aal1556.

Tóth, E., Varga, E., Kulcsár, P. I., Kocsis, J., Utka, V., Krausz, S. L., ... Welker, Z. (2020). Improved LbCas12a variants with altered pam specicities further broaden the genome targeting range of Cas12a nucleases. *Nucleic Acids Research*. Available from https://doi.org/10.1093/nar/gkaa110.

Truong, D. J., Kühner, K., Kühn, R., Werfel, S., Engelhardt, S., Wurst, W., & Ortiz, O. (2015). Development of an intein-mediated split-Cas9 system for gene therapy. *Nucleic Acids Research*, *43*(13), 6450–6458. Available from https://doi.org/10.1093/nar/gkv601.

Vanegas, K. G., Jarczynska, Z. D., Strucko, T., & Mortensen, U. H. (2019). Cpf1 enables fast and efficient genome editing in *Aspergilli*. *Fungal Biology and Biotechnology*, *6*, 6.

Vats, S., Kumawat, S., Kumar, V., Patil, G. B., Joshi, T., Sonah, H., ... Deshmukh, R. (2019). Genome editing in plants: Exploration of technological advancements and challenges. *Cells*, *8*, 1386. Available from https://doi.org/10.3390/cells8111386.

Vishwakarma, K., Mishra, M., Patil, G., Mulkey, S., Ramawat, N., Pratap Singh, V., ... Sharma, S. (2019). Avenues of the membrane transport system in adaptation of plants to abiotic stresses. *Critical Reviews in Biotechnology*, *39*, 861–883.

Vu, T. V., Sivankalyani, V., Kim, E. J., Doan, D. T. H., Tran, M. T., Kim, J., ... Kim, J. Y. (2020). Highly efficient homology-directed repair using CRISPR/Cpf1-geminiviral replicon in tomato. *Plant Biotechnology Journal*, *18*, 2133–2143. Available from https://doi.org/10.1111/pbi.13373.

Waltz, E. (2016). Gene-edited CRISPR mushroom escapes US regulation. *Nature*, *532*, 293. Available from https://doi.org/10.1038/nature.2016.19754.

Wang, F., Wang, C., Liu, P., Lei, C., Hao, W., Gao, Y., ... Zhao, K. (2016). Enhanced rice blast resistance by CRISPR/Cas9-targeted mutagenesis of the ERF transcription factor gene OSERF922. *PLoS One*, *11*, e0154027.

Wang, L., Chen, L., Li, R., Zhao, R., Yang, M., Sheng, J., & Shen, L. (2017). Reduced drought tolerance by CRISPR/Cas9-mediated SlMAPK3 mutagenesis in tomato plants. *Journal of Agricultural and Food Chemistry*, *65*, 8674–8682. Available from https://doi.org/10.1021/acs.jafc.7b027.45.

Wang, M., Mao, Y., Lu, Y., Tao, X., & Zhu, J. K. (2017). Multiplex gene editing in rice using the CRISPR-Cpf1 system. *Molecular Plant*, *10*, 1011–1013. Available from https://doi.org/10.1016/j.molp.2017.03.001.

Wang, X., Tu, M., Wang, D., Liu, J., Li, Y., Li, Z., ... Wang, X. (2018). CRISPR/Cas9-mediated efficient targeted mutagenesis in grape in the first generation. *Plant Biotechnology Journal*, *16*, 844–855.

Wang, Y., Cheng, X., Shan, Q., Zhang, Y., Liu, J., Gao, C., & Qiu, J. L. (2014). Simultaneous editing of three homoeoalleles in hexaploid bread wheat confers heritable resistance to powdery mildew. *Nature Biotechnology*, *32*, 947–951. Available from https://doi.org/10.1038/nbt.2969.

Xie, K., & Yang, Y. (2013). RNA-guided genome editing in plants using a CRISPR-Cas system. *Molecular Plant*, *6*, 1975–1983.

Xu, J., Soni, V., Chopra, M., & Chan, O. (2020). Genetic modification of the HSP90 Gene using CRISPR-Cas9 to enhance thermotolerance in *T. Suecica*. *URNCST Journal*, *21*(4), 4. Available from https://doi.org/10.26685/urncst.178.

Xu, R., Qin, R., Li, H., Li, D., Li, L., Wei, P., & Yang, J. (2017). Generation of targeted mutant rice using a CRISPR-Cpf1 system. *Plant Biotechnology Journal*, *15*(6), 713–717. Available from https://doi.org/10.1111/pbi.12669, Epub 2017 Feb 19. PMID: 27875019; PMCID: PMC5425385.

Xu, Y., Wang, F., Chen, Z., Wang, J., Li, W. Q., Fan, F., ... Zhu, Q. H. (2019). Intron-targeted gene insertion in rice using CRISPR/Cas9: A case study of the Pi-ta gene. *The Crop Journal*, *8*, 424–431.

Yamano, T., Nishimasu, H., Zetsche, B., Hirano, H., Slaymaker, I. M., Li, Y., ... Nureki, O. (2016). Crystal structure of Cpf1 in complex with guide RNA and target DNA. *Cell*, *165*(4), 949–962. Available from https://doi.org/10.1016/j.cell.2016.04.003.

Yan, W. X., Hunnewell, P., Alfonse, L. E., Carte, J. M., Keston-Smith, E., Sothiselvam, S., ... Scott, D. A. (2019). Functionally diverse type V CRISPR-Cas systems. *Science (New York, N.Y.)*, *363*(6422), 88–91. Available from https://doi.org/10.1126/science.aav7271.

Yang, H., Gao, P., Rajashankar, K. R., & Patel, D. J. (2016). PAM-dependent target DNA recognition and cleavage by C2c1 CRISPR-Cas endonuclease. *Cell*, *167*, 1814–1828.

Yin, X., Anand, A., Quick, P., & Bandyopadhyay, A. (2019). Editing a stomatal developmental gene in rice with CRISPR/Cpf1. In Y. Qi (Ed.), *Plant genome editing with CRISPR systems. Methods in molecular biology* (1917). New York, NY: Humana Press.

Zafar, K., Khan, M. Z., Amin, I., Mukhtar, Z., Yasmin, S., Arif, M., ... Mansoor, S. (2020). Precise CRISPR-Cas9 mediated genome editing in super basmati rice for resistance against bacterial blight by targeting the major susceptibility gene. *Frontiers in Plant Science*, *11*, 575. Available from https://doi.org/10.3389/fpls.2020.00575.

Zargar, S. M., Nagar, P., Deshmukh, R., Nazir, M., Wani, A. A., Masoodi, K. Z., ... Rakwal, R. (2017). Aquaporins as potential drought tolerance inducing proteins: Towards instigating stress tolerance. *Journal of Proteomics*, *169*, 233–238.

Zeng, D. D., Yang, C. C., Qin, R., Alamin, M., Yue, E. K., Jin, X. L., & Shi, C. H. (2018). A guanine insert in OsBBS1 leads to early leaf senescence and salt stress sensitivity in rice (*Oryza sativa* L.). *Plant Cell Reports*, *37*(6), 933–946. Available from https://doi.org/10.1007/s00299-018-2280-y.

Zeng, Y., Wen, J., Zhao, W., Wang, Q., & Huang, W. (2020). Rational improvement of rice yield and cold tolerance by editing the three genes OsPIN5b, GS3, and OsMYB30 with the CRISPR-Cas9 system. *Frontiers in Plant Science*, *10*, 1663. Available from https://doi.org/10.3389/fpls.2019.01663.

Zetsche, B., Gootenberg, J. S., Abudayyeh, O. O., Slaymaker, I. M., Makarova, K. S., Essletzbichler, P., ... Zhang, F. (2015). Cpf1 is a single RNA-guided endonuclease of a class 2 CRISPR-Cas system. *Cell*, *163*(3), 759–771. Available from https://doi.org/10.1016/j.cell.2015.09.038.

Zhang, A., Liu, Y., Wang, F., Li, T., Chen, Z., Kong, D., ... Wang, J. (2019). Enhanced rice salinity tolerance via CRISPR/Cas9-targeted mutagenesis of the OsRR22 gene. *Molecular Breeding.*, *39*, 47.

Zhang, H. X., Zhang, Y., & Yin, H. (2019). Genome editing with mRNA encoding ZFN, TALEN, and Cas9. *Molecular Therapy: The Journal of the American Society of Gene Therapy*, *27*, 735–746.

Zhang, M., Cao, Y., Wang, Z., Wang, Z. Q., Shi, J., Liang, X., ... Jiang, C. (2018). A retrotransposon in an HKT1 family sodium transporter causes variation of leaf Na + exclusion and salt tolerance in maize. *The New Phytologist*, *217*, 1161–1176.

Zhang, Y., Bai, Y., Wu, G., Zou, S., Chen, Y., Gao, C., & Tang, D. (2017). Simultaneous modification of three homoeologs of TaEDR1 by genome editing enhances powdery mildew resistance in wheat. *The Plant Journal: For Cell and Molecular Biology*, *91*, 714–724.

Zhang, Y., Massel, K., Godwin, I. D., & Gao, C. (2018). Applications and potential of genome editing in crop improvement. *Genome Biology*, *19*, 210.

Zhang, Y., Massel, K., Godwin, I. D., & Gao, C. (2019). Correction to: Applications and potential of genome editing in crop improvement. *Genome Biology*, *20*, 13.

Zhang, Y. M., Zhang, Y., & Xie, K. (2020). Evaluation of CRISPR/Cas12a-Based DNA detection for fast pathogen diagnosis and GMO test in rice. *Molecular Breeding*, *40*(1), 11. Available from https://doi.org/10.1007/s11032-0191092-2.

Zhang, Z., Ge, X., Luo, X., Wang, P., Fan, Q., Hu, G., ... Wu, J. (2018). Simultaneous editing of two copies of GH14-3-3D confers enhanced transgene-clean plant defense against *Verticillium dahliae* in allotetraploid upland cotton. *Frontiers in Plant Science*, *9*, 842.

Zhang, Z., Li, J., Pan, Y., Li, J., Zhou, L., Shi, H., et al. (2017). Natural variation in CTB4a enhances rice adaptation to cold habitats. *Nature Communications*, *8*, 14788.

Zheng, X., Yang, S., Zhang, D., Zhong, Z., Tang, X., Deng, K., ... Zhang, Y. (2016). Effective screen of CRISPR/Cas9-induced mutants in rice by single-strand conformation polymorphism. *Plant Cell Reports*, *35*, 1545–1554.

Zhong, Z., Zhang, Y., You, Q., Tang, X., Ren, Q., Liu, S., et al. (2018). Plant genome editing using FnCpf1 and LbCpf1 nucleases at redefined and altered PAM sites. *Molecular Plant*, *11*, 999–1002. Available from https://doi.org/10.1016/j.molp.2018.03.008.

Zhou, J., Peng, Z., Long, J., Sosso, D., Liu, B., Eom, J. S., ... Frommer, W. B. (2015). Gene targeting by the TAL effector PthXo2 reveals cryptic resistance gene for bacterial blight of rice. *The Plant Journal: For Cell and Molecular Biology*, *82*(4), 632–643.

Zuo, Z., & Liu, J. (2016). Cas9-catalyzed DNA cleavage generates staggered ends: Evidence from molecular dynamics simulations. *Scientific Reports*, *6*, 37584.

Chapter 12

Photosynthetic and cellular responses in plants under saline conditions

Khan Bilal Mukhtar Ahmed[1], Sarika Singh[1], Yawar Sadiq[1], M. Masroor A. Khan[1], Moin Uddin[2], M. Naeem[1] and Tariq Aftab[3]

[1]Plant Physiology Section, Department of Botany, Faculty of Life Sciences, Aligarh Muslim University, Aligarh, India, [2]Botany Section, Women's College, Aligarh Muslim University, Aligarh, India, [3]Department of Botany, Aligarh Muslim University, Aligarh, India

12.1 Introduction

Global food security heavily relies on agricultural produce, which requires a substantial increase in reducing the gap between production and demand (Majeed & Muhammad, 2019). Owing to the tremendous increase in the human population which is projected to cross 9.7 billion by 2050, the need for escalating crop productivity has been an unavoidable issue in the previous few years (Abdelaal et al., 2020). There is an immediate requirement to substantially improve food production to quench the demand for such a rising population. However, a variety of abiotic stresses such as heavy metal stress, heat stress, cold stress, and salinity stress severely affect the growth of food crops (Siddiqui, Hayat, & Bajguz, 2018). Salinity stress is a critical factor suppressing plant productivity (Hura, Szewczyk-Taranek, Hura, Nowak, & Pawłowska, 2017) especially in areas with an arid and semiarid climate, where the shortage of water and poor drainage of irrigated lands adversely decreases plant productivity (Hanin, Ebel, Ngom, Laplaze, & Masmoudi, 2016). It is expected that about 1 billion hectares (ha) of land covering over 100 countries face salinity-related issues (Food and Agriculture Organization of the United Nations, FAOUN, 2015). Furthermore, soil salinization affects approximately three hectares of arable land per minute, leading to an increase of 10% in the salinity-affected area on yearly basis (Bose et al., 2017; Shrivastava & Kumar, 2015). The vast majority of calories taken by humans are derived from the chief staple crops (e.g., wheat, rice, and corn), which are glycophytes, meaning that when

Frontiers in Plant–Soil Interaction. DOI: https://doi.org/10.1016/B978-0-323-90943-3.00007-9

subjected to NaCl concentrations of more than 200 mM they fail to complete their life cycle (Flowers & Colmer, 2008; Munns & Tester, 2008). By the year 2050, salinity could cover over 50% of the fertile land (Jamil, Riaz, Ashraf, & Foolad, 2011). Thus two major problems that affect agricultural sustainability are the exploding human population and declining cultivable land (Shahbaz & Ashraf, 2013). With this concern, improving tolerance in crops to salinity stress is of paramount importance for worldwide food security.

Due to the buildup of salt, either by natural or anthropogenic processes, soils are rendered saline. The weathering of parent rocks, seawater deposition, and atmospheric deposition are natural processes, whereas anthropogenic processes include using brackish groundwater for irrigation, poor drainage, incessant irrigation for extensive intervals, and inappropriate handling of water in irrigated agriculture system. A high amount of soluble salts found in salinity affected areas mainly consists of NaCl, but Na_2SO_4, $CaSO_4$, and KCl are also present (Hachicha, 2007). A soil is considered to be saline when its electric conductivity (EC) becomes 4 dS m^{-1} (equal to 40 mM NaCl), producing an osmotic pressure of nearly 0.2 MPa, (Munns & Tester, 2008). Using nonselective cation channels, cyclic nucleotide-gated channels along with high-affinity potassium transporter (HKT1) Na^+ enters the plant cells (Krishnamurthy, Qingsong, & Kumar, 2018). Plants can be categorized into two major groups on the basis of their response to salinity: the glycophytes (that cannot tolerate salinity and ultimately die) and the halophytes (that can tolerate salinity). The growth in glycophytes is repressed or even totally stopped by NaCl concentrations of 100–200 mM, causing plant death, whereas halophytes can grow in 300–500 mM NaCl concentrations, as they possess several salt resilience mechanisms (Acosta-Motos et al., 2017).

According to Negrão et al. (2017) plant responses to salinity can be dissected into two major stages, early responses (shoot ion-independent) and late responses (ion-dependent). An early response (a couple of minutes or a day following stress generation) is considered as a quick response, in which the plant senses the osmotic effect of the salt in the soil. Early responses are also referred to as the water-deficit phase as water accessibility is harshly affected (Roy et al., 2014). In the second phase accretion of Na^+ in photosynthetic tissues takes place which disturbs nutrient availability in the cytosol and is a delayed response developed after numerous days to weeks (Roy et al., 2014). Salinity stress inhibits plant growth, which is facilitated by a range of physiological, biochemical, and molecular alterations (Husen, Iqbal, Sohrab, & Ansari, 2018). Additionally, membrane rupture, hampered nutrient uptake, repression of enzymes function, and metabolic breakdown, can occur and may eventually prove lethal for plant survival (Hasanuzzaman & Fujita, 2012; Mahajan & Tuteja, 2005). Salinity restricts CO_2 uptake and ultimately photosynthesis through a decline of stomatal conductances

(Flexas, Bota, Loreto, Cornic, & Sharkey, 2004). Acosta-Motos et al. (2015) revealed that salinity stress affects the functioning of photosystem II (PSII). Furthermore, it increases the production of reactive oxygen species (ROS), which impair the cellular machinery (Costa, Martins, Agacka-Mołdoch, Czubacka, & de Sousa Araújo, 2018) by inducing oxidation of lipids, protein, and nucleic acid (Vaishnav, Varma, Tuteja, & Choudhary, 2016).

The strategies for salinity tolerance in plants are getting significant attention due to the augmented need to cultivate and preserve salt-afflicted areas to enhance food supply. Plants have developed several adaptive mechanisms to avert the salinity-induced damage. Salinity-mediated adaptations in plants are multifarious and involve the exclusion, compartmentalization, and selective uptake of ions to sustain an appropriate Na^+, K^+, and Cl^- content and also to produce compatible osmoprotectants, such as proline, glycine betaine, sucrose, etc. (Gil et al., 2013). Furthermore, to cope with excessive ROS generation during salinity stress, enzymatic antioxidant machinery are triggered including superoxide dismutase (SOD), catalase (CAT), and peroxidase (POX) (Gharsallah, Fakhfakh, Grubb, & Gorsane, 2016). Differential responses to salinity can be seen in the same genus. For example, among the three species in the genus *Brassica*, the salinity tolerance trait followed the order *Brassica napus* > *Brassica juncea* > *Brassica oleracea* (Chakraborty, Bose, Shabala, & Shabala, 2016). To detect and comprehend salt tolerance elements and mechanisms at the molecular level the various omics tools, that is, genomics, transcriptomics, proteomics, are now being used. Plant biologists recognize sets of genes and proteins that promote stress tolerance and incorporate them into various species to produce salt-tolerant varieties (Gupta, Stoimenova, & Kaiser, 2005).

In this chapter, we mainly emphasize the ill effects of soil salinity on plant growth, physiology, nutrient balance, etc., and the tolerance mechanisms that allow some plants to resist salinity stress, with specific emphasis on ion homeostasis, the role of osmolytes and antioxidant machinery. Lastly, a detailed section about the roles of transcriptomics and proteomics in plant salt stress tolerance has been discussed.

12.2 Negative effects of soil salinity in plants

12.2.1 Growth and yield parameters

The plant grown in excess saline condition, suffers through various processes like ionic imbalance, osmotic effect, mineral imbalance, etc. that more or less decreases its growth and yield (Arbona et al., 2005; Ashraf & Wu, 1994; Gorham, Jones, & McDonnell, 1985; Ruiz, Martínez, & Cerdá, 1997). In salinity stress, an increase in Na^+ uptake creates disturbances in metabolic processes, in which low K^+ plays an important role (Marschner, 1995). During salinity stress, the accumulation of high sodium (Na^+) and chloride

(Cl^-) ions inside the plant tissues causes detrimental effects including physiological and molecular disorders. Plant responses to osmotic stress are suppression of cell growth and development, poor chlorophyll content, and earlier senescence (Shannon & Grieve, 1998). Several pieces of research have shown that the plant biomass (fresh and dry biomass), leaf area, and the number of leaves were severely decreased by salinity levels up to 8 dS m^{-1} (Kamran et al., 2020). In tomato high soil salinity was accompanied by a significant decline in various growth parameters such as plant height, leaves number, shoot weight, root length, and root weight (Mohammad, Shibli, Ajlouni, & Nimri, 1998). In addition to this, in barley even high exogenous Cl^- concentrations show detrimental effects like Na^+ toxicity (Tavakkoli, Rengasamy, & McDonald, 2010a). In faba bean, high Cl^- concentrations accumulate in the chloroplast region and degrade the chlorophyll as well as reduces the activity of PSII whereas high Na^+ blocks the calcium and potassium uptake and reduces the stomatal conductance, thereby results in photosynthesis constraints and reduced growth of the plant (Tavakkoli, Rengasamy, & McDonald, 2010b). In germinating seeds of broccoli and cauliflower, salinity suppresses the water potential, protein content and reserve food material that more or less shows detrimental effects on the growth and development of developing plants (Wu, Huo, Yao, & Li, 2019). In a study on Ajwain (*Trachyspermum ammi*), salinity stress shows more clear effects on the seed yield as compared to the fresh and dry weight of the plant (Ashraf & Orooj, 2006). In wheat, higher salinity decreases the germination percentage, growth, and yield (Kumar, Singh, & Kumar, 2012). An increase in Na^+ concentrations and decrease of K^+ uptake is one of the physiological disorders that leads to a decrease in the growth and development of plants and productivity (James, Blake, Byrt, & Munns, 2011). A decrease in photosynthetic pigments is found to be an important criterion in the reduction of growth in *Pistacia vera* L. Rootstocks (Rahneshan, Nasibi, & Moghadam, 2018). In another experiment, five cultivars of canola grow at different salinity conditions, result shows that an increase in salinity levels adversely affects various variables like germination rate, plant height, and growth and accumulates high sodium and proline concentrations (Bybordi, 2010). Salinity condition affects plant growth in different stages of development. Rice is reported as one of the salt-sensitive crop (Maas & Hoffman, 1977; Shannon, Rhoades, Draper, Scardaci, & Spyres, 1998). An experiment by Zeng and Shannon (2000) in rice found that the harvest index gets decreased on a salinity level of 3.40 dS m^{-1} or higher. Another experiment was conducted by Zeng, Shannon, and Lesch in (2001), in which plant sensitivity toward salinity was recorded in different growth stages like the day of seedling, 1-leaf, 3-leaf, panicle initiation, and booting initiation. In *Aloe vera* (a medicinal plant), growth, and yield parameters (plant height, leaf area, and total gel content) get affected in all salinity levels (Moghbeli et al., 2012). Plant growth reduction under salinity stress is attributed to the inhibition of cell elongation (Shahverdi, Omidi, & Tabatabaei, 2018). Moreover, the detrimental influence of salinity stress on photosynthesis is attributed to plant stomatal closure and the resulting decreased CO_2 assimilation, ultimately limit plant and leaf growth (Netondo, Onyango, & Beck, 2004; Shahverdi et al., 2018). Seed germination plays a significant role in seedling growth and development. In rice, salinity reduces gibberellic acid content which further inhibits rice seed germination by a decrease in α-amylase activity (Liu et al., 2018). In alfalfa, TIR (Thermal infrared remote) technique reveals that reduction in leaf area index and transpiration rate increases the osmotic stress in plants that further decreases the plant growth and canopy under salinity stress (Tian, Hou, Qiu, Zhang, & Yuan, 2020).

12.2.2 Photosynthesis and allied attributes

Photosynthesis is an important physio-chemical mechanism in which diffusion of CO_2 from the atmosphere to the carboxylation site in the chloroplast stroma of the leaf takes place (Long & Bernacchi, 2003). However, salinity disrupts the photosynthesis mechanism by decreasing the availability of CO_2 (Flexas et al., 2007). Salinity affects the photosynthetic machinery either by affecting the stomatal aperture size thereby regulating the gas exchange into or out of the leaf tissues or by accumulating salt in the mesophyll tissues that further limits the CO_2 assimilation during photosynthesis (Pan et al., 2020). However, in various plant species, salinity stress also reduces the activity of photosynthetic-related pigments. In Indian mustard (*B. juncea* L), NaCl content degrades the photosynthetic pigment and decreases the Chl a content by 52% (Ahmad et al., 2015). In another experiment on *B. juncea* by Mittal, Kumari, and Sharma (2012), salt stress decreases the net photosynthetic rate, chlorophyll fluorescence, and photosynthetic pigments.

However, electron transport and CO_2 reduction play an important role in photosynthesis machinery, and dismantle of these two due to high salinity, leads to impairment in the photosynthesis, process (Ashraf & Harris, 2013). In maize salt stress generally dismantle the oxygen-evolving complex, inactivates the PSII reaction centre, and also blocks the electron flow from QA to QB (Kan et al., 2017). High salinity generally reduces the net photosynthetic rate and chlorophyll content of *Licum ruthenicum* by inhibiting the PSII activity, water splitting complex, and electron transport mechanism (Li, Zhang, Zhang, & He, 2019). *Posidonia australis* (seagrass) shows highly reduction in photosynthetic rates, electron transport rate, and quantum yield of PSII under salinity stress (Cambridge, Zavala-Perez, Cawthray, Mondon, & Kendrick, 2017). In *Medicago truncatula*, high salinity disrupts the PSII and inhibits the assimilation of CO_2, and reduces the photosynthesis mechanism (Najar, Aydi, Sassi-Aydi, Zarai, & Abdelly, 2019). According to Huihui et al. (2020) high salinity and pH show a detrimental effect on the PSII photochemical efficiency (Fv/Fm) in mulberry. On the PSII donor side, a decrease in the activity of OEC (Oxygen evolving complex) is also shown under salinity stress that further declines the electron transport rate whereas on the PSII acceptor site there is blockage of electron flow from QA to QB. Hence salinity leads to photoinhibition by the degradation of the PSII reaction centre (Huihui et al., 2020).

In *Arabidopsis*, hypertonic salinity stress increases the extracellular ATP levels due to damage to the plasma membrane and cell wall. In addition to this, high NaCl conditions also decrease the φPSII, qP, Fv/Fm, and intracellular ATP production (Hou et al., 2018). However, in tomato, proteomic analysis reveals that the presence of high salinity creates a negative impact on the multiprotein complex proteins like PSI and PSII, reduces the levels of cytochrome b_6f and ATP synthase, and destroy the trimers and monomers of light-harvesting complex (LHCs) (Muneer, Park, Manivannan, Soundararajan, & Jeong, 2014). D1 and D2 proteins are the essential components of PSII reaction centre. Expression of the psbA gene, which codes for D1 protein get reduced in response to salinity stress in tomato. In addition to this, the *Porphobilinogen deaminase* gene (PBGD) is a chlorophyll biosynthesis gene, also gets downregulated under salinity stress whereas the expression of chlorophyllase (*Chlase*) a chlorophyll degrading gene gets upregulated in the case of tomato (Hu, Xiang, Li, Zou, & Hu, 2016). In wheat, a combination of abiotic environmental stresses like light, osmotic, and salt phosphorylates the D1 and D2 protein of PSII machinery (Wu et al., 2020). Research work by He et al. (2014) on the soybean plant, revealed that 150 mM of NaCl decreased the net photosynthetic rate, carboxylation efficiency, and RuBisCO content. Apart from this, degradation of RuBisCO also occurred via a mechanism in which CPs (chloroplast protrusions) form more RCBs (rubisco containing bodies) that excludes RuBisCO from the chloroplast to the cytoplasm and unite with vacuole and degrade the RuBisCO (He et al., 2014). Proteomic analysis identified several photosynthetic related proteins associated with the PSII (OEE1 and OEE2) and Calvin cycle-related key enzymes expression and ATP-production decreased under salinity stress (Bai et al., 2017). Therefore salinity resulted in the decline in photosynthesis, chlorophyll content, net photosynthetic rate, PSII activity, stomatal conductance by disrupting the enzymatic cascade and downregulation of stress-responsive genes and proteins related to the normal functioning of photosynthesis machinery.

12.2.3 Salinity perturb ion homeostasis, water uptake, and reactive oxygen species metabolism

Maintaining persistent intracellular ion homeostasis, particularly that of K^+ and Na^+ is critical for various metabolic processes in living cells, and is especially crucial for plant adjusting in a high salt environment (Pan et al., 2016). Since there is antagonism between Na^+ and K^+, elevated Na^+ level in the root zone can upset K^+ uptake (Abdel Latef, Srivastava, El-sadek, Kordrostami, & Tran, 2018; Põldma, Moor, Tõnutare, Herodes, & Rebane, 2013). Under a saline environment, Na^+ uptake increases, while K^+ efflux increases that elicit K^+ loss from the plant cells, eventually causing deficiency of K^+ in the cell (Chokshi, Pancha, Ghosh, & Mishra, 2017). In saline conditions, the elevated Na^+ content leads to nutrient imbalance, particularly the Na^+/K^+ ratio (Rahman, Nahar, Hasanuzzaman, & Fujita, 2016). According to Munns and Tester (2008) elevated Na^+ levels in the cytoplasm not only induce K^+ ion deficit and thus interrupt several enzymatic reactions but also imposed an extra demand of energy on the cell required for the synthesis of organic solutes to be used as osmolytes. Potassium is an activator for about 50 enzymes, which cannot be replaced with Na^+ (Wu, 2018). Moreover due to the vital function of K^+ in cell turgor, and membrane potential its deficit will reduce plant growth (Jouyban, 2012). Salt stress disturbs cytoplasmic Na^+/K^+ homeostasis in the cytosol of wheat, chickpea causing a rise in Na^+/K^+ ratio (Garg & Bhandari, 2016; Gul, Wakeel, Steffens, & Lindberg, 2019). As compared to the salt-tolerant varieties of wheat and barley the salt-sensitive varieties displayed higher salt-induced K^+ efflux in their mesophyll cells (Wu, Shabala, et al., 2015; Wu, Shabala, Barry, Zhou, & Shabala, 2013; Wu, Zhu, Shabala, Zhou, & Shabala, 2015).

High soil salinity cause toxicities of ions such as Na^+, Cl^-, or sulfate (SO_4^{2-}) that interrupt the availability of essential elements like nitrogen (N), phosphorus (P), potassium (K^+), and calcium (Ca^{2+}) (Zhu, 2001). Lavandula plants grown under 100 mM NaCl showed a reduction in leaf N and K content by 24% and 37%, respectively (Chrysargyris, Michailidi, & Tzortzakis, 2018). High salinity in soybean causes a decrease in K^+/Na^+ ratio (Dogan, 2011). El-Esawi et al. (2018) showed the elevated levels of Na^+ and Cl^- ions and decreased levels of K^+ and Ca^{2+} ions under salinity stress in maize. Excess of Na^+ has been observed to reduce seed germination, growth, flowering, and fruiting, etc. (Parihar, Singh, Singh, Singh, & Prasad, 2015; Singh, Kumar, Singh, Singh, & Prasad, 2015). Besides Na^+ bind to the active site of the enzymes which are normally bound by K^+ causing their inhibition and affect protein synthesis and ribosome functioning (Munns & Tester, 2008). However, in comparison to Na^+ ion, Cl^- is more toxic as it induces chlorosis due to reduced chlorophyll synthesis in addition to dismantling nutrient equilibrium (Hasanuzzaman, Nahar, & Fujita, 2013; Tavakkoli et al., 2010a, 2010b).

Plant water status can be estimated by a physiological parameter known as water potential (Parida & Das, 2005). Under the nonsaline condition, in comparison to the soil, the plant cells possess greater osmotic potential as a result of which they can pick up water and essential nutrients proficiently from the soil. However, under salinity stress reverse condition occurs *viz.* the soil solution possesses higher osmotic pressure than the plant cells as a result the plant cannot extract water from the soil (Kader & Lindberg, 2010).

Moreover, as a saline environment induces osmotic stress, the drop in relative water content (RWC) is a common problem in plant growth under salinity (Hasanuzzaman et al., 2014). In coleus, salinity causes a decline in leaf water potential and RWC and enhanced electrolyte leakage and ROS formation (Kotagiri & Kolluru, 2017). A low RWC was registered in pomegranate (Soori et al., 2019), mung bean (Ahmad et al., 2019), and sweet pepper (Abdelaal et al., 2020) under salinity stress. This decrease may be attributed to poor water uptake (Parvin et al., 2019). RWC was significantly reduced at both levels of NaCl *viz.* 200 and 300 mmol L^{-1} in Eruca (Hniličková, Hnilička, Martinkova, & Kraus, 2017). In okra, salinity decreased leaf turgor, leaf RWC, and turgor potential (Abbas et al., 2017). The low RWC results in loss of turgor and hydrostatic pressure gradient, which restricts the water availability for cell growth (Vysotskaya et al., 2010). Leaf water and osmotic potentials in *P. australis* became more negative in both low and high salinity levels (Cambridge et al., 2017). It was found that water potentials declined from -3.04 to -3.74 MPa, and osmotic potentials declined from -3.61 to -4.14 MPa, when *Posidonia oceanica* was treated to salinities from 37 to 43 psu over 47 days (Sandoval-Gil, Marín-Guirao, & Ruiz, 2012). In the case of citrus, exposed to saline conditions witnessed a decrease in water potential (Romero-Trigueros et al., 2017). The availability of soil water and water use efficiency (WUE = yield/water use) determine plant productivity (Xu & Hsiao, 2004). Increased WUE can increase productivity and mitigate environmental stresses (Han et al., 2013). Salinity also reduces WUE due to the ion-specific toxicity and the inability to extract water, as well as the photosynthesis and crop yield (Khataar, Mohammadi, & Shabani, 2018). WUE sharply reduced under high salinity levels in pepper (Semiz, Suarez, Ünlükara, & Yurtseven, 2014).

There is a delicate balance between ROS generation and ROS removal which is perturbed by various stresses including salinity (Das & Roychoudhury, 2014). Under salinity stress, there is excess-reduction of electron transport chain during photosynthesis resulting in the excess generation of ROS due to the limited supply of CO_2 (Türkan & Demiral, 2009). The various ROS species produced include superoxide (O_2^-), hydroxyl radical (OH^-), and hydrogen peroxide (H_2O_2) (Yamamoto et al., 2015). Chloroplasts, mitochondria, peroxisomes, the endoplasmic reticulum, and plasma membranes are the multiple sites in the plant cells where ROS production takes place (Foyer & Noctor, 2009; Gechev, Van Breusegem, Stone, Denev, & Laloi, 2006). Under salt stress, the production of ROS exceeds their removal by the antioxidant system, which initiates oxidative damage to lipids, proteins, and DNA, RNA, etc. (Ahanger & Agarwal, 2017; Tang, Mu, Shao, Wang, & Brestic, 2015) ultimately leading to cell death. Salinity stress adversely affects membrane lipids by promoting their peroxidation and reduce the ROS detoxifying enzymes such as SOD, CAT, and ascorbic peroxidase (APX) in salt-exposed plants (Halo et al., 2015). Salinity causes the peroxidation of membrane lipid that led to the formation of Malondialdehyde (MDA), and its content reflects the amount of membrane damage (Liang, Ma, Wan, & Liu, 2018). Salt-tolerant phaseolus plants have a low degree of lipid peroxidation (Taïbi et al., 2016). Salinity dose of 150 mM increased the MDA and H_2O_2 contents in wheat leaves (Mohamed et al., 2017). Salinity induced ROS generation causes aberrations in the ultrastructure of organelles and ion leakage (Ahanger, Tomar, Tittal, Argal, & Agarwal, 2017). Mung bean under salinity stress, showed elevated H_2O_2 and MDA contents which led to enhanced electrolyte leakage (Ahmad et al., 2019). Higher MDA content was reported in pistachio rootstock and citrus (Goharrizi, Baghizadeh, Kalantar, & Fatehi, 2019; Tanou et al., 2009). Salt stress increased MDA and membrane damage due to excess ROS generation in watermelon (Li et al., 2017); *B. napus* seedling (Hasanuzzaman & Fujita, 2011). Salt-treated maize had significantly higher electrolyte leakage and lipid peroxidation (Jiang, Zu, et al., 2017). Under salinity levels (80 and 100 mM NaCl) a significant rise in the contents of H_2O_2, MDA, and electrolyte leakage in maize was observed by El-Esawi et al. (2018).

12.2.4 Mineral uptake

Salinity induced nutritional imbalance negatively influence crop productivity (Hasanuzzaman et al., 2013). Salinity restricts mineral uptake and causes its deficits that further hampers growth, root morphology, and physiological functions (Sirohi, Pandey, Deveshwar, & Giri, 2016). It disturbs the mineral contents of plants by diminishing N and other mineral nutrients uptake due to competition with Na^+ and Cl^- (Ahmad, Azooz, & Prasad, 2013). At high NaCl concentration, sodium content increases in plant tissues, and leaf N, K^+, and Ca^{2+} decreases (Tuna et al., 2007). According to Hussin, Geissler, and Koyro (2013), Ca^{2+} and Mg^{2+} contents of all organs dropped in *Atriplex nummularia* due to salinity stress. Salinity reduces N, K^+, and Zn^{2+} content in the leaves, whereas K^+, P, Mg^{2+}, and Ca^{2+} content were reduced in the roots of mentha (Chrysargyris, Papakyriakou, Petropoulos, & Tzortzakis, 2019). It was observed that elevated NaCl levels cause calcium unavailability in plants such as *Solanum lycopersicum* (Tuna et al., 2007), and *Vigna unguiculata* (Murillo-Amador et al., 2006). According to Karimi, Ghabooli, Rahimi, and Amerian (2020), NaCl at 75 mM reduced the leaf contents of K, Zn, Fe, and Mn in leaves of *Vitis vinifera*. In green pepper, saline soil decreases N, P, and K in the plant during various developmental stages of the plant because roots cannot absorb them (Al-Karaki, 2017). High salinity level has been reported to disrupt mineral uptake in various plants *viz. T. ammi* (Ashraf & Orooj, 2006), fennel (Abd El-Wahab, 2006), *Matricaria recutita* (Baghalian, Haghiry, Naghavi, & Mohammadi, 2008), and *Achillea fragratissima* (Abd EL-Azim & Ahmed, 2009). In tomato, salinity hampers uptake of N, P, Mn, B, and Zn (Chaichi, Keshavarz-Afshar, Lu, & Rostamza, 2017). Elevated salinity levels reduce the concentrations of Ca, K, Mg which play a key role during photosynthesis in the case of mungbean (Torabian, Farhangi-Abriz, & Rathjen, 2018). Likewise, supplementation of 100 mM NaCl caused a considerable decline in N, P, K^+, and Ca^{2+} content in maize (Kaya, Akram, Ashraf, & Sonmez, 2018). Leaf K^+ and Ca^{2+} accumulation decreased under 100 mM NaCl stress in mung bean (Ahmad et al., 2019). In response to high salinity K^+ and Ca^{2+} contents decreased and Na^+ and Cl^- increased in a sensitive cultivar of Lisianthus; however, its salt-tolerant cultivar was able to maintain a persistent K^+ content despite Na^+ and Cl^- accumulation, signifying effective salt compartmentalization (Ashrafi & Nejad, 2018). The content of leaf and root K^+, Ca^{2+} Mg, and Fe decreased in fenugreek subjected to both the level of salinity stress *viz.* 100 and 200 mM (Belmecheri-Cherifi, Albacete, Martínez-Andújar, Pérez-Alfocea, & Abrous-Belbachir, 2019). Salinity induced alterations in various plants processes are shown in Fig. 12.1.

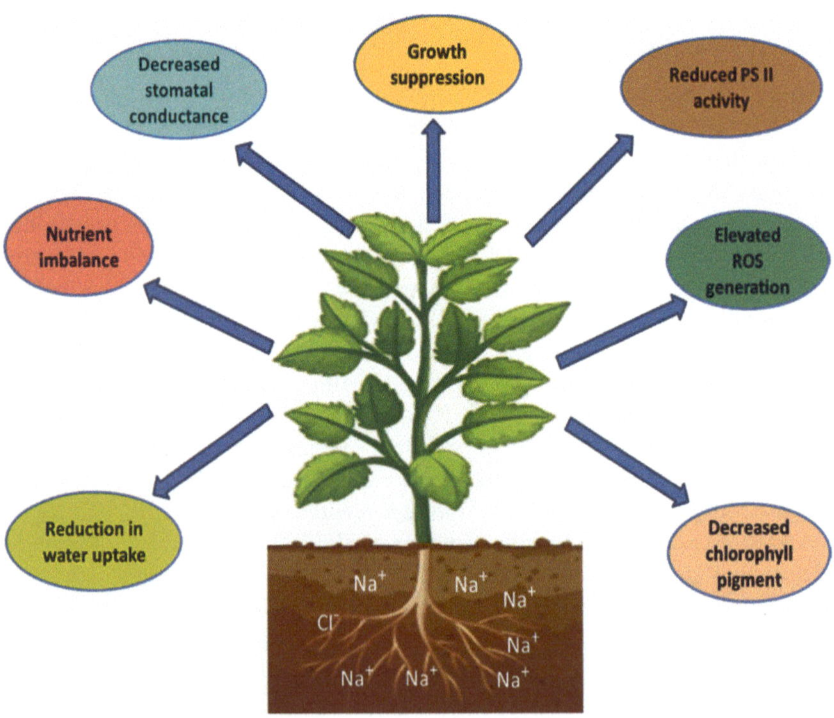

FIGURE 12.1 Negative effect of soil salinity on plant processes - Soil salinity adversely affect water uptake, nutrient balance, stomatal conductance, PS II activity, chlorophyll content which ultimately inhibit plant growth.

12.3 Adaptation of plants toward salinity

Tolerance to soil salinity is a physiologically multidimensional trait ascribed to multiple mechanisms. To facilitate crop production in the saline area the understanding of various adaptive mechanisms and identification of salt-tolerant genes/proteins is critical, which would help in crop improvements. Therefore a well-defined strategy linking the physiological, biochemical, metabolic, and molecular facets of salinity tolerance is necessary to develop salt-tolerant crop varieties. In this section, we will unravel various tolerance mechanisms employed by plants to cope with salinity problems such as ion homeostasis, compatible osmolytes production, and antioxidant regulation. Lastly, we assess the recent development of using some omics approaches (transcriptomics, proteomics) in developing salinity tolerance in crop plants.

12.3.1 Ionic homeostasis and compartmentalization: role of ion transporters

As mentioned above due to excessive uptake of Na^+ by plants under the saline condition the K^+/Na^+ ratio decreases (Almeida, Oliveira, & Saibo, 2017; Kader & Lindberg, 2005). This phenomenon hampers the supply of essential nutrients elements, particularly K^+, through its poor absorption by plants (Wakeel, 2013). To maintain ions homeostasis under the saline condition and to sustain photosynthesis and overall plant growth, a low cytosolic Na^+/K^+ ratio is needed (Hoang et al., 2016). Ion homeostasis is a key mechanism of cells under salt stress as it maintains lower content of Na^+ and higher content of K^+ ion (Nadeem et al., 2019). Ion homeostasis and thus salt tolerance are mediated by the following major approaches: Na^+ exclusion, K^+ retention, and tissue tolerance (Na^+ sequestration) (Munns & Tester, 2008; Wu, Shabala, et al., 2013; Wu, Wang, Feng, Li, & Wang, 2018).

Na^+ exclusion may occur during its loading in the xylem to prevent the transport of Na^+ to shoots in some genotypes (Chakraborty, Basak, & Bhaduri, 2018; Negrão, Courtois, & Ahmadi, 2011). High-affinity K^+ transporters (HKTs), are important Na^+-permeable transporters in plants, categorized as class I and class II types (Horie, Hauser, & Schroeder, 2009). HKT transporters play a vital role in salinity tolerance in several plant species and frequently occur in the xylem parenchyma and root epidermal cells (Møller et al., 2009; Munns et al., 2012). They also maintain a higher K^+ concentration in active plant tissues (Munns et al., 2012). Horie et al. (2005) showed that the AtHKT1 transporter has an important function of Na^+ unloading from xylem vessels to parenchyma cells in the leaves. The AtHKT1 transporter has also been involved in Na^+ transport from shoots back to the roots (Berthomieu et al., 2003) thereby maintaining a low Na^+/K^+ ratio in the cytosol of the shoot. Thus the main function of the HKT1 transporter is to pick up the Na^+ ion from the xylem and to protect the photosynthetically active tissues from damage (Ren et al., 2005). In plants such as lupine, clover, and maize Na^+ recirculation from shoot to roots through phloem promoted salinity tolerance (Tester & Davenport, 2003). It was observed in bread wheat that TaHKT1 localized in its root stele, was involved in salt tolerance by restricting Na^+ buildup in the shoots (Byrt et al., 2014). Likewise, in rice, OsHKT1, facilitates Na^+ exclusion from the leaf by excluding Na^+ from the xylem sap (Davenport et al., 2007; Horie et al., 2009).

Besides Na^+ exclusion, its sequestration also contributes to tissue tolerance to salinity stress. This trait shield cytosol from toxic Na^+ as these ions are compartmentalized in vacuoles, an approach usually seen in halophytes (Bonales-Alatorre, Shabala, Chen, & Pottosin, 2013). Sodium sequestration into vacuoles is carried out by the tonoplast-bound NHX proteins which function as Na^+/H^+ antiporters (Kumar & Mosa, 2015). The Na^+/H^+ EXCHANGER 1 (AtNHX1) obtained from *Arabidopsis thaliana* was overexpressed in transgenic plants that increased salt tolerance confirms that vacuolar Na^+ sequestration is mediated by tonoplast bound transporters such as AtNHX1 in *Arabidopsis* (Apse, Aharon, Snedden, & Blumwald, 1999; Blumwald, Aharon, & Apse, 2000). NHX transporters can be classified into two subtypes, class I and class II, based on protein sequence and their localization (Pardo, Cubero, Leidi, & Quintero, 2006; Rodriguez-Rosales et al., 2009). The vacuolar membrane *viz.* tonoplast bears the class I NHX (Venema, Quintero, Pardo, & Donaire, 2002), while the class II NHX proteins are present in endosomal vesicles of plants (Bassil et al., 2011). These proteins retain optimum K^+ concentration and help in supporting normal plant growth as well as endure salinity stress (Pardo et al., 2006). Overexpression of NHX1 transporters in wheat rice, and maize, improved their salt tolerance, further backing the concept that NHX transporters mediate the Na^+/H^+ exchange activity in the tonoplast (Yamaguchi & Blumwald, 2005). Overexpression of NHX1 enhances salt tolerance in several plants such as tomato (Zhang, Hodson, Williams, & Blumwald, 2001), rice (Chen, An, & Tang, 2007; Chen, Pottosin, et al., 2007), and tobacco (Gouiaa, Khoudi, Leidi, Pardo, & Masmoudi, 2012), revealing the significance of vacuolar Na^+ sequestration in salinity tolerance. According to Huertas et al. (2013) salinity tolerance in transgenic plants with overexpressed NHXs was due to a high K^+/Na^+ ratio which was credited to K^+ retention rather than Na^+ exclusion. Increased salt-tolerance in rice resulted from vacuolar Na^+ compartmentalization was facilitated by OsNHX1 (Prusty, Kim, & Vinarao, 2018). The overexpression of the *Arabidopsis* derived vacuolar NHX antiporter (AtNHX1) in wheat improved its salt tolerance (Xue, Zhi, Xue, Zhang, & Zhao, 2004). The vacuolar Na^+ fluorescence intensity in the mature root of wheat was significantly higher in the salt-tolerant varieties, as compared to sensitive ones (Cuin et al., 2011; Wu, Shabala, et al., 2015; Wu, Zhu, et al., 2015). Improved salinity tolerance was reported in transgenic tobacco by the overexpression of wheat antiporter gene TaNHX3 (Lu et al., 2014). Likewise, transgenic chili pepper plants exhibited increase salt tolerance due to wheat antiporter TaNHX2 gene (Bulle, Yarra, & Abbagani, 2016). This vacuolar sequestration of Na^+ not only avoids the toxic influence of Na^+ in the cytosol but also permits the plants to maintain the osmotic potential by using Na^+ ions as an osmolyte, that helps the plants to absorb more water into the cell. Thus NHX antiporters have central roles in the plant's response to salt stress (Rodriguez-Rosales et al., 2009).

Intra- and inter-cellular K^+ retention also contribute to plant salinity tolerance (Wu, Shabala, et al., 2013). Mesophyll tissue vacuolar storage of K^+ can prevent salt-induced injury and reduce K^+ content in the cytosol (Bassil et al., 2011). OsHAK1 and OsHAK5 (K^+ uptake transporter) were involved in the uptake of K^+ from potassium limited condition and its consequent allocation in various plant tissues (Chen, Hu, & Luo, 2015). The expression of HAK1 increases under K^+ deficient condition, which assists in root K^+ uptake, while HAK5 functions in K^+ allocation in upper plant tissues and maintains a higher K^+/Na^+ ratio (Chen et al., 2015). Chakraborty et al. (2019) found that there was an elevated expression of both HAK1 and HAK5 in the roots of tolerant genotypes of rice. Likewise, other approaches like the presence of specialized salt glands can work as a Na^+ storage organ under high-salinity conditions, and thus alleviate the rise of cytosolic Na^+, and improve survival (Agarie et al., 2007). Barhoumi, Djebali, Smaoui, Chaïbi, and Abdelly (2007) observed that *Aeluropus littoralis* possess special salt glands that excreted salts consisting mainly of NaCl. In Rhodes grass, salt glands secrete both Na^+ and K^+, but its Na^+ secreting ability was greater than K^+ (Kobayashi, Masaoka, Takahashi, Ide, & Sato, 2007). Bonales-Alatorre et al. (2013) studied quinoa under salt stress and found that in contrast to young leaves, the old leaves had considerably elevated Na^+ level in their leaf sap.

The salt overly sensitive (SOS) pathway represents an evolutionary primordial salt tolerance mechanism in the plant lineage (Edel & Kudla, 2015). Increasing research reveals the importance of the SOS stress signaling pathway in ion homeostasis and salinity resistance (Ji et al., 2013; Yang et al., 2009). SOS pathway includes proteins SOS1, SOS2, SOS3, and a Ca^{2+} sensor for the removal of Na^+ from the cell (Khan, 2011). SOS proteins detect the calcium signal that switches on the process for Na^+ exclusion (Zhu, 2007). In particular, SOS1, which encodes a plasma membrane $Na^+/H+$ antiporter, is a key player in Na^+ removal and in regulating Na^+ movement from the root to shoot (Shi, Ishitani, Kim, & Zhu, 2000; Shi, Quintero, Pardo, & Zhu, 2002). SOS1 protein operates as a Na^+/H^+ antiport to regulate the long-distance Na^+ transport in plants (Shi, Quintero, et al., 2002; Shi, Wu, & Zhu, 2002). It is well documented that salinity affects the levels of proteins such as calmodulin, calreticulin, annexin, and calcineurin-like phosphoesterases in various plants (Krishnamurthy et al., 2018). These changes may control the Ca^{2+} concentration in the intracellular stores, thereby triggering various phosphatases and kinases (Jiang, Yang, Harris, & Deyholos, 2007). Ca^{2+} signaling stimulates the SOS pathway which is imperative for controlling the cellular Na^+/K^+ ratio (Zhu, 2002). It begins with the sensing of extracellular salt stress by SOS3 through the Ca^{2+} signals which in turn triggers SOS2, a serine/threonine-protein kinase. The activated SOS2 then catalyzes phosphorylation of SOS1, a plasma membrane Na^+/H^+ antiporter for removing Na^+ out of the cytosol. SOS2 is a key protein in SOS-mediated salinity tolerance mechanism as it also phosphorylates the vacuolar Na^+/H^+ antiporter (NHX) for directing Na^+ into the vacuole (Chinnusamy, Jagendorf, & Zhu, 2005). Thus SOS2 activates both SOS1 and NHX, but actual Na^+ removal from the cell or storage inside the vacuole was mediated by the two latter proteins (Fig. 12.2). The role of SOS1 in effluxing Na^+ from leaf to xylem could be understood by the less buildup of Na^+ in stems of tomato (Olías, Eljakaoui, & Li, 2009). Augmented salt tolerance was observed in tobacco with an overexpressed SOS1 gene isolated from the extreme halophyte *Salicornia brachiate* (Yadav, Shukla, Jha, Agarwal, & Jha, 2012). SOS1 overexpression can also improve salinity tolerance in various transgenic plants (Yang et al., 2009; Yue, Zhang, Zhang,

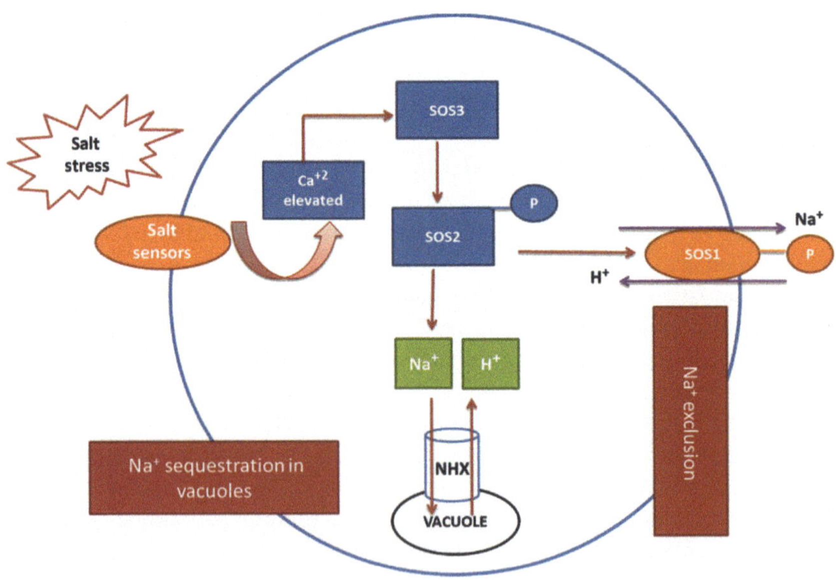

FIGURE 12.2 Salt overly sensitive (SOS) pathway for relieving salinity stress - Upon sensing salt stress, calcium ion (Ca^{+2}) concentration increased that triggers SOS pathway. The elevated Ca^{+2} first stimulates SOS3 protein, which phosphorylates SOS2 protein. The phosphorylated SOS2 protein directs both Na^+ exclusion from the cell and Na^+ sequestration in the vacuole by activating both cell membrane-bound SOS1 and tonoplast-bound NHX protein. SOS1 and NHX are Na^+/H^+ antiporters that mediate the salinity tolerance response.

Duan, & Li, 2012). A hyper-salt-sensitive phenotype was noticed in the halophytic *Arabidopsis* relative *Thellungiella salsuginea* due to the loss of SOS1 function (Oh et al., 2010). These findings established the important role of the SOS1 in Na^+ exclusion and overall plant salt tolerance.

12.3.2 Role of osmolytes in salinity tolerance

Plants also alleviate salt stress by the synthesis and accumulation of organic solutes known as osmolytes or compatible solutes (Per et al., 2017). The augmented levels of osmolytes in the cytosol decrease water loss and increase turgor pressure and cell expansion (Apse & Blumwald, 2002). Literature suggests that an escalation in osmolytes content resulted in reduced ROS generation, thus improving tolerance to salt stress (Behr et al., 2017). Various osmolytes like glycine betaine, proline, sucrose, trehalose, and myo-inositol are involved in osmotic regulation in plants under high salinity or other environmental stress conditions (Kumari & Parida, 2018).

12.3.2.1 Role of proline

Proline is a cyclic amino acid and is known to act as an osmolyte in plants under stressful conditions (Per et al., 2017). Proline accumulation is a general instant response of plants to salt stress (Lokhande & Suprasanna, 2012), showing its possible role in ROS detoxification (Behr et al., 2017). In addition to its role in osmotic regulation, the proline also inhibits NaCl-mediated K^+ leakage from the cell leading to K^+ homeostasis (Cuin & Shabala, 2005). As proline scavenge ROS it protects DNA, proteins, and membranes from damaging effects of NaCl (Kaur & Asthir, 2015). Compared to glycophytes halophytes are more proficient in modulating the metabolism of the proline and prominently accumulate it under saline condition (Slama, Abdelly, Bouchereau, Flowers, & Savoure, 2015). *Salvadora persica* a facultative halophytic plant endure salinity stress by accumulating proline (Kumari & Parida, 2018) which was in line with the finding of Kiani-Pouya et al. (2017) in *Chenopodium quinoa* where elevated proline content was reported under salinity. Lokhande, Nikam, and Penna (2010) reported that *Sesuvium portulacastrum* calli accumulated the highest level of proline grown at 200 and 400 mM NaCl. According to Wu, Cai, et al. (2013) wild barley, which demonstrated better salinity tolerance compared to cultivated barley, accumulates more proline. Similarly, *Atriplex halimus* registered high proline content under salinity stress (Belkheiri & Mulas, 2013). The accumulation of free proline under salinity stress was observed in *Pisum sativum* (Najafi, Khavari, Rastgar, & Sticklen, 2006), *B. juncea* (Rais, Masood, Inam, & Khan, 2013) *Triticum aestivum* (Ashfaque, Khan, & Khan, 2014) *Oryza sativa* (Hasanuzzaman et al., 2014) and *Kosteletzkya virginica* (Wang, Tang, Wang, & Shao, 2015). Amini and Ehsanpour (2005) have shown that increased proline content in tomato under salinity stress was attributed to the increased expression of genes concerned with the proline biosynthesis enzymes such as Pyrroline-5-carboxylate synthetase (P5CS) or repression in enzymes of proline catabolism such as proline dehydrogenase (PDH). Similarly, overexpression of P5CS1 can mitigate salinity stress in crops like rice and potato (Varghese, Alyammahi, Nasreddine, Alhassani, & Gururani, 2019). In *Arabidopsis*, mutation of the P5CS1 (proline biosynthesis gene), inhibits proline synthesis thereby, producing hyper- salt sensitivity (Mazars et al., 2008). When another gene concerned with proline biosynthesis *viz.* pyrroline-5-carboxylate reductase (P5CR) was overexpressed in transgenic tobacco, it conferred tolerance toward salinity (Kavi Kishor, Hong, Miao, Hu, & Verma, 1995). Under salt stress, Thellungiella displayed a low expression of the enzyme catalyzing proline catabolism, while a higher level of enzymes involved in proline biosynthesis (Pang et al., 2010). Naidu (2003) has reported that *Melanleuca bracteata* produced proline analog, 4-hydroxyN-methyl proline (MHP) under saline condition.

12.3.2.2 Role of sugars

Under salinity stress, sugar, and sugar alcohol derivatives are mostly accumulated to regulate cellular osmolarity, energy metabolism, and ROS scavenging inside plant cells (Kiani-Pouya et al., 2017; Kumari, Das, Parida, & Agarwal, 2015). Besides this, these compounds have also been reported to regulate the expressions of genes concerned with physiological processes (photosynthesis and respiration) for adaptation of the plants to various environmental stresses (Conde, Chaves, & Gerós, 2011). It was observed that maize plant growing under 50 mM NaCl, accumulates sugars like glucose, fructose, and sucrose (Richter, Erban, Kopka, & Zörb, 2015). Besides, *Juncus acutus* and *Juncus maritimus* plants also accumulate sucrose, glucose, and fructose in response to salinity stress (Ricardo et al., 2011). In durum wheat seedlings hexoses and sucrose acted as osmolytes in shoot and root tissues under salt conditions (Annunziata et al., 2017; Woodrow et al., 2017). Boriboonkaset et al. (2013) noted escalated content of starch and soluble sugar in the flag leaf of salt-tolerant genotype of rice which played a role in the salt tolerance mechanism. Similarly, ryegrass plants treated with 200 mM NaCl showed increased sucrose and glucose content (Hu et al., 2013). Garg and Bharti (2018) reported that salinity tolerance observed in the salt-tolerant *Cicer arietinum* cultivar (PBG 5) was attributed to a higher accumulation of sugars as compared to its salt-sensitive counterparts (BG256). Some carbohydrates such as sucrose, maltose, maltotriitol, and cellobiose were involved in the salinity resistance of *Prosopis strombulifera* (Llanes, Arbona, Gómez-Cadenas, & Luna, 2016). Trehalose mitigates salinity stress by detoxifying ROS, sustaining the K^+/Na^+ ratio, and enhancing the concentrations of soluble sugars in plants (Evelin, Devi, Gupta, & Kapoor, 2019).

Li, Zang, Deng, and Wang (2011) reported overexpression of OsTPS1 (trehalose-6-phosphate synthase) in transgenic rice that improved its salinity tolerance. Overexpression of the trehalose gene (OsTRE1) in transgenic rice plants have shown improved salt stress tolerance (Islam et al., 2019). Therefore salinity-induced carbohydrate accumulation might be crucial for plant growth.

Sugar alcohols could work as ROS scavengers and avert peroxidation of cell lipids (Keunen, Peshev, Vangronsveld, Van Den Ende, & Cuypers, 2013). The sugar alcohol, myo-inositol has been proposed to maintain the osmotic balance and protect the cellular structures from ROS and contribute to salt tolerance (Sengupta et al., 2012). Under salinity stress, increased levels of myo-inositol in *S. persica* were ascribed to the increased activity of myo-inositol1-phosphate synthase (MIPS), which is a key rate-limiting enzyme for the biosynthesis of myo-inositol (Kumari & Parida, 2018). In transgenic crops, augmented production of myo-inositol because of the overexpression of MIPS gene reported the enhancement of the salt, drought, and chilling stresses (Das-Chatterjee et al., 2006; Goswami et al., 2014; Zhai et al., 2016).

12.3.2.3 Role of glycine betaine

Glycine betaine (GB) acts as a nitrogen containing osmoregulator that protects the membrane integrity and scavenges ROS under environmental constraints (Annunziata, Ciarmiello, Woodrow, Dell'Aversana, & Carillo, 2019). GB is located in the chloroplast where it protects the thylakoid membrane and holds the photosynthetic machinery in the active state (Ashraf & Foolad, 2007; Genard, Le Saos, Billard, Tremolieres, & Boucaud, 1991). It helps in salt tolerance by reducing lipid peroxidation during and after the onset of salinity stress and stabilizes the protein quaternary structure (Kumari et al., 2015). It also maintains PSII and its allied proteins, manganese clusters, and maintains the activities of CO_2 fixing enzymes such as RuBisCO and rubisco activase (Pan et al., 2020; Talaat & Shawky, 2014). GB has been proposed as the organic component that plays the central role in the cytosol osmotic adjustment in *A. nummularia* (Silveira, Araújo, Lima, & Viégas, 2009). The sequential action of two enzymes CDH (choline dehydrogenase) and BADH (betaine aldehyde dehydrogenase) on choline produces GB. Increased tolerance to salt and drought stress was observed in transgenic crops via overexpression of the enzyme BADH (Fitzgerald, Waters, & Henry, 2009). Interestingly, even in salt-sensitive carrot, the BADH overexpression in the plastid resulted in a >50-fold increased GB generation compared to its wild relatives and resists salinity stress up to 400 mM NaCl, similar to halophytes (Kumar, Dhingra, & Daniell, 2004). In transgenic wheat, overexpression of the BADH gene increases the glycine betaine levels that protect the lipid and fatty acids of the thylakoid membrane, thereby improving photosynthesis

(Tian et al., 2017). Other enzymes related to the biosynthesis of GB such as SAMS (S adenosyl methionine synthetase), choline monooxygenase, and betaine aldehyde dehydrogenase were found in abundance in salt-treated *Suaeda aegyptiaca* (Askari, Edqvist, Hajheidari, Kafi, & Salekdeh, 2006) and *Setaria italica* (Veeranagamallaiah et al., 2008). Similarly, in the case of *Arabidopsis*, the overexpression of GOLS2, a GB synthase increased tolerance to elevated salinity levels (Sun et al., 2013). Under salinity stress in maize the exogenous application of GB led to an increase in growth (Yang & Lu, 2005). According to Tsutsumi, Yamada, Cha-um, Tanaka, and Takabe (2015), *Atriplex gmelini* registered enhanced GB accumulation in lamina leaves and bladder hairs in saline conditions. Salinity-induced GB accumulation in common beans enhanced stomatal conductance and relative leaf water content (Lopez, Takahashi, & Yamazaki, 2002). Tobacco plant overexpressing *Escherichia coli*-derived betA and betB genes coding for CDH and BADH enzymes resulted in increased biomass and easing of photoinhibition under salt stress (Holmström, Somersalo, Mandal, Palva, & Welin, 2000).

12.3.3 Role of antioxidant machinery

Under high salinity stress, overproduction of ROS such as superoxide (O_2^-), hydroxyl radical (OH^-), and hydrogen peroxide (H_2O_2) takes place (Yamamoto et al., 2015). The rapid increase in ROS concentration is called an "oxidative burst." The distinct subcellular localization, differential gene expression of antioxidant enzymes, and the variety of nonenzymatic scavengers make the antioxidant system a very handy unit that mitigates ROS accumulation both temporally and spatially (Tanveer & Shabala, 2018). Enzymatic ROS detoxifying enzymes include SOD, CAT, APX, and GPX (glutathione peroxidase). Additionally, nonenzymatic antioxidants include compounds such as ascorbate, glutathione (GSH), tocopherol, etc.

12.3.3.1 Role of enzymatic antioxidants

Activities of enzymatic antioxidants are associated to plant salinity tolerance as these antioxidants help detoxify the salt-induced ROS (Ashraf, 2009). The enhanced activity of antioxidant enzymes due to the salinity shields biomolecules from superoxide anion (O_2^-) (Kannan, Deepa, Kanth, & Rengasamy, 2013). SOD (Superoxide dismutase) is a major decomposer of O_2^- and its enzymatic action results in the production of H_2O_2 and O_2 (Lee et al., 2007). The produced H_2O_2 is then neutralized by various antioxidant enzymes including CAT (catalase), a variety of peroxidase such as GPX (glutathione peroxidase), and APX (ascorbate peroxidase). While CAT largely converts H_2O_2 into the water and molecular oxygen in the peroxisomes, the same function in the cytosol and the chloroplast is performed by APX and GPX is active intracellularly (cytosol, vacuole), as well as in the cell wall (Das & Roychoudhury, 2014).

To cope up with the oxidative stress in plants, the action of SODs have been considered as the "first line of defense." The universal role of SODs is scavenging O_2^- in plants (Bose, Rodrigo-Moreno, & Shabala, 2014) can be understood from its diverse distribution, viz. chloroplast localized Fe-SOD, mitochondria, and peroxisomes localized Mn-SOD and Cu/Zn SOD localized in the cytosol and chloroplast. The activities of Fe-SOD and Mn-SOD associated well with the salt tolerance of numerous halophytes (Jithesh, Prashanth, Sivaprakash, & Parida, 2006), however, the activity of Cu/Zn SOD showed no clear pattern in halophytes. Overexpression of Cu/Zn SOD derived from halophyte *Avicennia marina* in glycophyte indicate rice salinity tolerance, suggesting that Cu/Zn SOD does play a noteworthy role in stress tolerance (Prashanth, Sadhasivam, & Parida, 2008). Interestingly, some halophytes produce new isoforms of SOD with their appearance being dependent on the plant developmental stage and the time of exposure to salt stress (Ozgur, Uzilday, Sekmen, & Turkan, 2013). The expression of MnSOD in *A. thaliana* derived from *Triticum turgidum* increased tolerance to various abiotic stresses by boosting proline synthesis and reducing H_2O_2 level (Kaouthar et al., 2016). Kaya et al. (2018) reported the increased activities of enzymatic antioxidants such as CAT, POD, and SOD in the maize plants under salt stress. Under salinity stress, GPX, CAT, and SOD activities increased in pistachio rootstock (Goharrizi et al., 2019). Tobacco plants that overexpress the APX gene show enhanced tolerance to salt and drought stresses (Badawi et al., 2004).

Under high salinity conditions, the activity of CAT enzyme decreases in salt-sensitive soybean genotypes, (Dogan, 2011), whereas in salt-tolerant soybean genotypes, an increase in enzymatic activity of CAT was noted (Weisany, Sohrabi, Heidari, Siosemardeh, & Ghassemi-Golezani, 2012). Pea plants are sensitive to salt stress, which was linked with decreased CAT activity with an increased level of NaCl stress (Ozturk et al., 2012). Similarly, in chickpea, salt stress increased CAT activity (Eyidogan & Öz, 2007; Kukreja et al., 2005). From the research, it can be seen that CAT activities were lower in the salt-sensitive one than the salt-tolerant cultivar and high CAT activity helps plants to mitigate soil salinity (El-Shabrawi et al., 2010; Hasanuzzaman et al., 2014). GPX activity increased at both reproductive and vegetative stages in pearl millet under salinity (Heidari & Jamshidi, 2011).

12.3.3.2 Role of nonenzymatic antioxidants

The antioxidant machinery also consists of nonenzymatic antioxidants comprising ascorbic acids, flavonoids, glutathione, α-tocopherol, etc. They not only shield different cellular components from impairment, but also play a significant role in the growth and development of the plant by tuning several cellular processes like cell elongation, mitosis, senescence, and cell death (de Pinto & DeGara, 2004).

There are several nonenzymatic antioxidants like GSH, ascorbate (ASC), glycine betaine, proline, polyamines, tocopherols, flavonoids, and their derivates that increased during salinity stress. Ascorbate and glutathione are the common nonenzymatic antioxidant that scavenge ROS through the ascorbate–glutathione cycle in halophytes (Ozgur et al., 2013). Ascorbate and glutathione pathways contain AsA (Ascorbate), GSH together with four enzymes (ascorbate peroxidase, monodehydroascorbate reductase, dehydroascorbate reductase, and glutathione reductase) that detoxifies the toxic ROS (Hasanuzzaman et al., 2019). The most plentiful, potent, and water-soluble antioxidant is ASC that restricts the damage caused by ROS in plants (Tanveer & Shabala, 2018). ASC helps in dissipating excess excitation energy by acting as a cofactor of violaxanthin de-epoxidase in chloroplasts (Smirnoff, 2005). In celery, overexpression of AgAPX1 gene upregulates the production of ascorbate and various antioxidant enzymes that further scavenge ROS and close the stomatal aperture (Liu, Feng, et al., 2019; Liu, Xiao, et al., 2019). In maize, the exogenous application of ascorbic acid alleviated the salinity stress by escalating the antioxidant machinery (Tuna, Kaya, Altunlu, & Ashraf, 2013). In addition, exogenous application of glutathione on chickpea plant decreases the peroxide and lipid peroxidation levels and enhances the osmolytes and membrane stability as well as increases the level of several antioxidants (ascorbate peroxidase, superoxide dismutase, glutathione reductase, peroxidase) (Sadak, Abd Elhamid, & Mahmoud, 2017). Proline also works as an antioxidant and its accumulation under salinity stress protects the physiochemical properties of a plant cell in tomato; the accumulation of proline via the ornithine pathway in Grand Brix variety provides tolerance to the plant (De la Torre-González, Montesinos-Pereira, Blasco, & Ruiz, 2018). In *Panicum virgatum*, the expression of the proline biosynthesis gene (PuP5CS) increases the expression of genes related to ROS scavengers and increases the area of bulliform cells and bundle sheath cells in leaves (Guan et al., 2018). In *M. truncatula*, exogenous application of polyamines decreases the oxidative stress by scavenging the peroxides and reduces the lipid peroxidation, and also increases the expression of genes related to brassinosteroid biosynthesis pathways (López-Gómez et al., 2017). Tocopherols are other nonenzymatic lipophilic antioxidants that protect the cellular mechanism of plant cells facing osmotic stress. The tocopherols prevent lipid peroxidation during environmental stresses as they possess a superior capacity to detoxify lipid peroxyl radicals (Munné-Bosch & Alegre, 2002; Szarka, Tomasskovics, & Bánhegyi, 2012). In plant green tissues, α-tocopherol is the predominant form among its various isoforms which exhibits antioxidant activity in both glycophytes and halophytes and a close association was found with the abiotic stress tolerance (Bose et al., 2014). A comparison between halophyte species of *Cakile maritima* and glycophyte *A. thaliana* revealed that the α-tocopherol level in salt-grown plants was twofold higher in *C. maritima* (Ellouzi, Ben Hamed, Cela, Munné-Bosch, & Abdelly, 2011).

In transgenic sweet potato, the overexpression of IpTC (*Ipomea batatus* Tocopherol Cyclase) gene leads to the production of α-tocopherol that mitigates various abiotic stress (salt, drought, osmotic) by increasing the chlorophyll content and improves the activity of PSII (Kim et al., 2019). In addition to this, α-tocopherols strongly regulate the physiochemical composition of the plasma membrane and cell wall as well as increase the cell turgidity and elastic modulus (Surówka et al., 2020). Flavonoids are plant secondary metabolites that serve as ROS scavengers by neutralizing radicals before they damage the cell (Løvdal, Olsen, Slimestad, Verheul, & Lillo, 2010). Most flavonoids have higher antioxidant potential than the well-known antioxidants, such as α-tocopherol and ASC (Hernández et al., 2009). Under salinity stress, several enzymes related to flavonoid compounds metabolisms such as isoflavone reductase and chalcone isomerase were induced in soybean (Ji et al., 2016). The elevated synthesis of flavonoids has suggested that these molecules could play important roles in soybean under salinity stress (Qu et al., 2016). Rutin is a nonenzymatic flavonoid antioxidant that accumulates in the cytosolic region and scavenges the hydroxyl radical formed in response to salinity stress. In addition to this, rutin also exports Na^+ and improves K^+ retention inside the cell (Ismail et al., 2016). Moreover, under salinity stress, there are various flavonoid genes like phenylalanine ammonia-lyase, chalcone synthase, and flavanol synthase, and carotenoid genes like phytoene synthase 2 and β-lycopene cyclase get expressed and show antioxidant properties that protect the plant from oxidative damage (Abdallah et al., 2016). Thus plants increase the levels of various antioxidants to mitigate the stress during saline conditions.

12.3.4 Transcriptional regulation of salinity tolerance

Genomics study itself is a complex mechanism consisting of the central dogma that comprises transcription of genes and posttranscription modification of mRNA, translation of mRNA, and posttranslation modification of the protein. As discussed above, salinity stress is a major abiotic factor that affects the growth and development of plants. During salinity stress, upregulation or downregulation of stress-related genes products (mRNA/protein) takes place. Transcriptomic analysis reveals the identification of a large number of salt-responsive TFs (Transcription Factors) and genes which are either upregulated or downregulated in response to salinity stress. A large number of salt responsive genes have been identified in various plant species like *Arabidopsis*, rice, maize, *Medicago*, *Mesembryanthemum*, *Thellungiella*, and *Avicennia* with the help of transcriptomics and microarray technologies that might be further used in developing salt-tolerant crops (Krishnamurthy et al., 2018).

There has been various research undertaken on different plants to understand the mechanism by which a gene or genes participates in the salinity tolerance mechanism. Several transcriptomic studies have been done on salt stress with different species. In a study on Alfalfa root transcriptome, 604 genes have been found that changed their expression under salinity stress and their main function is to activate the TFs of several families, stress-related secondary metabolites, stabilization of plasma membrane, and nitrogen uptake (Postnikova, Shao, & Nemchinov, 2013). Salinity stress also affects nitrate assimilation (Dluzniewska et al., 2007). In a transcriptomic study on two varieties of *O. sativa*—IR64 (salt-sensitive) and Pokkali (salt-tolerant)—there are a special set of genes in the Pokkali variety like CaMBP (Calmodulin Binding Protein), GST (Glutathione S-transferase), LEA (Late Embryogenesis-Abundant), V-ATPase (Vacuolar-ATPase), OSAP1 zinc finger protein, and a transcription factor HBP1B which can prove beneficial by a transgenic method in developing a tolerance mechanism toward salinity in the IR64 variety (Kumari et al., 2009).

Plants use carbohydrates derivates like hexose and sucrose as a substrate for cellular metabolism, growth, and development, but also as osmoticum that maintains cellular homeostasis under salinity stress. To alleviate the salt stress, plants upregulate the expression of SWEET15 gene, which further regulates vacuolar storage and transport of sugar (Jithesh et al., 2018). Under salt stress plant also upregulate the expression of various genes related to enzymes such as SUS1 (invertase and sucrose synthase), TPS (trehalose-6 phosphate synthase), TPP (trehalose phosphate phosphatase), fructose 1,6-bisphosphatase, starch synthase, and ADP-glucose pyrophosphorylase that are involved in pathways related to plant metabolism (Ghaffari et al., 2016). In wheat, upregulation of genes encoded for hexokinase, cell wall invertase, and sucrose synthase take place in response to salinity stress (Amirbakhtiar, Ismaili, Ghaffari, Nazarian Firouzabadi, & Shobbar, 2019). In a recent study on root transcriptomic of Kharchia (a local variety of wheat), there have been various genes found, like V-ATPase, that provide energy to export cytosolic Na^+ into vacuoles by forming a proton gradient. ROS scavengers, ATP synthase that is engaged in energy metabolism, and Cbl-interacting protein kinase (gene related to signaling mechanism) are revealed as the most important genes that are involved in the salt tolerance mechanism (Goyal et al., 2016). In different plant species, such as tomato, tobacco, and soybean, an upregulation of Ca^{2+} ATPase has been reported that might help in dropping the levels of cytosolic calcium, and also maintain the Ca^{2+} homeostasis under salinity stress (Singh et al., 2014). Under osmotic stress, LEA (Late-embryogenesis-abundant) protein is induced in vegetative tissues that provide tolerance to dehydration (Chinnusamy et al., 2005). Twenty-seven genes coding for LEA proteins were discovered in the DEGs (Differentially Expressed Genes) identified in wheat under salinity stress (Amirbakhtiar et al., 2019). The production of LEA proteins attenuates the detrimental effects of salinity stress.

In transgenic sweet potato, downregulation of gene β-carotene hydroxylase increases the β-carotene content and total carotenoids that together regulate tolerance mechanism toward salinity (Kim, Ahn, Ahn, Lee, & Kwak,

2012). In tomato, res (restored cell structure) mutant is the first tomato mutant that accumulates jasmonic acid in roots and leaves. Under salinity stress, res shows upregulation of photosynthetic genes like RUBISCO ACTIVASE 1 and ALTERNATIVE OXIDASE 1 as well as genes related to TFs and protease inhibitor like METALLOCARBOXYPEPTIDASE INHIBITOR that contribute to the recovery of the plant (Albaladejo et al., 2018). Another study on res mutant discloses the role of DEAD-Box RNA helicase in developmental and stress responses in tomato. SlDEAD39 is a chloroplast targeted DEAD Box RNA helicase that plays a crucial role by increasing the sugar content and accumulation of photoassimilates in chloroplast under salinity stress (Capel et al., 2020). Expression of TaZNF (*T. aestivum* predicted Dof zinc-finger protein) gene (belongs to C2H2 zinc finger motif) in wheat, increases the export of Na^+ from roots and shrink the stomatal aperture in leaves under salt stress (Ma, Liang, Gu, & Huang, 2016). In halophytes, various salt responsive genes like antiporters (NHX, HKT, SOS, V-ATPase), antioxidant (APX, CAT, SOD, BADH, SOD), ion channels (Cl^-, Ca^{+2}, aquaporins) are isolated and can be used in transgenic plants to mitigate salinity stress (Fig. 12.3) (Mishra & Tanna, 2017).

In *Chlamydomonas reinhardtti* (green algae), increase in intracellular peroxide levels and inhibition of photosynthetic-electron flow takes place in response to salinity stress (200 mM NaCl for 24 h). A transcriptomic study on *C. reinhardtti*, reveals the identification of 10,635 unigenes (5920 upregulated and 4715 downregulated) which regulates the maintenance of lipid homeostasis and levels of phosphatidic acid acetate levels to cope up with the salinity stress (Wang et al., 2018). Another recent study by Wu, Zhao, et al. (2019) on flax transcriptome revealed the presence of 33,774 Differentially- Expressed Unigenes (DEUs) out of which 18,040 are upregulated and 15,734 are downregulated genes under NaCl stress. The functions of these DEUs in the biosynthesis of amino acids, photosynthesis-antenna proteins, and signal transduction of phytohormones that helps the plant to survive in NaCl stress. Many DEUs homologs are identified with the help of plant TFs likc bZIP, IID-ZIP, NAC, MYB, GATA, CAMTA. During seed germination in rice, a genome-wide association study (GWAS) reveals the presence of 17 genes that contribute to the salt tolerance mechanism. OSMAD31 coded for floral organ specification whereas, under salt stress and K^+ starvation in plant, OSHAK11 is upregulated for high-affinity K^+ transporter (Yu et al., 2018).

MicroRNAs (miRNAs) are the posttranscriptional gene regulators in eukaryotes. In plants, miRNAs regulate various physiological traits and environmental stress responses are among them (Zhang & Wang, 2015). In alfalfa, microRNA156 (MiR156) increases the biomass, decreases plant height, increases no. of branches, and delay the developmental process, and decreases Na^+ uptake under saline conditions. Another study by Shinde et al. (2020) on pearl millet reveals the role of 81 conserved and 14 novel

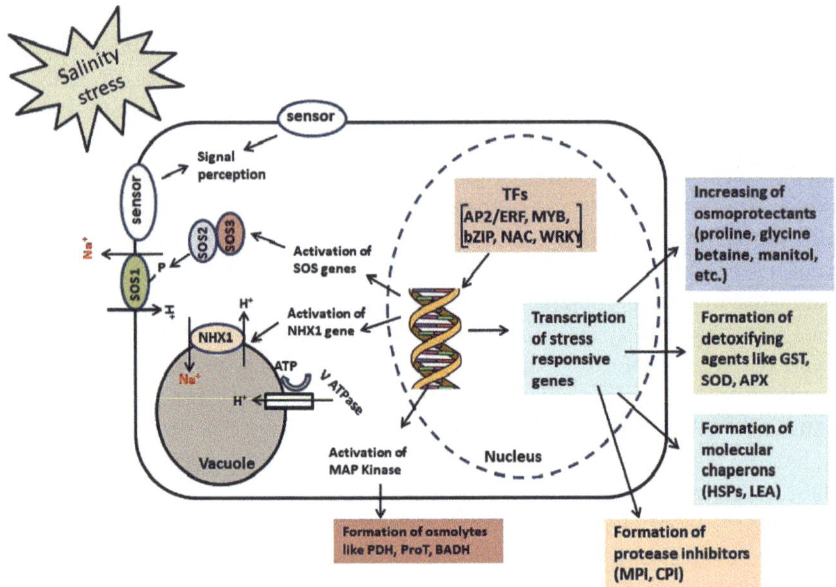

FIGURE 12.3 Generalized representation of Salt-responsive genes in mitigation of salinity stress - Transcription factors activate the expression of salt responsive genes like SOS, NHX1. Activation of SOS genes activate the SOS (SOS1, SOS2, and SOS3) proteins. SOS2 and SOS3 together form a complex that further activate SOS1 which export cytosolic Na^+ from the cell. In addition activation of NHX1 gene activates vacuolar NHX1 protein that import excessive cytosolic Na^+ inside the vacuole. However, transcription of salt responsive genes leads to formation of various osmoprotectants, detoxifying agents, molecular chaperons, osmolytes and protease inhibitors that more or less mitigate salinity stress. *Abbreviations: AP2/ERF, APETALA 2/ethylene responsive element binding factor; APX, ascorbate peroxidase; BADH, Betaine aldehyde dehydrogenase; bZIP, Basic leucine zipper; CPI, cysteine protease inhibitors; GST, glutathione S- transferase; HSPs, heat shock proteins; LEA, late embryogenesis abundant proteins; MPI, maize proteases inhibitors; NAC, NAM-ATAF1, 2-CUC2; NHX1, Na^+/H^+ exchanger; PDH, proline dehydrogenase; 1SOS1, salt overly sensitive 1; SOS2, salt overly sensitive 2; SOS3, salt overly sensitive 3; SOD, superoxide dismutase.*

miRNAs that shows a prominent role in salt stress responses. Apart from this, 448 pearl millet mRNAs were also identified, among them, 25% encode the TFs and the rest of the function as laccase, protein kinase, ion transporters, and so on. These miRNAs further regulates auxin-signaling pathways and show tolerance toward salinity in pearl millet (Shinde et al., 2020).

Histone modification plays a crucial role in gene regulation and expression under salinity stress. A study by Song et al. (2012) on soybean plant reveals that DNA methylation and histone modification regulates the expression of genes related to four TFs. ChIP analysis discloses that out of the four DNA methylated TFs, three of them show an increase in H3K4 trimethylation, H3K9 acetylation, or decrease in levels of H3K9 demethylation in various part of promoter or coding regions that provide tolerance toward salinity stress (Song et al., 2012). In *Arabidopsis*, ky-2 (histone deacetylase inhibitor) treatment reveals the acetylation of histone at H4 of the AtSOS1 gene (encodes a Na^{+}/K^{+} antiporter) through transcriptomic and metabolome analysis. Upregulated expression of AtSOS1 and AtSOS3 genes decreased the intracellular Na^{+} accumulation and increased high salinity stress tolerance in plant (Sako et al., 2016). Another study on cassava root indicates the enhancement of tolerance toward salinity can be done by using SAHA (suberoylanilide hydroxamic acid), a histone deacetylase inhibitor. The immune-blotting technique discloses that SAHA treatment induces H3 and H4 histone acetylation. Apart from this, the transcriptomic analysis revealed the upregulation of genes related to phytohormone biosynthesis mechanisms like ABA, jasmonic acid, ethylene, gibberellic acid by using SAHA treatment (Patanun et al., 2017). In castor bean, H3K4me3 (H3 Lysine 4 trimethylation) and H3K27me3 (H3 tri-methylated lysine 27) histone markers regulate the expression of a salt responsive MYB related TF, RMS1 (RADIALIS LIKE SANT) (Han, Du, et al., 2020; Han, Xu, et al., 2020). In rice, two different genotypic salt-tolerant cultivars, Nonabokra and Pokalli were used to study the comparative effects of genetic and epigenetic factors against IR64 variety. Results suggest that OsBZ8 gene is highly induced in Nonabokra variety as compared to the IR64 and histone H3K9 and H3K27 found the upstream region of Nonabokra were hyper-acetylated and show dynamic arrangements of nucleosomes as compared to IR64. DNA methylation was low in OsBZ8 locus of Nonabokra variety with low levels of H3K27me3 histone marker and high H3K4me3 whereas IR64 shows high DNA methylation and high H3K27me3. Taken together these results shows a differential modification in the chromatin region of different varieties of rice cultivars to mitigate salinity stress (Paul, Dasgupta, Roy, & Chaudhuri, 2017). Therefore histone modification and DNA methylation are tools of epigenetic regulation of genes that play an important role in salt stress tolerance.

12.3.4.1 Transcription factors

TFs are modulators of genes that play an important role in responses toward salinity stress in regulating the transcription of downstream genes that code for proteins that regulates salinity tolerance in plants. There are various TFs like, AP2/ERF, bZIP, NAC, MYB, WRKY have been identified which have a great potential in improving salinity tolerance (Table 12.1) (Kumar, Kalita, Srivastava, & Sahoo, 2017; Kumar, Singh, et al., 2017). These TFs interact with the cis-acting elements that are present in the promoter region of salt-tolerance genes and thus activate the transcription leading to an enhanced tolerance mechanism to multiple stresses (Lindemose, O'Shea, Jensen, & Skriver, 2013). The roles of these TFs in salinity stress are discussed below.

TABLE 12.1 Role of transcriptions factors in mitigating salinity stress.

Plant name	Salinity level	Transcription factors (TFs)	Family	Function of TFs	Authors
Fortunella crassifolia	200 mM	FcWRKY 40	WRKY	Gene regulation of SOS2, proline biosynthesis, abscisic acid (ABA) signaling pathway	Dai, Wang, Gong, and Liu (2018)
Arabidopsis thaliana	150 mM	AtbZIP62	bZIP	Negatively regulates the expression of SOS genes, and modulates antioxidant system and chloroplast pigments.	Rolly, Imran, Lee, and Yun (2020)
Lycopersicon esculentum	200 mM	SlWRKY 39 SlMYB14	WRKY	Downregulation of circadian gene GIGANTEA1 and shows a prominent role in photosynthetic efficiency, protease inhibitors further contributes to salt tolerance in roots.	Albaladejo et al. (2018)
Oryza sativa	100 mM	$OsHBP_1b$	bZIP	Higher fold expression of these genes leads to enhancement in antioxidant enzyme activity and photosynthetic efficiency under salinity stress	Das et al. (2019)
Triticum aestivum	200 mM	TaMYB73	MYB	Improved ionic resistance and participate in salinity tolerance mechanism	He et al. (2012)
A. thaliana	250 mM	PgNAC21 (TF in Pearl millet)	NAC	Induce by NaCl and ABA, further regulates salinity stress and ABA	Shinde et al. (2019)
O. sativa	150 mM	OsMYB6	MYB	Improving tolerance mechanism via over-expression of stress responsive genes.	Tang et al. (2019)
Pennisetum glaucum	250 mM	PgWRKY33, PgWRKY62, PgWRKY65	WRKY	Shows responses toward salinity stress	Chanwala et al. (2020)
A. thaliana	200 mM	MbNAC25	NAC	Increase in chlorophyll content, proline content and antioxidant levels that further provide tolerance to *Arabidopsis* plant.	Han, Du, et al. (2020), Han, Xu, et al. (2020)

Medicago sativa	150 mM and 200 mM	MsMYB4	MYB	Increase tolerance toward salinity in a ABA-dependent manner.	Dong, Liu, Li, Gao, and Song (2018)
Vitis venifera L.	150 mM	VvWRKY30	WRKY	Regulates ROS-scavenging and osmoticum accumulation	Zhu et al. (2019)
Musa paradisiaca	250 mM	MusaNAC042	NAC	Increases chlorophyll content and proline content and decreases MDA content. Modulates various abiotic stress-related genes further codes for CBF/DREB, LEA, WRKY.	Tak, Negi, and Ganapathi (2017)
Chrysanthemum	100,200 and 400 mM	DgNAC1	NAC	Decreases the content of MDA and ROS (H_2O_2 and O_2^-) and increases the accumulation of CAT, peroxidase and SOD and proline.	Wang et al. (2017)
T. aestivum	250 mM	TaSIM	MYB	Activated the transcription of genes RD22 (ABA- dependent) and RD29A (ABA-independent)	Yu, Ni, Chen, and Qu (2017)
L. esculentum	125 and 200 mM	JUB1	NAC	Regulating peroxide levels and increases proline content.	Alshareef et al. (2019)
Gossypium hirsutum L.	400 mM	GhERF4L GhERF54L	ERF	Function in ethylene signaling pathway	Long et al. (2019)
L. esculentum	300 mM	SlDof22	Dof	Controls ascorbate accumulation and regulate the expression of SlSOS1 gene	Cai et al. (2016)
Glycine max	200 mM	GmbZIP110	bZIP	Decreases Na^+ uptake, might be involved in ABA dependent pathway and ion transport	Xu et al. (2016)
Brassica napus	150 mM	BnNAC2 BnNAC5	NAC	BnNAC2 act as a transcriptional regulator and BnNAC5 controls salt-hypersensitivity and leaf senescence by increasing chlorophyll content in leaf.	Zhong et al. (2012)
T. aestivum	200 mM	TaMYB73	MYB	Binds to promoter region of salt-stress signaling genes AtCBF3 and AtABF3 and regulates ionic resistance	He et al. (2012)

12.3.4.1.1 Role of AP2/ERF

AP2/ERF belongs to the largest family of TFs comprising four transcriptional subfamilies that are: AP2, ERF, DREB, and RAV that are involved in responses toward salt stress (Das, Behera, Chatterjee, Das, & Mohapatra, 2020; Holmes, Hall, Gendall, Boon, & James, 2016; Jithesh et al., 2018). Induction of ERFs has been reported in response to high salinity, osmotic stress, and drought (Xu, Chen, Li, & Ma, 2011). A study by Kaashyap et al. (2018), reveals that AP2/ERF TFs was 10-fold more induced in the tolerant genotype as compared to the sensitive genotype of chickpea. The AP2/ERF is thought to bind DREs (dehydration responsive elements) at the promoter site of target genes and instigate salt stress responses (Sakuma et al., 2006). However, Kaashyap et al. (2018) also found that DREs/CRT was significantly induced in the tolerant genotype while being repressed in the sensitive genotypes. The expression of CBF4 (a member of AP2-EREB TF family) in *Medicago polymorpha* was greater than in *Medicago lacinata* genotypes under salinity (Mokhtari, Rafiei, Shabani, & Shiran, 2017). Interestingly, overexpression of CBF4 in *M. truncatula* leads to lower leakage of electrolytes and provides tolerance to salinity stress (Li et al., 2011).

12.3.4.1.2 Role of bZIP

bZIP TFs also play an important role in salinity tolerance in plants. In transgenic plants, overexpression of bZIP genes mitigate salinity induced dehydration of cells (Hossain et al., 2010; Huang, Liu, & Chen, 2010). In *A. thaliana*, translocation of AtbZIP17 to the nucleus followed with the upregulation of genes related to salinity stress to cope up with the proteolytic damages caused by high salinity (Liu, Srivastava, & Howell, 2008). Besides, the insertion of rice OsbZIP71 gene in transgenic plants can significantly improve the tolerance mechanism under high salt conditions (Liu, Chang, et al., 2014; Liu, Mao, et al., 2014). OrbHLH2 (Basic helix-loop-helix (bHLH) encoding gene), act as a transcription factor and positively regulates salt-stress signals, and overexpression of OrbHLH2 in *Arabidopsis* conferred increased tolerance to salt and osmotic stress (Zhou et al., 2009). However, in rice, OsHBP1b (Histone gene binding protein-1b), member of bZIP TF, shows a higher level expression of stress-responsive genes like APX (5-fold), NAC (2.5-fold), DREB (7-fold), SOD (4-fold), bZIP (1.5-fold) and HSP (4-fold) as compared to control which further leads to improvement in better shoot and root growth, enhanced photosynthetic pigments and antioxidant enzyme activity. However, a decrease in CAT transcript up to 3.5-fold is also seen under OsHBP1b over-expressing lines (Das, Lakra, Nutan, Singla-Pareek, & Pareek, 2019).

12.3.4.1.3 Role of NAC

NAC TFs play important role in salinity tolerance in plants. There are many members of NAC TFs like NAM, ATAF1/2, and CUC2, which are specifically expressed at the time of stress responses and development of plants (Kumar, Kalita, et al., 2017; Kumar, Singh, et al., 2017). NACs TFs have been reported to impart salt tolerance in *O. sativa* (Hu et al., 2008), *A. thaliana* (Xu, Wang, Xue, Zhang, & Ji, 2015), and tomato (Alshareef et al., 2019). In *O. sativa*, over-expression of specific transcription factor gene OsNAC04 increases the tolerance mechanism in plants against drought and salinity stress (Zheng, Chen, Lu, & Han, 2009). Liu, Sun, and Wu (2016) found that ATAF1 not only functions as a response to salinity stress but also may help in improving the salinity tolerance in *O. sativa*. Improved drought and salt tolerance could also be achieved by transgenic over-expression of diverse NAC factors in species ranging from *Arabidiopsis*, rice to chickpea, wheat, and tomato (Golldack, Lüking, & Yang, 2011). The seeds of ONAC063-expressing transgenic *Arabidopsis* showed enhanced tolerance to high-salinity and osmotic pressure indicating a possible role of ONAC063 in eliciting responses to high-salinity stress (Yokotani et al., 2009). OsNAC5 and ZFP179 show an upregulation under salinity stress, which may regulate the synthesis and accumulation of proline, sugar, and LEA proteins that in turn play an integral role in stress tolerance (Song, Chen, Chen, Dai, & Zhang, 2011). Overexpression of a NAC TF in both rice and wheat confers salt tolerance, thereby predicting their role in stress mitigation (Nakashima et al., 2007). In another study by Shinde et al. (2019), expression of a pearl millet transcription factor gene, PgNAC21 is induced by the treatment of salinity stress and abscisic acid (ABA). In *Arabidopsis*, gene PgNAC21 regulates the expression of various stress responsive genes like GSTF6 (GLUTATHIONES-TRANSFERASE 6), COR47 (COLD-REGULATED 47), RD20 (RESPONSIVE TO DEHYDRATION 20). The promoter of these genes has NAC core binding site CACG to which PgNAC21 directly binds and regulates the expression of these genes (Shinde et al., 2019).

12.3.4.1.4 Role of MYB transcriptions factors

Many MYB proteins function as TFs and have been proved to be engaged in regulating different cellular processes, including biotic and abiotic stress response (Stracke, Werber, & Weisshaar, 2001). Under salinity stress, a novel MYB TF in wheat, TaMYB73 is induced and bind with the promoter region of AtCBF3 and AtABF3 genes, regulates salinity tolerance mechanism by improving ionic resistance (He et al., 2012). MYB112 was expressed in *M. polymorpha* exhibiting upregulation in the roots and leaves under salinity stress (Mokhtari et al., 2017). Similarly, Li et al. (2011) identified an over-expression of MYB112 in the root of *M. truncatula* under salinity condition. According to yeast one-hybrid assay, PgMYB1 transcription factor of pearl millet binds to the promoter region of gene PgNAC21 and regulates its expression (Shinde et al., 2019).

12.3.4.1.5 Role of WRKY

There is another important family of TFs, WRKY that plays a crucial role in plant development, signaling of phytohormones, and developing tolerance against various biotic and abiotic stresses. Different WRKY TFs are highly tissue-specific and reported to be critical for salt stress tolerance (Konda, Farmer, Soren, PS, & Setti, 2018; Kumar et al., 2016). WRKY25 and WRKY33 are the WRKY TFs factors that are involved in plant salt adaptation also increased salt tolerance and ABA sensitivity when over-expressed in *Arabidiopsis thaliana* (Jiang & Deyholos, 2009). Gene expression profiling in soybean under salinity stress depicted the upregulation of 65 GmWRKY genes and downregulation of WRKY71 gene (Yu et al., 2016). In *Pennisetum glaucum*, there are approximately 97 PgWRKY proteins were identified, out of which expression of PgWRKY33, PgWRKY62, and PgWRKY65 shows tolerance toward drought and salinity stress (Chanwala et al., 2020) (Table 12.1). In chickpea, differential expression of WRKY-TF genes was also observed under salinity stress (Garg et al., 2016). The potential of WRKY-type TFs to confer increased salt tolerance by the transgenic expression is further supported by the different salt-induced regulation of a WRKY protein in salt-sensitive rice and a halophytic rice-relative model species (Diédhiou, Popova, & Golldack, 2009a; Diédhiou, Popova, & Golldack, 2009b).

Genes associated with salinity tolerance mechanism in plants:

1. SOS family genes: they maintain salt tolerance with a very important role in ion homeostasis. SOS family comprises three members, that is, SOS1, SOS2, and SOS3 that together regulate ion homeostasis. SOS1 is an antitransporter that export Na^+ from the cytoplasm of the cell. SOS1 gene is upregulated in plants facing salinity. SOS1 gene is activated by SOS2, SOS3, and the complex of SOS2-SOS3, H^+-ATPase. The function of H^+-ATPase is to export protons from the cell to maintain pH and ion balance. However, SOS2 encodes a serine/threonine type protein kinase of 446 amino acids and acts upstream of the SOS1 gene. Loss in the function of the SOS2 gene causes Na^+ and K^+ imbalance in plants and is required for salt tolerance mechanisms in plants (Liu, Ishitani, Halfter, Kim, & Zhu, 2000). SOS3 encodes a Ca-binding protein that binds with cytosolic Ca^{2+} molecule when levels of calcium become high inside the cell in response to salinity stress. This further activates SOS2 protein. SOS2 and SOS3 together form a complex, SOS3-SOS2 kinase complex that induces the expression of the SOS1 gene inside the nucleus. The Antiporter of SOS1 exports excess levels of Na^+ outside the cell and regulates ion homeostasis (Zhu, 2000). Gene expression study in *Brassica* species discloses that expression of more SOS genes in CS 52 and CS 54 varieties makes the plant more salinity tolerant (Chakraborty, Sairam, & Bhattacharya, 2012).

2. NHX1: NHX1 is a vacuolar Na^+/H^+ antiporter that exports Na + from the cytoplasm to vacuole in the exchange of protons (H^+). A transgenic mungbean plant is made by insertion of the AtNHX1 gene of *Arabidopsis*, to mungbean plant via Agrobacterium-mediated transformation. Expression of the AtNHX1 gene improved the tolerance mechanism in the plant via regulating ion homeostasis, osmoregulation, reduction in damage of plasma membrane, as well as improved photosynthetic capacity and transpiration rate (Kumar, Kalita, et al., 2017; Kumar, Singh, et al., 2017).
3. HKTs: HKT transporters play an important role in mediating salt tolerance in plants. Maximum expression of this gene is found in leaf vasculature and root stele and controls Na^+ distribution in leaves (Hamamoto et al., 2015; Mäser et al., 2002). In AtHKT1 promoter-GUS reporter gene analysis confirms that expression of AtHKT1 is also found in xylem parenchyma cells and it reduces the level of Na^+ in xylem vessels and leaves (Horie et al., 2005).
4. APX: during salt stress, an increase in ROS levels harms thc plant in different ways like the disruption of various physiological and biochemical processes. To cope up with this situation, plants possess a key antioxidant enzyme, APX involved in the Ascorbate-Glutathione pathway that mitigates the negative impact of H_2O_2 in the cell. In a recent experiment by Saxena et al. (2020), a transgenic plant (*B. juncea*), which is made with a gene Apx1 gene from *Arabidopsis* is artificially introduced. Expression of AtApx1 gene reduces the H_2O_2 levels and membrane damage; strengthen the antioxidant defense system and stress tolerance mechanism of the brassica plant. Furthermore higher proline accumulation, lower chlorophyll a/b ratio, higher chlorophyll stability index, and better antioxidant enzyme activities are shown by the leaf disc assay of the transgenic plant (Saxena et al., 2020).

Genetic engineering is now a day, provide new opportunities to crop improvement facing abiotic stresses like drought, salinity, high temperature, cold, etc. However, the selection of the desired gene is crucial for achieving the desired traits. There are two sets of genes found, one is a structural gene and the other one is regulatory genes. Structural gene encoding proteins provide direct tolerance to plants whereas the regulatory genes control the expression of other genes that function in different cellular processes (Zafar et al., 2020). Apart from this, there are cis-regulatory genes also, that provides tolerance to plants. With the help of the CRISPR/CRISPR-associated protein-9 (Cas9) approach, these genes (structural/regulatory, cis-regulatory) can be artificially induced to form transgenic plants that can deal better with the environmental stresses (Zafar et al., 2020). In a bioinformatic study, there are various cis-regulatory motifs, GT1, MBS, LTR, and ARE that function as stress-responsive genes. A halophytic plant (*Salicornia brachiata*) has an SbNHX1 promoter containing a 624 bp regulatory region comprises two-leaf structure motifs and GT-1 and MBS motifs that function as salt inducible motifs. This promoter can be used as an important tool in genome engineering to better deal with salinity stress tolerance (Tiwari, Patel, Chaturvedi, Mishra, & Jha, 2019).

12.3.5 Proteomics

The exact picture of the gene expression can only be obtained through proteomics (the study of gene expression at protein level), as some immature/unprocessed mRNA may not be translated into proteins in certain tissues/cells. Additionally, posttranscriptional regulation will be missed if we only depend on transcriptomics. In contrast to transcripts, proteins are the key elements that produce the plant responses (growth and development, physiological attributes, as well as response to various stress environments), so analysis of changes in the plant proteome is highly important. Hence for elucidating the mechanism of salt tolerance, a quantitative study of gene expression at the protein level is essential (Sobhanian, Aghaei, & Komatsu, 2011). Proteomics helps us to unravel the various strategies of plants whether at a cellular, metabolic, or whole plant level to adapt themselves under salt stress conditions (Jha, Bohra, Jha, & Parida, 2019).

Two-dimensional gel electrophoresis (2-DE) was usually employed for quantitative proteomics of biological samples for many years. However, its various drawbacks such as the low quantity of protein detection, low reproducibility, and challenging isolation of hydrophobic proteins limited the use of 2-DE in the investigation of proteomic alterations (Van den Bergh & Arckens, 2005). A new gel-free method for quantitative protein investigation containing isotope labeling, known as Isobaric tags for relative and absolute quantification (iTRAQ) analysis, has developed into an important proteomic tool. Due to its unique combinations of great sensitivity and more precise quantification, iTRAQ has been effectively employed in the study of proteins under several abiotic stresses including salinity (Sun et al., 2017).

Recently, proteins responsible for salinity tolerance mechanisms have been identified in many crop species, such as cotton (Li et al., 2015), cucumber (Fan, Xu, Du, & Wu, 2015), and wheat (Jiang, Li, et al., 2017), using proteomics-based technologies. Furthermore, proteomic analyses for salt stress-related proteins were conducted in many halophytes, such as *Aeluropus lagopoides*, *Halogeton glomeratus*, *Salicornia europaea*, and *Thellungiella halophile* (Kumari et al., 2015). These studies have identified plentiful proteins modulated in response to salinity stress which play different roles such as in photosynthesis, ROS scavenging systems, ionic homeostasis, signal transduction, etc. (Zhang et al., 2012).

Proteins associated with several important functions in plants (photosynthesis, carbohydrate, and energy metabolism, antioxidant/ROS scavenging, etc.) that are modulated under salinity stress are discussed below in the text and Table 12.2.

TABLE 12.2 Modulation of various proteins under salinity stress studied using proteomics.

Plant species	Salinity level	Proteins modulated	Functions	Authors
Haloxylon salicornicum	400 mM	RuBisCo, D1 protein, photosystem II-CP47, TCA cycle enzymes (citrate synthase, succinate dehydrogenase, and malate dehydrogenase) were upregulated	Improved photosynthesis, provides energy for salinity tolerance	Panda, Rangani, and Parida (2020)
Oryza sativa	50–200 mM	PS II CP47, chlorophyll a-b binding (CAB) protein, glutamine synthetase (GS), NADH dehydrogenase and mitochondrial cytochrome c oxidase were upregulated in tolerant cultivar	Maintenance of PS II, promotes photosynthesis, improved N metabolism and increased energy production	Frukh, Siddiqi, Khan, and Ahmad (2020)
Hordeum vulgare	150 mM	PsB protein, cyt b_6f iron-sulfur subunit, fructose 1,6-biphosphate aldolase (FBA) were in abundance	PS II repair, cyt b_6f increase ETR and ultimately increase production of ATP, NADPH, FBA	Zhu et al. (2020)
Beta vulgaris	100 and 200 mM	OEE, PSB27- H1, 14-3-3 proteins, MAP kinases, BHLH transcription factor (TF), superoxide dismutase (SOD), aspartic protease, L ascorbate oxidase, phosphoethanolamine N-methyltransferase and choline monooxygenase were increased	Oxygen-evolving enhancer (OEE), PSB27-H1 maintain PSII and improve photosynthesis, aspartic protease induces stomatal closure, ascorbate oxidase decreased reactive oxygen species (ROS) content, last two enzymes involved in glycine betaine synthesis.	Rasouli et al. (2020)
Sesamum indicum (salt tolerant variety-G441and salt sensitive variety-G358)	150 mM	SOD, peroxidase (POD) were upregulated in G441 and downregulated in G358. HSP-70, HSP- 83 were higher in G441 than G358. Amino acid metabolism enzymes (alanine and aspartate aminotransferase, glutamate	Mitigate ROS, enhanced proline synthesis, maintenance of cellular ATP levels, and carbohydrate metabolism	Zhang et al. (2019)

(*Continued*)

TABLE 12.2 (Continued)

Plant species	Salinity level	Proteins modulated	Functions	Authors
		decarboxylase were increased in G441 only), carbohydrate metabolism [glyceraldehyde-3-phosphate dehydrogenase (GAPDH), pyruvate decarboxylase etc.]		
Musa paradisiaca	60 mM	Catalase (CAT), POD, glutathione S-transferase (GST), Chaperone, GAPDH and enolase, RuBisCO, OEE, CAB and calreticulin were increased	ROS clearance, protein folding, energy production, PSII stability, calcium homeostasis	Ji et al. (2019)
Kandelia candel	600 mmol L^{-1}	SOD, POD, ascorbic peroxidase (APX), and CAT, heat shock proteins (HSPs), pectin lyase and pectin methylesterase downregulated, deoxyketose pentose phosphate (DXS) and 2-C-methyl-D-erythritol cyclophosphate synthase (MEP) were upregulated	ROS homeostasis, pectin accumulation in the cell wall to resist salinity, and increased terpenoids synthesis	Xing et al. (2019)
B. vulgaris	50 mM	Phosphofructokinase increased; malate dehydrogenase downregulated, vacuolar membrane proton pump (V-H^+-PPase), NADH dehydrogenase subunit were upregulated, PsbQ and CAB downregulated	Enhanced glycolysis, but TCA downregulated. Increased ATP synthesis; but photosynthesis was downregulated	Wu et al. (2018)
Festuca arundinacea	250 mM	OEE1 and OEE2, fructose-bisphosphate aldolase and Lipocalin were abundant in tolerant variety.	Increased PSII photoactivity and the Calvin cycle efficiency. Protection of photosystem from salinity	Pawłowicz et al. (2018)
Glycine max	50 mM	Dehydrins and late embryogenesis abundant (LEA) upregulated, pyruvate dehydrogenase, malate dehydrogenase. ATP synthase were downregulated	Improved resistance to salinity, repression of glycolysis and TCA cycle	Yin et al. (2018)

O. sativa	150 mM	Glucose-1-phosphate adenylyltransferase large subunit protein (GAS1 and GAS8), argininosuccinate synthase and pyruvate phosphate dikinase protein were found in abundance	Improved seed germination and energy production	Xu et al. (2017)
Triticum aestivum	350 mM	BZIP-type TFs, SOD, POD, CAT, Polyphenol oxidase were upregulated	LEA expression increased which led to salinity tolerance, ROS mitigation	Jiang, Li, et al. (2017)
Gossypium hirsutum	200 mM	PSI P700 chlorophyll a apoprotein A2 (PSIP700), calmodulin	Improved efficiency of electron transfer, calcium homeostasis	Gong et al. (2017)
Raphanus sativus	0–300 mM	Calcium signaling proteins (14-3-3-like proteins, annexin, calreticulin, phospholipase C), hexokinase, phosphofructokinase, pyruvate kinases, enolase), SOD, POD, CAT, dehydrins	Calcium homeostasis, salinity tolerance, ROS alleviation, desiccation tolerance	Sun et al. (2017)
Brassica juncea	50, 100, 150 and 200 mM	OEE, Ribulose activase increased, fructose kinase, GAPDH, malate and isocitrate dehydrogenase were downregulated, sucrose-phosphate synthase, sucrose synthase were upregulated, SOS2 overexpressed	Efficient functioning of PS II, stabilization of RuBiSCo, Repression of glycolysis and TCA cycle. Increased energy production and ions homeostasis	Yousuf et al. (2017)
G. max	150 mM	Chalcone synthase, cytochrome P450 monooxygenase, ERF, MYB TFs, APX, SOD, GST and POD were upregulated	Increased flavonoid synthesis, activation of MAPK and SOS pathways to eliminate ROS	Pi et al. (2016)
G. max	200 mM	Calreticulin, annexins, 14-3-3 proteins, MAPK, ATPase, glutamine synthetase, cysteine synthase, aspartate aminotransferase	Calcium homeostasis, signal transduction, energy production, and improved amino acid metabolism	Ji et al. (2016)

(*Continued*)

TABLE 12.2 (Continued)

Plant species	Salinity level	Proteins modulated	Functions	Authors
Tangut Nitraria	500 mM	CAB, PSI P700 apoprotein A1, RuBisCo LSU downregulated, ATP synthase upregulated, SOD, CAT, POD, GPX were upregulated	Increased efficiency of light and dark reactions of photosynthesis, increased energy production, and ROS homeostasis	Cheng et al. (2015)
G. hirsutum	200 mM	H^+ -ATPase, ABC transporters, SOD, POD, glutathione S-transferase were upregulated	ion homeostasis, and ROS detoxification	Li et al. (2015)
T. aestivum	100 and 200 mM	RuBisCO small subunit, phosphoenolpyruvate carboxykinases, TCA enzymes (citrate synthase)	Calvin cycle enhanced, provides substrate for amino acid synthesis	Capriotti et al. (2014)
O. sativa	120 mM	2-Cys peroxiredoxin, glutathione S-transferase, RuBisCo activase were increased	ROS detoxification, maintenance of RuBisCO activity	Ghaffari, Gharechahi, Nakhoda, and Salekdeh (2014)
Lycopersicon esculentum	50 mM	SOD, APX, CAT, POD, H- ATPase were over-expressed	ROS detoxification, ion homeostasis	Gong et al. (2014)
O. sativa	250 mM	Cysteine synthase, ATP synthase subunit β fructokinase 2, GAPDH, RuBisCO large subunit, lipoxygenase (LOX)	Provide energy needed to repair salt-induced damage, ROS alleviation	Liu, Chang, et al. (2014), Liu, Mao, et al. (2014)
K. candel	300, 450 and 600 mM	OEE2, CAB2 and RuBisCo upregulated at 300 and 450 mM and downregulated at 600 mM. EMP- TCA cycles enzymes were in abundance	OEE2 maintain PSII activity, CAB 2 promotes reduction of NADP at PSI. RuBisCo promotes CO_2 assimilation and EMP- TCA cycles enzymes energy production	Wang et al. (2014)

H. vulgare	100 and 150 mM	ATP synthase β subunit, peroxidase 1, glutathione transferase were increased	Energy production, ROS mitigation	Witzel et al. (2014)
Thellungiella halophila	200, 400 and 600 mM	Glyceraldehyde dehydrogenase, fructose bisphosphate dehydrogenase, ATPase, amylase, pyrroline-5-carboxylate dehydrogenase (P5CD) were upregulated	Energy production, starch and sucrose metabolism, proline biosynthesis	Wang et al. (2013)
Cakile maritima	75 mM	OEE, RuBisCo LSU and SSU, GAPDH, fructose bisphosphate aldolase, ATP synthase were upregulated	Improved photosynthesis and energy production	Debez et al. (2012)
Bruguiera gymnorrhiza	200 and 500 mM	RuBisCo activase, Glutathione transferase, HSP were upregulated	Improve photosynthesis, prevent oxidative damage, maintenance of protein structure	Zhu, Chen, and Zheng (2012)
Salicornia europaea	200 and 800 mM	CAB, Cyt b_6f Rubisco, RuBisCo activase, phosphoglycerate kinase and GAPDH were increased at 200 mM	Improved light harvesting, electron transfer from PSII to PSI and increased CO_2 assimilation	Fan et al. (2011)
T. aestivum	1.5%, 2.0%, and 2.5% NaCl	ATP synthase, Glutathione S-transferase, ferritin, Triosephosphate isomerase were upregulated	Energy production, ROS alleviation	Gao et al. (2011)
L. esculentum	100 mM	ATPase synthase, HSP-90, glutamine synthase, APX, thioredoxin were over-expressed	Energy production, amino acid synthesis, ROS mitigation	Manaa et al. (2011)
Puccinellia tenuiflora	50 and 150 mM	CAB, OEE, RuBisCO, RuBisCO activase were down- regulated. Tocopherol cyclase (VTE1), ferredoxin-NADP(H) oxidoreductases (FNR) caffeic acid O-methyltransferase (COMT) and	Regulate light harvesting by PSII and dark reaction of photosynthesis. Increased tocopherol content. FNR prevent oxidative damage. COMT and PPDK involve in lignin synthesis	Yu et al. (2011)

(Continued)

TABLE 12.2 (Continued)

Plant species	Salinity level	Proteins modulated	Functions	Authors
		pyruvate orthophosphate dikinase (PPDK) were upregulated		
Cucumis sativus	50 mM	Glyceraldehyde-3 phosphate dehydrogenase, enolase, S-adenosyl methionine synthase (SAMS), HSP-70 were in abundance	Carbohydrate metabolism, lignin synthesis, cellular homeostasis	Du, Fan, Guo, Tezuka, and Li (2010)
T. aestivum	200 mM	SOD, alternative oxidase (AOX) were upregulated	Superoxide mitigation, prevent over-reduction of the respiratory chain	Jacoby et al. (2010)
S. europaea	200 mM	Choline monooxygenase (CMO), betaine aldehyde dehydrogenase (BADH) were upregulated, vacuolar H^+-ATPases upregulated, OEE, RuBisCo, RuBisCo activase were in abundance	Increased glycine betaine content, ion homeostasis, improved photosynthesis	Wang et al. (2009)

12.3.5.1 Protein concerned with the metabolic processes (photosynthesis, respiration, etc.)

Leaf photosynthesis is the main source of plant biomass influencing potential crop yield. It is one of the principal processes affected by salt stress (Ghassemi-Golezani, Taifeh-Noori, Oustan, Moghaddam, & Rahmani, 2011). Photosynthesis-linked proteins were differentially expressed under salinity stress, as revealed by several proteomic analysis. Several proteomics investigations have shown that proteins associated with the first phase of photosynthesis (Oxygen-evolving enhancer protein 1 (OEE1), Oxygen-evolving enhancer protein 2 (OEE2), chloroplast manganese stabilizing protein), cytochrome b6f complex, etc.) were present in differential abundance in the presence of salt stress (El Rabey, Al-Malki, & Abulnaja, 2016; Silveira & Carvalho, 2016; Zhu et al., 2020). Besides this, the proteins are linked with the second phase of photosynthesis, that is, Calvin cycle-related proteins, such as Ribulose bisphosphate carboxylase/oxygenase (RuBisCO), RuBisCO activase, and phosphoglycerate kinase, were observed in abundance (Gong et al., 2017). However, a reduced abundance of RuBisCO activase, RuBisCO large and small subunits, and carbonic anhydrase were witnessed in various glycophytes exposed to salt (Parihar et al., 2015), whereas in the case of halophytic plants salt-stress increased the level of RuBisCO subunits, chlorophyll a/b binding proteins (CAB), and D2 protein (a core protein of photosystem II) (Cheng et al., 2015; Wang et al., 2008b; Wang, Tang, et al., 2015). This shows that unlike glycophytes, halophytes are effective in sustaining the plant growth and various metabolic processes under saline conditions.

For maintaining the optimum PS II activity under NaCl stress the presence of OEEs is required (Gururani, Venkatesh, & Tran, 2015; Wang, Liang, et al., 2013). Plants such as *Kandelia candel* (Wang, Pan, et al., 2015), *Thellungiella halophila* (Chang et al., 2015), and *S. portulacastrum* (Peng et al., 2019) enhanced the accumulation of OEEs to sustain the PS II stability during salt stress. Another protein concerned with the light reaction of photosynthesis is chlorophyll a/b binding protein (CAB), a component of PSI, which assists in light-harvesting and the allocation of the excitation energy to reaction centers for the reduction of NADP to NADPH (Wan & Liu, 2008). This protein accumulation increased under salinity stress in *A. lagopoides* (Sobhanian, Motamed, Jazii, Nakamura, & Komatsu, 2010). On the contrary, its downregulation has been observed in *Physcomitrella patens* (Wang et al., 2008a) and *Puccinellia tenuiflora* (Yu et al., 2011). The cytochrome b6f complex is the protein that forms a link between the PS I and PS II reaction centers. The major subunit of the cytochrome b6f complex is the iron−sulfur subunit (ISP) (de Vitry, Ouyang, Finazzi, Wollman, & Kallas, 2004) and these ISP proteins in *S. portulacastrum* were enhanced under differential salt conditions (Peng et al., 2019), which would encourage plastoquinone to release more electrons to the cytochrome b6f complex (Alric, 2015). Peroxiredoxin (Prx) is a protein that eliminates ROS in organisms through the oxidation of the conserved cysteine residue (-Cys) (Pena-Ahumada, Kahmann, Dietz, & Baier, 2006). 2-Cys Prx protein is located in the thylakoid membrane of the chloroplast, and its sequence is highly conserved (Bhatt & Tripathi, 2011). Elevated 2-Cys Prx expression can efficiently enhance tolerance to abiotic stress (Broin, Cuiné, Eymery, & Rey, 2002; König et al., 2002). Overexpression of 2-Cys Prx in tobacco can effectively alleviate H_2O_2-induced membrane lipid peroxidation and the extent of photoinhibition of PS II during salinity stress (Zhang, Xu, Li, & Gu, 2016; Zhang, Zhu, et al., 2016). Similarly, 2-Cys Prx protein was increased in barley (Rasoulnia, Bihamta, Peyghambari, Alizadeh, & Rahnama, 2011) and sugarbeet (Hossain, ElSayed, Moore, & Dietz, 2017) during salinity stress.

Moreover, the upregulation of the C3 cycle results in elevated production of photosynthates such as sucrose and starch, which confer tolerance against salt stress in plants (Wang, Chang, et al., 2013). Exposure to a saline environment enhanced the abundance of RuBisCO large subunit (LSU), RuBisCO small subunit (SSU) in gametophyte, or leaves of halophytes such as *P. patens* and *Porteresia coarctata* (Sengupta & Majumder, 2010; Wang et al., 2008a). However, a downregulation in the expression of the eight RuBisCO subunits has been reported by Zhang, Xu, et al. (2016), Zhang, Zhu, et al. (2016) in cotton. The main role of RuBisCO activase lies with the protection of the catalytic activity of RuBisCO. An increase of 1.7- and 2.5-fold in the content of RuBisCO activase has been reported under salinity stress (Parker, Flowers, Moore, & Harpham, 2006), however, some proteomic studies suggest that RuBisCO activase content decreases under a similar environment (Wang, Wang, Liu, & Meng, 2013). Any decrease in RuBisCO activase might reduce the activity of the Calvin- Benson cycle (Goharrizi et al., 2019). The transketolase (TK) protein, which is found in chloroplast and participates in the regeneration phase of the Calvin-Benson cycle, was found to be upregulated under salinity stress (Wang, Liang, et al., 2013; Yu et al., 2011). Proteins such as rubisco activase, ribulose-5-phosphate kinase (Ru5PK), and OEE protein were noted under salinity stress from two contrasting genotypes of cowpea that played vital roles in energy metabolism and photosynthesis, and these proteins showed abundance in salt-tolerant genotypes, in contrast to salt-sensitive cultivars (de Abreu et al., 2014).

Upon encountering salinity stress, the plant increases the demand for higher energy production as a stress response. The EMP-TCA pathway is regarded as the most important biological process in plants regulating carbohydrate and energy metabolism (Xing, Pan, Wang, Tan, & Chen, 2019). An elevation in the relative abundance of various proteins linked with energy production processes, for instance, the EMP-TCA pathway, pentose phosphate pathway (PPP), and photorespiration, has been found in both halophytes and glycophytes (Kosová, Vítámvás, Urban, & Prášil, 2013).

Enhancement in the carbon catabolism could be considered as a strategy to produce energy that is required for the operation of protective mechanisms (Chaves, Flexas, & Pinheiro, 2009). The regulation of proteins involved in carbohydrate and energy metabolisms in roots of *H. glomeratus* and *Gossypium hirsutum* played crucial roles in maintaining their tolerance capacity against salt (Li et al., 2015; Wang et al., 2016). In maize, the roots of the salt-tolerant variety improved salt tolerance by increasing energy supply (Luo et al., 2018). In *A. thaliana* and *T. salsuginea*, a comparative proteomic study revealed that a greater relative abundance of respiration-associated enzymes was found in the former as compared to the latter (Pang et al., 2010). EMP pathway (Glycolysis)-related enzymes such as pyruvate kinase isozyme A (PK), 6-phosphofructokinase 3 (PFK), and pyruvate decarboxylase 2 (PDC); and TCA pathway (Kreb's cycle)-related enzymes such as pyruvate dehydrogenase complex (PDC), isocitrate dehydrogenase (IDH), malate dehydrogenase (MDH), aconitate hydratase (AH), succinate dehydrogenase (SDH), and ATP-citrate synthase, were highly augmented under salinity stress (Kosová, Vítámvás, Urban, Prášil, & Renaut, 2018). The rise in levels of aconitate hydratase (AH) and (transketolase) TK in *P. tenuiflora* in the presence of salinity stress inferred that it is directly correlated to the TCA cycle and pentose phosphate pathway (Yu et al., 2011). Augmentation of the PPP is an obvious phenomenon that has been reported in several species as an adaptive response against various stress conditions (Henkes, Sonnewald, Badur, Flachmann, & Stitt, 2001; Nemoto & Sasakuma, 2000). Enhanced EMP-TCA activity along with raised ATP synthesis indicates that salt stress induces the remobilization of energy in plants to cope with the toxicity. Therefore this escalation in the abundance of these important enzymes of the EMP-TCA pathway ensured the supply of adequate energy to improve different metabolic processes under a saline environment. Moreover, along with the enhanced photosynthetic light reactions, this might ensure the uninterrupted supply of ATP and reducing power of NAD(P)H to mediate the salt resistance in halophytes (Kumari et al., 2015).

A significant increase in rice ATP synthase was noticed in rice (Kim et al., 2005), and also in the roots as well as shoots of *Zea mays* growing under salinity stress (Zuther et al., 2004). Moreover, a higher abundance subunit β of ATP synthase, playing a key role in ATP biosynthesis has been reported in several salt-tolerant plant species (Geissler, Hussin, & Koyro, 2010; Li, Zhang, et al., 2011). Remarkably, overexpression of a rice mitochondrial ATP synthase in tobacco imparted tolerance against salt stress (Zhang, Takano, & Liu, 2006). Improved ATP synthesis induction helps in conferring resistance toward various abiotic stresses (Zhang, Liu, & Takano, 2008). The elevation in ATP generation suggests that regulation of ATP formation and its utilization for the synthesis of compatible solutes is one of the crucial strategies that plants adapt to cope with salt stress. The ATP synthesized during cyclic electron flow around PSI has been reported to restrict the accumulation of Na^+ in the chloroplasts of soybean (He et al., 2015). Furthermore, it has been reported that salt stress triggered electron transport in some halophytes (Stepien & Johnson, 2009), and frequently these excess electrons get dissipated by alternative electron sinks such as the plastid terminal oxidase (PTOX) (Stepien & Johnson, 2009; Uzilday, Ozgur, Sekmen, Yildiztugay, & Turkan, 2015). PTOX inhibition raised the H_2O_2 levels inside thylakoids of *Eutrema* (Wiciarz, Gubernator, Kruk, & Niewiadomska, 2015), indicating the role of alternative oxidase (AOX) in controlling ROS production. The halophytic *Populus euphratica* exposed to 200 mM salt showed (3.1- and 1.8-fold) increase in AOX expression (Ottow et al., 2005). Overexpression of alternative oxidase (Ataox1a) in *Arabidopsis* lowered the ROS outburst and amended growth by 30%–40% during salt stress (Smith, Melino, Sweetman, & Soole, 2009). Upregulation of the AOX enzyme associated with supposed "cyanide-resistant respiration," that is, respiration occurring without the use of cytochromes resulting in reduced ROS formation, has been observed in mitochondria of salt-treated wheat (Jacoby, Millar, & Taylor, 2010). Thus it suggests that AOX plays an important role in ROS mitigation during salinity stress in both glycophytes and halophytes.

12.3.5.2 Protein concerned with antioxidant/ reactive oxygen species scavenging activity

It has been widely reported that scavenging of ROS by increasing the activation of antioxidant enzymes can improve salt tolerance (Moghaddam, Farhadi, Panjtandoust, & Ghanati, 2019). Under saline conditions, accumulation of SOD is known to play a protective role in *S. europaea* and *Solanum chilense* (Wang et al., 2009; Zhou et al., 2011). The three different SOD isoforms—copper-zinc SOD (Cu-Zn SOD), iron-SOD (Fe-SOD), and manganese-SOD (Mn-SOD)—were increased in different organelles during salinity stress (Javed, Zafar, & Ashraf, 2019). Previous studies established that SOD and POD proteins accumulate to high levels in wheat, tomato, and cotton under salt stress (Gong et al., 2014; Guo et al., 2012; Li et al., 2015). Kannan et al. (2013) while working on *S. portulacastrum* L. reported that antioxidant enzymes such as polyphenol oxidase (PPO), CAT, and SOD increased under saline condition. Overexpression of CAT gene (katE) in rice conferred tolerance against oxidative stress induced by salinity (Nagamiya et al., 2007). Tocopherol cyclase is an important enzyme required for the biosynthesis of α-tocopherol which is a nonenzymatic ROS scavenging molecule. An increase in its abundance has been reported *P. tenuiflora* (Yu et al., 2011).

Salt stress induces the accumulation of several proteins possessing protective functions, for instance, chaperones belonging to the heat shock proteins 90 (HSP90) family in tomato roots (Manaa et al., 2011) and HSP 70 family, Hsc70 (heat-shock cognate) proteins in *P. patens* (Wang et al., 2008a).

Osmoprotectants like late embryogenesis abundant (LEA) proteins, dehydrins, and omotions also showed higher expression in response to salinity (Jyothi-Prakash et al., 2014; Kosová et al., 2013; Pang et al., 2010). LEA proteins protect cells from cellular dehydration by countering any sort of water stress faced by the cells. They act as molecular chaperones, impart antioxidants, and membrane protective function against cold and saline stress (Bhardwaj et al., 2013). Salinity-induced accumulation of LEA proteins was observed that resulted in the growth arrest of *O. sativa* seedlings (Chourey, Ramani, & Apte, 2003). Similarly, expression of proteins belonging to LEA Group 1 and 2 of wheat have been reported in two transgenic rice plants under saline conditions (Cheng, Targolli, Huang, & Wu, 2002). Dehydrin (30 kDa) accumulated in *C. quinoa* embryos in presence of 300 and 500 mM of NaCl (Burrieza, Koyro, Tosar, Kobayashi, & Maldonado, 2012).

12.3.5.3 Protein concerned with the signaling processes

Proteomic analysis reveals that Ca^{2+} signaling pathway is involved in conferring salt tolerance. Calcium-binding proteins are modulated by salinity in plants like *A. thaliana, Suaeda salsa, and Avicennia officinalis* (Krishnamurthy et al., 2014; Li et al., 2011; Pang et al., 2010). Moreover, levels of calmodulin, annexin, calcineurin-like phosphoesterases, and calreticulin were also affected by salt (Jiang et al., 2007; Li et al., 2010; Zörb, Schmitt, & Mühling, 2010). These modifications might alter the levels of intracellular stores of Ca^{2+}. Reports suggest that Ca^{2+} signaling switches on the SOS pathway which is crucial for the maintenance of cellular Na^+/K^+ homeostasis (Zhu, 2002). Calnexin and calmodulin (calcium-binding proteins) detect the Ca concentration and activate NHK (Al Murad, Khan, & Muneer, 2020), thus induces salinity tolerance by translocating the Na^+ ions into the vacuole. Calreticulin (CRT) a calcium-binding protein also participates in calcium signaling in the endoplasmic reticulum during salt stress response in plants (Persson et al., 2003). CRT upregulation might play a role in signal transduction in halophytes growing in extreme saline conditions (Cheng et al., 2015; Sobhanian et al., 2011). A protein named translationally controlled tumor protein (TCTP) possesses a high affinity for calcium (Saraf, Saingar, Chaudhary, & Chakraborty, 2018). In a proteomic study conducted on barley, it was reported that TCTP was detected at varying levels in saline sensitive and tolerant genotypes during germination (Witzel et al., 2010).

Proteins like 14-3-3 are known to be associated with stress-induced responses including salinity in plants (Zhang et al., 2012). Besides, 14-3-3 proteins are involved in different signal transduction pathways through controlling the kinases and phosphatases activities, indicating its importance in multiple pathways involved in salt stress responses (Kumari et al., 2015). An increased relative abundance of 14-3-3-like proteins was reported in *P. patens* gametophyte, maize root, and grass pea leaves (Chattopadhyay et al., 2011;

Wang et al., 2008a; Zörb et al., 2010). These proteins interacted with the MAPK cascade to regulate the activity of H^+-ATPases present on the plasma membrane for maintaining intracellular ion homeostasis.

Thus the integration of proteomics with transcriptomic data could unravel the underlying candidate gene(s)/proteins and complex signaling pathways involved in generating responses and tolerance toward salinity (Ngara & Ndimba, 2014).

12.4 Conclusions

Considering the ever-increasing size of the human population, it is the immediate need to bring land under cultivation that is no longer in use due to poor soil health, viz., presence of excess salt in the soil. This would only be possible if the crop plants possess salt resilience traits that will allow their growth in such a harsh environment. Most of the crop plants (rice, wheat) are glycophytes (salt-sensitive plants), which cannot survive with excessive salt concentration in the soil. Halophytes, however, are bestowed with certain traits that enable them to survive under saline conditions. To increase the plant salt-tolerance, there is a need for understanding the mechanisms of salt limitations on plant growth and the mechanism of salt tolerance at the whole-plant, organelle, and molecular levels. Over the years, a plethora of research has been carried out in several halophytic plants that have identified several salinity mitigation traits by identifying key genes/proteins involved in stress response. These traits include ions homeostasis, ion compartmentalization, ions exclusion, expression of genes concerned with antioxidant machinery, photosynthesis, and energy production. Salinity inhibits water uptake, causes nutrient imbalance, and limits photosynthesis by decreasing chlorophyll content, PS II activity, and stomatal conductance. All these processes ultimately decrease plant growth. Some plants resist these salinity-induced problems through the synthesis of osmolytes that improve water uptake, ions homeostasis (exclusion of toxic Na^+ ions or compartmentalization) that maintain normal ionic equilibrium, or activation of antioxidant machinery that scavenges ROS. Additionally, with the advent of omics approaches, such as transcriptomics and proteomics, the key genes/proteins that mediate tolerance mechanisms are now being identified and used for the development of resistance in several salt sensitive crops.

References

Abbas, T., Sattar, A., Ijaz, M., Aatif, M., Khalid, S., & Sher, A. (2017). Exogenous silicon application alleviates salt stress in Okra. *Horticulture, Environment, and Biotechnology*, *58*(4), 342–349.

Abd EL-Azim, W. M., & Ahmed, S. T. H. (2009). Effect of salinity and cutting date on growth and chemical constituents of *Achillea fragratissima* Forssk, under Ras Sudr conditions. *Research Journal of Agriculture and Biological Sciences*, *5*(6), 1121–1129.

Abd El-Wahab, M. A. (2006). The efficiency of using saline and fresh water irrigation as alternating methods of irrigation on the productivity of *Foeniculum vulgare* Mill subsp. vulgare var. vulgare under North Sinai conditions. *Research Journal of Agriculture and Biological Sciences*, *2*(6), 571–577.

Abdallah, S. B., Aung, B., Amyot, L., Lalin, I., Lachâal, M., Karray-Bouraoui, N., & Hannoufa, A. (2016). Salt stress (NaCl) affects plant growth and branch pathways of carotenoid and flavonoid biosyntheses in *Solanum nigrum*. *Acta Physiologiae Plantarum*, *38*(3), 72.

Abdel Latef, A. A. H., Srivastava, A. K., El-sadek, M. S. A., Kordrostami, M., & Tran, L. S. P. (2018). Titanium dioxide nanoparticles improve growth and enhance tolerance of broad bean plants under saline soil conditions. *Land Degradation & Development*, *29*(4), 1065–1073.

Abdelaal, K. A., EL-Maghraby, L. M., Elansary, H., Hafez, Y. M., Ibrahim, E. I., El-Banna, M., & Elkelish, A. (2020). Treatment of sweet pepper with stress tolerance-inducing compounds alleviates salinity stress oxidative damage by mediating the physio-biochemical activities and antioxidant systems. *Agronomy*, *10*(1), 26.

Acosta-Motos, J. R., Diaz-Vivancos, P., Álvarez, S., Fernández-García, N., Sánchez-Blanco, M. J., & Hernández, J. A. (2015). NaCl-induced physiological and biochemical adaptative mechanisms in the ornamental *Myrtus communis* L. plants. *Journal of Plant Physiology*, *183*, 41–51.

Acosta-Motos, J. R., Ortuño, M. F., Bernal-Vicente, A., Diaz-Vivancos, P., Sanchez-Blanco, M. J., & Hernandez, J. A. (2017). Plant responses to salt stress: Adaptive mechanisms. *Agronomy*, *7*(1), 18.

Agarie, S., Shimoda, T., Shimizu, Y., Baumann, K., Sunagawa, H., Kondo, A., & Cushman, J. C. (2007). Salt tolerance, salt accumulation, and ionic homeostasis in an epidermal bladder-cell-less mutant of the common ice plant *Mesembryanthemum crystallinum*. *Journal of Experimental Botany*, *58*(8), 1957–1967.

Ahanger, M. A., & Agarwal, R. M. (2017). Salinity stress induced alterations in antioxidant metabolism and nitrogen assimilation in wheat (*Triticum aestivum* L) as influenced by potassium supplementation. *Plant Physiology and Biochemistry*, *115*, 449–460.

Ahanger, M. A., Tomar, N. S., Tittal, M., Argal, S., & Agarwal, R. M. (2017). Plant growth under water/saltstress: ROS production; antioxidants and significance of added potassium under such conditions. *Physiology and Molecular Biology of Plants*, *23*, 731–744.

Ahmad, P., Ahanger, M. A., Alam, P., Alyemeni, M. N., Wijaya, L., Ali, S., & Ashraf, M. (2019). Silicon (Si) supplementation alleviates NaCl toxicity in mung bean [*Vigna radiata* (L.) Wilczek] through the modifications of physio-biochemical attributes and key antioxidant enzymes. *Journal of Plant Growth Regulation*, *38*(1), 70–82.

Ahmad, P., Azooz, M. M., & Prasad, M. N. V. (2013). *Salt stress in plants* (pp. 1–14). Heidelberg: Springer.

Ahmad, P., Hashem, A., Abd-Allah, E. F., Alqarawi, A. A., John, R., Egamberdieva, D., & Gucel, S. (2015). Role of *Trichoderma harzianum* in mitigating NaCl stress in Indian mustard (*Brassica juncea* L) through antioxidative defense system. *Frontiers in Plant Science*, *6*, 868.

Al Murad, M., Khan, A. L., & Muneer, S. (2020). Silicon in horticultural crops: Cross-talk, signaling, and tolerance mechanism under salinity stress. *Plants*, *9*(4), 460.

Albaladejo, I., Egea, I., Morales, B., Flores, F. B., Capel, C., Lozano, R., & Bolarin, M. C. (2018). Identification of key genes involved in the phenotypic alterations of RES (restored cell structure by salinity) tomato mutant and its recovery induced by salt stress through transcriptomic analysis. *BMC Plant Biology*, *18*(1), 213.

Al-Karaki, G. N. (2017). Effects of mycorrhizal fungi inoculation on green pepper yield and mineral uptake under irrigation with saline water. *Advances in Plants and Agriculture Research*, *6*(5), 00231.

Almeida, D. M., Oliveira, M. M., & Saibo, N. J. (2017). Regulation of Na^+ and K^+ homeostasis in plants: Towards improved salt stress tolerance in crop plants. *Genetics and Molecular Biology*, *40*, 326–345.

Alric, J. (2015). The plastoquinone pool, poised for cyclic electron flow? *Frontiers in Plant Science*, *6*, 540.

Alshareef, N. O., Wang, J. Y., Ali, S., Al-Babili, S., Tester, M., & Schmöckel, S. M. (2019). Overexpression of the NAC transcription factor JUNGBRUNNEN1 (JUB1) increases salinity tolerance in tomato. *Plant Physiology and Biochemistry*, *140*, 113–121.

Amini, F., & Ehsanpour, A. A. (2005). Soluble proteins, proline, carbohydrates and Na + /K + changes in two tomato (*Lycopersicon esculentum* Mill.) cultivars under in vitro salt stress. *American Journal of Biochemistry and Biotechnology*, *1*(4), 204–208.

Amirbakhtiar, N., Ismaili, A., Ghaffari, M. R., Nazarian Firouzabadi, F., & Shobbar, Z. S. (2019). Transcriptome response of roots to salt stress in a salinity-tolerant bread wheat cultivar. *PLoS One*, *14*(3), e0213305.

Annunziata, M. G., Ciarmiello, L. F., Woodrow, P., Dell'Aversana, E., & Carillo, P. (2019). Spatial and temporal profile of glycine betaine accumulation in plants under abiotic stresses. *Frontiers in Plant Science*, *10*, 230.

Annunziata, M. G., Ciarmiello, L. F., Woodrow, P., Maximova, E., Fuggi, A., & Carillo, P. (2017). Durum wheat roots adapt to salinity remodeling the cellular content of nitrogen metabolites and sucrose. *Frontiers in Plant Science*, *7*, 2035.

Apse, M. P., Aharon, G. S., Snedden, W. A., & Blumwald, E. (1999). Salt tolerance conferred by overexpression of a vacuolar Na + /H + antiport in *Arabidopsis*. *Science (New York, N. Y.)*, *285*, 1256–1258.

Apse, M. P., & Blumwald, E. (2002). Engineering salt tolerance in plants. *Current Opinion in Biotechnology*, *13*(2), 146–150.

Arbona, V., Marco, A. J., Iglesias, D. J., López-Climent, M. F., Talon, M., & Gómez-Cadenas, A. (2005). Carbohydrate depletion in roots and leaves of salt-stressed potted *Citrus clementina* L. *Plant Growth Regulation*, *46*(2), 153–160.

Ashfaque, F., Khan, M. I. R., & Khan, N. A. (2014). Exogenously applied H_2O_2 promotes proline accumulation, water relations, photosynthetic efficiency and growth of wheat (*Triticum aestivum* L.) under salt stress. *Annual Research & Review in Biology*, *4*, 105–120.

Ashraf, M. (2009). Biotechnological approach of improving plant salt tolerance using antioxidants as markers. *Biotechnology Advances*, *27*, 84–93.

Ashraf, M., & Orooj, A. (2006). Salt stress effects on growth, ion accumulation and seed oil concentration in an arid zone traditional medicinal plant ajwain (*Trachyspermum ammi* [L.] Sprague). *Journal of Arid Environments*, *64*(2), 209–220.

Ashraf, M. F. M. R., & Foolad, M. R. (2007). Roles of glycine betaine and proline in improving plant abiotic stress resistance. *Environmental and Experimental Botany*, *59*(2), 206–216.

Ashraf, M. H. P. J. C., & Harris, P. J. (2013). Photosynthesis under stressful environments: An overview. *Photosynthetica*, *51*(2), 163–190.

Ashraf, M. Y., & Wu, L. (1994). Breeding for salinity tolerance in plants. *Critical Reviews in Plant Sciences*, *13*(1), 17–42.

Ashrafi, N., & Nejad, A. R. (2018). Lisianthus response to salinity stress. *Photosynthetica*, *56* (2), 487–494.

Askari, H., Edqvist, J., Hajheidari, M., Kafi, M., & Salekdeh, G. H. (2006). Effects of salinity levels on proteome of *Suaeda aegyptiaca* leaves. *Proteomics*, *6*(8), 2542–2554.

Badawi, G. H., Kawano, N., Yamauchi, Y., Shimada, E., Sasaki, R., Kubo, A., & Tanaka, K. (2004). Over-expression of ascorbate peroxidase in tobacco chloroplasts enhances the tolerance to salt stress and water deficit. *Physiologia Plantarum*, *121*(2), 231–238.

Baghalian, K., Haghiry, A., Naghavi, M. R., & Mohammadi, A. (2008). Effect of saline irrigation water on agronomical and phytochemical characters of chamomile (*Matricaria recutita* L.). *Scientia Horticulturae*, *116*(4), 437–441.

Bai, J., Qin, Y., Liu, J., Wang, Y., Sa, R., Zhang, N., & Jia, R. (2017). Proteomic response of oat leaves to long-term salinity stress. *Environmental Science and Pollution Research*, *24* (4), 3387–3399.

Barhoumi, Z., Djebali, W., Smaoui, A., Chaïbi, W., & Abdelly, C. (2007). Contribution of NaCl excretion to salt resistance of *Aeluropus littoralis* (Willd) Parl. *Journal of Plant Physiology*, *164*(7), 842–850.

Bassil, E., Tajima, H., Liang, Y. C., Ohto, M., Ushijima, K., Nakano, R., ... Blumwald, E. (2011). The *Arabidopsis* Na + /H + antiporters NHX1 and NHX2 control vacuolar pH and K + homeostasis to regulate growth, flower development, and reproduction. *The Plant Cell*, *23*, 3482–3497.

Behr, J. H., Bouchereau, A., Berardocco, S., Seal, C. E., Flowers, T. J., & Zörb, C. (2017). Metabolic and physiological adjustment of *Suaeda maritima* to combined salinity and hypoxia. *Annals of Botany*, *119*(6), 965–976.

Belkheiri, O., & Mulas, M. (2013). The effects of salt stress on growth, water relations and ion accumulation in two halophyte *Atriplex* species. *Environmental and Experimental Botany*, *86*, 17–28.

Belmecheri-Cherifi, H., Albacete, A., Martínez-Andújar, C., Pérez-Alfocea, F., & Abrous-Belbachir, O. (2019). The growth impairment of salinized fenugreek (*Trigonella foenum-graecum* L.) plants is associated to changes in the hormonal balance. *Journal of Plant Physiology*, *232*, 311–319.

Berthomieu, P., Conejéro, G., Nublat, A., Brackenbury, W., Lambert, C., Savió, C., ... Hand-Casse, F. S. (2003). Functional analysis of AtHKT1 in *Arabidopsis* shows that Na^+ recirculation by the phloem is crucial for salt tolerance. *The EMBO Journal*, *22*, 2004–2014.

Bhardwaj, R., Sharma, I., Kanwar, M., Sharma, R., Handa, N., Kaur, H., & Kapoor, D. (2013). *LEA proteins in salt stress tolerance. Salt stress in plants* (pp. 79–112). New York, NY: Springer.

Bhatt, I., & Tripathi, B. N. (2011). Plant peroxiredoxins: Catalytic mechanisms, functional significance and future perspectives. *Biotechnology Advances*, *29*(6), 850–859.

Blumwald, E., Aharon, G. S., & Apse, M. P. (2000). Sodium transport in plant cells. *Biochimica et Biophysica Acta*, *1465*, 140–151.

Bonales-Alatorre, E., Shabala, S., Chen, Z. H., & Pottosin, I. (2013). Reduced tonoplast fast-activating and slow-activating channel activity is essential for conferring salinity tolerance in a facultative halophyte, quinoa. *Plant Physiology*, *162*, 940–952.

Boriboonkaset, T., Theerawitaya, C., Yamada, N., Pichakum, A., Supaibulwatana, K., Cha-um, S., & Kirdmanee, C. (2013). Regulation of some carbohydrate metabolism-related genes, starch and soluble sugar contents, photosynthetic activities and yield attributes of two contrasting rice genotypes subjected to salt stress. *Protoplasma*, *250*(5), 1157–1167.

Bose, J., Munns, R., Shabala, S., Gilliham, M., Pogson, B., & Tyerman, S. D. (2017). Chloroplast function and ion regulation in plants growing on saline soils: Lessons from halophytes. *Journal of Experimental Botany*, *68*(12), 3129–3143.

Bose, J., Rodrigo-Moreno, A., & Shabala, S. (2014). ROS homeostasis in halophytes in the context of salinity stress tolerance. *Journal of Experimental Botany*, *65*(5), 1241–1257.

Broin, M., Cuiné, S., Eymery, F., & Rey, P. (2002). The plastidic 2-cysteine peroxiredoxin is a target for a thioredoxin involved in the protection of the photosynthetic apparatus against oxidative damage. *The Plant Cell*, *14*(6), 1417–1432.

Bulle, M., Yarra, R., & Abbagani, S. (2016). Enhanced salinity stress tolerance in transgenic chilli pepper (*Capsicum annuum* L.) plants overexpressing the wheat antiporter (TaNHX2) gene. *Molecular Breeding*, *36*(4), 36.

Burrieza, H. P., Koyro, H. W., Tosar, L. M., Kobayashi, K., & Maldonado, S. (2012). High salinity induces dehydrin accumulation in *Chenopodium quinoa* Willd. cv. Hualhuas embryos. *Plant and Soil*, *354*(1-2), 69–79.

Bybordi, A. (2010). The influence of salt stress on seed germination, growth and yield of canola cultivars. *Notulae Botanicae Horti Agrobotanici Cluj-Napoca*, *38*(1), 128–133.

Byrt, C. S., Xu, B., Krishnan, M., Lightfoot, D. J., Athman, A., Jacobs, A. K., & Gilliham, M. (2014). The Na^+ transporter, Ta HKT 1; 5-D, limits shoot Na^+ accumulation in bread wheat. *The Plant Journal*, *80*(3), 516–526.

Cai, X., Zhang, C., Shu, W., Ye, Z., Li, H., & Zhang, Y. (2016). The transcription factor SlDof22 involved in ascorbate accumulation and salinity stress in tomato. *Biochemical and Biophysical Research Communications*, *474*(4), 736–741.

Cambridge, M. L., Zavala-Perez, A., Cawthray, G. R., Mondon, J., & Kendrick, G. A. (2017). Effects of high salinity from desalination brine on growth, photosynthesis, water relations and osmolyte concentrations of seagrass *Posidonia australis*. *Marine Pollution Bulletin*, *115*(1-2), 252–260.

Capel, C., Albaladejo, I., Egea, I., Massaretto, I. L., Yuste-Lisbona, F. J., Pineda, B., & Lozano, R. (2020). The RES (restored cell structure by salinity) tomato mutant reveals the role of the DEAD-box RNA helicase SlDEAD39 in plant development and salt response. *Plant, Cell & Environment*, *43*, 1722–1739.

Capriotti, A. L., Borrelli, G. M., Colapicchioni, V., Papa, R., Piovesana, S., Samperi, R., & Laganà, A. (2014). Proteomic study of a tolerant genotype of durum wheat under salt-stress conditions. *Analytical and Bioanalytical Chemistry*, *406*(5), 1423–1435.

Chaichi, M. R., Keshavarz-Afshar, R., Lu, B., & Rostamza, M. (2017). Growth and nutrient uptake of tomato in response to application of saline water, biological fertilizer, and surfactant. *Journal of Plant Nutrition*, *40*(4), 457–466.

Chakraborty, K., Basak, N., Bhaduri, D., et al. (2018). Ionic basis of salt tolerance in plants: Nutrient homeostasis and oxidative stress tolerance. In M. Hasanuzzaman, M. Fujita, H. Oku, K. Nahar, & B. Hawrylak- Nowak (Eds.), *Plant nutrients and abiotic stress tolerance* (pp. 325–362). New York: Springer.

Chakraborty, K., Bose, J., Shabala, L., & Shabala, S. (2016). Difference in root $K+$ retention ability and reduced sensitivity of $K+$-permeable channels to reactive oxygen species confer differential.

Chakraborty, K., Chattaopadhyay, K., Nayak, L., Ray, S., Yeasmin, L., Jena, P., & Sarkar, R. K. (2019). Ionic selectivity and coordinated transport of $Na+$ and $K+$ in flag leaves render differential salt tolerance in rice at the reproductive stage. *Planta*, *250*(5), 1637–1653.

Chakraborty, K., Sairam, R. K., & Bhattacharya, R. C. (2012). Differential expression of salt overly sensitive pathway genes determines salinity stress tolerance in *Brassica* genotypes. *Plant Physiology and Biochemistry*, *51*, 90–101.

Chang, L., Guo, A., Jin, X., Yang, Q., Wang, D., Sun, Y., & Wang, X. (2015). The beta subunit of glyceraldehyde 3-phosphate dehydrogenase is an important factor for maintaining photosynthesis and plant development under salt stress—Based on an integrative analysis of the structural, physiological and proteomic changes in chloroplasts in *Thellungiella halophila*. *Plant Science*, *236*, 223–238.

Chanwala, J., Satpati, S., Dixit, A., Parida, A., Giri, M. K., & Dey, N. (2020). Genome-wide identification and expression analysis of WRKY transcription factors in pearl millet (*Pennisetum glaucum*) under dehydration and salinity stress. *BMC Genomics*, *21*(1), 1–16.

Chattopadhyay, A., Subba, P., Pandey, A., Bhushan, D., Kumar, R., Datta, A., & Chakraborty, N. (2011). Analysis of the grasspea proteome and identification of stress-responsive proteins upon exposure to high salinity, low temperature, and abscisic acid treatment. *Phytochemistry*, *72*(10), 1293–1307.

Chaves, M. M., Flexas, J., & Pinheiro, C. (2009). Photosynthesis under drought and salt stress: Regulation mechanisms from whole plant to cell. *Annals of Botany*, *103*(4), 551–560.

Chen, G., Hu, Q., Luo, L. E., et al. (2015). Rice potassium transporter OsHAK1 is essential for maintaining potassium mediated growth and functions in salt tolerance over low and high potassium concentration ranges. *Plant, Cell & Environment*, *38*, 2747–2765.

Chen, H., An, R., Tang, J. H., et al. (2007). Overexpression of a vacuolar Na + /H + antiporter gene improves salt tolerance in an upland rice. *Molecular Breeding*, *19*, 215–225.

Chen, Z., Pottosin, I. I., Cuin, T. A., Fuglsang, A. T., Tester, M., Jha, D., & Shabala, S. (2007). Root plasma membrane transporters controlling K + /Na + homeostasis in salt-stressed barley. *Plant Physiology*, *145*(4), 1714–1725.

Cheng, T., Chen, J., Zhang, J., Shi, S., Zhou, Y., Lu, L., & Shi, J. (2015). Physiological and proteomic analyses of leaves from the halophyte *Tangut Nitraria* reveals diverse response pathways critical for high salinity tolerance. *Frontiers in Plant Science*, *6*, 30.

Cheng, Z., Targolli, J., Huang, X., & Wu, R. (2002). Wheat LEA genes, PMA80 and PMA1959, enhance dehydration tolerance of transgenic rice (*Oryza sativa* L.). *Molecular Breeding*, *10* (1-2), 71–82.

Chinnusamy, V., Jagendorf, A., & Zhu, J. K. (2005). Understanding and improving salt tolerance in plants. *Crop Science*, *45*(2), 437–448.

Chokshi, K., Pancha, I., Ghosh, A., & Mishra, S. (2017). Salinity induced oxidative stress alters the physiological responses and improves the biofuel potential of green microalgae *Acutodesmus dimorphus*. *Bioresource Technology*, *244*, 1376–1383.

Chourey, K., Ramani, S., & Apte, S. K. (2003). Accumulation of LEA proteins in salt (NaCl) stressed young seedlings of rice (*Oryza sativa* L.) cultivar Bura Rata and their degradation during recovery from salinity stress. *Journal of Plant Physiology*, *160*(10), 1165–1174.

Chrysargyris, A., Michailidi, E., & Tzortzakis, N. (2018). Physiological and biochemical responses of *Lavandula angustifolia* to salinity under mineral foliar application. *Frontiers in Plant Science*, *9*, 489.

Chrysargyris, A., Papakyriakou, E., Petropoulos, S. A., & Tzortzakis, N. (2019). The combined and single effect of salinity and copper stress on growth and quality of *Mentha spicata* plants. *Journal of Hazardous Materials*, *368*, 584–593.

Conde, A., Chaves, M. M., & Gerós, H. (2011). Membrane transport, sensing and signaling in plant adaptation to environmental stress. *Plant and Cell Physiology*, *52*(9), 1583–1602.

Costa, S. F., Martins, D., Agacka-Mołdoch, M., Czubacka, A., & de Sousa Araújo, S. (2018). *Strategies to alleviate salinity stress in plants*, . *Salinity responses and tolerance in plants* (Volume 1, pp. 307–337). Cham: Springer.

Cuin, T. A., Bose, J., Stefano, G., Jha, D., Tester, M., Mancuso, S., & Shabala, S. (2011). Assessing the role of root plasma membrane and tonoplast Na + /H + exchangers in salinity tolerance in wheat: In planta quantification methods. *Plant, Cell & Environment*, *34*(6), 947–961.

Cuin, T. A., & Shabala, S. (2005). Exogenously supplied compatible solutes rapidly ameliorate NaCl-induced potassium efflux from barley roots. *Plant and Cell Physiology*, *46*(12), 1924–1933.

Dai, W., Wang, M., Gong, X., & Liu, J. H. (2018). The transcription factor Fc WRKY 40 of *Fortunella crassifolia* functions positively in salt tolerance through modulation of ion homeostasis and proline biosynthesis by directly regulating SOS_2 and P_5CS_1 homologs. *New Phytologist*, *219*(3), 972–989.

Das, K., & Roychoudhury, A. (2014). Reactive oxygen species (ROS) and response of antioxidants as ROS-scavengers during environmental stress in plants. *Frontiers in Environmental Science*, *2*, 53.

Das, P., Behera, B. K., Chatterjee, S., Das, B. K., & Mohapatra, T. (2020). De novo transcriptome analysis of halotolerant bacterium *Staphylococcus* sp. strain P-TSB-70 isolated from East coast of India: In search of salt stress tolerant genes. *PLoS One*, *15*(2), e0228199.

Das, P., Lakra, N., Nutan, K. K., Singla-Pareek, S. L., & Pareek, A. (2019). A unique bZIP transcription factor imparting multiple stress tolerance in Rice. *Rice*, *12*(1), 58.

Das-Chatterjee, A., Goswami, L., Maitra, S., Dastidar, K. G., Ray, S., & Majumder, A. L. (2006). Introgression of a novel salt-tolerant L-myo-inositol 1-phosphate synthase from *Porteresia coarctata* (Roxb.) Tateoka (PcINO1) confers salt tolerance to evolutionary diverse organisms. *FEBS Letters*, *580*(16), 3980–3988.

Davenport, R. J., Muñoz-Mayor, A. L. I. C. I. A., Jha, D., Essah, P. A., Rus, A. N. A., & Tester, M. (2007). The Na + transporter AtHKT1; 1 controls retrieval of Na + from the xylem in *Arabidopsis*. *Plant, Cell & Environment*, *30*(4), 497–507.

de Abreu, C. E., Araújo Gdos, S., Monteiro-Moreira, A. C., Costa, J. H., Leite Hde, B., Moreno, F. B., . . . Gomes-Filho, E. (2014). Proteomic analysis of salt stress and recovery in leaves of *Vigna unguiculata* cultivars differing in salt tolerance. *Plant Cell Reports*, *33*, 1289–1306.

De la Torre-González, A., Montesinos-Pereira, D., Blasco, B., & Ruiz, J. M. (2018). Influence of the proline metabolism and glycine betaine on tolerance to salt stress in tomato (*Solanum lycopersicum* L.) commercial genotypes. *Journal of Plant Physiology*, *231*, 329–336.

de Pinto, M. C., & De Gara, L. (2004). Changes in the ascorbate metabolism of apoplastic and symplastic spaces are associated with cell differentiation. *Journal of Experimental Botany*, *55*(408), 2559–2569.

de Vitry, C., Ouyang, Y., Finazzi, G., Wollman, F. A., & Kallas, T. (2004). The chloroplast Rieske iron-sulfur protein at the crossroad of electron transport and signal transduction. *Journal of Biological Chemistry*, *279*(43), 44621–44627.

Debez, A., Braun, H. P., Pich, A., Taamalli, W., Koyro, H. W., Abdelly, C., & Huchzermeyer, B. (2012). Proteomic and physiological responses of the halophyte *Cakile maritima* to moderate salinity at the germinative and vegetative stages. *Journal of Proteomics*, *75*(18), 5667–5694.

Diédhiou, C. J., Popova, O. V., & Golldack, D. (2009a). Comparison of salt responsive gene regulation in rice and in the salt-tolerant *Festucarubra* ssp. litoralis. *Plant Signaling & Behavior*, *4*, 533–535.

Diédhiou, C. J., Popova, O. V., & Golldack, D. (2009b). Transcript profiling of the salt-tolerant *Festucarubra* ssp. litoralis reveals a regulatorynetwork controlling salt acclimatization. *Journal of Plant Physiology*, *166*, 697–711.

Dluzniewska, P., Gessler, A., Dietrich, H., Schnitzler, J. P., Teuber, M., & Rennenberg, H. (2007). Nitrogen uptake and metabolism in *Populus canescens* as affected by salinity. *New Phytologist*, *173*(2), 279–293.

Dogan, M. (2011). Antioxidative and proline potentials as a protective mechanism in soybean plants under salinity stress. *African Journal of Biotechnology*, *10*(32), 5972–5978.

Dong, W., Liu, X., Li, D., Gao, T., & Song, Y. (2018). Transcriptional profiling reveals that a MYB transcription factor MsMYB4 contributes to the salinity stress response of alfalfa. *PLoS One*, *13*(9), e0204033.

Du, C. X., Fan, H. F., Guo, S. R., Tezuka, T., & Li, J. (2010). Proteomic analysis of cucumber seedling roots subjected to salt stress. *Phytochemistry*, *71*(13), 1450–1459.

Edel, K. H., & Kudla, J. (2015). Increasing complexity and versatility: How the calcium signaling toolkit was shaped during plant land colonization. *Cell Calcium*, *57*, 231–246.

El Rabey, H. A., Al-Malki, A. L., & Abulnaja, K. O. (2016). Proteome analysis of date palm (*Phoenix dactylifera* L.) under severe drought and salt stress. *International Journal of Genomics*, *2016*, 7840759.

El-Esawi, M. A., Alaraidh, I. A., Alsahli, A. A., Alzahrani, S. M., Ali, H. M., Alayafi, A. A., & Ahmad, M. (2018). *Serratia liquefaciens* KM4 improves salt stress tolerance in maize by regulating redox potential, ion homeostasis, leaf gas exchange and stress-related gene expression. *International Journal of Molecular Sciences*, *19*(11), 3310.

Ellouzi, H., Ben Hamed, K., Cela, J., Munné-Bosch, S., & Abdelly, C. (2011). Early effects of salt stress on the physiological and oxidative status of *Cakile maritima* (halophyte) and *Arabidopsis thaliana* (glycophyte). *Physiologia Plantarum*, *142*(2), 128–143.

El-Shabrawi, H., Kumar, B., Kaul, T., Reddy, M. K., Singla-Pareek, S. L., & Sopory, S. K. (2010). Redox homeostasis, antioxidant defense, and methylglyoxal detoxification as markers for salt tolerance in Pokkali rice. *Protoplasma*, *245*(1-4), 85–96.

Evelin, H., Devi, T. S., Gupta, S., & Kapoor, R. (2019). Mitigation of salinity stress in plants by arbuscular mycorrhizal symbiosis: Current understanding and new challenges. *Frontiers in Plant Science*, *10*, 470.

Eyidogan, F., & Öz, M. T. (2007). Effect of salinity on antioxidant responses of chickpea seedlings. *Acta Physiologiae Plantarum*, *29*(5), 485.

Fan, H., Xu, Y., Du, C., & Wu, X. (2015). Phloem sap proteome studied by iTRAQ provides integrated insight into salinity response mechanisms in cucumber plants. *Journal of Proteomics*, *125*, 54–67.

Fan, P., Feng, J., Jiang, P., Chen, X., Bao, H., Nie, L., & Li, Y. (2011). Coordination of carbon fixation and nitrogen metabolism in *Salicornia europaea* under salinity: Comparative proteomic analysis on chloroplast proteins. *Proteomics*, *11*(22), 4346–4367.

Food and Agriculture Organization of the United Nations, FAOUN. (2015). Status of the world's soil resources (SWSR)–main report. *Food and agriculture organization of the United Nations and intergovernmental technical panel on soils*, Rome, Italy, 650.

Fitzgerald, T. L., Waters, D. L., & Henry, R. J. (2009). Betaine aldehyde dehydrogenase in plants. *Plant Biology*, *11*(2), 119–130.

Flexas, J., Bota, J., Loreto, F., Cornic, G., & Sharkey, T. D. (2004). Diffusive and metabolic limitations to photosynthesis under drought and salinity in C3 plants. *Plant Biology*, *6*(3), 269–279.

Flexas, J., Diaz-espejo, A., Galmes, J., Kaldenhoff, R., Medrano, H., & Ribas-Carbo, M. (2007). Rapid variations of mesophyll conductance in response to changes in CO_2 concentration around leaves. *Plant, Cell & Environment*, *30*(10), 1284–1298.

Flowers, T. J., & Colmer, T. D. (2008). Salinity tolerance in halophytes. *New Phytologist*, 945–963.

Foyer, C. H., & Noctor, G. (2009). Redox regulation in photosynthetic organisms: Signaling, acclimation, and practical implications. *Antioxidants & Redox Signaling*, *11*, 861–905.

Frukh, A., Siddiqi, T. O., Khan, M. I. R., & Ahmad, A. (2020). Modulation in growth, biochemical attributes and proteome profile of rice cultivars under salt stress. *Plant Physiology and Biochemistry*, *146*, 55–70.

Gao, L., Yan, X., Li, X., Guo, G., Hu, Y., Ma, W., & Yan, Y. (2011). Proteome analysis of wheat leaf under salt stress by two-dimensional difference gel electrophoresis (2D-DIGE). *Phytochemistry*, *72*(10), 1180–1191.

Garg, N., & Bhandari, P. (2016). Silicon nutrition and mycorrhizal inoculations improve growth, nutrient status, K + /Na + ratio and yield of *Cicer arietinum* L. genotypes under salinity stress. *Plant Growth Regulation*, *78*(3), 371–387.

Garg, N., & Bharti, A. (2018). Salicylic acid improves arbuscular mycorrhizal symbiosis, and chickpea growth and yield by modulating carbohydrate metabolism under salt stress. *Mycorrhiza*, *28*(8), 727–746.

Garg, R., Shankar, R., Thakkar, B., Kudapa, H., Krishnamurthy, L., Mantri, N., & Jain, M. (2016). Transcriptome analyses reveal genotype-and developmental stage-specific molecular responses to drought and salinity stresses in chickpea. *Scientific Reports*, *6*, 19228.

Gechev, T. S., Van Breusegem, F., Stone, J. M., Denev, I., & Laloi, C. (2006). Reactive oxygen species as signals that modulate plant stress responses and programmed cell death. *Bioessays: News and Reviews in Molecular, Cellular and Developmental Biology*, *28*(11), 1091–1101.

Geissler, N., Hussin, S., & Koyro, H. W. (2010). Elevated atmospheric CO_2 concentration enhances salinity tolerance in *Aster tripolium* L. *Planta*, *231*(3), 583–594.

Genard, H., Le Saos, J., Billard, J. P., Tremolieres, A., & Boucaud, J. (1991). Effect of salinity on lipid composition, glycine betaine content and photosynthetic activity in chloroplasts of *Suaeda maritima*. *Plant Physiology and Biochemistry*, *29*(5), 421–427.

Ghaffari, A., Gharechahi, J., Nakhoda, B., & Salekdeh, G. H. (2014). Physiology and proteome responses of two contrasting rice mutants and their wild type parent under salt stress conditions at the vegetative stage. *Journal of Plant Physiology*, *171*(1), 31–44.

Ghaffari, M. R., Ghabooli, M., Khatabi, B., Hajirezaei, M. R., Schweizer, P., & Salekdeh, G. H. (2016). Metabolic and transcriptional response of central metabolism affected by root endophytic fungus *Piriformospora indica* under salinity in barley. *Plant Molecular Biology*, *90* (6), 699–717.

Gharsallah, C., Fakhfakh, H., Grubb, D., & Gorsane, F. (2016). Effect of salt stress on ion concentration, proline content, antioxidant enzyme activities and gene expression in tomato cultivars. *AoB Plants*, *8*.

Ghassemi-Golezani, K., Taifeh-Noori, M., Oustan, S. H., Moghaddam, M., & Rahmani, S. S. (2011). Physiological performance of soybean cultivars under salinity stress. *Journal of Plant Physiology and Breeding*, *1*(1), 1–7.

Gil, R., Boscaiu, M., Lull, C., Bautista, I., Lidón, A., & Vicente, O. (2013). Are soluble carbohydrates ecologically relevant for salt tolerance in halophytes? *Functional Plant Biology: FPB*, *40*, 805–818.

Goharrizi, K. J., Baghizadeh, A., Kalantar, M., & Fatehi, F. (2019). Assessment of changes in some biochemical traits and proteomic profile of UCB-1 pistachio rootstock leaf under salinity stress. *Journal of Plant Growth Regulation*, *39*, 608–630.

Golldack, D., Lüking, I., & Yang, O. (2011). Plant tolerance to drought and salinity: Stress regulating transcription factors and their functional significance in the cellular transcriptional network. *Plant Cell Reports*, *30*(8), 1383–1391.

Gong, B., Zhang, C., Li, X., Wen, D., Wang, S., Shi, Q., & Wang, X. (2014). Identification of NaCl and $NaHCO_3$ stress responsive proteins in tomato roots using iTRAQ-based analysis. *Biochemical and Biophysical Research Communications*, *446*(1), 417–422.

Gong, W., Xu, F., Sun, J., Peng, Z., He, S., Pan, Z., & Du, X. (2017). iTRAQ-based comparative proteomic analysis of seedling leaves of two upland cotton genotypes differing in salt tolerance. *Frontiers in Plant Science*, *8*, 2113.

Gorham, J., Jones, R. W., & McDonnell, E. (1985). *Some mechanisms of salt tolerance in crop plants. Biosalinity in action: Bioproduction with saline water* (pp. 15–40). Dordrecht: Springer.

Goswami, L., Sengupta, S., Mukherjee, S., Ray, S., Mukherjee, R., & Majumder, A. L. (2014). Targeted expression of L-myo-inositol 1-phosphate synthase from *Porteresia coarctata* (Roxb.) Tateoka confers multiple stress tolerance in transgenic crop plants. *Journal of Plant Biochemistry and Biotechnology*, *23*(3), 316–330.

Gouiaa, S., Khoudi, H., Leidi, E. O., Pardo, J. M., & Masmoudi, K. (2012). Expression of wheat Na + /H + antiporter TNHXS1 and H + -pyrophosphatase TVP1 genes in tobacco from a bicistronic transcriptional unit improves salt tolerance. *Plant Molecular Biology*, *79*(1-2), 137–155.

Goyal, E., Amit, S. K., Singh, R. S., Mahato, A. K., Chand, S., & Kanika, K. (2016). Transcriptome profiling of the salt-stressresponse in *Triticum aestivum* cv. Kharchia Local. *Scientific Reports*, *6*, 27752.

Guan, C., Huang, Y. H., Cui, X., Liu, S. J., Zhou, Y. Z., & Zhang, Y. W. (2018). Overexpression of gene encoding the key enzyme involved in proline-biosynthesis (PuP5CS) to improve salt tolerance in switchgrass (*Panicum virgatum* L.). *Plant Cell Reports*, *37*(8), 1187–1199.

Gul, M., Wakeel, A., Steffens, D., & Lindberg, S. (2019). Potassium-induced decrease in cytosolic Na + alleviates deleterious effects of salt stress on wheat (*Triticum aestivum* L.). *Plant Biology*, *21*(5), 825–831.

Guo, G., Ge, P., Ma, C., Li, X., Lv, D., Wang, S., & Yan, Y. (2012). Comparative proteomic analysis of salt response proteins in seedling roots of two wheat varieties. *Journal of Proteomics*, *75*(6), 1867–1885.

Gupta, K. J., Stoimenova, M., & Kaiser, W. M. (2005). In higher plants, only root mitochondria, but not leaf mitochondria reduce nitrite to NO, in vitro and in situ. *Journal of Experimental Botany*, *56*(420), 2601–2609.

Gururani, M. A., Venkatesh, J., & Tran, L. S. P. (2015). Regulation of photosynthesis during abiotic stress-induced photoinhibition. *Molecular Plant*, *8*(9), 1304–1320.

Hachicha, M. (2007). The salty soils and their development in Tunisia. *Science and Planetary Changes/Drought*, *18*(1), 45–50.

Halo, B. A., Khan, A. L., Waqas, M., Al-Harrasi, A., Hussain, J., Ali, L., … Lee, I. J. (2015). Endophytic bacteria (*Sphingomonas* sp. LK11) and gibberellin can improve *Solanum lycopersicum* growth and oxidative stress under salinity. *Journal of Plant Interactions*, *10*, 117–125.

Hamamoto, S., Horie, T., Hauser, F., Deinlein, U., Schroeder, J. I., & Uozumi, N. (2015). HKT transporters mediate salt stress resistance in plants: From structure and function to the field. *Current Opinion in Biotechnology*, *32*, 113–120.

Han, B., Xu, W., Ahmed, N., Yu, A., Wang, Z., & Liu, A. (2020). Changes and associations of genomic transcriptions and histone methylations with salt stress in castor bean. *Plant and Cell Physiology*, *61*, 1120–1133.

Han, D., Du, M., Zhou, Z., Wang, S., Li, T., Han, J., & Yang, G. (2020). Overexpression of a *Malus baccata* NAC transcription factor gene MbNAC25 increases cold and salinity tolerance in *Arabidopsis*. *International Journal of Molecular Sciences*, *21*(4), 1198.

Han, X., Tang, S., An, Y., Zheng, D. C., Xia, X. L., & Yin, W. L. (2013). Overexpression of the poplar NF-YB7 transcription factor confers drought tolerance and improves water-use efficiency in *Arabidopsis*. *Journal of Experimental Botany*, *64*(14), 4589–4601.

Hanin, M., Ebel, C., Ngom, M., Laplaze, L., & Masmoudi, K. (2016). New insights on plant salt tolerance mechanisms and their potential use for breeding. *Frontiers in Plant Science*, *7*, 1787.

Hasanuzzaman, M., Alam, M., Rahman, A., Hasanuzzaman, M., Nahar, K., & Fujita, M. (2014). Exogenous proline and glycine betaine mediated upregulation of antioxidant defense and glyoxalase systems provides better protection against salt-induced oxidative stress in two rice (*Oryza sativa* L.) varieties. *BioMed Research International*, 2014.

Hasanuzzaman, M., Bhuyan, M. H. M., Anee, T. I., Parvin, K., Nahar, K., Mahmud, J. A., & Fujita, M. (2019). Regulation of ascorbate-glutathione pathway in mitigating oxidative damage in plants under abiotic stress. *Antioxidants*, *8*(9), 384.

Hasanuzzaman, M., & Fujita, M. (2011). Exogenous silicon treatment alleviates salinity-induced damage in *Brassica napus* L. seedlings by upregulating the antioxidant defense and methylglyoxal detoxification system.

Hasanuzzaman, M., & Fujita, M. (2012). Selenium and plants' health: The physiological role of selenium. Selenium: Sources, functions and health effects. (pp. 101–122).

Hasanuzzaman, M., Nahar, K., & Fujita, M. (2013). *Plant response to salt stress and role of exogenous protectants to mitigate salt-induced damages. Ecophysiology and responses of plants under salt stress* (pp. 25–87). New York, NY: Springer.

He, Y., Fu, J., Yu, C., Wang, X., Jiang, Q., Hong, J., & Xu, L. (2015). Increasing cyclic electron flow is related to Na^+ sequestration into vacuoles for salt tolerance in soybean. *Journal of Experimental Botany*, *66*(21), 6877–6889.

He, Y., Li, W., Lv, J., Jia, Y., Wang, M., & Xia, G. (2012). Ectopic expression of a wheat MYB transcription factor gene, TaMYB73, improves salinity stress tolerance in *Arabidopsis thaliana*. *Journal of Experimental Botany*, *63*(3), 1511–1522.

He, Y., Yu, C., Zhou, L., Chen, Y., Liu, A., Jin, J., & Jiang, D. (2014). Rubisco decrease is involved in chloroplast protrusion and Rubisco-containing body formation in soybean (*Glycine max.*) under salt stress. *Plant Physiology and Biochemistry*, *74*, 118–124.

Heidari, M., & Jamshidi, P. (2011). Effects of salinity and potassium application on antioxidant enzyme activities and physiological parameters in pearl millet. *Agricultural Sciences in China*, *10*(2), 228–237.

Henkes, S., Sonnewald, U., Badur, R., Flachmann, R., & Stitt, M. (2001). A small decrease of plastid transketolase activity in antisense tobacco transformants has dramatic effects on photosynthesis and phenylpropanoid metabolism. *The Plant Cell*, *13*(3), 535–551.

Hernández, I., Chacón, O., Rodriguez, R., Portieles, R., López, Y., Pujol, M., & Borrás-Hidalgo, O. (2009). Black shank resistant tobacco by silencing of glutathione S-transferase. *Biochemical and Biophysical Research Communications*, *387*(2), 300–304.

Hniličková, H., Hnilička, F., Martinkova, J., & Kraus, K. (2017). Effects of salt stress on water status, photosynthesis and chlorophyll fluorescence of rocket. *Plant, Soil and Environment*, *63*(8), 362–367.

Hoang, T. M. L., Tran, T. N., Nguyen, T. K. T., Williams, B., Wurm, P., Bellairs, S., & Mundree, S. (2016). Improvement of salinity stress tolerance in rice: Challenges and opportunities. *Agronomy*, *6*(4), 54.

Holmes, G. D., Hall, N. E., Gendall, A. R., Boon, P. I., & James, E. A. (2016). Using transcriptomics to identify differential gene expression in response to salinity among Australian *Phragmites australis* clones. *Frontiers in Plant Science*, *7*, 432.

Holmström, K. O., Somersalo, S., Mandal, A., Palva, T. E., & Welin, B. (2000). Improved tolerance to salinity and low temperature in transgenic tobacco producing glycine betaine. *Journal of Experimental Botany*, *51*(343), 177–185.

Horie, T., Hauser, F., & Schroeder, J. I. (2009). HKT transporter-mediated salinity resistance mechanisms in *Arabidopsis* and monocot crop plants. *Trends in Plant Science*, *14*, 660–668.

Horie, T., Motoda, J., Kubo, M., Yang, H., Yoda, K., Horie, R., & Osumi, M. (2005). Enhanced salt tolerance mediated by AtHKT1 transporter-induced Na + unloading from xylem vessels to xylem parenchyma cells. *The Plant Journal*, *44*(6), 928–938.

Hossain, M. A., Cho, J. I., Han, M., Ahn, C. H., Jeon, J. S., An, G., & Park, P. B. (2010). The ABRE-binding bZIP transcription factor $OsABF_2$ is a positive regulator of abiotic stress and ABA signaling in rice. *Journal of Plant Physiology*, *167*(17), 1512–1520.

Hossain, M. S., ElSayed, A. I., Moore, M., & Dietz, K. J. (2017). Redox and reactive oxygen species network in acclimation for salinity tolerance in sugar beet. *Journal of Experimental Botany*, *68*(5), 1283–1298.

Hou, Q. Z., Sun, K., Zhang, H., Su, X., Fan, B. Q., & Feng, H. Q. (2018). The responses of photosystem II and intracellular ATP production of *Arabidopsis* leaves to salt stress are affected by extracellular ATP. *Journal of Plant Research*, *131*(2), 331–339.

Hu, H., You, J., Fang, Y., Zhu, X., Qi, Z., & Xiong, L. (2008). Characterization of transcription factor gene $SNAC_2$ conferring cold and salt tolerance in rice. *Plant Molecular Biology*, *67* (1-2), 169–181.

Hu, L., Xiang, L., Li, S., Zou, Z., & Hu, X. H. (2016). Beneficial role of spermidine in chlorophyll metabolism and D1 protein content in tomato seedlings under salinity–alkalinity stress. *Physiologia Plantarum*, *156*(4), 468–477.

Hu, Y., Chen, L., Wang, H., Zhang, L., Wang, F., & Yu, D. (2013). Arabidopsis transcription factor WRKY 8 functions antagonistically with its interacting partner VQ 9 to modulate salinity stress tolerance. *The Plant Journal*, *74*(5), 730–745.

Huang, X. S., Liu, J. H., & Chen, X. J. (2010). Overexpression of PtrABF gene, a bZIP transcription factor isolated from *Poncirus trifoliata*, enhances dehydration and drought tolerance in tobacco via scavenging ROS and modulating expression of stress-responsive genes. *BMC Plant Biology*, *10*(1), 230.

Huertas, R., Rubio, L., Cagnac, O., García-Sánchez, M. J., Alché, J. D. D., Venema, K., & Rodríguez-Rosales, M. P. (2013). The K + /H + antiporter $LeNHX_2$ increases salt tolerance by improving K + homeostasis in transgenic tomato. *Plant, Cell & Environment*, *36*(12), 2135–2149.

Huihui, Z., Xin, L., Yan-hui, C., Yue, W., Ma-bo, L., Rong-yi, Y., & Guang-yu, S. (2020). A study on the effects of salinity and pH on PSII function in mulberry seedling leaves under saline–alkali mixed stress. *Trees*, 1–14.

Hura, T., Szewczyk-Taranek, B., Hura, K., Nowak, K., & Pawłowska, B. (2017). Physiological responses of *Rosa rubiginosa* to saline environment. *Water, Air, and Soil Pollution*, *228*, 81.

Husen, A., Iqbal, M., Sohrab, S. S., & Ansari, M. K. A. (2018). Salicylic acid alleviates salinity-caused damage to foliar functions, plant growth and antioxidant system in Ethiopian mustard (*Brassica carinata* A. Br.). *Agriculture & Food Security*, *7*(1), 44.

Hussin, S., Geissler, N., & Koyro, H. W. (2013). Effect of NaCl salinity on *Atriplex nummularia* (L.) with special emphasis on carbon and nitrogen metabolism. *Acta Physiologiae Plantarum / Polish Academy of Sciences, Committee of Plant Physiology Genetics and Breeding*, *35*, 1025–1038.

Islam, M. O., Kato, H., Shima, S., Tezuka, D., Matsui, H., & Imai, R. (2019). Functional identification of a rice trehalase gene involved in salt stress tolerance. *Gene*, *685*, 42–49.

Ismail, H., Maksimović, J. D., Maksimović, V., Shabala, L., Živanović, B. D., Tian, Y., & Shabala, S. (2016). Rutin, a flavonoid with antioxidant activity, improves plant salinity tolerance by regulating K + retention and Na + exclusion from leaf mesophyll in quinoa and broad beans. *Functional Plant Biology*, *43*(1), 75–86.

Jacoby, R. P., Millar, A. H., & Taylor, N. L. (2010). Wheat mitochondrial proteomes provide new links between antioxidant defense and plant salinity tolerance. *Journal of Proteome Research*, *9*(12), 6595–6604.

James, R. A., Blake, C., Byrt, C. S., & Munns, R. (2011). Major genes for Na + exclusion, Nax1 and Nax2 (wheat HKT1; 4 and HKT1; 5), decrease Na + accumulation in bread wheat leaves under saline and waterlogged conditions. *Journal of Experimental Botany*, *62* (8), 2939–2947.

Jamil, A., Riaz, S., Ashraf, M., & Foolad, M. R. (2011). Gene expression profiling of plants under salt stress. *Critical Reviews in Plant Sciences*, *30*, 435–458.

Javed, M., Zafar, Z. U., & Ashraf, M. (2019). Leaf proteome analysis signified that photosynthesis and antioxidants are key indicators of salinity tolerance in canola (*Brassica napus* L.). *Pakistan Journal of Botany*, *51*(6), 1955–1968.

Jha, U. C., Bohra, A., Jha, R., & Parida, S. K. (2019). Salinity stress response and 'omics' approaches for improving salinity stress tolerance in major grain legumes. *Plant Cell Reports*, *38*(3), 255–277.

Ji, F. S., Tang, L., Li, Y. Y., Wang, W. C., Yang, Z., Li, X. G., & Zeng, C. (2019). Differential proteomic analysis reveals the mechanism of *Musa paradisiaca* responding to salt stress. *Molecular Biology Reports*, *46*(1), 1057–1068.

Ji, H., Pardo, J. M., Batelli, G., Van Oosten, M. J., Bressan, R. A., & Li, X. (2013). The salt overly sensitive (SOS) pathway: Established and emerging roles. *Molecular Plant*, *6*(2), 275–286.

Ji, W., Cong, R., Li, S., Li, R., Qin, Z., Li, Y., & Li, J. (2016). Comparative proteomic analysis of soybean leaves and roots by iTRAQ provides insights into response mechanisms to short-term salt stress. *Frontiers in Plant Science*, *7*, 573.

Jiang, C., Zu, C., Lu, D., Zheng, Q., Shen, J., Wang, H., & Li, D. (2017). Effect of exogenous selenium supply on photosynthesis, Na^+ accumulation and antioxidative capacity of maize (*Zea mays* L.) under salinity stress. *Scientific Reports*, *7*, 42039.

Jiang, Q., Li, X., Niu, F., Sun, X., Hu, Z., & Zhang, H. (2017). iTRAQ-based quantitative proteomic analysis of wheat roots in response to salt stress. *Proteomics*, *17*(8), 1600265.

Jiang, Y., & Deyholos, M. K. (2009). Functional characterization of *Arabidopsis* NaCl-inducible WRKY25 and WRKY33 transcription factors in abiotic stresses. *Plant Molecular Biology*, *69*(1-2), 91–105.

Jiang, Y., Yang, B., Harris, N. S., & Deyholos, M. K. (2007). Comparative proteomic analysis of NaCl stress-responsive proteins in *Arabidopsis* roots. *Journal of Experimental Botany*, *58* (13), 3591–3607.

Jithesh, M. N., Prashanth, S. R., Sivaprakash, K. R., & Parida, A. K. (2006). Antioxidative response mechanisms in halophytes: Their role in stress defence. *Journal of Genetics*, *85*(3), 237.

Jithesh, M. N., Shukla, P. S., Kant, P., Joshi, J., Critchley, A. T., & Prithiviraj, B. (2018). Physiological and transcriptomics analyses reveal that *Ascophyllum nodosum* extracts induce salinity tolerance in *Arabidopsis* by regulating the expression of stress responsive genes. *Journal of Plant Growth Regulation*, *38*(2), 463–478.

Jouyban, Z. (2012). The effects of salt stress on plant growth. *Technical Journal of Engineering and Applied Sciences*, *2*(1), 7–10.

Jyothi-Prakash, P. A., Mohanty, B., Wijaya, E., Lim, T. M., Lin, Q., Loh, C. S., & Kumar, P. P. (2014). Identification of salt gland-associated genes and characterization of a dehydrin from the salt secretor mangrove *Avicennia officinalis*. *BMC Plant Biology*, *14*(1), 291.

Kaashyap, M., Ford, R., Kudapa, H., Jain, M., Edwards, D., Varshney, R., & Mantri, N. (2018). Differential regulation of genes involved in root morphogenesis and cell wall modification is associated with salinity tolerance in chickpea. *Scientific Reports*, *8*(1), 1–19.

Kader, M. A., & Lindberg, S. (2005). Uptake of sodium in protoplasts of salt-sensitive and salt-tolerant cultivars of rice, *Oryza sativa* L. determined by the fluorescent dye SBFI. *Journal of Experimental Botany*, *56*, 3149–3158.

Kader, M. A., & Lindberg, S. (2010). Cytosolic calcium and pH signaling in plants under salinity stress. *Plant Signaling & Behavior*, *5*, 233–238.

Kamran, M., Parveen, A., Ahmar, S., Malik, Z., Hussain, S., Chattha, M. S., & Chen, J. T. (2020). An overview of hazardous impacts of soil salinity in crops, tolerance mechanisms, and amelioration through selenium supplementation. *International Journal of Molecular Sciences*, *21*(1), 148.

Kan, X., Ren, J., Chen, T., Cui, M., Li, C., Zhou, R., & Yin, Z. (2017). Effects of salinity on photosynthesis in maize probed by prompt fluorescence, delayed fluorescence and P700 signals. *Environmental and Experimental Botany*, *140*, 56–64.

Kannan, P. R., Deepa, S., Kanth, S. V., & Rengasamy, R. (2013). Growth, osmolyte concentration and antioxidant enzymes in the leaves of *Sesuvium portulacastrum* L. under salinity stress. *Applied Biochemistry and Biotechnology*, *171*(8), 1925–1932.

Kaouthar, F., Ameny, F. K., Yosra, K., Walid, S., Ali, G., & Faical, B. (2016). Responses of transgenic *Arabidopsis* plants and recombinant yeast cells expressing a novel durum wheat manganese superoxide dismutase TdMnSOD to various abiotic stresses. *Journal of Plant Physiology*, *198*, 56–68.

Karimi, R., Ghabooli, M., Rahimi, J., & Amerian, M. (2020). Effects of foliar selenium application on some physiological and phytochemical parameters of *Vitis vinifera* L. cv. Sultana under salt stress. *Journal of Plant Nutrition*, *43*, 2226–2242.

Kaur, G., & Asthir, B. J. B. P. (2015). Proline: A key player in plant abiotic stress tolerance. *Biologia Plantarum*, *59*(4), 609–619.

Kavi Kishor, P. B., Hong, Z., Miao, G. H., Hu, C. A., & Verma, D. P. S. (1995). Overexpression of delta 1-pyrroline-5-carboxylate synthetase increases proline production and confers osmotolerance in transgenic plants. *Plant Physiology (Lancaster, PA)*, *108*, 1387–1394.

Kaya, C., Akram, N. A., Ashraf, M., & Sonmez, O. (2018). Exogenous application of humic acid mitigates salinity stress in maize (*Zea mays* L.) plants by improving some key physico-biochemical attributes. *Cereal Research Communications*, *46*(1), 67–78.

Keunen, E. L. S., Peshev, D., Vangronsveld, J., Van Den Ende, W. I. M., & Cuypers, A. N. N. (2013). Plant sugars are crucial players in the oxidative challenge during abiotic stress: Extending the traditional concept. *Plant, Cell & Environment*, *36*(7), 1242–1255.

Khan, M. S. (2011). Role of sodium and hydrogen (Na + /H +) antiporters in salt tolerance of plants: Present and future challenges. *African Journal of Biotechnology*, *10*(63), 13693–13704.

Khataar, M., Mohammadi, M. H., & Shabani, F. (2018). Soil salinity and matric potential interaction on water use, water use efficiency and yield response factor of bean and wheat. *Scientific Reports*, *8*(1), 1–13.

Kiani-Pouya, A., Roessner, U., Jayasinghe, N. S., Lutz, A., Rupasinghe, T., Bazihizina, N., & Shabala, S. (2017). Epidermal bladder cells confer salinity stress tolerance in the halophyte quinoa and *Atriplex* species. *Plant, Cell & Environment*, *40*(9), 1900–1915.

Kim, D. W., Rakwal, R., Agrawal, G. K., Jung, Y. H., Shibato, J., Jwa, N. S., & Usui, K. (2005). A hydroponic rice seedling culture model system for investigating proteome of salt stress in rice leaf. *Electrophoresis*, *26*(23), 4521–4539.

Kim, S. E., Lee, C. J., Ji, C. Y., Kim, H. S., Park, S. U., Lim, Y. H., & Guo, X. (2019). Transgenic sweetpotato plants overexpressing tocopherol cyclase display enhanced α-tocopherol content and abiotic stress tolerance. *Plant Physiology and Biochemistry*, *144*, 436–444.

Kim, S. H., Ahn, Y. O., Ahn, M. J., Lee, H. S., & Kwak, S. S. (2012). Down-regulation of β-carotene hydroxylase increases β-carotene and total carotenoids enhancing salt stress tolerance in transgenic cultured cells of sweetpotato. *Phytochemistry*, *74*, 69–78.

Kobayashi, H., Masaoka, Y., Takahashi, Y., Ide, Y., & Sato, S. (2007). Ability of salt glands in Rhodes grass (*Chloris gayana* Kunth) to secrete Na + and K + . *Soil Science and Plant Nutrition*, *53*(6), 764–771.

Konda, A. K., Farmer, R., Soren, K. R., PS, S., & Setti, A. (2018). Structural modeling and molecular dynamics of a multi-stress responsive WRKY TF-DNA complex towards elucidating its role in stress signaling mechanisms in chickpea. *Journal of Biomolecular Structure and Dynamics*, *36*(9), 2279–2291.

König, J., Baier, M., Horling, F., Kahmann, U., Harris, G., Schürmann, P., & Dietz, K. J. (2002). The plant-specific function of 2-Cys peroxiredoxin-mediated detoxification of peroxides in the redox-hierarchy of photosynthetic electron flux. *Proceedings of the National Academy of Sciences*, *99*(8), 5738–5743.

Kosová, K., Vítámvás, P., Urban, M. O., & Prášil, I. T. (2013). Plant proteome responses to salinity stress–comparison of glycophytes and halophytes. *Functional Plant Biology*, *40*(9), 775–786.

Kosová, K., Vítámvás, P., Urban, M. O., Prášil, I. T., & Renaut, J. (2018). Plant abiotic stress proteomics: The major factors determining alterations in cellular proteome. *Frontiers in Plant Science*, *9*, 122.

Kotagiri, D., & Kolluru, V. C. (2017). Effect of salinity stress on the morphology and physiology of five different *Coleus* species. *Biomedical and Pharmacology Journal*, *10*(4), 1639–1649.

Krishnamurthy, P., Qingsong, L., & Kumar, P. P. (2018). *Proteomics perspectives in post-genomic era for producing salinity stress-tolerant crops*, . *Salinity responses and tolerance in plants* (Volume 2, pp. 239–266). Cham: Springer.

Krishnamurthy, P., Tan, X. F., Lim, T. K., Lim, T. M., Kumar, P. P., Loh, C. S., & Lin, Q. (2014). Proteomic analysis of plasma membrane and tonoplast from the leaves of mangrove plant *Avicennia officinalis*. *Proteomics*, *14*(21-22), 2545–2557.

Kukreja, S., Nandwal, A. S., Kumar, N., Sharma, S. K., Unvi, V., & Sharma, P. K. (2005). Plant water status, H_2O_2 scavenging enzymes, ethylene evolution and membrane integrity of *Cicer arietinum* roots as affected by salinity. *Biologia Plantarum*, *49*(2), 305–308.

Kumar, J., Singh, S., Singh, M., Srivastava, P. K., Mishra, R. K., Singh, V. P., & Prasad, S. M. (2017). Transcriptional regulation of salinity stress in plants: A short review. *Plant Gene*, *11*, 160–169.

Kumar, K., & Mosa, K.A. (2015). Ion transporters: a decisive component of salt stress tolerance in plants. In *Managing salt tolerance in plants: Molecular and genomic perspectives*. (pp. 373–390).

Kumar, K., Srivastava, V., Purayannur, S., Kaladhar, V. C., Cheruvu, P. J., & Verma, P. K. (2016). WRKY domain-encoding genes of a crop legume chickpea (*Cicer arietinum*): comparative analysis with *Medicago truncatula* WRKY family and characterization of group-III gene (s). *DNA Research*, *23*(3), 225–239.

Kumar, R. A. J. E. E. V., Singh, M. P., & Kumar, S. (2012). Effect of salinity on germination, growth, yield and yield attributes of wheat. *International Journal of Scientific & Technology Research*, *1*(6), 19–23.

Kumar, S., Dhingra, A., & Daniell, H. (2004). Plastid-expressed betaine aldehyde dehydrogenase gene in carrot cultured cells, roots, and leaves confers enhanced salt tolerance. *Plant Physiology*, *136*(1), 2843–2854.

Kumar, S., Kalita, A., Srivastava, R., & Sahoo, L. (2017). Co-expression of *Arabidopsis* NHX1 and bar improves the tolerance to salinity, oxidative stress, and herbicide in transgenic mungbean. *Frontiers in Plant Science*, *8*, 1896.

Kumari, A., Das, P., Parida, A. K., & Agarwal, P. K. (2015). Proteomics, metabolomics, and ionomics perspectives of salinity tolerance in halophytes. *Frontiers in Plant Science*, *6*, 537.

Kumari, A., & Parida, A. K. (2018). Metabolomics and network analysis reveal the potential metabolites and biological pathways involved in salinity tolerance of the halophyte *Salvadora persica*. *Environmental and Experimental Botany*, *148*, 85–99.

Kumari, S., Nee Sabharwal, V. P., Kushwaha, H. R., Sopory, S. K., Singla-Pareek, S. L., & Pareek, A. (2009). Transcriptome map for seedling stage specific salinity stress response indicates a specific set of genes as candidate for saline tolerance in *Oryza sativa* L. *Functional & Integrative Genomics*, *9*(1), 109.

Lee, S. H., Ahsan, N., Lee, K. W., Kim, D. H., Lee, D. G., Kwak, S. S., & Lee, B. H. (2007). Simultaneous overexpression of both CuZn superoxide dismutase and ascorbate peroxidase in transgenic tall fescue plants confers increased tolerance to a wide range of abiotic stresses. *Journal of Plant Physiology*, *164*(12), 1626–1638.

Li, D., Zhang, Y., Hu, X., Shen, X., Ma, L., Su, Z., & Dong, J. (2011). Transcriptional profiling of *Medicago truncatula* under salt stress identified a novel CBF transcription factor MtCBF4 that plays an important role in abiotic stress responses. *BMC Plant Biology*, *11*(1), 109.

Li, H., Chang, J., Chen, H., Wang, Z., Gu, X., Wei, C., & Zhang, X. (2017). Exogenous melatonin confers salt stress tolerance to watermelon by improving photosynthesis and redox homeostasis. *Frontiers in Plant Science*, *8*, 295.

Li, H. W., Zang, B. S., Deng, X. W., & Wang, X. P. (2011). Overexpression of the trehalose-6-phosphate synthase gene OsTPS1 enhances abiotic stress tolerance in rice. *Planta*, *234*(5), 1007–1018.

Li, W., Zhao, F. A., Fang, W., Xie, D., Hou, J., Yang, X., & Lv, S. (2015). Identification of early salt stress responsive proteins in seedling roots of upland cotton (*Gossypium hirsutum* L.) employing iTRAQ-based proteomic technique. *Frontiers in Plant Science*, *6*, 732.

Li, X. J., Yang, M. F., Chen, H., Qu, L. Q., Chen, F., & Shen, S. H. (2010). Abscisic acid pretreatment enhances salt tolerance of rice seedlings: Proteomic evidence. *Biochimica et Biophysica Acta (BBA)-Proteins and Proteomics*, *1804*(4), 929–940.

Li, Y., Zhang, T., Zhang, Z., & He, K. (2019). The physiological and biochemical photosynthetic properties of *Lycium ruthenicum* Murr in response to salinity and drought. *Scientia Horticulturae*, *256*, 108530.

Liang, W., Ma, X., Wan, P., & Liu, L. (2018). Plant salt-tolerance mechanism: A review. *Biochemical and Biophysical Research Communications*, *495*(1), 286–291.

Lindemose, S., O'Shea, C., Jensen, M. K., & Skriver, K. (2013). Structure, function and networks of transcription factors involved in abiotic stress responses. *International Journal of Molecular Sciences*, *14*(3), 5842–5878.

Liu, A., Xiao, Z., Li, M. W., Wong, F. L., Yung, W. S., Ku, Y. S., & Chan, T. F. (2019). Transcriptomic reprogramming in soybean seedlings under salt stress. *Plant, Cell & Environment*, *42*(1), 98–114.

Liu, C., Mao, B., Ou, S., Wang, W., Liu, L., Wu, Y., & Wang, X. (2014). OsbZIP71, a bZIP transcription factor, confers salinity and drought tolerance in rice. *Plant Molecular Biology*, *84*(1-2), 19–36.

Liu, C. W., Chang, T. S., Hsu, Y. K., Wang, A. Z., Yen, H. C., Wu, Y. P., & Lai, C. C. (2014). Comparative proteomic analysis of early salt stress responsive proteins in roots and leaves of rice. *Proteomics*, *14*(15), 1759–1775.

Liu, J., Ishitani, M., Halfter, U., Kim, C. S., & Zhu, J. K. (2000). The Arabidopsis thaliana SOS2 gene encodes a protein kinase that is required for salt tolerance. *Proceedings of the National Academy of Sciences*, *97*(7), 3730–3734.

Liu, J. X., Feng, K., Duan, A. Q., Li, H., Yang, Q. Q., Xu, Z. S., & Xiong, A. S. (2019). Isolation, purification and characterization of an ascorbate peroxidase from celery and overexpression of the AgAPX1 gene enhanced ascorbate content and drought tolerance in *Arabidopsis*. *BMC Plant Biology*, *19*(1), 1–13.

Liu, J. X., Srivastava, R., & Howell, S. H. (2008). Stress-induced expression of an activated form of AtbZIP17 provides protection from salt stress in *Arabidopsis*. *Plant, Cell & Environment*, *31*(12), 1735–1743.

Liu, L., Xia, W., Li, H., Zeng, H., Wei, B., Han, S., & Yin, C. (2018). Salinity inhibits rice seed germination by reducing α-amylase activity via decreased bioactive gibberellin content. *Frontiers in Plant Science*, *9*, 275.

Liu, Y., Sun, J., & Wu, Y. (2016). *Arabidopsis* ATAF1 enhances the tolerance to salt stress and ABA in transgenic rice. *Journal of Plant Research*, *129*(5), 955–962.

Llanes, A., Arbona, V., Gómez-Cadenas, A., & Luna, V. (2016). Metabolomic profiling of the halophyte *Prosopis strombulifera* shows sodium salt-specific response. *Plant Physiology and Biochemistry*, *108*, 145–157.

Lokhande, V. H., Nikam, T. D., & Penna, S. (2010). Biochemical, physiological and growth changes in response to salinity in callus cultures of *Sesuvium portulacastrum* L. Plant Cell. *Tissue and Organ Culture (PCTOC)*, *102*(1), 17–25.

Lokhande, V. H., & Suprasanna, P. (2012). *Prospects of halophytes in understanding and managing abiotic stress tolerance. Environmental adaptations and stress tolerance of plants in the era of climate change* (pp. 29–56). New York, NY: Springer.

Long, L., Yang, W. W., Liao, P., Guo, Y. W., Kumar, A., & Gao, W. (2019). Transcriptome analysis reveals differentially expressed ERF transcription factors associated with salt response in cotton. *Plant Science*, *281*, 72–81.

Long, S. P., & Bernacchi, C. J. (2003). Gas exchange measurements, what can they tell us about the underlying limitations to photosynthesis? Procedures and sources of error. *Journal of Experimental Botany*, *54*(392), 2393–2401.

Lopez, C. M. L., Takahashi, H., & Yamazaki, S. (2002). Plant–water relations of kidney bean plants treated with NaCl and foliarly applied glycinebetaine. *Journal of Agronomy and Crop Science*, *188*(2), 73–80.

López-Gómez, M., Hidalgo-Castellanos, J., Muñoz-Sánchez, J. R., Marín-Peña, A. J., Lluch, C., & Herrera-Cervera, J. A. (2017). Polyamines contribute to salinity tolerance in the symbiosis *Medicago truncatula-Sinorhizobium meliloti* by preventing oxidative damage. *Plant Physiology and Biochemistry*, *116*, 9–17.

Løvdal, T., Olsen, K. M., Slimestad, R., Verheul, M., & Lillo, C. (2010). Synergetic effects of nitrogen depletion, temperature, and light on the content of phenolic compounds and gene expression in leaves of tomato. *Phytochemistry*, *71*(5-6), 605–613.

Lu, W., Guo, C., Li, X., Duan, W., Ma, C., Zhao, M., & Xiao, K. (2014). Overexpression of TaNHX3, a vacuolar Na + /H + antiporter gene in wheat, enhances salt stress tolerance in tobacco by improving related physiological processes. *Plant Physiology and Biochemistry*, *76*, 17–28.

Luo, M., Zhao, Y., Wang, Y., Shi, Z., Zhang, P., Zhang, Y., & Zhao, J. (2018). Comparative proteomics of contrasting maize genotypes provides insights into salt-stress tolerance mechanisms. *Journal of Proteome Research*, *17*(1), 141–153.

Ma, X., Liang, W., Gu, P., & Huang, Z. (2016). Salt tolerance function of the novel C_2H_2-type zinc finger protein TaZNF in wheat. *Plant Physiology and Biochemistry*, *106*, 129–140.

Maas, E. V., & Hoffman, G. J. (1977). Crop salt tolerance–current assessment. *Journal of the Irrigation and Drainage Division*, *103*(2), 115–134.

Mahajan, S., & Tuteja, N. (2005). Cold, salinity and drought stresses: An overview. *Archives of Biochemistry and Biophysics*, *444*(2), 139–158.

Majeed, A., & Muhammad, Z. (2019). *Salinity: A major agricultural problem—causes, impacts on crop productivity and management strategies. Plant abiotic stress tolerance* (pp. 83–99). Cham: Springer.

Manaa, A., Ben Ahmed, H., Valot, B., Bouchet, J. P., Aschi-Smiti, S., Causse, M., & Faurobert, M. (2011). Salt and genotype impact on plant physiology and root proteome variations in tomato. *Journal of Experimental Botany*, *62*(8), 2797–2813.

Marschner, H. (1995). *Mineral nutrition of higher plants* (p. 889) London: Academic Press.

Mäser, P., Eckelman, B., Vaidyanathan, R., Horie, T., Fairbairn, D. J., Kubo, M., & Robertson, W. (2002). Altered shoot/root Na + distribution and bifurcating salt sensitivity in *Arabidopsis* by genetic disruption of the Na + transporter AtHKT1. *FEBS Letters*, *531*(2), 157–161.

Mazars, C., Abdelly, C., Savouré, A. 1, Székely, G., Abrahám, E., Cséplo, A., ... Szabados, L. (2008). Duplicated P5CS genes of *Arabidopsis* play distinct roles in stress regulation and developmental control of proline biosynthesis. *The Plant Journal: For Cell and Molecular Biology*, *53*(1), 11–28.

Mishra, A., & Tanna, B. (2017). Halophytes: Potential resources for salt stress tolerance genes and promoters. *Frontiers in Plant Science*, *8*, 829.

Mittal, S., Kumari, N., & Sharma, V. (2012). Differential response of salt stress on *Brassica juncea*: Photosynthetic performance, pigment, proline, D1 and antioxidant enzymes. *Plant Physiology and Biochemistry*, *54*, 17–26.

Moghaddam, M., Farhadi, N., Panjtandoust, M., & Ghanati, F. (2019). Seed germination, antioxidant enzymes activity and proline content in medicinal plant *Tagetes minuta* under salinity stress. *Plant Biosystems – An International Journal Dealing With all Aspects of Plant Biology*, *154*, 835–842.

Moghbeli, E., Fathollahi, S., Salari, H., Ahmadi, G., Saliqehdar, F., Safari, A., & Grouh, M. S. H. (2012). Effects of salinity stress on growth and yield of *Aloe vera* L. *Journal of Medicinal Plants Research*, *6*(16), 3272–3277.

Mohamed, A. K. S., Qayyum, M. F., Abdel-Hadi, A. M., Rehman, R. A., Ali, S., & Rizwan, M. (2017). Interactive effect of salinity and silver nanoparticles on photosynthetic and biochemical parameters of wheat. *Archives of Agronomy and Soil Science*, *63*(12), 1736–1747.

Mohammad, M., Shibli, R., Ajlouni, M., & Nimri, L. (1998). Tomato root and shoot responses to salt stress under different levels of phosphorus nutrition. *Journal of Plant Nutrition*, *21* (8), 1667–1680.

Mokhtari, F., Rafiei, F., Shabani, L., & Shiran, B. (2017). Differential expression pattern of transcription factors across annual *Medicago* genotypes in response to salinity stress. *Biologia Plantarum*, *61*(2), 227–234.

Møller, I. S., Gilliham, M., Jha, D., Mayo, G. M., Roy, S. J., Coates, J. C., ... Tester, M. (2009). Shoot Na + exclusion and increased salinity tolerance engineered by cell type–specific alteration of Na^+ transport in *Arabidopsis*. *The Plant Cell*, *21*, 2163–2178.

Muneer, S., Park, Y. G., Manivannan, A., Soundararajan, P., & Jeong, B. R. (2014). Physiological and proteomic analysis in chloroplasts of *Solanum lycopersicum* L. under silicon efficiency and salinity stress. *International Journal of Molecular Sciences*, *15*(12), 21803–21824.

Munné-Bosch, S., & Alegre, L. (2002). The function of tocopherols and tocotrienols in plants. *Critical Reviews in Plant Sciences*, *21*(1), 31–57.

Munns, R., James, R. A., Xu, B., Athman, A., Conn, S. J., Jordans, C., ... Plett, D. (2012). Wheat grain yield on saline soils is improved by an ancestral Na + transporter gene. *Nature Biotechnology*, *30*, 360–364.

Munns, R., & Tester, M. (2008). Mechanisms of salinity tolerance. *Annual Review of Plant Biology*, *59*, 651–681.

Murillo-Amador, B., Jones, H. G., Kaya, C., Aguilar, R. L., García-Hernández, J. L., Troyo-Diéguez, E., ... Rueda-Puente, E. (2006). Effects of foliar application of calcium nitrate on growth and physiological attributes of cowpea (*Vigna unguiculata* L. Walp.) grown under salt stress. *Environmental and Experimental Botany*, *58*, 188–196.

Nadeem, M., Li, J., Yahya, M., Wang, M., Ali, A., Cheng, A., & Ma, C. (2019). Grain legumes and fear of salt stress: Focus on mechanisms and management strategies. *International Journal of Molecular Sciences*, *20*(4), 799.

Nagamiya, K., Motohashi, T., Nakao, K., Prodhan, S. H., Hattori, E., Hirose, S., & Komamine, A. (2007). Enhancement of salt tolerance in transgenic rice expressing an *Escherichia coli* catalase gene, katE. *Plant Biotechnology Reports*, *1*(1), 49–55.

Naidu, B. P. (2003). Production of betaine from Australian *Melaleuca* spp. for use in agriculture to reduce plant stress. *Australian Journal of Experimental Agriculture*, *43*(9), 1163–1170.

Najafi, F., Khavari, R. A., Rastgar, F., & Sticklen, M. (2006). Physiological changes in pea (*Pisum sativum* L. cv. Green Arrow) under NaCl salinity. *Pakistan Journal of Biological Sciences*, *9*(5), 974–978.

Najar, R., Aydi, S., Sassi-Aydi, S., Zarai, A., & Abdelly, C. (2019). Effect of salt stress on photosynthesis and chlorophyll fluorescence in *Medicago truncatula*. *Plant Biosystems-An International Journal Dealing With all Aspects of Plant Biology*, *153*(1), 88–97.

Nakashima, K., Tran, L. S. P., Van Nguyen, D., Fujita, M., Maruyama, K., Todaka, D., et al. (2007). Functional analysis of a NAC-type transcription factor OsNAC6 involved in abiotic and biotic stress-responsive gene expression in rice. *The Plant Journal*, *51*(4), 617–630.

Negrão, S., Courtois, B., Ahmadi, N., et al. (2011). Recent updates on salinity stress in rice: From physiological to molecular responses. *Critical Reviews in Plant Sciences*, *30*, 329–377.

Negrão, S., Schmöckel, S. M., & Tester, M. (2017). Evaluating physiological responses of plants to salinity stress. *Annals of Botany*, *119*(1), 1–11.

Nemoto, Y., & Sasakuma, T. (2000). Specific expression of glucose-6-phosphate dehydrogenase (G6PDH) gene by salt stress in wheat (*Triticum aestivum* L.). *Plant Science*, *158*(1-2), 53–60.

Netondo, G. W., Onyango, J. C., & Beck, E. (2004). Sorghum and salinity: II. Gas exchange and chlorophyll fluorescence of sorghum under salt stress. *Crop Science*, *44*(3), 806–811.

Ngara, R., & Ndimba, B. K. (2014). Understanding the complex nature of salinity and drought-stress response in cereals using proteomics technologies. *Proteomics*, *14*(4-5), 611–621.

Oh, D. H., Dassanayake, M., Haas, J. S., Kropornika, A., Wright, C., d'Urzo, M. P., & Inan, G. (2010). Genome structures and halophyte-specific gene expression of the extremophile *Thellungiella parvula* in comparison with *Thellungiella salsuginea* (*Thellungiella halophila*) and *Arabidopsis*. *Plant Physiology*, *154*(3), 1040–1052.

Olías, R., Eljakaoui, Z., Li, J., et al. (2009). The plasma membrane Na + /H + antiporter SOS1 is essential for salt tolerance in tomato and affects the partitioning of Na + between plant organs. *Plant Cell and Environment*, *32*, 904–916.

Ottow, E. A., Brinker, M., Teichmann, T., Fritz, E., Kaiser, W., Brosché, M., & Polle, A. (2005). *Populus euphratica* displays apoplastic sodium accumulation, osmotic adjustment by decreases in calcium and soluble carbohydrates, and develops leaf succulence under salt stress. *Plant Physiology*, *139*(4), 1762–1772.

Ozgur, R., Uzilday, B., Sekmen, A. H., & Turkan, I. (2013). Reactive oxygen species regulation and antioxidant defence in halophytes. *Functional Plant Biology*, *40*(9), 832–847.

Ozturk, L., Demir, Y., Unlukara, A., Karatas, I., Kurunc, A., & Duzdemir, O. (2012). Effects of long-term salt stress on antioxidant system, chlorophyll and proline contents in pea leaves. *Romanian Biotechnological Letters*, *17*(3), 7227–7236.

Pan, T., Liu, M., Kreslavski, V. D., Zharmukhamedov, S. K., Nie, C., Yu, M., & Shabala, S. (2020). Non-stomatal limitation of photosynthesis by soil salinity. *Critical Reviews in Environmental Science and Technology*. Available from https://doi.org/10.1080/10643389.2020.1735231.

Pan, Y. Q., Guo, H., Wang, S. M., Zhao, B., Zhang, J. L., Ma, Q., & Bao, A. K. (2016). The photosynthesis, Na + /K + homeostasis and osmotic adjustment of *Atriplex canescens* in response to salinity. *Frontiers in Plant Science*, *7*, 848.

Panda, A., Rangani, J., & Parida, A. K. (2020). Comprehensive proteomic analysis revealing multifaceted regulatory network of the xero-halophyte *Haloxylon salicornicum* involved in salt tolerance. *Journal of Biotechnology*, *324*, 143–161.

Pang, Q., Chen, S., Dai, S., Chen, Y., Wang, Y., & Yan, X. (2010). Comparative proteomics of salt tolerance in *Arabidopsis thaliana* and *Thellungiella halophila*. *Journal of Proteome Research*, *9*(5), 2584–2599.

Pardo, J. M., Cubero, B., Leidi, E. O., & Quintero, F. J. (2006). Alkali cation exchangers: Roles in cellular homeostasis and stress tolerance. *Journal of Experimental Botany*, *57*(5), 1181–1199.

Parida, A. K., & Das, A. B. (2005). Salt tolerance and salinity effects on plants: A review. *Ecotoxicology and Environmental Safety*, *60*(3), 324–349.

Parihar, P., Singh, S., Singh, R., Singh, V. P., & Prasad, S. M. (2015). Effect of salinity stress on plants and its tolerance strategies: A review. *Environmental Science and Pollution Research*, *22*(6), 4056–4075.

Parker, R., Flowers, T. J., Moore, A. L., & Harpham, N. V. (2006). An accurate and reproducible method for proteome profiling of the effects of salt stress in the rice leaf lamina. *Journal of Experimental Botany*, *57*(5), 1109–1118.

Parvin, K., Hasanuzzaman, M., Bhuyan, M. H. M. B., Nahar, K., Mohsin, S. M., & Fujita, M. (2019). Comparative physiological and biochemical changes in tomato (*Solanum lycopersicum* L.) under Salt Stress and Recovery: Role of antioxidant defense and glyoxalase systems. *Antioxidants*, *8*, 350.

Patanun, O., Ueda, M., Itouga, M., Kato, Y., Utsumi, Y., Matsui, A., & Narangajavana, J. (2017). The histone deacetylase inhibitor suberoylanilide hydroxamic acid alleviates salinity stress in cassava. *Frontiers in Plant Science*, *7*, 2039.

Paul, A., Dasgupta, P., Roy, D., & Chaudhuri, S. (2017). Comparative analysis of Histone modifications and DNA methylation at OsBZ8 locus under salinity stress in IR64 and Nonabokra rice varieties. *Plant Molecular Biology*, *95*(1-2), 63–88.

Pawłowicz, I., Waśkiewicz, A., Perlikowski, D., Rapacz, M., Ratajczak, D., & Kosmala, A. (2018). Remodeling of chloroplast proteome under salinity affects salt tolerance of *Festuca arundinacea*. *Photosynthesis Research*, *137*(3), 475–492.

Pena-Ahumada, A., Kahmann, U., Dietz, K. J., & Baier, M. (2006). Regulation of peroxiredoxin expression versus expression of Halliwell-Asada-Cycle enzymes during early seedling development of *Arabidopsis thaliana*. *Photosynthesis Research*, *89*(2-3), 99–112.

Peng, C., Chang, L., Yang, Q., Tong, Z., Wang, D., Tan, Y., & Zhang, Y. (2019). Comparative physiological and proteomic analyses of the chloroplasts in halophyte *Sesuvium portulacastrum* under differential salt conditions. *Journal of Plant Physiology*, *232*, 141–150.

Per, T. S., Khan, N. A., Reddy, P. S., Masood, A., Hasanuzzaman, M., Khan, M. I. R., & Anjum, N. A. (2017). Approaches in modulating proline metabolism in plants for salt and drought stress tolerance: Phytohormones, mineral nutrients and transgenics. *Plant Physiology and Biochemistry*, *115*, 126–140.

Persson, S., Rosenquist, M., Svensson, K., Galvão, R., Boss, W. F., & Sommarin, M. (2003). Phylogenetic analyses and expression studies reveal two distinct groups of calreticulin isoforms in higher plants. *Plant Physiology*, *133*(3), 1385–1396.

Pi, E., Qu, L., Hu, J., Huang, Y., Qiu, L., Lu, H., & Wang, H. (2016). Mechanisms of soybean roots' tolerances to salinity revealed by proteomic and phosphoproteomic comparisons between two cultivars. *Molecular & Cellular Proteomics*, *15*(1), 266–288.

Põldma, P., Moor, U., Tõnutare, T., Herodes, K., & Rebane, R. (2013). Selenium treatment under field conditions affects mineral nutrition, yield and antioxidant properties of bulb onion (*Allium cepa* L.). *Acta Scientiarum Polonorum Hortorum Cultus*, *12*(6), 167–181.

Postnikova, O. A., Shao, J., & Nemchinov, L. G. (2013). Analysis of the alfalfa root transcriptome in response to salinity stress. *Plant and Cell Physiology*, *54*(7), 1041–1055.

Prashanth, S. R., Sadhasivam, V., & Parida, A. (2008). Over expression of cytosolic copper/zinc superoxide dismutase from a mangrove plant *Avicennia marina* in indica rice var Pusa Basmati-1 confers abiotic stress tolerance. *Transgenic Research*, *17*(2), 281–291.

Prusty, M. R., Kim, S. R., Vinarao, R., et al. (2018). Newly identified wild rice accessions conferring high salt tolerance might use a tissue tolerance mechanism in leaf. *Frontiers in Plant Science*, *9*, 417.

Qu, L., Huang, Y., Zhu, C., Zeng, H., Shen, C., Liu, C., & Pi, E. (2016). Rhizobia-inoculation enhances the soybean's tolerance to salt stress. *Plant and Soil*, *400*(1-2), 209–222.

Rahman, A., Nahar, K., Hasanuzzaman, M., & Fujita, M. (2016). Calcium supplementation improves Na + /K + ratio, antioxidant defense and glyoxalase systems in salt-stressed rice seedlings. *Frontiers in Plant Science*, *7*, 609.

Rahneshan, Z., Nasibi, F., & Moghadam, A. A. (2018). Effects of salinity stress on some growth, physiological, biochemical parameters and nutrients in two pistachio (*Pistacia vera* L.) rootstocks. *Journal of Plant Interactions*, *13*(1), 73–82.

Rais, L., Masood, A., Inam, A., & Khan, N. (2013). Sulfur and nitrogen co-ordinately improve photosynthetic efficiency, growth and proline accumulation in two cultivars of mustard under salt stress. *Journal of Plant Biochemistry & Physiology*, *1*.

Rasouli, F., Kiani-Pouya, A., Li, L., Zhang, H., Chen, Z., Hedrich, R., & Shabala, S. (2020). Sugar beet (*Beta vulgaris*) guard cells responses to salinity stress: A proteomic analysis. *International Journal of Molecular Sciences*, *21*(7), 2331.

Rasoulnia, A., Bihamta, M. R., Peyghambari, S. A., Alizadeh, H., & Rahnama, A. (2011). Proteomic response of barley leaves to salinity. *Molecular Biology Reports*, *38*(8), 5055–5063.

Ren, Z. H., Gao, J. P., Li, L. G., Cai, X. L., Wei, H., Chao, D. Y., ... Lin, H. X. (2005). A rice quantitative trait locus for salt tolerance encodes a sodium transporter. *Nature Genetics*, *37*, 1141–1146.

Ricardo, G. I. L., Cristina, L. U. L. L., Boscaiu, M., Bautista, I., Lidón, A., & Vicente, O. (2011). Soluble carbohydrates as osmolytes in several halophytes from a Mediterranean salt marsh. *Notulae Botanicae Horti Agrobotanici Cluj-Napoca*, *39*(2), 09–17.

Richter, J. A., Erban, A., Kopka, J., & Zörb, C. (2015). Metabolic contribution to salt stress in two maize hybrids with contrasting resistance. *Plant Science*, *233*, 107–115.

Rodriguez-Rosales, M. P., Galvez, F. J., Huertas, R., Aranda, M. N., Baghour, M., Cagnac, O., & Venema, K. (2009). Plant NHX cation/proton antiporters. *Plant Signal Behavior*, *4*, 265–276.

Rolly, N. K., Imran, Q. M., Lee, I. J., & Yun, B. W. (2020). Salinity stress-mediated suppression of expression of salt overly sensitive signaling pathway genes suggests negative regulation by AtbZIP62 transcription factor in *Arabidopsis thaliana*. *International journal of Molecular Sciences*, *21*(5), 1726.

Romero-Trigueros, C., Nortes, P. A., Alarcón, J. J., Hunink, J. E., Parra, M., Contreras, S., & Nicolás, E. (2017). Effects of saline reclaimed waters and deficit irrigation on Citrus physiology assessed by UAV remote sensing. *Agricultural Water Management*, *183*, 60–69.

Roy, S. J., Negrão, S., & Tester, M. (2014). Salt resistant crop plants. *Current Opinion in Biotechnology*, *26*, 115–124.

Ruiz, D., Martínez, V., & Cerdá, A. (1997). Citrus response to salinity: Growth and nutrient uptake. *Tree Physiology*, *17*(3), 141–150.

Sadak, M., Abd Elhamid, E., & Mahmoud, M. (2017). Glutathione induced antioxidant protection against salinity stress in chickpea (*Cicer arietinum* L.) Plant. *Egyptian Journal of Botany*, *57*(2), 293–302.

Sako, K., Kim, J. M., Matsui, A., Nakamura, K., Tanaka, M., Kobayashi, M., & Yoshida, M. (2016). Ky-2, a histone deacetylase inhibitor, enhances high-salinity stress tolerance in *Arabidopsis thaliana*. *Plant and Cell Physiology*, *57*(4), 776–783.

Sakuma, Y., Maruyama, K., Osakabe, Y., Qin, F., Seki, M., Shinozaki, K., & Yamaguchi-Shinozaki, K. (2006). Functional analysis of an *Arabidopsis* transcription factor, DREB2A, involved in drought-responsive gene expression. *The Plant Cell*, *18*(5), 1292–1309.

Sandoval-Gil, J. M., Marín-Guirao, L., & Ruiz, J. M. (2012). Tolerance of Mediterranean seagrasses (*Posidonia oceanica and Cymodocea nodosa*) to hypersaline stress: Water relations and osmolyte concentrations. *Marine Biology*, *159*(5), 1129–1141.

Saraf, R., Saingar, S., Chaudhary, S., & Chakraborty, D. (2018). *Response of plants to salinity stress and the role of salicylic acid in modulating tolerance mechanisms: Physiological and proteomic approach. Biotic and abiotic stress tolerance in plants* (pp. 103–136). Singapore: Springer.

Saxena, S. C., Salvi, P., Kamble, N. U., Joshi, P. K., Majee, M., & Arora, S. (2020). Ectopic overexpression of cytosolic ascorbate peroxidase gene (Apx1) improves salinity stress tolerance in *Brassica juncea* by strengthening antioxidativedefense mechanism. *Acta Physiologiae Plantarum*, *42*(4), 1–14.

Semiz, G. D., Suarez, D. L., Ünlükara, A., & Yurtseven, E. (2014). Interactive effects of salinity and N on pepper (*Capsicum annuum* L.) yield, water use efficiency and root zone and drainage salinity. *Journal of Plant Nutrition*, *37*(4), 595–610.

Sengupta, S., & Majumder, A. L. (2010). *Porteresia coarctata* (Roxb.) Tateoka, a wild rice: A potential model for studying salt-stress biology in rice. *Plant, Cell & Environment*, *33*(4), 526–542.

Sengupta, S., Mukherjee, S., Goswami, L., Sangma, S., Mukherjee, A., Mukherjee, R., & Majumder, A. L. (2012). Manipulation of inositol metabolism for improved plant survival under stress: A "network engineering approach. *Journal of Plant Biochemistry and Biotechnology*, *21*(1), 15–23.

Shahbaz, M., & Ashraf, M. (2013). Improving salinity tolerance in cereals. *Critical Reviews in Plant Sciences*, *32*(4), 237–249.

Shahverdi, M. A., Omidi, H., & Tabatabaei, S. J. (2018). Plant growth and steviol glycosides as affected by foliar application of selenium, boron, and iron under NaCl stress in *Stevia rebaudiana* Bertoni. *Industrial Crops and Products*, *125*, 408–415.

Shannon, M. C., & Grieve, C. M. (1998). Tolerance of vegetable crops to salinity. *Scientia Horticulturae*, *78*(1-4), 5–38.

Shannon, M. C., Rhoades, J. D., Draper, J. H., Scardaci, S. C., & Spyres, M. D. (1998). Assessment of salt tolerance in rice cultivars in response to salinity problems in California. *Crop Science*, *38*(2), 394–398.

Shi, H., Ishitani, M., Kim, C., & Zhu, J.-K. (2000). The *Arabidopsis thaliana* salt tolerance gene SOS1 encodes a putative Na + /H + antiporter. *Proceedings of the National Academy of Sciences*, *97*(12), 6896–6901.

Shi, H., Quintero, F. J., Pardo, J. M., & Zhu, J.-K. (2002). The putative plasma membrane Na + /H + antiporter SOS1 controls long-distance Na + transport in plants. *The Plant Cell Online*, *14*(2), 465–477.

Shi, H., Wu, S. J., & Zhu, J. K. (2002). Role of SOS1 as a plasma membrane Na + /H + antiporter improves salt tolerance in *Arabidopsis*. *Nature Biotechnology*, *21*, 81–85.

Shinde, H., Dudhate, A., Anand, L., Tsugama, D., Gupta, S. K., Liu, S., & Takano, T. (2020). Small RNA sequencing reveals the role of pearl millet miRNAs and their targets in salinity stress responses. *South African Journal of Botany*, *132*, 395–402.

Shinde, H., Dudhate, A., Tsugama, D., Gupta, S. K., Liu, S., & Takano, T. (2019). Pearl millet stress-responsive NAC transcription factor PgNAC21 enhances salinity stress tolerance in *Arabidopsis*. *Plant Physiology and Biochemistry*, *135*, 546–553.

Shrivastava, P., & Kumar, R. (2015). Soil salinity: A serious environmental issue and plant growth promoting bacteria as one of the tools for its alleviation. *Saudi Journal of Biological Sciences*, *22*(2), 123–131.

Siddiqui, H., Hayat, S., & Bajguz, A. (2018). Regulation of photosynthesis by brassinosteroids in plants. *Acta Physiologiae Plantarum*, *40*(3), 59.

Silveira, J. A., & Carvalho, F. E. (2016). Proteomics, photosynthesis and salt resistance in crops: An integrative view. *Journal of Proteomics*, *143*, 24–35.

Silveira, J. A. G., Araújo, S. A. M., Lima, J. P. M. S., & Viégas, R. A. (2009). Roots and leaves display contrasting osmotic adjustment mechanisms in response to NaCl-salinity in *Atriplex nummularia*. *Environmental and Experimental Botany*, *66*(1), 1–8.

Singh, A., Kanwar, P., Yadav, A. K., Mishra, M., Jha, S. K., Baranwal, V., et al. (2014). Genome-wide expressional andfunctional analysis of calcium transport elements during abiotic stress and development in rice. *The FEBS Journal*, *281*(3), 894–915.

Singh, M., Kumar, J., Singh, S., Singh, V. P., & Prasad, S. M. (2015). Roles of osmoprotectants in improving salinity and drought tolerance in plants: A review. *Reviews in Environmental Science and Bio/Technology*, *14*(3), 407–426.

Sirohi, G., Pandey, B. K., Deveshwar, P., & Giri, J. (2016). Emerging trends in epigenetic regulation of nutrient deficiency response in plants. *Molecular Biotechnology*, *58*(3), 159–171.

Slama, I., Abdelly, C., Bouchereau, A., Flowers, T., & Savoure, A. (2015). Diversity, distribution and roles of osmoprotective compounds accumulated in halophytes under abiotic stress. *Annals of Botany*, *115*(3), 433–447.

Smirnoff, N. (2005). Ascorbate, tocopherol and carotenoids: Metabolism, pathway engineering and functions. Antioxidants and reactive oxygen species in plants, (pp. 53–86).

Smith, C. A., Melino, V. J., Sweetman, C., & Soole, K. L. (2009). Manipulation of alternative oxidase can influence salt tolerance in *Arabidopsis thaliana*. *Physiologia Plantarum*, *137*(4), 459–472.

Sobhanian, H., Aghaei, K., & Komatsu, S. (2011). Changes in the plant proteome resulting from salt stress: Toward the creation of salt-tolerant crops? *Journal of Proteomics*, *74*(8), 1323–1337.

Sobhanian, H., Motamed, N., Jazii, F. R., Nakamura, T., & Komatsu, S. (2010). Salt stress induced differential proteome and metabolome response in the shoots of *Aeluropus lagopoides* (Poaceae), a halophyte C4 plant. *Journal of Proteome Research*, *9*(6), 2882–2897.

Song, S. Y., Chen, Y., Chen, J., Dai, X. Y., & Zhang, W. H. (2011). Physiological mechanisms underlying $OsNAC_5$-dependent tolerance of rice plant to abiotic stress. *Planta*, *234*(2), 331–345.

Song, Y., Ji, D., Li, S., Wang, P., Li, Q., & Xiang, F. (2012). The dynamic changes of DNA methylation and histone modifications of salt responsive transcription factor genes in soybean. *PLoS One*, *7*(7), e41274.

Soori, N., Bakhshi, D., Rezaei Nejad, A., & Faizian, M. (2019). Effect of salinity stress on some physiological characteristics and photosynthetic parameters of several Iranian commercial pomegranate genotypes. *Journal of Plant Process and Function*, *8*(30), 155–170.

Stepien, P., & Johnson, G. N. (2009). Contrasting responses of photosynthesis to salt stress in the glycophyte Arabidopsis and the halophyte *Thellungiella*: Role of the plastid terminal oxidase as an alternative electron sink. *Plant Physiology*, *149*(2), 1154–1165.

Stracke, R., Werber, M., & Weisshaar, B. (2001). The R2R3-MYB gene family in *Arabidopsis thaliana*. *Current Opinion in Plant Biology*, *4*(5), 447–456.

Sun, X., Wang, Y., Xu, L., Li, C., Zhang, W., Luo, X., & Liu, L. (2017). Unraveling the root proteome changes and its relationship to molecular mechanism underlying salt stress response in radish (*Raphanus sativus* L.). *Frontiers in Plant Science*, *8*, 1192.

Sun, Z., Qi, X., Wang, Z., Li, P., Wu, C., Zhang, H., & Zhao, Y. (2013). Overexpression of $TsGOLS_2$, a galactinol synthase, in *Arabidopsis thaliana* enhances tolerance to high salinity and osmotic stresses. *Plant Physiology and Biochemistry*, *69*, 82–89.

Surówka, E., Potocka, I., Dziurka, M., Wróbel-Marek, J., Kurczyńska, E., Żur, I., & Miszalski, Z. (2020). Tocopherols mutual balance is a key player for maintaining *Arabidopsis thaliana* growth under salt stress. *Plant Physiology and Biochemistry*.

Szarka, A., Tomasskovics, B., & Bánhegyi, G. (2012). The ascorbate-glutathione-α-tocopherol triad in abiotic stress response. *International Journal of Molecular Sciences*, *13*(4), 4458–4483.

Taïbi, K., Taïbi, F., Abderrahim, L. A., Ennajah, A., Belkhodja, M., & Mulet, J. M. (2016). Effect of salt stress on growth, chlorophyll content, lipid peroxidation and antioxidant defence systems in *Phaseolus vulgaris* L. *South African Journal of Botany*, *105*, 306–312.

Tak, H., Negi, S., & Ganapathi, T. R. (2017). Banana NAC transcription factor MusaNAC042 is positively associated with drought and salinity tolerance. *Protoplasma*, *254*(2), 803–816.

Talaat, N. B., & Shawky, B. T. (2014). Protective effects of arbuscular mycorrhizal fungi on wheat (*Triticum aestivum* L.) plants exposed to salinity. *Environmental and Experimental Botany*, *98*, 20–31.

Tang, X., Mu, X., Shao, H., Wang, H., & Brestic, M. (2015). Global plant-responding mechanisms to salt stress: Physiological and molecular levels and implications in biotechnology. *Critical Reviews in Biotechnology*, *35*(4), 425–437.

Tang, Y., Bao, X., Zhi, Y., Wu, Q., Guo, Y., Yin, X., & Liu, K. (2019). Overexpression of a MYB family gene, OsMYB6, increases drought and salinity stress tolerance in transgenic rice. *Frontiers in Plant Science*, *10*, 168.

Tanou, G., Job, C., Rajjou, L., Arc, E., Belghazi, M., Diamantidis, G., & Job, D. (2009). Proteomics reveals the overlapping roles of hydrogen peroxide and nitric oxide in the acclimation of citrus plants to salinity. *The Plant Journal*, *60*(5), 795–804.

Tanveer, M., & Shabala, S. (2018). *Targeting redox regulatory mechanisms for salinity stress tolerance in crops, . Salinity responses and tolerance in plants* (Volume 1, pp. 213–234). Cham: Springer.

Tavakkoli, E., Rengasamy, P., & McDonald, G. K. (2010a). The response of barley to salinity stress differs between hydroponic and soil systems. *Functional Plant Biology*, *37*(7), 621–633.

Tavakkoli, E., Rengasamy, P., & McDonald, G. K. (2010b). High concentrations of Na + and Cl–ions in soil solution have simultaneous detrimental effects on growth of faba bean under salinity stress. *Journal of Experimental Botany*, *61*(15), 4449–4459.

Tester, M., & Davenport, R. (2003). Na^+ tolerance and Na^+ transport in higher plants. *Annals of Botany*, *91*(5), 503–527.

Tian, F., Hou, M., Qiu, Y., Zhang, T., & Yuan, Y. (2020). Salinity stress effects on transpiration and plant growth under different salinity soil levels based on thermal infrared remote (TIR) technique. *Geoderma*, *357*, 113961.

Tian, F., Wang, W., Liang, C., Wang, X., Wang, G., & Wang, W. (2017). Overaccumulation of glycine betaine makes the function of the thylakoid membrane better in wheat under salt stress. *The Crop Journal*, *5*(1), 73–82.

Tiwari, V., Patel, M. K., Chaturvedi, A. K., Mishra, A., & Jha, B. (2019). Cloning and functional characterization of the Na + /H + antiporter (NHX1) gene promoter from an extreme halophyte *Salicornia brachiata*. *Gene*, *683*, 233–242.

Torabian, S., Farhangi-Abriz, S., & Rathjen, J. (2018). Biochar and lignite affect H^+-ATPase and H^+-PPase activities in root tonoplast and nutrient contents of mung bean under salt stress. *Plant Physiology and Biochemistry*, *129*, 141–149.

Tsutsumi, K., Yamada, N., Cha-um, S., Tanaka, Y., & Takabe, T. (2015). Differential accumulation of glycinebetaine and choline monooxygenase in bladder hairs and lamina leaves of *Atriplex gmelini* under high salinity. *Journal of Plant Physiology*, *176*, 101–107.

Tuna, A. L., Kaya, C., Altunlu, H., & Ashraf, M. (2013). Mitigation effects of non-enzymatic antioxidants in maize (*'Zea mays'* L.) plants under salinity stress. *Australian Journal of Crop Science*, *7*(8), 1181.

Tuna, A. L., Kaya, C., Ashraf, M., Altunlu, H., Yokas, I., & Yagmur, B. (2007). The effects of calcium sulphate on growth, membrane stability and nutrient uptake of tomato plants grown under salt stress. *Environmental and Experimental Botany*, *59*, 173–178.

Türkan, I., & Demiral, T. (2009). Recent developments in understanding salinity tolerance. *Environmental and Experimental Botany*, *67*(1), 2–9.

Uzilday, B., Ozgur, R., Sekmen, A. H., Yildiztugay, E., & Turkan, I. (2015). Changes in the alternative electron sinks and antioxidant defence in chloroplasts of the extreme halophyte *Eutrema parvulum* (*Thellungiella parvula*) under salinity. *Annals of Botany*, *115*(3), 449–463.

Vaishnav, A., Varma, A., Tuteja, N., & Choudhary, D. K. (2016). *PGPR-mediated amelioration of crops under salt stress. Plant-microbe interaction: An approach to sustainable agriculture* (pp. 205–226). Singapore: Springer.

Van den Bergh, G., & Arckens, L. (2005). Recent advances in 2D electrophoresis: An array of possibilities. *Expert Review of Proteomics*, *2*(2), 243–252.

Varghese, N., Alyammahi, O., Nasreddine, S., Alhassani, A., & Gururani, M. A. (2019). Melatonin positively influences the photosynthetic machinery and antioxidant system of *Avena sativa* during salinity stress. *Plants*, *8*(12), 610.

Veeranagamallaiah, G., Jyothsnakumari, G., Thippeswamy, M., Reddy, P. C. O., Surabhi, G. K., Sriranganayakulu, G., & Sudhakar, C. (2008). Proteomic analysis of salt stress responses in foxtail millet (*Setaria italica* L. cv. Prasad) seedlings. *Plant Science*, *175*(5), 631–641.

Venema, K., Quintero, F. J., Pardo, J. M., & Donaire, J. P. (2002). The *Arabidopsis* Na + / H + exchanger AtNHX1 catalyzes low affinity Na + and K + transport in reconstituted liposomes. *The Journal of Biological Chemistry*, *277*, 2413–2418.

Vysotskaya, L., Hedley, P. E., Sharipova, G., Veselov, D., Kudoyarova, G., Morris, J., & Jones, H. G. (2010). Effect of salinity on water relations of wild barley plants differing in salt tolerance. *AoB Plants*, *2010*, plq006.

Wakeel, A. (2013). Potassium–sodium interactions in soil and plant under saline-sodic conditions. *Journal of Plant Nutrition and Soil Science*, *176*, 344–354.

Wan, X. Y., & Liu, J. Y. (2008). Comparative proteomics analysis reveals an intimate protein network provoked by hydrogen peroxide stress in rice seedling leaves. *Molecular & Cellular Proteomics*, *7*(8), 1469–1488.

Wang, F., Yang, H., Gao, F., & Liu, G. (2008b). Effects of acute temperature or salinity stress on the immune response in sea cucumber. *Apostichopus japonicus. Comparative Biochemistry and Physiology Part A: Molecular & Integrative Physiology*, *151*(4), 491–498.

Wang, H., Tang, X., Wang, H., & Shao, H. B. (2015). Proline accumulation and metabolism-related genes expression profiles in *Kosteletzkya virginica* seedlings under salt stress. *Frontiers in Plant Science*, *6*, 792.

Wang, J., Yao, L., Li, B., Meng, Y., Ma, X., Lai, Y., & Wang, H. (2016). Comparative proteomic analysis of cultured suspension cells of the halophyte *Halogeton glomeratus* by iTRAQ provides insights into response mechanisms to salt stress. *Frontiers in Plant Science*, *7*, 110.

Wang, K., Zhong, M., Wu, Y. H., Bai, Z. Y., Liang, Q. Y., Liu, Q. L., & Liu, G. L. (2017). Overexpression of a chrysanthemum transcription factor gene DgNAC1 improves the salinity tolerance in chrysanthemum. *Plant Cell Reports*, *36*(4), 571–581.

Wang, L., Liang, W., Xing, J., Tan, F., Chen, Y., Huang, L., & Chen, W. (2013). Dynamics of chloroplast proteome in salt-stressed mangrove *Kandelia candel* (L.) Druce. *Journal of Proteome Research*, *12*(11), 5124–5136.

Wang, L., Liu, X., Liang, M., Tan, F., Liang, W., Chen, Y., & Chen, W. (2014). Proteomic analysis of salt-responsive proteins in the leaves of mangrove *Kandelia candel* during short-term stress. *PLoS One*, *9*(1), e83141.

Wang, L., Pan, D., Li, J., Tan, F., Hoffmann-Benning, S., Liang, W., & Chen, W. (2015). Proteomic analysis of changes in the *Kandelia candel* chloroplast proteins reveals pathways associated with salt tolerance. *Plant Science*, *231*, 159–172.

Wang, N., Qian, Z., Luo, M., Fan, S., Zhang, X., & Zhang, L. (2018). Identification of salt stress responding genes using transcriptome analysis in green alga *Chlamydomonas reinhardtii*. *International Journal of Molecular Sciences*, *19*(11), 3359.

Wang, X., Chang, L., Wang, B., Wang, D., Li, P., Wang, L., & Guo, A. (2013). Comparative proteomics of *Thellungiella halophila* leaves from plants subjected to salinity reveals the importance of chloroplastic starch and soluble sugars in halophyte salt tolerance. *Molecular & Cellular Proteomics*, *12*(8), 2174–2195.

Wang, X., Fan, P., Song, H., Chen, X., Li, X., & Li, Y. (2009). Comparative proteomic analysis of differentially expressed proteins in shoots of *Salicornia europaea* under different salinity. *Journal of Proteome Research*, *8*(7), 3331–3345.

Wang, X., Yang, P., Gao, Q., Liu, X., Kuang, T., Shen, S., & He, Y. (2008a). Proteomic analysis of the response to high-salinity stress in *Physcomitrella patens*. *Planta*, *228*(1), 167–177.

Wang, Z., Wang, M., Liu, L., & Meng, F. (2013). Physiological and proteomic responses of diploid and tetraploid black locust (*Robinia pseudoacacia* L.) subjected to salt stress. *International Journal of Molecular Sciences*, *14*(10), 20299–20325.

Weisany, W., Sohrabi, Y., Heidari, G., Siosemardeh, A., & Ghassemi-Golezani, K. (2012). Changes in antioxidant enzymes activity and plant performance by salinity stress and zinc application in soybean (*'Glycine max'*L.). *Plant Omics*, *5*(2), 60.

Wiciarz, M., Gubernator, B., Kruk, J., & Niewiadomska, E. (2015). Enhanced chloroplastic generation of H_2O_2 in stress-resistant *Thellungiella salsuginea* in comparison to *Arabidopsis thaliana*. *Physiologia Plantarum*, *153*(3), 467–476.

Witzel, K., Matros, A., Strickert, M., Kaspar, S., Peukert, M., Mühling, K. H., & Mock, H. P. (2014). Salinity stress in roots of contrasting barley genotypes reveals time-distinct and genotype-specific patterns for defined proteins. *Molecular Plant*, *7*(2), 336–355.

Witzel, K., Weidner, A., Surabhi, G. K., Varshney, R. K., Kunze, G., Buck-sorlin, G. H., & Mock, H. P. (2010). Comparative analysis of the grain proteome fraction in barley genotypes with contrasting salinity tolerance during germination. *Plant, Cell & Environment*, *33*(2), 211–222.

Woodrow, P., Ciarmiello, L. F., Annunziata, M. G., Pacifico, S., Iannuzzi, F., Mirto, A., & Carillo, P. (2017). Durum wheat seedling responses to simultaneous high light and salinity involve a fine reconfiguration of amino acids and carbohydrate metabolism. *Physiologia Plantarum*, *159*(3), 290–312.

Wu, D., Cai, S., Chen, M., Ye, L., Chen, Z., Zhang, H., & Zhang, G. (2013). Tissue metabolic responses to salt stress in wild and cultivated barley. *PLoS One*, *8*(1), e55431.

Wu, G. Q., Wang, J. L., Feng, R. J., Li, S. J., & Wang, C. M. (2018). iTRAQ-based comparative proteomic analysis provides insights into molecular mechanisms of salt tolerance in Sugar Beet (*Beta vulgaris* L.). *International Journal of Molecular Sciences*, *19*(12), 3866.

Wu, H. (2018). Plant salt tolerance and Na^+ sensing and transport. *The Crop Journal*, *6*(3), 215–225.

Wu, H., Shabala, L., Barry, K., Zhou, M., & Shabala, S. (2013). Ability of leaf mesophyll to retain potassium correlates with salinity tolerance in wheat and barley. *Physiologia Plantarum*, *149*(4), 515–527.

Wu, H., Shabala, L., Liu, X., Azzarello, E., Zhou, M., Pandolfi, C., … Shabala, S. (2015). Linking salinity stress tolerance with tissue-specific Na + sequestration in wheat roots. *Frontiers in Plant Science*, *6*, 71.

Wu, H., Zhu, M., Shabala, L., Zhou, M., & Shabala, S. (2015). K + retention in leaf mesophyll, an overlooked component of salinity tolerance mechanism: A case study for barley. *Journal of Integrative Plant Biology*, *57*(2), 171–185.

Wu, J., Zhao, Q., Wu, G., Yuan, H., Ma, Y., Lin, H., & Sun, D. (2019). Comprehensive analysis of differentially expressed unigenes under NaCl stress in flax (*Linum usitatissimum* L.) using RNA-Seq. *International Journal of Molecular Sciences*, *20*(2), 369.

Wu, L., Huo, W., Yao, D., & Li, M. (2019). Effects of solid matrix priming (SMP) and salt stress on broccoli and cauliflower seed germination and early seedling growth. *Scientia Horticulturae*, *255*, 161–168, c.

Wu, N., Mao, H. T., Chen, M. Y., Dong, J., Yuan, M., Zhang, Z. W., & Chen, Y. E. (2020). Different responses of photosystem and antioxidant defense system to three environmental stresses in wheat seedlings. *Photosynthetica*, *58*(1), 87–99.

Xing, J., Pan, D., Wang, L., Tan, F., & Chen, W. (2019). Proteomic and physiological responses in mangrove *Kandelia candel* roots under short-term high-salinity stress. *Turkish Journal of Biology*, *43*(5), 314–325.

Xu, E., Chen, M., He, H., Zhan, C., Cheng, Y., Zhang, H., & Wang, Z. (2017). Proteomic analysis reveals proteins involved in seed imbibition under salt stress in rice. *Frontiers in Plant Science*, *7*, 2006.

Xu, L. K., & Hsiao, T. C. (2004). Predicted versus measured photosynthetic water-use efficiency of crop stands under dynamically changing field environments. *Journal of Experimental Botany*, *55*(407), 2395–2411.

Xu, Z., Ali, Z., Xu, L., He, X., Huang, Y., Yi, J., & Zhang, D. (2016). The nuclear protein GmbZIP110 has transcription activation activity and plays important roles in the response to salinity stress in soybean. *Scientific Reports*, *6*, 20366.

Xu, Z., Wang, C., Xue, F., Zhang, H., & Ji, W. (2015). Wheat NAC transcription factor TaNAC29 is involved in response to salt stress. *Plant Physiology and Biochemistry*, *96*, 356–363.

Xu, Z. S., Chen, M., Li, L. C., & Ma, Y. Z. (2011). Functions and application of the AP2/ERF transcription factor family in crop improvement F. *Journal of Integrative Plant Biology*, *53* (7), 570–585.

Xue, Z. Y., Zhi, D. Y., Xue, G. P., Zhang, H., Zhao, Y. X., et al. (2004). Enhanced salt tolerance of transgenic wheat (*Triticum aestivum* L.) expressing a vacuolar Na + /H + antiporter gene with improved grain yields in saline soils in the field and a reduced level of leaf Na +. *Plant Science (Shannon, Ireland)*, *167*, 849–859.

Yadav, N. S., Shukla, P. S., Jha, A., Agarwal, P. K., & Jha, B. (2012). The SbSOS1 gene from the extreme halophyte *Salicornia brachiata* enhances Na + loading in xylem and confers salt tolerance in transgenic tobacco. *BMC Plant Biology*, *12*, 188.

Yamaguchi, T., & Blumwald, E. (2005). Developing salt-tolerant crop plants: Challenges and opportunities. *Trends in Plant Science*, *10*(12), 615–620.

Yamamoto, N., Takano, T., Tanaka, K., Ishige, T., Terashima, S., Endo, C., & Tada, Y. (2015). Comprehensive analysis of transcriptome response to salinity stress in the halophytic turf grass *Sporobolus virginicus*. *Frontiers in Plant Science*, *6*, 241.

Yang, Q., Chen, Z. Z., Zhou, X. F., Yin, H. B., Li, X., Xin, X. F., & Gong, Z. (2009). Overexpression of SOS (salt overly sensitive) genes increases salt tolerance in transgenic *Arabidopsis*. *Molecular Plant*, *2*(1), 22–31.

Yang, X., & Lu, C. (2005). Photosynthesis is improved by exogenous glycinebetaine in salt-stressed maize plants. *Physiologia Plantarum*, *124*(3), 343–352.

Yin, Y., Qi, F., Gao, L., Rao, S., Yang, Z., & Fang, W. (2018). iTRAQ-based quantitative proteomic analysis of dark-germinated soybeans in response to salt stress. *RSC Advances*, *8*(32), 17905–17913.

Yokotani, N., Ichikawa, T., Kondou, Y., Matsui, M., Hirochika, H., Iwabuchi, M., & Oda, K. (2009). Tolerance to various environmental stresses conferred by the salt-responsive rice gene ONAC063 in transgenic *Arabidopsis*. *Planta*, *229*(5), 1065–1075.

Yousuf, P. Y., Ahmad, A., Ganie, A. H., Sareer, O., Krishnapriya, V., Aref, I. M., & Iqbal, M. (2017). Antioxidant response and proteomic modulations in Indian mustard grown under salt stress. *Plant Growth Regulation*, *81*(1), 31–50.

Yu, J., Chen, S., Zhao, Q., Wang, T., Yang, C., Diaz, C., & Dai, S. (2011). Physiological and proteomic analysis of salinity tolerance in *Puccinellia tenuiflora*. *Journal of Proteome Research*, *10*(9), 3852–3870.

Yu, J., Zhao, W., Tong, W., He, Q., Yoon, M. Y., Li, F. P., & Park, Y. J. (2018). A Genome-wide association study reveals candidate genes related to salt tolerance in rice (*Oryza sativa*) at the germination stage. *International Journal of Molecular Sciences*, *19*(10), 3145.

Yu, Y., Liu, Z., Wang, L., Kim, S. G., Seo, P. J., Qiao, M., & Xiang, F. (2016). WRKY 71 accelerates flowering via the direct activation of FLOWERING LOCUS T and LEAFY in *Arabidopsis thaliana*. *The Plant Journal*, *85*(1), 96–106.

Yu, Y., Ni, Z., Chen, Q., & Qu, Y. (2017). The wheat salinity-induced R2R3-MYB transcription factor TaSIM confers salt stress tolerance in *Arabidopsis thaliana*. *Biochemical and Biophysical Research Communications*, *491*(3), 642–648.

Yue, Y., Zhang, M., Zhang, J., Duan, L., & Li, Z. (2012). SOS1 gene overexpression increased salt tolerance in transgenic tobacco by maintaining a higher K + /Na + ratio. *Journal of Plant Physiology*, *169*(3), 255–261.

Zafar, S. A., Zaidi, S. S. E. A., Gaba, Y., Singla-Pareek, S. L., Dhankher, O. P., Li, X., & Pareek, A. (2020). Engineering abiotic stress tolerance via CRISPR/Cas-mediated genome editing. *Journal of Experimental Botany*, *71*(2), 470–479.

Zeng, L., & Shannon, M. C. (2000). Effects of salinity on grain yield and yield components of rice at different seeding densities. *Agronomy Journal*, *92*(3), 418–423.

Zeng, L., Shannon, M. C., & Lesch, S. M. (2001). Timing of salinity stress affects rice growth and yield components. *Agricultural Water Management*, *48*(3), 191–206.

Zhai, H., Wang, F., Si, Z., Huo, J., Xing, L., An, Y., & Liu, Q. (2016). A myo-inositol-1-phosphate synthase gene, Ib MIPS 1, enhances salt and drought tolerance and stem nematode resistance in transgenic sweet potato. *Plant Biotechnology Journal*, *14*(2), 592–602.

Zhang, B., & Wang, Q. (2015). MicroRNA-based biotechnology for plant improvement. *Journal of Cellular Physiology*, *230*(1), 1–15.

Zhang, F., Zhu, G., Du, L., Shang, X., Cheng, C., Yang, B., & Guo, W. (2016). Genetic regulation of salt stress tolerance revealed by RNA-Seq in cotton diploid wild species, *Gossypium davidsonii*. *Scientific Reports*, *6*, 20582.

Zhang, H., Han, B., Wang, T., Chen, S., Li, H., Zhang, Y., & Dai, S. (2012). Mechanisms of plant salt response: Insights from proteomics. *Journal of Proteome Research*, *11*(1), 49–67.

Zhang, H. H., Hodson, J. N., Williams, J. P., & Blumwald, E. (2001). Engineering salt tolerant Brassica plants: Characterization of yield and seed oil quality in transgenic plants with increased vacuolar sodium accumulation. *Proceedings of the National Academy of Sciences of the United States of America*, *98*, 12832–12836.

Zhang, H. H., Xu, N., Li, X., & Gu, S. Y. (2016). Overexpression of 2-Cys Prx increased salt tolerance of photosystem II (PSII) in tobacco. *International Journal of Agriculture And Biology*, *19*, 735–745.

Zhang, X., Liu, S., & Takano, T. (2008). Overexpression of a mitochondrial ATP synthase small subunit gene (AtMtATP6) confers tolerance to several abiotic stresses in *Saccharomyces cerevisiae* and *Arabidopsis thaliana*. *Biotechnology Letters*, *30*(7), 1289–1294.

Zhang, X., Takano, T., & Liu, S. (2006). Identification of a mitochondrial ATP synthase small subunit gene (RMtATP6) expressed in response to salts and osmotic stresses in rice (*Oryza sativa* L.). *Journal of Experimental Botany*, *57*(1), 193–200.

Zhang, Y., Wei, M., Liu, A., Zhou, R., Li, D., Dossa, K., & You, J. (2019). Comparative proteomic analysis of two sesame genotypes with contrasting salinity tolerance in response to salt stress. *Journal of Proteomics*, *201*, 73–83.

Zheng, X., Chen, B., Lu, G., & Han, B. (2009). Overexpression of a NAC transcription factor enhances rice drought and salt tolerance. *Biochemical and Biophysical Research Communications*, *379*(4), 985–989.

Zhong, H., Guo, Q. Q., Chen, L., Ren, F., Wang, Q. Q., Zheng, Y., & Li, X. B. (2012). Two *Brassica napus* genes encoding NAC transcription factors are involved in response to high-salinity stress. *Plant Cell Reports*, *31*(11), 1991–2003.

Zhou, J., Li, F., Wang, J. L., Ma, Y., Chong, K., & Xu, Y. Y. (2009). Basic helix-loop-helix transcription factor from wild rice (OrbHLH2) improves tolerance to salt-and osmotic stress in *Arabidopsis*. *Journal of Plant Physiology*, *166*(12), 1296–1306.

Zhou, S., Sauvé, R. J., Liu, Z., Reddy, S., Bhatti, S., Hucko, S. D., & Thannhauser, T. W. (2011). Identification of salt-induced changes in leaf and root proteomes of the wild tomato, *Solanum chilense*. *Journal of the American Society for Horticultural Science*, *136*(4), 288–302.

Zhu, D., Hou, L., Xiao, P., Guo, Y., Deyholos, M. K., & Liu, X. (2019). VvWRKY30, a grape WRKY transcription factor, plays a positive regulatory role under salinity stress. *Plant Science*, *280*, 132–142.

Zhu, J., Fan, Y., Shabala, S., Li, C., Lv, C., Guo, B., & Zhou, M. (2020). Understanding mechanisms of salinity tolerance in barley by proteomic and biochemical analysis of near-isogenic lines. *International Journal of Molecular Sciences*, *21*(4), 1516.

Zhu, J. K. (2000). Genetic analysis of plant salt tolerance using *Arabidopsis*. *Plant Physiology*, *124*(3), 941–948.

Zhu, J. K. (2001). Plant salt tolerance. *Trends in Plant Science*, *6*, 66–71.

Zhu, J. K. (2002). Salt and drought stress signal transduction in plants. *Annual Review of Plant Biology*, *53*(1), 247–273.

Zhu, J.-K. (2007). *Plant salt stress*. John Wiley & Sons, Ltd.

Zhu, Z., Chen, J., & Zheng, H. L. (2012). Physiological and proteomic characterization of salt tolerance in a mangrove plant, *Bruguiera gymnorrhiza* (L.) Lam. *Tree Physiology*, *32*(11), 1378–1388.

Zörb, C., Schmitt, S., & Mühling, K. H. (2010). Proteomic changes in maize roots after short-term adjustment to saline growth conditions. *Proteomics*, *10*(24), 4441–4449.

Zuther, E., Huang, S., Jelenska, J., Eilenberg, H., Arnold, E. M., Su, X., & Gornicki, P. (2004). Complex nested promoters control tissue-specific expression of acetyl-CoA carboxylase genes in wheat. *Proceedings of the National Academy of Sciences*, *101*(5), 1403–1408.

Chapter 13

Physiological and molecular strategies of plant adaptation in phosphorus-deficient soils

Muhammad Khashi u Rahman[1] and Faisal Siddique[2]
[1]*College of Horticulture and Landscape Architecture, Northeast Agricultural University, Harbin, P.R. China,* [2]*Department of Plant Pathology, Northeast Agricultural University, Harbin, P.R. China*

13.1 Introduction

The growth of plant is subjected to several cellular, biochemical, physiological, and developmental processes which involve a variety of mineral nutrients. Among those, phosphorus (P) is an essential element that plays key roles in multipartite functioning of plant physiology. Numerous plant molecules such as DNA, RNA, proteins, sugars, lipids, ADP, ATP, and NDPH, all contain P as a central constituent that make it the second most important nutrient after nitrogen for plant growth and crop production. Despite its importance, the availability of P is a big challenge for global food security because of its poor recovery from applied fertilizers and it is a given fact that the natural P reservoirs are being depleted so quickly that P could be scarce by the year 2050 (Cordell, Rosemarie, Schröder, & Smit, 2011). When P is applied to agricultural soils in the form of fertilizers, it is readily sorbed on to cations such as iron and aluminum in acidic soils and magnesium and calcium in alkaline soils, making it sparingly available to plants. Another form of P present in the soil is organic P which is not utilized by plants. Since all living organisms are blessed with survival ability, plants use numerous adaptive strategies for P acquisition in a P-deficient soil environment.

Phosphorus-use efficiency, which is the ability of a crop to produce high yield under P limiting condition, can be attained through improved P uptake efficiency (the ability to take more P from the soil under P limiting condition) and/or through improved P utilization efficiency (the ability to produce higher dry matter yield per unit of P taken up) (Balemi & Negisho, 2012). Plants possess several physiological, biochemical, and molecular strategies

Frontiers in Plant–Soil Interaction. DOI: https://doi.org/10.1016/B978-0-323-90943-3.00005-5

to cope with P starvation. Changes in belowground traits such as increased lateral root biomass, root to shoot ratio, and root hair density are the usual adaptive abilities of plants in the hunt for nutrients. Likewise, plants alter the amount of root exudates and the exudation pattern amid the mobilization of barely available inorganic P. Besides the direct mobilization of inorganic P, several facilitative and cooperative dialogs take place between plants and soil microbiota in the rhizosphere (Fig. 13.1). Soil microbes use components of root exudates and their energy source and in return secrete different hormones to desorbed plant unavailable P in the soil (u Rahman, Tan, Ma, Wu, & Zhou, 2020). Moreover, inter- and intraspecific interactions between plants consequently influence soil available nutrients and their acquisition. All these processes involve several complex mechanisms at the molecular

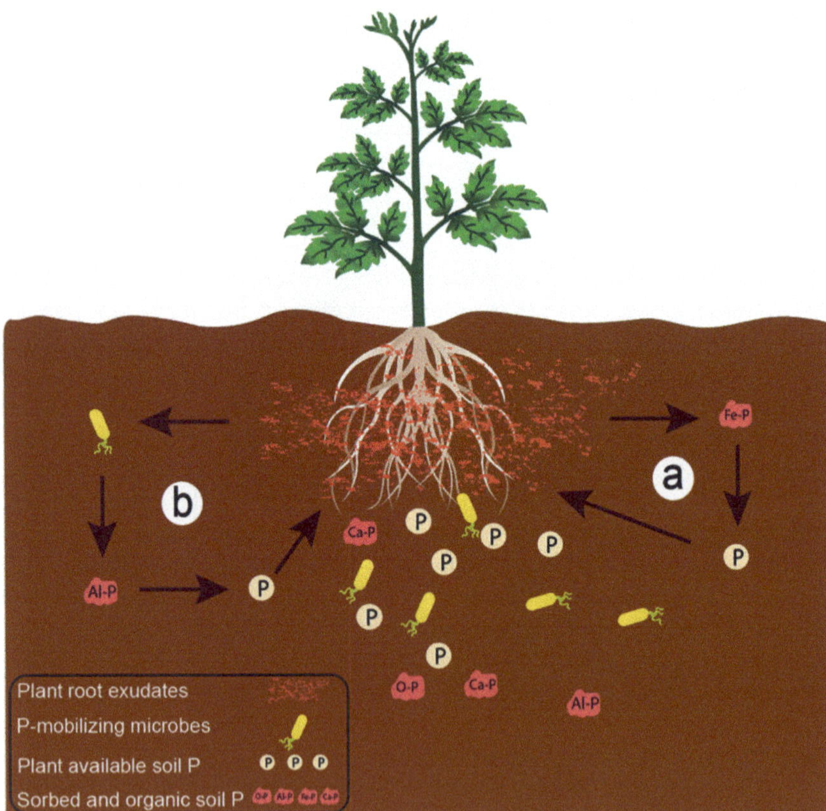

FIGURE 13.1 Speculated predominant mechanisms of plant acquisition of phosphorus (P) in P-deficit soils. (A) plant secret various compounds in its root exudates that can directly mobilize plant partially available P and improve P acquisition. (B) Plants also shape P-mobilizing bacteria through its root exudation that further mobilize sparingly available P and benefit roots to improve P acquisition.

level like sensing of P deficiency, signaling through xylem and phloem, and systemic reprogramming, which are recently well explored (Table 13.1).

Modern agricultural practices are proved to overcome the issue of poor P availability and substantially contribute to increased crop yield (u Rahman, Ma, Wu, & Zhou, 2021). However, due to the limited global P reservoirs and increased demand due to an increase in world population, the future security of a high level of agricultural production is required. Hence, it is necessary to develop a better understanding of the molecular proceedings of controlling P homeostasis in plants. Here, we evaluate the current advances in the field of plant physiology and discuss molecular strategies that plants adopt in limiting P growth conditions.

13.2 Physiological and morphological responses to phosphorus stress

Plants adopt physiological and morphological strategies to acquire P from the soil environment. This includes the search for a new P source, reaching out to an existing one, and sources that exist but are sparingly available to plant. In order to respond to the limiting P in the soil, plants modify their root systems and undergo an array of adoptive responses including change in root morphology, root exudation pattern, composition of root exudates, induction of phosphate transporters, and changes in the recruitment of soil microbiota (Gao et al., 2020). In an adequate P environment, plants may exert energy on axial roots elongation, higher root to shoot rations, dense root hairs, and even development of cluster roots in some plant species to increase foraging (Lambers et al., 2011). In addition, the ability of mobilization and acquisition of P could be genotype and species dependent. Some P-efficient crops such as soybean and maize use shallow root growth angles that allow them to obtain more of the available P in the soil (Lynch, 2011). Similarly, barley plants develop long root hairs to increase the root surface area, more root–soil contact, and thereby increase P acquisition in P-limiting conditions (Haling et al., 2013). In the model plant *Arabidopsis*, researchers have found a reduction in primary root growth in response to P stress (Chevalier, Pata, Nacry, Doumas, & Rossignol, 2003).

Most of the plants develop larger root systems to get more access to P available in the soil, hence the root to shoot ratio of plant biomass is usually higher in P-stressed conditions compared to P-sufficient conditions (Jungk, 2001). Moreover, root surface area may not always be a sufficient trait to fulfill the required P acquisition, therefore the presence of specific characterized root hairs is of importance for adequate P acquisition. Root hairs grow perpendicular to roots to allow better reach to nutrients resource as well as to increase the root absorbance area. The increased root diameter is another strategy that a plant uses during P deficiency, because thinner roots allow a contact with a large amount of soil, thus increasing the proportion of

TABLE 13.1 A list of highlighted studies from recent five years demonstrating plant strategies and systematic mechanisms involved in phosphorus acquisition during low phosphorus availability.

Crop (s)	Main finding (s)	Reference
Phaeodactylum tricornutum	PtPSR is a key transcription factor influencing phosphorus (P) scavenging in plant adaption to P stress	Sharma et al. (2020)
Zea mays L.	Maize allocated more photosynthates to roots as sucrose, which acts as a low-P signal in P stress adoption	Zhou et al. (2020)
Z. mays L.	Phosphate-solubilizing bacteria plus P application as organic manures is an ecofriendly option to improve crop growth and P nutrition in a calcareous soil under changing climate	Adnan et al. (2020)
Sorghum bicolour L.	Plant growth promoting drought-adaptive phosphorus solubilizers could be used as bioinoculants for crops under water scarcity ecosystems	Kour et al. (2020)
Oryza sativa L.	Reduced water availability induces root modifications which, apart from enhancing drought resilience, also affect P uptake efficiency	De Bauw et al. (2019)
Gossypium hirsutum L.	Root morphology, especially surface fine root length and middle root length, played an important role in P uptake under P-deficient conditions	Chen et al. (2019)
Zea mays L.	Large CN improves plant P acquisition from low-P soils by reducing rooting depth and increasing topsoil foraging	Sun, Gao, and Lynch (2018)
Nannochloropsis oceanica	An *N. oceanica* strain has a unique repertoire of genes that facilitate P acquisition and the degradation of phospholipids compared with other stramenopiles	Mühlroth et al. (2017)
Different soybean genotypes	Cross-talk between various metabolic pathways implicated in conferring superior P acquisition efficiency under stress	Vengavasi, Pandey, Abraham, and Yadav (2017)
Hybrid *Eucalyptus*	*Eucalyptus* relies on changes to morphological characteristics of the root system to increase accessibility to phosphorus resources	Zhou et al. (2017)

nutrients absorption. Some of the plant species of *Fabaceae*, *Cyperaceae*, and *Proteaceae* are known for the development of clustering roots in response to P deficiency. Clustering roots are characterized with a greater number of rootlets, dense root branches, longer root hairs, and higher density, with an ultimate goal to get more contact with the soil available P.

Roots secrete a variety of organic compounds that play versatile functions in the rhizosphere. Some organic anions and H^+ could solubilize mineral sorbed inorganic P and make it available to plants. Similarly, citrate secreted by white lupin cluster roots solubilizes insoluble P and enhances P acquisition (Cheng et al., 2011). In response to P stress, roots secrete a higher quantity of acid phosphatases that hydrolyze insoluble P and increase the availability of soil P (Robinson, Carson, Ying, Ellis, & Plaxton, 2012). Besides a direct effect, roots also form symbiotic associations with soil microorganisms, such as arbuscular mycorrhizal fungi (AMF). Increased P acquisition by hyphae in arbuscular networks is related to both modification of the root environment as well as exploration of soil. Organic P, which is not available to plants, can also be hydrolyzed by acid phosphatase released by AMF hyphae. The polyphosphates stored in hyphal vacuoles are also hydrolyzed to inorganic P and transported to the host plant. According to Gashaw Deressa and Schenk (2008), AMF association with onion accounted for almost all of the predicted P uptake.

P-solubilizing microorganisms in the soil play an important role in P availability to plants. These microbes, including plant growth-promoting rhizobacteria (PGPR), hydrolyze soil available organic P by secreting phosphatases such as alkaline/acidic phosphatases (Singh & Satyanarayana, 2011). For example, most of the *Pseudomonas* and *Bacillus* spp. have the ability to solubilize and mineralize P and also enhance its absorption by plant roots (Alaylar, Güllüce, Karadayi, & Isaoglu, 2019; Tan et al., 2020). Similarly, some fungi, such as *Trichoderma* spp., a group of mostly biological control species, have the ability to dissolve the fixed phosphate existing in the soil, so as to improve the soil fertility and plant growth (Kapri & Tewari, 2010). During P starvation, plants could alter root exudation to change the microbial community composition and to promote/attract these P-solubilizing microbes to attain their P requirements (Li, Tian, u Rahman, & Wu, 2020).

13.3 Molecular mechanisms of senescing, signaling, and communication during phosphorus starvation

Plants have a wide range of adaptive mechanisms under P-deficient conditions to absorb sufficient P to maintain metabolic activities and growth. Molecular studies have revealed alterations in several biochemical and signaling pathways resulting in changes at the gene, protein, and metabolite levels. The expression of genes causing alternations in phenotype is under complex regulation and is generally upregulated by P starvation, resulting in transcript accumulation over longer periods (Hammond et al., 2003). The sensing of P takes place either locally, that is, within tissues and organs, or systemically in plants. In general, root tissues perceive nutrient stress and send signals/warning the aboveground plant via xylem about the limitations of specific nutrients. In response, plant shoots send signals to the root system via phloem to adapt to stress through developmental strategies (Zhang, Liao, & Lucas, 2014). This whole process involves ethylene, auxin, cytokinins, abscisic acid, gibberellins, strigolactones, sugars, and P itself as dominant signaling pathway components (Chiou & Lin, 2011). In local P sensing, it has been shown that when root tips interact with P deficit soil patches, a stimulus is perceived by a plasma membrane localized sensor. Once the perception of P deficiency is finally made by root tip P sensors, the signaling pathway in P sensing cells is activated that further transfers signal to plant shoots through xylem by different signaling components.

Among the several signaling components taking part in the signaling pathway, Ca^{2+}, reactive oxygen species, and inositol polyphosphates are the main players and are considered as universal secondary messengers in systemic plant communication. In a study, when tomato plant was subjected to P-deficit soil, the level of Ca^{2+}-ATPase was increased (Muchhal, Liu, & Raghothama, 1997), indicating the possible involvement of Ca^{2+} in P stress signaling. Reactive oxygen species are sensitive to nutrient stress that affects their distribution. It has been shown that during P stress, reactive oxygen species were abundant in the proximal area of lateral roots as compared to the primary root meristem and elongation zones when an adequate or higher amount of P was available (Tyburski, Dunajska, & Tretyn, 2009). These findings indicate that reactive oxygen species play a key role as P stress causes the elongation of lateral roots and the inhibition of primary roots.

Plant secreted hormones also play a significant role during P starvation. The P stress can affect the production of hormones and their transport to regulate P stress gene expression. In *Arabidopsis*, plants change the root architecture by inhibiting primary roots and promoting lateral roots during P deprivation. Similarly, when exogenously auxin was applied at different concentrations, the plant promoted lateral roots at a higher rate but not the primary roots, suggesting that the root system changed the production of auxin and thus the architecture in response to P stress. P stress induced the expression of Transport Inhibitor Response 1 (TIR1), an auxin receptor that degrades the expression of Auxin Response Repressor (ARR) protein. This mechanism allows the Auxin Response Factor 19 (ARF19), along with other ARF, to regulate specific auxin responsive genes responsible for lateral root growth. Similarly, OsARF12 and OsARF19 in rice have been shown to be associated with auxin signaling and communication in P starvation response (Wang et al., 2014).

Ethylene is another important hormone that has been shown to respond to P stress by enhancing its production (Li, Mao, Tian, Li, & Zhang, 2009).

According to Borch, Bouma, Lynch, and Brown (1999), exogenously applied AVG (2-aminoethoxyvinyl glycine), an ethylene inhibitor, can promote lateral roots as compared to primary roots in P deficit conditions, however, the process can be reversed by applying ethylene. Similarly, application of another ethylene inhibitor MCP (1-methylcyclopropen) could increase primary root growth under adequate to high P concentration, while decreasing in P-deficit soils. Likewise, the process can be reversed by the application of ACC (1-aminocyclopropane-1-carboxylic acid), an ethylene precursor (Ma, Baskin, Brown, & Lynch, 2003). Moreover, when ethylene is applied to plants in sufficient P condition, the functioning of the root meristem is disrupted to simulate the effects of P stress on root meristem development (Chacón-López, Ibarra-Laclette, Sánchez-Calderón, Gutiérrez-Alanís, & Herrera-Estrella, 2011). In these situations, the effects of P deficiency on root meristem development are overcome by the application of Ag^+ or AVG, which are ethylene signaling inhibitors. Another role of ethylene during plant adaptation of P stress is the growth of root hairs that are governed by the process in which ethylene signaling leads to the expression of ethylene responsive genes that are responsible for epidermal cell differentiation (Nagarajan & Smith, 2012).

Cytokinins are negatively associated with plant response to P stress as their levels increase in higher P availability and decrease in P deficiency. Manually applied cytokinins suppress several phosphorus stress-induced genes such as AtPT1, AtPS1, and At4 (Karthikeyan et al., 2002). In addition, exogenously applied cytokinin do not suppress the growth of the root meristem as well as root hairs during P stress. Strigolactones are another important plant hormone that play a key role in seed germination and root colonization with soil microbiota such as AMF and hyphal branching. The production of strigolactones governs the shoot branching (Umehara et al., 2008), root development, and regulates several other systemic processes (Koltai, 2013). According to Czarnecki, Yang, Weston, Tuskan, and Chen (2013), strigolactones function both locally and systemically to facilitate the acclimation of a plant during the P-deficit period. Studies have demonstrated that the application of GR24, a synthetic strigolactone, inhibits adventitious, primary, and lateral roots and promotes growth and density of root hairs, which is an efficient strategy that plants use for their survival in nutrient stress (Rasmussen et al., 2012).

After sensing P stress through the root tip, root–shoot and shoot–root signaling and communication takes place via the xylem and phloem, respectively. The main agents involved in root–shoot signaling are P itself along with cytokinins and strigolactones, while shoot–root signaling is usually carried out by phloem translocation of mineral nutrients, some hormones, proteins, and RNAs (Zhang et al., 2014). Due to recent advances in transcriptomic analysis, there are more than 1000 transcripts known in phloem that perform several functions including aboveground–belowground signaling (Guo et al., 2011). However, studies are required to understand their role and functioning, especially as signaling components during nutrient stress conditions.

13.4 Transcriptional response to phosphorus starvation

The mechanism of adaption to P stress involves the expression of hundreds of genes that regulate the whole process in terms of P uptake, recycling, and overall stress protection. Recent transcriptional techniques, such as subtractive hybridization and deep sequencing have provided fundamental knowledge of P stress assimilation. On the basis of responsive characteristics, P stress response genes are categorized into early and late responsive genes (Misson et al., 2005). As the name shows, early P stress-responsive genes (ePSR) respond quickly (within few hours) to P deficiency while late responsive genes (lPSR) respond approximately over 1 day. The list of ePSR genes is vast including general stress-related genes such as jasmonic acid, ethylene, and jasmonic acid-induced genes, cytochrome P450 and peroxidase genes, and single transduction-related genes, such as genes encoding MAP kinase kinases, MAP kinases, Ca^{+}, and CDPK-related protein genes (Lin et al., 2011). The WRKY transcription factors, such as WRK40 and WRK18, are typical examples of ePSR genes that are induced within an hour of roots' interaction with P stress.

The early developmental response of roots to P-stress is an alteration in induction of ePSR genes that are usually involved in cell wall remodeling, that is, genes encoding for members of the extension, expansin, chitinase, pectinesterase, and endochitinase families during P sensing in stress conditions (Misson et al., 2005). Similarly, some of the genes specific to P deficiency, such as those encoding for ribonucleases, acid phosphatases, and phosphate transporters, are also induced quickly in response to P stress (Lin et al., 2011). However, Hammond et al. (2003) found late induction of some ePSR genes in *Arabidopsis*, about 6 hours after introducing a P-stress treatment. Instead of those genes that are encoded for the cellular function, lPSR genes are usually related to downstream functions such as P uptake, recycling, remobilization, metabolism, protein synthesis, and photosynthesis (Amtmann, Hammond, Armengaud, & White, 2005). During P stress, the genes in root and leaves especially those encoding for phosphatases, P transporters, and RNAs, and enzymes involved in lipid metabolism and glycolysis are greatly induced to overcome P homeostasis (Lan, Li, & Schmidt, 2012; O'Rourke et al., 2013). At the same time, expression of genes encoding of protein degradation is upregulated while the expression of genes encoding protein synthesis regulators is downregulated (Müller, Morant, Jarmer, Nilsson, & Nielsen, 2007).

A P-deficit environment strongly affects exudation pattern as plants increase the secretion of organic acid as a P-stress adaptation strategy

(Khashi u Rahman, Zhou, & Wu, 2019). When roots observe P deficiency, the genes encoding enzymes associated with exudation and secretion of organic acids are strongly upregulated. In this process, another class of genes involved in glycolysis is also upregulated because the increased carbon supply is required for organic acid synthesis. Furthermore, the expression of genes involved in sensing, signaling, hormones synthesis and metabolism change according to their specific roles in signaling and communication during long-lasting P deficiency. For example, the expression of several genes from the auxin family such as auxin synthesis genes, auxin transporter genes, auxin transcription factors, and degradation genes have been shown to be affected by P starvation (O'Rourke et al., 2013). The genes involved in ethylene synthesis, that is, ACC synthesis and oxidase, and signaling (i.e., EIN 3, ERF/AP2 are also induced during P-stress). The degradation of cytokinin is catalyzed by cytokinin oxidases, the expression of which is also induced during P stress and plays a negative role in P stress response.

After the root, the shoot is the second medium in which the process of signaling takes place during stress response. In this matter, genes involved in different metabolisms are greatly involved and their expression is affected by long-term P deficiency. The expression of predominant genes involved in photosynthesis such as PSI, PSII, small subunits, Rubisco, chlorophyll-binding proteins, and Calvin cycle enzymes are suppressed during P starvation, while the expression of genes associated with glycolysis, phosphoenolpyruvate, carboxylase, sucrose transporters, and those that have functions in sucrose and starch synthesis are induced (Wasaki et al., 2006). The induction of sugar synthesis gene expression is positively correlated with the P deficiency, as many sugar responsive genes are regulated by P deficiency. According to Zhang et al. (2014), there are almost 150 genes that are antagonistically or synergistically regulated by both sugar and P deficiency at the same time.

13.5 Metabolic adaptations: role of sugar and lipid

Adaptations through remodeling of lipid and sugar metabolism are the two main metabolic strategies that plants use to combat P stress. According to Park, Baek, Benildo, Yun, and Hasenstein (2012), the deficiency of P in soil can lead to an increase in starch and sucrose level, and a decrease in chlorophyll, pyruvate, and glucose level. This suggests that the change in availability of P could strongly influence metabolic pathways associated with sucrose, starch, chlorophyll, glucose, and pyruvate. Moreover, deficiency of P also affects phosphorylated sugar levels in roots and leaves because of the lower activities of hexokinase and fructokinase. The increase in sugar level, particularly sucrose, during P deficiency is reported in several agronomic crops such as barley, beans, and soybean, as well as in the model plant *Arabidopsis* (Hammond & White, 2008).When plant roots encounter P deficiency, the level of accumulated sugars start increasing, which results in a decline in respiration rate, that further affects plant growth and development (Wanke, Ciereszko, Podbielkowska, & Rychter, 1998).

The sucrose and starch synthesis is greatly handled by the availability of P as P governs the distribution and movement of fixed carbon within chloroplasts and the movement of triose phosphate to cytoplasm that is used in sucrose synthesis (Zhang et al., 2014). Moreover, the production and accumulation of ADP glucose pyrophosphorylase in chloroplast is also induced in P-stressed plants due to low P and high amount of sugars. Here, limited P plays another significant role in increasing starch production, that is when carbon is fixed in the form of triose-P, that is required to be transported to chloroplast by triose-P antiporter, poor availability of P blocks the antiporter system and redirects it to starch production. At last, the production of starch increases and its accumulation hinders the photosynthesis rate through reducing the stromal volume available to CO_2 fixation. As Hermans, Hammond, White, and Verbruggen (2006) stated, the whole mechanism increases movement and translocation of carbohydrate within phloem to the root, hence promoting root growth for better and extensive soil exploration which leads the way to ameliorating the deficiency of P.

Phospholipids, another important constituent of biological membranes, also play key role in plant adaptations to P. Lipid metabolism is mainly driven by the availability of P and during P stress conditions, phospholipids act as substitutes of phosphorus-free galactolipids, as increased concentration of these are reported in *Arabidopsis* in P stress conditions (Gaude, Nakamura, Scheible, Ohta, & Dörmann, 2008). Moreover, the expression of sulfoquinovosyldiacylglycerol (SQDG), sulfoquinovosyldiacyl 1 (SQD1), SQD2, and genes that are encoding enzymes related to sulfolipid biosynthesis are induced in rice and *Arabidopsis* when P stress conditions were imposed (Misson et al., 2005; Wang et al., 2006). According to Okazaki et al. (2013), the levels of glucuronosyldiacylglycerol in rice and *Arabidopsis* were significantly increased during P starvation. Taken together, these studies suggest the lipids play a major part in plant response to P stress.

13.6 Conclusions

When starvation occurs, living organisms hunt for food for their survival. In spite of being immobile creature, plants do search for nutrients using different strategies, specifically in the belowground world. Among the essential plant nutrients, P is an important element in the plant life cycle. Recent advances in research techniques have made it easier to understand the molecular insights of strategies that plants use to face P starvation. Several physiomorphological adaptations, such as alteration in root architecture, involve the expression of numerous genes that regulate an array of physiological functions. From sensing P stress to signaling and systemic communication, several components, including ethylene, auxin, cytokinins, abscisic acid, gibberellins, strigolactones, sugars, proteins, lipids, RNAs, and P itself are involved in signaling or being a responsive trait. In this regard, xylem and phloem are the main channels where systemic or local signaling and communication takes place. These molecular mechanistic insights have broadened our knowledge to understand the systemic response of plants to P stress, which will eventually help us to develop and manage the agricultural sector accordingly to increase crop yield per capita.

References

Adnan, M., Fahad, S., Zamin, M., Shah, S., Mian, I. A., Danish, S., ... Saud, S. (2020). Coupling phosphate-solubilizing bacteria with phosphorus supplements improve maize phosphorus acquisition and growth under lime induced salinity stress. *Plants*, *9*(7), 900.

Alaylar, B., Güllüce, M., Karadayi, M., & Isaoglu, M. (2019). Rapid detection of phosphate-solubilizing bacteria from agricultural areas in Erzurum. *Current Microbiology*, *76*, 804–809.

Amtmann, A., Hammond, J. P., Armengaud, P., & White, P. J. (2005). Nutrient sensing and signalling in plants: Potassium and phosphorus. *Advances in Botanical Research*, *43*, 209–257.

Balemi, T., & Negisho, K. (2012). Management of soil phosphorus and plant adaptation mechanisms to phosphorus stress for sustainable crop production: A review. *Journal of Soil Science and Plant Nutrition*, *12*(3), 547–562.

Borch, K., Bouma, T. J., Lynch, J. P., & Brown, K. M. (1999). Ethylene: A regulator of root architectural responses to soil phosphorus availability. *Plant, Cell & Environment*, *22*(4), 425–431.

Chacón-López, A., Ibarra-Laclette, E., Sánchez-Calderón, L., Gutiérrez-Alanís, D., & Herrera-Estrella, L. (2011). Global expression pattern comparison between low phosphorus insensitive 4 and WT *Arabidopsis* reveals an important role of reactive oxygen species and jasmonic acid in the root tip response to phosphate starvation. *Plant Signaling & Behavior*, *6*(3), 382–392.

Chen, B., Wang, Q., Bücking, H., Sheng, J., Luo, J., Chai, Z., ... Feng, G. (2019). Genotypic differences in phosphorus acquisition efficiency and root performance of cotton (*Gossypium hirsutum*) under low-phosphorus stress. *Crop and Pasture Science*, *70*(4), 344–358.

Cheng, L., Bucciarelli, B., Liu, J., Zinn, K., Miller, S., Patton-Vogt, J., ... Vance, C. P. (2011). White lupin cluster root acclimation to phosphorus deficiency and root hair development involve unique glycerophosphodiester phosphodiesterases. *Plant Physiology*, *156*(3), 1131–1148.

Chevalier, F., Pata, M., Nacry, P., Doumas, P., & Rossignol, M. (2003). Effects of phosphate availability on the root system architecture: Large-scale analysis of the natural variation between *Arabidopsis* accessions. *Plant, Cell & Environment*, *26*(11), 1839–1850.

Chiou, T. J., & Lin, S. I. (2011). Signaling network in sensing phosphate availability in plants. *Annual Review of Plant Biology*, *62*, 185–206.

Cordell, D., Rosemarin, A., Schröder, J. J., & Smit, A. L. (2011). Towards global phosphorus security: A systems framework for phosphorus recovery and reuse options. *Chemosphere*, *84*(6), 747–758.

Czarnecki, O., Yang, J., Weston, D. J., Tuskan, G. A., & Chen, J. G. (2013). A dual role of strigolactones in phosphate acquisition and utilization in plants. *International Journal of Molecular Sciences*, *14*(4), 7681–7701.

De Bauw, P., Vandamme, E., Lupembe, A., Mwakasege, L., Senthilkumar, K., & Merckx, R. (2019). Architectural root responses of rice to reduced water availability can overcome phosphorus stress. *Agronomy*, *9*(1), 11.

Gao, H., Yang, W., Li, C., Zhou, X., Gao, D., Khashi u Rahman, M., ... Wu, F. (2020). Gene expression and K + uptake of two tomato cultivars in response to sub-optimal temperature. *Plants*, *9*(1), 65.

Gashaw Deressa, T., & Schenk, M. K. (2008). Contribution of roots and hyphae to phosphorus uptake of mycorrhizal onion (*Allium cepa* L.)—A mechanistic modeling approach. *Journal of Plant Nutrition and Soil Science*, *171*(5), 810–820.

Gaude, N., Nakamura, Y., Scheible, W. R., Ohta, H., & Dörmann, P. (2008). Phospholipase C5 (NPC5) is involved in galactolipid accumulation during phosphate limitation in leaves of *Arabidopsis*. *The Plant Journal*, *56*(1), 28–39.

Guo, W., Zhao, J., Li, X., Qin, L., Yan, X., & Liao, H. (2011). A soybean β-expansin gene GmEXPB2 intrinsically involved in root system architecture responses to abiotic stresses. *The Plant Journal*, *66*(3), 541–552.

Haling, R. E., Brown, L. K., Bengough, A. G., Young, I. M., Hallett, P. D., White, P. J., & George, T. S. (2013). Root hairs improve root penetration, root–soil contact, and phosphorus acquisition in soils of different strength. *Journal of Experimental Botany*, *64*(12), 3711–3721.

Hammond, J. P., & White, P. J. (2008). Sucrose transport in the phloem: Integrating root responses to phosphorus starvation. *Journal of Experimental Botany*, *59*(1), 93–109.

Hammond, J. P., Bennett, M. J., Bowen, H. C., Broadley, M. R., Eastwood, D. C., May, S. T., ... White, P. J. (2003). Changes in gene expression in *Arabidopsis* shoots during phosphate starvation and the potential for developing smart plants. *Plant Physiology*, *132*(2), 578–596.

Hermans, C., Hammond, J. P., White, P. J., & Verbruggen, N. (2006). How do plants respond to nutrient shortage by biomass allocation? *Trends in Plant Science*, *11*(12), 610–617.

Jungk, A. (2001). Root hairs and the acquisition of plant nutrients from soil. *Journal of Plant Nutrition and Soil Science*, *164*(2), 121–129.

Kapri, A., & Tewari, L. (2010). Phosphate solubilization potential and phosphatase activity of rhizospheric *Trichoderma* spp. *Brazilian Journal of Microbiology*, *41*, 787–795.

Karthikeyan, A. S., Varadarajan, D. K., Mukatira, U. T., D'Urzo, M. P., Damsz, B., & Raghothama, K. G. (2002). Regulated expression of *Arabidopsis* phosphate transporters. *Plant Physiology*, *130*(1), 221–233.

Khashi u Rahman, M., Zhou, X., & Wu, F. (2019). The role of root exudates, CMNs, and VOCs in plant–plant interaction. *Journal of Plant Interactions*, *14*(1), 630–636.

Koltai, H. (2013). Strigolactones activate different hormonal pathways for regulation of root development in response to phosphate growth conditions. *Annals of Botany*, *112*(2), 409–415.

Kour, D., Rana, K. L., Kaur, T., Sheikh, I., Yadav, A. N., Kumar, V., ... Saxena, A. K. (2020). Microbe-mediated alleviation of drought stress and acquisition of phosphorus in great millet (*Sorghum bicolour* L.) by drought-adaptive and phosphorus-solubilizing microbes. *Biocatalysis and Agricultural Biotechnology*, *23*, p. 101501.

Lambers, H., Finnegan, P. M., Laliberté, E., Pearse, S. J., Ryan, M. H., Shane, M. W., & Veneklaas, E. J. (2011). Phosphorus nutrition of *Proteaceae* in severely phosphorus-impoverished soils: Are there lessons to be learned for future crops? *Plant Physiology*, *156* (3), 1058–1066.

Lan, P., Li, W., & Schmidt, W. (2012). Complementary proteome and transcriptome profiling in phosphate-deficient *Arabidopsis* roots reveals multiple levels of gene regulation. *Molecular & Cellular Proteomics*, *11*(11), 1156–1166.

Li, C., Tian, Q., u Rahman, M. K., & Wu, F. (2020). Effect of anti-fungal compound phytosphingosine in wheat root exudates on the rhizosphere soil microbial community of watermelon. *Plant and Soil*, 1–18.

Li, Y. S., Mao, X. T., Tian, Q. Y., Li, L. H., & Zhang, W. H. (2009). Phosphorus deficiency-induced reduction in root hydraulic conductivity in *Medicago falcata* is associated with ethylene production. *Environmental and Experimental Botany*, *67*(1), 172–177.

Lin, W. D., Liao, Y. Y., Yang, T. J., Pan, C. Y., Buckhout, T. J., & Schmidt, W. (2011). Coexpression-based clustering of *Arabidopsis* root genes predicts functional modules in early phosphate deficiency signaling. *Plant Physiology*, *155*(3), 1383–1402.

Lynch, J. P. (2011). Root phenes for enhanced soil exploration and phosphorus acquisition: Tools for future crops. *Plant Physiology*, *156*(3), 1041–1049.

Ma, Z., Baskin, T. I., Brown, K. M., & Lynch, J. P. (2003). Regulation of root elongation under phosphorus stress involves changes in ethylene responsiveness. *Plant Physiology*, *131*(3), 1381–1390.

Misson, J., Raghothama, K. G., Jain, A., Jouhet, J., Block, M. A., Bligny, R., ... Doumas, P. (2005). A genome-wide transcriptional analysis using *Arabidopsis thaliana* affymetrix gene chips determined plant responses to phosphate deprivation. *Proceedings of the National Academy of Sciences*, *102*(33), 11934–11939.

Muchhal, U. S., Liu, C., & Raghothama, K. G. (1997). Ca2 + -ATPase is expressed differentially in phosphate-starved roots of tomato. *Physiologia Plantarum*, *101*(3), 540–544.

Mühlroth, A., Winge, P., El Assimi, A., Jouhet, J., Maréchal, E., Hohmann-Marriott, M. F., ... Bones, A. M. (2017). Mechanisms of phosphorus acquisition and lipid class remodeling under P limitation in a marine microalga. *Plant Physiology*, *175*(4), 1543–1559.

Müller, R., Morant, M., Jarmer, H., Nilsson, L., & Nielsen, T. H. (2007). Genome-wide analysis of the *Arabidopsis* leaf transcriptome reveals interaction of phosphate and sugar metabolism. *Plant Physiology*, *143*(1), 156–171.

Nagarajan, V. K., & Smith, A. P. (2012). Ethylene's role in phosphate starvation signaling: More than just a root growth regulator. *Plant and Cell Physiology*, *53*(2), 277–286.

O'Rourke, J. A., Yang, S. S., Miller, S. S., Bucciarelli, B., Liu, J., Rydeen, A., ... Gronwald, J. W. (2013). An RNA-Seq transcriptome analysis of orthophosphate-deficient white lupin reveals novel insights into phosphorus acclimation in plants. *Plant Physiology*, *161*(2), 705–724.

Okazaki, Y., Otsuki, H., Narisawa, T., Kobayashi, M., Sawai, S., Kamide, Y., ... Saito, K. (2013). A new class of plant lipid is essential for protection against phosphorus depletion. *Nature Communications*, *4*(1), 1–10.

Park, M. R., Baek, S. H., Benildo, G., Yun, S. J., & Hasenstein, K. H. (2012). Transcriptome profiling characterizes phosphate deficiency effects on carbohydrate metabolism in rice leaves. *Journal of Plant Physiology*, *169*(2), 193–205.

Rasmussen, A., Mason, M. G., De Cuyper, C., Brewer, P. B., Herold, S., Agusti, J., ... Beveridge, C. A. (2012). Strigolactones suppress adventitious rooting in *Arabidopsis* and pea. *Plant Physiology*, *158*(4), 1976–1987.

Robinson, W. D., Carson, I., Ying, S., Ellis, K., & Plaxton, W. C. (2012). Eliminating the purple acid phosphatase At PAP 26 in *Arabidopsis thaliana* delays leaf senescence and impairs phosphorus remobilization. *New Phytologist*, *196*(4), 1024–1029.

Sharma, A. K., Mühlroth, A., Jouhet, J., Maréchal, E., Alipanah, L., Kissen, R., ... Winge, P. (2020). The Myb-like transcription factor phosphorus starvation response (PtPSR) controls conditional P acquisition and remodeling in marine microalgae. *New Phytologist*, *225*, 2380–2395.

Singh, B., & Satyanarayana, T. (2011). Microbial phytases in phosphorus acquisition and plant growth promotion. *Physiology and Molecular Biology of Plants*, *17*(2), 93–103.

Sun, B., Gao, Y., & Lynch, J. P. (2018). Large crown root number improves topsoil foraging and phosphorus acquisition. *Plant Physiology*, *177*(1), 90–104.

Tan, S. C., Liu, J. Y., Rahman, M., Ma, C. L., Wu, F. Z., & Zhou, X. G. (2020). Effects of selected root exudates components on soil *Pseudomonas* spp. community structures and abundances. *Allelopathy Journal*, *50*(1), 85–93.

Tyburski, J., Dunajska, K., & Tretyn, A. (2009). Reactive oxygen species localization in roots of *Arabidopsis thaliana* seedlings grown under phosphate deficiency. *Plant Growth Regulation*, *59*(1), 27–36.

u Rahman, M. K., Tan, S., Ma, C., Wu, F., & Zhou, X. (2020). Exogenously applied ferulic acid and p-coumaric acid differentially affect cucumber rhizosphere *Trichoderma* spp. community structure and *abundance. Plant, Soil and Environment*, *66*(9), 461–467.

u Rahman, M. K., Ma, C., Wu, F., & Zhou, X. (2021). Changes in cucumber rhizosphere *Bacillus* spp. abundance and community structure during long-term continuous monocropping. *Polish Journal of Environmental Studies*, *30*(2), 1795–1800.

Umehara, M., Hanada, A., Yoshida, S., Akiyama, K., Arite, T., Takeda-Kamiya, N., ... Kyozuka, J. (2008). Inhibition of shoot branching by new terpenoid plant hormones. *Nature*, *455*(7210), 195–200.

Vengavasi, K., Pandey, R., Abraham, G., & Yadav, R. K. (2017). Comparative analysis of soybean root proteome reveals molecular basis of differential carboxylate efflux under low phosphorus stress. *Genes*, *8*(12), 341.

Wang, X., Yi, K., Tao, Y., Wang, F., Wu, Z., Jiang, D., ... Wu, P. (2006). Cytokinin represses phosphate-starvation response through increasing of intracellular phosphate level. *Plant, Cell & Environment*, *29*(10), 1924–1935.

Wang, L., Chu, H., Li, Z., Wang, J., Li, J., Qiao, Y., ... Xu, J. (2014). Origin and development of the root cap in rice. *Plant physiology*, *166*(2), 603–613.

Wanke, M., Ciereszko, I., Podbielkowska, M., & Rychter, A. M. (1998). Response to phosphate deficiency in bean (*Phaseolus vulgaris* L.) roots. Respiratory metabolism, sugar localization and changes in ultrastructure of bean root cells. *Annals of Botany*, *82*(6), 809–819.

Wasaki, J., Shinano, T., Onishi, K., Yonetani, R., Yazaki, J., Fujii, F., ... Hashimoto, A. (2006). Transcriptomic analysis indicates putative metabolic changes caused by manipulation of phosphorus availability in rice leaves. *Journal of Experimental Botany*, *57*(9), 2049–2059.

Zhang, Z., Liao, H., & Lucas, W. J. (2014). Molecular mechanisms underlying phosphate sensing, signaling, and adaptation in plants. *Journal of Integrative Plant Biology*, *56*(3), 192–220.

Zhou, C., Jiang, W., Li, Y., Hou, X., Liu, A., & Cai, L. (2017). Morphological plasticity and phosphorus uptake mechanisms of hybrid Eucalyptus roots under spatially heterogeneous phosphorus stress. *Journal of Forestry Research*, *28*(4), 713–724.

Zhou, T., Wang, L., Sun, X., Wang, X., Chen, Y., Rengel, Z., ... Yang, W. (2020). Light intensity influence maize adaptation to low P stress by altering root morphology. *Plant and Soil*, *447*(1), 183–197.

Chapter 14

Biotic stresses on plants: reactive oxygen species generation and antioxidant mechanism

Saurav Suman[1,*], Diksha Bagal[2,*], Deepti Jain[3], Ragini Singh[1], Indrakant K. Singh[4] and Archana Singh[1]

[1]Department of Botany, Hansraj College, University of Delhi, New Delhi, India, [2]Department of Botany, University of Delhi, New Delhi, India, [3]Interdisciplinary Centre for Plant Genomics, Department of Plant Molecular Biology, Delhi University South Campus, New Delhi, India, [4]Molecular Biology Research Lab, Department of Zoology, Deshbandhu College, University of Delhi, New Delhi, India

14.1 Introduction

Plants often encounter the menace of infection by pathogens (like bacteria, fungi, viruses, and nematodes), invasion by pests, and attack by herbivores during their lifetime, which impedes their growth and development. Biotic stress negatively influences biomass production and crop yield. Due to their sessile nature, plants need to evolve certain mechanisms to combat these adverse environmental conditions (Kissoudis, van de Wiel, Visser, & van der Linden, 2014). Biotic stress leads to various physiological, biochemical, molecular, and metabolic alterations in plants (Singh & Singh, 2018). One of the crucial changes, which is observed in response to herbivore and pathogen attack, is reactive oxygen species (ROS) generation (Tripathy & Oelmüller, 2012). The major types of ROS that have been reported so far include hydrogen peroxide (H_2O_2), superoxide anion (O_2), hydroxyl radicals (OH), and singlet oxygen ($^1O_2$) (Dismukes et al., 2001; Gill & Tuteja, 2010; Karuppanapandian, Moon, Kim, Manoharan, & Kim, 2011; Vellosillo, Vicente, Kulasekaran, Hamberg, & Castresana, 2010). ROS are either partially reduced or activated oxygen derivatives. ROS are extremely reactive

* Equally contributed.

Frontiers in Plant–Soil Interaction. DOI: https://doi.org/10.1016/B978-0-323-90943-3.00014-6

and toxic, so they can easily react with other biomolecules in a cell and often cause oxidative stress and cell death (Overmyer, Brosché, & Kangasjärvi, 2003; Balaban, Nemoto & Finkel, 2005; Van Aken & Van Breusegem, 2015). ROS generation also causes posttranslational modifications of proteins, reduced photosynthetic capability, change in protein turnover; as well as transcriptional, translational, and epigenetic changes (Bailey-Serres & Mittler, 2006; Choudhury, Panda, Sahoo, & Panda, 2013; Dias et al., 2011; Golldack, Li, Mohan, & Probst, 2014; Rentel et al., 2004; Singh, Kumar, Yadav, & Singh, 2019; Soares, Wojtkowska, & Jackson, 2011)

Under normal conditions, ROS are generated as the by-product of aerobic respiration in cell organelles like chloroplast, mitochondria, and peroxisome, which are the primary sites for the production of ROS. But under biotic stress, ROS production is upregulated even at the secondary sites like endoplasmic reticulum, cell membrane, cell wall, and the apoplast (Das & Roychoudhury, 2014). Thus biotic stress leads to oxidative stress, which can cause lipid peroxidation, protein and nucleic acid damage, inhibition of enzyme activities, and activation of the programmed cell death (PCD) pathway that will eventually lead to cell death (Foyer & Noctor, 2016; Sharma, Jha, Dubey, & Pessarakli, 2012). Therefore for their protection from damage caused by higher levels of ROS, plants have devised various mechanisms for redox homeostasis, in order to seek balance between the ROS generated and scavenged. ROS level is primarily maintained by quenching activity of enzymatic antioxidants such as superoxide dismutase (SOD), catalase (CAT), phenyl ammonia lyase (PAL), glutathione reductase (GR), and nonenzymatic antioxidants such as flavonoids, phenols, ascorbic acid (AA), and reduced glutathione (GSH). (Das & Roychoudhury, 2014). At the low/moderate levels, ROS can act as the secondary messenger for various signaling pathways including tolerance to biotic stresses so as to regulate the antioxidant pathway against the oxidative stress. ROS are detected by receptors that activate the signal transduction pathway, which carry the signal to the nucleus via the mitogen-activated protein kinase (MAPK) pathway and result in transcription and translation of new proteins and modifications in already existing proteins by binding of stress-responsive transcription factors to their respective cis-acting elements (Choudhury et al., 2013; Dias et al., 2011; Golldack et al., 2014; Kumar, Ebel, & Roberts, 2011; Mittler, Vanderauwera, Gollery, & Van Breusegem, 2004; Rentel et al., 2004; Soares et al., 2011). Here, in this chapter, we summarize the impact of ROS generated during biotic stress, the mechanism by which a low/moderate level of ROS is maintained and how ROS serve as signaling molecules to induce plant defense against biotic stress.

14.2 Different forms of reactive oxygen species generated during biotic stress

The first report on production of ROS during pathogenic attack was demonstrated in *Solanum tuberosum* on infection with *Phythophthora infestans*.

In this study, it was reported that ROS are associated with plant defense against pathogens and they play a significant role in host protection. Many other plants also exhibited production of ROS during biotic stress and their involvement in different modes of plant defense such as hypersensitive reactions (HR), systemic acquired resistance (SAR), and phytohormone-mediated defense were also noted (Apel & Hirt, 2004; Tripathy & Oelmüller, 2012). When a pathogen attacks, ROS generation is mainly mediated by NADPH oxidases/respiratory burst oxidase (Fridovich, 1995; Li, Jia, & Trush, 2016; Bailey-Serres & Mittler, 2006; Torres & Dangl, 2005), cell wall peroxidases (Bolwell, Davies, Gerrish, Auh, & Murphy, 1998; Grant, Yun, & Loake, 2000), and copper containing amine oxidase (Jasso-Robles et al., 2020). This ROS production in the apoplast causes oxidative burst and is considered to be the first line of inducible defense. NADPH oxidases are located in the plasma membrane and synthesize superoxide from oxygen molecules by transferring e^- from NAPDH (Giesbert, Schuerg, Scheele, & Tudzynski, 2008; Marino, Dunand, Puppo, & Pauly, 2012). This superoxide can be converted to H_2O_2 by SOD present on the cell wall (Kar, 2015). The apoplastic H_2O_2 can also be generated by peroxidases (Bolwell et al., 1998; Wojtaszek, 1997). ROS production from molecular oxygen occurs either by absorbing energy that reverses the spin of one of the unpaired electrons, which leads to the formation of singlet oxygen $^1O_2$ or by stepwise monovalent reduction in which OH•, H_2O_2, and $O_2•^-$ can be formed (Sharma et al., 2012). Different forms of ROS have a particular half-life and they vary in their capacity to oxidize cellular compartments. ROS produced during adverse environmental conditions can be either free radicals (superoxide $O_2•^-$, Hydroxyl ions OH•) or nonradicals (H_2O_2, $^1O_2$) type.

14.2.1 Singlet oxygen ($^1O_2$)

$^1O_2$ is generated during biotic stress due to closing of the stomata and lower concentration of CO_2 inside the cell. In the antennae system and reaction center of photosystem II of chloroplasts, $^1O_2$ is produced by the triplet chlorophyll, ^{3}Chl. Inadequate energy depletion during photosynthesis in the antenna system forms ^{3}Chl, while in the reaction center it is formed by the charge recombination of the light-induced charge pair. In the antennae system, O_2 reacts with the ^{3}Chl to produce a profoundly reactive ROS $^1O_2$ (Krieger-Liszkay, 2005). The half-life of the $^1O_2$ is approximately 3 μs (Hatz, Lambert, & Ogilby, 2007) and it is able to diffuse over 100 nm distance, which may lead to the destruction of various cellular components (Krieger-Liszkay, 2005; Wagner et al., 2004). The $^1O_2$ produced can oxidize the polyunsaturated fatty acids (PUFA), guanine, and several other amino acids. It can also cause nucleic acid modification by reacting with the deoxyguanosine (Kasai, 1997). The $^1O_2$ produced can also react to D1 protein of the PSII that may result in loss of PSII activity (Krieger-Liszkay, 2005).

But $^1O_2$ can be scavenged by β-carotene and α-tocopherol (Hasanuzzaman et al., 2020).

14.2.2 Superoxide radical ($O_2\bullet^-$)

When O_2 gains an electron and is left with only one unpaired electron, superoxide radicals are formed (Halliwell, 2006). $O_2\bullet^-$ is the primary ROS and furthermore it can form other secondary ROS directly or indirectly by enzyme- or metal-catalyzed reactions (Valko, Morris, & Cronin, 2005). During the noncyclic electron transport chain (ETC), a superoxide radical $O_2\bullet^-$ is formed in the PSII localized in the thylakoid, and also in other cellular compartments like chloroplast and plasma membrane (Das & Roychoudhury, 2014). Superoxide radicals have the half-life of approximately 1 μs (Halliwell, 1977). They are considered to be moderately reactive, so they undergo various reactions to form highly reactive ROS like hydroxyl ion OH•, singlet oxygen $^1O_2$, or H_2O_2 (Halliwell, 2006). Superoxide radicals react with the proteins having Fe-S center and can either cause protein degradation or activation of various signaling pathways (Imlay, 2003).

14.2.3 Hydrogen peroxide (H_2O_2)

Hydrogen peroxide is generated by the dismutation of oxygen superoxide at low pH either nonenzymatically or usually by the enzyme SOD. H_2O_2 is also produced during normal conditions as it regulates various physiological processes like stomatal closure, PCD, SAR, growth regulation, regulatory signals for processes like senescence, photorespiration, and photosynthesis (Ślesak, Libik, Karpinska, Karpinski, & Miszalski, 2007). The half-life of H_2O_2 is 1 ms (Bienert et al., 2007). It is a much more stable ROS. It can cross membranes and its diffusion is mediated by the aquaporins. H_2O_2 can be produced in the chloroplast, peroxisome, mitochondria, cytoplasm, extracellular matrix, and cell membrane (Halliwell, 2006). At lower concentration it can work as a signaling molecule for various processes, but at higher concentration it leads to the oxidation of sulfur containing amino acids, like cysteine (-SH) or methionine (-SCH3) and the enzymes of the Calvin cycle, and inactivates Cu/Zn-SOD and Fe-SOD by their thiol group oxidation (Halliwell, 2006). It also hampers other processes by oxidizing the protein kinases, phosphatases, and transcription factors containing thiolate residue which can ultimately lead to cell mortality due to higher oxidative stress. H_2O_2 can be scavenged by various antioxidants like ascorbate, glutathione, α-tocopherol, GSH, and catalase (Halliwell & Gutteridge, 2015).

14.2.4 Hydroxyl radical (OH•)

Among all the ROS produced, the hydroxyl radical is the most reactive and toxic due to its extremely unstable single unpaired electron (Sharma et al.,

2012). It is generated by the Fenton reaction in which H_2O_2 is converted into OH• and transition metals like (Fe^{+2}, Fe^{+3}) work as the catalyst. OH•, being most reactive, causes oxidative stress to the great extent and damages the cell by initiating lipid peroxidation and protein, nucleic acid, and membrane destruction. Since no enzymatic system to scavenge the hydroxyl radical is known as yet, its excess amount in the cell can lead to cell death (Pinto et al., 2003). OH• are scavenged only through nonenzymatic antioxidants like proline, ascorbate, etc. (Hasanuzzaman et al., 2020).

14.3 Generation of reactive oxygen species in different cellular compartments

In the plant cell, ROS are produced under both normal and stressed conditions. Normally, ROS are formed primarily in the chloroplast, mitochondria, and peroxisomes but when plants undergo biotic stress, ROS are also produced at the secondary sites such as endoplasmic reticulum, cell wall, cell membrane, and the apoplast (Fig. 14.1). In the chloroplast, mitochondria, and plasma membranes ROS are formed by the unavoidable leakage of electrons in the ETS to the O_2 or as the by-product of various processes in other cellular compartments(Sharma et al., 2012).

14.3.1 Chloroplast

The PSI and PSII are responsible for the ROS production in the chloroplast. PSI and PSII form the center of the light harvesting system and are incorporated into the thylakoid membrane of the chloroplast. Under normal conditions, PS gets excited on receiving light and electrons flowing from it reduce $NADP^+$ to NADPH, which further enters the C_3 cycle and reduces CO_2 into glucose (Elstner & Osswald, 1994; Khorobrykh, Havurinne, Mattila, & Tyystjärvi, 2020). Under stressed conditions, CO_2 concentration is limiting and leads to the formation of superoxide ($O_2^{\bullet-}$) by the Mehler reaction in which electrons leaked from ferredoxin reduce O_2 to $O_2^{\bullet-}$. In ETC of PSI leakage of electron from 2Fe-2S and in PSII flowing out of electrons from Q^A and Q^B acceptors are responsible for the formation of $O_2\bullet^-$ (Cleland & Grace, 1999). Leakage of electrons in the PSII also leads to the formation of the singlet oxygen ($^1O_2$) or it can also be formed by the reaction of triplet chlorophyll ($^3Chl^*$) and O_2 (Karuppanapandian et al., 2011). The conversion of O_2 to $O_2^{\bullet-}$ is the rate-limiting step. $O_2^{\bullet-}$ can further undergo reactions to convert into more reactive ROS. It may be dismutated to H_2O_2 by the enzyme SOD on the external stromal membrane by the membrane bound Cu/Zn SOD at PSI (Miller, Suzuki, Ciftci-Yilmaz, & Mittler, 2010), while in the lumen $O_2^{\bullet-}$ is protonated to form the most reactive hydroxyl form (OH) at the Fe-S center via the Fenton reaction. The 'O_2 which gets accumulated in the chloroplast is dangerous because it causes lipid peroxidation, mainly

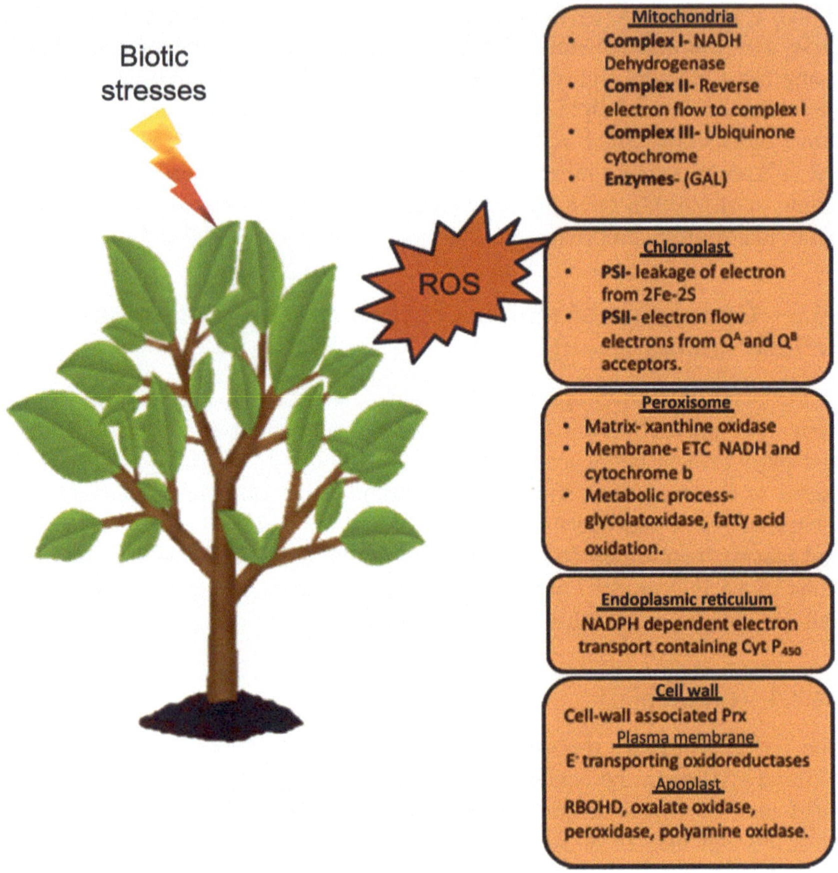

FIGURE 14.1 ROS producing cellular compartments. *RBOHD*, respiratory burst oxidase homology D; *ROS*, reactive oxygen species.

of PUFA. It also disrupts membrane proteins. Consequently, the P_{680} reaction center of PSII becomes at risk, which can also directly lead to cell mortality (Møller, Jensen, & Hansson, 2007; Triantaphylidès et al., 2008).

14.3.2 Mitochondria

In mitochondria, ROS like H_2O_2 and $O_2\bullet^-$ are produced under normal conditions but when the plant undergoes biotic stress then electron leakage and overreduction of electron carriers lead to an increase in the amount of ROS produced (Murphy, 2009). ROS formation takes place in the complex I, complex II, and complex III of ETC (Møller et al., 2007; Noctor, De Paepe, & Foyer, 2007). In the complex I (NADH dehydrogenase) flavoprotein region it leads to the reduction of oxygen to superoxide anions ($O_2\bullet^-$). The ROS formation at NADH dehydrogenase further increases during the reverse electron transport from the complex III to complex I due to limiting NAD^+-linked substrate. The electron backflow can be regulated by ATP hydrolysis (Turrens, 2003). In the complex III (ubiquinone cytochrome) region an electron is donated by the fully reduced ubiquinone to the cytochrome c_1 that forms a highly unstable ubisemiquinone radical that in turn leaks electron to O_2 and thus forms $O_2\bullet^-$. The complex III releases $O_2\bullet^-$ to both the intermembrane and matrix side (Rasmusson, Geisler, & Møller, 2008). ROS can also be produced by some enzymes in the mitochondria, e.g., aconitase produces ROS directly while 1-galactono-γ-lactone dehydrogenase (GAL) feeds electrons directly to the ETC. $O_2\bullet^-$, being unstable, is converted quickly to H_2O_2 by Mn-SOD (mitochondrial SOD) or into H_2O_2 by Ascorbate peroxidase (APX), which is more stable and membrane permeable (Sharma et al., 2012). It has been reported that further H_2O_2 is converted to OH• by Fenton reaction. To restrain the oxidative stress two enzymes, alternative oxidase (AOX) and Mn-SOD, play a vital role in mitochondria (Ho et al., 2008; Mittova, Tal, Volokita, & Guy, 2003).

14.3.3 Peroxisome

Peroxisomes are the foremost site for H_2O_2 formation due to its innate oxidative metabolism. When a plant undergoes any biotic stress, the reduction in CO_2 availability causes stomata to close and leads to photorespiration that produces glycolate. The oxidation of glycolate into glycolate oxidase causes H_2O_2 production inside the peroxisome (Noctor, Veljovic-Jovanovic, Driscoll, Novitskaya, & Foyer, 2002). Similar to chloroplast and mitochondria, $O_2\bullet^-$ production also takes place in peroxisome at two different sites. One, in the organelle matrix, xanthine oxidase produces $O_2\bullet^-$ as the by-product while converting xanthine and hypoxanthine into uric acid. Another site for $O_2\bullet^-$ production is a small ETC that is NADPH dependent, composed of NADH and Cyt b confined in the membrane of the peroxisome. The Cyt b present in the membrane uses oxygen molecule as the acceptor of electron and releases $O_2\bullet^-$ into the cytosol. For $O_2\bullet^-$ production three integrated peroxisomal membrane proteins (pmps) of molecular masses 18, 24, and 32 kDa have been found to be responsible. For the pmps of molecular mass 18 and 32 kDa, NADH act as the electron donor while NADPH act as the electron donor for the 29 kDa PMP (Del Río, Sandalio, Corpas, Palma, & Barroso, 2006).

14.3.4 Endoplasmic reticulum

The major ROS produced in the ER is superoxide. The NADPH-dependent electron transport containing Cyt P_{450} in the ER produces $O_2\bullet^-$. When organic substrate comes in contact with the Cyt P_{450} then it is reduced to a radical intermediate (Cyt $P_{450}R^-$) that is reduced by Cyt b or decomposed to produce $O_2\bullet^-$ as its by-product(Mittler, 2002).

14.3.5 Apoplast

Apoplastic ROS is mainly produced in response to pathogen-associated molecular patterns (PAMPs) that are wholly dependent on the NADPH-regulated respiratory burst oxidase homology D (RBOHD) (Smith & Heese, 2014). Also, while exposed to biotic stress, phytohormone biosynthesis is coupled with ROS generation, primarily H_2O_2. In the apoplast ROS are produced by the enzymes present at the cell wall, like oxalate oxidase (also known as germin), which on reacting with oxalic acid produces H_2O_2 and CO_2. Other enzymes like pH-dependent peroxidase (poxs) and polyamine oxidase also produce H_2O_2 (Wojtaszek, 1997). Enzymes like amine oxidase may help in defense mainly due to the production of H_2O_2 in the apoplast. Amine oxidase uses FAD as a cofactor and catalyzes the oxidative deamination of polyamine (Cona, Rea, Angelini, Federico, & Tavladoraki, 2006). The extracellular alkalization due to Ca^{2+} influx in cytoplasm and K^+ influx due to the pathogen infection is also responsible for the oxidative burst in the apoplast. In the apoplast H_2O_2 can also undergo various reactions to produce more reactive radicals like OH• (Heyno, Mary, Schopfer, & Krieger-Liszkay, 2011).

14.3.6 Plasma membrane

The plasma membrane has an important role in cell survival when the plant is undergoing biotic stress. There is an extensive presence of the electron transporting oxidoreductase at the plasma membrane that is responsible for ROS production during biotic stress. The NADPH oxidase forms $O_2\bullet^-$ by catalyzing the electron transferred from the cystolic NADPH to O_2, which is then instantaneously dismutated to H_2O_2 or with the activity of enzyme SOD (Apel & Hirt, 2004). NADPH oxidase is the key player in the plant defense against biotic stressors (Kwak et al., 2003).

14.3.7 Cell wall

The cell wall is actively engaged in the ROS production. In the presence of NADH, peroxidase associated with the cell wall catalyzes H_2O_2 production. NADH is provided mainly by malate dehydrogenase, and the reaction is triggered by monophenols like coniferyl alcohol l. Lipoxygenase present in the cell wall caused PUFA hydroperoxidation releasing various ROS like $^1O_2$, OH•, H_2O_2, and $O_2\bullet^-$. ROS can also be produced by the diamine oxidase present in the cell wall by utilizing diamine or polyamine. On pathogen attack, H_2O_2 present in the cell wall mediated the substantial lignin precursor cross-links (Higuchi, 2006)

14.4 Deleterious effects of high levels of reactive oxygen species

At low/medium concentration ROS act as the secondary messenger molecules of various signal transduction pathways but when the ROS concentration increases due to biotic stress, it leads to oxidative stress and eventually can cause cell death. When the balance between the ROS formed and scavenged is disturbed, it can instigate oxidative stress, which can damage biomolecules like lipid, proteins, DNA (Fig. 14.2), and alter various other cell properties (Sharma et al., 2012)

14.4.1 Effect of reactive oxygen species on lipid

The major portion of the plasma membrane is constituted by lipid. Under stressed conditions, increased ROS level leads to oxidation of the lipid in cellular and organellar membranes. After lipid peroxidation lipid radicals are formed which further destruct protein and nucleic acid by starting chain reactions, ROS can target two sites on the phospholipid molecule: one at the unsaturated C atom (double bond C atom), and another at the ester bond formed between the glycerol and the fatty acid. In the membranes, many

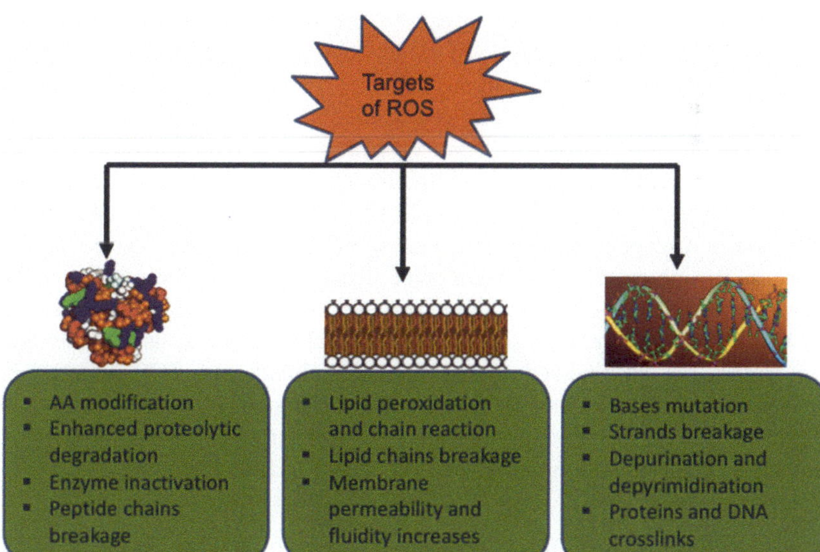

FIGURE 14.2 Main targets of ROS: DNA, lipids and proteins. *AA*, ascorbic acid; *ROS*, reactive oxygen species.

PUFAs such as linoleic acid and linolenic acid are present, which are the sites for the attack by ROS such as $^1O_2$ and OH• (Das & Roychoudhury, 2014). PUFA oxidation can produce malondialdehyde (MDA) and 4-hydroxynonenal. Among all the ROS produced, the hydroxyl ion (OH•) is the most reactive and initiates chain reactions that lead to further damage of other PUFA present (Tsikas, 2017). Lipid peroxidation occurs in three steps: initiation, propagation, and termination (Girotti, 1998). Initially ROS like OH• and $O_2{\bullet}^-$ are produced by activating O_2 which is the rate-limiting step. ROS formed then reacts with the methylene of PUFA to yield conjugate dienes, lipid peroxyl radicals, and hydroperoxide (Smirnoff, 2000). The peroxyl radical (PUFA − OO•) is highly reactive and continues the lipid peroxidation by further withdrawing H-atoms from the side chain of nearby PUFA. The lipid hydroperoxide formed is cleaved by the reduced metals like Fe^{2+}. The PUFA alkoxy radical can again start another reaction (Buettner, 1993). Conjugated dienes are formed when the rearrangement of the double bond in the methylene group occurs by the free radical attack on its hydrogen. The termination of lipid peroxidation occurs when the lipid-derived radicals react with each other to form lipid dimers. Lipid peroxidation breaks the cell membrane unsaturated fatty acid that can damage the receptors present in membrane, enzymes, and ion-channels and increases the fluidity and permeability of the membrane.

14.4.2 Effect of high dose of reactive oxygen species on proteins

During stress conditions, ROS can cause oxidation of proteins which may be direct or indirect. By direct mode of oxidation, proteins are modified via nitrosylation, carbonylation, glutathionylation, and disulfide bond formation (Yamauchi et al., 2008). Cells exposed to biotic stress enhance the concentration of carbonylated proteins that are very often used as the markers for protein oxidation (Møller et al., 2007). Proteins are indirectly modified when they interact with the lipid peroxidation products. In the protein, all the amino acids differ in the susceptibility to be attacked by ROS. Thiol- and sulfur-containing amino acids like cysteine and methionine, respectively, are more prone to the attack of ROS such as $^1O_2$ and OH• (Brot & Weissbach, 1982). ROS extracts the H-atom from cysteine to form a thiyl radical to create a disulfide bridge., and methionine is oxidized to methionine sulfoxide on ROS attack. The enzymes with a Fe-S center are inactivated irreversibly by $O_2{\bullet}^-$ (Gardner & Fridovich, 1991). Oxidized proteins are the better targets for proteolytic digestion and undergo ubiquitination-mediated proteasomal degradation.

14.4.3 Effect of high dose of reactive oxygen species on DNA

During oxidative stress, ROS produced can damage nuclear, mitochondrial, and chloroplast DNA. Oxidative damage on the DNA can cause double-strand breakage, deoxyribose sugar oxidation, DNA–protein cross-linking, or nucleotide base modification and their removal. To protect nuclear DNA, it has histones and nonhistone proteins, but mitochondrial and chloroplast DNA lacks histone, so they are more prone to ROS attack. OH• damages deoxyribose sugar by scavenging its hydrogen atoms and also reacts with the double bond of purine and pyrimidine (Das & Roychoudhury, 2014). ROS can also attack DNA indirectly through the highly reactive radicals generated by the lipid peroxidation. ROS causes single-strand breaks in DNA by extracting H-atoms from the fourth position carbon atom (C_4) of deoxyribose sugar and forms deoxyribose radicals (Evans, Dizdaroglu, & Cooke, 2004). H_2O_2 and $O_2{\bullet}^-$ alone cannot cause much damage. But OH• produced via Fenton reaction can cause DNA–protein cross-linking. If not repaired before the onset replication or transcription, these cross-links can be highly deleterious for the cell (Oleinick, Chiu, Ramakrishnan, & Xue, 1987).

14.5 Reactive oxygen species scavenging by antioxidants

To decline the harmful effects of the ROS, plants develop ROS scavenging machinery which consists of enzymatic and nonenzymatic antioxidants. The ROS scavenging pathways from several organelles like the mitochondria, chloroplasts, and peroxisomes are coordinated.

14.5.1 Enzymatic antioxidants

The enzymatic antioxidants are present in different subcellular compartments that respond on the onset of oxidative stress. The enzymatic antioxidants are sodium dismutase (SOD), CAT, phenyl ammonia lyase (PAL), and GR.

14.5.1.1 Superoxide dismutase (EC: 1.15.1.1)

SOD are the members of metalloenzymes and give prime protection to the plant when it undergoes oxidative stress due to ROS in all aerobic organisms. The dismutation of $O_2{\bullet}^-$ into H_2O_2, and O_2 is catalyzed by SOD. This avoids the production of more reactive OH• via the Haber–Weiss reaction. On the basis of SOD binding on metal it is classified into three isoenzymes; Mn-SOD (in mitochondria), Fe-SOD (chloroplast), and Cu/Zn-SOD (cytosol, chloroplast and peroxisome) (Mittler, 2002). Increased production of SOD also corresponds to the increased tolerance of plants against oxidative stress caused due to biotic invaders.

14.5.1.2 Catalase (E.C:1.11.1.6)

Catalase was the first discovered antioxidant enzyme. It is an ubiquitous heme-containing tetrameric enzyme. It catalyzes the dismutation of H_2O_2 into H_2O and O_2. It has a very high enzyme activity and CAT can dismutate approximately 6×10^6 molecules of H_2O_2 per minute. Catalases are peculiar as there is no requirement for a cellular reducing equivalent in its case. They are predominately found in the peroxisome as it is the hub of the H_2O_2 production due to fatty acid β-oxidation, photorespiration, oxidative stress, and purine catabolism (Mittler, 2002). CATs are also found in other cellular parts like cytosol, chloroplast, and mitochondria (Mhamdi et al., 2010). In angiosperms three CAT genes exist: class I CAT (in photosynthetic tissues), class II CAT (in vascular tissue), and class III CAT (in seeds and young seedlings) (Willekens et al., 1997). Under stressed conditions and oxidative stress, H_2O_2 is removed efficiently by CATs.

14.5.1.3 Guaiacol peroxidase (EC:1.11.1.7)

GPX is a heme-containing protein which is able to scavenge H_2O_2 under normal and stress conditions. The GPX enzyme is a monomer of 40–50 kDa with four conserved disulfide and two structural calcium ions (Schuller, Ban, Van Huystee, McPherson, & Poulos, 1996). GPX plays an essential role in the lignification of the cell wall, IAA degradation, ethylene biosynthesis, wound healing, and H_2O_2 scavenging during biotic stress (Kobayashi, Kumazawa, Miwa, & Yamanaka, 1996). Some of the isoenzymes of GPX are also found in vacuoles, the cell wall, and cytosol (Asada, 1992). GPX mostly oxidizes aromatic compounds such as guaiacol and pyragallol at the expense of H_2O_2 (Asada, 1999). GPX catalyzes the H_2O_2 conversion into H_2O and reduced glutathione (GSH), which is further oxidized to glutathione disulfide (GSSH).

14.5.1.4 Ascorbate peroxidase (E.C: 1.1.11.1)

It is the crucial enzyme for the ascorbate–glutathione (ASC-GSH) cycle where it catalyzes H_2O_2 conversion into water molecule by utilizing AA as an electron donor, which is then converted into DHA. APX is mostly found in chloroplasts, but can also scavenge ROS in the cytosol, mitochondria, and peroxisome (Caverzan et al., 2012). Based on the different amino acid sequences and location, the APX family has five distinct isoforms: cytosolic, mitochondrial, peroxisomal, stromal, and thylakoidal (Sharma & Dubey, 2004). APX is broadly distributed in most of the cellular compartments and is more efficient for scavenging H_2O_2 than the CATs under stress conditions (J. Wang, Zhang, & Allen, 1999)

14.5.1.5 Monodehydroascorbate reductase (E.C: 1.6.5.4)

Monodehydroascorbate reductase (MDHAR) is a FAD (flavin adenine dinucleotide) enzyme that catalyzes the conversion of MDHA radical to AA using NADPH as the electron donor (Hossain & Asada, 1985). This is the only enzyme that can utilize organic radicals, i.e., MDA, and can also reduce phenoxyl radicals. As the MDHAR generated AA, so it is present in the peroxisome and mitochondria with APX as it can oxidize the AA produced by MDHAR. There are various isoenzymes of MDHAR present in the chloroplast, mitochondria, peroxisome, cytosol, and glyoxysome.

14.5.1.6 Dehydroascorbate reductase (EC:1.8.5.1)

Dehydroascorbate reductase (DHAR) is the monomeric enzyme and it catalyzes the reduction of DHA to AA by utilizing reduced glutathione as an electron donor (Eltayeb et al., 2007). Like MDHAR, DHAR also contributes AA production. This is important to sustain the redox state by modulating the AA in both symplast and apoplast (Z. Chen & Gallie, 2006). DHAR is found in seeds, roots, and green and etiolated seedlings.

14.5.1.7 Glutathione reductase (E.C: 1.6.4.2)

GR is a flavoenzyme with prime disulfide group (Gill et al., 2013). It reduces GSSH (glutathione disulfide) to GSH (reduced glutathione) by utilizing NADPH. In the GSH cycle, GSH is used to rejuvenate AA from MDHA and DHA and then converted to GSSR. Thus GR is a key enzyme to catalyze the reduction of GSSR to GSH so as to maintain a high cellular GSH:GSSR ratio. GR is mostly found in the chloroplast and some amount is also present in the mitochondria and cytosol.

14.5.2 Nonenzymatic antioxidants

They are present in many cellular compartments and also play an important part in plant growth, development, and stress management by regulating processes like meiosis, cell elongation, senescence, and cell death (De Pinto & De Gara, 2004). They include flavonoids, phenols, AA, and reduced glutathione (GSH).

14.5.2.1 Ascorbic acid

AA is a low-molecular-weight compound, which is a widely found antioxidant in many cellular compartments. It is an efficient antioxidant as it can donate electrons to enzymatic as well as nonenzymatic reactions. It plays a major role in several physiological processes like growth, differentiation, and metabolism. The AA pool is mostly formed by the D-mannose/L-galactose via the Smirnoff–Wheeler pathway (Wheeler, Jones, & Smirnoff, 1998) and also synthesized from uronic acid intermediates like D-galacturonic acid (Isherwood, Chen, & Mapson, 1954). AA is the principal nonenzymatic antioxidant as apart from the cytoplasm, it is also found in apoplasts (Barnes, Zheng, & Lyons, 2002). AA removes H_2O_2 via the ASA-GSH cycle which occurs in two consecutive steps: first MDHA production and eventually DHA. DHA is reactive at a pH of more than 6 and breaks down into tartarate and oxalate (Noctor & Foyer, 1998). To avoid this DHA is reduced by the action of enzyme DHAR. AA protects the membrane from oxidative stress by reacting with ROS like OH•, H_2O_2, and O_2•$^-$ and producing α-tocopherol from tocopheroxyl radical., AA also prevents photooxidation by regulating pH-dependent PSII activity and its downregulation which is connected with zeaxanthin formation.

14.5.2.2 *Reduced glutathione (GSH)*

Glutathione or γ-glutamyl-cysteinyl-glycine is a ubiquitous important low-molecular-weight thiol tripeptide localized in almost all cellular compartments like chloroplast, mitochondria, vacuoles, cytosol, peroxisome, and apoplast. Apart from being a potential reducing agent, due to its high reproductive potential, GSH plays a crucial role in processes like cell differentiation, growth and division, senescence and cell death, xenobiotics detoxification, sulfate transport regulation, metabolite conjugation, enzyme activity regulation, and synthesis of proteins, nucleotides, and phytochelatins (Mullineaux & Rausch, 2005). GSH reacts with ROS such as OH•, H_2O_2, and O_2•$^-$ and protects macromolecules like proteins, lipids, and DNA by directly reacting with electrophiles like glutathionate and forming adducts or by forming GSSH (oxidized glutathione) by donating electrons to ROS or free organic radicals. The GSH pool is refilled by converting GSSH produced into GSH again either de novo or by the enzymatic action of GR. Heavy metal ions are one of the major sources of the ROS production, GSH helps in the formation of phytochelatins, which can chelate these heavy metal ions, and ceases the ROS production (Roychoudhury, Pradhan, Chaudhuri, & Das, 2012). So, the ratio of GSH: GSSG should be maintained to sustain the cellular redox state.

14.5.2.3 *Tocopherol*

Tocopherol is a lipophilic antioxidant which has the potential to scavenge oxygen and lipid radicals. Among the four isomers of tocopherol (α, β, γ, and δ), antioxidant activity varies according to the methylation pattern and attached methyl group amount (Fukuzawa, Tokumura, Ouchi, & Tsukatani, 1982). α-Tocopherol has the highest antioxidant activity due to the presence of three methyl groups (Kamal-Eldin & Appelqvist, 1996). Tocopherol is present only in the green parts of the plant as it is synthesized only by the photosynthetic regions. By physically quenching ROS tocopherols protect membrane lipids and in chloroplasts they react with O_2 to protect the PSII structure. Tocopherols are also known to cease the lipid free radical chain reaction by reacting with the lipid radicals RO•, ROO•, and RO* and reduces them into TOH• which recycle by interacting with the GSH and AA.

14.5.2.4 *Carotenoids*

Carotenoids are the lipophilic antioxidants which are found in both photosynthetic and nonphotosynthetic plants. Apart from plants the presence of carotenoids can be seen in the microorganisms. They form the antennae molecules and can absorb light of wavelength 450–570 nm length (Sharma et al., 2012). They are known to protect the photosystem in several ways: they react with lipid free radicals and inhibit chain reaction; they scavenge singlet oxygen $^1O_2$ and produce heat as the by-product; they react with the triplet and excited chlorophyll ($^3Chl^*/Chl^*$) to avoid the $^1O_2$ formation; and they are involved in the dissipation of extra energy with the help of xanthophyll cycle (Das & Roychoudhury, 2014).

14.5.2.5 Flavonoids

Flavonoids are the phenolic compounds present in leaves, floral parts, and pollen grains. On the basis of their structure flavonoids are classified into four major classes: flavones, isoflavones, flavanols, and anthocyanin. Apart from providing pigmentation and their role in plant fertility and pollen germination, they act as the defense against the plant pathogens. They require excess excitation energy, so are considered as the secondary ROS scavenging system (Fini, Brunetti, Di Ferdinando, Ferrini, & Tattini, 2011).

14.5.2.6 Proline

Proline is an osmolyte, which is also considered to be a strong antioxidant. It is a nonenzymatic antioxidant that can neutralize the ROS effects on the membrane. It is synthesized using glutamic acid as a substrate, via a pyrroline 5-carboxylate (P5C) intermediate. This process in plants is catalyzed by two enzymes: 8-pyrroline-5-carboxylate synthetase (P5CS) and ð1-Pyrroline-5- carboxylate reductase (P5CR). It can avoid lipid damage by effectively removing OH• and $^1O_2$. Accumulation of proline occurs during stress conditions in plants which may be due to an increase in synthesis or a reduction in degradation (Verbruggen & Hermans, 2008).

14.6 Role of reactive oxygen species in signaling and instigation of plant defense

Plants have developed complex defense mechanisms to defend against invading pathogens and pests. When pathogens approach their host, they release MAMPS/PAMPs that are perceived by membrane-bound pattern recognition receptors (PRRs), that trigger basal defense, i.e., PAMP-triggered immunity (PTI) (Jones and Dangl, 2006). However, a few pathogens can overcome this defense mechanism, therefore plants have an alternative mechanism to protect themselves: R-gene-mediated effector triggered immunity (ETI) (Jones and Dangl, 2006; Dangl, Horvath, & Staskawicz, 2013). Both defense mechanisms involve ROS generation (Grant et al., 2000; Torres, Jones, & Dangl, 2006; Kadota et al., 2014). Pathogen-induced ROS production plays a significant role in signal transduction during plant immune responses. During

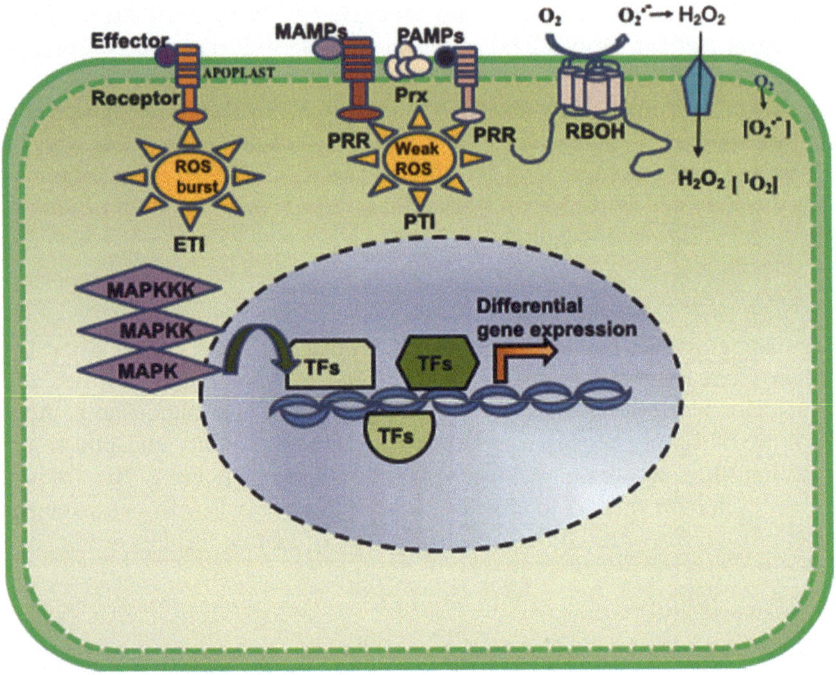

FIGURE 14.3 Generation of ROS and signaling during biotic stress. *MAMPs*, Microbe-associated molecular patterns; *Prx*, peroxiredoxin; *TFs*, transcription factors. Upon pathogen attack, Pattern recognition receptors/receptors (R) perceive MAMPs, PAMPs or effectors, and RBOH and peroxidases are activated to produce ROS. Interaction between PRR and MAMPs/PAMPs causes a weak ROS burst leading to PTI, whereas interaction between receptor/R and effectors lead to a strong ROS burst, which results in ETI. High levels of ROS activate the MAPK cascade, which further stimulates transcription factors leading to differential gene expression. *ETI*, effector triggered immunity; *MAPK*, mitogen-activated protein kinase; *MAPKK*, MAPK kinase; *MAPKKK*, MAPKK kinase; *PAMPs*, pathogen-associated molecular patterns; *PRR*, pattern recognition receptor; *PTI*, PAMP-triggered immunity; *RBOH*, respiratory burst oxidase homolog; *ROS*, reactive oxygen species.

pathogen attack, ROS are generated in the apoplast mainly by peroxidases and NADPH-oxidase/RBOHD with electrons delivered by the NADP-malic enzyme. When plants perceive MAMPs/PAMPs, either a weak or strong ROS bursts is witnessed that leads to PTI and ETI, respectively (Fig. 14.3). Amongst all the ROS generated, H_2O_2 is present at maximum concentration and is stable as well. Moreover, it can be transported from extracellular region to cytosol and also to neighboring cells. H_2O_2 and other ROS are considered as efficient weapons against invading pathogens. They can inhibit pathogen growth themselves and can also instigate signal transduction for further activation of efficient defense mechanisms.

During biotic stress, an extra load of ROS are generated that are polarized and delivered to other cellular compartments and can contribute to pathogen growth inhibition (Hückelhoven & Kogel, 2003). ROS, along with SA (salicylic acid), were found to regulate the establishment of SAR (Durrant, & Dong, 2004). The rapidity of ROS generation and the potential for free diffusion of H_2O_2 across the membranes gives an indication that ROS could function as an intercellular or intracellular secondary messenger; (Lamb & Dixon, 1997; Levine, Tenhaken, Dixon, & Lamb, 1994). Also, ROS metabolism could affect the NPR1 functioning, a significant mediator/regulator of these systemic responses, by controlling the redox state of NPR1 (Mou, Fan, & Dong, 2003). In tomato plants, ROS also acts as a second runner for incorporation of defense genes in reaction to cuts and wounds (Orozco-Cárdenas, Narváez-Vásquez, & Ryan, 2001). ROS are produced nearby the cell walls of vascular bundles and the resultant H_2O_2 produced by wound-triggered polygalacturonate acts as a reduplicative messenger for induction of defense genes in cells of mesophyll tissues of tomato leaves; however, they don't mediate pathways of signaling genes in cells of stele, particularly vascular bundles (Orozco-Cárdenas et al., 2001).

Additionally, ROS also act as signaling molecules regulating various plant responses in cells, such as apoptosis (Bethke & Jones, 2001; Mittler, 2002), and endurance to various biotic as well as abiotic stress, particularly at optimum or limited concentration (Torres, Dangl, & Jones, 2002). Various redox-sensitive proteins, calcium mobilization, protein phosphorylation, and, through altered gene expression, ROS assist plants in sensing and transmitting ROS signals into reasonable cellular activities. Various signaling biomolecules, namely protein signaling, i.e., tyrosine phosphate can recognize ROS by oxidation of cysteine residues (Xiong, Schumaker, & Zhu, 2002). ROS may often regulate the functioning of various signaling molecules, namely phosphatases, kinases, and transcription factors (Z. Chen & Gallie, 2006). ROS also interact with other molecules constituting footsteps for signal transduction network regulating ROS downstream response (Neill, Desikan, & Hancock, 2002). Equilibrium between the production of an oxidant and its eviction by antioxidant determines the efficiency and longevity of ROS signaling networks. Plants have developed an intricate regulatory network to intervene in biotic and abiotic stress reaction by regulating ROS synthesis, scavenging, and signaling (Nanda, Andrio, Marino, Pauly, & Dunand, 2010). SA is considered essential for dynamic ROS signaling to establish HR (Klessig et al., 2000).

14.7 Reactive oxygen species sensing and stimulation of mitogen-activated protein kinase signaling

Cells need to sense the change in levels of ROS in response to environmental stresses. An accelerated as well as an efficient sensing mechanism is needed for ROS sensing as every ROS has a very short life span. Receptors (Liang & Zhou, 2018), redox sensitive compounds, such as transcription factors, heat shock factors, NPR1, and phosphatase inhibitors are the main factors involved in ROS sensing. The foremost sensing of ROS is encountered at the apoplast, which causes activation of intra- and intercellular signaling resulting in changes in the expression of genes. ROS exists in four distinct forms, which are produced in various cell organelle/compartments (Mhamdi & Van Breusegem, 2018; Mittler, 2017). Interaction with particular cellular targets for various forms of ROS concerned with signaling pathway is the prerequisite for signal transmission. Once ROS sensing is done, the pathway for signal transduction resulting in altered gene expression is activated within the cell organelles or in its cytoplasm; this pathway of signal transduction also plays a significant role in maintaining the homeostasis of ROS. Signal transductions involves variation in the redox state of the factors (after ROS sensing is perceived); these factors further involve signaling of phospholipids, Ca^{2+} signaling, calmodulins (binding proteins for Ca^{2+}), and G protein activation. During biotic stress, H_2O_2 (most common ROS) and Ca^{2+} signatures are produced triggering CDPKs and CBLs-CIPKs responsible for respiratory burst oxidase homologs (RBOHs) activation. RBOHs are membrane-bound proteins with an N-terminal cytoplasmic domain having Ca-binding EF-hands and with a domain of extracellular oxidases. The superoxide production in the extracellular region is governed by the apoplastic RBOHs domain; whereas phosphorylation is governed by its cytoplasmic counterpart (Ogasawara et al., 2008; Takeda et al., 2008). RBOHs contribute toward cytosolic NADPH catalytic oxidation from superoxide for H_2O_2 production. The synchronous Ca^{2+} activation and H_2O_2 generate ROS waves; which may go across distantly located tissues resulting into programmed/regulated ROS signaling, which in turn leads to activated tolerance mechanisms in nearby as well as distantly located cells (Raja, Majeed, Kang, Andrabi, & John, 2017). Environmental stress causes plants to stimulate the MAPK pathway. The activated MAPK is mainly specific, however sometimes cross-talk is also observed in the signaling pathway. MAPK cascades activation are induced once various forms of environmental stress stimuli (H_2O_2-generated) are perceived (J. Zhou et al., 2014). The MAPK cascade includes MAPKK kinase (MAPKKK), MAPK kinase (MAPKK), and MAPK and their activation source through phosphorylation. Various downstream targets, namely transcription factors, other kinases, phosphatases, and cytoskeleton-associated proteins are activated by MAPK through phosphorylation.

Under environmental stress, ROS production and activity of RBOHs is regulated by MAPKs (Asai, Ohta, & Yoshioka, 2008). There are two main MAPK cascades: NPK1-MEK1-NTF6 and MEK2-SIPK, that are known to be involved in the regulation of oxidative burst mediated by RBOHs and ROS production for mediation of disease resistance (Asai et al., 2008). According to recent reports, MEKK1-MKK5-MPK6 regulates expression of Fe SOD gene which in turn induces production of ROS suggesting the downstream of MAPK by ROS (Xing, Chen, Jia, & Zhang, 2015) Although ROS production under variety of stress responses have also been observed to enforce their effect on MAPKs, resulting in upstream of MAPKs. In *Arabidopsis* MPK3, MPK4 and MPK6; are activated by ROS produced in response to attack by pathogens. MEKK1-MKK4/5-MPK3/6 is involved as MAPK cascade functioning in *Arabidopsis* against pathogen attack involving ROS downstream. MEKK1-MKK2-MPK4/6 is another MAPK cascade working as ROS downstream, involved in signaling for both biotic as well as abiotic stress (Furuya, Matsuoka, & Nanmori, 2014; Teige et al., 2004). In response to specific stimuli, ROS-activated MAPK cascades mainly govern the regulation of ROS production through a feedback mechanism. Various reports have suggested a positively regulated feedback mechanism of ROS production by MAPK cascades. ABA-activated 46KDa MAPK acts downstream to H_2O_2 and further there is positive RBOH regulation for production of H_2O_2 (Lin et al., 2009). Another positively regulated ROS production by MAPK cascade is OXI1-MPK6 (oxidative signal-induced kinase 1 activated by ROS itself). OXI1 is a serine/threonine MAPKKK (Asai et al., 2008). MEKK1-MKK4-MPK3/6 acts in stimulation of ROS production through upstream NADPH oxidase in response to pathogen attack, and H_2O_2 production in turn activates MPK3 and MPK6 (Kovtun, Chiu, Tena, & Sheen, 2000). Apart from positive regulation of ROS generation, certain MAPK cascades, namely NDPK2-MPK3/6, cause negative regulation of ROS production; thereby providing tolerance to cold, salinity, and oxidative stress (Moon et al., 2003).

Although it is evident that both ROS and MAPK control each other's functioning, the exact mechanism of connections and grounds of positive and negative regulation of feedback (mechanism) still remains unanswered. Even though activation of MAPKs by ROS play key role in refereeing cellular activities in response to stressful conditions; however currently there exist no clue on how interaction of ROS signaling and activation of MAPKs operates. Also, the knowledge about various kind of proteins phosphorylated by MAPKs, are largely unidentified. In recent studies, it was also observed that MAPK phosphatase inactivation by ROS activates MAPKs (Y. Liu & He, 2017).

14.8 Influence of reactive oxygen species on transcriptional, translational and epigenetic changes

During environmental stress, gene expression is influenced by ROS and ROS-derived signals. transcript profiling through microarray, sage analysis, QPCR as well as RNA Seq has rebuilt our sightedness on gene expression. In total, approximately 1%–10% of the transcriptome is altered by enhanced ROS levels. ROS-detoxification, protein folding and degradation, organellar function, transport, etc. involves a plethora of genes, which alter their expression upon ROS-generation. In *Arabidopsis*, approximately 175 genes encode proteins that are responsible for antioxidant functioning, defense mechanisms against various types of environmental stress, and events involving proteins accountable for signaling upregulated by H_2O_2. Additionally, ROS- induced receptor kinase, annexins, and peroxisomes biogenesis-related genes have also been identified (Apel & Hirt, 2004). It has also been observed that of the total genes, 2% exhibited altered gene expression while treating tobacco plants with a sublethal dose of ROS-generating chemicals. During oxidative stress, accumulated proteins involved in antioxidant defense, peroxiredoxins, disulfide isomerase proteins, and TCA cycle-associated proteins exhibited downregulation as revealed in a study of mitochondrial proteomics performed in *Arabidopsis*. Also, transcriptomic studies conducted in antisense lines revealed lower antioxidant activity by catalases and APX but upregulated SOD level, and GR has been observed during increased oxidative stress. Inhibited or scavenged production of H_2O_2 results in upregulated levels of PR proteins, APX, GST, and PAL upon ROS signaling activation. In *Arabidopsis*, APETALA2/ETHYLENE RESPONSE FACTOR (AP2/ERF) transcription factor and Redox Responsive Transcription Factor 1 (RRTF1) act as core redox signaling network components and expression is stimulated by ROS (produced by various biotic as well as abiotic signals) frequently and transiently. Increased RRTF1 levels (plants) result in accumulation of ROS, suggesting that ROS generation is amplified by RRTF1 upon interaction with a variety of environmental stresses (Matsuo et al., 2015). The invasion of pathogens stimulates ROS generation in plants, which in turn triggers the activation of HR (hypersensitive response) (Delledonne, Zeier, Marocco, & Lamb, 2001). In resistance to plant disease, ROS play a key role in killing the pathogens such as bacteria (Paiva & Bozza, 2014), fungi, etc.; along with simultaneous enhancement in neighboring cell walls thickening in order to prevent the spread of infections caused by pathogens (Wang and Higgins, 2005). For protection against pathogen infections, plants have evolved PRRs that are utilized to perceive MAMPs/PAMPs, such as chitin perception, triggering the activation of intracellular MAPK cascades for brisk production of ROS (Kawasaki, Yamada, Yoshimura, & Yamaguchi, 2017). In *Arabidopsis*, PRR-associated kinase Bik1 directly phosphorylates NADPH oxidase RBOHD resulting in PAMP-induced ROS burst and antibacterial immunity (Kadota et al., 2014).

Epigenetic modifications include both DNA modifications and posttranslational modifications, which play a significant role in the regulation of gene expression in distinct redox pathways. Hence, ROS also show their impact on the regulation of genes via epigenetic mechanisms (Shimazu et al., 2013). Additionally, chromatin remodeling is involved in epigenetic changes through methylation, acetylation, phosphorylation, ADP-ribosylation, SUMOylating, deamination, and ubiquitylation of histones, thus influencing gene activation in response to stresses that generate ROS. Some evidence has revealed a close link between the metabolism of ROS and epigenetic regulation during the course of environment acclimation and plant growth. Removal of 5-methylcytosine from a molecule of DNA is actively catalyzed by four specific DNA demethylases, repressor of silencing 1, demeter (DME), DME-like 2 (DML 2), and DML 3. A recent study indicates that direct interaction of ROS1 and DME occurs with Fe-S cluster assembly machinery; which is largely sensitive to oxidation caused by ROS. Consequently, their functioning can be influenced by oxidative conditions resulting from stress. Hence, the outcome establishes connections between DNA methylation and ROS metabolism.

During plant growth, development, and in response to environmental stresses, histone acetylation changes (Z. Wang, Cao, Chen, & Liu, 2014; Zhang, Zhu, Song, Harmon, & Chen, 2015). The existence of links between alterations in redox status and chromatin remodeling has been reported in yeasts, animals, and also in the plant kingdom (Shen, Issakidis-Bourguet, & Zhou, 2016). Overexpression of OsSRT1 and OsSRT2 (sirtuin-like genes) in rice plants incorporate increased tolerance in the transformants during oxidative stress but the silencing of the aforesaid genes results in overproduction of ROS (Huang et al., 2007), establishing the significance of the role of OsSRT during epigenetic changes resulting from oxidative stress (Huang et al., 2007).

ROS-generating stressful conditions also influence histone methylases (Z. J. Chen & Tian, 2007). At high concentration of H_2O_2, the activity of JmjC (jumonji C) demethylase crucial for the demethylation of histone is inhibited. Fluctuations in ROS levels oxidize histone deacetylases, namely HDA19 and HDA9, resulting in lower activity, and hence less deacetylation; thereby increasing histone acetylation and eventually causing differential expression of genes (J. Liu, Feng, Li, & He, 2015). This gives the implications that ROS and epigenetic regulation of expression of genes work in close cooperation. However, furthermore research and investigation regarding detailed insights of the interaction of ROS and epigenetic modifications regulating gene expressions are required.

14.9 Influence of reactive oxygen species on posttranslational modifications

In order to instigate alteration in the activity of transcriptome, proteome, and metabolome to develop tolerance against oxidative stress, transport of some of the transcriptional regulators from one location to other for activation is required, in addition to proteolysis and increased binding affinity toward CREs. Normally, NPR1, an oligomer, is a resident of the cytoplasm under optimal conditions. However, SA-induced oxidative stress causes reduction of a disulfide bond mediated by thioredoxin. As a result of which monomeric NPR1is produced, which in turn is fetched to the nucleus in order to bring about transcriptional modifications. Additionally, other posttranslational modifications, namely, phosphorylation and proteasome-mediated degradation, also activate NPR1. In response to H_2O_2, HSF transcription factors in a similar way get activated and transported to the nucleus from cytoplasm following trimerization (Giesguth, Sahm, Simon, & Dietz, 2015; Y. Liu et al., 2013). During ROS, whirly 1 also reaches the nucleus from the chloroplast (Foyer, Karpinska, & Krupinska, 2014). Also, ANAC089 and ANAC013 are a few transcription factors, which are proteolytically activated and translocated from the membrane to the nucleus in response to high ROS levels. A recent study revealed that in rice leaves an alteration in ROS levels results in apparent epigenetic modifications, namely acylation; as a result of which, the ROS related proteins' activity in turn gets regulated (H. Zhou et al., 2018). This implies that the interaction between ROS and acylation might play a significant role in the posttranscriptional modifications (PTM) of proteins of leaf having crucial metabolic functions. These observations give an indication that the process of epigenetic control and redox metabolism are interrelated mechanisms (Locato, Cimini, & De Gara, 2018).

14.10 Conclusion and future perspectives

Currently, it is well-known that ROS are generated at many primary and secondary sites during biotic stresses and they assist plants in the development of PTI and ETI. However, information related to the participation of ROS in signal transduction, how it interacts with MAPK signaling components, their interplay with other plant hormones, their impact on miRNA and protein modifications leading to cell death are largely unknown. Therefore further investigation in this direction will improve our knowledge on the role of ROS in plant defense against biotic stresses.

References

Apel, K., & Hirt, H. (2004). Reactive oxygen species: Metabolism, oxidative stress, and signal transduction. *Annual Review of Plant Biology*. Available from https://doi.org/10.1146/annurev.arplant.55.031903.141701.

Asada, K. (1992). Ascorbate peroxidase – A hydrogen peroxide-scavenging enzyme in plants. *Physiologia Plantarum*. Available from https://doi.org/10.1111/j.1399-3054.1992.tb04728.x.

Asada, K. (1999). The water-water cycle in chloroplasts: Scavenging of active oxygens and dissipation of excess photons. *Annual Review of Plant Biology*. Available from https://doi.org/10.1146/annurev.arplant.50.1.601.

Asai, S., Ohta, K., & Yoshioka, H. (2008). MAPK signaling regulates nitric oxide and NADPH oxidase-dependent oxidative bursts in *Nicotiana benthamiana*. *The Plant Cell*. Available from https://doi.org/10.1105/tpc.107.055855.

Bailey-Serres, J., & Mittler, R. (2006). The roles of reactive oxygen species in plant cells. *Plant Physiology*. Available from https://doi.org/10.1104/pp.104.900191.

Balaban, R. S., Nemoto, S., & Finkel, T. (2005). Mitochondria, oxidants, and aging. *Cell*, *120* (4), 483–495.

Barnes, J., Zheng, Y., & Lyons, T. (2002). *Plant resistance to ozone: The role of ascorbate. Air pollution and plant biotechnology* (pp. 235–252). Springer.

Bethke, P. C., & Jones, R. L. (2001). Cell death of barley aleurone protoplasts is mediated by reactive oxygen species. *The Plant Journal: For Cell and Molecular Biology*. Available from https://doi.org/10.1046/j.1365-313X.2001.00930.x.

Bienert, G. P., Møller, A. L. B., Kristiansen, K. A., Schulz, A., Møller, I. M., Schjoerring, J. K., & Jahn, T. P. (2007). Specific aquaporins facilitate the diffusion of hydrogen peroxide across membranes. *The Journal of Biological Chemistry*. Available from https://doi.org/10.1074/jbc.M603761200.

Bolwell, G. P., Davies, D. R., Gerrish, C., Auh, C. K., & Murphy, T. M. (1998). Comparative biochemistry of the oxidative burst produced by rose and French bean cells reveals two distinct mechanisms. *Plant Physiology*. Available from https://doi.org/10.1104/pp.116.4.1379.

Brot, N., & Weissbach, H. (1982). The biochemistry of methionine sulfoxide residues in proteins. *Trends in Biochemical Sciences*. Available from https://doi.org/10.1016/0968-0004(82)90204-3.

Buettner, G. R. (1993). The pecking order of free radicals and antioxidants: Lipid peroxidation, α-tocopherol, and ascorbate. *Archives of Biochemistry and Biophysics*. Available from https://doi.org/10.1006/abbi.1993.1074.

Caverzan, A., Passaia, G., Rosa, S. B., Ribeiro, C. W., Lazzarotto, F., & Margis-Pinheiro, M. (2012). Plant responses to stresses: Role of ascorbate peroxidase in the antioxidant protection. *Genetics and Molecular Biology*. Available from https://doi.org/10.1590/S1415-47572012000600016.

Chen, Z., & Gallie, D. R. (2006). Dehydroascorbate reductase affects leaf growth, development, and function. *Plant Physiology*. Available from https://doi.org/10.1104/pp.106.085506.

Chen, Z. J., & Tian, L. (2007). Roles of dynamic and reversible histone acetylation in plant development and polyploidy. *Biochimica et Biophysica Acta—Gene Structure and Expression*. Available from https://doi.org/10.1016/j.bbaexp.2007.04.007.

Choudhury, S., Panda, P., Sahoo, L., & Panda, S. K. (2013). Reactive oxygen species signaling in plants under abiotic stress. *Plant Signaling & Behavior*. Available from https://doi.org/10.4161/psb.23681.

Cleland, R. E., & Grace, S. C. (1999). Voltammetric detection of superoxide production by photosystem II. *FEBS Letters*. Available from https://doi.org/10.1016/S0014-5793(99)01067-4.

Cona, A., Rea, G., Angelini, R., Federico, R., & Tavladoraki, P. (2006). Functions of amine oxidases in plant development and defence. *Trends in Plant Science*. Available from https://doi.org/10.1016/j.tplants.2005.12.009.

Das, K., & Roychoudhury, A. (2014). Reactive oxygen species (ROS) and response of antioxidants as ROS-scavengers during environmental stress in plants. *Frontiers in Environmental Science*. Available from https://doi.org/10.3389/fenvs.2014.00053.

Dangl, J. L., Horvath, D. M., & Staskawicz, B. J. (2013). Pivoting the plant immune system from dissection to deployment. *Science*, *341*(6147), 746–751.

De Pinto, M. C., & De Gara, L. (2004). Changes in the ascorbate metabolism of apoplastic and symplastic spaces are associated with cell differentiation. *Journal of Experimental Botany*. Available from https://doi.org/10.1093/jxb/erh253.

Del Río, L. A., Sandalio, L. M., Corpas, F. J., Palma, J. M., & Barroso, J. B. (2006). Reactive oxygen species and reactive nitrogen species in peroxisomes. Production, scavenging, and role in cell signaling. *Plant Physiology*. Available from https://doi.org/10.1104/pp.106.078204.

Delledonne, M., Zeier, J., Marocco, A., & Lamb, C. (2001). Signal interactions between nitric oxide and reactive oxygen intermediates in the plant hypersensitive disease resistance response. *Proceedings of the National Academy of Sciences of the United States of America*. Available from https://doi.org/10.1073/pnas.231178298.

Dias, C. V., Mendes, J. S., dos Santos, A. C., Pirovani, C. P., da Silva Gesteira, A., Micheli, F., ... de Mattos Cascardo, J. C. (2011). Hydrogen peroxide formation in cacao tissues infected by the hemibiotrophic fungus Moniliophthora perniciosa. *Plant Physiology and Biochemistry*. Available from https://doi.org/10.1016/j.plaphy.2011.05.004.

Dismukes, G. C., Klimov, V. V., Baranov, S. V., Kozlov, Y. N., DasGupta, J., & Tyryshkin, A. (2001). The origin of atmospheric oxygen on Earth: The innovation of oxygenic photosynthesis. *Proceedings of the National Academy of Sciences of the United States of America*. Available from https://doi.org/10.1073/pnas.061514798.

Durrant, W. E., & Dong, X. (2004). Systemic acquired resistance. *Annual Review of Phytopathology*. Available from https://doi.org/10.1146/annurev.phyto.42.040803.140421.

Elstner, E. F., & Osswald, W. (1994). Mechanisms of oxygen activation during plant stress. *Proceedings of the Royal Society of Edinburgh. Section B: Biological Sciences*. Available from https://doi.org/10.1017/s0269727000014068.

Eltayeb, A. E., Kawano, N., Badawi, G. H., Kaminaka, H., Sanekata, T., Shibahara, T., ... Tanaka, K. (2007). Overexpression of monodehydroascorbate reductase in transgenic tobacco confers enhanced tolerance to ozone, salt and polyethylene glycol stresses. *Planta*. Available from https://doi.org/10.1007/s00425-006-0417-7.

Evans, M. D., Dizdaroglu, M., & Cooke, M. S. (2004). Oxidative DNA damage and disease: Induction, repair and significance. *Mutation Research—Reviews in Mutation Research*. Available from https://doi.org/10.1016/j.mrrev.2003.11.001.

Fini, A., Brunetti, C., Di Ferdinando, M., Ferrini, F., & Tattini, M. (2011). Stress-induced flavonoid biosynthesis and the antioxidant machinery of plants. *Plant Signaling & Behavior*. Available from https://doi.org/10.4161/psb.6.5.15069.

Foyer, C. H., Karpinska, B., & Krupinska, K. (2014). The functions of WHIRLY1 and REDOXRESPONSIVE TRANSCRIPTION FACTOR 1 in cross tolerance responses in plants: A hypothesis. *Philosophical Transactions of the Royal Society B: Biological Sciences*. Available from https://doi.org/10.1098/rstb.2013.0226.

Foyer, C. H., & Noctor, G. (2016). Stress-triggered redox signaling: What's in pROSpect? *Plant, Cell & Environment*. Available from https://doi.org/10.1111/pce.12621.

Fridovich, I. (1995). Superoxide radical and superoxide dismutases. *Annual Review of Biochemistry*, *64*, 97–112.

Fukuzawa, K., Tokumura, A., Ouchi, S., & Tsukatani, H. (1982). Antioxidant activities of tocopherols on Fe2 + -ascorbate-induced lipid peroxidation in lecithin liposomes. *Lipids*. Available from https://doi.org/10.1007/BF02535334.

Furuya, T., Matsuoka, D., & Nanmori, T. (2014). Membrane rigidification functions upstream of the MEKK1-MKK2-MPK4 cascade during cold acclimation in *Arabidopsis thaliana*. *FEBS Letters*. Available from https://doi.org/10.1016/j.febslet.2014.04.032.

Gardner, P. R., & Fridovich, I. (1991). Superoxide sensitivity of the *Escherichia coli* 6-phosphogluconate dehydratase. *The Journal of Biological Chemistry*.

Giesbert, S., Schuerg, T., Scheele, S., & Tudzynski, P. (2008). The NADPH oxidase Cpnox1 is required for full pathogenicity of the ergot fungus *Claviceps purpurea*. *Molecular Plant Pathology*, *9*(3), 317–327.

Giesguth, M., Sahm, A., Simon, S., & Dietz, K. J. (2015). Redox-dependent translocation of the heat shock transcription factor AtHSFA8 from the cytosol to the nucleus in *Arabidopsis thaliana*. *FEBS Letters*. Available from https://doi.org/10.1016/j.febslet.2015.01.039.

Gill, S. S., Anjum, N. A., Hasanuzzaman, M., Gill, R., Trivedi, D. K., Ahmad, I., . . . Tuteja, N. (2013). Glutathione and glutathione reductase: A boon in disguise for plant abiotic stress defense operations. *Plant Physiology and Biochemistry*. Available from https://doi.org/10.1016/j.plaphy.2013.05.032.

Gill, S. S., & Tuteja, N. (2010). Reactive oxygen species and antioxidant machinery in abiotic stress tolerance in crop plants. *Plant Physiology and Biochemistry*. Available from https://doi.org/10.1016/j.plaphy.2010.08.016.

Girotti, A. W. (1998). Lipid hydroperoxide generation, turnover, and effector action in biological systems. *Journal of Lipid Research*. Available from https://doi.org/10.1016/S0022-2275(20)32182-9.

Golldack, D., Li, C., Mohan, H., & Probst, N. (2014). Tolerance to drought and salt stress in plants: Unraveling the signaling networks. *Frontiers in Plant Science*. Available from https://doi.org/10.3389/fpls.2014.00151.

Grant, J. J., Yun, B. W., & Loake, G. J. (2000). Oxidative burst and cognate redox signaling reported by luciferase imaging: Identification of a signal network that functions independently of ethylene, SA and Me-JA but is dependent on MAPKK activity. *The Plant Journal: For Cell and Molecular Biology*. Available from https://doi.org/10.1046/j.1365-313X.2000.00902.x.

Halliwell, B. (1977). Generation of hydrogen peroxide, superoxide and hydroxyl radicals during the oxidation of dihydroxyfumaric acid by peroxidase. *The Biochemical Journal*. Available from https://doi.org/10.1042/bj1630441.

Halliwell, B. (2006). Reactive species and antioxidants. Redox biology is a fundamental theme of aerobic life. *Plant Physiology*. Available from https://doi.org/10.1104/pp.106.077073.

Halliwell, B., Gutteridge, J.M.C. (2015). Free radicals in biology and medicine, free radicals in biology and medicine. https://doi.org/10.1093/acprof:oso/9780198717478.001.0001.

Hasanuzzaman, M., Bhuyan, M. H. M. B., Zulfiqar, F., Raza, A., Mohsin, S. M., Al Mahmud, J., . . . Fotopoulos, V. (2020). Reactive oxygen species and antioxidant defense in plants under abiotic stress: Revisiting the crucial role of a universal defense regulator. *Antioxidants*. Available from https://doi.org/10.3390/antiox9080681.

Hatz, S., Lambert, J. D. C., & Ogilby, P. R. (2007). Measuring the lifetime of singlet oxygen in a single cell: Addressing the issue of cell viability. *Photochemical & Photobiological Sciences: Official Journal of the European Photochemistry Association and the European Society for Photobiology*. Available from https://doi.org/10.1039/b707313e.

Heyno, E., Mary, V., Schopfer, P., & Krieger-Liszkay, A. (2011). Oxygen activation at the plasma membrane: Relation between superoxide and hydroxyl radical production by isolated membranes. *Planta*. Available from https://doi.org/10.1007/s00425-011-1379-y.

Higuchi, T. (2006). Look back over the studies of lignin biochemistry. *Journal of Wood Science*. Available from https://doi.org/10.1007/s10086-005-0790-z.

Ho, L. H. M., Giraud, E., Uggalla, V., Lister, R., Clifton, R., Glen, A., . . . Whelan, J. (2008). Identification of regulatory pathways controlling gene expression of stress-responsive mitochondrial proteins in arabidopsis. *Plant Physiology*. Available from https://doi.org/10.1104/pp.108.121384.

Hossain, M. A., & Asada, K. (1985). Monodehydroascorbate reductase from cucumber is a flavin adenine dinucleotide enzyme. *The Journal of Biological Chemistry*.

Huang, L., Sun, Q., Qin, F., Li, C., Zhao, Y., & Zhou, D. X. (2007). Down-regulation of a Silent Information Regulator2-related histone deacetylase gene, OsSRT1, induces DNA fragmentation and cell death in rice. *Plant Physiology*. Available from https://doi.org/10.1104/pp.107.099473.

Hückelhoven, R., & Kogel, K. H. (2003). Reactive oxygen intermediates in plant-microbe interactions: Who is who in powdery mildew resistance? *Planta*. Available from https://doi.org/10.1007/s00425-003-0973-z.

Imlay, J. A. (2003). Pathways of oxidative damage. *Annual Review of Microbiology*. Available from https://doi.org/10.1146/annurev.micro.57.030502.090938.

Isherwood, F. A., Chen, Y. T., & Mapson, L. W. (1954). Synthesis of L-ascorbic acid in plants and animals. *The Biochemical Journal*, *56*, 1–15.

Jasso-Robles, F. I., Gonzalez, M. E., Pieckenstain, F. L., Ramírez-García, J. M., Guerrero-González, M., de la, L., ... Rodríguez-Kessler, M. (2020). Decrease of Arabidopsis PAO activity entails increased RBOH activity, ROS content and altered responses to Pseudomonas. *Plant Science (Shannon, Ireland)*. Available from https://doi.org/10.1016/j.plantsci.2019.110372.

Jones, J. D., & Dangl, J. L. (2006). The plant immune system. *Nature*, *444*(7117), 323–329.

Kadota, Y., Sklenar, J., Derbyshire, P., Stransfeld, L., Asai, S., Ntoukakis, V., ... Zipfel, C. (2014). Direct regulation of the NADPH oxidase RBOHD by the PRR-associated kinase BIK1 during plant immunity. *Molecular Cell*. Available from https://doi.org/10.1016/j.molcel.2014.02.021.

Kamal-Eldin, A., & Appelqvist, L. Å. (1996). The chemistry and antioxidant properties of tocopherols and tocotrienols. *Lipids*. Available from https://doi.org/10.1007/BF02522884.

Kar, R. K. (2015). *ROS signaling: Relevance with site of production and metabolism of ROS. Reactive oxygen species and oxidative damage in plants under stress* (pp. 115–125). Cham: Springer.

Karuppanapandian, T., Moon, J. C., Kim, C., Manoharan, K., & Kim, W. (2011). Reactive oxygen species in plants: Their generation, signal transduction, and scavenging mechanisms. *Australian Journal of Crop Science*.

Kasai, H. (1997). Analysis of a form of oxidative DNA damage, 8-hydroxy-2'-deoxyguanosine, as a marker of cellular oxidative stress during carcinogenesis. *Mutation Research—Reviews in Mutation Research*. Available from https://doi.org/10.1016/S1383-5742(97)00035-5.

Kawasaki, T., Yamada, K., Yoshimura, S., & Yamaguchi, K. (2017). Chitin receptor-mediated activation of MAP kinases and ROS production in rice and Arabidopsis. *Plant Signaling & Behavior*. Available from https://doi.org/10.1080/15592324.2017.1361076.

Khorobrykh, S., Havurinne, V., Mattila, H., & Tyystjärvi, E. (2020). Oxygen and ROS in photosynthesis. *Plants*. Available from https://doi.org/10.3390/plants9010091.

Kissoudis, C., van de Wiel, C., Visser, R. G. F., & van der Linden, G. (2014). Enhancing crop resilience to combined abiotic and biotic stress through the dissection of physiological and molecular crosstalk. *Frontiers in Plant Science*. Available from https://doi.org/10.3389/fpls.2014.00207.

Klessig, D. F., Durner, J., Noad, R., Navarre, D. A., Wendehenne, D., Kumar, D., ... Silva, H. (2000). Nitric oxide and salicylic acid signaling in plant defense. *Proceedings of the National Academy of Sciences of the United States of America*. Available from https://doi.org/10.1073/pnas.97.16.8849.

Kobayashi, K., Kumazawa, Y., Miwa, K., & Yamanaka, S. (1996). ε-(γ-Glutamyl)lysine cross-links of spore coat proteins and transglutaminase activity in *Bacillus subtilis*. *FEMS Microbiology Letters*. Available from https://doi.org/10.1016/0378-1097(96)00353-9.

Kovtun, Y., Chiu, W. L., Tena, G., & Sheen, J. (2000). Functional analysis of oxidative stress-activated mitogen-activated protein kinase cascade in plants. *Proceedings of the National Academy of Sciences of the United States of America*. Available from https://doi.org/10.1073/pnas.97.6.2940.

Krieger-Liszkay, A. (2005). Singlet oxygen production in photosynthesis, in*Journal of Experimental Botany*. Available from https://doi.org/10.1093/jxb/erh237.

Kumar, N., Ebel, R. C., & Roberts, P. D. (2011). H_2O_2 degradation is suppressed in kumquat leaves infected with *Xanthomonas axonopodis* pv. *citri*. *Scientia Horticulturae*. Available from https://doi.org/10.1016/j.scienta.2011.07.005.

Kwak, J. M., Mori, I. C., Pei, Z. M., Leonhard, N., Angel Torres, M., Dangl, J. L., ... Schroeder, J. I. (2003). NADPH oxidase AtrbohD and AtrbohF genes function in ROS-dependent ABA signaling in arabidopsis. *The EMBO Journal*. Available from https://doi.org/10.1093/emboj/cdg277.

Lamb, C., & Dixon, R. A. (1997). The oxidative burst in plant disease resistance. *Annual Review of Plant Biology*. Available from https://doi.org/10.1146/annurev.arplant.48.1.251.

Levine, A., Tenhaken, R., Dixon, R., & Lamb, C. (1994). H_2O_2 from the oxidative burst orchestrates the plant hypersensitive disease resistance response. *Cell*. Available from https://doi.org/10.1016/0092-8674(94)90544-4.

Li, R., Jia, Z., & Trush, M. A. (2016). Defining ROS in biology and medicine. *Reactive Oxygen Species (Apex, NC)*, *1*(1), 9.

Liang, X., & Zhou, J. M. (2018). Receptor-like cytoplasmic kinases: Central players in plant receptor kinase-mediated signaling. *Annual Review of Plant Biology*. Available from https://doi.org/10.1146/annurev-arplant-042817-040540.

Lin, F., Ding, H., Wang, J., Zhang, H., Zhang, A., Zhang, Y., ... Jiang, M. (2009). Positive feedback regulation of maize NADPH oxidase by mitogen-activated protein kinase cascade in abscisic acid signaling. *Journal of Experimental Botany*. Available from https://doi.org/10.1093/jxb/erp157.

Liu, J., Feng, L., Li, J., & He, Z. (2015). Genetic and epigenetic control of plant heat responses. *Frontiers in Plant Science*. Available from https://doi.org/10.3389/fpls.2015.00267.

Liu, Y., & He, C. (2017). A review of redox signaling and the control of MAP kinase pathway in plants. *Redox Biology*. Available from https://doi.org/10.1016/j.redox.2016.12.009.

Liu, Y., Zhang, C., Chen, J., Guo, L., Li, X., Li, W., ... Zhang, L. (2013). Arabidopsis heat shock factor HsfA1a directly senses heat stress, pH changes, and hydrogen peroxide via the engagement of redox state. *Plant Physiology and Biochemistry*. Available from https://doi.org/10.1016/j.plaphy.2012.12.013.

Locato, V., Cimini, S., & De Gara, L. (2018). ROS and redox balance as multifaceted players of cross-tolerance: Epigenetic and retrograde control of gene expression. *Journal of Experimental Botany*. Available from https://doi.org/10.1093/jxb/ery168.

Marino, D., Dunand, C., Puppo, A., & Pauly, N. (2012). A burst of plant NADPH oxidases. *Trends in Plant Science*, *17*(1), 9–15.

Matsuo, M., Johnson, J. M., Hieno, A., Tokizawa, M., Nomoto, M., Tada, Y., ... Oelmüller, R. (2015). High REDOX RESPONSIVE TRANSCRIPTION FACTOR1 levels result in accumulation of reactive oxygen species in *Arabidopsis thaliana* shoots and roots. *Molecular Plant*. Available from https://doi.org/10.1016/j.molp.2015.03.011.

Mhamdi, A., Queval, G., Chaouch, S., Vanderauwera, S., Van Breusegem, F., & Noctor, G. (2010). Catalase function in plants: A focus on Arabidopsis mutants as stress-mimic models. *Journal of Experimental Botany*. Available from https://doi.org/10.1093/jxb/erq282.

Mhamdi, A., & Van Breusegem, F. (2018). Reactive oxygen species in plant development. *Development*. Available from https://doi.org/10.1242/dev.164376.

Miller, G., Suzuki, N., Ciftci-Yilmaz, S., & Mittler, R. (2010). Reactive oxygen species homeostasis and signaling during drought and salinity stresses. *Plant, Cell & Environment*. Available from https://doi.org/10.1111/j.1365-3040.2009.02041.x.

Mittler, R. (2002). Oxidative stress, antioxidants and stress tolerance. *Trends in Plant Science*. Available from https://doi.org/10.1016/S1360-1385(02)02312-9.

Mittler, R. (2017). ROS are good. *Trends in Plant Science*. Available from https://doi.org/10.1016/j.tplants.2016.08.002.

Mittler, R., Vanderauwera, S., Gollery, M., & Van Breusegem, F. (2004). Reactive oxygen gene network of plants. *Trends in Plant Science*. Available from https://doi.org/10.1016/j.tplants.2004.08.009.

Mittova, V., Tal, M., Volokita, M., & Guy, M. (2003). Up-regulation of the leaf mitochondrial and peroxisomal antioxidative systems in response to salt-induced oxidative stress in the wild salt-tolerant tomato species *Lycopersicon pennellii*. *Plant, Cell & Environment*. Available from https://doi.org/10.1046/j.1365-3040.2003.01016.x.

Møller, I. M., Jensen, P. E., & Hansson, A. (2007). Oxidative modifications to cellular components in plants. *Annual Review of Plant Biology*. Available from https://doi.org/10.1146/annurev.arplant.58.032806.103946.

Moon, H., Lee, B., Choi, G., Shin, D., Theertha Prasad, D., Lee, O., ... Yun, D. J. (2003). NDP kinase 2 interacts with two oxidative stress-activated MAPKs to regulate cellular redox state and enhances multiple stress tolerance in transgenic plants. *Proceedings of the National Academy of Sciences of the United States of America*. Available from https://doi.org/10.1073/pnas.252641899.

Mou, Z., Fan, W., & Dong, X. (2003). Inducers of plant systemic acquired resistance regulate NPR1 function through redox changes. *Cell*. Available from https://doi.org/10.1016/S0092-8674(03)00429-X.

Mullineaux, P. M., & Rausch, T. (2005). Glutathione, photosynthesis and the redox regulation of stress-responsive gene expression. *Photosynthesis Research*. Available from https://doi.org/10.1007/s11120-005-8811-8.

Murphy, M. P. (2009). How mitochondria produce reactive oxygen species. *The Biochemical Journal*. Available from https://doi.org/10.1042/BJ20081386.

Nanda, A. K., Andrio, E., Marino, D., Pauly, N., & Dunand, C. (2010). Reactive oxygen species during plant-microorganism early interactions. *Journal of Integrative Plant Biology*. Available from https://doi.org/10.1111/j.1744-7909.2010.00933.x.

Neill, S., Desikan, R., & Hancock, J. (2002). Hydrogen peroxide signaling. *Current Opinion in Plant Biology*. Available from https://doi.org/10.1016/S1369-5266(02)00282-0.

Noctor, G., De Paepe, R., & Foyer, C. H. (2007). Mitochondrial redox biology and homeostasis in plants. *Trends in Plant Science*. Available from https://doi.org/10.1016/j.tplants.2007.01.005.

Noctor, G., & Foyer, C. H. (1998). Ascorbate and glutathione: Keeping active oxygen under control. *Annual Review of Plant Biology*. Available from https://doi.org/10.1146/annurev.arplant.49.1.249.

Noctor, G., Veljovic-Jovanovic, S., Driscoll, S., Novitskaya, L., & Foyer, C. H. (2002). Drought and oxidative load in the leaves of C3 plants: A predominant role for photorespiration? *Annals of Botany*. Available from https://doi.org/10.1093/aob/mcf096.

Ogasawara, Y., Kaya, H., Hiraoka, G., Yumoto, F., Kimura, S., Kadota, Y., ... Kuchitsu, K. (2008). Synergistic activation of the arabidopsis NADPH oxidase AtrbohD by Ca 2 + and phosphorylation. *The Journal of Biological Chemistry*. Available from https://doi.org/10.1074/jbc.M708106200.

Oleinick, N. L., Chiu, S.-M., Ramakrishnan, N., & Xue, L. Y. (1987). The formation, identification, and significance of DNA-protein cross-links in mammalian cells. *British Journal of Cancer*.

Orozco-Cárdenas, M. L., Narváez-Vásquez, J., & Ryan, C. A. (2001). Hydrogen peroxide acts as a second messenger for the induction of defense genes in tomato plants in response to wounding, systemin, and methyl jasmonate. *The Plant Cell*. Available from https://doi.org/10.1105/tpc.13.1.179.

Overmyer, K., Brosché, M., & Kangasjärvi, J. (2003). Reactive oxygen species and hormonal control of cell death. *Trends in Plant Science*, *8*(7), 335–342.

Paiva, C. N., & Bozza, M. T. (2014). Are reactive oxygen species always detrimental to pathogens? *Antioxidants & Redox Signaling*. Available from https://doi.org/10.1089/ars.2013.5447.

Pinto, E., Sigaud-Kutner, T. C. S., Leitão, M. A. S., Okamoto, O. K., Morse, D., & Colepicolo, P. (2003). Heavy metal-induced oxidative stress in algae. *Journal of Phycology*. Available from https://doi.org/10.1111/j.0022-3646.2003.02-193.x.

Raja, V., Majeed, U., Kang, H., Andrabi, K. I., & John, R. (2017). Abiotic stress: Interplay between ROS, hormones and MAPKs. *Environmental and Experimental Botany*. Available from https://doi.org/10.1016/j.envexpbot.2017.02.010.

Rasmusson, A. G., Geisler, D. A., & Møller, I. M. (2008). The multiplicity of dehydrogenases in the electron transport chain of plant mitochondria. *Mitochondrion*. Available from https://doi.org/10.1016/j.mito.2007.10.004.

Rentel, M. C., Lecourieux, D., Ouaked, F., Usher, S. L., Petersen, L., Okamoto, H., ... Knight, M. R. (2004). OXI1 kinase is necessary for oxidative burst-mediated signalling in Arabidopsis. *Nature*. Available from https://doi.org/10.1038/nature02353.

Roychoudhury, A., Pradhan, S., Chaudhuri, B., Das, K. (2012). 11 Phytoremediation of toxic metals and the involvement of Brassica species. In: Anjum N.A., Pereira M.E., Ahmad I., Duarte A.C., Umar S. and Khan N.A. (eds.), *Phytotechnologies: remediation of environmental contaminants*. CRC Press.

Schuller, D. J., Ban, N., Van Huystee, R. B., McPherson, A., & Poulos, T. L. (1996). The crystal structure of peanut peroxidase. *Structure*. Available from https://doi.org/10.1016/S0969-2126(96)00035-4.

Sharma, P., & Dubey, R. S. (2004). Ascorbate peroxidase from rice seedlings: Properties of enzyme isoforms, effects of stresses and protective roles of osmolytes. *Plant Science*. Available from https://doi.org/10.1016/j.plantsci.2004.04.028.

Sharma, P., Jha, A. B., Dubey, R. S., & Pessarakli, M. (2012). Reactive oxygen species, oxidative damage, and antioxidative defense mechanism in plants under stressful conditions. *Journal of Botany*. Available from https://doi.org/10.1155/2012/217037.

Shen, Y., Issakidis-Bourguet, E., & Zhou, D. X. (2016). Perspectives on the interactions between metabolism, redox, and epigenetics in plants. *Journal of Experimental Botany*. Available from https://doi.org/10.1093/jxb/erw310.

Shimazu, T., Hirschey, M. D., Newman, J., He, W., Shirakawa, K., Le Moan, N., ... Verdin, E. (2013). Suppression of oxidative stress by β-hydroxybutyrate, an endogenous histone deacetylase inhibitor. *Science (80-.)*. Available from https://doi.org/10.1126/science.1227166.

Singh, A., Kumar, A., Yadav, S., & Singh, I. K. (2019). Reactive oxygen species-mediated signaling during abiotic stress. *Plant Gene*, *18*, 100173.

Singh, A., Singh, I.K. (2018). *Molecular aspects of plant-pathogen interaction*. https://doi.org/10.1007/978-981-10-7371-7.

Ślesak, I., Libik, M., Karpinska, B., Karpinski, S., & Miszalski, Z. (2007). The role of hydrogen peroxide in regulation of plant metabolism and cellular signaling in response to environmental stresses. *Acta Biochimica Polonica*. Available from https://doi.org/10.18388/abp.2007_3267.

Smirnoff, N. (2000). Ascorbic acid: Metabolism and functions of a multi-facetted molecule. *Current Opinion in Plant Biology*. Available from https://doi.org/10.1016/S1369-5266(00)00069-8.

Smith, J. M., & Heese, A. (2014). Rapid bioassay to measure early reactive oxygen species production in Arabidopsis leave tissue in response to living *Pseudomonas syringae*. *Plant Methods*. Available from https://doi.org/10.1186/1746-4811-10-6.

Soares, N. C., Wojtkowska, J., & Jackson, P. A. (2011). A proteomic analysis of the wound response in Medicago leaves reveals the early activation of a ROS-sensitive signal pathway. *Journal of Proteomics*. Available from https://doi.org/10.1016/j.jprot.2011.03.017.

Takeda, S., Gapper, C., Kaya, H., Bell, E., Kuchitsu, K., & Dolan, L. (2008). Local positive feedback regulation determines cell shape in root hair cells. *Science (80-.)*. Available from https://doi.org/10.1126/science.1152505.

Teige, M., Scheikl, E., Eulgem, T., Dóczi, R., Ichimura, K., Shinozaki, K., … Hirt, H. (2004). The MKK2 pathway mediates cold and salt stress signaling in Arabidopsis. *Molecular Cell*. Available from https://doi.org/10.1016/j.molcel.2004.06.023.

Torres, M. A., & Dangl, J. L. (2005). Functions of the respiratory burst oxidase in biotic interactions, abiotic stress and development. *Current Opinion in Plant Biology*. Available from https://doi.org/10.1016/j.pbi.2005.05.014.

Torres, M. A., Dangl, J. L., & Jones, J. D. G. (2002). Arabidopsis gp91phox homologues Atrbohd and Atrbohf are required for accumulation of reactive oxygen intermediates in the plant defense response. *Proceedings of the National Academy of Sciences of the United States of America*. Available from https://doi.org/10.1073/pnas.012452499.

Torres, M. A., Jones, J. D., & Dangl, J. L. (2006). Reactive oxygen species signaling in response to pathogens. *Plant Physiology*, *141*(2), 373–378.

Triantaphylidès, C., Krischke, M., Hoeberichts, F. A., Ksas, B., Gresser, G., Havaux, M., … Mueller, M. J. (2008). Singlet oxygen is the major reactive oxygen species involved in photooxidative damage to plants. *Plant Physiology*. Available from https://doi.org/10.1104/pp.108.125690.

Tripathy, B. C., & Oelmüller, R. (2012). Reactive oxygen species generation and signaling in plants. *Plant Signaling & Behavior*. Available from https://doi.org/10.4161/psb.22455.

Tsikas, D. (2017). Assessment of lipid peroxidation by measuring malondialdehyde (MDA) and relatives in biological samples: Analytical and biological challenges. *Analytical Biochemistry*. Available from https://doi.org/10.1016/j.ab.2016.10.021.

Turrens, J. F. (2003). Mitochondrial formation of reactive oxygen species. *The Journal of Physiology*. Available from https://doi.org/10.1113/jphysiol.2003.049478.

Valko, M., Morris, H., & Cronin, M. (2005). Metals, toxicity and oxidative stress. *Current Medicinal Chemistry*. Available from https://doi.org/10.2174/0929867053764635.

Van Aken, O., & Van Breusegem, F. (2015). Licensed to kill: Mitochondria, chloroplasts, and cell death. *Trends in Plant Science*, *20*(11), 754–766.

Vellosillo, T., Vicente, J., Kulasekaran, S., Hamberg, M., & Castresana, C. (2010). Emerging complexity in reactive oxygen species production and signaling during the response of plants to pathogens. *Plant Physiology*. Available from https://doi.org/10.1104/pp.110.161273.

Verbruggen, N., & Hermans, C. (2008). Proline accumulation in plants: A review. *Amino Acids*. Available from https://doi.org/10.1007/s00726-008-0061-6.

Wagner, D., Przybyla, D., Op Den Camp, R., Kim, C., Landgraf, F., Keun, P. L., … Apel, K. (2004). The genetic basis of singlet oxygen-induced stress response of *Arabidopsis thaliana*. *Science (80-.)*. Available from https://doi.org/10.1126/science.1103178.

Wang, J., Zhang, H., & Allen, R. D. (1999). Overexpression of an Arabidopsis peroxisomal ascorbate peroxidase gene in tobacco increases protection against oxidative stress. *Plant & Cell Physiology*. Available from https://doi.org/10.1093/oxfordjournals.pcp.a029599.

Wang, J., & Higgins, V. J. (2005). Nitric oxide has a regulatory effect in the germination of conidia of *Colletotrichum coccodes*. *Fungal Genetics and Biology*, *42*(4), 284–292.

Wang, Z., Cao, H., Chen, F., & Liu, Y. (2014). The roles of histone acetylation in seed performance and plant development. *Plant Physiology and Biochemistry*. Available from https://doi.org/10.1016/j.plaphy.2014.09.010.

Wheeler, G. L., Jones, M. A., & Smirnoff, N. (1998). The biosynthetic pathway of vitamin C in higher plants. *Nature*. Available from https://doi.org/10.1038/30728.

Willekens, H., Chamnongpol, S., Davey, M., Schraudner, M., Langebartels, C., Van Montagu, M., ... Van Camp, W. (1997). Catalase is a sink for H_2O_2 and is indispensable for stress defence in C3 plants. *The EMBO Journal*. Available from https://doi.org/10.1093/emboj/16.16.4806.

Wojtaszek, P. (1997). Oxidative burst: An early plant response to pathogen infection. *The Biochemical Journal*. Available from https://doi.org/10.1042/bj3220681.

Xing, Y., Chen, W. H., Jia, W., & Zhang, J. (2015). Mitogen-activated protein kinase kinase 5 (MKK5)-mediated signaling cascade regulates expression of iron superoxide dismutase gene in Arabidopsis under salinity stress. *Journal of Experimental Botany*. Available from https://doi.org/10.1093/jxb/erv305.

Xiong, L., Schumaker, K. S., & Zhu, J. K. (2002). Cell signaling during cold, drought, and salt stress. *The Plant Cell*. Available from https://doi.org/10.1105/tpc.000596.

Yamauchi, Y., Furutera, A., Seki, K., Toyoda, Y., Tanaka, K., & Sugimoto, Y. (2008). Malondialdehyde generated from peroxidized linolenic acid causes protein modification in heat-stressed plants. *Plant Physiology and Biochemistry*. Available from https://doi.org/10.1016/j.plaphy.2008.04.018.

Zhang, T., Zhu, M., Song, W. Y., Harmon, A. C., & Chen, S. (2015). Oxidation and phosphorylation of MAP kinase 4 cause protein aggregation. *Biochimica et Biophysica Acta—Proteins and Proteomics*. Available from https://doi.org/10.1016/j.bbapap.2014.11.006.

Zhou, H., Finkemeier, I., Guan, W., Tossounian, M. A., Wei, B., Young, D., ... Foyer, C. H. (2018). Oxidative stress-triggered interactions between the succinyl- and acetyl-proteomes of rice leaves. *Plant, Cell & Environment*. Available from https://doi.org/10.1111/pce.13100.

Zhou, J., Wang, J., Li, X., Xia, X. J., Zhou, Y. H., Shi, K., ... Yu, J. Q. (2014). H2O2 mediates the crosstalk of brassinosteroid and abscisic acid in tomato responses to heat and oxidative stresses. *Journal of Experimental Botany*. Available from https://doi.org/10.1093/jxb/eru217.

Chapter 15

Physiological mechanisms and adaptation strategies of plants under heavy metal micronutrient deficiency/toxicity conditions

Muhammad Ashar Ayub[1,2], Muhammad Zia ur Rehman[1], Wajid Umar[3], Muhammad Adnan[4], Zia Ur Rahman Farooqi[1], Maira Naveed[1], Muhammad Zohaib Aslam[1] and Hamaad Raza Ahmad[1]

[1]Institute of Soil and Environmental Sciences, University of Agriculture Faisalabad, Faisalabad, Pakistan, [2]Indian River Research and Education Centre, Institute of Food and Agricultural Sciences, University of Florida (UF-IFAS), Fort Pierce, FL, United States, [3]Doctoral School of Environmental Sciences, Szent Istvan University, Godollo, Hungary, [4]College of Agriculture, University of Sargodha, Sargodha, Pakistan

15.1 Introduction

Heavy metal is a vast term and a wide range of plant pollutants like Cd, Pb, Cr (Ayub et al., 2020; Naeem et al., 2019) and nutrients are classified in this category called heavy metal plant micro nutrients (HMmNs) (Riaz et al., 2020). There are 17 essential plant elements (plant nutrients) required for the normal functioning of plants (Brown, Welch, & Cary, 1987; Taiz, Zeiger, Møller, & Murphy, 2015), of which six are heavy metals (Fe, Mn, Zn, Ni, Cu, and Mo) (Reeves & Baker, 2000). These metallic elements can be taken up by plants from soil and can be toxic if taken up in surplus amount (Blaylock & Huang, 2000; Marschner, 2012; Monni, Salemma, & Millar, 2000). Plant micronutrients are present in the range of 0.1–200 ppm making less than 0.02% of a plant dry biomass (Marschner, 2012). Plants take up these nutrients from the soil and accumulate in their body. Eight micronutrients (Fe, Mn, B, Cu, Mn, Zn, Ni, Cl) perform various important functions in plants like reproductive growth, nutrient regulation, seed and fruit development, chlorophyll synthesis, and plant metabolism. Moreover, the deficiency

Frontiers in Plant–Soil Interaction. DOI: https://doi.org/10.1016/B978-0-323-90943-3.00016-X

of these nutrients can cause several diseases, thus reducing the quality as well as quantity of food. Out of these eight, five elements are classified as heavy metals (Fe, Mn, Zn, Ni, and Cu). These plant heavy metal micronutrients (HMmNs) are necessary for plants' normal growth and development (Riaz et al., 2020; Taiz et al., 2015). In the past decade, a lot of research has been executed concluding the significance of HMmNs in the plant body under normal and stressed conditions, a brief overview of which is shown in Table 15.1.

15.2 Specific role of HMmNs in plants

15.2.1 Zinc

Zinc uptake from soil and its accumulation and translocation in various plant tissues is highly dependent upon plant species type. The ionic form of Zn (Zn^{+2}) is present in soil solution and its up-take by plants is assisted in two ways: (1) release of reductants like organic acids and H^+; and (2) release of phytosiderophores which make chelates with Zn and improve its uptake via root epidermal cells (Gupta, Ram, & Kumar, 2016). Soil bioavailable Zn content is highly dependent upon soil parent material and extent of weathering (Chesworth, 1991) as its concentration varies from 40 to 120 mg kg^{-1} in magmatic rocks, 80–120 mg kg^{-1} in argillaceous sediments, only 15–30 mg kg^{-1} in shales, and 10–25 mg kg^{-1} in limestones and dolomites (Kabata-Pendias, 2010).

Zinc is an essential plant and human nutrient with a significant role in plant physiology (Riaz et al., 2020; Taiz et al., 2015) and human health (Frassinetti, Bronzetti, Caltavuturo, Cini, & Della Croce, 2006). In plants it is involved in enzymes activation, metabolic processes, redox reactions, synthesis of cytochromes, protein, and carbohydrates (Broadley, White, Hammond, Zelko, & Lux, 2007; Taiz et al., 2015), phytohormones regulation, and improving crop yield and produce quality (Chattha et al., 2017; Hassan et al., 2019; Mengel, Kosegarten, Kirkby, & Appel, 2001; Mousavi, Galavi, & Ahmadvand, 2007). Plants under stress use Zn as a weapon against abiotic stress as it is involved in uplifting plants' carbon assimilation (Tsoney & CebolaLidon, 2012). Though it is highly beneficial for plant growth and development, higher Zn content in soil act as a pollutant affecting crop growth significantly. Worldwide a large population is suffering from Fe and Zn deficiency due to consumption of poor Fe and Zn content-containing food crops (Graham, Welch, & Bouis, 2001; Naveed et al., 2020).

15.2.2 Iron

Iron is the most abundant element in the universe and the fourth most abundant in the Earth's crust (Morgan & Anders, 1980). It is essential for most

TABLE 15.1 Heavy metals micronutrients and their deficiency symptoms in plants.

Sr. no.	Micronutrients	Deficiency symptoms	Plant adaptation to deficiency	References
1	Zn	• Reduced stem elongation in tomato • Necrosis of root apex known as dieback commonly • Mottling of leaves and interveinal chlorosis • Bronzing of leaves (reddish-brown color) • Resetting (shortening of internodes) • Inward curl of plant leaves termed as epinasty • Reduction in size of leaves termed as little leaf and "Goblet" leaves	• Carbonic anhydrase regulation for their fixation to the carbohydrates in the plants • Increase in carbohydrate metabolism and protein metabolism • Formation of pollen and auxins • Performs defense mechanism and governs the biological membranes against harmful pathogens • Protects the plants from oxidative stress • Presence of Zn as a cofactor in CAT and SOD	Alloway (2004a, b); Disante, Fuentes, and Cortina (2010)
2	Fe	• Chlorosis of young leaves due to less chlorophyll synthesis • Reduced growth and hill reaction activity	• Chlorophyll biosynthesis • Bioactivation of enzymes • Part of protein ferredoxin, required in nitrate and sulfate reduction • Involved in protein metabolism	Nikolic and Kastori (2000), Goos, Johnson, Jackson, and Hargrove (2004)

(Continued)

TABLE 15.1 (Continued)

Sr. no.	Micronutrients	Deficiency symptoms	Plant adaptation to deficiency	References
3	Mn	• Interveinal chlorosis • Premature falling of leaves • Delayed maturity and formation of whitish-gray spots on leaves • Development of necrotic spots on cotyledon of leguminous crops • Presence of greenish-gray specks at the lower base of monocots	• Plays role in synthesis of proteins, fatty acids and ATP acyl lipids • Participates in redox reactions and photosynthesis of plants • Bioactivation of enzymes • Photolysis of water at photosystem-II of photosynthesis	Schulte and Kelling (1999)
4	Ni	• High concentration of Ni consequently decreased a potential inhibitor of photosynthesis • Chlorophyll content decrease in young leaves • Chlorophyll contents and the number of leaves decreased with 24% and 47% • Decrease in photosynthetic pigments • Burning of leaf tip • Chlorosis and necrosis of plants • ultimately caused decrease in crop yield	• Plays role in plant metabolism • Part of many enzymes like glyoxalase, urease, superoxide dismutase and hydrogenase	Dubey and Pandey (2011)

5	Mo	• Leaf cupping • Mottling • Gray tinting • And flaccid leaves • The formation of necrotic spots on margins of leaves resulting in decreased plant growth • Leathery leaves and meristem necrosis • Formation of whip-tale (loss in proper development of lamina)	• Part of various molybdoenzymes which broke down to compounds used in assimilation and nitrogen reduction • Serves as nitrogenase in nitrogen fixation • Take part in IAA (indole-3 acetic acid), AO (aldehyde oxidase) and abscisic acid synthesis • Sulfur metabolism • Serves as xanthine dehydrogenase (XDH) or oxidase in purine catabolism	Mendel and Haensch (2002).

organisms, but its availability is limited in oxidized environments. Being abundant in the Earth's crust does not make Fe a surplus element available to plants, especially in aerobic and neutral pH/alkaline conditions as in these conditions it exists as Fe^{+3}. Among all micronutrients, Fe is the third most important limiting factor for plant growth due to its lower mobility/solubility in soil and oxidized status in aerobic conditions (Samaranayke, Peiris, & Dssanayake, 2012; Zuo & Zhang, 2011). Its availability is very important for plants as Fe being an essential plant nutrient, plays vital role in plant physiology and body functioning (Römheld & Nikolic, 2006) like DNA synthesis, respiration, chlorophyll synthesis, and nitrogen fixation (Kim & Rees, 1992). Many cellular components of plants also have significant Fe content like mitochondria, chloroplast, and vacuoles (Vigani, Zocchi, Bashir, Philippar, & Briat, 2013). Worldwide Fe deficiency affects around 2 billion people (McLean, Cogswell, Egli, Wojdyla, & De Benoist, 2009) as most of them are eating poor Fe-containing food (Abadía et al., 2011). Iron plays many vital roles in plants, as Fe cofactors are involved in oxygen transport and regulation of proteins stability. The major portion of chloroplasts comprises Fe-S proteins, such as ferredoxins present in PS-I. Mitochondria is another hotspot for Fe enzymes as they are involved in respiratory complexes, such as in complex I and II, and in a mix of FeS and heme (complex III) or heme and Cu (complex IV). In endoplasmic reticulum heme proteins like peroxidases and cytochrome P450 are present (Connorton, Balk, & Rodríguez-Celma, 2017).

15.2.3 Manganese

Manganese, another HMmN, is the most important micronutrient after Fe (Millaleo et al., 2010) and its availability in soil is also related with soil redox potential development (Porter, Bajita-Locke, Hue, & Strand, 2004). Among various roles of Mn in plants, enzyme activation and producing metalloenzyme cluster in photosystem are important (Andresen, Peiter, & Küpper, 2018). It is required in a very minute concentration by plants for normal growth and development (Millaleo et al., 2010) as it is involved in various metabolic processes, acts as a cofactor of various plant enzymes (Burnell, 1988), and plays a role in carbon assimilation (Millaleo et al., 2010). Manganese (Mn) is an important plant HMmN involved in plant metabolism via acting as a cofactor of enzymes and activating biological activity (Andresen et al., 2018). It is an important part of the metalloenzyme cluster of oxygen evolving complex of PS-II (Broadley, 2012). It helps the plant in photosynthesis, respiration, and scavenging reactive oxygen species (ROS) under stressed conditions as well as acts as a signal for hormone production. In many enzymes, Mn is interchangeable with other cations (Ca, Co, Cu, Mg, or Zn) while in plants only the OES, in PS-II, MnSOD, and oxalate oxidase require exclusively Mn (Alejandro, Höller, Meier, & Peiter, 2020).

15.2.4 Nickel

The concept of Ni being an essential plant nutrient is quite old (started in the 1970s), as reported by Dixon, Gazzola, Blakeley, and Zerner (1975), but very little studied (Bai, Reilly, & Wood, 2006). In the 1980s, Eskew, Welch, and Norvell (1984) reported that Ni is an essential element for leguminous plants but still its role for other plants was not clear until Brown et al. (1987) proposed its essentiality for nonleguminous plants. A wide range of studies have proved its essentiality and plants can't complete their life cycle without it (Azcon-Bieto & Talón, 2008; Bai et al., 2006; Epstein & Bloom, 2005), which is the main criteria for being an essential element (Arnon & Stout, 1939). It is a part of various enzymes like glyoxalases, urease, superoxide dismutase, and hydrogenase making up to 10 mg kg^{-1} concentration in plant dry body mass (Küpper & Kroneck, 2007). Nickel absorption from soil is governed by various processes, for example, its solubility and soil properties like soil solution chemistry, organic matter composition, etc. (Chen, Huang, & Liu, 2009), and is transported into the plant body via active and passive movement (Antić-Mladenović et al., 2011). Nickel availability is higher in neutral to acidic pH, as with increasing pH more complexes of Ni are formed, thus decreasing its availability in soil (Nkrumah et al., 2019).

15.2.5 Molybdenum

Molybdenum (Mo), a transition element, is an essential plant nutrient taken up from soil in the form of molybdate (Bittner & Mendel, 2010; Bittner, 2014). The fate of Mo in soil varies greatly depending on its oxidation state, as it exists in 0–VI oxidation states with Mo^{+6} being most prominent in soil solution. Once taken up, Mo plays an integral role in plant physiology acting as a metallic cofactor for various enzymes and acting as a co-factor called Moco (Mendel & Kruse, 2012; Williams & Frausto da Silva, 2002). Bioavailability of Mo in soil is highly dependent upon soil REDOX potential, soil pH, and organic matter contents. In alkaline pH its availability increases due to the formation of anionic compounds of Mo (Reddy, Munn, & Wang, 1997).

15.2.6 Copper

Copper (Cu) is another metal element essential for the normal function and growth of plants and is required in very minute quantity but can be toxic if up taken in higher quantity (Zhou et al., 2019). It is involved in various physiological and biochemical processes in plants and its deficiency causes decreased carbon assimilation and photosynthetic activities (Yruela, 2005).

15.3 Effect of HMmNs stress on plants

15.3.1 Zinc deficiency

Zinc deficiency in alkaline and calcareous soils is a concern as its availability is compromised under high soil pH conditions (Humayan Kabir, Swaraz, & Stangoulis, 2014). Deficiency of this important nutrient in plants causes leaf browning, growth retardation, biochemical dysfunction, and is even responsible for yield loss (Fageria, Baligar, & Clark, 2002; Naveed et al., 2020). Zn-deficiency symptoms are very much common in younger and middle-aged leaves (Ozturk et al., 2006; Riaz et al., 2020). High bicarbonate contents, intense cropping, ratio of Mg and Ca in the soil >1, use of alkaline water for irrigation, and the use of high yielding crop varieties also affect the content of Zn in the soil (Das, Avasthe, & Gopi, 2014).

Zinc is an important plant nutrient absorbed from the soil as described in above section. The absorption of Zn from soil by plants is done via two ways: first via direct absorption as Zn^{+2} ions along with water, and second via the production of Zn-PS (phytosiderophores), which occurs in some special cases (Welch & Shuman, 1995). Plants' responses under Zn-deficient conditions vary drastically from species to species and crop variety to variety and contradictory results on Zn-PS production under Zn-deficient conditions have been reported (Gries, Brunn, Crowley, & Parker, 1995; Pedler, Parker, & Crowley, 2000; Rengel & Romheld, 2000). A major physiological response of plants under Zn deficiency is the secretion of these Zn-PS which are highly dependent upon the cultivar's Zn harvesting capacity. Zn-efficient plant are more tolerant to Zn-deficiency due to their higher ability of Zn-quenching from soil via PS production. Furthermore, these PS secretions from plants are also highly dependent on soil type and condition. For inter- and intracellular translocation of Zn, Zn-transporter proteins zinc/iron permease (ZIP) play a vital role. More than 100 ZIP members have been discovered in various organisms, out of which 16 have been discovered in *Arabidopsis*. In higher plants, the messenger RNA (mRNA) responsible for ZIPs is predominantly activated under Zn deficiency. Various kinds of these ZIP proteins have been classified in different plants like for soybean and *Bradyrhizobium japonicum* symbiosis controlling GmZIP1 (Moreau et al., 2002), OsZIP4 in rice (Ishimaru et al., 2005), OsZIP1 and OsZIP2 in *Arabidopsis* (Ramesh, Shin, Eide, & Schachtman, 2003), mRNA which are primarily targeted in plant tissues where Zn is meant to be more translocated. Another type of facilitator family of protein, cation diffusion facilitator (CDF) is also found to be responsible for Zn translocation under deficient conditions. A type of metal translocator protein (MTP) is also crucially involved in Zn homeostasis in plants. Another class of Zn transporter, heavy metal ATPase (HMA) is also activated in soil plants under Zn deficiency (Cobbett, Hussain, & Haydon, 2003).

Deficiency of Zn in soil also triggers gene edits in various crop plants. Significant genetic variability has been observed in crops like rice (Höller, Meyer, & Frei, 2014), barley (Genc, Huang, & Langridge, 2007), bean (Hacisalihoglu, Ozturk, Cakmak, Welch, & Kochian, 2004), and wheat (Cakmak, Gülüt, Marschner, & Graham, 1994) for effective absorption of Zn under deficient conditions (Impa & Johnson-Beebout, 2012; Kabir, Hossain, Khatun, Sarkar, & Haider, 2017). Besides these intercellular responses, plants tend to modulate and overcome soil-mediated hindrance in Zn availability via exudating H^+ into the rhizosphere, thus decreasing rhizosphere pH (Santi & Schmidt, 2009). Lower soil pH is a key participant in increasing Zn bioavailability (Palmer & Guerinot, 2009). Other important molecular responses are upregulation of Zn sensors and chelators [like deoxymugineic acid (DMA) and nicotianamine (NA)] that are helpful in net sequestration of Zn by various plants (Takahashi et al., 2003; 2012). Genetic engineering for Zn-transported incorporation into plant cell can be an effective strategy in escaping Zn deficiency stress (Kabir et al., 2017).

15.3.2 Zinc toxicity

Zinc, being an essential plant nutrient, can be taken up by plants in adequate concentrations. The typical Zn concentration required for proper growth is approximately between 15 and 20 $mg\ kg^{-1}$ dry weight (Cakmak, Öztürk, Karanlik, Marschner, & Ekiz, 1996). Higher contents of Zn once up taken into the plant's body can be toxic to plants. The potential toxicity of Zn increases as its bioavailable concentration surpasses the permissible limit. The bioavailable concentration of Zn in soil is predominantly controlled by soil physicochemical properties (Kwon et al., 2017; Wyszkowski & Modrzewska, 2016). Upon experiencing toxic Zn concentrations, plant's physiology and molecular aspects get disturbed and it tends to either survive this toxicity via adaptation or permanently develop tolerance against Zn toxicity. A wide range of cellular and molecular phenomena are involved in plants which help them to withstand Zn toxicity.

The Zn toxicity response of plants varies as hyperaccumulator Zn-tolerant plant species can accumulate significantly more Zn compared to other plants. The Zn hyperaccumulator plants are characterized based upon their physiological and morphological as well as Zn translocation behavior which makes them distinct (Clemens, 2017; Moreira, Moraes, & dos Reis, 2018). Upon absorption from the root, Zn is translocated in plants body via mass flow following two routes (symplastic and apoplastic paths) and its way is only gets checked at Casparian strips. Under Zn toxicity/excess conditions plants develop a specialized cell layer as well as lignification of the cell wall, which is a distinct feature of Zn-tolerant plants (Gupta et al., 2016). Specialized genes present in plants are activated for potential heavy metal tolerance which can also help in Zn homeostasis (Merlot, de la Torre, & Hanikenne, 2018).

Zinc translocation in plants is primarily governed by Zn-transporters like ZRT-IRT or ZIP family proteins. These transporters are responsible for transportation of Zn radially to stele as well as contribute to Zn translocation to endodermal cells (Van de Mortel et al., 2006). Overexpression of NcZNT1 (another Zn transported located on cell membrane) is linked with excessive Zn exposure in *A. thaliana*, which is responsible for enhanced Zn translocation, accumulation, and tolerance compared to wild plant species (Lin, Hassan, Talukdar, Schat, & Aarts, 2016). Another two important transporters, NcZNT 2, 5, iron-transported (IRT3), and ZIP5 are involved in the upregulation of more Zn absorption from roots (Fasani, 2012; Van de Mortel et al., 2006) from which IRT3 is also reported in the cell-membrane of *A. halleri* root cells upon exposure of Zn (Caldelas & Weiss, 2017; Talke, Hanikenne, & Krämer, 2006). The IRT3 Zn transporter is the main characteristic of Zn hyperaccumulator plants (Caldelas & Weiss, 2017; Van de Mortel et al., 2006), which can also be present at subcellular levels assisting in potential translocation and assimilation of metal ions. Beside this, the expression of ZIP-19/23 was also found to be uplifted in *A. halleri* and *N. caerulescens* roots upon higher Zn levels (Kabir et al., 2017). For aboveground translocation of Zn, ATPase4 (HMA4) is reported to play potential role in Zn homeostasis (Guerinot, 2000; Merlot et al., 2018; Van de Mortel et al., 2006).

15.3.3 Iron deficiency

Iron is a crucial part of various enzymes involved in electron transfer, such as the cytochromes. It is also a part of non-heme Fe proteins that take part in photosynthesis, nitrogen fixation, respiration, etc. (Taiz & Zeiger, 1991). Iron is an integral part of a plant's body and once it becomes a limiting element then plants tend to adopt an integrated approach to manage its deficiency (Grotz & Guerinot, 2006). It is a co-factor of a variety of enzymes and processes like the electron transport chain and carbon assimilation and its limitation compromises plant's yield and growth significantly. It is also is an integral part of the electron transport chain and chlorophyll, that is why its deficiency can decrease photosynthetic rate, stomatal conductance, and yield of plants (Molassiotis, Tanou, Diamantidis, Patakas, & Therios, 2006). Several enzymes also have Fe as its integral part and Fe deficiency can significantly influence metabolic activities being controlled by these enzymes (Borlotti, Vigani, & Zocchi, 2012). The Fe deficiency is also reported to be involved in mitochondrial distortion and poor light harvesting (Vigani et al., 2013).

15.3.4 Iron toxicity

Once taken up in excess, Fe becomes potentially toxic for plants with the primary effects being evident on chloroplasts as 80% of cellular Fe is present there (Marschner, 1995). The Fe toxicity in cells brings oxidative stress disturbing the photosystem and light harvesting complex (Suh, Kim, Lee, & Jung, 2002). Tiny brown necrotic spots on plant leaves followed by reddish-color and leaf dying are prominent symptoms of Fe toxicity, which is called leaf bronzing (Becker & Asch, 2005).

15.3.5 Manganese deficiency

Manganese, being an important plant nutrient, plays many vital roles in plants' physiology and cellular mechanisms, like regulation of various biosynthesis pathways, help in light harvesting by photosystem, and providing defense against diseases (Eaton, 2015; Yano & Yachandra, 2014). The deficiency of Mn is a problem for crop productivity, and widely affects winter cereals like wheat and barley (Hebbern, Pedas, Schjoerring, Knudsen, & Husted, 2005; Jiang, 2006). This phenomenon is termed as "winterkill." It further persuade the scientists to explore chlorophyll using fluorescence tools for the diagnosis of Mn deficiency. The Mn deficiency does not show clear symptoms, so the intensity of its deficiency is still unknown. However, Mn deficiency affects the thylakoid structure and promotes the degradation of chlorophyll (Papadakis et al., 2007), further leading to the formation of interveinal chlorosis. The symptoms of Mn deficiency in the fields are scattered due to the uneven field soil conditions that favor oxidation of Mn. This process is known as "light spot disease." Manganese deficiency is widespread in soil like in saline soils (George et al., 2014; Hernandez-Apaolaza, 2014) as its deficiency symptoms do not appear at most levels of stress (Schmid et al., 2014). The Mn deficiency in plants is coupled with extreme oxidative stress (Zhao et al., 2017) and decreased plant growth (Moradtalab et al., 2018; Saidi et al., 2012), increased susceptibility to pathogen (Heine et al., 2011), imbalance in water relation (Hebbern et al., 2009), and decreased resistance to cold stress (Ihnatowicz et al., 2014). Higher Mn concentration is toxic to plants, but this depends upon plant type (Broadley, 2012; Husted, Thomsen, Mattsson, & Schjoerring, 2009) as Mn inter- and intracellular accumulation in plants vary highly. The Mn toxicity in plants can disturb nutrient acquisition (St. Clair & Lynch, 2005), cause imbalance in chlorophyll biosynthesis (Subrahmanyam & Rathore, 2001), decrease in cell division and elongation (Zhao et al., 2017), and oxidative damage to cells (Fecht-Christoffers, Führs, Braun, & Horst, 2006)

15.3.6 Manganese toxicity

Though Mn is an essential plant nutrient needed for various biochemical processes in plants, its higher concentration can be lethal for plant physiology and cellular homeostasis and can also cause damage to the human food chain (Yang, Deng, & Li, 2008). Manganese is required for effective working of various enzymes involved in ROS quenching, thus upon its deficiency a plant cell experiences severe oxidative stress (Hohle & O'Brian, 2009). A decrease in the bioavailable Mn content can severely influence plant growth and secondary metabolites biosynthesis making plants prone to various disorders (Wang et al., 2012). The Mn limit of deficiency is around 10–20 mg kg^{-1} body weight of young plants which widely varies plant to plant and even among cultivars of the same species (Shao, Yamaji, Shen, & Ma, 2017).

As mentioned earlier, even a minute concentration of Mn is beneficial for physicochemical processes of plants cell, but its higher concentrations are toxic to plants. The most evident toxicity is observed on cellular light harvesting complex, ROS quenching abilities, and disturbed nutritional acquisition (Fernando & Lynch, 2015). Molecular and cellular responses of plants toward Mn toxicity vary widely based upon the genetic makeup and environmental conditions of growing plants. For examples, the Australian seaboard with Mn-rich acidic soil in which Mn contents surpass 5000 mg kg^{-1} soil have predominantly a native flora which has significant tolerance toward Mn toxicity (Fernando, Baker, & Woodrow, 2009). A major population of Mn tolerant plants have this capability owing to their genetic makeup given by years of generation of evolution under specific geographical and climatic conditions. Toxic levels of Mn concentration in plants also vary widely and are crop dependent, as maize is reported to show toxicity symptoms upon Mn accumulation of 200 mg kg^{-1} while sunflower showed toxicity at 5300 mg kg^{-1}. Some plants which are hyperaccumulators of Mn can uptake and accumulate significantly higher contents of Mn as they have specific transporters for effective uptake, distribution, and help in detoxification of Mn, for example, OsNram3 in rice (Edwards & Asher, 1981; Shao et al., 2017). A wide range of transporters present in plants are involved in net homeostasis of Mn such as AtNram1, AtZIP1/2, HvNramp5, and HvIRT1, which help plants in the efficient management of both Mn deficiency and toxicity stresses (Shao et al., 2017).

15.3.7 Nickel deficiency

Nickel (Ni) was announced to be an essential plant nutrient in 1987 and its deficiency can lead to severe physiological and molecular disturbances in plants. Ni plays a vital role in embryonic development and its deficiency can lead to hindrances in embryonic growth. The critical Ni concentration for barley is reported to be 90 ± 10 ng g^{-1} dry mass, which reduced its yield by 15%. Ni is a key component of urease biosynthesis and function in plants and its binding is controlled by two important genes, EU2 and EU3, coding for auxiliary proteins involved in activation of urease (Seregin & Kozhevnikova, 2006).

15.3.8 Nickel toxicity

Nickel, although classified as a plant nutrient, can become toxic for plants if taken up in surplus amount. The major mode of toxicity on plants is via disturbed mineral nutrient acquisition and induction of oxidative stress (Sreekanth, Nagajyothi, Lee, & Prasad, 2013). Among various HMmNs, Ni, if available in toxic concentrations, can lead to imbalanced mineral acquisition (Ahmad, Hussain, Saddiq, & Alvi, 2007), subsequently affecting plant growth and disrupting its cellular biochemical processes (Gajewska, Skłodowska, Słaba, & Mazur, 2006). Exogenous application of other micronutrients can lead to net cancelation of Ni toxicity in plants. The antagonistic effect of Ni on Fe (Pandey & Sharma, 2002), Zn and Cu (Parida, Chhibba, & Nayyar, 2003) nutrient ions also disrupts biosynthesis of antioxidant enzymes like superoxide dismutase and catalase (SOD, CAT), which are important metalloenzymes (Gajewska et al., 2006). Other molecular toxicity of Ni involves its removal of Ca from Ca-binding sites in the photosystem O-evolution complex and swapping of Mg from chlorophyll (Solymosi et al., 2004). Direct onset of oxidative stress is also associated with Ni toxicity (Boominathan & Doran, 2002) which is due to the potential increase of OH, O^-, NO, and H_2O_2 (Gajewska & Skłodowska, 2005). As Ni is not a redox-active metal, its role in direct onset of oxidative stress is less evident. The potential increase of ROS observed under Ni toxicity (Hao, Wang, & Chen, 2006) is due to its direct interaction with many antioxidants (antioxidant disruption like of SOD, CAT, glutathione GSH, peroxidase POD, and ascorbate peroxidase) which are involved in potential scavenging of these ROS (Gomes-Junior et al., 2006). All these effects of Ni toxicity are responsible for the net decrease of photosynthesis, thus ultimately decreasing crop yield (Dubey & Pandey, 2011).

15.3.9 Molybdenum deficiency

The metabolism of plants is mainly affected by Mo-deficiency at different stages. Various types of molybdoenzymes present in the plants strongly require molybdenum, like nitrate reductase, sulfite oxidase, xanthine dehydrogenase, aldehyde oxidase, and mitochondrial amidoxime reductase (Rana et al., 2020). These molybdoenzymes are further classified on the basis of their interaction with molybdenum, like Mo-Co. NR and dioxi-Mo cofactor in SO, which further activated the proteins inserted in the protein complex (Mendel & Haensch, 2002). Furthermore, Mo-redox behavior, growth promotion effect, biosynthesis of Mo-Co (Mo-containing enzymes), as well as its role in N-fixation makes it a very crucial limiting element.

The responses of plants toward Mo deficiency is intricate as it is very difficult to determine the specific enzyme system involved. The involvement of molybdoenzymes in the metabolism of nitrogen makes it clear that a decrease in plant growth and physiology affect the grain and fruit development which are easily susceptible to pest damage (Graham & Stangoulis, 2005). The deficiency of Mo is evident from several plant species whose phenotypes differ in appearance and severity (Hewitt & Bolle-Jones, 1952a).

The Mo deficiency is clearly noticeable in many members of the Brassicaceae family. The Mo-deficiency symptoms in young plants include mottling, cupping of leaf, gray colored tinting, and flaccid leaves found mostly on the seedlings which remain dwarf leading to the death of plant (Hewitt & Bolle-Jones, 1952a). Whereas in the older plants, the Mo deficiency can be liberated or at the modest levels of Mo deficiency, the younger leaves show the deficiency symptoms like meristem necrosis, loss of leaf lamina development (Whip-tail), and leathering of leaves (Hewitt & Bolle-Jones, 1952b). By the indirect effects of Mo on the nitrogen assimilating enzymes, these deficiency symptoms are masked. Most legume crops, cereal crops, and horticultural crops grown in the soil (deficient in Mo) with the addition of nitrogen fertilizer show the symptoms of necrosis at the margins of leaves, pale green leaves, and reduction in the growth of plants (Chatterjee & Nautiyal, 2001). Moreover, oat and wheat crops deficient in Mo show symptoms like necrotic regions on plant leaves and poor development of seeds (Chatterjee & Nautiyal, 2001). Exogenous application of Mo can be a possible solution to tackle the issue of Mo-deficiency-mediated symptoms (Rana et al., 2020)

15.3.10 Molybdenum toxicity

A very limited information is available for Mo toxicity in terrestrial plants and its toxic limits in soil are not well defined yet due to the lack of sufficient literature. The potential toxicity of Mo in plants is highly dependent upon soil type and properties (McGrath et al., 2010) with very distinct physiological and molecular responses.

15.3.11 Copper deficiency

Copper is an important HMmNs involved in the proper functioning of PS II, and its deficiency causes chlorosis, dieback of fruit, paleness of leaves, quenching and inhibited growth, and PS II activity. Furthermore, Cu deficiency causes disturbances in biostimulation and regulation of a variety of genes involved in biochemical processes like light harvesting (Thomas, Andresen, Mattusch, Hubáček, & Küpper, 2016; Yruela, 2005). The Cu is a crucial element for proper carbohydrates metabolism (Ameh & Sayes, 2019), thus it is required for bioproduction of lignin (critical to provide an escape against cell wall wilting) which helps in cell wall strengthening (Raldugina, Krasavina, Lunkova, & Burmistrova, 2016). Cu-mediated disturbance in photosynthesis is attributed toward Cu-initiated oxidative stress (due to light diversion) which damages the photosystem, thus decreasing light harvesting and food production. Another possible reason is damage to Cu/Zn-SOD under Cu deficiency as its bioproduction is severely decreased. An investigation by Thomas et al. (2016) has shown that the absence of Cu from the system can severely influence the electrotransport chain under actinic light between PS-I and PS-II, decrease plastocyanin production, and increase production of ROS in *Ceratophyllum demersum* fragile stem.

15.3.12 Copper toxicity

Excessive availability, uptake, and accumulation of Cu in plants significantly hinders its growth and metabolism (Gong et al., 2019; Marques, da Silva, Mantovani, Magalhães, & de Souza, 2019). The Cu toxicity has direct as well indirect toxicity toward plants, as reported by published work (Jaime-Pérez et al., 2019; Marques et al., 2019). The Cu toxicity-mediated growth retardation in plants (Buapet et al., 2019; Michaud, Chappellaz, & Hinsinger, 2008; Mwamba et al., 2016) but proper mechanism of Cu toxicity in plants is still under investigation. The first organs of plants to be affected by Cu toxicity are plant roots (Cambrollé, García, Ocete, Figueroa, & Cantos, 2013) in which higher Cu accumulation causes damage that decreases its growth and water/nutrient absorption capacity (Sheldon & Menzies, 2005). The Cu toxicity results in damage and rupturing of root epidermis and cortex (Kopittke, Blamey, McKenna, Wang, & Menzies, 2011) resulting in less branching, decreased growth and elongation, dark coloration, and less secondary growth (Marques et al., 2019). It is reported by Marques et al. (2019) that higher application of Cu (1000 ppm solution) can result in disturbed root morphology.

Among the molecular aspects of Cu toxicity are decreased plant pigment formation, decreased photosynthesis (Jaime-Pérez et al., 2019), altered photosynthesis pigment, and protein composition (Silva, Echeveste, & Lombardi, 2018). An investigation by Nazir, Hussain, and Fariduddin (2019) reported that application of 100 ppm Cu for 30 days significantly reduced photosynthesis, disturbed stomatal opening and closing, as well as activity of antioxidants in tomato plants. It is also reported that a high level of Cu can affect light harvesting, PS-II functioning, ROS formation, the electron transport chain (Gong et al., 2019), and photorespiration (Ameh & Sayes, 2019; Marcec, Gilroy, Poovaiah, & Tanaka, 2019) (Fig. 15.1).

15.4 Management of HMmNs deficiency and toxicity

15.4.1 Physiological and molecular responses of plants

The majority of the response in relation to environmental stresses in plants is linked with growth, differentiation, and physiological aspects such as photosynthesis, ions uptake, and transport (Seh, 2002). Movement of metals from soil to roots and their absorption and translocation is concerned with the physiology of specific metal toxicity. Availability of metal to the plant mainly depends upon the soil absorption strength and the plants effectors

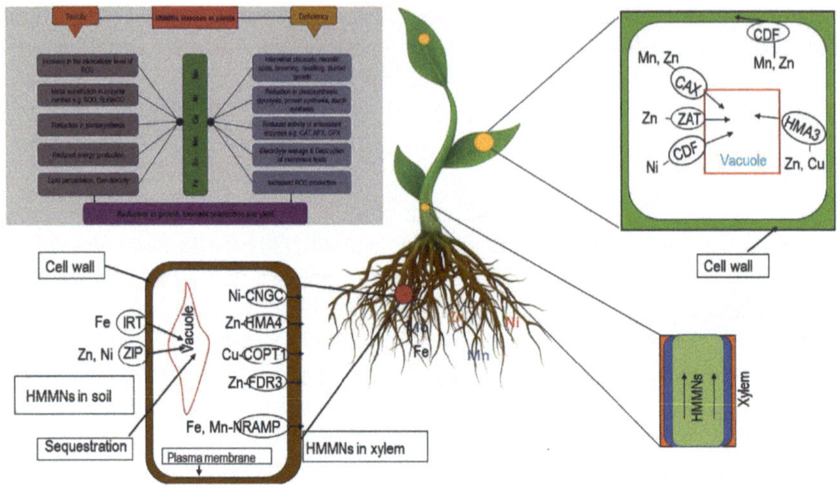

FIGURE 15.1 Response of plants to the excess of heavy metal micronutrients (HMmNs).

(root exudates for chelation and reduction). There are very few studies on the biochemical and physiological metal toxicity. The chelators like amino acids and carboxylic acids affect the activity of heavy metals in the cytoplasm (Rout & Das, 2009). The sensitivity of plants to heavy metals depends on an inter-related network of molecular and physiological processes, including the uptake and accumulation of heavy metals by binding with the cell wall and extracellular exudates and the complexity of ions with various substances present in the cells such as amino acids, organic acids, metallothioneins, ferritins, and phytochelatins. Moreover, the general defense responses toward biochemical stress, like modification and activation of metabolism in the plants and introduction of antioxidant enzymes, permit the proper functioning of metabolic pathway in plants and fast repair of damaged cell structure (Cho, Chardonnens, & Dietz, 2003; Hall, 2002).

It was observed in a study on *Elodea densa* that Ni and Zn have significantly caused a decrease in plant chlorophyll levels, thereby decreasing the photochemical efficiency of PS-II (Maleva, Nekrasova, Borisova, Chukina, & Ushakova, 2012). In another study conducted on *Thalassia hemprichii*, it was reported that Zn and Cu have reduced the levels of carotenoids and chlorophyll contents as well as the quantum yield of PS-II which indicates that these heavy metals have negative effects on plant photosynthetic activity. Heavy metals also reduce CO_2 assimilation by decreasing RUBP carboxylase activity or by reaction with the thiol group pf RUBISCO. The RUBISCO activity has been reduced by Zn in *Phaseolus vulgaris* after replacement of Zn^{2+} by Mg^{2+} (Monnet et al., 2001).

Heavy metals like Cu and Mn also accumulate in the root and shoot of plants due to Zn toxicity (Nagajyoti, Lee, & Sreekanth, 2010). In a study, Zn toxicity inhibited the PS-I and PS-II in *P. vulgaris*, negatively affecting the synthesis of ATP and photosynthesis in the plant. These effects can be reversed but can be permanent if stress persists (Assche and Clijsters, 1990). During Zn toxicity, Zn ion replaces Mg^{2+} ion in RuBisCO (ribulose-1,5-bisphosphate-carboxylase/oxygenase), thus inhibiting its activity and carbon fixation. Todeschini et al. (2011) observed the effect of Zn toxicity in leaves of poplar tree. In this study, it was observed that Zn accumulated in the parenchyma and xylem cells. Moreover, the accumulation of calcium oxalate crystals was observed in the leaves due to Zn toxicity. The growth and photosynthetic activity of the plant was also affected by Ni toxicity. It also induced oxidative stress, nitrogen metabolism, mitotic and enzymatic activities in the plant and also interfered with the uptake of other metal elements. The uptake of metal ions happens through active or passive transport and diffusion depending upon the soil pH, type of plant, concentration and form of nickel ion, and the availability of other metals (Ghori et al., 2019).

15.4.1.1 Zinc homeostasis

Zinc homeostasis is a key factor in sustainable agriculture and its deficiency and toxicity must be addressed properly. Plants have developed (evolved) to manage Zn deficiency and toxicity in the most distinct ways possible (Henriques, Chalfun-Junior, & Aarts, 2012). For effective management of Zn homeostasis, a wide and complex network of cellular and tissues in plants have been discovered controlling metal uptake, accumulation, assimilation, and detoxification. The potential control of Zn deficiency/toxicity starts from absorption control from the soil, which is affected by several soil factors, among which Zn solubility is most important and is affected by soil physicochemical properties (Cakmak, Pfeiffer, & Mcclafferty, 2011). Once dissolved in soil (and reaching soil solution) it is absorbed by roots and moves into the plant via xylem in the water-soluble Zn^{+2} form, passing through epidermal and cortical cells, and finally gets loaded into the xylem following the symplastic pathway (Broadley et al., 2007). Potential Zn dissolution, absorption, transport/accumulation, and assimilation in plants is highly dependent upon its bioavailable fraction as well as plant type (Marschner, 1995) as under deficiency plants tend to absorb more, while under toxicity avoidance is opted. Under limited Zn concentration, inter-plant distribution of Zn is managed, ultimately giving the surplus Zn to the root, vegetative portion, and reproductive tissues of young plants. Under Zn toxicity plants tend to absorb and accumulate most of the Zn in cortical cells (Robson, 1994). Plant roots also manage to release many metal chelators involved in the enhancement of Zn under Zn-deficient conditions, production of which is controlled by nicotinamide-synthase genes (Klatte et al., 2009). Among these genes, AtNAS2, -4 are expressed under Zn-deficient conditions (van de Mortel et al., 2006). Furthermore, release of phytosiderophores, like 2-deoxymugenic acid, by crops like wheat also help in Zn management (Rengel, 1997). In Zn-deficient soils, plants tend to release low-molecular-weight compounds like malate and mugineic acid acting as siderophores (Gao, Zhang, & Hoffland, 2009). Plants experiencing Zn deficiency also tend to modulate root morphology for effective Zn absorption (Impa & Johnson-Beebout, 2012).

Upon uptake, the accumulation and partitioning of Zn in plants is controlled by a variety of selective transporters (Clemens, Palmgren, & Kramer, 2002), like natural resistance associated macrophage protein (NRAMP). Research has shown that NRAMPs have little metal specificity thus helping in translocation of a variety of metal elements. Among the various NARMPs, the AtNARMP3, -4 are present in tonoplast and helps in intracellular translocation of metals like Fe, Zn, and Mn (Thomine et al., 2000; 2003) with similar levels in various crop plants (Grotz & Guerinot, 2006) under HMmNs deficiency conditions. Under Zn toxic conditions, Zn efflux from a plant cell is an acceptable approach and Zn-efflux-transporters like P1B-ATPase play a significant role in this regard, including HMA1,2,3,4 which is involved in Zn transport (Eren & Argüello, 2004). The AtHMA1 is present in the chloroplast envelope and helps in the detoxification of Zn toxicity to organelles via active efflux of Zn (Kim et al, 2009). The AtHMA2 gene encodes an Zn-ATPase present on plasma proteins with expression induced by Zn and Cd (Eren & Argüello, 2004). AtHMA3 is present in potential hyperaccumulator plants helping to accumulate higher Zn than other plants (Hassan & Aarts, 2011). The last type of Zn-ATPase, AtHMA4 works like AtHMA2 and acts in potential retranslocation of Zn into the xylem (Waters & Sankaran, 2011). Another class of transporters, CDF is also involved in the homeostasis of HMmNs in plants (Gustin, Zanis, & Salt, 2011). Among these, MTP 1, -3 are involved in Zn-sequestration in root vacuoles thus limiting translocation in aerial tissues (Arrivault, Senger, & Kramer, 2006). The MTP2 is involved in Zn homeostasis under Zn deficiency conditions and facilitates Zn influx. While MTP8 is involved in the hindrance to Mn transport while facilitating Zn uptake (van de Mortel et al., 2006). The last class of transporters, iron-regulated transporters-like proteins (IRT-ZIP) are involved in potential translocation of Zn into cytosol across plasma membrane which is an important task in Zn-uptake (Palmer & Guerinot, 2009; Song et al., 2010). There are eight transmembrane domains of ZIP and a histidine-rich domain is present in III and IV members of the ZIP family (Colangelo & Guerinot, 2004), and in *Arabidopsis* plant around 15 ZIP gene members have been found, among which ZIP1, -3, -5, -6, -10, and IRT3 are involved in the management of Zn deficiency (Assunção et al., 2010; van de Mortel et al., 2006).

15.4.1.2 Iron homeostasis

Although Fe is present in a surplus quantity in the Earth's crust, its bioavailability is much lower and thus plants tend to adopt various strategies to tackle Fe deficiency. To cope with this issue, plants tend to adopt various molecular mechanisms, the first of which is the reduction approach (Hindt & Guerinot, 2012). In this approach, plants modulate root cells to release H^+ ions into the rhizosphere, thus lowering soil pH which ultimately helps in increasing bioavailable Fe concentrations (Santi, Cesco, Varanini, & Pinton, 2005; Santi & Schmidt, 2008). One unit decrease in soil solution pH can increase Fe concentration (bioavailable contents) up to 1000 times (Guerinot & Yi 1994). After that, the plant uses membrane-bound Fe-chelate-reductase-enzyme (AtFRO2 in *Arabidopsis* or PsFRO1 in pea) to convert available Fe^{+3} to Fe^{+2} via an active process of reduction and later form is more available for plants (Robinson, Procter, Connolly, & Guerinot, 1999; Waters, Blevins, & Eide, 2002). The process of Fe conversion (from oxidized to reduced state) is an event of prime importance as transgenic expression of Fe-chelate-reductase-enzyme can be visible in many crops under Fe-deficit conditions (Ishimaru et al., 2007; Vasconcelos et al., 2006). Once uptaken, Fe^{+2} is transported into roots via IRT1 transporters, an essential gene which predominantly activates under Fe-deficit conditions on the epidermal layer (Dinneny et al., 2008; Vert et al., 2002).

Besides modifying rhizosphere chemistry and reducing rhizosphere, plants also tend to release some chelating agents into the rhizosphere to tackle Fe deficiency. Among the various crop plants, grasses release siderophores (phytosiderophores sauce as mutinies acid or Mas) which have high affinity for Fe-chelation, making it available for longer time periods (Conte & Walker, 2011). The bioprecursor of these siderophores is NA which is an important amino acid which can be synthesized by almost all plants and acts as a metal chelator, but only in grasses it is converted to phytosiderophores (Hindt & Guerinot 2012). The Fe-siderophore complex (Fe-Ps) can be directly taken up by plant roots via yellow strip family transporters like YS1 in maize (Curie et al., 2009; Curie et al., 2001) and OsYSL15 in rice (Inoue et al., 2009; Lee et al., 2009). Specific genes in various plants have also reported to be involved in Fe stress management like 21 NAS and TaNAAT and Ta DMAS1 present in wheat (Bonneau, Baumann, Beasley, Li, & Johnson, 2016), while HuNAAT and HuDMAS1 in barley (Beasley, Bonneau, & Johnson, 2017). Beasley et al. (2017) reported several genes (TaDMAS1/2, TaNAAT1) involved in potential uplifting of Fe translocation under Fe-deficit conditions. The siderophores which act as Fe chelates are also transported via phytosiderophores-efflux-transporter-TOM1 (Nozoye et al., 2011). Once Fe toxicity is observed, plant molecular response includes (1) active exclusion of Fe^{+2} by roots into rhizosphere (Deng, Wu, Wu, Deng, & Wong, 2010); (2) potential assimilation of Fe in vacuole of cell making it harmless for plant's cellular machinery (Briat, Duc, Ravet, & Gaymard, 2010); and (3) Fe tolerance enhancement via detoxification of ROS with excessive production of antioxidants like glutathione and ascorbic acids (Gallie 2013). These molecular responses are evident in the management of Fe stress in crops like wheat (Wang, Kawakami, & Bhullar, 2019).

15.4.1.3 Manganese homeostasis

As explained in earlier sections, Mn is an important HMmN required for a variety of functions for plants but at higher concentration is toxic to plants (Goulding, 2016). Depending upon Mn availability plants either acquire it more (under Mn deficiency) or detoxify its effects (under superfluous availability). Various transporters are reported in plants for effective management of Mn homeostasis (Shao et al., 2017). The deficiency of Mn in plants occur as a latent disorder with very few clear visual symptoms. To effectively manage Mn deficiency and toxicity, a wide range of transporters are present in plants which are multielement transporters, for example, Mn/Fe transporter AtMTP8 (Chu et al., 2017). Importer transporters translocate Mn into cytosol from the extracellular matrix while exporter transporters are involved in the exclusion of Mn from cytosol. The NRAMP family, Zn-Regulated-Transporter/Fe-Regulated transporters (ZRT/IRT)-related protein family, and yellow-stripe like (YSL) family are the most famous importers of Mn in cells. In contrast, the Cation-Diffusion-Facilitator/Metal-Transport-Protein (CDF/MTP) family, the Vacuolar-Fe-Transporter (VIT) family, Ca^{+2}/Cation antiporter (CaCA) superfamily, Bivalent-Cation-Transporter (VIT) family, and the P2A-type-ATPase family are active exporters of Mn for effective exclusion from plant cells. These transporters work antagonistically for the effective management of Mn deficiency and toxicity conditions (Alejandro et al., 2020).

15.4.1.4 Nickel homeostasis

There is less information about the uptake mechanism of Ni by plants. Ni/Co transporters play an important role in the uptake of Ni in prokaryotes. The genome of *A. thaliana* also contains the genes of the Ni/Co protein family (Eitinger, Suhr, Moore, & Smith, 2005). There is evidence that suggests that uptake of Ni also takes place with the help of Fe transporters due to a broad efficacy for divalent cations (Morrissey and Guerinot, 2009). This uptake of Ni by Fe transporters showed a strong relation between the transport of these two metals. Nishida et al. (2011) reported that in a hydroponic experiment with higher Ni concentration *A. thaliana* showed higher uptake of Fe as well as increased expression of IRT1. After uptake metals accumulated in different plant parts. A study conducted by Yamaguchi, Tomioka, & Takenaka (2015) showed that in *Clethra barbinervis* plants the concentration of Ni and cobalt was the same in roots, while the concentration of Co in leaves was higher as compared to Ni, which showed that long distance transport of Ni and Co occurred through different transporters. It was reported that long distance transport of Ni in *Noccaea caerulescens* occurred through the ligand NA. In hyperaccumulators the accumulation of Ni mostly happened in the vacuole of epidermal cells to avoid the interference of Ni with the photosynthetic mechanism (Küpper and Kroneck, 2007). Histidine acts as a chelator of Ni and a higher concentration of free histidine is present in hyperaccumulators (Ingle, Mugford, Rees, Campbell, & Smith, 2005).

15.4.1.5 Molybdenum homeostasis

The soluble and plant available form of Mo in soil is molybdate. It was reported that the uptake of Mo in *A. thaliana* occurred through molybdate transporters known as MOT1 (Tomatsu et al., 2007). MOT1 are found in the mitochondrial envelop, vesicles, and also are localized partially in the plasma membrane (Baxter et al., 2008). Gasber et al. (2011) reported that *A. thaliana* lacks another member of the MOT1 family known as AtMOT1;2 which is involved in inter organ and intercellular translocation of Mo and showed different patterns of Mo accumulation. Another Mo transporter MOT2 is involved in the translocation of Mo which is stored in the vacuole (Carter et al., 2004). It is an interesting fact that the activity and transcription of CrMOT1 depends on nitrate instead of Mo and the sufficiency and toxicity of Mo does not affect it (Tejada-Jiménez, Chamizo-Ampudia, Galván, Fernández, & Llamas, 2013). On the other hand, the activity of CrMOT2 depends on Mo availability and the deficiency of Mo leads to the upregulation of its transcription (Tejada-Jiménez, Galván, & Fernández, 2011). There are two types of Mo transporters that depend on the affinity toward Mo have been discovered in *C. reinhardtii* strain 21gr (Llamas & Kalakoutskii, 2000). Fitzpatrick, Tyerman, & Kaiser (2008) also discussed the transport of Mo by sulfate transporters.

15.4.1.6 Copper homeostasis

Eukaryotes possess a widespread family of Cu transporters (Ctrs). Additional Cu transporters was also discovered in *A. thaliana* (COPT1-COPT6) (Puig, 2014). These transporters showed different levels of expressions in different parts of the plants, for example, COPT4 have higher expression level in roots, COPT1, 2, and 6 in leaves, and COPT3 and 5 in stem. COPT5 is involved in the translocation of already stored Cu in plants and is mainly localized in the vacuolar membrane (Gracia-Molina et al., 2011). Deficiency of Cu significantly reduced the level of expression of transporters (Gracia-Molina et al., 2013). It was reported that the long-distance transport of Cu occurred through various ways, such as by complex formation with histidine, nicotinamine, or Cu(II), or binding with metallothioneins as Cu(I) (Liao, Hedley, Woolley, Brooks, & Nichols, 2000). According to Zheng, Yamaji, Yokosho, & Ma (2012), YSL proteins lead to the transport of Cu to vasculature. It was also reported that transporters like HMA5 which depends on ATP are also involved in the efflux of excessive Cu in apoplasts (Pilon, Ravet, & Tapken, 2011). In *A. thaliana* P-type ATPase is involved in the transport of Cu to chloroplasts where Cu is needed for the synthesis of plastocyanin and for Cu/Zn SOD (Ravet and Pilon, 2013).

15.5 Management of HMmNs toxicity

15.5.1 Physiological and molecular responses of plants

When HMs acts as a toxic metal in plants, it increases the oxidative stress in the plants by generating ROS that further damage the proteins and photosynthetic pigments and cause lipid peroxidation when the ROS is not properly scavenged (Liu et al., 2019). Some more changes caused by high toxicity of HMs in plants include activation of the antioxidant system through the antioxidant enzymes [e.g., CAT, SOD, APX (ascorbate peroxidase), POD (peroxidase), and glutathione reductase] and nonenzymatic antioxidant components (glutathione and ascorbate) (Ribera-Fonseca, Inostroza-Blancheteau, Cartes, Rengel, & Mora, 2013; Ribera-Fonseca et al., 2013) detailed review of physiological and molecular responses are given in Table 15.2.

15.6 Hyperaccumulators

The capability of a hyperaccumulating plant to gather HMs in its vegetative parts ranges at 100–1000-fold higher than in other non-hyperaccumulating plant species (Baker, 1981). This ability lies in (1) the decrease in the HMs sequestration within root-cell vacuoles, (2) the improvement in HMs loading in the xylem for their transport to the shoot part, (3) the increased uptake of heavy metals from the rhizosphere across the plant's root cell plasma membrane, and (4) the stimulation of metal influx across the plasma membrane of leaf cell and sequestration in the leaf (mesophyll cells) vacuoles (Milner and Kochian, 2008). The use of genetic engineering to modify plants for metal uptake, transport, and sequestration opens up new avenues for enhancing the efficiency of phytoremediation. The studies have revealed better knowledge about the mechanism of hyperaccumulation of metals in the plants that increased the metal loading in the xylem part and increased the detoxification in the shoot part of plant (Verbruggen, Hermans, & Schat, 2009; Xing et al., 2008). Metal chelator, metal transporter, MT, and PC genes have been transferred to plants for improved metal uptake and sequestration (Eapen & D'souza, 2005; Yeargan, Maiti, Nielsen, Hunt, & Wagner, 1992).

Analyzing trace metal tolerance and accumulation has been greatly enhanced using high-tech molecular technologies, in particular microarray, which has enhanced understanding the difficulty in the hyperaccumulation phenomenon. These studies have supported the idea that the genes which are thought to be tangled in the phenomenon of hypertolerance and hyperaccumulation are not species-specific, but they are expressed differently in comparison to the other hyperaccumulator species. However, complete genome sequence of hyperaccumulator plants are not available yet, thus this assumption cannot be confirmed completely. Comparative transcriptomics studies on hyperaccumulators and related non-accumulating non-tolerant species have identified a large array of genes that are constitutively (in the absence

TABLE 15.2 Effect of HMmNs deficiency and toxicity on molecular and physiological aspects of plants.

Metal	Deficiency				Toxicity			
	Effect of plant physiology	**Reference/s**	**Plant cellular and molecular response**	**Reference/s**	**Effect of plant physiology**	**Reference/s**	**Plant cellular and molecular response**	**Reference/s**
Cu	Concentration of plastocyanin reduced and leads to reduced electron transport to PSI	Shikanai, Müller-Moulé, Munekage, Niyogi, and Pilon (2003)	Laccase (LAC transcripts was reduced)	Bernal et al. (2012)	Reduced chlorophyll contents, increased proline accumulation and increase the activity of guaiacol peroxidase in *Colobanthus quitensis*	Cuba-Díaz, Marín, Castel, Machuca, and Rifo (2017)	Overexpression of ATX1 starts to regulate the plant growth	Shin, Lo, & Yeh (2012)
Cu	Downregulation of SOD expression	Yamasaki, Pilon, and Shikanai (2008)	Cu deficiency triggered the activity of Cu transporters HMA1 in leaves and COPT2 in roots of *Brassica napus* L.	Billard et al. (2014)	Lignin and phenolics concentration were increased in pepper	Diaz, Bernal, Pomar, & Merino (2001)	Activity of CsHMA5.2 increased to sequester the excessive amount of Cu	Migocka, Posyniak, Maciaszczyk-Dziubinska, Papierniak, & Kosieradzaka (2015)

(Continued)

TABLE 15.2 (Continued)

Metal	Deficiency				Toxicity			
	Effect of plant physiology	Reference/s	Plant cellular and molecular response	Reference/s	Effect of plant physiology	Reference/s	Plant cellular and molecular response	Reference/s
Cu	Increased lipid peroxidation and H_2O_2 concentration was noted in wheat plants	Sarika, Sairam, Meena, Aruna, and Srivastava (2010)	Cu deficiency increased the activation of CPX1 and CYC6	Eriksson et al. (2004)	Increased the production of ROS and increase the activity of antioxidant enzymes	Liu et al. (2014)	Efflux of Cu ions initiated with the help of H^+/Cu^{+2} antiporters in cucumber	Parrotta, Guerriero, Sergeant, Cai, & Hausman (2015)
Cu	Cu deficiency triggered the biosynthesis of lignin	Van Maarschalkerweerd, Bro, Egebo, and Husted (2013)	SPL7 expression was increased in plants by Cu deficiency	Gracia-Molina, Xing, & Huijser (2014)	Significant reduction in chlorophyll and ETR was observed in *Arundo donax* L. plants	Pietrini et al. (2019)	Significantly affect the root to shoot signaling in plants	Hippler, Mattos-Jr, Boaretto, & Williams (2018)
Zn	Activities of SOD, POD, CAT increased in mandarin plants	Subba et al. (2014)	Activity of bZIP23 and bZIP19 was increased in *Arabidopsis*	Assunção et al. (2010)	Reduce the chlorophyll contents, reduce the photosynthesis and transpiration	Subba et al. (2014)	Zn toxicity affect the Calvin cycle and photosystem activities	Paunov, Koleva, Vassilev, Vangronsveld, & Goltsev (2018); Vassilev, Nikolova, Koleva, & Lidon (2011)

Zn	Increased the production of ROS	Cakmak (2000)	ZIP family genes expression was induced	Fei et al. (2016)	Antioxidant enzyme activity, H_2O_2, lipid peroxidation and ascorbate were increased in coffee plants	dos Santos et al. (2019)	Activities of MTP1, MTP3, CDF, HMA3 and ZIF1 were increased in *A. thaliana*	Haydon et al. (2012); Shahzad et al. (2018)
Zn	Net photosynthesis, chlorophyll content was reduced and antioxidant enzyme activity was increased	Fei et al. (2016)	Upregulation of OsDMAS1, OsPCS1, OsZIP4 and OsNAS2 was reported in rice	Kabir et al. (2017)	Increased the ROS production and disturbed the redox homeostasis in mulberry plants	Kumar Tewari, Kumar, and Nand Sharma (2008)	Natural resistance-associated macrophage proteins (AtNRAMP4) were induced in the roots of *S. cerevisiae*	Van de Mortel et al. (2006)
Zn	Malonaldehyde was increased and SOD activity was decreased in wheat plants	Hong and Jin (2007)	Genes expression of ZIP4, ZIP3 and ZIP1 was induced in *A. thaliana*	Grotz and Guerinot (2006)	Activity of photosystem II was reduced in *Orphyra haitanesis*	Li et al. (2010)	In Arabidopsis gene expression of ZIF1 was upregulated	Haydon et al. (2012)
Fe	Under Fe deficiency different organic solutes such as flavonoids, phenolics, flavins and organic acids were released to acidify the root	Connorton et al. (2017)	Expressions of genes associated with Fe uptake such as AtAHA7 and AtAHA2 were increased	Yuan et al. (2017)	In response to excess Fe, excessive Fe retained in the roots and Fe transporters have been inhibited in rice	Aung and Masuda (2020)	Excessive Fe induced OsNAS3 OsVIT2 and OsFers in the roots of rice plants. Expression of P450	Aung and Masuda (2020)

(Continued)

TABLE 15.2 (Continued)

Metal	Deficiency				Toxicity			
	Effect of plant physiology	Reference/s	Plant cellular and molecular response	Reference/s	Effect of plant physiology	Reference/s	Plant cellular and molecular response	Reference/s
	zone for increased Fe uptake						cytochrome proteins, NAC type transcription factor was increased	
Fe	Coumarins have been secreted in alkaline soil under Fe deficiency by *A. thaliana*	Clemens and Weber (2016)	Expression of cytochrome B5 reductase 1 (CBR1) was induced under Fe deficiency	Oh et al. (2016)	Under Fe toxicity Ferritin expression was increased in rice leaves	Müller et al. (2015)	OsNAC4,5,6, S nitroso glutathione reductase, and transcription factor OsWRKY were induced to detoxify ROS produced due to Fe toxicity in rice	Aung and Masuda (2020)
Fe	Complex auxin-ethylene signaling was reported in tomato plants under Fe deficiency	Celletti et al. (2020)	Expressions of PDR9 and feruloyl coenzyme A 6′-hydrozylase 1 (F6′H1) were upregulated in *A. thaliana*	Schmid et al. (2014); Fourcroy, Tissot, Gaymard, Briat, & Dubos (2016)	Due to Fe toxicity chlorophyll contents in rice plants decreased while stomatal resistance was increased	Dufey, Hakizimana, Draye, Lutts, and Bertin (2009)	Qualitative trait locus effect was confirmed on chromosome 1 and 3 in rice to increase the tolerance to Fe under excessive Fe application	Wu et al. (2014)

Fe	Increase in Phytosiderophore excretion was reported in barley plants under Fe deficiency	Yousfi, Houmani, Zribi, Abdelly, and Gharsalli (2012)	Overexpression of ferric reductase oxidase 2 (FRO2) was induced by Fe deficiency	Connolly, Campbell, Grotz, Prichard, and Guerinot (2003)	Overproduction of ROS was reported due to Fe toxicity	Tripathi et al. (2018)	Fe toxicity alters the Fenton reaction and increase the production of ROS	Becana, Moran, & Iturbe-Ormaetxe (1998); Wu et al. (2017)
Mn	Mn deficiency reduced the number of chloroplasts and photosynthetic efficiency in plants	Alejandro et al (2017)	Expression of HvIRT1 was upregulated under Mn deficiency	Wu et al. (2016)	Inhibit the chlorophyll biosynthesis	Subrahmanyam and Rathore (2000)	Inhibit the meristematic cell division	Zhao et al. (2017)
Mn	Mn deficiency disintegrates the complexes of PSII and reduced the net photosynthesis	Schmidt, Jensen, and Husted (2016)	OsMTP8.1 was inhibited under Mn toxicity in rice	Chen et al. (2013)	Reduce the rate of photosynthesis	Amao and Ohashi (2008)	AtMTP8 expression was enhanced under Mn toxicity in *Arabidopsis*	Eroglu, Meier, von Wirén, & Peiter (2016)
Mn	Decrease the activity of MnSOD and increase the free oxygen radicals	Broadley (2012)	Mn deficiency leads to changes in chromatin structure in *Arabidopsis*	Rodríguez-Celma et al. (2016)	Mn toxicity cause thylakoid swelling in plants	Doncheva et al. (2009)	OsMTP8.1 was induced under Mn toxicity in rice	Chen et al. (2013)
Ni	Increase in the activity of GOPX, POD and SOD was reported	Gautam and Pandey (2008)	Ni deficiency disturbed the citric acid cycle	Bai et al. (2006)	Increase in MDA concentration	Gonnelli, Galardi, and Gabbrielli (2001)	Ni toxicity leads to the inhibition of	Robertson and Meakin (1980)

(*Continued*)

TABLE 15.2 (Continued)

Metal	Deficiency				Toxicity			
	Effect of plant physiology	Reference/s	Plant cellular and molecular response	Reference/s	Effect of plant physiology	Reference/s	Plant cellular and molecular response	Reference/s
					was reported in plants		cell division at root meristems	
Ni	Ni deficiency reduced the scavenging of ROS	Ahmad and Ashraf (2012)	Ni deficiency increase the accumulation of oxalic acid and lactic acid in plant cells	Bai et al. (2006)	Ni toxicity reduced the efficiency of antioxidative machinery	Gajewska, Bernat, Długoński, and Skłodowska (2012)	Ni toxicity affects PIN2 and the orientation of microtubules and disturbance in integrity was also reported in *Arabidopsis*	Lešková, Zvarík, Araya, & Giehl (2020)
Ni	Ni plays important role in the functioning of Glyoxalase-I and Urease	Shahzad et al. (2018)			Ni toxicity leads to the overproduction of ROS	Gill and Tuteja (2010)	Ni toxicity cause reduction in chromosome integrity and leads to abnormalities during mitosis	Pavlova (2017)
Mo			Activity of nitrate reductase reduced die to Mo deficiency	Rana et al. (2020)				

of excess of metallic ions) highly expressed (Becher, Talke, Krall, & Krämer, 2004; Chakrabarty et al., 2009; Weber, Harada, Vess, von Roepenack-Lahaye, & Clemens, 2004). The availability of full-genome sequences will allow the development of microarrays in other plant families. A wide array of genes is involved in metal uptake, translocation, sequestration, chemical modification, and tolerance (Eapen & D'souza, 2005; Verbruggen et al., 2009).

The overexpression of any or combination of these genes is a possible strategy for genetic engineering, for example, more efficient sequestration of metals in plant storage compartments, overproduction of metal chelating molecules, or increasing activity of enzymes involved in general (oxidative) stress resistance. The metal accumulation, tolerance, and plant productivity are not necessarily correlated (Macnair, Tilstone, & Smith, 2000).

15.7 Summary

Plants tend to show deficiency and toxicity symptoms of HMmNs and its management is very necessary for proper growth and functioning of plants. These HMnNs show a variety of deficiency and toxicity symptoms which have been described above.

References

Abadía, J., Vázquez, S., Rellán-Álvarez, R., El-Jendoubi, H., Abadía, A., Álvarez-Fernández, A., et al. (2011). Towards a knowledge-based correction of iron chlorosis. *Plant Physiology and Biochemistry: PPB/Societe Francaise de Physiologie Vegetale*, *49*, 471–482. Available from https://doi.org/10.1016/j.plaphy.2011.01.026.

Ahmad, M. S. A., & Ashraf, M. (2012). *Essential roles and hazardous effects of nickel in plants. Reviews of environmental contamination and toxicology* (pp. 125–167). New York, NY: Springer.

Ahmad, M. S. A., Hussain, M., Saddiq, R., & Alvi, A. K. (2007). Mungbean: A nickel indicator, accumulator or excluder? *Bulletin of Environmental Contamination and Toxicology*, *78*(5), 319–324.

Alejandro, S., Cailliatte, R., Alcon, C., Dirick, L., Domergue, F., Correia, D., & Curie, C. (2017). Intracellular distribution of manganese by the trans-Golgi network transporter NRAMP2 is critical for photosynthesis and cellular redox homeostasis. *The Plant Cell*, *29* (12), 3068–3084.

Alejandro, S., Höller, S., Meier, B., & Peiter, E. (2020). Manganese in plants: From acquisition to subcellular allocation. *Frontiers in Plant Science*, *11*, 300.

Alloway, B. J. (2004a). *Zinc in soils and crop nutrition*. Brussels, Belgium: International Zinc Association.

Alloway, B. J. (2004b). Contamination of soils in domestic gardens and allotments: A brief review. *Land Contamination and Reclamation*, *12*(3), 179–187.

Amao, Y., & Ohashi, A. (2008). Effect of Mn ion on the visible light induced water oxidation activity of photosynthetic organ grana from spinach. *Catalysis Communications*, *10*(2), 217–220.

Ameh, T., & Sayes, C. M. (2019). The potential exposure and hazards of copper nanoparticles: A review. *Environmental Toxicology and Pharmacology*, *71*, 103220.

Andresen, E., Peiter, E., & Küpper, H. (2018). Trace metal metabolism in plants. *Journal of Experimental Botany*, *69*, 909–954. Available from https://doi.org/10.1093/jxb/erx465.

Antić-Mladenović, S., Rinklebe, J., Frohne, T., Stärk, H. J., Wennrich, R., Tomić, Z., & Ličina, V. (2011). Impact of controlled redox conditions on nickel in a serpentine soil. *Journal of Soils and Sediments*, *11*(3), 406–415.

Arnon, D. I., & Stout, P. R. (1939). The essentiality of certain elements in minute quantity for plants with special reference to copper. *Plant Physiology*, *14*, 371–375.

Arrivault, S., Senger, T., & Kramer, U. (2006). The Arabidopsis metal tolerance protein AtMTP3 maintains metal homeostasis by mediating Zn exclusion from the shoot under Fe deficiency and Zn oversupply. *The Plant Journal: For Cell and Molecular Biology*, *46*, 861–879.

Assche, F. V., & Clijsters, H. (1990). Effects of metals on enzyme activity in plants. Plant Cell Environ*ment*, *13*(3), 195–206.

Assunção, A. G., Herrero, E., Lin, Y. F., Huettel, B., Talukdar, S., Smaczniak, C., & Aarts, M. G. (2010). Arabidopsis thaliana transcription factors bZIP19 and bZIP23 regulate the adaptation to zinc deficiency. *Proceedings of the National Academy of Sciences of the United States of America*, *107*(22), 10296–10301.

Aung, M. S., & Masuda, H. (2020). How does rice defend against excess iron?: Physiological and molecular mechanisms. *Frontiers in Plant Science*, *11*, 1102.

Ayub, M. A., Usman, M., Faiz, T., Umair, M., Ul Haq, M. A., Rizwan, M., ... Ur Rehman, M. Z. (2020). *Restoration of degraded soil for sustainable agriculture. Soil health restoration and management* (pp. 31–81). Singapore: Springer.

Azcon-Bieto, J., & Talón, M. (2008). *Fundamentos de fisiología vegetal* (2nd ed., pp. 83–97). Barcelona, Spain: Mc Graw Hill Interamericana.

Bai, C., Reilly, C., & Wood, B. W. (2006). Ni deficiency disrupts metabolism of ureides, amino acids, and organic acids of young pecan foliage. *Plant Physiology*, *140*, 433–443.

Baker, A. J. (1981). Accumulators and excluders-strategies in the response of plants to heavy metals. *Journal of Plant Nutrition*, *3*(1–4), 643–654.

Baxter, I., Muthukumar, B., Park, H. C., Buchner, P., Lahner, B., Danku, J., ... Salt, D. E. (2008). Variation in molybdenum content across broadly distributed populations of *Arabidopsis thaliana* is controlled by a mitochondrial molybdenum transporter (MOT1). *PLoS Genetics*, *4*(2), e1000004.

Beasley, J. T., Bonneau, J. P., & Johnson, A. A. T. (2017). Characterisation of the nicotianamine aminotransferase and deoxymugineic acid synthase genes essential to strategy II iron uptake in bread wheat (*Triticum aestivum* L.). *PLoS One 12*, e0177061. Available from https://doi.org/10.1371/journal.pone.0177061.

Becana, M., Moran, J. F., & Iturbe-Ormaetxe, I. (1998). Iron-dependent oxygen free radical generation in plants subjected to environmental stress: Toxicity and antioxidant protection. *Plant and Soil*, *201*(1), 137–147.

Becher, M., Talke, I. N., Krall, L., & Krämer, U. (2004). Cross-species microarray transcript profiling reveals high constitutive expression of metal homeostasis genes in shoots of the zinc hyperaccumulator *Arabidopsis halleri*. *Plant Journal*, *37*, 251–268.

Becker, M., & Asch, F. (2005). Iron toxicity in rice—Conditions and management concepts. *Journal of Plant Nutrition and Soil Science*, *168*, 558–573. Available from https://doi.org/10.1002/(ISSN)1522-2624.

Bernal, M., Casero, D., Singh, V., Wilson, G. T., Grande, A., Yang, H., & Merchant, S. S. (2012). Transcriptome sequencing identifies SPL7-regulated copper acquisition genes FRO4/

FRO5 and the copper dependence of iron homeostasis in Arabidopsis. *The Plant Cell*, *24*(2), 738–761.

Billard, V., Ourry, A., Maillard, A., Garnica, M., Coquet, L., Jouenne, T., & Etienne, P. (2014). Copper-deficiency in *Brassica napus* induces copper remobilization, molybdenum accumulation and modification of the expression of chloroplastic proteins. *PLoS One*, *9*(10), e109889.

Bittner, F. (2014). Molybdenum metabolism in plants and crosstalks to iron. *Frontiers in Plant Science*, *5*, 28.

Bittner, F., & Mendel, R. R. (2010). Cell biology of molybdenum. In R. Hell, & R. R. Mendel (Eds.), *Cell biology of metals and nutrients. Plant Cell Monographs, 17* (pp. 119–143). Berlin, Heidelberg: Springer-Verlag.

Blaylock, M. J., & Huang, J. W. (2000). Phytoextraction of metals. In I. Raskin, & B. D. Ensley (Eds.), *Phytoremediation of toxic metals-using plants to clean up the environment* (pp. 53–70). New York: Wiley.

Bonneau, J., Baumann, U., Beasley, J., Li, Y., & Johnson, A. A. (2016). Identification and molecular characterization of the nicotianamine synthase gene family in bread wheat. *Plant Biotechnology Journal*, *14*(12), 2228–2239.

Boominathan, R., & Doran, P. M. (2002). Ni-induced oxidative stress in roots of the Ni hyperaccumulator, *Alyssum bertolonii*. *New Phytologist*, *156*(2), 205–215.

Borlotti, A., Vigani, G., & Zocchi, G. (2012). Iron deficiency affects nitrogen metabolism in cucumber (*Cucumis sativus* L.) plants. *BMC Plant Biology*, *12*(1), 1–15.

Briat, J. F., Duc, C., Ravet, K., & Gaymard, F. (2010). Ferritins and iron storage in plants. *Biochimica Et Biophysica Acta-General Subjects.*, *1800*, 806–814. Available from https://doi.org/10.1016/j.bbagen.2009.12.003.

Broadley, M. R. (2012). *Marschner's mineral nutrition of higher plants*. Elsevier/Academic Press.

Broadley, M. R., White, P. J., Hammond, J. P., Zelko, I., & Lux, A. (2007). Zinc in plants. *The New Phytologist*, *173*, 677–702.

Brown, P. H., Welch, R. M., & Cary, E. E. (1987). Ni: A micronutrient essential for higher plants. *Plant Physiology*, *85*, 801–803.

Buapet, P., Mohammadi, N. S., Pernice, M., Kumar, M., Kuzhiumparambil, U., & Ralph, P. J. (2019). Excess copper promotes photoinhibition and modulates the expression of antioxidant-related genes in *Zostera muelleri*. *Aquatic Toxicology*, *207*, 91–100.

Burnell, J. N. (1988). *The biochemistry of manganese in plants. Manganese in soils and plants* (pp. 125–137). Netherlands: Springer.

Cakmak, I. (2000). Possible roles of Zinc in protecting plant cells from damage by reactive oxygen species. *New Phytologist*, *146*, 185–205.

Cakmak, I., Pfeiffer, W., & Mcclafferty, B. (2011). Biofortification of durum wheat with zinc and iron. *Cereal Chemistry*, *87*, 10–20.

Cakmak, I., Öztürk, L., Karanlik, S., Marschner, H., & Ekiz, H. (1996). Zinc-efficient wild grasses enhance release of phytosiderophores under zinc deficiency. *Journal of Plant Nutrition*, *19*, 551–563.

Cakmak, S., Gülüt, K. Y., Marschner, H., & Graham, R. D. (1994). Effect of zinc and iron deficiency on phytos1derophore release in wheat genotypes differing in zinc efficiency. *Journal of Plant Nutrition*, *17*(1), 1–17.

Caldelas, C., & Weiss, D. J. (2017). Zinc homeostasis and isotopic fractionation in plants: A review. *Plant and Soil*, *411*, 17–46.

Cambrollé, J., García, J. L., Ocete, R., Figueroa, M. E., & Cantos, M. (2013). Growth and photosynthetic responses to copper in wild grapevine. *Chemosphere*, *93*(2), 294–301.

Carter, C., Songqin, P., Zouhar, J., Avila, E. L., Girke, T., & Raikhel, N. V. (2004). The vegetative vacuole proteome of Arabidopsis thaliana reveals predicted and unexpected proteins. *Plant Cell*, *16*, 3285–3303.

Celletti, S., Pii, Y., Valentinuzzi, F., Tiziani, R., Fontanella, M. C., Beone, G. M., & Astolfi, S. (2020). Physiological responses to Fe deficiency in split-root tomato plants: Possible roles of auxin and ethylene? *Agronomy*, *10*(7), 1000.

Chakrabarty, D., Trivedi, P. K., Misra, P., Tiwari, M., Shri, M., Shukla, D., ... Rai, A. (2009). Comparative transcriptome analysis of arsenate and arsenite stresses in rice seedlings. *Chemosphere*, *74*, 688–702.

Chatterjee, C., & Nautiyal, N. (2001). Molybdenum stress affects viability and vigour of wheat seeds. *Journal of Plant Nutrition*, *24*, 1377–1386.

Chattha, M. U., Hassan, M. U., Khan, I., Chattha, M. B., Mahmood, A., Chattha, M. U., ... Khan, S. (2017). Biofortification of wheat cultivars to combat zinc deficiency. *Frontiers in Plant Science*, *8*, 281.

Chen, C., Huang, D., & Liu, J. (2009). Functions and toxicity of nickel in plants: Recent advances and future prospects. *Clean*, *37*, 304–313.

Chen, Z., Fujii, Y., Yamaji, N., Masuda, S., Takemoto, Y., Kamiya, T., & Ma, J. F. (2013). Mn tolerance in rice is mediated by MTP8. 1, a member of the cation diffusion facilitator family. *Journal of Experimental Botany*, *64*(14), 4375–4387.

Chesworth, W. (1991). Geochemistry of micronutrients. *Micronutrients in Agriculture*, *4*, 1–30.

Cho, M., Chardonnens, A. N., & Dietz, K. J. (2003). Differential heavy metal tolerance of *Arabidopsis halleri* and *Arabidopsis thaliana*: A leaf slice test. *The New Phytologist*, *158*, 287–293.

Chu, H.-H., Car, S., Socha, A. L., Hindt, M. N., Punshon, T., & Guerinot, M. L. (2017). The *Arabidopsis* MTP8 transporter determines the localization of manganese and iron in seeds. *Scientific Reports*, *7*, 11024. Available from https://doi.org/10.1038/s41598-017-11250-9.

Clemens, S. (2017). How metal hyperaccumulating plants can advance Zn biofortification. *Plant and Soil*, *411*, 111–120.

Clemens, S., Palmgren, M., & Kramer, U. (2002). A long way ahead: Understanding and engineering plant metal accumulation. *Trends in Plant Science*, *7*, 309–315.

Clemens, S., & Weber, M. (2016). The essential role of coumarin secretion for Fe acquisition from alkaline soil. *Plant Signaling & Behavior*, *11*(2), e1114197.

Cobbett, C. S., Hussain, D., & Haydon, M. J. (2003). Structural and functional relationships between type 1B heavy metal-transporting P-type ATPases in Arabidopsis. *New Phytologist*, *159*(2), 315–321.

Colangelo, E., & Guerinot, M. (2004). Put the metal to the petal: Metal uptake and transport throughout plants. *Current Opinion in Plant Biology*, *9*, 322–330.

Connolly, E. L., Campbell, N. H., Grotz, N., Prichard, C. L., & Guerinot, M. L. (2003). Overexpression of the FRO2 ferric chelate reductase confers tolerance to growth on low iron and uncovers posttranscriptional control. *Plant Physiology*, *133*(3), 1102–1110.

Connorton, J. M., Balk, J., & Rodríguez-Celma, J. (2017). Iron homeostasis in plants—A brief overview. *Metallomics*, *9*(7), 813–823.

Conte, S. S., & Walker, E. L. (2011). Transporters contributing to iron trafficking in plants. *Molecular Plant*, *4*, 464–476.

Cuba-Díaz, M., Marín, C., Castel, K., Machuca, Á., & Rifo, S. (2017). Effect of copper (II) ions on morpho-physiological and biochemical variables in Colobanthus quitensis. *Journal of Soil Science and Plant Nutrition*, *17*(2), 429–440.

Curie, C., Cassin, G., Counch, D., Divol, F., Higuchi, K., Le Jean, M., ... Mari, S. (2009). Metal movement within the plant: Contribution of nictotianamine and yellow stripe 1-like transporters. *Annals of Botany*, *103*, 1–11.

Curie, C., Panaviene, Z., Loulergue, C., Dellaporta, S. L., Briat, J.-F., & Walker, E. L. (2001). Maize yellow stripe1 encodes a membrane protein directly involved in Fe(III) uptake. *Nature*, *409*, 346–349.

Das, S. K., Avasthe, R. K., & Gopi, R. (2014). Vermiwash: Use in organic agriculture for improved crop production. *Popular Kheti*, *2*, 45–46.

Deng, D., Wu, S. C., Wu, F. Y., Deng, H., & Wong, M. H. (2010). Effects of root anatomy and Fe plaque on arsenic uptake by rice seedlings grown in solution culture. *Environmental Pollution*, *158*, 2589–2595. Available from https://doi.org/10.1016/j.envpol.2010.05.015.

Diaz, J., Bernal, A., Pomar, F., & Merino, F. (2001). Induction of shikimate dehydrogenase and peroxidase in pepper (*Capsicum annuum* L.) seedlings in response to copper stress and its relation to lignification. *Plant Science*, *161*, 179–188.

Dinneny, J. R., Long, T. A., Wang, J. Y., Jung, J. W., Mace, D., Pointer, S., ... Benfey, P. N. (2008). Cell identity mediates the response of Arabidopsis roots to abiotic stress. *Science (New York, NY)*, *320*, 942–945.

Disante, K. B., Fuentes, D., & Cortina, J. (2010). Response to drought of Zn-stressed *Quercus suber* L. seedlings. *Environmental and Experimental Botany*, *70*, 96–103.

Dixon, N. E., Gazzola, C., Blakeley, R. L., & Zerner, R. (1975). Jack bean urease. A metalloenzyme. A simple biological role for Ni. *Journal of the American Chemical Society*, *97*, 4131–4133.

Doncheva, S. N., Poschenrieder, C., Stoyanova, Z., Georgieva, K., Velichkova, M., & Barceló, J. (2009). Silicon amelioration of manganese toxicity in Mn-sensitive and Mn-tolerant maize varieties. *Environmental and Experimental Botany*, *65*(2-3), 189–197.

dos Santos, J. O., Andrade, C. A., de Souza, K. R. D., de Oliveira Santos, M., Brandão, I. R., Alves, J. D., & Santos, I. S. (2019). Impact of zinc stress on biochemical and biophysical parameters in *Coffea arabica* seedlings. *Journal of Crop Science and Biotechnology*, *22*(3), 253–264.

Dubey, D., & Pandey, A. (2011). Effect of nickel (Ni) on chlopophyll, lipid peroxidation andantioxidant enzymes activities in black gram (*Vigna mungo*) leaves. *International Journal of Science and Nature*, *2*(2), 395–401.

Dufey, I., Hakizimana, P., Draye, X., Lutts, S., & Bertin, P. (2009). QTL mapping for biomass and physiological parameters linked to resistance mechanisms to ferrous iron toxicity in rice. *Euphytica*, *167*(2), 143–160.

Eapen, S., & D'souza, S. F. (2005). Prospects of genetic engineering of plants for phytoremediation of toxic metals. *Biotechnology Advances*, *23*, 97–114.

Eaton, E. T. (2015). Manganese. In A. V. Barker, & D. J. Pilbeam (Eds.), *Handbook of plant nutrition* (pp. 427–485). CRC Press.

Edwards, D. G., & Asher, C. J. (1981). Tolerance of crop and pasture species to manganese toxicity. In A. Scaife (Ed.), *Proceedings of the ninth international plant nutrition coloquium* (pp. 145–150). Warwick University.

Eitinger, T., Suhr, J., Moore, L., & Smith, J. A. (2005). Secondary transporters for nickel and cobalt ions: Theme and variations. *Biometals*, *18*, 399–405. Available from https://doi.org/10.1007/s10534-005-3714-x.

Epstein, E., & Bloom, A. (2005). *Mineral nutrition of plant: Principles and perspectives*. Sunderland, MA: Sinauer Associates, Inc. Publishers.

Eren, E., & Argüello, J. (2004). Arabidopsis HMA2, a divalent heavy metal-transporting P_{IB}-Type ATPase, is involved in cytoplasmic Zn^{2+} homeostasis. *Plant Physiology*, *136*, 3712–3723.

Eriksson, M., Moseley, J. L., Tottey, S., Del Campo, J. A., Quinn, J., Kim, Y., & Merchant, S. (2004). Genetic dissection of nutritional copper signaling in Chlamydomonas distinguishes regulatory and target genes. *Genetics*, *168*(2), 795–807.

Eroglu, S., Meier, B., von Wirén, N., & Peiter, E. (2016). The vacuolar manganese transporter MTP8 determines tolerance to iron deficiency-induced chlorosis in *Arabidopsis*. *Plant Physiology*, *170*(2), 1030–1045.

Eskew, D. L., Welch, R. M., & Norvell, W. A. (1984). Ni in higher plants: Further evidence for an essential role. *Plant Physiology*, *76*, 691–693.

Fageria, N. K., Baligar, V. C., & Clark, R. B. (2002). *Micronutrients in crop production*, . *Advances in agronomy* (77, pp. 185–268). Academic Press.

Fasani, E. (2012). *Plants that hyperaccumulate heavy metals*. *Plants and heavy metals* (pp. 55–74). Berlin, Germany: Springer.

Fecht-Christoffers, M. M., Führs, H., Braun, H. P., & Horst, W. J. (2006). The role of hydrogen peroxide-producing and hydrogen peroxide-consuming peroxidases in the leaf apoplast of cowpea in manganese tolerance. *Plant Physiology*, *140*, 1451–1463. Available from https://doi.org/10.1104/pp.105.070474.

Fei, X., Fu, X. Z., Wang, N. Q., Xi, J. L., Huang, Y., Wei, Z., & Peng, L. Z. (2016). Physiological changes and expression characteristics of ZIP family genes under zinc deficiency in navel orange (*Citrus sinensis*). *Journal of Integrative Agriculture*, *15*(4), 803–811.

Fernando, D. R., Baker, A. J. M., & Woodrow, I. E. (2009). Physiological responses in *Macadamia integrifolia* on exposure to manganese treatment. *Australian Journal of Botany*, *57*(5), 406–413.

Fernando, D. R., & Lynch, J. P. (2015). Manganese phytotoxicity: New light on an old problem. *Annals of Botany*, *116*(3), 313–319.

Fitzpatrick, K. L., Tyerman, S. D., & Kaiser, B. N. (2008). Molybdate transport through the plant sulfate transporter SHST1. *FEBS Letters*, *582*(10), 1508–1513.

Fourcroy, P., Tissot, N., Gaymard, F., Briat, J. F., & Dubos, C. (2016). Facilitated Fe nutrition by phenolic compounds excreted by the *Arabidopsis* ABCG37/PDR9 transporter requires the IRT1/FRO2 high-affinity root Fe^{2+} transport system. *Molecular Plant*, *9*(3), 485–488.

Frassinetti, S., Bronzetti, G. L., Caltavuturo, L., Cini, M., & Della Croce, C. (2006). The role of zinc in life: A review. *Journal of Environmental Pathology, Toxicology and Oncology*, *25* (3), 597–610.

Gajewska, E., Bernat, P., Długoński, J., & Skłodowska, M. (2012). Effect of nickel on membrane integrity, lipid peroxidation and fatty acid composition in wheat seedlings. *Journal of Agronomy and Crop Science*, *198*(4), 286–294.

Gajewska, E., & Skłodowska, M. (2005). Antioxidative responses and proline level in leaves and roots of pea plants subjected to nickel stress. *Acta Physiologiae Plantarum*, *27*(3), 329–340.

Gajewska, E., Skłodowska, M., Słaba, M., & Mazur, J. (2006). Effect of nickel on antioxidative enzyme activities, proline and chlorophyll contents in wheat shoots. *Biologia Plantarum*, *50* (4), 653–659.

Gallie, D. R. (2013). The role of l-ascorbic acid recycling in responding to environmental stress and in promoting plant growth. *Journal of Experimental Botany*, *64*, 433–443. Available from https://doi.org/10.1093/jxb/ers330.

Gao, X., Zhang, F., & Hoffland, E. (2009). Malate exudation by six aerobic rice genotypes varying in zinc uptake efficiency. *Journal of Environmental Quality*, *38*(6), 2315–2321.

Gasber, A., Klaumann, S., Trentmann, O., Trampczynska, A., Clemens, S., Schneider, S., ... Neuhaus, H. E. (2011). Identification of an *Arabidopsis* solute carrier critical for intracellular transport and inter-organ allocation of molybdate. *Plant Biology*, *13*(5), 710–718.

Gautam, S., & Pandey, S. N. (2008). Growth and biochemical responses of nickel toxicity on leguminous crop (*Lens esculentum*) grown in alluvial soil. *Research in Environment and Life Sciences*, *1*, 25–28.

Genc, Y., Huang, C. Y., & Langridge, P. (2007). A study of the role of root morphological traits in growth of barley in zinc-deficient soil. *Journal of Experimental Botany*, *58*(11), 2775–2784.

George, T. S., French, A. S., Brown, L. K., Karley, A. J., White, P. J., Ramsay, L., & Daniell, T. J. (2014). Genotypic variation in the ability of landraces and commercial cereal varieties to avoid manganese deficiency in soils with limited manganese availability: Is there a role for rootexuded phytases? *Physiologia Plantarum*, *151*, 243–256.

Ghori, N. H., Ghori, T., Hayat, M. Q., Imadi, S. R., Gul, A., Altay, V., & Ozturk, M. (2019). Heavy metal stress and responses in plants. *International Journal of Environmental Science and Technology*, *16*(3), 1807–1828.

Gill, S. S., & Tuteja, N. (2010). Reactive oxygen species and antioxidant machinery in abiotic stress tolerance in crop plants. *Plant Physiology and Biochemistry*, *48*(12), 909–930.

Gomes-Junior, R. A., Moldes, C. A., Delite, F. S., Gratão, P. L., Mazzafera, P., Lea, P. J., & Azevedo, R. A. (2006). Nickel elicits a fast antioxidant response in *Coffea arabica* cells. *Plant Physiology and Biochemistry*, *44*(5-6), 420–429.

Gong, Q., Wang, L., Dai, T., Zhou, J., Kang, Q., Chen, H., & Li, Z. (2019). Effects of copper on the growth, antioxidant enzymes and photosynthesis of spinach seedlings. *Ecotoxicology and Environmental Safety*, *171*, 771–780.

Gonnelli, C., Galardi, F., & Gabbrielli, R. (2001). Nickel and copper tolerance and toxicity in three Tuscan populations of Silene paradoxa. *Physiologia Plantarum*, *113*(4), 507–514.

Goos, R. J., Johnson, B., Jackson, G., & Hargrove, G. (2004). Greenhouse evaluation of controlled-release iron fertilizers for soybean. *Journal of Plant Nutrition*, *27*, 43–55.

Goulding, K. W. T. (2016). Soil acidification and the importance of liming agricultural soils with particular reference to the United Kingdom. *Soil Use and Management*, *32*(3), 390–399.

Gracia-Molina, A., Andrés-Colás, N., Perea-García, A., Del Valle-Tascón, S., Peñarrubia, L., & Puig, S. (2011). The intracellular *Arabidopsis* COPT5 transport protein is required for photosynthetic electron transport under severe copper deficiency. *Plant Journal*, *65*, 848–860. Available from https://doi.org/10.1111/j.1365-313X.2010.04472.x.

Gracia-Molina, A., Andrés-Colás, N., Perea-García, A., Neumann, U., Dodani, S. C., Huijser, P., Puig, S., et al. (2013). The *Arabidopsis* COPT6 transport protein functions in copper distribution under copper-deficient conditions. *Plant Cell Physiology*, *54*, 1378–1390. Available from https://doi.org/10.1093/pcp/pct088.

Gracia-Molina, A., Xing, S., & Huijser, P. (2014). A conserved KIN17 curved DNA-binding domain protein assembles with SQUAMOSA PROMOTER-BINDING PROTEIN-LIKE7 to adapt *Arabidopsis* growth and development to limiting copper availability. *Plant Physiology*, *164*, 828–840. Available from https://doi.org/10.1104/pp.113.228239.

Graham, R. D., & Stangoulis, J. R. C. (2005). Molybdenum and disease. In L. Dantoff, W. Elmer, & D. Huber (Eds.), *Mineral nutrition and plant diseases*. St Paul, MN: APS Press.

Graham, R. D., Welch, R. M., & Bouis, H. E. (2001). Addressing micronutrients malnutrition through enhancing the nutritional quality of staple foods principles, perspectives and knowledge gaps. *Advanced Agronomy*, *70*, 77–142.

Gries, D., Brunn, S., Crowley, D. E., & Parker, D. R. (1995). Phytosiderophore release in relation to micronutrient metal deficiencies in barley. *Plant and Soil*, *172*, 299–308.

Grotz, N., & Guerinot, M. L. (2006). Molecular aspects of Cu, Fe and Zn homeostasis in plants. *Biochimica et Biophysica Acta*, *1763*, 595–608.

Guerinot, M. L. (2000). The ZIP family of metal transporters. *Biochimica et Biophysica Acta*, *1465*, 190–198.

Guerinot, M. L., & Yi, Y. (1994). Iron: Nutritious, noxious, and not readily available. *Plant Physiology*, *104*, 815–820.

Gupta, N., Ram, H., & Kumar, B. (2016). Mechanism of Zinc absorption in plants: Uptake, transport, translocation and accumulation. *Reviews in Environmental Science and Bio/ Technology*, *15*, 89–109.

Gustin, J. L., Zanis, M. J., & Salt, D. E. (2011). Structure and evolution of the plant cation diffusion facilitator family of ion transporters. *BMC Evolutionary Biology*, *11*, 76.

Hacisalihoglu, G., Ozturk, L., Cakmak, I., Welch, R. M., & Kochian, L. (2004). Genotypic variation in common bean in response to zinc deficiency in calcareous soil. *Plant and Soil*, *259* (1-2), 71–83.

Hall, J. L. (2002). Cellular mechanisms for heavy metal detoxification and tolerance. *Journal of Experimental Botany*, *53*, 1–11.

Hao, F., Wang, X., & Chen, J. (2006). Involvement of plasma-membrane NADPH oxidase in nickel-induced oxidative stress in roots of wheat seedlings. *Plant Science*, *170*(1), 151–158.

Hassan, M. U., Chattha, M. U., Ullah, A., Khan, I., Qadeer, A., Aamer, M., … Khan, T. A. (2019). Agronomic biofortification to improve productivity and grain Zn concentration of bread wheat. *International Journal of Agriculture And Biology*, *21*, 615–620.

Hassan, Z., & Aarts, M. (2011). Opportunities and feasibilities for biotechnological improvement of Zn, Cd or Ni tolerance and accumulation in plants. *Environmental and Experimental Botany*, *72*, 53–63.

Haydon, M. J., Kawachi, M., Wirtz, M., Hillmer, S., Hell, R., & Krämer, U. (2012). Vacuolar nicotianamine has critical and distinct roles under iron deficiency and for zinc sequestration in *Arabidopsis*. *The Plant Cell*, *24*(2), 724–737.

Hebbern, C. A., Laursen, K. H., Ladegaard, A. H., Schmidt, S. B., Pedas, P., Bruhn, D., et al. (2009). Latent manganese deficiency increases transpiration in barley (*Hordeum vulgare*). *Physiologia Plantarum*, *135*, 307–316. Available from https://doi.org/10.1111/j.1399-3054.2008.01188.x.

Hebbern, C. A., Pedas, P., Schjoerring, J. K., Knudsen, L., & Husted, S. (2005). Genotypic differences in manganese efficiency: Field experiments with winter barley (*Hordeum vulgare*). *Plant and Soil*, *272*, 233–244.

Heine, G., Max, J. F. J., Führs, H., Moran-Puente, D. W., Heintz, D., & Horst, W. J. (2011). Effect of manganese on the resistance of tomato to *Pseudocercospora fuligena*. *Journal of Plant Nutrition and Soil Science*, *174*, 827–836. Available from https://doi.org/10.1002/jpln.201000440.

Henriques, A. R., Chalfun-Junior, A., & Aarts, M. (2012). Strategies to increase zinc deficiency tolerance and homeostasis in plants. *Brazilian Journal of Plant Physiology*, *24*(1), 3–8.

Hernandez-Apaolaza, L. (2014). Can silicon partially alleviate micronutrient deficiency in plants? A review. *Planta*, *240*, 447–458. Available from https://doi.org/10.1007/s00425-014-2119-.

Hewitt, E. J., & Bolle-Jones, E. W. (1952a). Molybdenumas a plant nutrient. II. The effects of molybdenum deficiency on some horticultural and agricultural crop plants in sand culture. *Journal of Horticultural Science*, *27*, 257–265.

Hewitt, E. J., & Bolle-Jones, E. W. (1952b). Molybdenum as a plant nutrient. I. The influence of molybdenumon the growth of some Brassica crops in sand culture. *Journal of Horticultural Science*, *27*, 245–256.

Hindt, M. N., & Guerinot, M. L. (2012). Getting a sense for signals: Regulation of the plant iron deficiency response. *Biochimica et Biophysica Acta (BBA)-Molecular Cell Research*, *1823* (9), 1521–1530.

Hippler, F. W., Mattos-Jr, D., Boaretto, R. M., & Williams, L. E. (2018). Copper excess reduces nitrate uptake by *Arabidopsis* roots with specific effects on gene expression. *Journal of Plant Physiology*, *228*, 158–165.

Hohle, T. H., & O'Brian, M. R. (2009). The mntH gene encodes the major Mn^{2+} transporter in *Bradyrhizobium japonicum* and is regulated by manganese via the Fur protein. *Molecular Microbiology*, *72*(2), 399–409.

Höller, S., Meyer, A., & Frei, M. (2014). Zinc deficiency differentially affects redox homeostasis of rice genotypes contrasting in ascorbate level. *Journal of Plant Physiology*, *171*(18), 1748–1756.

Hong, W., & Jin, J. Y. (2007). Effects of zinc deficiency and drought on plant growth and metabolism of reactive oxygen species in maize (*Zea mays* L). *Agricultural Sciences in China*, *6*(8), 988–995.

Humayan Kabir, A., Swaraz, A. M., & Stangoulis, J. (2014). Zinc-deficiency resistance and biofortification in plants. *Journal of Plant Nutrition and Soil Science*, *177*(3), 311–319.

Husted, S., Thomsen, M. U., Mattsson, M., & Schjoerring, J. (2009). Influence of nitrogen and sulfur form on manganese acquisition by barley (*Hordeum vulgare*). *Plant and Soil*, *268*, 309–317.

Ihnatowicz, A., Siwinska, J., Meharg, A. A., Carey, M., Koornneef, M., & Reymond, M. (2014). Conserved histidine of metal transporter AtNRAMP1 is crucial for optimal plant growth under manganese deficiency at chilling temperatures. *The New Phytologist*, *202*, 1173–1183. Available from https://doi.org/10.1111/nph.12737.

Impa, S. M., & Johnson-Beebout, S. E. (2012). Mitigating zinc deficiency and achieving high grain Zn in rice through integration of soil chemistry and plant physiology research. *Plant and Soil*, *361*(1-2), 3–41.

Ingle, R. A., Mugford, S. T., Rees, J. D., Campbell, M. M., & Smith, J. A. C. (2005). Constitutively high expression of the histidine biosynthetic pathway contributes to nickel tolerance in hyperaccumulator plants. *The Plant Cell*, *17*(7), 2089–2106.

Inoue, H., Kobayashi, T., Nozoye, T., Takahashi, M., Kakei, Y., Suzuki, K., … Nishizawa, N. K. (2009). Rice OsYSL15 is an iron-regulated iron(III)-deoxymugineic acid transporter expressed in the roots and is essential for iron uptake in early growth of the seedlings. *The Journal of Biological Chemistry*, *284*, 3470–3479.

Ishimaru, Y., Kim, S. A., Tsukamoto, T., Oki, H., Kobayashi, T., Watanabe, S., … Nishizawa, N. K. (2007). Mutational reconstructed ferric chelate reductase confers enhanced tolerance in rice to iron deficiency in calcareous soil. *Proceedings of the National Academy of Sciences of the United States of America*, *104*, 7373–7378.

Ishimaru, Y., Suzuki, M., Kobayashi, T., Takahashi, M., Nakanishi, H., Mori, S., & Nishizawa, N. K. (2005). OsZIP4, a novel zinc-regulated zinc transporter in rice. *Journal of Experimental Botany*, *56*(422), 3207–3214.

Jaime-Pérez, N., Kaftan, D., Bina, D., Bokhari, S. N. H., Shreedhar, S., & Küpper, H. (2019). Mechanisms of sublethal copper toxicity damage to the photosynthetic apparatus of *Rhodospirillum rubrum*. *Biochimica et Biophysica Acta (BBA)-Bioenergetics*, *1860*(8), 640–650.

Jiang, W. Z. (2006). Mn use efficiency in different wheat cultivars. *Environmental and Experimental Botany*, *57*, 41–50.

Kabata-Pendias, A. (2010). *Trace elements in soils and plants*. CRC Press.

Kabir, A. H., Hossain, M. M., Khatun, M. A., Sarkar, M. R., & Haider, S. A. (2017). Biochemical and molecular mechanisms associated with Zn deficiency tolerance and signaling in rice (*Oryza sativa* L.). *Journal of Plant Interactions*, *12*(1), 447–456.

Kim, J., & Rees, D. C. (1992). Structural models for the metal centers in the nitrogenous molybdenum-iron protein. *Science (New York, NY)*, *257*, 1677–1682.

Kim, Y. Y., Choi, H., Segami, S., Cho, H. T., Martinoia, E., Maeshima, M., et al. (2009). AtHMA1 contributes to detoxification cation of excess Zn(II) in Arabidopsis. *The Plant Journal: for Cell and Molecular Biology*, *58*, 737–753.

Klatte, M., Schuler, M., Wirtz, M., Fink-Straube, C., Hell, R., & Bauer, P. (2009). The analysis of Arabidopsis nicotianamine synthase mutants reveals functions for nicotianamine in seed iron loading and iron deficiency responses. *Plant Physiology, 150*, 257–271.

Kopittke, P. M., Blamey, F. P. C., McKenna, B. A., Wang, P., & Menzies, N. W. (2011). Toxicity of metals to roots of cowpea in relation to their binding strength. *Environmental Toxicology and Chemistry, 30*(8), 1827–1833.

Kumar Tewari, R., Kumar, P., & Nand Sharma, P. (2008). Morphology and physiology of zinc-stressed mulberry plants. *Journal of Plant Nutrition and Soil Science, 171*(2), 286–294.

Küpper, H., & Kroneck, P. M. H. (2007). Nickel in the environment and its role in the metabolism of plants and cyanobacteria. In A. Sigel, H. Sigel, & R. K. O. Sigel (Eds.), *Metal ions in life sciences* (2, pp. 31–62). Wiley Ltd.

Kwon, M. J., Boyanov, M. I., Yang, J.-S., Lee, S., Hwang, Y. H., Lee, J. Y., ... Kemner, K. M. (2017). Transformation of zinc-concentrate in surface and subsurface environments: Implications for assessing zinc mobility/toxicity and choosing an optimal remediation strategy. *Environmental Pollution (Barking, Essex: 1987), 226*, 346–355.

Lee, S., Chiecko, J. C., Kim, S. A., Walker, E. L., Lee, Y., Guerinot, M. L., & An, G. (2009). Disruption of OsYSL15 leads to iron inefficiency in rice plants. *Plant Physiology, 150*, 786–800.

Lešková, A., Zvarík, M., Araya, T., & Giehl, R. F. (2020). Nickel toxicity targets cell wall-related processes and PIN2-mediated auxin transport to inhibit root elongation and gravitropic responses in *Arabidopsis*. *Plant and Cell Physiology, 61*(3), 519–535.

Li, Y. X., Zhou, S., Zhao, F. J., Liu, Y., Fan, P. P., & Wang, G. C. (2010). Physiological responses of *Porphyra haitanesis* to different copper and zinc concentrations. *Brazilian Journal of Oceanography, 58*(4), 261–267.

Liao, M. T., Hedley, M. J., Woolley, D. J., Brooks, R. R., & Nichols, M. A. (2000). Copper uptake and translocation in chicory (*Cichorium intybus* L. cv. Grasslands Puna) and tomato (*Lycopersicon esculentum* Mill. cv. Rondy) plants grown in NFT system. I. Copper uptake and distribution in plants. *Plant Soil, 223*, 243–252. Available from https://doi.org/10.1023/A:1004843505053.

Lin, Y.-F., Hassan, Z., Talukdar, S., Schat, H., & Aarts, M. G. (2016). Expression of the ZNT1 zinc transporter from the metal hyperaccumulator *Noccaea caerulescens* confers enhanced zinc and cadmium tolerance and accumulation to *Arabidopsis thaliana*. *PLoS One, 11*, e0149750.

Liu, L., Rehman, M., Wang, Q., Saleem, M. H., Bashir, S., Ullah, S., & Peng, D. (2019). Copper environmental toxicology, recent advances, and future outlook: A review. *Environmental Science and Pollution Research, 26*, 18003–18016.

Liu, S. L., Yang, R. J., Ma, M. D., Dan, F., Zhao, Y., Jiang, P., & Wang, M. H. (2014). Effects of exogenous NO on the growth, mineral nutrient content, antioxidant system, and ATPase activities of *Trifolium repens* L. plants under cadmium stress. *Acta Physiologiae Plantarum/Polish Academy of Sciences, Committee of Plant Physiology Genetics and Breeding, 37*(1), 1–16.

Llamas, A., & Kalakoutskii, K. L. (2000). Molybdenum cofactor amounts in *Chlamydomonas reinhardtii* depend on the Nit5 gene function related to molybdate transport, *Plant, Cell &. Environment, 23*(11), 1247–1255.

Macnair, M. R., Tilstone, G. H., & Smith, S. E. (2000). The genetics of metal tolerance and accumulation in higher plants. In N. Terry, & G. Banuelos (Eds.), *Phytoremediation of contaminated soil and water* (pp. 235–250). Boca Raton: Lewis Publishers.

Maleva, M. G., Nekrasova, G. F., Borisova, G. G., Chukina, N. V., & Ushakova, O. S. (2012). Effect of heavy metal on photosynthetic apparatus and antioxidant status of elodea. *Russian*

Journal of Plant Physiology, *59*, 190–197. Available from https://doi.org/10.1134/S1021443712020069.

Marcec, M. J., Gilroy, S., Poovaiah, B. W., & Tanaka, K. (2019). Mutual interplay of Ca^{2+} and ROS signaling in plant immune response. *Plant Science*, *283*, 343–354.

Marques, D. M., da Silva, A. B., Mantovani, J. R., Magalhães, P. C., & de Souza, T. C. (2019). Root morphology and leaf gas exchange in *Peltophorum dubium* (Spreng.) Taub. (Caesalpinioideae) exposed to copper-induced toxicity. *South African Journal of Botany*, *121*, 186–192.

Marschner, H. (1995). *Mineral nutrition of higher plants* (2nd ed.). London: Academic Press.

Marschner, H. (2012). In P. Marschner (Ed.), *Marschner's mineral nutrition of higher plants* (89). Academic Press.

McGrath, S. P., Micó, C., Zhao, F. J., Stroud, J. L., Zhang, H., & Fozard, S. (2010). Predicting molybdenum toxicity to higher plants: Estimation of toxicity threshold values. *Environmental Pollution*, *158*(10), 3085–3094.

McLean, E., Cogswell, M., Egli, I., Wojdyla, D., & De Benoist, B. (2009). Worldwide prevalence of anaemia, WHO vitamin and mineral nutrition information system, 1993–2005. *Public Health Nutrition*, *12*(4), 444–454.

Mendel, R. R., & Haensch, R. (2002). Molybdoenzymes and molybdenum cofactor in plants. *Journal of Experimental Botany*, *53*, 1689–1698.

Mendel, R. R., & Kruse, T. (2012). Cell biology of molybdenum in plants and humans. *Biochimica et Biophysica Acta (BBA)-Molecular Cell Research*, *1823*(9), 1568–1579.

Mengel, K., Kosegarten, H., Kirkby, E. A., & Appel, T. (2001). *Principles of plant nutrition*. Berlin, Germany: Springer.

Merlot, S., de la Torre, V. S. G., & Hanikenne, M. (2018). *Physiology and molecular biology of trace element hyperaccumulation. Agromining: Farming for metals* (pp. 93–116). Berlin, Germany: Springer.

Michaud, A. M., Chappellaz, C., & Hinsinger, P. (2008).). Copper phytotoxicity affects root elongation and iron nutrition in durum wheat (*Triticum turgidum* durum L.). *Plant and Soil*, *310*(1-2), 151–165.

Migocka, M., Posyniak, E., Maciaszczyk-Dziubinska, E., Papierniak, A., & Kosieradzaka, A. (2015). Functional and biochemical characterization of cucumber genes encoding two copper ATPases CsHMA5. 1 and CsHMA5. 2. *The Journal of Biological Chemistry*, *290*(25), 15717–15729.

Milner, M. J., & Kochian, L. V. (2008). Investigating heavy-metal hyperaccumulation using *Thlaspi caerulescens* as a model system. *Annals of Botany*, *102*(1), 3–13.

Millaleo, R., Reyes-Díaz, M., Ivanov, A. G., Mora, M. L., & Alberdi, M. (2010). Manganese as essential and toxic element for plants: Transport, accumulation and resistance mechanisms. *Journal of Soil Science and Plant Nutrition*, *10*(4), 470–481.

Molassiotis, A., Tanou, G., Diamantidis, G., Patakas, A., & Therios, I. (2006). Effects of 4-month Fe deficiency exposure on Fe reduction mechanism, photosynthetic gas exchange, chlorophyll fluorescence and antioxidant defense in two peach rootstocks differing in Fe deficiency tolerance. *Journal of Plant Physiology*, *163*(2), 176–185.

Monnet, F., Vaillant, N., Vernay, P., Coudret, A., Sallanon, H., & Hitmi, A. (2001). Relationship between PSII activity, CO_2 fixation, and Zn, Mn and Mg contents of Lolium perenne under zinc stress. *Journal of Plant Physiology*, *158*, 1137–1144. Available from https://doi.org/10.1078/S0176-1617(04)70140-6.

Monni, S., Salemma, M., & Millar, N. (2000). The tolerance of *Empetrum nigrum* to copper and nickel. *Environmental Pollution (Barking, Essex: 1987)*, *109*, 221–229.

Moradtalab, N., Weinmann, M., Walker, F., Höglinger, B., Ludewig, U., & Neumann, G. (2018). Silicon improves chilling tolerance during early growth of maize by effects on micronutrient homeostasis and hormonal balances. *Frontiers in Plant Science*, *9*, 1–17. Available from https://doi.org/10.3389/fpls.2018.00420.

Moreau, S., Thomson, R. M., Kaiser, B. N., Trevaskis, B., Guerinot, M. L., Udvardi, M. K., ... Day, D. A. (2002). GmZIP1 encodes a symbiosis-specific zinc transporter in soybean. *Journal of Biological Chemistry*, *277*(7), 4738–4746.

Moreira, A., Moraes, L. A., & dos Reis, A. R. (2018). *The molecular genetics of zinc uptake and utilization efficiency in crop plants. Plant micronutrient use efficiency* (pp. 87–108). Amsterdam, The Netherlands: Elsevier.

Morgan, J. W., & Anders, E. (1980). Chemical composition of Earth, Venus, and Mercury. *Proceedings of the National Academy of Sciences of the United States of America*, *77*, 6973–6977. Available from https://doi.org/10.1073/pnas.77.12.6973.

Morrissey, J., & Guerinot, M. L. (2009). Iron uptake and transport in plants: The good, the bad, and the ionome. *Chemical Reviews*, *109*(10), 4553–4567.

Mousavi, S. R., Galavi, M., & Ahmadvand, G. (2007). Effect of zinc and manganese foliar application on yield, quality and enrichment on potato (*Solanum tuberosum* L.). *Asian Journal of Plant Sciences*, *6*, 1256–1260.

Müller, C., Kuki, K. N., Pinheiro, D. T., de Souza, L. R., Silva, A. I. S., Loureiro, M. E., & Almeida, A. M. (2015). Differential physiological responses in rice upon exposure to excess distinct iron forms. *Plant and Soil*, *391*(1-2), 123–138.

Mwamba, T. M., Li, L., Gill, R. A., Islam, F., Nawaz, A., Ali, B., & Zhou, W. (2016). Differential subcellular distribution and chemical forms of cadmium and copper in *Brassica napus*. *Ecotoxicology and Environmental Safety*, *134*, 239–249.

Naeem, A., Zafar, M., Khalid, H., Zia-ur-Rehman, M., Ahmad, Z., Ayub, M. A., ... Qayyum, M. F. (2019). *Cadmium-induced imbalance in nutrient and water uptake by plants. Cadmium toxicity and tolerance in plants* (pp. 299–326). Academic Press.

Nagajyoti, P. C., Lee, K. D., & Sreekanth, T. V. M. (2010). Heavy metals, occurrence and toxicity for plants: A review. *Environmental Chemistry Letters*, *8*(3), 199–216.

Naveed, M., Khalid, H., Ayub, M. A., ur Rehman, M. Z., Rizwan, M., Rasul, A., & ul Haq, M. A. (2020). *Biofortification of cereals with zinc and iron: Recent advances and future perspectives. Resources use efficiency in agriculture* (pp. 615–646). Springer Nature Singapore Pte Ltd.

Nazir, F., Hussain, A., & Fariduddin, Q. (2019). Hydrogen peroxide modulate photosynthesis and antioxidant systems in tomato (*Solanum lycopersicum* L.) plants under copper stress. *Chemosphere*, *230*, 544–558.

Nikolic, M., & Kastori, R. (2000). Effect of bicarbonate and Fe supply on Fe nutrition of grapevine. *Journal of Plant Nutrition*, *23*, 1619–1627.

Nishida, S., Tsuzuki, C., Kato, A., Aisu, A., & Yoshida, J. (2011). and Mizuno T. AtIRT1, the primary iron uptake transporter in the root, mediates excess nickel accumulation in Arabidopsis thaliana. *Plant Cell Physiology*, *52*, 1433–1442.

Nkrumah, P. N., Echevarria, G., Erskine, P. D., Chaney, R. L., Sumail, S., & van der Ent, A. (2019). Effect of nickel concentration and soil pH on metal accumulation and growth in tropical agromining 'metal crops'. *Plant and Soil*, *443*(1-2), 27–39.

Nozoye, T., Nagasaka, S., Kobayashi, T., Sato, Y., Uozumi, N., Nakanishi, H., et al. (2011). The phytosiderophore efflux transporter TOM2 is involved in metal transport in rice. *The Journal of Biological Chemistry*, *290*, 27688–27699. Available from https://doi.org/10.1074/jbc.M114.635193.

Oh, Y. J., Kim, H., Seo, S. H., Hwang, B. G., Chang, Y. S., Lee, J., & Hwang, I. (2016). Cytochrome b5 reductase 1 triggers serial reactions that lead to iron uptake in plants. *Molecular Plant*, *9*(4), 501–513.

Ozturk, L., Yazici, M. A., Yucel, C., Torun, A., Cekic, C., Bagci, A., ... Cakmak, I. (2006). Concentration and localization of zinc during seed development and germination in wheat. *Physiologia Plantarum*, *128*, 144–152.

Palmer, C., & Guerinot, M. (2009). Facing the challenges of Cu, Fe and Zn homeostasis in plants. *Nature Chemical Biology*, *5*, 333–340.

Pandey, N., & Sharma, C. P. (2002). Effect of heavy metals Co^{2+}, Ni^{2+} And Cd^{2+} on growth and metabolism of cabbage. *Plant Science (Shannon, Ireland)*, *163*, 753–758.

Papadakis, I. E., Giannakoula, A., Therios, I. N., Bosabalidis, A. M., Moustakas, M., & Nastou, A. (2007). Mn-induced changes in leaf structure and chloroplast ultrastructure of *Citrus volkameriana* (L.) plants. *Journal of Plant Physiology*, *164*, 100–103.

Parida, B. K., Chhibba, I. M., & Nayyar, V. K. (2003). Influence of nickel-contaminated soils on fenugreek (*Trigonella corniculata* L.) growth and mineral composition. *Scientia Horticulturae*, *98*(2), 113–119.

Parrotta, L., Guerriero, G., Sergeant, K., Cai, G., & Hausman, J. F. (2015). Target or barrier? The cell wall of early-and later-diverging plants vs cadmium toxicity: Differences in the response mechanisms. *Frontiers in Plant Science*, *6*, 133.

Paunov, M., Koleva, L., Vassilev, A., Vangronsveld, J., & Goltsev, V. (2018). Effects of different metals on photosynthesis: Cadmium and zinc affect chlorophyll fluorescence in durum wheat. *International Journal of Molecular Sciences*, *19*(3), 787.

Pavlova, D. (2017). Nickel effect on root-meristem cell division in *Plantago lanceolata* (Plantaginaceae) seedlings. *Australian Journal of Botany*, *65*(5), 446–452.

Pedler, J. F., Parker, D. R., & Crowley, D. E. (2000). Zinc deficiency-induced phytosiderophore release by the Triticaceae is not consistently expressed in solution culture. *Planta*, *211*, 120–126.

Pietrini, F., Carnevale, M., Beni, C., Zacchini, M., Gallucci, F., & Santangelo, E. (2019). Effect of different copper levels on growth and morpho-physiological parameters in giant reed (*Arundo donax* L.) in semi-hydroponic mesocosm experiment. *Water*, *11*(9), 1837.

Pilon, M., Ravet, K., & Tapken, W. (2011). The biogenesis and physiological function of chloroplast superoxide dismutases. *Biochimica et Biophysica Acta (BBA)-Bioenergetics*, *1807*(8), 989–998.

Porter, G. S., Bajita-Locke, J. B., Hue, N. V., & Strand, D. (2004). Manganese solubility and phytotoxicity affected by soil moisture, oxygen levels, and green manure additions. *Communications in Soil Science and Plant Analysis*, *35*(1–2), 99–116.

Puig, S. (2014). Function and regulation of the plant COPT family of high-affinity copper transport proteins. *Advances in Botany*.

Raldugina, G. N., Krasavina, M. S., Lunkova, N. F., & Burmistrova, N. A. (2016). *Resistance of plants to Cu stress: Transgenesis. Plant metal interaction* (pp. 69–114). Elsevier.

Ramesh, S. A., Shin, R., Eide, D. J., & Schachtman, D. P. (2003). Differential metal selectivity and gene expression of two zinc transporters from rice. *Plant Physiology*, *133*(1), 126–134.

Rana, M., Bhantana, P., Sun, X. C., Imran, M., Shaaban, M., Moussa, M., & Afzal, J. (2020). Molybdenum as an essential element for crops: An overview. *Biomedical Journal of Scientific and Technical Research*, *24*, 18535.

Ravet, K., & Pilon, M. (2013). Copper and iron homeostasis in plants: The challenges of oxidative stress. *Antioxidants & redox signaling*, *19*(9), 919–932.

Reddy, K. J., Munn, L. C., & Wang, L. (1997). Chemistry and mineralogy of molybdenum in soils. In U. C. Gupta (Ed.), *Molybdenum in agriculture*. Cambridge: Cambridge University Press.

Reeves, R. D., & Baker, A. J. M. (2000). Metal-accumulating plants. In I. Raskin, & B. D. Ensley (Eds.), *Phytoremediation of toxic metals: using plants to clean up the environment* (pp. 193–229). New York: Wiley.

Rengel, Z. (1997). *Root exudation and microflora populations in rhizosphere of crop genotypes differing in tolerance to micronutrient deficiency. Plant nutrition for sustainable food production and environment* (pp. 243–248). Dordrecht: Springer.

Rengel, Z., & Romheld, V. (2000). Root exudation and Fe uptake and transport in wheat genotypes differing in tolerance to Zn deficiency. *Plant and Soil, 222*, 25–34.

Riaz, M. U., Ayub, M. A., Khalid, H., ul Haq, M. A., Rasul, A., ur Rehman, M. Z., & Ali, S. (2020). *Fate of micronutrients in alkaline soils. Resources use efficiency in agriculture* (pp. 577–613). Singapore: Springer.

Ribera-Fonseca, A., Inostroza-Blancheteau, C., Cartes, P., Rengel, Z., & Mora, M. L. (2013). Early induction of Fe-SOD gene expression is involved in tolerance to Mn toxicity in perennial ryegrass. *Plant Physiology and Biochemistry, 73*, 77–82. Available from https://doi.org/10.1016/j.plaphy.2013.08.012.

Robertson, A. I., & Meakin, M. E. R. (1980). The effect of nickel on cell division and growth of *Brachystegia spiciformis* seedlings. *Kirkia*, 115–125.

Robinson, N. J., Procter, C. M., Connolly, E. L., & Guerinot, M. L. (1999). A ferric-chelate reductase for iron uptake from soils. *Nature, 397*, 694–697.

Robson, A. D. (1994). *Zinc in soils and plants*. New York: Springer.

Rodríguez-Celma, J., Tsai, Y. H., Wen, T. N., Wu, Y. C., Curie, C., & Schmidt, W. (2016). Systems-wide analysis of manganese deficiency-induced changes in gene activity of Arabidopsis roots. *Scientific Reports, 6*, 35846.

Römheld, V., & Nikolic, M. (2006). Iron. In A. V. Barker, & D. J. Pilbeam (Eds.), *Handbook of plant nutrition* (pp. 329–350). Boca Raton: CRC Press.

Rout, G. R., & Das, P. (2009). *Effect of metal toxicity on plant growth and metabolism: I. Zinc. Sustainable agriculture* (pp. 873–884). Dordrecht: Springer.

Saidi, M. N., Jbir, R., Ghorbel, I., Namsi, A., Drira, N., & Gargouri-Bouzid, R. (2012). Brittle leaf disease induces an oxidative stress and decreases the expression of manganese-related genes in date palm (*Phoenix dactylifera* L.). *Plant Physiology and Biochemistry: PPB/Societe Francaise de Physiologie Vegetale, 50*, 1–7. Available from https://doi.org/10.1016/j.plaphy.2011.09.016.

Samaranayke, P., Peiris, B. D., & Dssanayake, S. (2012). Effect of excessive ferrous (Fe^{2+}) on growth and iron content in rice (*Oryza sativa. International Journal of Agriculture And Biology, 14*, 296–298.

Santi, S., Cesco, S., Varanini, Z., & Pinton, R. (2005). Two plasma membrane H + -ATPase genes are differentially expressed in iron-deficient cucumber plants. *Plant Physiology and Biochemistry: PPB/Societe Francaise de Physiologie Vegetale, 43*, 287–292.

Santi, S., & Schmidt, W. (2008). Laser microdissection-assisted analysis of the functional fate of iron deficiency-induced root hairs in cucumber. *Journal of Experimental Botany, 59*, 697–704.

Santi, S., & Schmidt, W. (2009). Dissecting iron deficiency-induced proton extrusion in Arabidopsis roots. *New Phytologist, 183*(4), 1072–1084.

Sarika, A., Sairam, R. K., Meena, R. C., Aruna, T., & Srivastava, G. C. (2010). Effect of excess and deficient levels of iron and copper on oxidative stress and antioxidant enzymes activity in wheat. *Journal of Plant Sciences, 5*(1), 105–116.

Schmid, N. B., Giehl, R. F., Döll, S., Mock, H. P., Strehmel, N., Scheel, D., & von Wirén, N. (2014). Feruloyl-CoA 6′-hydroxylase1-dependent coumarins mediate iron acquisition from alkaline substrates in Arabidopsis. *Plant Physiology, 164*(1), 160–172.

Schmidt, S. B., Jensen, P. E., & Husted, S. (2016). Manganese deficiency in plants: The impact on photosystem II. *Trends in Plant Science, 21*(7), 622–632.

Schulte, E. E., & Kelling, K. A. (1999). *Soil and applied manganese. Understanding plant nutrients* (p. A2526) Madison, WI: University of Wisconsin Cooperative Extension.

Seh, E. (2002). Metal permeability, transport and efflux in plants. In M. N. V. Prasad, & K. Strazalka (Eds.), *Physiology and biochemistry of metal toxicity and tolerance in plants* (p. 1). London: Kluwer Academic Publishers.

Seregin, I. V., & Kozhevnikova, A. D. (2006). Physiological role of nickel and its toxic effects on higher plants. *Russian Journal of Plant Physiology, 53*, 285–308.

Shahzad, B., Tanveer, M., Rehman, A., Cheema, S. A., Fahad, S., Rehman, S., & Sharma, A. (2018). Nickel; whether toxic or essential for plants and environment—A review. *Plant Physiology and Biochemistry, 132*, 641–651.

Shao, J. F., Yamaji, N., Shen, R. F., & Ma, J. F. (2017). The key to Mn homeostasis in plants: Regulation of Mn transporters. *Trends in Plant Science, 22*(3), 215–224.

Sheldon, A. R., & Menzies, N. W. (2005). The effect of copper toxicity on the growth and root morphology of Rhodes grass (*Chloris gayana* Knuth.) in resin buffered solution culture. *Plant and Soil, 278*(1-2), 341–349.

Shikanai, T., Müller-Moulé, P., Munekage, Y., Niyogi, K. K., & Pilon, M. (2003). PAA1, a P-type ATPase of Arabidopsis, functions in copper transport in chloroplasts. *The Plant Cell, 15*(6), 1333–1346.

Silva, J. C., Echeveste, P., & Lombardi, A. T. (2018). Higher biomolecules yield in phytoplankton under copper exposure. *Ecotoxicology and Environmental Safety, 161*, 57–63.

Shin, L. J., Lo, J. C., & Yeh, K. C. (2012). Copper chaperone antioxidant protein1 is essential for copper homeostasis. *Plant Physiology, 159*, 1099–1110.

Solymosi, K., Lenti, K., Myśliwa-Kurdziel, B., Fidy, J., Strzałka, K., & Böddi, B. (2004). Hg2 + reacts with different components of the NADPH: Protochlorophyllide oxidoreductase macrodomains. *Plant Biology, 6*(3), 358–368.

Song, W. Y., Choi, K. S., Kim do, Y., Geisler, M., Park, J., Vincenzetti, V., Schellenberg, M., et al. (2010). Arabidopsis PCR2 is a zinc exporter involved in both zinc extrusion and long-distance zinc transport. *The Plant Cell, 22*, 2237–2252.

Sreekanth, T. V. M., Nagajyothi, P. C., Lee, K. D., & Prasad, T. N. V. K. V. (2013). Occurrence, physiological responses and toxicity of nickel in plants. *International Journal of Environmental Science and Technology, 10*, 1129–1140.

St. Clair, S. B., & Lynch, J. P. (2005). Element accumulation patterns of deciduous and evergreen tree seedlings on acid soils: Implications for sensitivity to manganese toxicity. *Tree Physiology, 25*, 85–92. Available from https://doi.org/10.1093/treephys/25.1.85.

Subba, P., Mukhopadhyay, M., Mahato, S. K., Bhutia, K. D., Mondal, T. K., & Ghosh, S. K. (2014). Zinc stress induces physiological, ultra-structural and biochemical changes in mandarin orange (*Citrus reticulata* Blanco) seedlings. *Physiology and Molecular Biology of Plants, 20*(4), 461–473.

Subrahmanyam, D., & Rathore, V. S. (2000). Influence of manganese toxicity on photosynthesis in ricebean (*Vigna umbellata*) seedlings. *Photosynthetica, 38*(3), 449–453.

Subrahmanyam, D., & Rathore, V. S. (2001). Influence of manganese toxicity on photosynthesis in ricebean (*Vigna umbellata*) seedlings. *Photosynthetica, 38*, 449–453.

Suh, H. J., Kim, C. S., Lee, J. Y., & Jung, J. (2002). Photodynamic effect of iron excess on photosystem II function in pea plants. *Photochemistry and Photobiology, 75*, 513–518.

Taiz, L., & Zeiger, E. (1991). *Plant physiology* (p. 565) Redwood City, CA: The Benjamin/Cummings Publ. Co. Inc.

Taiz, L., Zeiger, E., Møller, I. M., & Murphy, A. (2015). *Plant physiology and development*. Sunderland, CT: Sinauer Associates.

Takahashi, M., Terada, Y., Nakai, I., Nakanishi, H., Yoshimura, E., Mori, S., & Nishizawa, N. K. (2003). Role of nicotianamine in the intracellular delivery of metals and plant reproductive development. *The Plant Cell*, *15*(6), 1263–1280.

Takahashi, R., Ishimaru, Y., Shimo, H., Ogo, Y., Senoura, T., Nishizawa, N. K., & Nakanishi, H. (2012). The OsHMA2 transporter is involved in root-to-shoot translocation of Zn and Cd in rice. *Plant, Cell & Environment*, *35*(11), 1948–1957.

Talke, I. N., Hanikenne, M., & Krämer, U. (2006). Zinc-dependent global transcriptional control, transcriptional deregulation, and higher gene copy number for genes in metal homeostasis of the hyperaccumulator *Arabidopsis halleri*. *Plant Physiology*, *142*, 148–167.

Tejada-Jiménez, M., Chamizo-Ampudia, A., Galván, A., Fernández, E., & Llamas, Á. (2013). Molybdenum metabolism in plants. *Metallomics*, *5*(9), 1191–1203.

Tejada-Jiménez, M., Galván, A., & Fernández, E. (2011). Algae and humans share a molybdate transporter. *Proceedings of the National Academy of Sciences*, *108*(16), 6420–6425.

Thomas, G., Andresen, E., Mattusch, J., Hubáček, T., & Küpper, H. (2016). Deficiency and toxicity of nanomolar copper in low irradiance—A physiological and metalloproteomic study in the aquatic plant *Ceratophyllum demersum*. *Aquatic Toxicology*, *177*, 226–236.

Thomine, S., Lelièvre, F., Debarbieux, E., Schroeder, J. I., Barbier-Brygoo, H., Thomine, S., et al. (2003). AtNRAMP3, a multispecific vacuolar metal transporter involved in plant responses to iron deficiency. *The Plant Journal: For Cell and Molecular Biology*, *34*, 685–695.

Thomine, S., Wang, R., Ward, J. M., Crawford, N. M., Schroeder, J. I., Thomine, S., et al. (2000). Cadmium and iron transport by members of a plant metal transporter family in Arabidopsis with homology to Nramp genes. *Proceedings of the National Academy of Sciences of the United States of America*, *97*, 4991–4996.

Todeschini, V., Lingua, G., D'Agostino, G., Carniato, F., Roccotiello, E., & Berta, G. (2011). Efects of high zinc concentration on poplar leaves: A morphological and biochemical study. *Environmental and Experimental Botany*, *71*(1), 50–56.

Tomatsu, H., Takano, J., Takahashi, H., Watanabe-Takahashi, A., Shibagaki, N., & Fujiwara, T. (2007). An *Arabidopsis thaliana* high-affinity molybdate transporter required for efficient uptake of molybdate from soil. *Proceedings of the National Academy of Sciences*, *104*(47), 18807–18812.

Tripathi, D. K., Singh, S., Gaur, S., Singh, S., Yadav, V., Liu, S., & Dubey, N. K. (2018). Acquisition and homeostasis of iron in higher plants and their probable role in abiotic stress tolerance. *Frontiers in Environmental Science*, *5*, 86.

Tsoney, T., & CebolaLidon, F. J. (2012). Zinc in plants—An overview. *Emirates Journal of Agricultural and Food Science*, *24*(4), 322–333.

Vasconcelos, M., Eckert, H., Arahana, V., Graef, G., Grusak, M. A., & Clemente, T. (2006). Molecular and phenotypic characterization of transgenic soybean expressing the Arabidopsis ferric chelate reductase gene, FRO2. *Planta*, *224*, 1116–1128.

van de Mortel, J. E., Almar Villanueva, L., Schat, H., Kwekkeboom, J., Coughlan, S., Moerland, P. D., et al. (2006). Large expression differences in genes for iron and zinc homeostasis, stress response, and lignin biosynthesis distinguish roots of *Arabidopsis thaliana* and the related metal hyperaccumulator *Thlaspi caerulescens*. *Plant Physiology*, *42*, 1127–1147.

Van Maarschalkerweerd, M., Bro, R., Egebo, M., & Husted, S. (2013). Diagnosing latent copper deficiency in intact barley leaves (*Hordeum vulgare* L.) using near infrared spectroscopy. *Journal of Agricultural and Food Chemistry*, *61*(46), 10901–10910.

Vassilev, A., Nikolova, A., Koleva, L., & Lidon, F. (2011). Effects of excess Zn on growth and photosynthetic performance of young bean plants. *Journal of Phytology*, *3*(6).

Verbruggen, N., Hermans, C., & Schat, H. (2009). Tansley review molecular mechanisms of metal hyperaccumulation in plants. *New Phytologist*, *181*, 759–776.

Vert, G., Grotz, N., Dedaldechamp, F., Gaymard, F., Guerinot, M. L., Briat, J.-F., & Curie, C. (2002). IRT1, an Arabidopsis transporter essential for iron uptake from the soil and plant growth. *The Plant Cell*, *14*, 1223–1233.

Vigani, G., Zocchi, G., Bashir, K., Philippar, K., & Briat, J. F. (2013). Signals from chloroplasts and mitochondria for iron homeostasis regulation. *Trends in Plant Science*, *18*(6), 305–311.

Wang, D., Wan, C. Y., Wang, W. Q., Hou, J. L., Li, W. D., Wei, S. L., & Yu, J. B. (2012). Effects of manganese deficiency on growth and contents of active constituents of *Glycyrrhiza uralensis* Fisch. *Communications in Soil Science and Plant Analysis*, *43*(17), 2218–2227.

Wang, M., Kawakami, Y., & Bhullar, N. K. (2019). Molecular analysis of iron deficiency response in hexaploid wheat. *Frontiers in Sustainable Food Systems*, *3*, 67.

Waters, B., & Sankaran, R. (2011). Moving micronutrients from the soil to the seeds: Genes and physiological processes from a biofortification perspective. *Plant Science (Shannon, Ireland)*, *180*, 562–574.

Waters, B. M., Blevins, D. G., & Eide, D. J. (2002). Characterization of FRO1, a pea ferric-chelate reductase involved in root iron acquisition. *Plant Physiology*, *129*, 85–94.

Weber, M., Harada, E., Vess, C., von Roepenack-Lahaye, E., & Clemens, S. (2004). Comparative microarray analysis of *Arabidopsis thaliana* and *Arabidopsis halleri* roots identifies nicotianamine synthase, a ZIP transporter and other genes as potential metal hyperaccumulation factors. *The Plant Journal*, *37*, 269–281.

Welch, R. M., & Shuman, L. (1995). Micronutrient nutrition of plants. *Critical Reviews in Plant Sciences*, *14*(1), 49–82.

Williams, R. J. P., & Frausto da Silva, J. J. R. (2002). The involvement of molybdenum in life. *Biochemical and Biophysical Research Communications*, *292*, 293–299.

Wu, D., Yamaji, N., Yamane, M., Kashino-Fujii, M., Sato, K., & Ma, J. F. (2016). The HvNramp5 transporter mediates uptake of cadmium and manganese, but not iron. *Plant Physiology*, *172*(3), 1899–1910.

Wu, G., Chen, W., Zheng, X., He, D., Luo, Y., Wang, X., … Li, Y. (2017). Hierarchical Fe-doped NiOx nanotubes assembled from ultrathin nanosheets containing trivalent nickel for oxygen evolution reaction. *Nano Energy*, *38*, 167–174.

Wu, L. B., Shhadi, M. Y., Gregorio, G., Matthus, E., Becker, M., & Frei, M. (2014). Genetic and physiological analysis of tolerance to acute iron toxicity in rice. *Rice*, *7*(1), 1–12.

Wyszkowski, M., & Modrzewska, B. (2016). Acidity and sorption properties of Zinc-contaminated soil following the application of neutralising substances. *Journal of Ecological Engineering*, 17.

Xing, J. P., Jiang, R. F., Ueno, D., Ma, J. F., Schat, H., McGrath, S. P., & Zhao, F. J. (2008). Variation in root-to-shoot translocation of cadmium and zinc among different accessions of the hyperaccumulators *Thlaspi caerulescens* and *Thlaspi prae-cox*. *New Phytologist*, *178*, 315–325.

Yamaguchi, T., Tomioka, R., & Takenaka, C. (2015). Can *Clethra barbinervis* distinguish nickel and cobalt in uptake and translocation? *International Journal of Molecular Sciences*, *16*(9), 21378–21391.

Yamasaki, H., Pilon, M., & Shikanai, T. (2008). How do plants respond to copper deficiency? *Plant Signaling & Behavior*, *3*(4), 231–232.

Yang, S. X., Deng, H., & Li, M. S. (2008). Manganese uptake and accumulation in a woody hyperaccumulator, *Schima superba*. *Plant, Soil and Environment*, *54*(10), 441–446.

Yano, J., & Yachandra, V. (2014). Mn4Ca cluster in photosynthesis: Where and how water is oxidized to dioxygen. *Chemical Reviews*, *114*, 4175–4205.

Yeargan, R., Maiti, I. B., Nielsen, M. T., Hunt, A. G., & Wagner, G. J. (1992). Tissue partitioning of cadmium in transgenic tobacco. Seedlings and field grown plants expressing mouse metallothionein-I gene. *Transgenic Research*, *1*, 261–267.

Yousfi, S., Houmani, H., Zribi, F., Abdelly, C., & Gharsalli, M. (2012). Physiological responses of wild and cultivated barley to the interactive effect of salinity and iron deficiency. *ISRN Agronomy*, *2012*, 121983.

Yruela, I. (2005). Copper in plants. *Brazilian Journal of Plant Physiology*, *17*(1), 145–156.

Yuan, W., Zhang, D., Song, T., Xu, F., Lin, S., Xu, W., & Zhang, J. (2017). Arabidopsis plasma membrane H + -ATPase genes AHA2 and AHA7 have distinct and overlapping roles in the modulation of root tip H + efflux in response to low-phosphorus stress. *Journal of Experimental Botany*, *68*(7), 1731–1741.

Zhao, J., Wang, W., Zhou, H., Wang, R., Zhang, P., Wang, H., & Xu, J. (2017). Manganese toxicity inhibited root growth by disrupting auxin biosynthesis and transport in Arabidopsis. *Frontiers in Plant Science*, *8*, 272.

Zheng, L., Yamaji, N., Yokosho, K., & Ma, J. F. (2012). YSL16 is a phloem-localized transporter of the copper–nicotianamine complex that is responsible for copper distribution in rice. *Plant Cell*, *24*, 3767–3782. Available from https://doi.org/10.1105/tpc.112.103820.

Zhou, Q., Li, X., Lin, Y., Yang, C., Tang, W., Wu, S., ... Lou, W. (2019). Effects of copper ions on removal of nutrients from swine wastewater and on release of dissolved organic matter in duckweed systems. *Water Research*, *158*, 171–181.

Zuo, Y., & Zhang, F. (2011). Soil and crop management strategies to prevent iron deficiency in crops. *Plant and Soil*, *339*, 83–95.

Chapter 16

Halophytes as effective tool for phytodesalination and land reclamation

Nair G. Sarath[1], Palliyath Sruthi[1], A.M. Shackira[2] and Jos T. Puthur[1]
[1]*Plant Physiology and Biochemistry Division, Department of Botany, University of Calicut, Malappuram, India,* [2]*Department of Botany, Sir Syed College, Taliparamba, Kannur, India*

16.1 Introduction

Soil salinization is one of the major abiotic environmental stress and global issues, adversely affecting growth and yield of the crops and thereby affecting the food security. Higher levels of soil salinity cause a broad range of alterations in plant metabolism and finally lead to death (Kamran et al., 2020). More than one billion hectares of soil around the world are degraded due to soil salinity. Increased anthropogenic activities and climate change are accelerating the process of soil salinization. Salinized soil contains more than 4 $dS\ m^{-1}$ electrical conductivity (EC) and higher amounts of cations like Na^+, K^+, Mg^{+2}, and Ca^{+2} and anions like HCO^{3-}, Cl^-, NO^{3-}, SO_4^{-2}, and CO_3^{-2} (Corwin & Yemoto, 2017; Ivushkin et al., 2019). Higher levels of salts in the soil affect its texture, decrease soil productivity, and become a major problem for global food production (Cuevas, Daliakopoulos, del Moral, Hueso, & Tsanis, 2019).

Increased human activities on natural resources have resulted in global warming and this in turn affects the melting of icebergs and the rise of sea level. This situation has resulted in the higher intrusion of saltwater from the sea into the coastal land areas, which creates degraded lands with high salt content (Dasgupta, Hossain, Huq, & Wheeler, 2015; Velautham, Ranney, & Brow, 2019). Nowadays the use of saltwater for irrigation purposes in agricultural fields due to the shortage of good quality freshwater is accelerating the soil salinity. The overuse of groundwater and the changing climatic pattern with reduced precipitation also increase soil salinization. The unpredictable changes in soil characteristics due to salinization create ecologically bare areas (Chandio, Lee, & Mirjat, 2013; Okur & Örçen, 2020).

Frontiers in Plant–Soil Interaction. DOI: https://doi.org/10.1016/B978-0-323-90943-3.00020-1

The increase in sea level rise is also a major factor for the reduction of good quality drinking water and the scarcity of fresh drinking water is a problem in many countries. An increase in water salinity makes it less palatable for drinking purposes (Vineis, Chan, & Khan, 2011). About 20% of cultivated and 50% of irrigated land is affected due to salinity (Zhou et al., 2018). By the end of 2050 about half of the fertile agricultural land will have become salinized. This conversion may negatively affect agricultural and food production across the entire world (Hasanuzzaman et al., 2014; Menason, Betty, Vijayan, & Anbudurai, 2015).

Salinized soils can be classified into three groups, namely saline soils, sodic soils, and saline sodic soils. The amount of soluble salt content is higher in saline soils but in sodic soil sodium ions are higher. The saline sodic soil is characterized by the higher content of both soluble salts and exchangeable sodium ions (Sastre-Conde, Lobo, Beltrán-Hernández, & Poggi-Varaldo, 2015). The parameters which indicate the intensity of salinity in these three types of soil are EC, exchangeable sodium percentage (ESP), and pH. In the case of sodic soil and saline sodic soil the pH is more than 8.5 and ESP is above 15% but there is variation in EC. In sodic soil EC is below 4 mmhos cm^{-1} and in saline sodic soil it is above 4 mmhos cm^{-1}. Saline soil is characterized with more than 4 mmhos cm^{-1} EC, ESP below 15% and with lower pH (<8.5). Coastal salt marshes and inland saline lands are two types of natural saline soils which occur all around the world. But another type of soil, called heavily irrigated soil, in arid regions becomes salinized due to higher evapotranspiration and wrong irrigation practices.

The plants have developed mechanisms in response to abiotic stresses like salinity and these include physiological and molecular mechanisms to avoid or tolerate the condition. The salt sensitive plants, called glycophytes, have less tolerance and the growth of these plants is retarded even under a small elevation of salinity in their growing habitats. But some other plants, which are highly tolerant to salinity, may flourish and complete their life cycle in harsh saline condition, these are called salt-tolerant plants or halophytes. Approximately 1% of the world's flora are halophytes (Colmer & Flowers, 2008). Various inbuilt tolerance mechanisms in halophytes, including effective compartmentalization of ions, osmotic regulation, regulation of the movement of ions, salt influx, and timely expression of salt tolerance genes, empower these plants to survive better in saline habitats as compared to glycophytes (Munns, 2002; Parida & Jha, 2010a, 2010b). An increase in salt content cannot affect halophytes and it acts as a positive modulator in the vegetative as well as reproductive growth of these plants. The NaCl content benefits the different phases of reproduction, such as pollen viability, anther development, and also seed development (Yuan, Guo, Shabala, & Wang, 2019, Yuan, Liang, Li, Yin, & Wang, 2019).

Due to the tremendous increase of salinization throughout the world, its management using lower cost techniques is of great importance.

Identification of halo tolerant plants and their salt tolerable limits and their potential to survive in salt-affected soil has got great significance for land reclamation. The conversion of salinized land into suitable agriculture land by utilizing saline tolerant plants is a process of sustainable agricultural practice. Saline agriculture is a possible strategy of the reclamation of land degraded with salinity, which utilizes the technique of cultivating salt-tolerant plants in such lands and it has significant roles from ecological and agricultural perspectives.

16.2 Phytodesalination; sustainable solution for soil salinity

Remediation strategies which were conventionally practiced for the recuperation of salinity affected land have several disadvantages, like high cost, harmful effects on soil microflora, complex implementation strategies, less efficiency, etc. Hence, the planning and implementation of a cost-effective technique for the proper and sustainable management of soil salinity and its recuperation by natural means has got great importance nowadays. For this purpose, halophytes are used and are regarded as effective bioremediating agents. Halophytes have effective physiological and molecular mechanisms for easy growth under harsh saline conditions. The halophyte-based technology for remediating soil salinity called phytodesalination is a means of sustainable management of soil salinity and of restoring the degraded land for cultivation purposes (Devi et al., 2016; Farzi, Borghei, & Vossoughi, 2017; Jesus, Danko, Fiúza, & Borges, 2015). Because of its cost-effectiveness, easy management, and ecofriendly nature, this technique has attracted considerable attention in the 21st century.

The capabilities of halophytes, such as their capacity to tolerate salt content and absorb and store high level of salts in their biomass, helps them to easily grow on the soil degraded with salinity. This also improves the soil quality and restores it for the growth of glycophytes. Due to the scarcity of freshwater, conventional methods such as leaching and flushing with freshwater have become impractical. Instead haloresistant plants are used for the successful desalination of saline lands (Kalanaki, Ritzema, Bamshad, Jones, & Fazilatnia, 2020). Three categories of halophytes exist and each has shown varied tolerance potential against high salinity and the mechanisms underlined also vary. The first category of halophytes are called euhalophytes, and they have the potential to accumulate ions in vacuoles and have succulent leaves and stems. Pseduohalophytes grow in saline soil by excluding salts by using their own inherent mechanisms and have a low level of Na^+ in the upper part of the biomass when compared with the underground part. Recretohalophytes avoid salinity stress by using specialized structures like salt glands and salt bladders. This helps to reduce the salt content in the plant's parts by excreting the absorbed salts (Song & Wang, 2015; Yuan, Leng, & Wang, 2016). The selection of the best suited candidate for the salinized soil is the initial and key step in the process of phytodesalination. The plants used for this purpose should have the following characters,

1. High salinity tolerance potential
2. Significant biomass production
3. Capacity to sequestrates sodium ions in the biomass
4. High economic value after the desalination process and harvest (like fodder, fuel, essential oils, etc.)

Sesuvium portulacastrum, an obligate halophyte, was utilized in desalination processes. It reduces the salinity and sodicity of the salinized soil and restores the soil quality. After phytodesalination, the biomass of *S. portulacastrum* contains about 4.36 g per pot and 872 mg per plant of Na^+ ions in shoots and roots, respectively (Rabhi et al., 2009). The reclamation of soil degraded with salinity using *Atriplex nummulariawas* was reported by Miranda et al. (2020). They have efficient capacity to reduce the salinity, sodium adsorption ratio, and sodium exchangeable percentage. The use of *Spinacia oleracea* for phytodesalination and its use in bioactive compounds production was studied in detail by Muchate, Nikalje, Rajurkar, Suprasanna, and Nikam (2016). The modulations in physiochemical processes help to withstand the salinity condition and this stress condition augments the production of the bioactive compounds like 20-hydroxyecdysone. There was increased production of osmolytes, enhanced antioxidant activity, and Na^+ was increasingly sequestered in the biomass of *S. oleracea* under high salinity than control plants. This type of desalination has dual importance, both in desalination and also in economic aspects. Due to the high potential to accumulate sodium and chloride ions in their plant biomass, Xu, Renault, and Yuan (2019) reported two plants for the purpose of desalination: alkali grass (*Puccinellia nuttalliana*) and cat tail (*Typha latifolia*). These two plants showed significant growth in saline conditions compared to the control. The supplementation of gypsum ($CaSO_4 \cdot 2H_2O$) and Polisul-C with the plant *Sesuvium verrucosum* enhances the phytodesalination potential of this halophyte. The studies on *Bacopa monnieri* and *S. verrucosum* have also proven that these are the best candidates for phytodesalination, because they improve soil characteristics, such as EC, pH, and soil porosity, and restore the goodness of the soil for normal cultivation (Lastiri-Hernández, Alvarez-Bernal, Bermúdez-Torres, Cárdenas, & Ceja-Torres, 2019; Lastiri-Hernández, Cruz-Cárdenas, Álvarez-Bernal, Vázquez-Sánchez, & Bermúdez-Torres, 2020). The luxuriantly growing salt-loving plants such as *Suaeda maritima, S. portulacastrum, Clerodendron inerme, Ipomoea pes-caprae, Heliotropium curassavicum*, and *Excoecaria agallocha* were used for the study of the bioaccumulation of salts in their biomass by Ravindran, Venkatesan, Balakrishnan, Chellappan, and Balasubramanian (2007). This field study revealed *S. maritima* and *S. portulacastrum* are the best among them to potentially accumulate and reduce the soil salinity.

The EC of the saline water reduced with an increase in time of the growth of aquatic plants (*Ipomoea aquatica*, *Alternanthera philoxeroides*, and *Ludwigia adscendens*) in the treatment system. The harvested root and leaf after the desalination study showed a higher amount of sodium. High phytodesalination potential was exhibited by *I. aquatica* (130 kg $Na^{+}ha^{-1}$) as its productivity range was high compared with *A. philoxeroides* (105 kg $Na^{+}ha^{-1}$) and *L. adscendens* (80 kg Na^{+} ha^{-1}). About 32.1% and 29.8% of sodium and chloride accumulation were showed by *Suaeda salsa* and *Kalidium folium* plants, respectively (Zhao et al., 2005). *Salsola soda*, a halophyte, can effectively reduce the soil salinity compared with *Portulaca oleracea* and it is recommended as a good agent for phytodesalination by Karakaş, Cullu, and Dikilitaş (2017). *P. nuttalliana* and *Poa juncifolia* are two halophytic grass species with the potential to tolerate 300 mM of NaCl without affecting the physiological processes such as photosynthesis, transpiration rate, gas exchange, and leaf water content (Vaziriyeganeh, Lee, & Zwiazek, 2018). The plants with these types of potentiality have got great significance for phytodesalination purposes. Moderately salinized soil can be desalinized using the halophyte *Sulla carnosa* (Jlassi et al., 2013). The 80-day-old plants showed 5.0 t $DWha^{-1}$ productivity and 0.3 t Na^{+}/ha phytodesalination capacity. Palliyath and Puthur (2018) conducted an experiment with the mangrove plant *Bruguiera cylindrica*, they showed this halophyte had NaCl tolerance up to 600 mM. The modulation of various physiochemical features including oxidation and antioxidation (enzymatic and nonenzymatic) and potential to accumulate Na^{+} help them to tolerate the high salinity. The overall working method of phytodesalination in a salt-affected land is depicted in Fig. 16.1 and promising candidates for the purpose of phytodesalination are listed in Table 16.1.

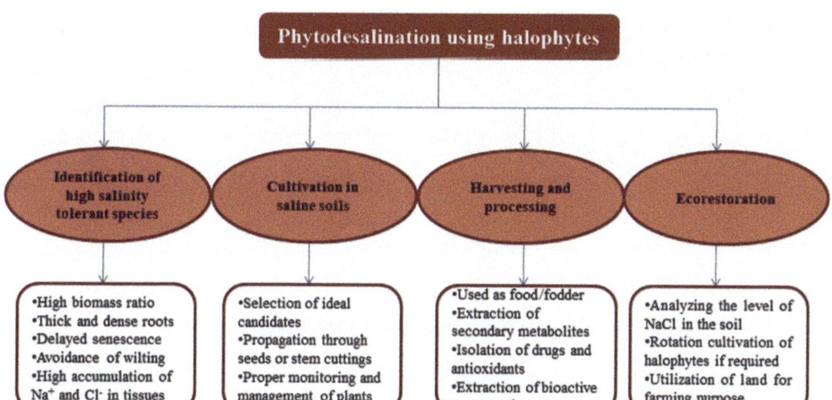

FIGURE 16.1 Overall working method of phytodesalination in a salt-affected land.

TABLE 16.1 List of promising candidates for phytodesalination.

Sl. no	Plant name	Salinity tolerance level	Reference
1	*Bruguiera cylindrica*	600 mM	Palliyath and Puthur (2018)
2	*Salsola drummondii*	500 mM	Elnaggar, Mosa, El-Keblawy, Tammam, and El-Naggar (2020)
3	*Suaeda rigida*	500 mM	Han, Sun, Zhang, and Zhai (2020)
4	*Paspalum vaginatum*	400 mM	Wu et al. (2020)
5	*Acanthus ilicifolius*	400 mM	Shackira and Puthur (2016)
6	*Haloxylon salicornicum*	400 mM	Panda, Rangani, and Parida (2019)
7	*Suaeda salsa*	400 mM	Lu, Qiu, Lu, Wang, and Kuang (2002)
8	*Bruguiera parviflora*	400 mM	Parida et al. (2004)
9	*Aegiceras corniculatum*	350 mM	Chen et al. (2014)
10	*Puccinellia tenuiflora*	300 mM	Niu et al. (2016)
11	*Chenopodium quinoa*	300 mM	Manaa et al. (2019)
12	*Puccinellia nuttalliana*	300 mM	Vaziriyeganeh et al. (2018)
13	*Salicornia dolichostachya*	300 mM	Katschnig, Broekman, and Rozema (2013)
14	*Aster tripolium*	300 mM	Wiszniewska, Koźmińska, Hanus-Fajerska, Dziurka, and Dziurka (2019)
15	*Suaeda maritima*	300 mM	Rajaravindran and Natarajan (2012)
16	*Poa juncifolia*	300 mM	Vaziriyeganeh et al. (2018)

16.3 Halophytes as a boon for desalination

Halophytes can grow and complete their life cycle in an environment that contains high concentrations of soluble salt, and based on their salt-tolerating capacity these are categorized as obligate or facultative. Obligate halophytes can grow up to moderate or high salinity and are incapable of growth at low salinity or in freshwater (Mitsch & Gosselink, 2000; Sabovljevic & Sabovljevic, 2007). Facultative halophytes can be established in comparatively less saline surroundings besides the border between saline and nonsaline areas and are characterized by enhanced physiological diversity enabling them to cope with the saline condition (Atreya, Vartak, & Bhargava, 2009). The capability of halophytes to thrive under extreme saline conditions has enabled the categorization of these plants as one of the finest germplasm used for saline agriculture. *Mesembryanthemum, Suaeda, Cakile, Aeluropus, Atriplex, Thellungiella*, and *Salicornia* are some halophytic genera serving as potential candidates for the selection and isolation of salt-responsive genes and promoters. A number of antiporters genes (HKT, SOS, NHX, VTPase), ion channels (aquaporins, Cl^-, Ca^{2+}), antioxidant encoding genes (BADH, SOD, APX, GST, CAT) etc. were isolated from halophytes and these genes were incorporated into glycophytes for increasing the stress tolerance of the latter (Mishra & Tanna, 2017).

Halophytes have evolved many strategies to mitigate the effects of salt stress and to permit growth in high-salt environments. Understanding how high salinity affects the morphological, physiological, biochemical, metabolic, and gene expression properties of plants will lead to revealing the mechanisms underlying the salt stress tolerance of halophytes. Earlier published literatures have discussed various adaptations or specific mechanisms for salinity tolerance in halophytes, a multiplex trait regulated by different mechanisms like several biochemical pathways leading to the synthesis of free radicals, osmotically active metabolites, specific proteins for maintenance of osmotic and ionic flux, and histoanatomical adaptations (Nawaz et al., 2020).

16.4 Inherent tolerance mechanisms in halophytes

16.4.1 Morphological and anatomical peculiarities

Morphoanatomical features are considered as adaptive features justifying the positioning of a plant in a specific ecological group (Grigore & Toma, 2006). Salinity tolerance of halophytes evolving adaptive responses to salt stress are characterized by various morphoanatomical adaptations associated with avoidance of water loss by evapotranspiration. The most important characters are reduced leaves or absence of leaves, large number of glandular trichomes, presence of salt glands, greater development of support tissues, and presence of aqueous tissue (Pérez Cuadra, Verolo, & Cambi, 2020). Leaf thickening, increased thickness of epidermal cells, variation in stomatal distribution, etc. are the anatomical adaptations acquired as a result of salinity (Flowers & Colmer, 2015).

In response to the regulation of leaf salt concentrations, the halophytic species possess some morphological adaptations like succulence of leaves and presence of salt glands. Also, an increase in the thickness of epidermal and spongy tissues of the leaves and reduction of xylem vessels of stem and root are other adaptations in halophytes (Grigore, Ivanescu, & Toma, 2014). Halophytes acquired more salinity tolerant genes than glycophytes, which are mostly involved in membrane transport, redox reactions, signal transduction, and other processes (Askari, Edqvist, Hajheidari, Kafi, & Salekdeh, 2006; Zhang et al., 2008). Tissue tolerance of salinity requires compartmentalization of toxic sodium ions at the cellular and intracellular levels to avoid toxic concentrations in the cytoplasm, especially in mesophyll cells of leaves (Munns & Tester, 2008; Wang et al., 2015). Increased leaf succulence is a typical adaptive response in halophytes and may be achieved by increasing the size of mesophyll cells and the relative size of their vacuoles (Shabala & Mackay, 2011).

This increase in epidermal thickness in salinity might be an adaptation to minimize the transpiration rate and hence to maintain the water content in mesophyll tissues of halophytes. Salinity induced increase in epidermal thickness not only improves the water use efficiency of plants but also provides additional space for efficient sequestration of Na^+ in the leaf epidermis (Shabala et al., 2012). Distribution of stomata on the leaves of several halophytes, stomatal aperture diameter, and pore size were also affected by salinity. The stomatal density and pore size decreased on both adaxial and abaxial surfaces in leaves of several halophytes, such as *Bruguiera parviflora* (Parida, Das, & Mittra, 2004); *Nitraria retusa*, *Atriplex halimus* (Boughalleb, Denden, & Tiba, 2009), and *Chenopodium quinoa* (Shabala, Hariadi, & Jacobsen, 2013).

Glandular trichomes are distinguishing features of halophytes toward salinity tolerance. Salt-secreting trichomes were seen in different species of *Atriplex* (Grigore & Toma, 2017). Salt bladders and glands are remarkable features found in halophytes and the secretion of salt via salt glands was found in more than 50 species belonging to 14 different angiosperm families (Dassanayake & Larkin, 2017) and it signifies a well-organized way of the removal of excessive salts from photosynthetically active mesophyll tissues. The direct role of epidermal bladder cells in salinity tolerance was first demonstrated in *C. quinoa*. The gentle mechanical removal of bladder cells from salt-grown *C. quinoa* plants resulted in a salt-sensitive phenotype, affecting ion transport and causing dramatic metabolic changes including in the accumulation of gamma-aminobutyric acid, proline, sucrose, and inositol (Kiani-Pouya et al., 2017). A recent screening of 140 quinoa accessions showed that the bladder density increased in most accessions when plants were grown under 400 mM NaCl conditions and showed a positive correlation between salinity tolerance and bladder diameter/volume (Kiani-Pouya et al., 2019). A high-quality genome assembly of *C. quinoa* reported the presence of 54,630 genes of which 99.3% have orthologous genes in glycophytes (Zou et al., 2017). In *M. crystallinum*, salt stress triggered an increase in the cell size and a ploidy level of the bladder (Barkla, Garibay-Hernández, Melzer, Rupasinghe, & Roessner, 2018).

Salt glands are usually more complex in their structure and can be roughly divided into bicellular (with a single basal and cap cells, e.g., in Poaceae (Liphschitz & Waisel, 1974) and multicellular-type salt glands, consisting of about 4–40 cells (Balsamo & Thomson, 1993). In dicotyledonous recretohalophyte *Limonium bicolor*, salt glands constituted of 16 cells including four sets of the outer cup cells, inner cup cells, accessory cells, and secretory cells, with the basal outer cup cells seen adjacent to mesophyll cells (Yuan et al., 2015). The glands on mangroves *Avicennia officinalis* and *Avicennia marina* possess 2–4 collecting cells and 8–12 secretory cells (Drennan, Berjak, Lawton, & Pammenter, 1987; Tan, Lin, Lim, Kumar, & Loh, 2013). Other species like *Tamarix, Aegiceras corniculatum*, and *Glaux maritima* are all equipped with salt glands with multiple secretory cells (Dassanayake & Larkin, 2017). In *Suaeda fruticosa* salt crystals were reported on the epidermis, confirming the existence of salt excretion mechanisms. One of the practically constant cellular contents in the mesophyll cells of plants is calcium oxalate crystals, which take different forms. According to the findings of recent research, the presence of these crystals has a connection with salt tolerance (Grigore & Toma, 2008). Also in a nonsecreting mangrove species *B. cylindrica*, the formation of abundant calcium oxalate crystals was found when these plants were subjected to high NaCl concentration (600 mM NaCl), and the presence of these crystals has revealed the NaCl tolerance potential of this potential (Sruthi & Puthur, 2019).

Plant roots facilitate the first line of damage or first line of defense due to the direct encounter with the saline condition (Rewald, Raveh, Gendler, Ephrath, & Rachmilevitch, 2012). Due to increased salinity, a decrease in the vessel diameter occur in the plants and thereby reduces the hydraulic conductivity and gives protection to the xylem vessels with substantial conductivity (Junghans et al., 2006). An increase in the thickness of root cortical cells under high saline condition was reported in the xerohalophyte shrubs *N. retusa* and *A. halimus* (Boughalleb et al., 2009). Salinity persuaded enhancement in root epidermal thickness thereby reduces salt ion diffusion in to the root. Also the process of root elongation and suberization of root cell walls helps to avoid an invasion of root cells by excess salts. Root elongation maintains the capacity for accumulating ions by providing new cells and suberization blocks the apoplasmic pathway (Stelzer, Kuo, & Koyro, 1988).

16.4.2 Physiological and biochemical peculiarities

16.4.2.1 Photosynthesis

In general, salinity induces a reduction in crop productivity indirectly by declining the photosynthetic yield as excess salt impairs the assimilation of CO_2, transpiration rate, and also stomatal conductance (Rabhi et al., 2012). However, in halophytes photosynthesis may be unaffected or in some cases may be enhanced under saline stress, especially under low concentrations of salt ions (Parida et al., 2004; Rabhi et al., 2012). For example, in *S. portulacastrum*, enhanced photosynthesis and stomatal conductance was recorded under increased NaCl concentration of about 600 mM (Venkatesalu & Chellappan, 1993). This indicates that halophytes are salinity resilient plants and was further evidenced by the increase in the photochemical reactions. For instance, in *Arthrocnemum macrostachyum*, *Salicornia europaea*, and *Atriplex portulacoides* the maximum quantum efficiency of PSII (Fv/Fm) seems to be enhanced under salinity (Benzarti, Ben Rejeb, Debez, Messedi, & Abdelly, 2012; Fan et al., 2011; Redondo-Gómez, Mateos-Naranjo, Figueroa, & Davy, 2010).

Moreover, in a halophyte *Suadea physophora*, there was no adverse effects on the morphology of chloroplasts and the ultrastructure of chloroplasts seems to be intact even when exposed to a higher concentration of 700 mM NaCl (Zhang, Song, Wang, & Feng, 2010). The increased tolerance of halophyte chloroplasts toward the excess salt ions may be attributed to the preferential exclusion of salt from the cytoplasm and subsequent sequestration in the vacuole there by lowering the toxicity of excess salt ions in the cytoplasm (Shabala, 2013). Chloroplast being the site of photosynthesis, an understanding of the molecular basis of the chloroplast genome may pave a new way for imparting salinity tolerance in crops. To date, more than 50 salt responsive genes have been characterized from the chloroplast genome, which include proteins involved in the organization of thylakoid membrane, PSII regulation, assimilation of CO_2, photorespiration, ABA metabolism and signaling, ion homeostasis, and scavenging of reactive oxygen species developed under excess salt ions (Rabhi et al., 2012).

Halophytes have the capacity to switch the CO_2 concentrating mechanism from the C3 pathway to other pathways like C4, crassulacean acid metabolism (CAM) pathways, and also employ photorespiration under increased concentrations of salt inside the mesophyll cells. Photorespiration is reported to be a positive regulation which acts as an energy sink during the occurrence of photoinhibition induced by salinity; but on the other hand photorespiration may impair the growth processes (Wingler, Lea, Quick, & Leegood, 2000). For example, halophytes like *Salvadora persica*, *A. macrostachyum*, and *Sarcocornia fruticosa* are reported to have a higher rate of photorespiration and slightly reduced growth rate when subjected to high NaCl concentration of 500 mM (Rangani, Parida, Panda, & Kumari, 2016; Redondo-Gómez et al., 2006, 2010). On the other hand, plants that fix CO_2 through the C4 and CAM pathways are reported to be naturally better adapted to salt stress when compared to C3 plants (Stepien & Klobus, 2005; Vasquez, Glenn, Guntenspergen, Brown, & Nelson, 2006). Halophytes like *P. nuttalliana, Aeluropus littoralis*, and *S. europaea* are able to switch from C3 to C4 pathways and similarly the halophytes *Mesembryanthemum crystallinum* and *Portulacaria afra* switch from C3 to CAM pathways under excess salt concentrations (Bose et al., 2017). Surprisingly, both C3 and C4 pathways were found to operate in the same cell in some halophytes like *Suaeda aralocaspica*, *Bienertia cycloptera*, and *Bienertia sinuspersici*, which is enabled by the sorting of chloroplasts that resemble mesophyll and bundle sheath cells (Chuong, Franceschi, & Edwards, 2006; von Caemmerer, Edwards, Koteyeva, & Cousins, 2014). In addition to this, about 30%–45% of halophytes are C4 plants, which indicates that salinity favors the C4 pathway as a CO_2 concentrating mechanism rather than the C3 pathway (Bromham & Bennett, 2014; Feldman, Bisaro, Biani, & Prado, 2008; Santos, Al-Azzawi, Aronson, & Flowers, 2016).

16.4.2.2 Oxidative stress and its management

Salinity stress may induce oxidative damage to cells by impairing the proteins, nucleic acid, and lipid metabolism and in response to these deleterious effects plant cell have evolved efficient antioxidant system. Halophytes in general are exposed to a multitude of stress factors naturally and hence have developed extensive antioxidant mechanisms so as to thrive in the harsh environmental conditions. For example, recently, a halophyte *Cakile maritima* was studied for salinity tolerance and it was found that the increased activities of antioxidant enzymes like SOD, POD, APX, and the accumulation of glutathione impart tolerance in the plants under salinity stress (Ben Amor, Jiménez, Boudabbous, Sevilla, & Abdelly, 2020).

To date several salt responsive genes have been cloned that impart stress tolerance in halophytes toward an excess amount of salt ions by efficiently scavenging the ROS produced inside the cell. The MDAR (monodehydroascorbate reductase) gene from the mangrove *A. marina* was isolated and incorporated in tobacco plants to help the transgenic lines to thrive better under saline stress through the reduction of monodehydroascorbate to ascorbate which ensures the proper functioning of the enzyme APX (Kavitha, George, Venkataraman, & Parida, 2010). Similarly, transgenic tobacco plants were developed by incorporating SbpAPX gene (located in the peroxisomal membrane) from the halophyte *Salicornia brachiata* and were found to be upregulated during salinity stress aiding tolerance to the plant. This enhanced tolerance was attributed to the potassium efflux as well as the scavenging of hydroxyl radicals through the activity of APX enzyme (Tiwari, Chaturvedi, Mishra, & Jha, 2014). *Arabidopsis thaliana* plants were imparted with NaCl tolerance up to a level of 150 mM through overexpression of the gene ThMSD from the halophyte *Thellungiella halophila* (Xu et al., 2014).

16.4.2.3 Osmotic adjustment

Accumulation of osmolytes under salt stress is a common defensive strategy in halophytes as well as glycophytes so as to cope with the adverse environmental condition. However, being saline tolerant by nature halophytes possess an array of osmolytes (low-molecular-weight compounds) like proline, glycine betaine, sorbitol, mannitol, myoinositol, glycerol, sucrose, glucose, polyphenols, etc. in increased amounts, which helps the cells to operate the metabolic pathways normally (Lokhande & Suprasanna, 2012; Patel, Mishra, & Jha, 2016). The increased concentration of Na^+ and Cl^- ions inside the cytosol may hamper the metabolic process in halophytes and was effectively overcome by the sequestration of ions into the vacuoles facilitated by specific transporter proteins in the tonoplast membrane. For the normal metabolic process only optimal levels of Na^+ and Cl^- (10–30 mM) are maintained in the cytoplasm (Kronzucker & Britto, 2011; Munns, Passioura, Colmer, & Byrt, 2020; Sreeshan, Meera, & Augustine, 2014). Osmolytes play a crucial role in halophytes by regulating the level of ROS and cellular redox status through activating the osmotic response network and thereby aiding tolerance to salinity. Moreover, they act as sources of carbon, nitrogen, and sulfur, which are essential for the metabolic pathways of the cell and also help to stabilize the enzymes/proteins and provide immunity to plants (Surówka & Hura, 2020).

Several genes from halophytes encoding osmolytes were characterized, isolated, and incorporated in the glycophytes so as to impart salinity tolerance. For example, the most prominent osmolyte proline was accumulated in a higher amount in the glycophyte *A. thaliana* through the integration of the proline transporter gene AhProT1 that was isolated from the halophyte *Atriplex hortensis*. The overexpression of AhProT1 gene in transgenic *A. thaliana* conferred salinity tolerance up to a level of 200 mM NaCl and also resulted in an increased biomass (Yi-Guo et al., 2002). Transgenic *A. thaliana* was also produced by incorporating the SalINO1 gene from the halophyte *Spartina alterniflora* which encodes for the osmolyte inositol. The resultant transgenic lines of *A. thaliana* exhibited tolerance to a level of 150 mM NaCl as reported by Joshi, Ramanarao, and Baisakh (2013). Likewise, the transgenic tobacco plants incorporated with DREB2A gene from the halophyte *S. brachiata* imparted salinity tolerance by reducing the level of Na^+ and Cl^- ions in the foliar tissues of tobacco plants (Gupta, Jha, & Agarwal, 2014). The precursor of glycine betaine, namely choline, was found to have role in lowering the functions of ion channels which help to minimize the leakage of sodium ions from the vacuole, thereby aiding the salinity tolerance in plants (Pottosin, Bonales-Alatorre, & Shabala, 2014). Thus overexpression of osmolyte-encoding genes from the halophytes into the crop plants may confer enhanced salt resistance, but may need extensive studies in the transcriptomic and metabolomic analysis to identify the most promising set of genes for the transgenic approach.

16.4.3 Uniqueness of halophytes

Halophytes deploy different strategies like avoidance, resistance, or tolerance to cope up with excess salt (Nikalje, Bhaskar, Yadav, & Penna, 2019). The salt tolerance mechanism may involve certain physiological and/or biochemical alterations in the plants that enable the plant to maintain capability of the protoplasm to accumulate the salt ions inside the cells. Salt tolerance in halophytes can be accomplished either by salt exclusion or salt inclusion and salt-tolerant plants utilize the energy for the exclusion of excess salt in order to protect themselves from toxic effects of excess salt (Ashraf, Akhtar, Hussain, & Iqbal, 2006). Salt avoidance may involve some structural and physiological adaptations so as to exclude salt through the root membrane or to minimize the salt concentrations in the cell. To cope with hypersaline condition, mangroves adopt various strategies, namely, salt exclusion, salt accumulation, and salt secretion.

16.4.3.1 Salt excluders

Some halophytes avoid salt by restricting or avoiding salt uptake and the subsequent adverse effects (Nikalje et al., 2019; Saddhe, Manuka, Nikalje, & Penna, 2020). The salt excluders can accumulate a larger amount of salt in the vacuoles of roots and xylem than in the shoot (Matsushita & Matoh, 1991). In these halophytes, most of the Na^+ and Cl^- are excluded into the soil solution or by accumulating salt ions in the roots and thus maintaining low concentrations of salt in the shoots, making the shoots free from salt damage, which is a characteristic feature of salt tolerance (Munns, 2005). The accumulation of excess Na^+ by halophytes occurs in the pericycle, xylem parenchyma cells, the root cortex (especially the casparian strip), and the phloem cells (Munns, James, & Läuchli, 2006). Salt-exclusion halophytes survive in saline habitats by controlling the uptake and the distribution of Na^+, for example, Euhalophytes such as *Suaeda sp., Aeluropus repens, A. littoralis, Poa littoralis, Dactylis littoralis, Kochia scoparia*. Moreover, the salt excluding mangrove species (e.g., species of *Rhizophora, Ceriops, Bruguiera*, etc.) have the adaptation for eliminating excess salts by ultrafiltration in root cell membranes (Aziz & Khan, 2001; Khan & Aziz, 2001; Sruthi, 2019; Takemura et al., 2000; Wang, Zheng, Ren, & Zhang, 2002; Zheng, Wang, & Lin, 1999).

16.4.3.2 Salt secretors

The salt resistant strategy of halophytes is also accomplished by the accumulation and secretion of excess salts through salt glands or bladders (Nikalje et al., 2019). *Spartina anglica, Aeluropus sinensis*, and *Distichlis spicata* are the halophytic species with high salt secretion ability (Semenova, Fomina, & Biel, 2010). Salt-secreting mangroves (*Acanthus* spp., *A. marina, A. officinalis, A. alba, A. corniculatum*, and *Aegialitis* spp.) regulate the internal level of salt by expelling the excess salt through glands present on the leaves and thus attain normal growth and development in saline soil environments (Selvam, 2003). Ye, Tam, Lu, and Wong (2005) reported that, in three salt-secreting mangrove species, that is, *Acanthus ilicifolius*, *A. corniculatum*, and *A. marina*, salt secretion increases with an increase in salinity of the growing environment. Na^+ and Cl^- may be unidirectionally transported into salt glands through plasmodesmata, while the cuticle around the salt gland prevents ion leakage, and thus plasmodesmata may play a crucial role in salt secretion from salt glands (Yuan & Wang, 2020). In addition, various ion transporters located in the membrane are also expected to participate in salt secretion. Salt secretion is considered to be an active process: the first ion transporter involved in energy generation is H^+-ATPase, the activity of which increases in *C. maritima* (Debez et al., 2006) and *A. marina* under increasing salinity (Chen et al., 2010). High-throughput sequencing of recretohalophytes under salt treatment also identified several ion transporters that are highly expressed under salt treatment, including Na^+/H^+ antiporters (SOS1 and NHX1), HKT1, and NSCC (Dang et al., 2013; Sanadhya, Agarwal, Khedia, & Agarwal, 2015; Yamamoto et al., 2015; Yuan, Guo et al., 2019, Yuan, Liang et al., 2019).

16.4.3.3 Salt accumulators

Salt accumulators accumulate high level of salts in the cells and tissues, while at the same time in order to avoid toxicity due to these salts by sequestering toxic ions in the vacuoles of leaves, they translocate the ions out of leaves by cuticular transpiration and also shed the leaves (species of *Lumnitzera* and *Excoecaria*) (Aziz & Khan, 2001; Tomlinson, 1986). The salt-accumulating halophytes compartmentalize ions in the vacuoles and accumulate the compatible solutes in the cytoplasm, and thus confer salt resistance (Gorham, 1995). *K. folium, S. salsa, S. maritima*, and *Suaeda portulacastrum* are significant examples of Na^+ and Cl^- hyperaccumulating halophytes (Ravindran et al., 2007; Zhao et al., 2005). Vacuolar sequestration of Na^+ is an important and cost-effective strategy for osmotic adjustment that also reduces the Na^+ concentration in the cytosol (Parida & Jha, 2010a, 2010b). Salt accumulation in halophytes occurs with the sequestration of Na^+ and Cl^- into the vacuoles of the hypodermal storage tissue of the leaves (Aziz & Khan, 2001; Kura-Hotta, Mimura, Tsujimura, Washitani-Nemoto, & Mimura, 2001; Mimura et al., 2003). Researchers suggest that if the harvest of the aerial plant portion of hypersalt accumulators is not added back to the same soil, it can significantly contribute to the removal of salts and amelioration processes (Owens, 2001). The mechanisms underlined in halophytes for salinity tolerance are depicted in Fig. 16.2.

16.5 Transgenic approaches for salinity tolerance

Halophytes adopt different strategies to cope with salinity and other stress factors prevailing in the wetland ecosystem. Hence, they can be used as

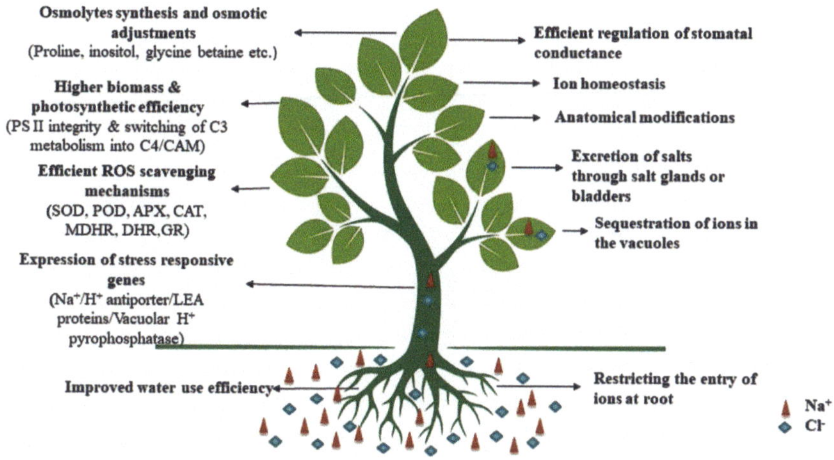

FIGURE 16.2 Inherent mechanisms underlined in halophytes for salt tolerance.

model plants for enlightening and elucidating the molecular mechanisms behind the abiotic stresses tolerance potential in plants, especially toward salinity stress. In addition, the characterization and isolation of novel salt responsive genes from halophytes may pave the way to the development of transgenics as it may help to produce crop plants that are better adapted to harsh environmental conditions like high salinity, flood, and low oxygen condition. To date, a number of halophytes (*Mesembryanthemum, Thellungiella, Salicornia, Aeluropus, Cakile, Atriplex, Suaeda*, etc.) have proven to be potential sources for salt responsive genes and promoters, aiding in the development of salt-resistant plants (Ben Amor et al., 2020; Gupta et al., 2014; Xu et al., 2014). Important genes isolated from halophytes to impart salinity tolerance in glycophytes include antioxidant encoding genes like APX, CAT, GST, SOD, antiporters like NHX, SOS, HKT, VTPase, and ion channels like Cl^-, Ca^{2+}, and aquaporins (Mishra & Tanna, 2017).

Even though transgenic attempts have been made to enhance the salt tolerance in crop plants, the genetic diversity of salinity tolerance in the glycophytes poses a major drawback. This was defeated to a great extent by halophytes which have the inherent ability to thrive in salinity stress. Hence, it is crucial to understand the mechanism of salt tolerance in halophyte, especially the molecular basis of ion transport, photosynthesis, and ROS scavenging mechanisms under increased concentration of NaCl, which will help in the breeding process to produce crop plants which are saline tolerant. Salinity tolerance imparting genes and their source halophytes for transgenic approaches are listed in Table 16.2.

TABLE 16.2 Salinity tolerance imparting genes isolated from halophytes for transgenic approaches.

Sl. no.	Halophyte	Gene	Specific protein	Glycophyte to which isolated gene is incorporated	Reference
1.	*Atriplex gmelini* C.A. Mey.	AgNHX1	Vacuolar-type Na^+/H^+ antiporter	*Oryza sativa* L.	Ohta et al. (2002)
2.	*Atriplex hortensis* L.	AhProT1	Proline transporter	*Arabidopsis thaliana* (L.) Heynh.	Yi-Guo et al. (2002)
3.	*Suaeda salsa* L.	SsNHX1	Vacuolar-type Na^+/H^+ antiporter	*O. sativa* L.	Zhao, Wang, Zhang, Zhao, and Zhang (2006)
4.	*Thellungiella halophila* (C.A. Mey) O.E. Schulz	TsVP	Vacuolar H^+ pyrophosphatase	*N. tabacum* L.	Gao et al. (2006)
5.	*S. salsa* L.	SsVP	Vacuolar H^+ pyrophosphatase	*A. thaliana* (L.) Heynh.	Guo et al. (2006)
6.	*Porteresia coarctata* T.	PcSrp	cDNA clone encoding serine-rich-protein	*Eleusine coracana* L.	Mahalakshmi, Christopher, Reddy, Rao, and Reddy (2006)
7.	*P. coarctata* (Roxb.)T.	PINO1orPcINO1	L-myo-inositol 1-phosphate synthase enzyme	*N. tabacum* L., *O. sativa* L. and *Brassica juncea* (L.)Czern.	Das-Chatterjee et al. (2006)
8.	*T. halophila* (C.A. Mey) O.E. Schulz	ThCYP1	Cyclophilin protein	*N. tabacum* L.	Chen et al. (2007)

9.	*Agropyron elongatum* (Host)P. Beauv.	AeNHX1	Vacuolar-type Na^+/H^+ antiporter	*A. thaliana* (L.) Heynh.	Qiao, Zhao, Li, Luo, and Zhang (2007)
10.	*T. halophila* (C.A. Mey) O.E. Schulz	TsVP	Vacuolar H^+ pyrophosphatase	*Gossypium hirsutum* L.	Lv et al. (2008)
11.	*Avicennia marina* (Forssk.) Vierh.	Sod1	Copper/zinc superoxide dismutase enzyme	*O. sativa* L. *var. Pusa Basmati-1*	Prashanth, Sadhasivam, and Parida (2008)
12.	*Chenopodium quinoa* Willd.	CqSOS1A and CqSOS1B	Plasma membrane Na^+/H^+ antiporter	*A. thaliana* (L.) Heynh.	Maughan et al. (2009)
13.	*T. halophila* (C.A. Mey) O.E. Schulz	ThNHX1	Vacuolar Na^+/H^+ antiporter	*A. thaliana* (L.) Heynh.	Wu, Gao, Kong, Zhao, and Zhang (2009)
14.	*Populus euphratica* Oliv.	PeNHX1–6	Vacuolar Na^+/H^+ antiporter	*Allium cepa* L.	Ye, Zhang, Chen, Xia, and Yin (2009)
15.	*Puccinellia tenuiflora* (Griesb.) Scribn. & Merr.	PutHKT2;1 PutAKT1	K^+ transporter	*A. thaliana* (L.) Heynh.	Ardie, Liu, and Takano (2010)
16.	*Tamarix androssowii* L.	TaMnSOD	Manganese superoxide dismutase enzyme	*Populus davidiana* × *P. bolleana*	Wang et al. (2010)
17.	*Aeluropus littoralis* (Gouan) Parl.	AlSAP	A20/AN1 zinc-finger protein	*N. tabacum* L.	Saad et al. (2010)
18.	*A. marina* (Forssk.) Vierh.	AmMDAR	Chloroplastic monodehydroascorbate reductase enzyme	*N. tabacum* L.	Kavitha et al. (2010)
19.	*Salicornia brachiata* Roxb.	SbGSTU	Tau class glutathione S-transferase enzyme	*N. tabacum* L.	Jha, Sharma, and Mishra (2011)

(Continued)

TABLE 16.2 (Continued)

Sl. no.	Halophyte	Gene	Specific protein	Glycophyte to which isolated gene is incorporated	Reference
20.	*P. tenuiflora* (Griesb.) Scribn. & Merr.	KPutB1 & KOB1	K^+ transporter	*A. thaliana* (L.) Heynh.	Ardie, Nishiuchi, Liu, and Takano (2011)
21.	*Spartina anglica* C.E. Hubb.	SaNHX1	Vacuolar Na^+/H^+ antiporter	*O. sativa* L.	Lan, Duan, Wang, Zhou, and Wu (2011)
22.	*Halostachys caspica* (Pall.) C.A. Mey. ex Schrenk	HcNHX1	Vacuolar Na^+/H^+ antiporter	*A. thaliana* (L.) Heynh.	Guan, Hu, Zeng, Wang, and Zhang (2011)
23.	*Salsola soda* L.	SsNHX1	Vacuolar Na^+/H^+ antiporter	*Medicago sativa* L.	Li et al. (2011)
24.	*Suaeda corniculata* Forssk. ex J.F. Gmel.	ScVP	Vacuolar H^+ pyrophosphatase	*A. thaliana* (L.) Heynh.	Liu et al. (2011)
25.	*A. littoralis* (Gouan) Parl.	Pr_{AISAP}	Zinc-finger protein	*N. tabacum* L.	Saad et al. (2011)
26.	*T. halophila* (C.A. Mey) O.E. Schulz	TsHKT1;2	K^+ transporter	*A. thaliana* (L.) Heynh.	Ali et al. (2012)
27.	*Kalidium foliatum* (Pall.) Moq.	KfVP1	Vacuolar H^+ pyrophosphatase	*A. thaliana* (L.) Heynh.	Yao et al. (2012)
28.	*P. tenuiflora* (Griesb.) Scribn. & Merr.	PtNHA1 or PtSOS1	Plasma membrane and vacuolar-type Na^+/H^+ antiporter	*A. thaliana* (L.) Heynh.	Kobayashi, Abe, Yoshida, Liu, and Takano (2012), Wang et al. (2011)

29.	*S. brachiata* Roxb.	SbSOS1	Plasma membrane Na^+/H^+ antiporter	*N. tabacum* L.	Yadav, Shukla, Jha, Agarwal, and Jha (2012)
30.	*P. tenuiflora* (Griesb.) Scribn. & Merr.	PtNHA1 and PutNHX	Plasma membrane and vacuolar Na^+/H^+ antiporter	*O. sativa* L.	Kobayashi et al. (2012)
31.	*Halostachys caspica* (Pall.) C.A. Mey. ex Schrenk	HcVP1 and HcVHA-B	Vacuolar H^+ pyrophosphatase and B subunit of H^+-ATPase	*A. thaliana* (L.) Heynh.	Hu, Zeng, Guan, and Zhang (2012)
32.	*Spartina alterniflora* Loisel.	SaVHAc1	Vacuolar H^+-ATPase subunit c1	*O. sativa* L.	Baisakh et al. (2012)
33.	*S. alterniflora* Loisel.	SaSce9	SUMO conjugating enzyme	*A. thaliana* (L.) Heynh.	Karan and Subudhi (2012)
34.	*Suaeda asparagoides* (Miq.) Makino	SaRBP1	cDNA clones encoding RNA binding proteins	*A. thaliana* (L.) Heynh.	Ayarpadikannan et al. (2012)
35.	*P. tenuiflora* (Griesb.) Scribn. & Merr.	PtNHA1 and PutNHX	Plasma membrane and vacuolar Na^+/H^+ antiporter	*O. sativa* L.	Kobayashi et al. (2012)
36.	*S. brachiata* Roxb.	SbASR-1	LEA protein	*N. tabacum* L.	Jha, Lal, Tiwari, Yadav, and Agarwal (2012)
37.	*T. salsuginea* (C.A. Mey) O.E. Schulz	TsLEA1	LEA protein	*A. thaliana* (L.) Heynh.	Zhang et al. (2012)
38.	*S. brachiata* Roxb.	SbNHX1	Vacuolar Na^+/H^+ antiporter	*N. tabacum* L. *and Jatropha curcas* L.	Jha, Mishra, Jha, and Joshi (2013)

(Continued)

TABLE 16.2 (Continued)

Sl. no.	Halophyte	Gene	Specific protein	Glycophyte to which isolated gene is incorporated	Reference
39.	*A. hortensis* L., *Suaeda maritime* (L.) Dumort. and *Suaeda liaotungensis* Kitag.	Badh	Betaine aldehyde dehydrogenase enzyme	*O. sativa* L. and *Solanum lycopersicum L. cv. Micro-Tom*	Wang, Tong, and Li (2013)
40.	*S. corniculata* Forssk. ex J.F. Gmel.	ScNHX1 and ScVP	Vacuolar H^+ pyrophosphatase	*M. sativa* L.	Liu et al. (2013)
41.	*S. brachiata* Roxb.	SbNHX1	Vacuolar Na^+/H^+ antiporter	*N. tabacum* L. *and J. curcas* L.	Jha et al. (2013)
42.	*S. alterniflora* Loisel.	SaINO1	Myo-inositol 1-phosphate synthase enzyme	*A. thaliana* (L.) Heynh.	Joshi et al. (2013)
43.	*S. brachiata* Roxb.	SbpAPX	Peroxisomal ascorbate peroxidase enzyme	*Arachis hypogaea* L. and *N. tabacum* L.	Singh, Mishra, and Jha (2014), Tiwari et al. (2014)
44.	*Leptochloa fusca* P. Beauv.	LfNHX1	Vacuolar Na^+/H^+ antiporter	*N. tabacum* L.	Rauf et al. (2014)
45.	*S. salsa* L.	SsHKT1;1	K^+ transporter	*A. thaliana* (L.) Heynh.	Shao, Han, Ding, Zhou, and Wang (2014)

46.	*A. littoralis* (Gouan) Parl.	AlNHX	Vacuolar Na^{+}/H^{+} antiporter	*N. tabacum* L. and *Glycine max* (L.) Merr.	Liu, Zhang, Dong, and Chu (2014)
47.	*Eutrema halophilum* (C. A. Mey.) Al-Shehbaz & Warwick	ThMSD	Manganese superoxide dismutase enzyme	*A. thaliana* (L.) Heynh.	Xu et al. (2014)
48.	*S. liaotungensis* Kitag.	SlNAC1 & SlNAC2	NAC transcription factor	*A. thaliana* (L.) Heynh.	Li, Hu, Yang, Yu, and Li (2014)
49.	*S. liaotungensis* Kitag.	SlPAPA1	Zinc-finger HIT protein	*A. thaliana* (L.) Heynh.	Li, Yang, Hu, Yu, and Li (2014)
50.	*S. alterniflora* Loisel.	SaARF	Adenosine diphosphate-ribosylation factor	*A. thaliana* (L.) Heynh.	Karan and Subudhi (2014)
51.	*A. hortensis* L., *Atriplex halimus* L. *and S. brachiata* Roxb.	AhDREB, AtDREB & SbDREB2A	Dehydration-responsive element binding transcription factor	*N. tabacum* L.	Gupta et al. (2014), Khedr et al. (2011), Shen et al. (2003)
52.	*Salicornia europaea* L.	SeNHX1	Vacuolar Na^{+}/H^{+} antiporter	*N. tabacum* L. and *M. sativa* L.	Chen, Bao, Guo, Jia, and Li (2015), Chen, Zong et al. (2015), Zhang et al. (2014), Zhou, Chen, Zhang, and Li (2008)
53.	*Zoysia matrella* (L.)Merr.	ZmVP1	Vacuolar H^{+} pyrophosphatase	*A. thaliana* (L.) Heynh.	Chen, Bao, et al. (2015), Chen, Zong et al. (2015)
54.	*S. salsa* L.	*SsGPAT*	Glycerol-3-phosphate acyltransferase	*A. thaliana* (L.) Heynh.	Sui, Tian, Wang, Wang, and Fan (2017)

(Continued)

TABLE 16.2 (Continued)

Sl. no.	Halophyte	Gene	Specific protein	Glycophyte to which isolated gene is incorporated	Reference
55.	*Aeluropus lagopoides*	AlNAC4	NAC transcription factor	*N. tabacum* L.	Khedia, Agarwal, and Agarwal (2018)
56.	*Kosteletzkya virginica* (L.) Presl.	*KvNHX1*	Vacuolar Na^+/H^+ antiporter	*N. tabacum* L.	Wang, Ding, and Wang (2018)
57.	*S. brachiata* Roxb.	*SbNHX1*	Vacuolar Na^+/H^+ antiporter	*N. tabacum* L.	Tiwari, Patel, Chaturvedi, Mishra, and Jha (2019)

16.6 Future prospects and conclusion

Salinization is becoming a global problem and is increasing day by day, negatively affecting the whole ecosystem and also threatening food security. In this scenario, phytodesalination has importance for the effective management of salinized land. Phytodesalination is an emerging ecofriendly sustainable solution for salinization which is focused on salt-loving plant (halophytes), and has been proven to be an effective remediation strategy for saline soil. Halophytes exhibit a vast array of physiological and biochemical mechanisms for tolerance in saline habitats. The expression of certain stress inducible genes also give more resistance to them for their successful survival under high salinity. The application of molecular biology with genomic and proteomic approaches for developing salt-tolerant crop plants and its use in saline agriculture lands could assure food security. Through these saline agriculture practices, it is possible to cope with the current salinization problem. The conversion of salinized agricultural land into suitable agricultural land by using saline-tolerant plants and desalination using proper candidates are the best strategies for the management of soil salinization.

The identification and molecular characterization of salinity hyperaccumulating halophyte species are essential for the successful transgenic approaches of crop plants to impart salinity tolerance. The future use of halophytes for the desalination technique largely depends on the characterization of salt tolerance genes and to check whether the genes are coping with the regulatory networks of glycophytes. Sequencing of the halophytic genome and elucidation of the function of stress-tolerant genes are crucial in engineering glycophytic crop plants that are salt resistant and with higher yield also. Recently, genome editing techniques like CRISPR/Cas9 have been applied for crop improvement program, and halophytes genome editing through this technique will definitely be a breakthrough for the ever-threatening salinity issues in the agricultural crops.

Acknowledgments

The authors greatly acknowledge the financial assistance provided by KSCSTE with funding through a research grant (KSCSTE/5179/2017-SRSLS).

References

Ali, Z., Park, H. C., Ali, A., Oh, D. H., Aman, R., Kropornicka, A., & Bressan, R. A. (2012). TsHKT1; 2, a HKT1 homolog from the extremophile *Arabidopsis* relative *Thellungiella salsuginea*, shows K^+ specificity in the presence of NaCl. *Plant Physiology*, *158*(3), 1463–1474. Available from http://doi.org/10.1104/pp.111.193110.

Ardie, S. W., Liu, S., & Takano, T. (2010). Expression of the AKT1-type K^+ channel gene from *Puccinellia tenuiflora*, PutAKT1, enhances salt tolerance in *Arabidopsis*. *Plant Cell Reports*, *29*(8), 865–874. Available from http://doi.org/10.1007/s00299-010-0872-2.

Ardie, S. W., Nishiuchi, S., Liu, S., & Takano, T. (2011). Ectopic expression of the K^+ channel β3 subunits from *Puccinellia tenuiflora* (KPutB1) and rice (KOB1) alters K^+ homeostasis of yeast and *Arabidopsis*. *Molecular Biotechnology*, *48*(1), 76–86. Available from http://doi.org/10.1007/s12033-010-9349-3.

Ashraf, M. Y., Akhtar, K., Hussain, F., & Iqbal, J. (2006). Screening of different accessions of three potential grass species from Cholistan desert for salt tolerance. *Pakistan Journal of Botany*, *38*(5), 1589–1597.

Askari, H., Edqvist, J., Hajheidari, M., Kafi, M., & Salekdeh, G. H. (2006). Effects of salinity levels on proteome of *Suaeda aegyptiaca* leaves. *Proteomics*, *6*(8), 2542–2554.

Atreya, A., Vartak, V., & Bhargava, S. (2009). Salt priming improves tolerance to dessication stress and to extreme salt stress in *Bruguiera cylindrica*. *International Journal of Integrative Biology*, *6*(2), 68–73.

Ayarpadikannan, S., Chung, E., Cho, C. W., So, H. A., Kim, S. O., Jeon, J. M., & Lee, J. H. (2012). Exploration for the salt stress tolerance genes from a salt-treated halophyte, *Suaeda asparagoides*. *Plant Cell Reports*, *31*(1), 35–48. Available from http://doi.org/10.1007/s00299-011-1137-4.

Aziz, I., & Khan, M. A. (2001). Experimental assessment of salinity tolerance of *Ceriops tagal* seedlings and saplings from the Indus delta, Pakistan. *Aquatic Botany*, *70*(3), 259–268.

Baisakh, N., RamanaRao, M. V., Rajasekaran, K., Subudhi, P., Janda, J., Galbraith, D., & Pereira, A. (2012). Enhanced salt stress tolerance of rice plants expressing a vacuolar H^+-ATPase subunit c1 (SaVHAc1) gene from the halophyte grass *Spartina alterniflora* Löisel. *Plant Biotechnology Journal*, *10*(4), 453–464. Available from http://doi.org/10.1111/j.1467-7652.2012.00678.x.

Balsamo, R. A., & Thomson, W. W. (1993). Ultrastructural features associated with secretion in the salt glands of *Frankenia grandifolia* (*Frankeniaceae*) and *Avicennia germinans* (*Avicenniaceae*). *American Journal of Botany*, *80*(11), 1276–1283.

Barkla, B. J., Garibay-Hernández, A., Melzer, M., Rupasinghe, T. W., & Roessner, U. (2018). Single cell-type analysis of cellular lipid remodelling in response to salinity in the epidermal bladder cells of the model halophyte *Mesembryanthemum crystallinum*. *Plant, Cell & Environment*, *41*(10), 2390–2403.

Ben Amor, N., Jiménez, A., Boudabbous, M., Sevilla, F., & Abdelly, C. (2020). Chloroplast implication in the tolerance to salinity of the halophyte *Cakile maritima*. *Russian Journal of Plant Physiology*, *67*, 507–514.

Benzarti, M., Ben Rejeb, K., Debez, A., Messedi, D., & Abdelly, C. (2012). Photosynthetic activity and leaf antioxidative responses of *Atriplex portulacoides* subjected to extreme salinity. *Acta Physiologiae Plantarum*, *34*(5), 1679–1688.

Bose, J., Munns, R., Shabala, S., Gilliham, M., Pogson, B., & Tyerman, S. D. (2017). Chloroplast function and ion regulation in plants growing on saline soils: Lessons from halophytes. *Journal of Experimental Botany*, *68*(12), 3129–3143.

Boughalleb, F., Denden, M., & Tiba, B. B. (2009). Anatomical changes induced by increasing NaCl salinity in three fodder shrubs, *Nitraria retusa, Atriplex halimus* and *Medicago arborea*. *Acta Physiologiae Plantarum*, *31*(5), 947–960.

Bromham, L., & Bennett, T. H. (2014). Salt tolerance evolves more frequently in C4 grass lineages. *Journal of Evolutionary Biology*, *27*(3), 653–659.

Chandio, A. S., Lee, T. S., & Mirjat, M. S. (2013). Simulation of horizontal and vertical drainage systems to combat waterlogging problems along the Rohri Canal in Khairpur District, Pakistan. *Journal of Irrigation and Drainage Engineering*, *139*(9), 710–717.

Chen, A. P., Wang, G. L., Qu, Z. L., Lu, C. X., Liu, N., Wang, F., & Xia, G. X. (2007). Ectopic expression of ThCYP1, a stress-responsive cyclophilin gene from *Thellungiella halophila*, confers salt tolerance in fission yeast and tobacco cells. *Plant Cell Reports*, *26*(2), 237–245. Available from http://doi.org/10.1007/s00299-006-0238-y.

Chen, J., Xiao, Q., Wang, C., Wang, W. H., Wu, F. H., He, B. Y., & Zheng, H. L. (2014). Nitric oxide alleviates oxidative stress caused by salt in leaves of a mangrove species, *Aegiceras corniculatum*. *Aquatic Botany*, *117*, 41–47.

Chen, J., Xiao, Q., Wu, F., Dong, X., He, J., Pei, Z., & Zheng, H. (2010). Nitric oxide enhances salt secretion and Na^+ sequestration in a mangrove plant, *Avicennia marina*, through increasing the expression of H^+-ATPase and Na^+/H^+ antiporter under high salinity. *Tree Physiology*, *30*(12), 1570–1585.

Chen, X., Bao, H., Guo, J., Jia, W., & Li, Y. (2015). Overexpression of SeNHX1 improves both salt tolerance and disease resistance in tobacco. *Plant Signaling & Behavior*, *10*(4), e993240. Available from http://doi.org/10.4161/15592324.2014.993240.

Chen, Y., Zong, J., Tan, Z., Li, L., Hu, B., Chen, C., & Liu, J. (2015). Systematic mining of salt-tolerant genes in halophyte-*Zoysia matrella* through cDNA expression library screening. *Plant Physiology and Biochemistry*, *89*, 44–52. Available from http://doi.org/10.1016/j.plaphy.2015.02.007.

Chuong, S. D., Franceschi, V. R., & Edwards, G. E. (2006). The cytoskeleton maintains organelle partitioning required for single-cell C4 photosynthesis in *Chenopodiaceae* species. *The Plant Cell*, *18*(9), 2207–2223.

Colmer, T. D., & Flowers, T. J. (2008). Flooding tolerance in halophytes. *New Phytologist*, *179*(4), 964–974.

Corwin, D. L., & Yemoto, K. (2017). Salinity: Electrical conductivity and total dissolved solids. *Methods of Soil Analysis*, *2*. (msaonline2017).

Cuevas, J., Daliakopoulos, I. N., del Moral, F., Hueso, J. J., & Tsanis, I. K. (2019). A review of soil-improving cropping systems for soil salinization. *Agronomy*, *9*(6), 295.

Dang, Z. H., Zheng, L. L., Wang, J., Gao, Z., Wu, S. B., Qi, Z., & Wang, Y. C. (2013). Transcriptomic profiling of the salt-stress response in the wild recretohalophyte *Reaumuria trigyna*. *BMC Genomics*, *14*(1), 29.

Das-Chatterjee, A., Goswami, L., Maitra, S., Dastidar, K. G., Ray, S., & Majumder, A. L. (2006). Introgression of a novel salt-tolerant L-myo-inositol 1-phosphate synthase from *Porteresia coarctata* (Roxb.) Tateoka (PcINO1) confers salt tolerance to evolutionary diverse organisms. *FEBS Letters*, *580*(16), 3980–3988. Available from http://doi.org/10.1016/j.febslet.2006.06.033.

Dasgupta, S., Hossain, M. M., Huq, M., & Wheeler, D. (2015). Climate change and soil salinity: The case of coastal Bangladesh. *Ambio*, *44*(8), 815–826.

Dassanayake, M., & Larkin, J. C. (2017). Making plants break a sweat: The structure, function, and evolution of plant salt glands. *Frontiers in Plant Science*, *8*, 406.

Debez, A., Saadaoui, D., Ramani, B., Ouerghi, Z., Koyro, H. W., Huchzermeyer, B., & Abdelly, C. (2006). Leaf H^+-ATPase activity and photosynthetic capacity of *Cakile maritima* under increasing salinity. *Environmental and Experimental Botany*, *57*(3), 285–295.

Devi, S., Nandwal, A. S., Angrish, R., Arya, S. S., Kumar, N., & Sharma, S. K. (2016). Phytoremediation potential of some halophytic species for soil salinity. *International Journal of Phytoremediation*, *18*(7), 693–696.

Drennan, P. M., Berjak, P., Lawton, J. R., & Pammenter, N. W. (1987). Ultrastructure of the salt glands of the mangrove, *Avicennia marina* (Forssk.) Vierh., as indicated by the use of selective membrane staining. *Planta*, *172*(2), 176–183.

Elnaggar, A., Mosa, K., El-Keblawy, A., Tammam, A., & El-Naggar, M. (2020). Physiological and biochemical insights of salt stress tolerance in the habitat-indifferent halophyte *Salsola drummondii* during the vegetative stage. *Botany*, *98*. (ja).

Fan, P., Feng, J., Jiang, P., Chen, X., Bao, H., Nie, L., & Li, Y. (2011). Coordination of carbon fixation and nitrogen metabolism in *Salicornia europaea* under salinity: Comparative proteomic analysis on chloroplast proteins. *Proteomics*, *11*(22), 4346–4367. Available from https://doi.org/10.1002/pmic.201100054.

Farzi, A., Borghei, S. M., & Vossoughi, M. (2017). The use of halophytic plants for salt phytoremediation in constructed wetlands. *International Journal of Phytoremediation*, *19*(7), 643–650.

Feldman, S. R., Bisaro, V., Biani, N. B., & Prado, D. E. (2008). Soil salinity determines the relative abundance of C3/C4 species in Argentinean grasslands. *Global Ecology and Biogeography*, *17*(6), 708–714.

Flowers, T. J., & Colmer, T. D. (2015). Plant salt tolerance: Adaptations in halophytes. *Annals of Botany*, *115*(3), 327–331.

Gao, F., Gao, Q., Duan, X., Yue, G., Yang, A., & Zhang, J. (2006). Cloning of an H^+-PPase gene from *Thellungiella halophila* and its heterologous expression to improve tobacco salt tolerance. *Journal of Experimental Botany*, *57*(12), 3259–3270. Available from http://doi.org/10.1093/jxb/erl090.

Gorham, J. (1995). Mechanism of salt tolerance of halophytes. Halophytes and biosaline agriculture (pp. 207–223).

Grigore, M. N., Ivanescu, L., & Toma, C. (2014). *Halophytes: An integrative anatomical study*. Springer.

Grigore, M. N., & Toma, C. (2008). Ecological anatomy investigations related to some halophyte species from Moldavia. *Romanian Journal of Biology-Plant Biology*, *53*(1), 27–31.

Grigore, M. N., & Toma, C. (2017). *Anatomical adaptations of halophytes. A review of classic literature and recent findings*. Cham: Springer.

Grigore, M. N., & Toma, C. O. N. S. T. A. N. T. I. N. (2006). Ecological anatomy elements related to *Asteraceae halophytes* species. *Studii Şil Comunicări, Complexul Muzeal de Ştiinţele Naturii "Ion Borcea" Bacău*, *21*, 94–98.

Guan, B., Hu, Y., Zeng, Y., Wang, Y., & Zhang, F. (2011). Molecular characterization and functional analysis of a vacuolar Na^+/H^+ antiporter gene (HcNHX1) from *Halostachys caspica*. *Molecular Biology Reports*, *38*(3), 1889–1899. Available from http://doi.org/10.1007/s11033-010-0307-8.

Guo, S., Yin, H., Zhang, X., Zhao, F., Li, P., Chen, S., & Zhang, H. (2006). Molecular cloning and characterization of a vacuolar H^+-pyrophos-phatase gene, SsVP, from the halophyte *Suaeda salsa* and its overexpression increases salt and drought tolerance of *Arabidopsis*. *Plant Molecular Biology*, *60*(1), 41–50. Available from http://doi.org/10.1007/s11103-005-2417-6.

Gupta, K., Jha, B., & Agarwal, P. K. (2014). A dehydration-responsive element binding (DREB) transcription factor from the succulent halophyte *Salicornia brachiata* enhances abiotic stress tolerance in transgenic tobacco. *Marine Biotechnology*, *16*(6), 657–673. Available from http://doi.org/10.1007/s10126-014-9582-z.

Han, Z. J., Sun, Y., Zhang, M., & Zhai, J. T. (2020). Transcriptomic profile analysis of the halophyte *Suaeda rigida* response and tolerance under NaCl stress. *Scientific Reports*, *10*(1), 1–10.

Hasanuzzaman, M., Nahar, K., Alam, M., Bhowmik, P. C., Hossain, M., Rahman, M. M., & Fujita, M. (2014). Potential use of halophytes to remediate saline soils. *BioMed Research International*, *2014*.

Hu, Y. Z., Zeng, Y. L., Guan, B., & Zhang, F. C. (2012). Overexpression of a vacuolar H^+-pyrophosphatase and a B subunit of H^+-ATPase cloned from the halophyte *Halostachys caspica* improves salt tolerance in *Arabidopsis thaliana*. *Plant Cell, Tissue and Organ Culture (PCTOC)*, *108*(1), 63–71. Available from http://doi.org/10.1007/s11240-011-0013-9.

Ivushkin, K., Bartholomeus, H., Bregt, A. K., Pulatov, A., Kempen, B., & De Sousa, L. (2019). Global mapping of soil salinity change. *Remote Sensing of Environment*, *231*, 111260.

Jesus, J. M., Danko, A. S., Fiúza, A., & Borges, M. T. (2015). Phytoremediation of salt-affected soils: A review of processes, applicability, and the impact of climate change. *Environmental Science and Pollution Research*, *22*(9), 6511–6525.

Jha, B., Lal, S., Tiwari, V., Yadav, S. K., & Agarwal, P. K. (2012). The SbASR-1 gene cloned from an extreme halophyte *Salicornia brachiata* enhances salt tolerance in transgenic tobacco. *Marine Biotechnology*, *14*(6), 782–792. Available from http://doi.org/10.1007/s10126-012-9442-7.

Jha, B., Mishra, A., Jha, A., & Joshi, M. (2013). Developing transgenic *Jatropha* using the SbNHX1 gene from an extreme halophyte for cultivation in saline wasteland. *PLoS One*, *8*(8), e71136. Available from http://doi.org/10.1371/journal.pone.0071136.

Jha, B., Sharma, A., & Mishra, A. (2011). Expression of SbGSTU (tau class glutathione S-transferase) gene isolated from *Salicornia brachiata* in tobacco for salt tolerance. *Molecular Biology Reports*, *38*(7), 4823–4832. Available from http://doi.org/10.1007/s11033-010-0625-x.

Jlassi, A., Zorrig, W., Khouni, A. E., Lakhdar, A., Smaoui, A., Abdelly, C., & Rabhi, M. (2013). Phytodesalination of a moderately-salt-affected soil by *Sulla carnosa*. *International Journal of Phytoremediation*, *15*(4), 398–404.

Joshi, R., Ramanarao, M. V., & Baisakh, N. (2013). *Arabidopsis* plants constitutively overexpressing a myo-inositol 1-phosphate synthase gene (SaINO1) from the halophyte smooth cordgrass exhibits enhanced level of tolerance to salt stress. *Plant Physiology and Biochemistry*, *65*, 61–66. Available from http://doi.org/10.1016/j.plaphy.2013.01.009.

Junghans, U., Polle, A., Düchting, P., Weiler, E., Kuhlman, B., Gruber, F., & Teichmann, T. (2006). Adaptation to high salinity in poplar involves changes in xylem anatomy and auxin physiology. *Plant, Cell & Environment*, *29*(8), 1519–1531.

Kalanaki, M., Ritzema, H., Bamshad, R., Jones, E., & Fazilatnia, M. (2020). Application of bio-desalinization for reclamation of salt-affected soil under composted cow manure and deficit irrigation with saline water. *Paddy and Water Environment*, 1–11.

Kamran, M., Parveen, A., Ahmar, S., Malik, Z., Hussain, S., Chattha, M. S., & Chen, J. T. (2020). An overview of hazardous impacts of soil salinity in crops, tolerance mechanisms, and amelioration through selenium supplementation. *International Journal of Molecular Sciences*, *21*(1), 148.

Karakaş, S., Cullu, M. A., & Dikilitaş, M. (2017). Comparison of two halophyte species (*Salsola soda* and *Portulaca oleracea*) for salt removal potential under different soil salinity conditions. *Turkish Journal of Agriculture and Forestry*, *41*(3), 183–190.

Karan, R., & Subudhi, P. K. (2012). A stress inducible SUMO conjugating enzyme gene (SaSce9) from a grass halophyte *Spartina alterniflora* enhances salinity and drought stress tolerance in *Arabidopsis*. *BMC Plant Biology*, *12*(1), 187. Available from http://doi.org/10.1186/1471-2229-12-187.

Karan, R., & Subudhi, P. K. (2014). Overexpression of an adenosine diphosphate-ribosylation factor gene from the halophytic grass *Spartina alterniflora* confers salinity and drought tolerance in transgenic *Arabidopsis*. *Plant Cell Reports*, *33*(2), 373–384. Available from http://doi.org/10.1007/s00299-013-1537-8.

Katschnig, D., Broekman, R., & Rozema, J. (2013). Salt tolerance in the halophyte *Salicornia dolichostachya* Moss: Growth, morphology and physiology. *Environmental and Experimental Botany*, *92*, 32–42.

Kavitha, K., George, S., Venkataraman, G., & Parida, A. (2010). A salt-inducible chloroplastic monodehydroascorbate reductase from halophyte *Avicennia marina* confers salt stress tolerance on transgenic plants. *Biochimie*, *92*(10), 1321–1329. Available from http://doi.org/10.1016/j.biochi.2010.06.009.

Khan, M. A., & Aziz, I. (2001). Salinity tolerance in some mangrove species from Pakistan. *Wetlands Ecology and Management*, *9*(3), 229–233.

Khedia, J., Agarwal, P., & Agarwal, P. K. (2018). AlNAC4 transcription factor from halophyte *Aeluropus lagopoides* mitigates oxidative stress by maintaining ROS homeostasis in transgenic tobacco. *Frontiers in Plant Science*, *9*, 1522. Available from https://doi.org/10.3389/fpls.2018.01522.

Khedr, A. H. A., Serag, M. S., Nemat-Alla, M. M., Abo-Elnaga, A. Z., Nada, R. M., Quick, W. P., & Abogadallah, G. M. (2011). A DREB gene from the xero-halophyte *Atriplex halimus* is induced by osmotic but not ionic stress and shows distinct differences from glycophytic homologues. *Plant Cell, Tissue and Organ Culture (PCTOC)*, *106*(2), 191–206. Available from http://doi.org/10.1007/s11240-010-9906-2.

Kiani-Pouya, A., Rasouli, F., Bazihizina, N., Zhang, H., Hedrich, R., & Shabala, S. (2019). A large-scale screening of *quinoa* accessions reveals an important role of epidermal bladder cells and stomatal patterning in salinity tolerance. *Environmental and Experimental Botany*, *168*, 103885.

Kiani-Pouya, A., Roessner, U., Jayasinghe, N. S., Lutz, A., Rupasinghe, T., Bazihizina, N., & Shabala, S. (2017). Epidermal bladder cells confer salinity stress tolerance in the halophyte *quinoa* and *Atriplex* species. *Plant, Cell & Environment*, *40*(9), 1900–1915.

Kobayashi, S., Abe, N., Yoshida, K. T., Liu, S., & Takano, T. (2012). Molecular cloning and characterization of plasma membrane-and vacuolar-type Na^+/H^+ antiporters of an alkaline-salt-tolerant monocot, *Puccinellia tenuiflora*. *Journal of Plant Research*, *125*(4), 587–594. Available from http://doi.org/10.1007/s10265-012-0475-9.

Kronzucker, H. J., & Britto, D. T. (2011). Sodium transport in plants: A critical review. *New Phytologist*, *189*(1), 54–81. Available from https://doi.org/10.1111/j.1469-8137.2010.03540.x.

Kura-Hotta, M., Mimura, M., Tsujimura, T., Washitani-Nemoto, S., & Mimura, T. (2001). High salt-treatment-induced Na^+ extrusion and low salt-treatment-induced Na^+ accumulation in suspension-cultured cells of the mangrove plant, *Bruguiera sexangula*. *Plant, Cell & Environment*, *24*(10), 1105–1112.

Lan, T., Duan, Y., Wang, B., Zhou, Y., & Wu, W. (2011). Molecular cloning and functional characterization of a Na^+/H^+ antiporter gene from halophyte *Spartina anglica*. *Turkish Journal of Agriculture and Forestry*, *35*(5), 535–543. Available from http://doi.org/10.3906/tar-1003-2.

Lastiri-Hernández, M. A., Alvarez-Bernal, D., Bermúdez-Torres, K., Cárdenas, G. C., & Ceja-Torres, L. F. (2019). Phytodesalination of a moderately saline soil combined with two inorganic amendments. *Bragantia*, *78*(4), 579–586.

Lastiri-Hernández, M. A., Cruz-Cárdenas, G., Álvarez-Bernal, D., Vázquez-Sánchez, M., & Bermúdez-Torres, K. (2020). Ecological niche modeling for halophyte species with possible anthropogenic use in agricultural saline soils. *Environmental Modeling & Assessment*, 1–12.

Li, W., Wang, D., Jin, T., Chang, Q., Yin, D., Xu, S., & Liu, L. (2011). The vacuolar Na^+/H^+ antiporter gene SsNHX1 from the halophyte *Salsola soda* confers salt tolerance in transgenic alfalfa (*Medicago sativa* L.). *Plant Molecular Biology Reporter*, *29*(2), 278–290. Available from http://doi.org/10.1007/s11105-010-0224-y.

Li, X. L., Hu, Y. X., Yang, X., Yu, X. D., & Li, Q. L. (2014). A novel zinc-finger HIT protein with an additional PAPA-1-like region from *Suaeda liaotungensis* K. enhanced transgenic *Arabidopsis* drought and salt stresses tolerance. *Molecular Biotechnology*, *56*(12), 1089–1099. Available from http://doi.org/10.1007/s12033-014-9789-2.

Li, X. L., Yang, X., Hu, Y. X., Yu, X. D., & Li, Q. L. (2014). A novel NAC transcription factor from *Suaeda liaotungensis* K. enhanced transgenic *Arabidopsis* drought, salt, and cold stress tolerance. *Plant Cell Reports*, *33*(5), 767–778. Available from http://doi.org/10.1007/s00299-014-1602-y.

Liphschitz, N., & Waisel, Y. (1974). Existence of salt glands in various genera of the Gramineae. *New Phytologist*, *73*(3), 507–513.

Liu, J., Zhang, S., Dong, L., & Chu, J. (2014).). Incorporation of Na^+/H^+ antiporter gene from *Aeluropus littoralis* confers salt tolerance in soybean (*Glycine max* L.). *Indian Journal of Biochemistry and Biophysics*, *51*, 58–65, 24791418.

Liu, L., Fan, X. D., Wang, F. W., Wang, N., Dong, Y. Y., Liu, X. M., & Li, H. Y. (2013). Coexpression of ScNHX1 and ScVP in transgenic hybrids improves salt and saline-alkali tolerance in alfalfa (*Medicago sativa* L.). *Journal of Plant Growth Regulation*, *32*(1), 1–8. Available from http://doi.org/10.1007/s00344-012-9270-z.

Liu, L., Wang, Y., Wang, N., Dong, Y. Y., Fan, X. D., Liu, X. M., & Li, H. Y. (2011). Cloning of a vacuolar H^+-pyrophosphatase gene from the halophyte *Suaeda corniculata* whose heterologous overexpression improves salt, saline-alkali and drought tolerance in *Arabidopsis*. *Journal of Integrative Plant Biology*, *53*(9), 731–742. Available from http://doi.org/10.1111/j.1744-7909.2011.01066.x.

Lokhande, V. H., & Suprasanna, P. (2012). *Prospects of halophytes in understanding and managing abiotic stress tolerance. Environmental adaptations and stress tolerance of plants in the era of climate change* (pp. 29–56). New York, NY: Springer.

Lu, C., Qiu, N., Lu, Q., Wang, B., & Kuang, T. (2002). Does salt stress lead to increased susceptibility of photosystem II to photoinhibition and changes in photosynthetic pigment composition in halophyte *Suaeda salsa* grown outdoors? *Plant Science*, *163*(5), 1063–1068.

Lv, S., Zhang, K., Gao, Q., Lian, L., Song, Y., & Zhang, J. (2008). Overexpression of an H + -PPase gene from *Thellungiella halophila* in cotton enhances salt tolerance and improves growth and photosynthetic performance. *Plant and Cell Physiology*, *49*(8), 1150–1164. Available from http://doi.org/10.1093/pcp/pcn090.

Mahalakshmi, S., Christopher, G. S. B., Reddy, T. P., Rao, K. V., & Reddy, V. D. (2006). Isolation of a cDNA clone (PcSrp) encoding serine-rich-protein from *Porteresia coarctata* T. and its expression in yeast and finger millet (*Eleusine coracana* L.) affording salt tolerance. *Planta*, *224*(2), 347. Available from http://doi.org/10.1007/s00425-005-0218-4.

Manaa, A., Goussi, R., Derbali, W., Cantamessa, S., Abdelly, C., & Barbato, R. (2019). Salinity tolerance of *quinoa* (*Chenopodium quinoa* Willd) as assessed by chloroplast ultrastructure and photosynthetic performance. *Environmental and Experimental Botany*, *162*, 103–114.

Matsushita, N., & Matoh, T. (1991). Characterization of Na^+ exclusion mechanisms of salt-tolerant reed plants in comparison with salt-sensitive rice plants. *Physiologia Plantarum*, *83*(1), 170–176.

Maughan, P. J., Turner, T. B., Coleman, C. E., Elzinga, D. B., Jellen, E. N., Morales, J. A., & Bonifacio, A. (2009). Characterization of salt overly sensitive 1 (SOS1) gene homoeologs in *quinoa* (*Chenopodium quinoa* Willd.). *Genome/National Research Council Canada = Genome/Conseil National de Recherches Canada*, *52*(7), 647–657. Available from http://doi.org/10.1139/G09-041.

Menason, E., Betty, T., Vijayan, K. K., & Anbudurai, P. R. (2015). Modification of fatty acid composition in salt adopted Synechocystis 6803 cells. *Annals of Biological Research*, *6*, 4–9.

Mimura, T., Kura-Hotta, M., Tsujimura, T., Ohnishi, M., Miura, M., Okazaki, Y., & Washitani-Nemoto, S. (2003). Rapid increase of vacuolar volume in response to salt stress. *Planta, 216*(3), 397–402.

Miranda, M. F. A., Freire, M. B. G. D. S., de Almeida, B. G., Freire, F. J., Pessoa, L. G. M., & Freire, A. G. (2020). Phytodesalination and chemical and organic conditioners to recover the chemical properties of saline-sodic soil. *Soil Science Society of America Journal*. Available from http://doi.org/10.1002/saj2.20173.

Mishra, A., & Tanna, B. (2017). Halophytes: Potential resources for salt stress tolerance genes and promoters. *Frontiers in Plant Science, 8*, 829.

Mitsch, W. J., & Gosselink, J. G. (2000). The value of wetlands: Importance of scale and landscape setting. *Ecological Economics, 35*(1), 25–33.

Muchate, N. S., Nikalje, G. C., Rajurkar, N. S., Suprasanna, P., & Nikam, T. D. (2016). Physiological responses of the halophyte *Sesuvium portulacastrum* to salt stress and their relevance for saline soil bio-reclamation. *Flora, 224*, 96–105.

Munns, R. (2002). Comparative physiology of salt and water stress. *Plant Cell and Environment, 25*, 239–250. Available from http://doi.org/10.1046/j.0016-8025.2001.00808.x.

Munns, R. (2005). Genes and salt tolerance: Bringing them together. *New Phytologist, 167*(3), 645–663.

Munns, R., James, R. A., & Läuchli, A. (2006). Approaches to increasing the salt tolerance of wheat and other cereals. *Journal of Experimental Botany, 57*(5), 1025–1043.

Munns, R., Passioura, J. B., Colmer, T. D., & Byrt, C. S. (2020). Osmotic adjustment and energy limitations to plant growth in saline soil. *New Phytologist, 225*(3), 1091–1096. Available from https://doi.org/10.1111/nph.15862.

Munns, R., & Tester, M. (2008). Mechanisms of salinity tolerance. *Annual Review of Plant Biology, 59*, 651–681.

Nawaz, A., Shahbaz, M., Imran, A., Marghoob, U., Imtiaz, M., Mubeen, F., & Khan, A. U. (2020). Potential of salt tolerant PGPR in growth and yield augmentation of wheat (*Triticum aestivum* L.) under saline conditions. *Frontiers in Microbiology, 11*, 2019.

Nikalje, G. C., Bhaskar, S. D., Yadav, K., & Penna, S. (2019). *Halophytes: Prospective plants for future. Ecophysiology, abiotic stress responses and utilization of halophytes* (pp. 221–234). Singapore: Springer.

Niu, S. Q., Li, H. R., Paré, P. W., Aziz, M., Wang, S. M., Shi, H., & Guo, Q. (2016). Induced growth promotion and higher salt tolerance in the halophyte grass *Puccinellia tenuiflora* by beneficial rhizobacteria. *Plant and Soil, 407*(1-2), 217–230.

Ohta, M., Hayashi, Y., Nakashima, A., Hamada, A., Tanaka, A., Nakamura, T., & Hayakawa, T. (2002). Introduction of a Na^+/H^+ antiporter gene from *Atriplex gmelini* confers salt tolerance to rice. *FEBS Letters, 532*(3), 279–282. Available from http://doi.org/10.1016/S0014-5793(02)03679-7.

Okur, B., & Örçen, N. (2020). *Soil salinization and climate change. Climate change and soil interactions* (pp. 331–350). Elsevier.

Owens, S. (2001). Salt of the earth. *EMBO Reports, 2*(10), 877–879.

Palliyath, S., & Puthur, J. T. (2018). The modulation of various physiochemical changes in *Bruguiera cylindrica* (L.) Blume affected by high concentrations of NaCl. *Acta Physiologiae Plantarum, 40*(9), 160.

Panda, A., Rangani, J., & Parida, A. K. (2019). Cross talk between ROS homeostasis and antioxidative machinery contributes to salt tolerance of the xero-halophyte *Haloxylon salicornicum. Environmental and Experimental Botany, 166*, 103799.

Parida, A. K., Das, A. B., & Mittra, B. (2004). Effects of salt on growth, ion accumulation, photosynthesis and leaf anatomy of the mangrove, *Bruguiera parviflora*. *Trees*, *18*(2), 167–174.

Parida, A. K., & Jha, B. (2010a). Antioxidative defense potential to salinity in the euhalophyte *Salicornia brachiata*. *Journal of Plant Growth Regulation*, *29*(2), 137–148.

Parida, A. K., & Jha, B. (2010b). Salt tolerance mechanisms in mangroves: A review. *Trees*, *24* (2), 199–217.

Patel, M. K., Mishra, A., & Jha, B. (2016). Untargeted metabolomics of halophytes. In S. Kim (Ed.), *Marine omics: Principles and applications* (pp. 309–325). Boca Raton, FL: CRC Press. Available from http://doi.org/10.1201/9781315372303-18.

Pérez Cuadra, V., Verolo, M., & Cambi, V. (2020). Morphological and anatomical traits of halophytes: Adaptive versus phylogenetic value. In Handbook of halophytes: From molecules to ecosystems towards biosaline agriculture, (pp. 1–20), Springer.

Pottosin, I., Bonales-Alatorre, E., & Shabala, S. (2014). Choline but not its derivative betaine blocks slow vacuolar channels in the halophyte *Chenopodium quinoa*: implications for salinity stress responses. *FEBS Letters*, *588*, 3918–3923.

Prashanth, S. R., Sadhasivam, V., & Parida, A. (2008). Over expression of cytosolic copper/zinc superoxide dismutase from a mangrove plant *Avicennia marina* in indica rice var Pusa Basmati-1 confers abiotic stress tolerance. *Transgenic Research*, *17*(2), 281–291. Available from http://doi.org/10.1007/s11248-007-9099-6.

Qiao, W. H., Zhao, X. Y., Li, W., Luo, Y., & Zhang, X. S. (2007). Overexpression of AeNHX1, a root-specific vacuolar Na^+/H^+ antiporter from *Agropyron elongatum*, confers salt tolerance to *Arabidopsis* and *Festuca* plants. *Plant Cell Reports*, *26*(9), 1663–1672. Available from http://doi.org/10.1007/s00299-007-0354-3.

Rabhi, M., Castagna, A., Remorini, D., Scattino, C., Smaoui, A., Ranieri, A., & Abdelly, C. (2012). Photosynthetic responses to salinity in two obligate halophytes: *Sesuvium portulacastrum* and *Tecticornia indica*. *South African Journal of Botany*, *79*, 39–47.

Rabhi, M., Hafsi, C., Lakhdar, A., Hajji, S., Barhoumi, Z., Hamrouni, M. H., & Smaoui, A. (2009). Evaluation of the capacity of three halophytes to desalinize their rhizosphere as grown on saline soils under nonleaching conditions. *African Journal of Ecology*, *47*(4), 463–468.

Rajaravindran, M., & Natarajan, S. (2012). Effects of salinity stress on growth and biochemical constituents of the halophyte *Sesuvium portulacastrum*. *International Journal of Biological Sciences*, *2*(1), 18–25.

Rangani, J., Parida, A. K., Panda, A., & Kumari, A. (2016). Coordinated changes in antioxidative enzymes protect the photosynthetic machinery from salinity induced oxidative damage and confer salt tolerance in an extreme halophyte *Salvadora persica* L. *Frontiers in Plant Science*, *7*, 50.

Rauf, M., Shahzad, K., Ali, R., Ahmad, M., Habib, I., Mansoor, S., & Saeed, N. A. (2014). Cloning and characterization of Na + /H + antiporter (LfNHX1) gene from a halophyte grass *Leptochloa fusca* for drought and salt tolerance. *Molecular Biology Reports*, *41*(3), 1669–1682. Available from http://doi.org/10.1007/s11033-013-3015-3.

Ravindran, K. C., Venkatesan, K., Balakrishnan, V., Chellappan, K. P., & Balasubramanian, T. (2007). Restoration of saline land by halophytes for Indian soils. *Soil Biology and Biochemistry*, *39*(10), 2661–2664.

Redondo-Gómez, S., Mateos-Naranjo, E., Figueroa, M. E., & Davy, A. J. (2010). Salt stimulation of growth and photosynthesis in an extreme halophyte, *Arthrocnemum macrostachyum*. *Plant Biology*, *12*(1), 79–87.

Redondo-Gómez, S., Wharmby, C., Castillo, J. M., Mateos-Naranjo, E., Luque, C. J., De Cires, A., & Enrique Figueroa, M. (2006). Growth and photosynthetic responses to salinity in an extreme halophyte, *Sarcocornia fruticosa*. *Physiologia Plantarum, 128*(1), 116–124.

Rewald, B., Raveh, E., Gendler, T., Ephrath, J. E., & Rachmilevitch, S. (2012). Phenotypic plasticity and water flux rates of Citrus root orders under salinity. *Journal of Experimental Botany, 63*(7), 2717–2727.

Saad, R. B., Romdhan, W. B., Zouari, N., Azaza, J., Mieulet, D., Verdeil, J. L., & Hassairi, A. (2011). Promoter of the AlSAP gene from the halophyte grass *Aeluropus littoralis* directs developmental-regulated, stress-inducible, and organ-specific gene expression in transgenic tobacco. *Transgenic Research, 20*(5), 1003–1018. Available from http://doi.org/10.1007/s11248-010-9474-6.

Saad, R. B., Zouari, N., Ramdhan, W. B., Azaza, J., Meynard, D., Guiderdoni, E., & Hassairi, A. (2010). Improved drought and salt stress tolerance in transgenic tobacco overexpressing a novel A20/AN1 zinc-finger "AlSAP" gene isolated from the halophyte grass *Aeluropus littoralis*. *Plant Molecular Biology, 72*(1-2), 171. Available from http://doi.org/10.1007/s11103-009-9560-4.

Sabovljevic, M., & Sabovljevic, A. (2007). Contribution to the coastal bryophytes of the Northern Mediterranean: Are there halophytes among bryophytes. *Phytologia Balcanica, 13* (2), 131–135.

Saddhe, A.A., Manuka, R., Nikalje, G.C., & Penna, S. (2020). Halophytes as a potential resource for phytodesalination. In Handbook of halophytes: From molecules to ecosystems towards biosaline agriculture (pp. 1-21).

Sanadhya, P., Agarwal, P., Khedia, J., & Agarwal, P. K. (2015). A low-affinity K^+ transporter AlHKT2; 1 from recretohalophyte *Aeluropus lagopoides* confers salt tolerance in yeast. *Molecular Biotechnology, 57*(6), 489–498.

Santos, J., Al-Azzawi, M., Aronson, J., & Flowers, T. J. (2016). eHALOPH a database of salt-tolerant plants: Helping put halophytes to work. *Plant and Cell Physiology, 57*(1), e10, e10.

Sastre-Conde, I., Lobo, M. C., Beltrán-Hernández, R. I., & Poggi-Varaldo, H. M. (2015). Remediation of saline soils by a two-step process: Washing and amendment with sludge. *Geoderma, 247*, 140–150.

Selvam, V. (2003). Environmental classification of mangrove wetlands of India. *Current Science, 84*(6), 757–765.

Semenova, G. A., Fomina, I. R., & Biel, K. Y. (2010). Structural features of the salt glands of the leaf of *Distichlis spicata* 'Yensen 4a'(Poaceae). *Protoplasma, 240*(1-4), 75–82.

Shabala, L., Mackay, A., Tian, Y., Jacobsen, S. E., Zhou, D., & Shabala, S. (2012). Oxidative stress protection and stomatal patterning as components of salinity tolerance mechanism in quinoa (*Chenopodium quinoa*). *Physiologia Plantarum, 146*(1), 26–38.

Shabala, S. (2013). Learning from halophytes: Physiological basis and strategies to improve abiotic stress tolerance in crops. *Annals of Botany, 112*(7), 1209–1221.

Shabala, S., Hariadi, Y., & Jacobsen, S. E. (2013). Genotypic difference in salinity tolerance in *quinoa* is determined by differential control of xylem Na^+ loading and stomatal density. *Journal of Plant Physiology, 170*(10), 906–914.

Shabala, S., & Mackay, A. (2011). *Ion transport in halophytes, . Advances in botanical research* (Vol. 57, pp. 151–199). Academic Press.

Shackira, A. M., & Puthur, J. T. (2016). Antioxidant defence mechanism operational in a mangrove – *Acanthus ilicifolius* L subjected to NaCl stress. *International Journal of Advanced Research, 4*, 1613–1623.

Shao, Q., Han, N., Ding, T., Zhou, F., & Wang, B. (2014). SsHKT1; 1 is a potassium transporter of the C3 halophyte *Suaeda salsa* that is involved in salt tolerance. *Functional Plant Biology*, *41*(8), 790–802. Available from http://doi.org/10.1071/FP13265.

Shen, Y. G., Zhang, W. K., Yan, D. Q., Du, B. X., Zhang, J. S., Liu, Q., & Chen, S. Y. (2003). Characterization of a DRE-binding transcription factor from a halophyte *Atriplex hortensis*. *Theoretical and Applied Genetics*, *107*(1), 155–161. Available from http://doi.org/10.1007/s00122-003-1226-z.

Singh, N., Mishra, A., & Jha, B. (2014). Ectopic over-expression of peroxisomal ascorbate peroxidase (SbpAPX) gene confers salt stress tolerance in transgenic peanut (*Arachis hypogaea*). *Gene*, *547*(1), 119–125. Available from http://doi.org/10.1016/j.gene.2014.06.037.

Song, J., & Wang, B. (2015). Using euhalophytes to understand salt tolerance and to develop saline agriculture: *Suaeda salsa* as a promising model. *Annals of Botany*, *115*(3), 541–553.

Sreeshan, A., Meera, S. P., & Augustine, A. (2014). A review on transporters in salt tolerant mangroves. *Trees*, *28*(4), 957–960.

Sruthi, P. (2019). Functional biology of *Bruguiera cylindrica* (L.) Blume in response to individual and combined stresses with NaCl and heavy metals. (Ph.D. dissertation), University of Calicut, Kerala, India.

Sruthi, P., & Puthur, J. T. (1990). Characterization of physiochemical and anatomical features associated with enhanced phytostabilization of copper in *Bruguiera cylindrica* (L.) Blume. *International Journal of Phytoremediation*, *21*, 1423–1441. Available from http://doi.org/10.1080/15226514.2019.1633263.

Stelzer, R., Kuo, J., & Koyro, H. W. (1988). Substitution of Na^+ by K^+ in tissues and root vacuoles of barley (*Hordeum vulgare* L. cv. Aramir). *Journal of Plant Physiology*, *132*(6), 671–677.

Stepien, P., & Klobus, G. (2005). Antioxidant defense in the leaves of C3 and C4 plants under salinity stress. *Physiologia Plantarum*, *125*(1), 31–40.

Sui, N., Tian, S., Wang, W., Wang, M., & Fan, H. (2017). Overexpression of glycerol-3-phosphate acyltransferase from *Suaeda salsa* improves salt tolerance in *Arabidopsis*. *Frontiers in Plant Science*, *8*, 1337. Available from https://doi.org/10.3389/fpls.2017.01337.

Surówka, E., & Hura, T. (2020). Osmoprotectants and nonenzymatic antioxidants in halophytes. In *Handbook of halophytes: From molecules to ecosystems towards biosaline agriculture*, (pp.1–31), Springer.

Takemura, T., Hanagata, N., Sugihara, K., Baba, S., Karube, I., & Dubinsky, Z. (2000). Physiological and biochemical responses to salt stress in the mangrove, *Bruguiera gymnorrhiza*. *Aquatic Botany*, *68*(1), 15–28.

Tan, W. K., Lin, Q., Lim, T. M., Kumar, P., & Loh, C. S. (2013). Dynamic secretion changes in the salt glands of the mangrove tree species *Avicennia officinalis* in response to a changing saline environment. *Plant, Cell & Environment*, *36*(8), 1410–1422.

Tiwari, V., Chaturvedi, A. K., Mishra, A., & Jha, B. (2014). The transcriptional regulatory mechanism of the peroxisomal ascorbate peroxidase (pAPX) gene cloned from an extreme halophyte, *Salicornia brachiata*. *Plant and Cell Physiology*, *55*(1), 201–217. Available from http://doi.org/10.1093/pcp/pct172.

Tiwari, V., Patel, M. K., Chaturvedi, A. K., Mishra, A., & Jha, B. (2019). Cloning and functional characterization of the Na^+/H^+ antiporter (NHX1) gene promoter from an extreme halophyte *Salicornia brachiata*. *Gene*, *683*, 233–242.

Tomlinson, P. B. (1986). *The botany of mangroves*. London: Cambridge University Press.

Vasquez, E. A., Glenn, E. P., Guntenspergen, G. R., Brown, J. J., & Nelson, S. G. (2006). Salt tolerance and osmotic adjustment of *Spartina alterniflora* (Poaceae) and the invasive M haplotype of *Phragmites australis* (Poaceae) along a salinity gradient. *American Journal of Botany*, *93*(12), 1784–1790.

Vaziriyeganeh, M., Lee, S. H., & Zwiazek, J. J. (2018). Water transport properties of root cells contribute to salt tolerance in halophytic grasses *Poa juncifolia* and *Puccinellia nuttalliana*. *Plant Science*, *276*, 54–62.

Velautham, L., Ranney, M. A., & Brow, Q. S. (2019). Communicating climate change oceanically: Sea level rise information increases mitigation, inundation, and global warming acceptance. *Frontiers in Communication*, *4*, 7.

Venkatesalu, V., & Chellappan, K. P. (1993). Photosynthetic characteristic of *Sesuvium portulacastrum* L. under salt stress. *Photosynthetica (Praha)*, *28*(2), 313–316.

Vineis, P., Chan, Q., & Khan, A. (2011). Climate change impacts on water salinity and health. *Journal of Epidemiology and Global Health*, *1*(1), 5–10.

von Caemmerer, S., Edwards, G. E., Koteyeva, N., & Cousins, A. B. (2014). Single cell C4 photosynthesis in aquatic and terrestrial plants: A gas exchange perspective. *Aquatic Botany*, *118*, 71–80.

Wang, H., Ding, Q., & Wang, H. (2018). A new Na^+/H^+ antiporter gene KvNHX1 isolated from the halophyte *Kosteletzkya virginica* improves salt tolerance in transgenic tobacco. *Biotechnology & Biotechnological Equipment*, *32*(6), 1378–1386.

Wang, J. Y., Tong, S. M., & Li, Q. L. (2013). Constitutive and salt-inducible expression of SlBADH gene in transgenic tomato (*Solanum lycopersicum* L. cv. Micro-Tom) enhances salt tolerance. *Biochemical and Biophysical Research Communications*, *432*(2), 262–267. Available from http://doi.org/10.1016/j.bbrc.2013.02.001.

Wang, L., Zhao, G., Li, M., Zhang, M., Zhang, L., Zhang, X., & Xu, S. (2015). C: N: P stoichiometry and leaf traits of halophytes in an arid saline environment, northwest China. *PLoS One*, *10*(3), e0119935.

Wang, S., Zheng, W., Ren, J., & Zhang, C. (2002). Selectivity of various types of salt-resistant plants for K^+ over Na^+. *Journal of Arid Environments*, *52*(4), 457–472.

Wang, X., Yang, R., Wang, B., Liu, G., Yang, C., & Cheng, Y. (2011). Functional characterization of a plasma membrane Na^+/H^+ antiporter from alkali grass (*Puccinellia tenuiflora*). *Molecular Biology Reports*, *38*(7), 4813–4822. Available from http://doi.org/10.1007/s11033-010-0624-y.

Wang, Y. C., Qu, G. Z., Li, H. Y., Wu, Y. J., Wang, C., Liu, G. F., & Yang, C. P. (2010). Enhanced salt tolerance of transgenic poplar plants expressing a manganese superoxide dismutase from *Tamarix androssowii*. *Molecular Biology Reports*, *37*(2), 1119. Available from http://doi.org/10.1007/s11033-009-9884-9.

Wingler, A., Lea, P. J., Quick, W. P., & Leegood, R. C. (2000). Photorespiration: Metabolic pathways and their role in stress protection. *Philosophical Transactions of the Royal Society of London. Series B: Biological Sciences*, *355*(1402), 1517–1529.

Wiszniewska, A., Koźmińska, A., Hanus-Fajerska, E., Dziurka, M., & Dziurka, K. (2019). Insight into mechanisms of multiple stresses tolerance in a halophyte *Aster tripolium* subjected to salinity and heavy metal stress. *Ecotoxicology and Environmental Safety*, *180*, 12–22.

Wu, C., Gao, X., Kong, X., Zhao, Y., & Zhang, H. (2009). Molecular cloning and functional analysis of a Na^+/H^+ antiporter gene ThNHX1 from a halophytic plant *Thellungiella halophila*. *Plant Molecular Biology Reporter*, *27*(1), 1. Available from http://doi.org/10.1007/s11105-008-0048-1.

Wu, P., Cogill, S., Qiu, Y., Li, Z., Zhou, M., Hu, Q., & Raymer, P. (2020). Comparative transcriptome profiling provides insights into plant salt tolerance in seashore paspalum (*Paspalum vaginatum*). *BMC Genomics*, *21*(1), 131.

Xu, Q., Renault, S., & Yuan, Q. (2019). Phytodesalination of landfill leachate using *Puccinellia nuttalliana* and *Typha latifolia*. *International Journal of Phytoremediation*, *21*(9), 831–839.

Xu, X. J., Zhou, Y. J., Ren, D. T., Bu, H. H., Feng, J. C., & Wang, G. Y. (2014). Cloning and characterization of gene encoding a Mn-containing superoxide dismutase in *Eutrema halophilum*. *Biologia Plantarum*, *58*(1), 105–113. Available from http://doi.org/10.1007/s10535-013-0363-8.

Yadav, N. S., Shukla, P. S., Jha, A., Agarwal, P. K., & Jha, B. (2012). The SbSOS1 gene from the extreme halophyte *Salicornia brachiata* enhances Na^+ loading in xylem and confers salt tolerance in transgenic tobacco. *BMC Plant Biology*, *12*(1), 188. Available from http://doi.org/10.1186/1471-2229-12-188.

Yamamoto, N., Takano, T., Tanaka, K., Ishige, T., Terashima, S., Endo, C., & Tada, Y. (2015). Comprehensive analysis of transcriptome response to salinity stress in the halophytic turf grass *Sporobolus virginicus*. *Frontiers in Plant Science*, *6*, 241.

Yao, M., Zeng, Y., Liu, L., Huang, Y., Zhao, E., & Zhang, F. (2012). Overexpression of the halophyte *Kalidium foliatum* H^+-pyrophosphatase gene confers salt and drought tolerance in *Arabidopsis thaliana*. *Molecular Biology Reports*, *39*(8), 7989–7996. Available from http://doi.org/10.1007/s11033-012-1645-5.

Ye, C. Y., Zhang, H. C., Chen, J. H., Xia, X. L., & Yin, W. L. (2009). Molecular characterization of putative vacuolar NHX-type Na^+/H^+ exchanger genes from the salt-resistant tree *Populus euphratica*. *Physiologia Plantarum*, *137*(2), 166–174. Available from http://doi.org/10.1111/j.1399-3054.2009.01269.x.

Ye, Y., Tam, N. F. Y., Lu, C. Y., & Wong, Y. S. (2005). Effects of salinity on germination, seedling growth and physiology of three salt-secreting mangrove species. *Aquatic Botany*, *83*(3), 193–205.

Yi-Guo, S. H. E. N., Wan-Ke, Z. H. A. N. G., Dong-Qing, Y. A. N., Bao-Xing, D. U., Jin-Song, Z. H. A. N. G., & Shou-Yi, C. N. (2002). Overexpression of proline transporter gene isolated from halophyte confers salt tolerance in Arabidopsis. *Journal of Integrative Plant Biology*, *44*(8), 956–962.

Yuan, F., Guo, J., Shabala, S., & Wang, B. (2019). Reproductive physiology of halophytes: Current standing. *Frontiers in Plant Science*, *9*, 1954.

Yuan, F., Leng, B., & Wang, B. (2016). Progress in studying salt secretion from the salt glands in recretohalophytes: How do plants secrete salt? *Frontiers in Plant Science*, *7*, 977.

Yuan, F., Liang, X., Li, Y., Yin, S., & Wang, B. (2019). Methyl jasmonate improves tolerance to high salt stress in the recretohalophyte *Limonium bicolor*. *Functional Plant Biology*, *46*(1), 82–92.

Yuan, F., Lyu, M. J. A., Leng, B. Y., Zheng, G. Y., Feng, Z. T., Li, P. H., & Wang, B. S. (2015). Comparative transcriptome analysis of developmental stages of the *Limonium bicolor* leaf generates insights into salt gland differentiation. *Plant, Cell & Environment*, *38*(8), 1637–1657.

Yuan, F., & Wang, B. (2020). Adaptation of recretohalophytes to salinity: Salt secretion and salt gland development. In *Handbook of halophytes: From molecules to ecosystems towards biosaline agriculture*, (pp. 1–21), Springer.

Zhang, L. Q., Niu, Y. D., Huridu, H., Hao, J. F., Qi, Z., & Hasi, A. (2014).). *Salicornia europaea* L. Na^+/H^+ antiporter gene improves salt tolerance in transgenic alfalfa (*Medicago sativa* L.). *Genetics and Molecular Research*, *13*(3), 5350–5360. Available from http://doi.org/10.4238/2014.

Zhang, S., Song, J., Wang, H., & Feng, G. (2010). Effect of salinity on seed germination, ion content and photosynthesis of cotyledons in halophytes or xerophyte growing in Central Asia. *Journal of Plant Ecology*, *3*(4), 259–267.

Zhang, Y., Lai, J., Sun, S., Li, Y., Liu, Y., Liang, L., & Xie, Q. (2008). Comparison analysis of transcripts from the halophyte *Thellungiella halophila*. *Journal of Integrative Plant Biology*, *50*(10), 1327–1335.

Zhang, Y., Li, Y., Lai, J., Zhang, H., Liu, Y., Liang, L., & Xie, Q. (2012). Ectopic expression of a LEA protein gene TsLEA1 from *Thellungiella salsuginea* confers salt-tolerance in yeast and *Arabidopsis*. *Molecular Biology Reports*, *39*(4), 4627–4633. Available from http://doi.org/10.1007/s11033-011-1254-8.

Zhao, F., Wang, Z., Zhang, Q., Zhao, Y., & Zhang, H. (2006). RETRACTED ARTICLE: Analysis of the physiological mechanism of salt-tolerant transgenic rice carrying a vacuolar Na^+/H^+ antiporter gene from *Suaeda salsa*. *Journal of Plant Research*, *119*(2), 95–104. Available from http://doi.org/10.1007/s10265-005-0250-2.

Zhao, K. F., Fan, H., Song, J., Sun, M. X., Wang, B. Z., Zhang, S. Q., & Ungar, I. A. (2005). Two Na^+ and Cl^- hyperaccumulators of the Chenopodiaceae. *Journal of Integrative Plant Biology*, *47*(3), 311–318.

Zheng, W. J., Wang, W. Q., & Lin, P. (1999). Dynamics of elemental contents during the development of hypocotyls and leaves of certain mangrove species. *Journal of Experimental Marine Biology and Ecology*, *233*, 247–257.

Zhou, S., Chen, X., Zhang, X., & Li, Y. (2008). Improved salt tolerance in tobacco plants by co-transformation of a betaine synthesis gene BADH and a vacuolar Na^+/H^+ antiporter gene SeNHX1. *Biotechnology Letters*, *30*(2), 369–376. Available from http://doi.org/10.1007/s10529-007-9548-6.

Zhou, Y., Tang, N., Huang, L., Zhao, Y., Tang, X., & Wang, K. (2018). Effects of salt stress on plant growth, antioxidant capacity, glandular trichome density, and volatile exudates of *Schizonepeta tenuifolia* Briq. *International Journal of Molecular Sciences*, *19*(1), 252.

Zou, C., Chen, A., Xiao, L., Muller, H. M., Ache, P., Haberer, G., & Lang, D. (2017). A high-quality genome assembly of *quinoa* provides insights into the molecular basis of salt bladder-based salinity tolerance and the exceptional nutritional value. *Cell Research*, *27*(11), 1327–1340.

Chapter 17

Role of nickel in regulation of nitrogen metabolism in legume–rhizobium symbiosis under critical conditions

Rukhsar Parwez[1], Aarifa Nabi[1], Mohammad Mukarram[1], Tariq Aftab[2], M. Masroor A. Khan[1] and M. Naeem[1]
[1]*Plant Physiology Section, Department of Botany, Faculty of Life Sciences, Aligarh Muslim University, Aligarh, India,* [2]*Department of Botany, Aligarh Muslim University, Aligarh, India*

17.1 Introduction

Although nickel (Ni) is an essential and irreplaceable micronutrient, it is a debatable element given its beneficial and toxic roles in the plants. Nickel forms an integral part of many metalloenzymes, like Ni-Fe dehydrogenase, acireductone dioxygenase, acetyl-CoA decarbonylase synthase, peptide deformylase, superoxide dismutase, methyl-coenzyme M reductase, lactate racemase, Ni-dependent glyoxylase, and carbon monoxide dehydrogenase in bacteria and lower plants (Boer, Mulrooney, & Hausinger, 2014). Moreover, it forms the structural component of urease and glyoxylase family I enzymes in higher plants (Siegbahn, Chen, & Liao, 2019). These enzymes play a significant role in maintaining accurate cellular redox status, improving overall growth, biochemical and physiological responses, stress tolerance, and optimum N use efficiency in plants (Alfano & Cavazza, 2020). Nickel is highly required in the case of leguminous plants as legumes formulate symbiotic association with different nitrogen-fixing bacteria and play an imperative role in facilitating symbiosis by forming the constituent of enzymes involved in biological nitrogen fixation. Nickel plays crucial roles in various enzymatic, developmental, symbiotic association, and fundamental processes in plants, as depicted in Fig. 17.1. Moreover, Ni also takes part in nitrogen acquisition, assimilation, and utilization controlling overall nitrogen metabolism processes in plants (de Macedo et al., 2016).

Some researchers suggested that Ni facilitates normal growth processes in minimal concentrations (0.01 to 5 $\mu g\ g^{-1}$ of plant dry weight) while its amount

Frontiers in Plant–Soil Interaction. DOI: https://doi.org/10.1016/B978-0-323-90943-3.00021-3

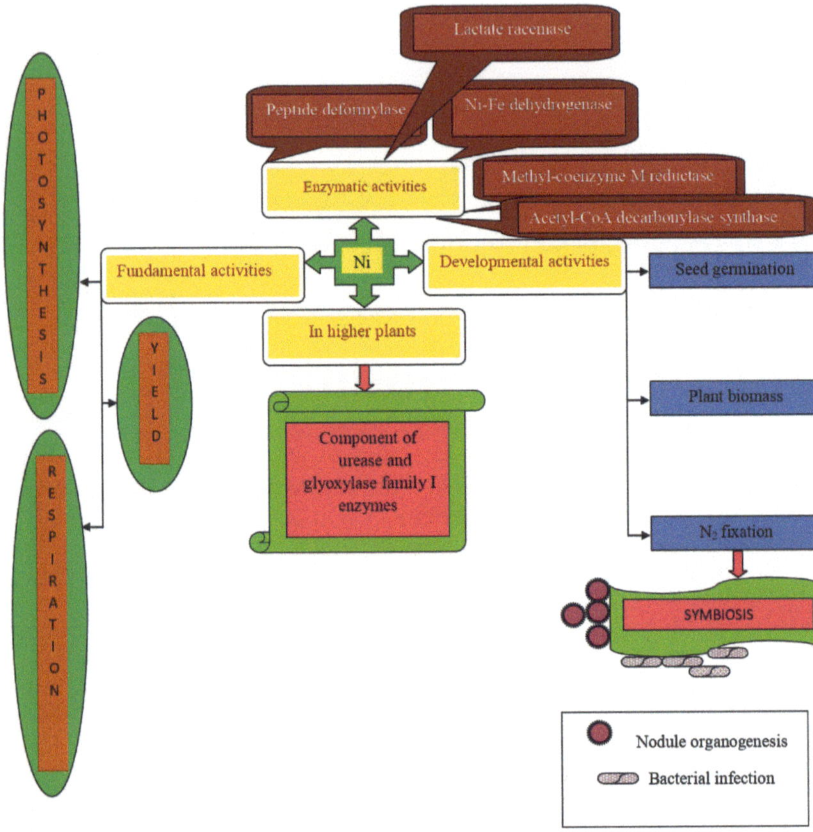

FIGURE 17.1 Involvement of Ni in various enzymatic, developmental, symbiotic association and some basic fundamental processes in plants.

in plants may vary from 0.05 to 5 mg kg^{-1} dry weight. However, an excess of Ni in soil affects various cellular and metabolic processes in plants, causing a reduction in seed germination, plant biomass, nutrient absorption by roots, N_2 fixation, photosynthesis and its pigments, respiration, yield, etc. (Nabi, Parwez, Aftab, Khan, & Naeem, 2020; Shahzad et al., 2018). Moreover, Ni stimulates a negative impact on legume–rhizobia symbiosis by affecting the process of nodulation and N_2 fixation (Athar & Ahmad, 2002; Polacco, Mazzafera, & Tezotto, 2013).

A general overview of the establishment of legume rhizobia symbiosis is required to understand the role of Ni in the modulation of the different processes involved. Major processes in symbiosis can be divided into two main events: bacterial infection and nodule organogenesis. Among the diverse group of bacteria, the most popular Gram-negative nodulating bacteria belong to genera *Rhizobium*, *Mesorhizobium*, *Sinorhizobium*, and *Bradyrhizobium*. Legumes release phenolic compounds, mainly flavonoids, into the rhizosphere, which result in the chemotactic movement of bacteria toward plant roots and the induction of nodulation (Nod) factors in specific rhizobial strain, ensuring high specificity between host and rhizobia. As a result, bacteria secrete Nod D, which gets activated on interaction with flavonoids and renders bacterial cell to induce transcription of other nod genes to release nod factors (Liu & Murray, 2016). These nod factors induce root hair curling. An invagination in the root cell at the interaction site progressively deepens, resulting in the formation and extension of infection thread through the cortex and allows rhizobial invasion into the rapidly dividing cortex cell forming a nodule primordium. Rhizobia released are grouped and enclosed in peribacteroid membranc-forming N_2-fixing symbiotic bacteroids due to the possession of oxygen-sensitive nitrogenase enzyme (Kumar & Mina, 2019). Moreover, the hosts' nodules synthesize a monomeric protein leghemoglobin (LHb) to maintain a low oxygen level to maintain nitrogenase activity (Singh & Varma, 2017).

The ability of the leguminous crops to fix atmospheric N in the soil through legume–rhizobia symbiosis makes it advantageous, reducing dependability on chemical fertilizers as nitrogen serves to be one of the key growth-limiting elements affecting crop productivity. Symbiotic nitrogen fixation is a complex process finely tuned by carbon and energy metabolism processes operating in plant and bacteria; therefore Ni serves to be one of the essential factors which affect the legume–rhizobia association. In this chapter, we have summarized the Ni-sensing process in legumes and *Rhizobium* sp. and how soil Ni may cause feedback control in the regulation of nodulation and other nitrogen metabolism activities operating in legumes, which may be of immense importance in developing a better sustainable agriculture.

17.2 Nickel uptake and transport

Nickel content in soil on an average may vary from 2 to 750 mg kg^{-1} soil with the highest content reported in serpentine soil (Seregin & Kozhevnikova, 2006). Ni concentration in soil has risen in recent years due to the use of fungicides, activities of electroplating, mining, metallurgical, and pharmaceutical industries (Iyaka, 2011; Tsadilas, Rinklebe, & Selim, 2018). Nickel uptake in plants is primarily from the soil. Nickel uptake may take place either through passive diffusion or an active transport process, depending on Ni levels in the soil. There are no specific nickel transporters disclosed so far in plants. The active transport mechanism operates at low Ni concentrations, while high Ni levels result in its passive transport. Ni uptake is favored at lower humus content, acidic pH, lighter soil granulometric composition, higher soil moisture levels, and at temperature between 23oC and 30oC (Sachan & Lal, 2017). Besides, several metal ions also play an important inhibitory role in Ni uptake and translocation from root to shoot, which follows the order: $Fe^{3+} > Co^{2+} > Ca^{2+} > Mg^{2+} > NH_4^+ > K^+ > Na^+$ (Seregin & Kozhevnikova, 2006). Cation transport systems like Fe^{2+}, Cu^{2+}, Zn^{2+}, and Mg^{2+} are responsible for the transport of solubilized Ni compounds through chelators like nicotinamide (NA), histidine (His), and citric acid, as well as through the involvement of proteins like metallothionein, YS-1 like (YSLs), permeases, and metallochaperons (Vatansever, Ozyigit, & Filiz, 2017).

In the majority of rhizobia, Ni transport proceeds through Nramp or ZIP transporters, NiCoT, ABC transporters, or HupE/UreJ homologs (Jose palaCios, 2017). HupN is another transmembrane protein involved in nickel transport in *Bradyrhizobium japonicum* (González-Guerrero et al., 2013). Nia (Nickel/Iron ATPase) is a specified P1B-5-ATPase family metal transporter responsible for Ni^{2+} and Fe^{2+} uptake in *Sinorhizobium meliloti* (Zielazinski et al., 2013).

17.3 Regulatory role of Ni in legume rhizobia symbiosis and N metabolism in plants

Nickel plays an important role to establish and maintain the efficiency of legume–*Rhizobium* symbiosis through modulating nodule structure and elicitor synthesis to establish symbiosis. It controls various enzyme activities in host and bacteria, affecting the nutritional status of the nodule and controls C-N balance in the host by regulating ureide and amino acid metabolism, urea synthesis, and Kreb's cycle intermediates. The following are important parameters affected by Ni for efficient legume–rhizobia symbiosis.

17.4 Nodule development

The nodule coordinates unique roles for the accommodation and proliferation of symbionts in metal-contaminated sites, thereby acting as microsites for limiting the toxic impact of metal ions. In a few legumes, low levels of Ni are essential for nodule growth. Interestingly, the optimum concentration of Ni promotes nodulation through an increase in nodule biomass in *Glycine max* and *Trigonella foenum-graecum* (Abd-Alla, Bashandy, Bagy, & El-enany, 2014; Bertrand & de Wolf, 1967), and its deficiency delays nodule formation in *G. max* (Eskew, Welch, & Cary, 1983; Zobiole et al., 2010). On the other hand, Ni-contaminated soils (above $50\,mg\,kg^{-1}$ of soil) show a decrease in nodulation in legumes like *Vigna radiate, Lens culinaris*, and *Cicer arientinum* manifested in the form of a decline in number and biomass of nodules (Athar & Ahmad, 2002; Saad, Kobaissi, Robin, Echevarria, & Benizri, 2016; Yusuf, Fariduddin, Ahmad, & Ahmad, 2014). Electron microscopic studies on root nodules of *Lotus corniculatus* reveal that Ni induces ultramorphological changes during nodules organogenesis reflected in the form of cell wall thickening, especially at corners and enhanced phenol accumulation (Sujkowska-Rybkowska, Kasowska, Gediga, Banasiewicz, & Stepkowski, 2020) as an adaptation to eliminate toxic metal effects and other abiotic stress effects on legume–rhizobia symbiosis (Krzesłowska, 2011; Le Gall et al., 2015). Studies revealed that 1 μg L^{-1} of Ni in culture promotes the growth of *Bradyrhizobia*, and the nodulation and growth of cowpea and pigeon pea were also enhanced (Wheeler, Hughes, Oldroyd, Pulford, 2001). Rhizobia are effective in infecting *Cajanus cajan* roots within the natural range of Ni concentration, but at a higher level Ni curbs the process of root hair formation affecting the overall infection, thereby resulting in a limited number of functional nodules in leguminous species (de Macedo et al., 2016; Dudeja, Sheokand, & Kumari, 2012; Ishtiaq & Mahmood, 2011). Further, Ni toxicity at 200 mg kg^{-1} lowers the membrane stability index of nodules in *C. cajan* genotypes, suggesting membrane damage as a major reason for growth reduction through restricting nutrient uptake (Garg & Saroy, 2020). Ni at 200 mg kg^{-1} also induces nodule senescence (Garg & Saroy, 2020), which may result from metal driven upregulation of a nodule specific NA synthase or oxidative damage due to free radical production (Curie et al., 2009).

17.5 Nodulation factor expression

The impact of Ni on the regulation of nod factors expression mainly arises due to its control over phenylalanine ammonia lyase (PAL) activity and nod gene expression. The PAL is one of the preliminary enzymes in the phenylpropanoid pathway whose upregulation in response to metal stress enhances phenol, flavanones, and isoflavanones production (Wang, Li, Zhang, & Ke, 2010; Yan et al., 2008). Flavones and isoflavones act as potent inducers of nodulation genes in *Rhizobia* (Cooper, 2007; Zoń, Amrhein, & Gancarz, 2002), while the phenolic compounds help the plant in coping up with harsh environmental conditions (Siddiqui, Siddiqui, Khan, & Bano, 2019). A recent study on *T. foenum-graecum* reveals that Ni at 75 and 100 mg L^{-1} showed a significant promontory effect on PAL activity in *Rhizobium* infected roots while flavonoid level is enhanced up to 25 mg L^{-1} Ni only; however, above this concentration, there is a significant reduction in flavonoid production under Ni toxicity (Abd-Alla et al., 2014). Ni generates osmotic stress (Rucińska-Sobkowiak, 2016), and osmotic stress results in increased expression of transcriptional regulators OnfD and NodD2 in *Rhizobium tropici* CIAT 899 infecting *Phaseolus vulgaris* and *L. corniculatus* enhancing overall nodulation factors synthesis under salt stress condition (Del Cerro et al., 2020). Besides, in the case of *T. foenum-graecum*, Ni concentration up to 25 mg L^{-1} does not affect nod gene expression while above this concentration, Ni has a pronounced deleterious effect on nod gene expression (Abd-Alla et al., 2014). Nod gene expression releases nod factors. These nod factors serve a dual role. Firstly, they act as signaling molecules for dedifferentiation and division of cortical cells of roots initiating nodule organogenesis (Geurts & Bisseling, 2002), and secondly, it facilitates rhizobial colonization through stimulating infection thread formation (Malolepszy et al., 2018; Oldroyd, Murray, Poole, & Downie, 2011). Thus it could be concluded that at a lower concentration, Ni enhances the symbiotic process, while at higher concentrations, it has a marked inhibitory effect on legume–rhizobia symbiosis.

17.6 Leghemoglobin content and nitrogenase activity

LHb is a monomeric oxygen scavenger hemoprotein responsible for maintaining anaerobic conditions inside the legume, protecting the activity of oxygen-sensitive nitrogenase (Taiz & Zeiger, 2006). Nitrogenase is the primary enzyme which catalyzes the process of biological nitrogen fixation in legumes:

$$N_2 + 16ATP + 8H^+ + 8e^- \xrightarrow{\text{Nitrogenase}} 2NH_3 + 16ADP + H_2 + 16P_i$$

The establishment of symbiosis with *Rhizobium* is the essential condition which confers the property to synthesize LHb in root nodules. At a concentration of 2.1 μM of $NiCl_2$ stimulated nitrogenase activity in *Nostoc muscorum* (Rai & Raizada, 1985) and at a concentration of 1 μg m L^{-1} Ni increased *Bradyrhizobium* growth, nodule number, and nitrogenase activity in *C. cajan* and *V. radiata* demonstrating that Ni is beneficial for legume–rhizobia symbiosis at low levels while at a higher concentration it proved toxic (Singh & Rao, 1997). Nickel stress significantly deteriorates the functional efficacy of nodules (ARA) and a subsequent decline in LHb content, suggesting a direct correlation between LHb concentration and ARA. It is reported that Ni stress in soybean interferes with the N_2 fixation system manifested in the form of LHb synthesis inhibition and a toxic effect on *Rhizobium* growth (de Macedo et al., 2016). Further, Ni showed a harmful effect on nodulation and nitrogenase activity in *Frankia–Alnus* symbiosis and *L. culinaris–Rhizobium leguminosarum* symbiosis (Saad et al., 2016; Wheeler et al., 2001). Other leguminous species like *G. max*, *Trifolium alexandrium*, and *Vicia faba* exhibited a significant dose-dependent reduction in N uptake and nodulation process with increasing Ni concentrations (Haddad, Tabatabai, Abdel-Moneim, & Loynachan, 2015). These results could be justified as Ni stress causes a reduction in O_2 flux in nodules due to a decline in LHb content (Layzell et al., 1993), inhibiting nitrogenase action and further impairing carbon metabolism, which proved to be deleterious for *Rhizobial* growth as bacteroid respiration is compromised (Soussi, Khadri, Lluch, & Ocana, 2001). A decline in nitrogenase activity due to the inhibition of LHb biosynthesis under Ni stress occurs. However, several other factors like ammonia accumulation curb nitrogenase biosynthesis (Vesper & Weidensaul, 1978), while the increase in nitrate level (Das, 1982) as a result of the decrease in nitrate reductase (NR) activity and limitation of ATP (a prerequisite for nitrogenase activity) due to inhibition of H^+-ATPase activity (Gill, 2014; Janicka-Russak, Kabala, Burzynski, & Klobus, 2008) further contribute to the decline in nitrogen fixation ability in legumes under Ni stress.

17.7 Nitrate reductase activity

Enzyme NR occupies a pivotal role in the integration of N metabolism in both bacteria and plants by governing nitrogen flux through several regulatory mechanisms. NR is known to catalyze NAD(P)H mediated reduction of nitrate to nitrite, and also NR is responsible for NO production and NO homeostasis in many plants (Chamizo-Ampudia, Sanz-Luque, Llamas, Galvan, & Fernandez, 2017; Tejada-Jimenez, Llamas, Galván, & Fernández, 2019). Both nitrate and nitric oxide play a pivotal role in the plant signaling process by initiating several reaction cascades that enable plant survival under critical conditions (Astier, Gross, & Durner, 2018; Konishi & Yanagisawa, 2013). Low Ni concentration favours NR expression, which encourages nitrate reduction and increases N acquisition and its incorporation into C-skeletons such as (2-ketoglutarate synthesis) maintaining C/N balance in plants for efficient symbiosis (Tejada-Jimenez et al., 2019). In contrast, NR-mediated NO production is responsible for posttranslational regulation of many plant proteins through nitrosylation and ROS reduction resulting in abiotic stress protection (Fancy, Bahlmann, & Loake, 2017).

Nickel at low concentrations, such as 0.4 μM, 25 μM, and 0.05 $\mu mol\ L^{-1}$, is reported to increase a reduction in nitrate level through enhancement of the activity of NR in lettuce roots (Matraszek, 2008), onion bulbs (Alibakhshi & Khoshgoftarmanesh, 2015), and soybean (dos Reis et al., 2017), respectively. The activity of NR is intended to be enhanced through Ni-dependent activation of malate dehydrogenase enzyme resulting in NADH synthesis, which acts as a source of metabolic energy for NR driven nitrate reduction (Pilbeam & Barker, 2007). However, a high concentration of Ni at 100 μM exhibited a marked decline in NR activity in soybean (El-Shintinawy & El-Ansary, 2000), *Brassica juncea* (Alam, Hayat, Ali, & Ahmad, 2007), and wheat (Gajewska, Wielanek, Bergier, & Skłodowska, 2009; Yusuf, Fariduddin, Hayat, & Ahmad, 2011; Yusuf, Fariduddin, Hayat, Hasan, & Ahmad, 2011), thereby suggesting a hindrance of the nitrate uptake mechanism by roots. Interference of Ni with the −SH group present in most nitrate uptake systems (Tan, Ikeda, & Oda, 2000), reduction in NADPH generation as a result of reduced rate of photosynthesis and respiration, interference in translocation of nitrate from roots to shoots (Sharma & Subhadra, 2010), and strong affinity of nickel for the −SH group present at the active site of NR are some of the chief suggested mechanisms of Ni-mediated NR inhibition (Gouia, Ghorbal, & Meyer, 2000). PII is a plant signaling protein playing an important role in plants' nitrogen regulation through mediating a complex signal transduction pathway, and its expression may be indirectly controlled by Ni as it variably affects nitrate, ammonium, and nitric oxide levels in plants. PII expression is upregulated in the presence of nitrate and nitrite, and its expression is downregulated in the presence of ammonium due to NO generation leading to the activation of NO-dependent guanylate cyclase (De Montaigu, Sanz-Luque, Galvan, & Fernandez, 2010; Tejada-Jimenez et al., 2019).

17.8 Hydrogenase activity

Nitrogen fixation in the presence of nitrogenase releases a hydrogen molecule as a by-product for each molecule of nitrogen fixed, making this process energetically inefficient in legume–rhizobium symbiosis (Flynn, Graham, Scott, Layzell, & Dong, 2014). However, a limited number of rhizobial strains of *Rhizobium* and *Bradyrhizobium* have procured an efficient Hup system responsible for uptake of the evolved hydrogen, which upon further oxidation yields ATP (González-Guerrero, Matthiadis, Sáez, & Long, 2014) that is required in several plant physiological processes including N reduction to ammonia. The Hup system is rarely found in *Sinorhizobium* and *Mesorhizobium* strains. This Hup system is responsible for hydrogenase expression in bacteroids (Annan, Golding, Zhao, & Dong, 2012). Ni induced hydrogenase activity has been reported in soybean where the application of 1 mM $NiCl_2$ increased hydrogenase activity up to 45% over the control at day 52 of plant growth while its promoting effect disappeared at day 100 as a result of increased accumulation of Ni with time in plant tissue (Dalton, Evans, & Hanus, 1985). Ni deficiency lowering the hydrogenase activity in root nodules of legumes is also supported by the instance of compromised hydrogenase activity in pea bacteroids in the absence of Ni. Furthermore, hydrogenase activity enhances with increasing Ni concentration in culture media during initial days, but after 100 days, a decline in the activity was observed under Ni toxicity (Brito, Palacios, Hidalgo, Imperial, & Ruiz-Argüeso, 1994). Reports revealed that Ni at 85 M increased the full potential of hydrogenase activity in the growing media of plants (Polacco et al., 2013).

Further transcriptomic studies revealed that hydrogenase expression is regulated by posttranscriptional activation of Ni-dependent heterologous regulators (Brito, Monza, Imperial, Ruiz-Argüeso, & Palacios, 2000). The Hup system comprises of several components in which the first component is a dimeric membrane-bound [NiFe] hydrogenase (Baginsky, Brito, Imperial, Palacios, & Ruiz-Argüeso, 2002; Ruiz-Argüeso, Palacios, & Imperial, 2000). It is composed of two polypeptides, which result from proteolytic processing of its precursors as a result of metal-cluster insertion. The other component of the Hup system includes several *hup* and *hyp* gene products. These gene products result from the expression of 17 common genes termed *hupSLCDFGHIJKhypABFCDEX*. These genes are clustered and organized in at least three operons with conserved gene composition and organization (Baginsky et al., 2002). HupS, and hupL encode the main structural subunit of hydrogenase while other hup and hyp gene products are concerned with the conscription and incorporation of different metallic groups, including nickel into the active site of hydrogenase (Casalot & Rousset, 2001; Forzi & Sawers, 2007). A study of the *Escherichia coli* model reveals that nickel incorporation is facilitated by the interaction of two nickel binding proteins HypA and HypB (Blokesch, Rohrmoser, Rode, & Böck, 2004). Hyp A coordinates single Ni^{2+} to N terminal motif of the enzyme while HypB possesses GTPase activity for the complete maturation of hydrogenase (Watanabe, Sasaki, Tominaga, & Miki, 2012; Werner, Romeis, & Witte, 2009). Besides another protein, SlyD interacts with HypB to release Ni into the active center of the hydrogenase enzyme (Khorasani-Motlagh, Lacasse, & Zamble, 2017). Finally, HupL maturation takes place by proteolytic processing of a large subunit at the C terminus carried out by nickel-dependent protease HupD (Lacasse & Zamble, 2016).

The hypX gene in the hydrogenase gene cluster of *Rhizobium leguminosarum viciae* is responsible for nickel-dependent processing and activation of hydrogenase (Muraki et al., 2019). Whereas in the case of *Rhizobium leguminosarum UPM791, R. leguminosarum viciae, Rhizobium tropici*, and *Mesorhizobium loti*, the *hup*E gene product has been reported to act as a Ni permease, thereby functioning as a nickel transporter (Brito et al., 2010; Ruiz-Argüeso et al., 2000), while hupNOP gene products in *B. japonicum* are involved in nickel metabolism (Fu & Maier, 1991; Fu, Javedan, Moshiri, & Maier, 1994). During posttranscriptional activation of hydrogenase in *R. leguminosarum*, the immature small HupS and large HupL polypeptides get converted to active mature form only in the presence of Ni. Analysis of the *hupSL* promoter expression revealed that the transcription of the *hup* gene is activated by NifA, the key regulator of the nitrogen fixation (Ruiz-Argüeso et al., 2000). Also, in a system highly homologous to *R. leguminosarum*, it is reported that a large subunit of precursor HycE homologous to HupD is made active through proteolytic cleavage by a hycl gene product whose expression depends on the presence of nickel (Sargent, 2016). These findings were supported by a marked increase in hydrogenase activity upon Ni supplementation during pea–*R. leguminosarum* UPM791 symbiosis rendering Ni to be a limiting factor during legume–rhizobia symbiosis (Brito et al., 1994).

17.9 Urease activity

Urease is an enzyme in the higher plants that stoichiometrically incorporates Ni in its active site as a metallocenter. Ni also forms an integral part of bacterial urease. The vital function of urease is to catabolize urea into ammonia and hydroxamic acid, the latter further hydrolyzed to ammonia and bicarbonate through the following reaction:

$$(NH_2)_2CO + 3H_2O \rightarrow 2NH^{4+} + HCO_3^- + OH^-$$

Urea is endogenously produced in plants as a result of an increase in arginase activity. Excessive urea results in leaf tip necrosis, and hampers overall plant productivity (Ali, Hayat, Fariduddin, & Ahmad, 2009; Gerendas & Sattelmacher, 1999). In addition, it releases NH^{4+}, which is used as an N source for plant growth and metabolite synthesis (Barcelos et al., 2018). Urease also plays an important role in seasonal recycling of arginine, which serves as a major nitrogen repository in the form of stored amino acids in seed (Polacco et al., 2013; Sirko & Brodzik, 2000). The activity of urease depends largely on Ni availability in the soil. Ni acts both as a cofactor of urease as well as an initiator of urease maturation. Though Ni is responsible for urease maturation, it does not induce urease biosynthesis (Winkler, Polacco, Eskew, and Welch, 1983). Deficiency of Ni lowers the activity of urease, leading to the accumulation of toxic levels of urea, upsetting N metabolism in plant tissues (Yusuf, Fariduddin, Hayat, & Ahmad, 2011; Yusuf, Fariduddin, Hayat, Hasan, & Ahmad, 2011). While the excess of Ni ranging from 50–300 mg kg^1 soil resulted in a decline in urease activity in *C. cajan* (Garg & Saroy, 2020).

Urease activation in the case of *Rhizobium* bacteria is also functionally dependent on Ni availability. Transcriptome analysis of *R. leguminosarum* bv. viciae urease gene cluster revealed that it consists of three structural genes (ureA, ureB, and ureC), four accessory genes (ureD, ureE, ureF, and ureG), and five additional ORFs (orf83, orf135, orf207, orf223, and orf287) encoding proteins of unknown function (Toffanin, Cadahia, Imperial, Ruiz-Argüeso, & Palacios, 2002). *S. meliloti* has the same three structural genes intervened by three ORFs designated as orf3, orf5, and orf6, which are functional homologs of orf83, orf135, and orf207 respectively (Miksch, Arnold, Lentzsch, Priefer, & Puhler, 1994; Toffanin et al., 2002). Urease regulation, as studied in the *Klebsiella aerogenes* model system, reveals that Ni-dependent urease maturation results from the formation of apourease complex constituted by UreD-UreF-UreG (Moncrief & Hausinger, 1997). This process starts by recruitment of a molecular chaperon UreD to the UreABC trimer (Park, Carr, & Hausinger, 1994), which is responsible for making urease competent for Ni metallocenter insertion (Park & Hausinger, 1996) and subsequent binding of UreF to form UreABC-UreDF complex covers all the epitopes of UreD (M. Moncrief & Hausinger, 1996). Finally, UreG binds to the above complex forming UreABC-UreDFG upon which UreF activates GTPase activity of UreG, leading to activation of urease activity (Lam et al., 2010; Salomone-Stagni, Zambelli, Musiani, & Ciurli, 2007). Among these, UreE is a Ni-binding protein involved in Ni delivery to various other enzymes and Ni storage (Lee, Mulrooney, Renner, Markowicz, & Hausinger, 1992), and UreG also binds to Ni (Boer, Quiroz-Valenzuela, Anderson, & Hausinger, 2010). The dimeric UreE-UreG sequentially delivers two Ni atoms to UreF, then UreD finally followed by carbamate at modified lysine residue of urease active site (Carter & Hausinger, 2010).

Urease activation in plants slightly differs from bacterial pathways as plant urease lacks UreE encoding genes (Cao et al., 2010; Witte, Tiller, Isidore, Davies, & Taylor, 2005). In plants, activation of urease is facilitated by three urease accessory proteins (UAPs): UreG, UreD, and UreF. N-terminus of UreG in plant resembles C-terminus of UreE which is responsible for Ni^{2+} coordination (Zambelli et al., 2013) and two copies of UreD and UreF are present in many legumes (Bailey & Boulter, 1971), soybean (Polacco, Hyten, Medeiros-Silva, Sleper, & Bilyeu, 2011), and cucurbits (Fahmy, Mohamed, & Kamel, 1993). Moreover, most legumes contain more than one urease; for example, in soybean, two functional ureases, namely embryo specific urease and ubiquitous urease are present (Carlini & Polacco, 2008), thereby generating the possibility of involvement of multiple accessory proteins for multiple apo-urease activations in plants. In soybean, attachment of Ni to urease active site is facilitated by gene products of Eu2 and Eu3, which are functional homologs of UreF and UreG of model bacteria (Polacco et al., 2011). The gene product of Eu3 is a 32-kD protein which, upon interaction with the product of Eu2, leads to Ni recruitment at the active site of urease facilitating embryonic urease activation. However, Ni deprivation at this stage leads to the inactivation of the gene product of Eu3 resulting in failure of urease activation (Polacco, Freyermuth, Gerendás, & Cianzio, 1999; Sirko & Brodzik, 2000). Recent investigations using strep tagged urease: UAPs provide a new insight into urease activation in the plant. Deletion studies in UreG exhibit that the hypervariable sites of its N-terminal domain containing His and Asp/Glu-rich residues play a minor role in urease activation efficiency. At the same time, the highly conserved first HFX motif proves to be an essential and second HFX motif compulsory for Ni binding and delivery, playing a crucial role in urease activation (Myrach, Zhu, & Witte, 2017). Nickel uptake, through rooting medium, reduced the urea concentration in the leaves along with a considerable decline in urease activity. Thus Ni plays an essential role in optimal N-use efficiency under various N methods by controlling urea metabolism in both legume and *Rhizobium* (Polacco et al., 2013).

17.10 Ureide metabolism

Ureides (allantoin and allantoate) are the characteristic main N-containing compounds occurring freely in the xylem of leguminous plants. Several investigations reveal that ureide breakdown occurs through a purine breakdown pathway in tropical legumes without the involvement of urea as intermediate, unlike in the case of many bacteria (Stahlhut & Widholm, 1989).

Ni is required for ureide transport and its catabolism in plant tissues (Siqueira Freitas et al., 2018). A study of a xanthine dehydrogenase (an enzyme responsible for purine breakdown) mutant of *Arabidopsis thaliana* suggested that ureides play a role in combating reactive oxygen species (ROS) mitigating oxidative stress under critical conditions (Brychkova, Alikulov, Fluhr, & Sagi, 2008). The translational product of the allantoate aminohydrolase gene causes the breakdown of allantoate in soybean, escaping the release of urea (Todd & Polacco, 2006; Werner, Sparkes, Romeis, & Witte, 2008). Degradation of allantoate by this pathway in three sequential enzymatic steps forms ureidoglycine, ureidoglycolate, and glyoxylate, releasing ammonia and bicarbonate as by-products (Werner et al., 2009). Deficiency of nickel is reported to disrupt ureide metabolism resulting in overaccumulation of allantoic acid, xanthine, and ureidoglycolate content in *Carya illinoinensis* foliage (Bai, Reilly, & Wood, 2006). Ni deprivation also results in increased urea accumulation formed due to the nonenzymatic breakdown of highly unstable ureidoglycine, and urease inactivity further elevates urea level during Ni deficiency (Polacco et al., 2013; Todd & Polacco, 2004). Ni at low concentration (0.5 mg kg^{-1} of soil) increased ureide concentration in soybean (Saif & Khan, 2018). In contrast, high concentrations of Ni (300 mg kg^{-1} of soil) significantly lowered total ureides content in nodules correlated with a decrease in N content in nodules of *C. cajan* (Garg & Saroy, 2020).

17.11 Amino acid metabolism

Nickel modulates the activities of several amino acid metabolizing enzymes that alter amino acid levels in the plant. Ni deprivation caused altered amino acid metabolism in *Vigna unguiculata* (Walker, Graham, Madison, Cary, & Welch, 1985). In *G. max*, Ni at 200 mM upregulates aspartate aminotransferase and alanine aminotransferase activities and leads to the accumulation of aspartic acid and alanine. Moreover, Ni also increases cysteine level constituted phytochelatin (El-Shintinawy & El-Ansary, 2000). The proline concentrations were enhanced in several legumes such as *G. max*, *V. radiata*, and *Pisum sativum* with elevated Ni concentrations (Gajewska & Sklodowska, 2005; Sirhindi, Mir, Abd-Allah, Ahmad, & Gucel, 2016; Yusuf et al., 2014). Proline combats various abiotic stresses through improved osmoregulation, photosynthesis, secondary metabolite accumulation, enhanced nitrogen fixation, and N-use efficiency in plants (Alyemeni, Hayat, Hayat, Faizan, & Faraz, 2016; Noreen, Akhter, Yaamin, & Arfan, 2018; Zali & Ehsanzadeh, 2018).

Further, a study on Ni deficient pelican (*C. illinoinensis*) elucidates that Ni deficiency enhances glycine, isoleucine, valine, arginine, tyrosine, tryptophan, and total free amino acid contents but decreases glutamic acid and histidine levels (Bai et al., 2006). However, elevated levels of Ni at 100 μM decrease activities of glutamine synthetase and NADH-GOGAT in wheat (Gajewska et al., 2009), which is responsible for the incorporation of ammonia generated as a result of nitrogen fixation into C-skeleton forming glutamate (Hanfrey, Sommer, Mayer, Burtin, & Michael, 2001). This decrease in glutamate levels is overcome by the upregulation of activities of aminating glutamate dehydrogenase (NADH-GDH) and aspartate aminotransferase (AspAT) in the presence of any concentration of nickel (Mishra & Dubey, 2011). Glutamine generated as result acts as an effector molecule and is sensed by a conserved PII plant protein (responsible for N regulation), which in turn enables the PII to interact with and activate other chloroplast proteins like N-acetyl glutamate kinase and results in arginine and other polyamines biosynthesis (Chellamuthu et al., 2014). Ni stress is also reported to upregulate the activities of S-adenosylmethionoine decarboxylase, ornithine decarboxylase, and arginine decarboxylase, increasing polyamine biosynthesis in *P. sativum* (Shahid et al., 2014). Thus enhanced production of polyamines like putrescine, spermidine, and spermine plays a regulatory role in overcoming abiotic stress conditions (Ali, Hassan, Shah, Sajjad, & Bhat, 2020).

17.12 Carbon metabolism

Carbon metabolites in plants mainly occur in the form of carbohydrate photosynthates and organic acids. Carbon compounds generated by the host serve as sources of energy to support the growth and respiration of *Rhizobium* in legume–rhizobia symbiosis, and organic acids further serve to overcome nutritional deficiencies, confer metal tolerance, and improve legume rhizobia interaction at root–soil interphase. Deficiency of Ni disrupted carbon metabolism via the accumulation of lactic and oxalic acid and reduction in citrate levels, causing the famous morphological "mouse-ear symptom" in pecan (Bai et al., 2006). Ni at a concentration of 20 μM significantly improved malic, oxalic, citric, and cis-aconitic acid accumulation in shoots of *Lolium perenne* L., whereas Ni levels ranging from 40 to 80 μM significantly enhanced oxalic, malic, and cis-aconitate levels in *Zea mays* L. (Yang, Baligar, Foster, & Martens, 1997). An increase in citrate levels in rhizobia plays an important role in normal nodulation through direct impact on iron uptake and other metabolic reactions enhancement in the cells (Pardo, Lagunez, Miranda, & Martinez, 1994). Citrate levels enhance Fe transport through MATE family citrate transporters in both *Rhizobi*um and plants (Roschzttardtz, Séguéla-Arnaud, Briat, Vert, & Curie, 2011; Takanashi et al., 2013) resulting in activation of several enzymes involved in symbiotic nitrogen fixation requiring iron as a cofactor (González-Guerrero et al., 2014). The presence of organic acids lowers Ni toxicity by reducing Ni mobility and bioavailability by forming stable chelation complexes with Ni and other heavy metals (Yang et al., 1997). Nickel hyperaccumulators have shown the efficiency of Ni^{2+} to form a complex with citrate correlated from an increase in citrate levels with increasing Ni^{2+} concentrations (Hossain, Piyatida, Silva, & Fujita, 2012).

17.13 Nutrient balance

Low nickel concentration promotes the overall growth of the plant and enhances legume rhizobia symbiosis, as Ni at low concentration increases N, P, Fe, and Cu uptake and translocation in several plants (Amjad et al., 2020; Kumar, Rouphael, Cardarelli, & Colla, 2015). Uptake of nitrogen improves the nitrogen fixation rate of nitrogenase, and an increase in P enhances the growth of both host and rhizobia, serving as a source of energy acquisition, utilization, and storage during various physiological and biochemical growth processes (Epstein & Bloom, 2005). Fe is important for enhanced efficiency of nitrogen fixation as it forms a structural part of nitrogenase NifH and NifDK components (Jasniewski, Sickerman, Hu, & Ribbe, 2018), required for LHb biosynthesis and antioxidant enzymes SOD and CAT activity (Jamet, Sigaud, Van de Sype, Puppo, & Hérouart, 2003; Rubio et al., 2007), and serves as an electron carrier responsible for various energy transduction processes by rhizobial cytochromes (Brear, Day, & Smith, 2013). Improvement in Cu levels results in enhanced nitrogen fixation per nodule resulting in increased N concentration in plant tissues, and it also acts as a cofactor of energy transducing cytochromes in bacteroids (González-Guerrero et al., 2014).

However, at higher concentrations of Ni, there is a reduction in P, Ca, Mg, Mn, Fe, K, Cu, and Zn content in higher plants, including legumes (Batool, 2018; dos Reis et al., 2017; Panday & Sharma, 2002; Parida, Chhibba, & Nayyar, 2003) either by competing with these metals in absorption or reducing their uptake by causing membrane alterations and damage resulting in hindrance on metal uptake and its translocation to sink (Chen, Haung, & Liu, 2009; A. Sharma & Dhiman, 2013). Particularly, Fe deficiency under excess Ni is reported to be caused by increased expression of IRT1 (an iron uptake transporter which can also transport Ni), FROS2 (a ferric reductase), and a transcription factor FIT involved in Fe homeostasis (Nishida, Aisu, & Mizuno, 2012). Ca and Mn are initially involved in rhizobial colonization through the facilitation of association of rhizobial lectin to root hair tips (González-Guerrero et al., 2014), and Ca spiking at later stages induces upregulation of transcription factors NIN or NSP1/2 activating nodulation genes (Andriankaja et al., 2007; Kalo et al., 2005). Further, K transport via putative channels like CASTOR, POLLUX, or DMI1 plays an essential role in Ca-mediated early signaling processes by acting as a counter ion compensating Ca in earlier nodulation signaling processes (Charpentier & Oldroyd, 2013; Peiter et al., 2007). Ni supply may also reduce Mo uptake, which acts as the NR cofactor (Akpinar et al., 2015). Therefore, a decrease in the aforementioned metals content under Ni stress causes a heavy toll on legume–rhizobia symbiosis.

17.14 Nickel and rhizobia-mediated molecular responses for the alleviation of critical conditions during legume–rhizobia symbiosis

Legume–rhizobia symbiosis faces a significant setback during environmental stress, posing a threat for host survival and rhizobial existence and a decline in its overall metabolic functions. Nickel within tolerable limits causes an increase in activities of various antioxidant enzymes resulting in detoxification of free radicals produced as a result of the generation of ROS, an indicator of oxidative stress. There are several instances which ensure the presence of low levels of Ni in soil has shown to increase the activities of superoxide dismutase, catalase, peroxidase, ascorbate peroxidase, glutathione reductase, and guaiacol peroxidase in plants, finally resulting in effective scavenging of ROS (Freeman et al., 2004; Gomes-Juniora, 2006). In order to gain more information on Ni metabolism, 2D gel electrophoresis in combination with LC-MS/MS analysis was performed, which showed that Ni tolerance is associated with enhanced synthesis of cysteine or downstream metabolite like GSH, resulting in the accumulation of various proteins involved in sulfur metabolism (Ingle, Smith, & Sweetlove, 2005).

Further, Ni also accumulates osmolytes like proline (Saif & Khan, 2018) and trehalose (Garg & Saroy, 2020). These osmolytes exhibit a myriad of roles under osmotic stress conditions as proline is involved in membrane stabilization and stabilization of proteins and subcellular structures (Kavi Kishor, Hima Kumari, Sunita, & Sreenivasulu, 2015), while trehalose acts as a C-source safeguarding the survival and tolerance of rhizobium toward heat, desiccation, and salinity stress conditions (Reina-Bueno et al., 2012). During extreme environmental conditions, these endogenously produced osmolytes in plants are utilized by the rhizobial partner resulting in the secretion of certain proteins to alleviate the problems of salinity, aridity, and toxicity (Karmakar, Rana, Rajwar, Sahgal, & Johri, 2015). Ni also plays an important role in the synthesis of metal binding proteins like phytochelatin and metallothioneins (MTs), responsible for the compartmentation and sequestration of toxic metal ions to overcome oxidative stress and confer heavy metal tolerance to plants (Ferraz, Fidalgo, Almeida, & Teixeira, 2012; Shahzad et al., 2018). Transcriptomic studies in *Solanum nigrum* revealed that MT-related transcripts like MT2a, MT2d, MT2c, and MT3 mRNA overaccumulated in response to Ni stress (Ferraz et al., 2012). Further, the accumulation of methyl glyoxime (MG) elevates under stressful conditions as a result of imbalances in glycolysis and photosynthetic pathways (Richard, 1993). These molecules (MG) generate oxidative stress as a result of ROS generation (Kalapos, 2008), glycation end-products accumulation (Thornalley, 2003), and compromisation of antioxidant machinery (Martins, Cordeiro, & Ponces Freire, 2001). Nickel acts as a cofactor and activator of different isoforms of glyoxylase I enzyme in plants (Mustafiz et al., 2014), responsible for the degradation of MG (Mustafiz, Singh, Pareek, Sopory, & Singla-Pareek, 2011). Moreover, nickel also regulates reduced glutathione homeostasis, which is consumed and regenerated during MG detoxification events (Fabiano, Tezotto, Favarin, Polacco, & Mazzafera, 2015) through modulation of glutathione reductase activity (Fabiano et al., 2015). This enhancement of stress regulatory pathways by nickel in plants ensures host survival, ensuring a sustainable legume–rhizobium symbiosis.

However, at higher concentration, Ni itself leads to oxidative stress as a combined result of the increase in lipid peroxidation mediated in the form of elevation of MDA (malondialdehyde) levels and a reduction in Fe uptake and translocation, which forms a major component of ROS scavenging SOD and CAT (Helaoui et al., 2020; Nishida et al., 2011; Seregin & Kozhevnikova, 2006) leading to inefficiency of enzymatic and nonenzymatic antioxidant modulation. Also, phytocystatins, a class of protease inhibitors performing a myriad of functions in plants such as seed maturation, seed germination, programmed cell death, and plant defense system toward various abiotic stress factors (Kunert, van Wyk, Cullis, Vorster, & Foyer, 2015), are affected by high Ni concentrations. Proteomic analysis (fluorescence spectroscopy, ANS binding, and caseinolytic activity assay) reveals excess nickel is responsible for causing a conformational change in phytocystatins leading to a loss in its activity (Siddiqui et al., 2019). In such cases, symbiotic rhizobial species also play an important role in facilitating better survival of host plant during different environmental stress conditions, including Ni stress.

Several N_2-fixing bacteria like *Ensifer, Bradyrhizobium*, and *Mesorhizobium* possess an active Ni efflux system to overcome Ni accumulation, causing toxicity (Chaintreuil et al., 2007; Zielazinski et al., 2013). DmeR is one such metalloregulator responsible for the efflux of Ni^{2+} and Co^{2+} in *R. leguminosarum* and *S. meliloti*, which helps to cope with nickel and cobalt toxicity (Higgins, 2019). *Rhizobium* strains L9 and L19 and *Mesorhizobium* strains L42 and L50 isolated from *L. corniculatus*–rhizobia symbiosis on Ni-rich ultramafic disclose the potential of these rhizobial strains to overcome Ni phytotoxicity through enhanced production of IAA, improved siderophore synthesis, and increased phosphate solubilization (Sujkowska-Rybkowska et al., 2020). A multiomic approach reflects upregulation of various stress response genes in different symbiotic *Rhizobium* species under critical conditions; for instance, upregulation of rpoH2 in *S. meliloti* and *Rhizobium etli* to face salinity stress (da-Silva, Alexandre, Brígido, & Oliveira, 2017) and enhanced induction of various genes encoding for molecular chaperones and heat shock proteins to combat extreme temperature stress (Alexandre & Oliveira, 2016). *S. melioti* strain establishing symbiosis with *Medicago truncatula* harbors an additional pathway for indole-3-acetic acid (IAA) synthesis, which increases its tolerance toward salinity, low pH, and UV radiations (Bianco & Defez, 2009).

In this regard, *R. etli–P. vulgaris* symbiosis easily bypasses drought due to enhanced production of cbb3 oxidase which is a cytochrome responsible for bacteroid survival under low O_2 concentrations in root nodules, arising as a result of LHb presence (Talbi, Sanchez, & Hidalgo-Garcia, 2012).

Metabolomic profiling of *Rhizobium*-inoculated *Medicago sativa* reflects a better response to alkali stress over noninoculated control through the better accumulation of osmolytes (proline, sugar, glycols), antioxidants (CAT, POX, GSH), organic acids (alpha-ketoglutaric acid, fumaric acid, and succinic acid), and other metabolites involved in nitrogen fixation (Song et al., 2017). The above evidence well describes that legume–rhizobia symbiosis reduces the dependence on nitrogen fertilizers for enhancing crop productivity and enables the host plant to resist adverse environmental conditions.

17.15 Conclusion

The essentiality of Ni has become a moot topic in legumes given its both essential and toxic roles. It is not only required for proper plant growth, but it

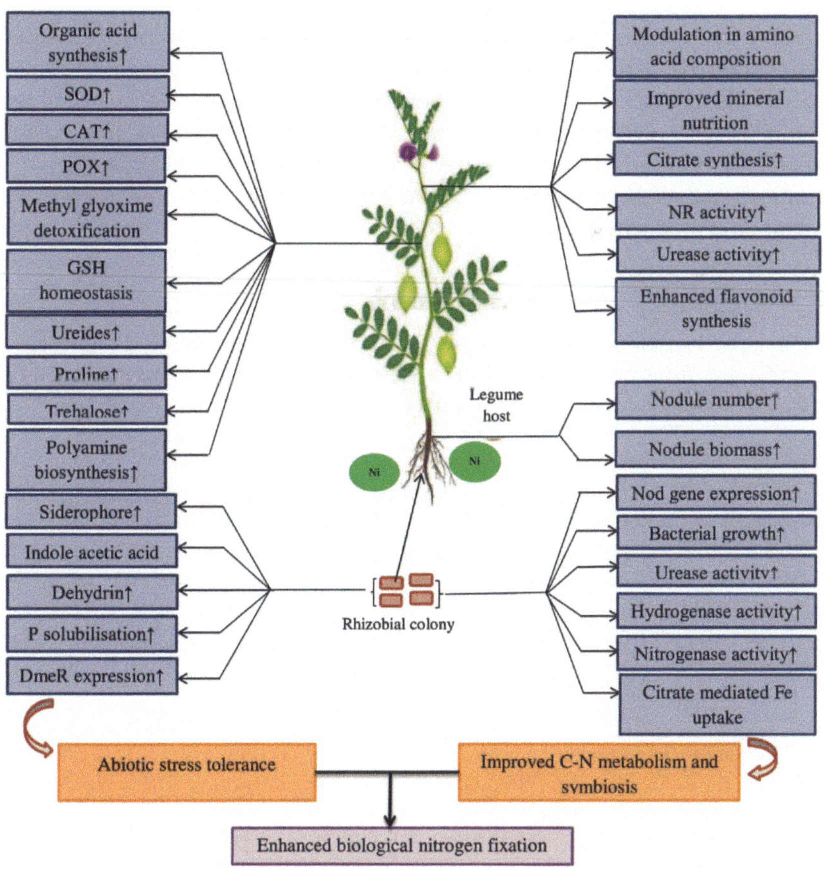

FIGURE 17.2 Beneficial role of Ni in legume–*Rhizobium* symbiosis under critical conditions.

also plays prodigious roles related to legume–rhizobium symbiosis, which is one of the essential characteristics of legumes from an ecological as well as nutritional point of view. Nickel content in soil regulates the expression of nitrogenase, hydrogenase and urease involved in nitrogen fixation, nitrogen assimilation and N-use efficiency during host–rhizobia interaction. Moreover, Ni level differentially regulated C:N ratio in the host, which forms the basis of the symbiotic system as the plant receives N fixed by rhizobia, while rhizobia rely on the photosynthates (carbon compounds) supplied by the plant for its growth and survival. Variation in Ni level varies with NO_3^-, NH_4^+, and NO levels in plants, which modulate signaling of PII proteins responsible for N regulation in plants, while C levels are mainly regulated through changes in the content of Kreb's cycle intermediates. Low concentration of Ni exhibits a promoting role on the above factors to enhance the efficiency of legume–rhizobia symbiosis. Besides, Ni also induces several stress strategies in the host to cope with critical conditions like an increase in osmoprotectant, antioxidant enzymes, and metal chelation proteins, which is further enhanced by the rhizobia species invading the legume. A generalized overview of the beneficial role of Ni during legume–rhizobium symbiosis is illustrated in Fig. 17.2. Though Ni plays a crucial role in various physiological and biochemical processes in plants, the molecular basis of Ni metabolism is still an enigma. Further molecular investigations on Ni-mediated responses would unravel the full potential of Ni in breeding and engineering efforts around Ni-dependent enzymes to enhance overall crop productivity.

Acknowledgment

Ms. Rukhsar Parwez would like to duly acknowledge the funding by the Council of Scientific and Industrial Research (CSIR), New Delhi, under the JRF scheme (09/112 (0651)/2019-EMR-1).

References

Abd-Alla, M. H., Bashandy, S. R., Bagy, M. K., & El-enany, A. W. E. (2014). *Rhizobium tibeticum* activated with a mixture of flavonoids alleviates nickel toxicity in symbiosis with fenugreek (*Trigonella foenum graecum* L.). *Ecotoxicology (London, England)*, *23*(5), 946–959.

Akpinar, A., Arslan, H., Güleryüz, G., Kırmızı, S., Erdemİr, Ü. S., & Güçer, Ş. (2015). ni-induced changes in nitrate assimilation and antioxidant metabolism of *Verbascum olympicum* Boiss.: Could the plant be useful for phytoremediation or/and restoration purposes? *International Journal of Phytoremediation*, *17*(6), 546–555.

Alam, M. M., Hayat, S., Ali, B., & Ahmad, A. (2007). Effect of 28-homobrassinolide treatment on nickel toxicity in *Brassica juncea*. *Photosynthetica*, *45*(1), 139–142.

Alexandre, A., & Oliveira, S. (2016). *Heat shock response in bacteria with large genomes: Lessons from rhizobia. Stress and environmental control of gene expression in bacteria* (pp. 737–746). New Jersey, United States: Wiley-Blackwell Publishers.

Alfano, M., & Cavazza, C. (2020). Structure, function, and biosynthesis of nickel-dependent enzymes. *Protein Science*, *29*(5), 1071–1089.

Ali, B., Hayat, S., Fariduddin, Q., & Ahmad, A. (2009). *Nickel: Essentiality, toxicity and tolerance in plants. Nickel in relation to plants* (pp. 73–80). New Delhi, India: Narosa Publishing House.

Ali, R., Hassan, S., Shah, D., Sajjad, N., & Bhat, E. A. (2020). *Role of polyamines in mitigating abiotic stress. Protective chemical agents in the amelioration of plant abiotic stress: Biochemical and molecular perspectives* (p. 291) .

Alibakhshi, M., & Khoshgoftarmanesh, A. H. (2015). Effects of nickel nutrition in the mineral form and complexed with histidine in the nitrogen metabolism of onion bulb. *Plant Growth Regulation, 75*, 733–740.

Alyemeni, M. N., Hayat, Q., Hayat, S., Faizan, M., & Faraz, A. (2016). Exogenous proline application enhances the efficiency of nitrogen fixation and assimilation in chickpea plants exposed to cadmium. *Legume Research-An International Journal, 39*(2), 221–227.

Amjad, M., Raza, H., Murtaza, B., Abbas, G., Imran, M., Shahid, M., & Iqbal, M. M. (2020). Nickel toxicity induced changes in nutrient dynamics and antioxidant profiling in two maize (*Zea mays* L.) hybrids. *Plants, 9*(1), 5.

Andriankaja, A., Boisson-Dernier, A., Frances, L., Sauviac, L., Jauneau, A., Barker, D. G., & de Carvalho-Niebel, F. (2007). AP2-ERF transcription factors mediate Nod factor–dependent Mt ENOD11 activation in root hairs via a novel cis-regulatory motif. *The Plant Cell, 19*(9), 2866–2885.

Annan, H., Golding, A. L., Zhao, Y., & Dong, Z. (2012). Choice of hydrogen uptake (Hup) status in legume-rhizobia symbioses. *Ecology and Evolution, 2*(9), 2285–2290.

Astier, J., Gross, I., & Durner, J. (2018). Nitric oxide production in plants: An update. *Journal of Experimental Botany, 69*(14), 3401–3411.

Athar, R., & Ahmad, M. (2002). Heavy metal toxicity in legume-microsymbiont system. *Journal of Plant Nutrition, 25*(2), 369–386.

Baginsky, C., Brito, B., Imperial, J., Palacios, J. M., & Ruiz-Argüeso, T. (2002). Diversity and evolution of hydrogenase systems in rhizobia. *Applied and Environmental Microbiology, 68* (10), 4915–4924.

Bai, C., Reilly, C. C., & Wood, B. W. (2006). Nickel deficiency disrupts metabolism of ureides, amino acids, and organic acids of young pecan foliage. *Plant Physiology, 140*(2), 433–443.

Bailey, C. J., & Boulter, D. (1971). *Urease, a typical seed protein of the Leguminosae. Chemotaxonomy of the leguminosae* (pp. 485–502). New York: Academic Press.

Barcelos, J. P. Q., Reis, H. P. G., Godoy, C. V., Gratão, P. L., Furlani Junior, E., Putti, F. F., & Reis, A. R. (2018). Impact of foliar nickel application on urease activity, antioxidant metabolism and control of powdery mildew (*Microsphaera diffusa*) in soybean plants. *Plant Pathology, 67*(7), 1502–1513.

Batool, S. (2018). Impact of bioaccumulation of nickel on growth, seed yield and mineral uptake of Chickpea (*Cicer arietinum* L.) varieties. *Pakistan Journal of Botany, 50*(6), 2147–2150.

Bertrand, D., & de Wolf, A. (1967). Nickel, a dynamic trace element for higher plants. *Comptes rendus de l'Académie des Sciences, 265*, 1053–1055.

Bianco, C., & Defez, R. (2009). *Medicago truncatula* improves salt tolerance when nodulated by an indole-3-acetic acid-overproducing *Sinorhizobium meliloti* strain. *Journal of Experimental Botany, 60*, 3097–3107.

Blokesch, M., Rohrmoser, M., Rode, S., & Böck, A. (2004). HybF, a zinc-containing protein involved in NiFe hydrogenase maturation. *Journal of Bacteriology, 186*(9), 2603–2611.

Boer, J. L., Mulrooney, S. B., & Hausinger, R. P. (2014). Nickel-dependent metalloenzymes. *Archives of Biochemistry and Biophysics, 544*, 142–152.

Boer, J. L., Quiroz-Valenzuela, S., Anderson, K. L., & Hausinger, R. P. (2010). Mutagenesis of *Klebsiella aerogenes* UreG to probe nickel binding and interactions with other urease-related proteins. *Biochemistry*, *49*(28), 5859–5869.

Brear, E. M., Day, D. A., & Smith, P. M. C. (2013). Iron: An essential micronutrient for the legume-rhizobium symbiosis. *Frontiers in Plant Science*, *4*, 359.

Brito, B., Monza, J., Imperial, J., Ruiz-Argüeso, T., & Palacios, J. M. (2000). Nickel availability and hupSL activation by heterologous regulators limit symbiotic expression of the *Rhizobium leguminosarum* bv. *viciae* hydrogenase system in Hup − rhizobia. *Applied and Environmental Microbiology*, *66*(3), 937–942.

Brito, B., Palacios, J. M., Hidalgo, E., Imperial, J., & Ruiz-Argüeso, T. (1994). Nickel availability to pea (*Pisum sativum* L.) plants limits hydrogenase activity of *Rhizobium leguminosarum* bv. *viciae* bacteroids by affecting the processing of the hydrogenase structural subunits. *Journal of Bacteriology*, *176*(17), 5297–5303.

Brito, B., Prieto, R. I., Cabrera, E., Mandrand-Berthelot, M. A., Imperial, J., Ruiz-Argüeso, T., & Palacios, J. M. (2010). *Rhizobium leguminosarum* hupE encodes a nickel transporter required for hydrogenase activity. *Journal of Bacteriology*, *192*(4), 925–935.

Brychkova, G., Alikulov, Z., Fluhr, R., & Sagi, M. (2008). A critical role for ureides in dark and senescence-induced purine remobilization is unmasked in the Atxdh1 *Arabidopsis* mutant. *The Plant Journal: For Cell and Molecular Biology*, *54*, 496–509.

Cao, F. Q., Werner, A. K., Dahncke, K., Romeis, T., Liu, L. H., & Witte, C. P. (2010). Identification and characterization of proteins involved in rice urea and arginine catabolism. *Plant Physiology*, *154*(1), 98–108.

Carlini, C. R., & Polacco, J. C. (2008). Toxic properties of urease. *Crop Science*, *48*(5), 1665–1672.

Carter, E. L., & Hausinger, R. P. (2010). Characterization of the *Klebsiella aerogenes* urease accessory protein UreD in fusion with the maltose binding protein. *Journal of Bacteriology*, *192*(9), 2294–2304.

Casalot, L., & Rousset, M. (2001). Maturation of the [NiFe] hydrogenases. *Trends in Microbiology*, *9*(5), 228–237.

Chaintreuil, C., Rigault, F., Moulin, L., Jaffre, T., Fardoux, J., & Giraud, E. (2007). Nickel resistance determinants in *Bradyrhizobium* strains from nodules of the endemic New Caledonia legume *Serianthes calycina*. *Applied and Environmental Microbiology*, *73*, 8018–8022.

Chamizo-Ampudia, A., Sanz-Luque, E., Llamas, A., Galvan, A., & Fernandez, E. (2017). Nitrate reductase regulates plant nitric oxide homeostasis. *Trends in Plant Science*, *22*(2), 163–174.

Charpentier, M., & Oldroyd, G. E. (2013). Nuclear calcium signaling in plants. *Plant Physiology*, *163*(2), 496–503.

Chellamuthu, V. R., Ermilova, E., Lapina, T., Lüddecke, J., Minaeva, E., Herrmann, C., & Forchhammer, K. (2014). A widespread glutamine-sensing mechanism in the plant kingdom. *Cell*, *159*(5), 1188–1199.

Chen, C., Haung, D., & Liu, J. (2009). Functions and toxicity of nickel in plants: Recent advances and future prospects. *Clean*, *37*, 304–313.

Cooper, J. E. (2007). Early interactions between legumes and rhizobia: Disclosing complexity in a molecular dialogue. *Journal of Applied Microbiology*, *103*(5), 1355–1365.

Curie, C., Cassin, G., Couch, D., Divol, F., Higuchi, K., & LeJean, M. (2009). Metal movement within the plant: Contribution of nicotianamine and yellowstripe1-like transporters. *Annals of Botany*, *103*, 1–11.

Dalton, D. A., Evans, H. J., & Hanus, F. J. (1985). Stimulation by nickel of soil microbial urease activity and urease and hydrogenase activities in soybeans grown in a low-nickel soil. *Plant and Soil*, *88*, 245–258.

Das, G. (1982). A stimulatory effect of low nitrate levels on nodulation, leghaemoglobin content, nitrogenase activity, and growth of mung-bean plants (*Vigna radiata*). *Canadian Journal of Botany*, *60*(10), 1907–1912.

da-Silva, J. R., Alexandre, A., Brígido, C., & Oliveira, S. (2017). Can stress response genes be used to improve the symbiotic performance of rhizobia? *AIMS Microbiology*, *3*(3), 365.

de Macedo, F. G., Bresolin, J. D., Santos, E. F., Furlan, F., Lopes da Silva, W. T., Polacco, J. C., & Lavres, J. (2016). Nickel availability in soil as influenced by liming and its role in soybean nitrogen metabolism. *Frontiers in Plant Science*, *7*, 1358.

De Montaigu, A., Sanz-Luque, E., Galvan, A., & Fernandez, E. (2010). A soluble guanylate cyclase mediates negative signaling by ammonium on expression of nitrate reductase in *Chlamydomonas*. *The Plant Cell*, *22*, 1532–1548.

Del Cerro, P., Ayala-García, P., Buzón, P., Castells-Graells, R., López-Baena, F. J., Ollero, F. J., & Pérez-Montaño, F. (2020). OnfD, an AraC-type transcriptional regulator encoded by *Rhizobium tropici* CIAT 899 and involved in Nod factor synthesis and symbiosis. *Applied and Environmental Microbiology*, *86*(19).

dos Reis, A. R., de Queiroz Barcelos, J. P., de Souza Osório, C. R. W., Santos, E. F., Lisboa, L. A. M., Santini, J. M. K., & Lavres, J. (2017). A glimpse into the physiological, biochemical and nutritional status of soybean plants under Ni-stress conditions. *Environmental and Experimental Botany*, *144*, 76–87.

Dudeja, S. S., Sheokand, S., & Kumari, S. (2012). Legume root nodule devel legume root nodule development and functioning under opment and functioning under tropics and subtropics: Perspectives and challenges. *Legume Research*, *35*, 85–103.

El-Shintinawy, F., & El-Ansary, A. (2000). Differential effect of Cd^{2+} and Ni^{2+} on amino acid metabolism in soybean seedlings. *Biologia Plantarum*, *43*(1), 79–84.

Epstein, E., & Bloom, A. J. (2005). *Mineral metabolism-mineral nutrition of plants: Principles and perspectives* (2nd edn). Saunderland, Massachusetts: Sinuar Associates. Inc. Publishers.

Eskew, D. L., Welch, R. M., & Cary, E. E. (1983). Nickel: An essential micronutrient for legumes and possibly all higher plants. *Science (New York, N.Y.)*, *222*(4624), 621–623.

Fabiano, C., Tezotto, T., Favarin, J. L., Polacco, J. C., & Mazzafera, P. (2015). Essentiality of nickel in plants: A role in plant stresses. *Frontiers in Plant Science*, *6*, 754.

Fahmy, A. S., Mohamed, M. A., & Kamel, M. Y. (1993). Ureases in the cucurbitaceae, distribution and properties. *Phytochemistry*, *35*(1), 151–154.

Fancy, N. N., Bahlmann, A. K., & Loake, G. J. (2017). Nitric oxide function in plant abiotic stress. *Plant, Cell & Environment*, *40*(4), 462–472.

Ferraz, P., Fidalgo, F., Almeida, A., & Teixeira, J. (2012). Phytostabilization of nickel by the zinc and cadmium hyperaccumulator *Solanum nigrum* L. Are metallothioneins involved? *Plant Physiology and Biochemistry*, *57*, 254–260.

Flynn, B., Graham, A., Scott, N., Layzell, D. B., & Dong, Z. (2014). Nitrogen fixation, hydrogen production and N_2O emissions. *Canadian Journal of Plant Science*, *94*(6), 1037–1041.

Forzi, L., & Sawers, R. G. (2007). Maturation of [NiFe]-hydrogenases in *Escherichia coli*. *Biometals: An International Journal on the Role of Metal Ions in Biology, Biochemistry, and Medicine*, *20*(3-4), 565.

Freeman, J. L., Persans, M. W., Nieman, K., Albrecht, C., Peer, W., Pickering, I. J., & Salt, D. E. (2004). Increased glutathione biosynthesis plays a role in nickel tolerance in *Thlaspi* nickel hyperaccumulators. *The Plant Cell*, *16*, 2176–2191.

Fu, C., Javedan, S., Moshiri, F., & Maier, R. J. (1994). Bacterial genes involved in incorporation of nickel into a hydrogenase enzyme. *Proceedings of the National Academy of Sciences*, *91* (11), 5099–5103.

Fu, C., & Maier, R. J. (1991). Competitive inhibition of an energy-dependent nickel transport system by divalent cations in *Bradyrhizobium japonicum* JH. *Applied and Environmental Microbiology*, *57*, 3511–3516.

Gajewska, E., & Sklodowska, M. (2005). Antioxidative responses and proline level in leaves and roots of pea plants subjected to nickel stress. *Acta Physiologiae Plantarum*, *27*, 329–339.

Gajewska, E., Wielanek, M., Bergier, K., & Skłodowska, M. (2009). Nickel-induced depression of nitrogen assimilation in wheat roots. *Acta Physiologiae Plantarum*, *31*(6), 1291.

Garg, N., & Saroy, K. (2020). Interactive effects of polyamines and arbuscular mycorrhiza in modulating plant biomass, N_2 fixation, ureide, and trehalose metabolism in *Cajanus cajan* (L.) Millsp. genotypes under nickel stress. *Environmental Science and Pollution Research*, *27*(3), 3043–3064.

Gerendas, J., & Sattelmacher, B. (1999). Influence of Ni Supply on Growth and Nitrogen Metabolism ofBrassica napusL. Grown with NH_4NO_3 or Urea as N Source. *Annals of Botany*, *83*(1), 65–71.

Geurts, R., & Bisseling, T. (2002). *Rhizobium* Nod factor perception and signalling. *The Plant Cell*, *14*(1), S239–S249.

Gill, M. (2014). Heavy metal stress in plants: A review. *International Journal of Advanced Research*, *2*, 1043–1055.

Gomes-Juniora, R. A. (2006). Nickel elicits a fast antioxidant response in *Coffea Arabica* cells. *Plant Physiology and Biochemistry*, *44*, 420–429.

González-Guerrero, M., Matthiadis, A., Sáez, Á., & Long, T. A. (2014). Fixating on metals: New insights into the role of metals in nodulation and symbiotic nitrogen fixation. *Frontiers in Plant Science*, *5*, 45.

González-Guerrero, M., Rubio-Sanz, L., Rodríguez-Haas, B., Albareda, M., Menéndez-Cerón, M., Brito, B., & Palacios, J. M. (2013). *Metal transport in the* Rhizobium–*legume symbiosis* (pp. 141–163). Boca Raton, FL: CRC Press.

Gouia, H., Ghorbal, M. H., & Meyer, C. (2000). Effects of cadmium on activity of nitrate reductase and on other enzymes of the nitrate assimilation pathway in bean. *Plant Physiology and Biochemistry*, *38*(7-8), 629–638.

Haddad, S. A., Tabatabai, M. A., Abdel-Moneim, A. M. A., & Loynachan, T. E. (2015). Inhibition of nodulation and nitrogen nutrition of leguminous crops by selected heavy metals. *Air, Soil and Water Research*, *8*, ASWR-S21098.

Hanfrey, C., Sommer, S., Mayer, M. J., Burtin, D., & Michael, A. J. (2001). *Arabidopsis* polyamine biosynthesis: Absence of ornithine decarboxylase and the mechanism of arginine decarboxylase activity. *The Plant Journal: For Cell and Molecular Biology*, *27*, 551–560.

Helaoui, S., Boughattas, I., Hattab, S., Mkhinini, M., Alphonse, V., Livet, A., & Banni, M. (2020). Physiological, biochemical and transcriptomic responses of *Medicago sativa* to nickel exposure. *Chemosphere*, *249*, 126121.

Higgins, K. (2019). Nickel metalloregulators and chaperones. *Inorganics*, *7*(8), 104.

Hossain, M. A., Piyatida, P., Silva, J. A. T. D., & Fujita, M. (2012). Molecular mechanism of heavy metal toxicity and tolerance in plants: Central role of glutathione in detoxification of reactive oxygen species and methylglyoxal and in heavy metal chelation. *Journal of Botany, 2012*.

Ingle, R. A., Smith, J. A., & Sweetlove, L. J. (2005). Responses to nickel in the proteome of the hyperaccumulator plant *Alyssum lesbiacum*. *Biometals: An International Journal on the Role of Metal Ions in Biology, Biochemistry, and Medicine*, *18*, 627–641.

Ishtiaq, S., & Mahmood, S. (2011). Phytotoxicity of nickel and its accumulation in tissues of three *Vigna* species at their early growth stages. *Journal of Applied Botany and Food Quality*, *84*, 223–228.

Iyaka, Y. A. (2011). Nickel in soils: A review of its distribution and impacts. *Scientific Research and Essays*, *6*(33), 6774–6777.

Jamet, A., Sigaud, S., Van de Sype, G., Puppo, A., & Hérouart, D. (2003). Expression of the bacterial catalase genes during *Sinorhizobium meliloti-Medicago sativa* symbiosis and their crucial role during the infection process. *Molecular Plant-Microbe Interactions*, *16*(3), 217–225.

Janicka-Russak, M., Kabala, K., Burzynski, M., & Klobus, G. (2008). Response of plasma membrane H^+-ATPase to heavy metal stress in *Cucumis sativus* roots. *Journal of Experimental Botany*, *59*, 3721–3728.

Jasniewski, A. J., Sickerman, N. S., Hu, Y., & Ribbe, M. W. (2018). The Fe protein: An unsung hero of nitrogenase. *Inorganics*, *6*(1), 25.

Jose palaCios, M. B. (2017). Nickel in microbial physiology—from single proteins to complex trafficking systems: Nickel import/export. *The Biological Chemistry of Nickel*, 237.

Kalapos, M. P. (2008). The tandem of free radicals and methylglyoxal. *Chemico-Biological Interactions*, *171*, 251–271.

Kalo, P., Gleason, C., Edwards, A., Marsh, J., Mitra, R. M., & Hirsch, S. (2005). Nodulation signaling in legumes requires NSP2, a member of the GRAS family of transcriptional regulators. *Science (New York, N.Y.)*, *308*, 1786–1789.

Karmakar, K., Rana, A., Rajwar, A., Sahgal, M., & Johri, B. N. (2015). Legume-Rhizobia Symbiosis Under Stress. In N. Arora (Ed.), *Plant microbes symbiosis: Applied facets*. New Delhi: Springer.

Kavi Kishor, P. B., Hima Kumari, P., Sunita, M. S. L., & Sreenivasulu, N. (2015). Role of proline in cell wall synthesis and plant development and its implications in plant ontogeny. *Frontiers in Plant Science*, *6*, 544.

Khorasani-Motlagh, M., Lacasse, M. J., & Zamble, D. B. (2017). High-affinity metal binding by the *Escherichia coli* [NiFe]-hydrogenase accessory protein HypB is selectively modulated by SlyD. *Metallomics*, *9*(5), 482–493.

Konishi, M., & Yanagisawa, S. (2013). Arabidopsis NIN-like transcription factors have a central role in nitrate signalling. *Nature Communications*, *4*(1), 1–9.

Krzesłowska, M. (2011). The cell wall in plant cell response to trace metals: Polysaccharide remodeling and its role in defense strategy. *Acta Physiologiae Plantarum*, *33*(1), 35–51.

Kumar, P., & Mina, U. (2019). *Life sciences: Fundamentals and practice. Biological nitrogen fixation* (6th ed., pp. 357–359). New Delhi, India: Pathfinder publication, ISBN: 978-81-906427-7-4.

Kumar, P., Rouphael, Y., Cardarelli, M., & Colla, G. (2015). Effect of nickel and grafting combination on yield, fruit quality, antioxidative enzyme activities, lipid peroxidation, and mineral composition of tomato. *Journal of Plant Nutrition and Soil Science*, *178*(6), 848–860.

Kunert, K. J., van Wyk, S. G., Cullis, C. A., Vorster, B. J., & Foyer, C. H. (2015). Potential use of phytocystatins in crop improvement, with a particular focus on legumes. *Journal of Experimental Botany*, *66*(12), 3559–3570.

Lacasse, M. J., & Zamble, D. B. (2016). [NiFe]-hydrogenase maturation. *Biochemistry*, *55*, 1689–1701.

Lam, R., Romanov, V., Johns, K., Battaile, K. P., Wu-Brown, J., Guthrie, J. L., & Chirgadze, N. Y. (2010). Crystal structure of a truncated urease accessory protein UreF from *Helicobacter pylori*. *Proteins: Structure, Function, and Bioinformatics*, *78*(13), 2839–2848.

Layzell, D. B., Del Castillo, L. D., Hunt, S., Kuzma, M., Van Cauwenberghe, O., & Oresnik, I. (1993). *The regulation of oxygen and its role in regulating nodule metabolism. New horizons in nitrogen fixation* (pp. 393–398). Dordrecht: Springer.

Le Gall, H., Philippe, F., Domon, J. M., Gillet, F., Pelloux, J., & Rayon, C. (2015). Cell wall metabolism in response to abiotic stress. *Plants*, *4*(1), 112–166.

Lee, M. H., Mulrooney, S. B., Renner, M. J., Markowicz, Y., & Hausinger, R. P. (1992). *Klebsiella aerogenes* urease gene cluster: Sequence of ureD and demonstration that four accessory genes (ureD, ureE, ureF, and ureG) are involved in nickel metallocenter biosynthesis. *Journal of Bacteriology*, *174*(13), 4324–4330.

Liu, C. W., & Murray, J. D. (2016). The role of flavonoids in nodulation host-range specificity: An update. *Plants*, *55*, 1689–1701.

Malolepszy, A., Kelly, S., Sørensen, K. K., James, E. K., Kalisch, C., Bozsoki, Z., & Jensen, D. B. (2018). A plant chitinase controls cortical infection thread progression and nitrogen-fixing symbiosis. *eLife*, *7*, e38874.

Martins, A. M., Cordeiro, C. A., & Ponces Freire, A. M. (2001). In situ analysis of methylglyoxal metabolism in *Saccharomyces cerevisiae*. *FEBS Letters*, *499*, 41–44.

Matraszek, R. (2008). Nitrate reductase activity of two leafy vegetables as affected by nickel and different nitrogen forms. *Acta Physiologiae Plantarum*, *30*(3), 361.

Miksch, G., Arnold, W., Lentzsch, P., Priefer, U., & Puhler, A. (1994). A 4.6 kb DNA region of *Rhizobium meliloti* involved in determining urease and hydrogenase activities carries he structural genes for urease (ureA, ureB, ureC) interrupted by other open reading frames. *Molecular & General Genetics*, *242*, 539–550.

Mishra, P., & Dubey, R. S. (2011). Nickel and Al-excess inhibit nitrate reductase but upregulate activities of aminating glutamate dehydrogenase and aminotransferases in growing rice seedlings. *Plant Growth Regulation*, *64*(3), 251–261.

Moncrief, M., & Hausinger, R. (1996). Purification and activation properties of UreD–UreF–urease apoprotein complexes. *Journal of Bacteriology*, *178*, 5417–5421.

Moncrief, M. B., & Hausinger, R. P. (1997). Characterization of UreG, identification of a UreD-UreF-UreG complex, and evidence suggesting that a nucleotide-binding site in UreG is required for in vivo metallocenter assembly of *Klebsiella aerogenes* urease. *Journal of Bacteriology*, *179*(13), 4081–4086.

Muraki, N., Ishii, K., Uchiyama, S., Itoh, S. G., Okumura, H., & Aono, S. (2019). Structural characterization of HypX responsible for CO biosynthesis in the maturation of NiFe-hydrogenase. *Communications Biology*, *2*(1), 1–12.

Mustafiz, A., Ghosh, A., Tripathi, A. K., Kaur, C., Ganguly, A. K., & Bhavesh, N. S. (2014). A unique Ni^{2+}-dependent and methylglyoxal-inducible rice glyoxalase I possesses a single active site and functions in abiotic stress response. *The Plant Journal: For Cell and Molecular Biology*, *78*, 951–963.

Mustafiz, A., Singh, A. K., Pareek, A., Sopory, S. K., & Singla-Pareek, S. L. (2011). Genome-wide analysis of rice and Arabidopsis identifies two glyoxalase genes that are highly expressed in abiotic stresses. *Functional & Integrative Genomics*, *11*, 293–305.

Myrach, T., Zhu, A., & Witte, C. P. (2017). The assembly of the plant urease activation complex and the essential role of the urease accessory protein G (UreG) in delivery of nickel to urease. *Journal of Biological Chemistry*, *292*(35), 14556–14565.

Nabi, A., Parwez, R., Aftab, T., Khan, M. M. A., & Naeem, M. (2020). Triacontanol protects *Mentha arvensis* L. from nickel-instigated repercussions by escalating antioxidant machinery, photosynthetic efficiency and maintaining leaf ultrastructure and root morphology. *Journal of Plant Growth Regulation*, 1–19.

Nishida, S., Aisu, A., & Mizuno, T. (2012). Induction of IRT1 by the nickel-induced iron-deficient response in Arabidopsis. *Plant Signaling & Behavior*, *7*(3), 329–331.

Nishida, S., Tsuzuki, C., Kato, A., Aisu, A., Yoshida, J., & Mizuno, T. (2011). AtIR1, the primary iron uptake transporter in the roots, mediates excess nickel accumulation in *Arabidopsis thaliana*. *Plant Cell Physiology*, *52*, 1433–3499.

Noreen, S., Akhter, M. S., Yaamin, T., & Arfan, M. (2018). The ameliorative effects of exogenously applied proline on physiological and biochemical parameters of wheat (*Triticum aestivum* L.) crop under copper stress condition. *Journal of Plant Interactions, 13*(1), 221–230.

Oldroyd, G. E., Murray, J. D., Poole, P. S., & Downie, J. A. (2011). The rules of engagement in the legume-rhizobial symbiosis. *Annual Review of Genetics, 45*, 119–144.

Panday, N., & Sharma, C. P. (2002). Effect of heavy metals Co^{2+}, Ni^{2+} and Cd^{2+} on growth and metabolism of cabbage. *Plant Science (Shannon, Ireland), 163*, 753–758.

Pardo, M. A., Lagunez, J., Miranda, J., & Martinez, E. (1994). Nodulating ability of *Rhizobium tropici* is conditioned by a plasmid-encoded citrate synthase. *Molecular Microbiology, 11*, 315–321.

Parida, B. K., Chhibba, J. M., & Nayyar, V. K. (2003). Effect of nickel contaminated soil on fenugreek (*Trigonella corniculata* L.) growth and mineral composition. *Scientia Horticulturae, 98*, 113–119.

Park, I. S., Carr, M. B., & Hausinger, R. P. (1994). In vitro activation of urease apoprotein and role of UreD as a chaperone required for nickel metallocenter assembly. *Proceedings of the National Academy of Sciences, 91*(8), 3233–3237.

Park, I. S., & Hausinger, R. P. (1996). Metal ion interactions with urease and UreD-urease apoproteins. *Biochemistry, 35*(16), 5345–5352.

Peiter, E., Sun, J., Heckmann, A. B., Venkateshwaran, M., Riely, B. K., & Otegui, M. S. (2007). The *Medicago truncatula* DMI1 protein modulates cytosolic calcium signaling. *Plant Physiology, 145*, 192–203.

Pilbeam, D. J., & Barker, A. V. (2007). *Handbook of plant nutrition*. Florida, US: CRC Press. (No. 631.81 H236h).

Polacco, J. C., Freyermuth, S. K., Gerendás, J., & Cianzio, S. R. (1999). Soybean genes involved in nickel insertion into urease. *Journal of Experimental Botany, 50*(336), 1149–1156.

Polacco, J. C., Hyten, D. L., Medeiros-Silva, M., Sleper, D. A., & Bilyeu, K. D. (2011). Mutational analysis of the major soybean UreF paralogue involved in urease activation. *Journal of Experimental Botany, 62*(10), 3599–3608.

Polacco, J. C., Mazzafera, P., & Tezotto, T. (2013). Opinion–nickel and urease in plants: Still many knowledge gaps. *Plant Science, 199*, 79–90.

Rai, L. C., & Raizada, M. (1985). Effect of nickel and silver ions on survival, growth, carbon fixation and nitrogenase activity in *Nostoc muscorum*: Regulation of toxicity by EDTA and calcium. *The Journal of General and Applied Microbiology, 31*(4), 329–337.

Reina-Bueno, M., Argandoña, M., Nieto, J. J., Hidalgo-García, A., Iglesias-Guerra, F., Delgado, M. J., & Vargas, C. (2012). Role of trehalose in heat and desiccation tolerance in the soil bacterium *Rhizobium etli*. *BMC Microbiology, 12*(1), 207.

Richard, J. P. (1993). Mechanism for the formation of methylglyoxal from triosephosphates. *Biochemical Society Transactions, 21*, 549–553.

Roschzttardtz, H., Séguéla-Arnaud, M., Briat, J. F., Vert, G., & Curie, C. (2011). The FRD3 citrate effluxer promotes iron nutrition between symplastically disconnected tissues throughout *Arabidopsis* development. *The Plant Cell, 23*, 2725–2737.

Rubio, M. C., Becana, M., Sato, S., James, E. K., Tabata, S., & Spaink, H. P. (2007). Characterization of genomic clones and expression analysis of the three types of superoxide dismutases during nodule development in *Lotus japonicus*. *Molecular Plant-Microbe Interactions, 20*(3), 262–275.

Rucińska-Sobkowiak, R. (2016). Water relations in plants subjected to heavy metal stresses. *Acta Physiologiae Plantarum, 38*(11), 257.

Ruiz-Argüeso, T., Palacios, J.M., & Imperial, J. (2000). Uptake hydrogenases in root nodule.

Saad, R., Kobaissi, A., Robin, C., Echevarria, G., & Benizri, E. (2016). Nitrogen fixation and growth of *Lens culinaris* as affected by nickel availability: A pre-requisite for optimization of agromining. *Environmental and Experimental Botany*, *131*, 1–9.

Sachan, P., & Lal, N. (2017). An overview of nickel (Ni2 +) essentiality, toxicity and tolerance strategies in plants. *Asian Journal of Biology*, 1–15.

Saif, S., & Khan, M. S. (2018). Assessment of toxic impact of metals on proline, antioxidant enzymes, and biological characteristics of *Pseudomonas aeruginosa* inoculated *Cicer arietinum* grown in chromium and nickel-stressed sandy clay loam soils. *Environmental Monitoring and Assessment*, *190*(5), 290.

Salomone-Stagni, M., Zambelli, B., Musiani, F., & Ciurli, S. (2007). A model-based proposal for the role of UreF as a GTPase-activating protein in the urease active site biosynthesis. *Proteins: Structure, Function, and Bioinformatics*, *68*(3), 749–761.

Sargent, F. (2016). The model [NiFe]-hydrogenases of *Escherichia coli*. *Advances in Microbial Physiology*, *68*, 433–507.

Seregin, I. V., & Kozhevnikova, A. D. (2006). Physiological role of nickel and its toxic effects on higher plants. *Russian Journal of Plant Physiology*, *53*, 257–277.

Shahid, M. A., Balal, R. M., Pervez, M. A., Abbas, T., Aqeel, M. A., Javaid, M. M., & Garcia-Sanchez, F. (2014). Exogenous proline and proline-enriched *Lolium perenne* leaf extract protects against phytotoxic effects of nickel and salinity in *Pisum sativum* by altering polyamine metabolism in leaves. *Turkish Journal of Botany*, *38*(5), 914–926.

Shahzad, B., Tanveer, M., Rehman, A., Cheema, S. A., Fahad, S., Rehman, S., & Sharma, A. (2018). Nickel; whether toxic or essential for plants and environment-a review. *Plant Physiology and Biochemistry*, *132*, 641–651.

Sharma, A., & Dhiman, A. (2013). Nickel and cadmium toxicity in plants. *Journal of Pharmaceutical and Scientific Innovation*, *2*, 20–24.

Sharma, J. O. T., & Subhadra, A. V. (2010). The effect of mercury on nitrate reductase activity in bean leaf segments (*Phaseolus vulgaris*) and its chelation by phytochelatin synthesis. *Life Sciences and Medicine Research*, *2010*, 1–8.

Siddiqui, S., Siddiqui, M. F., Khan, S., & Bano, B. (2019). Insight into the biochemical characterization of phytocystatin from *Glycine max* and its interaction with $Cd^{+\ 2}$ and Ni^{+2}. *Journal of Molecular Recognition*, *32*(10), e2787.

Siegbahn, P. E., Chen, S. L., & Liao, R. Z. (2019). Theoretical studies of nickel-dependent enzymes. *Inorganics*, *7*(8), 95.

Singh, R. K., & Rao, P. J. M. (1997). Biological significance of nickel on the nitrogen fixing ability of cowpea *Bradyrhizobium*. *Journal of Plant Nutrition*, *20*(10), 1449–1455.

Singh, S., & Varma, A. (2017). *Structure, function, and estimation of leghemoglobin*. Rhizobium *biology and biotechnology* (pp. 309–330). Cham: Springer.

Siqueira Freitas, D., Wurr Rodak, B., Rodrigues dos Reis, A., de Barros Reis, F., Soares de Carvalho, T., Schulze, J., & Guimaraes Guilherme, L. R. (2018). Hidden nickel deficiency? Nickel fertilization via soil improves nitrogen metabolism and grain yield in soybean genotypes. *Frontiers in Plant Science*, *9*, 614.

Sirhindi, G., Mir, M. A., Abd-Allah, E. F., Ahmad, P., & Gucel, S. (2016). Jasmonic acid modulates the physio-biochemical attributes, antioxidant enzyme activity, and gene expression in *Glycine max* under nickel toxicity. *Frontiers in Plant Science*, *7*, 591.

Sirko, A., & Brodzik, R. (2000). Plant ureases: Roles and regulation. *Acta Biochimica Polonica*, *47*(4), 1189–1195.

Song, T., Xu, H., Sun, N., Jiang, L., Tian, P., Yong, Y., & Cui, G. (2017). Metabolomic analysis of alfalfa (*Medicago sativa* L.) root-symbiotic rhizobia responses under alkali stress. *Frontiers in Plant Science*, *8*, 1208.

Soussi, M., Khadri, M., Lluch, C., & Ocana, A. (2001). Carbon metabolism and bacteroid respiration in nodules of chick-pea (*Cicer arietinum* L.) plants grown under saline conditions. *Plant Biosystems-An International Journal Dealing with all Aspects of Plant Biology*, *135* (2), 157–164.

Stahlhut, R. W., & Widholm, J. M. (1989). Ureide catabolism by soybean [*Glycine max* (L.) Merrill] cell suspension cultures I. Urea is not an intermediate in allantoin degradation. *Journal of Plant Physiology*, *134*(1), 85–89.

Sujkowska-Rybkowska, M., Kasowska, D., Gediga, K., Banasiewicz, J., & Stepkowski, T. (2020). *Lotus corniculatus*-rhizobia symbiosis under Ni, Co and Cr stress on ultramafic soil. *Plant and Soil*, *451*(459), 484.

Taiz, L., & Zeiger, E. (2006). *Plant physiology online* (3rd ed., p. 269)MA: Sinauwr Associates, Sunderland, ISBN:978-0-87893-856-8.

Takanashi, K., Yokosho, K., Saeki, K., Sugiyama, A., Sato, S., & Tabata, S. (2013). LjMATE1:a citrate transporter responsible for iron supply to the nodule infection zone of *Lotus japonicus*. *Plant & Cell Physiology*, *54*, 585–594.

Talbi, C., Sanchez, C., & Hidalgo-Garcia, A. (2012). Enhanced expression of *Rhizobium etli* cbb (3) oxidase improves drought tolerance of common bean symbiotic nitrogen fixation. *Journal of Experimental Botany*, *63*, 5035–5043.

Tan, X. W., Ikeda, H., & Oda, M. (2000). Effects of nickel concentration in the nutrient solution on the nitrogen assimilation and growth of tomato seedlings in hydroponic culture supplied with urea or nitrate as the sole nitrogen source. *Scientia Horticulturae*, *84*(3-4), 265–273.

Tejada-Jimenez, M., Llamas, A., Galván, A., & Fernández, E. (2019). Role of nitrate reductase in NO production in photosynthetic eukaryotes. *Plants*, *8*(3), 56.

Thornalley, P. J. (2003). Glyoxalase I—structure, function and a critical role in the enzymatic defence against glycation. *Biochemical Society Transactions*, *31*, 1343–1348.

Todd, C. D., & Polacco, J. C. (2004). Soybean cultivars 'Williams 82' and Maple Arrow produce both urea and ammonia using ureide degradation. *Journal of Experimental Botany*, *398*, 867–877.

Todd, C. D., & Polacco, J. C. (2006). AtAAH encodes a protein with allantoate amidohydro-lase activity from *Arabidopsis thaliana*. *Planta*, *223*, 1108–1113.

Toffanin, A., Cadahia, E., Imperial, J., Ruiz-Argüeso, T., & Palacios, J. (2002). Characterization of the urease gene cluster from *Rhizobium leguminosarum* bv. *viciae*. *Archives of Microbiology*, *177*(4), 290–298.

Tsadilas, C., Rinklebe, J., & Selim, M. (Eds.), (2018). *Nickel in soils and plants*. CRC Press.

Vatansever, R., Ozyigit, I. I., & Filiz, E. (2017). Essential and beneficial trace elements in plants, and their transport in roots: A review. *Applied Biochemistry and Biotechnology*, *181* (1), 464–482.

Vesper, S. J., & Weidensaul, T. C. (1978). Effects of cadmium, nickel, copper, and zinc on nitrogen fixation by soybeans. *Water, Air, and Soil Pollution*, *9*(4), 413–422.

Walker, D. W., Graham, R. D., Madison, J. T., Cary, E. E., & Welch, R. M. (1985). Effects of Ni deficiency on some nitrogen metabolites in cowpea (*Vigna unguiculata* L. Wap.). *Plant Physiology*, *79*, 474–479.

Wang, J., Li, W., Zhang, C., & Ke, S. (2010). Physiological responses and detoxific mechanisms to Pb, Zn, Cu and Cd in young seedlings of *Paulownia fortunei*. *Journal of Environmental Sciences*, *22*(12), 1916–1922.

Watanabe, S., Sasaki, D., Tominaga, T., & Miki, K. (2012). Structural basis of [NiFe] hydrogenase maturation by Hyp proteins. *Biological Chemistry*, *393*(10), 1089–1100.

Werner, A. K., Romeis, T., & Witte, C. P. (2009). Ureide catabolism in *Arabidopsis thaliana* and *Escherichia coli*. *Nature Chemical Biology*, *6*, 19–21.

Werner, A. K., Sparkes, I. A., Romeis, T., & Witte, C. P. (2008). Identification, biochemical characterization, and subcellular localization of allantoate amidohydrolases from *Arabidopsis* and soybean. *Plant Physiology*, *146*, 418–430.

Wheeler, C. T., Hughes, L. T., Oldroyd, J., & Pulford, I. D. (2001). Effects of nickel on Frankia and its symbiosis with *Alnus glutinosa* (L.). Gaertn. *Plant and Soil*, *231*, 81–90.

Winkler, R. G., Polacco, J. C., Eskew, D. L., & Welch, R. M. (1983). Nickel is not required for apourease synthesis in soybean seeds. *Plant Physiology*, *72*, 262–263.

Witte, C. P., Tiller, S., Isidore, E., Davies, H. V., & Taylor, M. A. (2005). Analysis of two alleles of the urease gene from potato: Polymorphisms, expression, and extensive alternative splicing of the corresponding mRNA. *Journal of Experimental Botany*, *56*(409), 91–99.

Yan, R., Gao, S., Yang, W., Cao, M., Wang, S., & Chen, F. (2008). Ni toxicity induced antioxidant enzyme and phenylalanine ammonia-lyase activities in *Jatropha curcas* L. cotyledons. *Plant, Soil and Environment*, *54*, 294–300.

Yang, X. E., Baligar, V. C., Foster, J. C., & Martens, D. C. (1997). Accumulation and transport of nickel in relation to organic acids in ryegrass and maize with different nickel levels. *Plant and Soil*, *196*, 271–276.

Yusuf, M., Fariduddin, Q., Ahmad, I., & Ahmad, A. (2014). Brassinosteroid-mediated evaluation of antioxidant system and nitrogen metabolism in two contrasting cultivars of *Vigna radiata* under different levels of nickel. *Physiology and Molecular Biology of Plants*, *20*(4), 449–460.

Yusuf, M., Fariduddin, Q., Hayat, S., & Ahmad, A. (2011). Nickel: An overview of uptake, essentiality and toxicity in plants. *Bulletin of Environmental Contamination and Toxicology*, *86*(1), 1–17.

Yusuf, M., Fariduddin, Q., Hayat, S., Hasan, S. A., & Ahmad, A. (2011). Protective response of 28-homobrassinolide in cultivars of *Triticum aestivum* with different levels of nickel. *Archives of Environmental Contamination and Toxicology*, *60*(1), 68–76.

Zali, A. G., & Ehsanzadeh, P. (2018). Exogenous proline improves osmoregulation, physiological functions, essential oil, and seed yield of fennel. *Industrial Crops and Products*, *111*, 133–140.

Zambelli, B., Banaszak, K., Merloni, A., Kiliszek, A., Rypniewski, W., & Ciurli, S. (2013). Selectivity of Ni(II) and Zn(II) binding to *Sporosarcina pasteurii* UreE, a metallochaperone in the urease assembly: A calorimetric and crystallographic study. *Journal of Biological Inorganic Chemistry*, *18*, 1005–1017.

Zielazinski, E. L., González-Guerrero, M., Subramanian, P., Stemmler, T. L., Argüello, J. M., & Rosenzweig, A. C. (2013). *Sinorhizobium meliloti* Nia is a P1B-5-ATPase expressed in the nodule during plant symbiosis and is involved in Ni and Fe transport. *Metallomics*, *5*(12), 1614.

Zobiole, L. H. S., Oliveira, R. S., Jr, Kremer, R. J., Constantin, J., Yamada, T., Castro, C., & Oliveira, A., Jr (2010). Effect of glyphosate on symbiotic N_2 fixation and nickel concentration in glyphosate-resistant soybeans. *Applied Soil Ecology*, *44*(2), 176–180.

Zoń, J., Amrhein, N., & Gancarz, R. (2002). Inhibitors of phenylalanine ammonia-lyase: 1-aminobenzylphosphonic acids substituted in the benzene ring. *Phytochemistry*, *59*(1), 9–21.

Chapter 18

Beneficial aspects of cobalt uptake in plants exposed to abiotic stresses

Aditya Banerjee and Aryadeep Roychoudhury
Department of Biotechnology, St. Xavier's College (Autonomous), Kolkata, India

18.1 Introduction

Cobalt (Co) has been widely regarded as a beneficial trace element in plants as well as humans, though it exerts extreme toxicity at high concentrations by inducing oxidative stress and associated molecular damages (Yamagata, Murata, & Torii, 1962). At trace levels, Co is systematically incorporated within cobalamin (vitamin B_{12}) moieties and is also utilized during the biosynthesis of several amino acids and crucial neurotransmitters (Bhattacharya, Misra, & Hussain, 2016). According to "Frieden's Categorical Classification of Elements," Co has been placed into Group III which comprises essential trace elements required in minor quantities for optimum metabolism, growth, and development (Frieden, 1974). Though the requirement of Co fertilization is usually rare, it has been reported to benefit crop growth under nutrient-deprived conditions. The most important reference of the beneficial role of Co in plants is during nitrogen (N) fixation in nodulated roots of leguminous plant species. During N fixation, Co is specifically utilized by the nodulating *Rhizobia* in symbiotic association with the legume roots (Ali, Hayat, Hayat, & Ahmad, 2010). Delwiche, Johnson, and Reisenauer (1961) reported that coapplication of exogenous Co and an inoculum of *Rhizobium* promoted efficient N fixation in *Medicago sativa* even under conditions in which the plants would have been unable to fix N. A quick search in the literature databases would clearly show that much of the research work regarding the presence of Co in plants has been performed from the point of Co toxicity. As a result, detailed knowledge regarding Co nutrition in plants and crops is unavailable. The present chapter briefly outlines the beneficial roles of Co as an important micronutrient in plants exposed to abiotic stresses like drought, senescence due to oxidative stress, and even nutrient deficiencies. These

Frontiers in Plant–Soil Interaction. DOI: https://doi.org/10.1016/B978-0-323-90943-3.00012-2

environmental stresses drastically affect the growth, development, and productivity of crops and vegetation, thus disturbing the overall ecological balance and food security (Banerjee & Roychoudhury, 2020a, 2020b, 2020c, 2020d, 2020e; Chandra, Banerjee, & Roychoudhury, 2020; Das, Banerjee, & Roychoudhury, 2019). Thus the chapter holistically illustrates the instances of Co fertilization for growing healthy plants even under suboptimal conditions.

18.2 Co in soil and its uptake by plants

Unlike macroelements like sodium (Na), potassium (K), etc., Co is detected in soil at trace levels usually between 15–25 mg kg^{-1} of soil and 0.04 mg L^{-1} of water (Palit, Sharma, & Talukder, 1994). It has been reported that plants accumulate Co at 0.1–10 mg kg^{-1} dry weight. Within soil, Co is usually present in the ionic form as Co^{2+}. It has been observed that soil with fine texture and high organic matter usually contains higher amounts of Co (Palit et al., 1994). Watanabe et al. (2007) showed that apart from Ericales, Euasterids, and Asparagales, foliar Co content is usually lower than 0.2 mg kg^{-1}. However Co hyperaccumulators belonging to Lamiaceae, Scrophulariaceae, Asteraceae, and Fabaceae accumulated greater than 1 g kg^{-1} Co in their leaves (Baker, McGrath, Reeves, & Smith, 2000). Like nickel (Ni), Co also has a tendency to bind strongly to roots. Co might passively enter into the root cells or it could be imported through the membrane carriers which are also involved in Ni transport, as already reported in *Lotus japonicus* (Chen et al., 2009). In another study, it was found that Co import into root cells might be mediated via the IRON-REGULATED TRANSPORTER IRT1 in *Arabidopsis* seedlings grown in iron (Fe)-deficient growth medium (Baxter et al., 2008). Overexpression of the gene encoding Co-Ni transporter (permease) from *Rhodopseudomonas palustris* increased Co uptake by fivefold and Ni acquisition by twofold in transgenic *Nicotiana tabacum* seedlings compared to those in wild-type plants (Nair et al., 2012).

18.3 Roles of Co in stimulating N metabolism in legumes

Application of optimum concentrations of Co stimulated the synthesis of leghemoglobin essential for N fixation in the symbiotic nodules (Yadav & Khanna, 1988). Highly efficient N fixation was observed in groundnut seedlings treated with 8 mg L^{-1} Co due to elevated activity of nitrogenase enzyme (Gad, 2012). Exogenous treatment with Co stimulated the growth, productivity, number of nodules, and nodule size in faba beans, due to which they could maintain optimum pod size even under minimal conditions (Minz et al., 2018). Yadav and Khanna (1983) also reported increased N fixation in cowpea and soybean varieties treated with very low doses of Co solution. Coapplication of phosphorus (P) and Co has been observed to increase the number of nodules significantly in legume plants (Yadav, Yadav, & Khanna, 1984). Soil fertilization with Co at 50 mg kg^{-1} led to increased biomass formation and seedling growth in cowpea seedlings (Vijyarengan, 2012).

18.4 Roles of Co in stimulating growth and yield in plants

Application of cobalt ferrite ($CoFe_2O_4$) nanoparticles at a concentration of 125 mg L^{-1} significantly increased Fe and magnesium (Mg) translocation in tomato seedlings, though percentage of germination and overall growth remained unaffected (Lopez-Moreno, Aviles, Perez, Irizarry, & Perales, 2016). Increased growth and productivity was observed in soybean seedlings fertilized with 12 mg kg^{-1} Co (Gad, El-Moez, Bekbayeva, Karabayeva, & Surif, 2013; Kandil, Ihab, & El-Maghraby, 2013). Aqueous solution of $CoSO_4$ at 20 mg L^{-1} concentration effectively increased the biomass and number of pods in the seedlings of faba beans (Kandil, 2007; Minz et al., 2018). Jaleel, Jayakumar, Chang-Xing, and Azooz (2008) also illustrated the growth-promoting roles of Co in groundnut seedlings. Apart from these instances, fertilization of growth medium with Co also stimulated growth and yield in lentil and chickpea seedlings (Khan & Khan, 2010; Sahay & Singh, 2012).

18.5 Roles of Co in overcoming oxidative stress and nutrient deficiency

It has been reported that Co could delay foliar senescence and chlorosis by inhibiting the biosynthesis and production of the gaseous phytohormone, ethylene (Piton-Smits, Quinn, Tapken, Malagoli, & Schiavon, 2009). Palit et al. (1994) also stated that exposure to optimum concentrations of Co promoted drought tolerance in seeds. This was possibly mediated via increased production of protective alkaloids like isoquinolines by utilizing precursor aromatic amino acids (Palit et al., 1994).

Application of Co has been closely associated with increased nutrient and mineral homeostasis in plants. Basu, Bhadoria, & Mahapatra (2006) reported that the seedlings of summer groundnut inoculated with *Rhizobium* and phosphobacterium and treated with Co could maximally translocate the macroelements like N, P, and K. Similar increase in N, P, and K uptake was also reported in the seedlings of broad beans exogenously treated with Co (Manal, Manal, & Allam, 2016). Soybean seedlings grown in soil fertilized with Co, also accumulated higher levels of micronutrients like copper (Cu), zinc (Zn), and manganese (Mn) which participate in several enzymatic and physiological processes as crucial co-factors (Kandil et al., 2013).

18.6 Conclusion

Growth and development of plants is highly dependent on optimum intracellular levels of beneficial mineral elements. Like several other micronutrients, Co also has been observed to stimulate seedling growth and nodule formation in seedlings at trace levels. It was observed that optimum doses of Co promoted the growth of symbiotic *Rhizobacterium* in the rhizosphere of legumes and increased N fixation and absorption. Co has been directly found to be involved in the synthesis of cobalamin and leghemoglobin. Exogenous application of Co in the form of solution or fertilization of soil with Co significantly triggered shoot, root length, biomass, yield, and productivity especially in leguminous plants due to increased N assimilation, efficient mobilization of macroelements like N, P, and K, and microelements like Cu, Zn, and Mn. Exogenous Co also promoted desiccation tolerance in seeds and overall can be used as a useful supplement in crop plants which are grown in nutrient-deficient conditions (Fig. 18.1).

18.7 Future perspectives

The use of Co to ameliorate abiotic stress-induced injuries in crop plants requires more exhaustive and elaborate investigations. Apart from legumes, the utility of low doses of Co should also be investigated in staple cereal crops like rice, wheat, maize, and sorghum. The roles of exogenous Co in ameliorating edaphic stresses (other than nutrient deficiency or drought) like

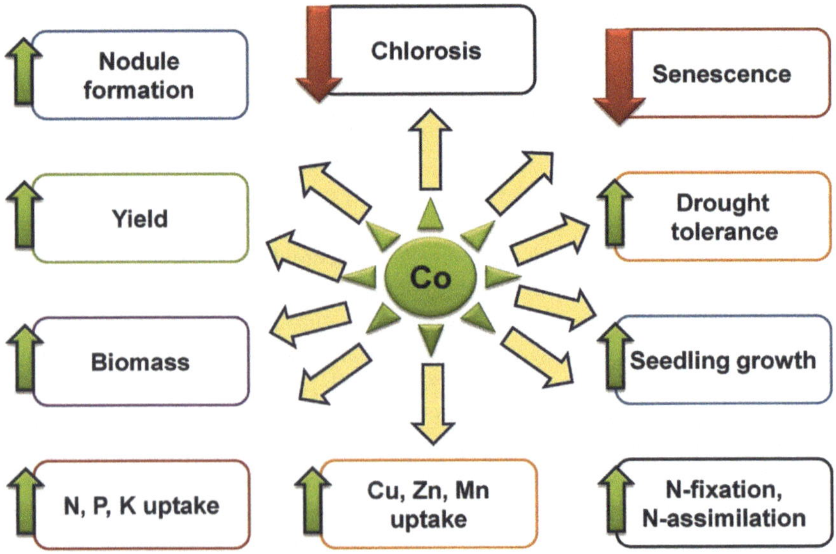

FIGURE 18.1 The beneficial roles of exogenous Co in plants.

salinity and heavy metal toxicity also require proper elucidation. The various molecular alterations at the genomic, epigenomic, transcriptomic, and metabolomic levels upon application of optimum doses of Co should also be characterized in important crop and plant species. Successful execution of these studies would definitely help in refining the scientific knowledge regarding the roles of Co as a protective agent against plant abiotic stress.

Acknowledgments

Financial assistance from the Science and Engineering Research Board, Government of India through the grant [EMR/2016/004799] and Department of Higher Education, Science and Technology and Biotechnology, Government of West Bengal, through the grant [264 (Sanc.)/ST/P/S&T/1G-80/2017] to Dr. Aryadeep Roychoudhury is gratefully acknowledged. Mr. Aditya Banerjee is thankful to University Grants Commission, Government of India for providing a Senior Research Fellowship in the course of this work.

References

Ali, B., Hayat, S., Hayat, Q., & Ahmad, A. (2010). Cobalt stress affects nitrogen metabolism, photosynthesis and antioxidant system in chickpea (*Cicer arietinum* L.). *Journal of Plant Interactions*, *5*, 223–231.

Baker, A. J. M., McGrath, S. P., Reeves, R. D., & Smith, J. A. C. (2000). Metal hyperaccumulator plants: A review of the ecology and physiology of a biological resource for phytoremediation of metal polluted soils. In N. Terry, & G. S. Banuelos (Eds.), *Phytoremediation of contaminated soil and water* (pp. 85–107). Boca Raton: CRC Press.

Banerjee, A., & Roychoudhury, A. (2020a). Plant responses to environmental nickel toxicity. In T. Aftab, & K. R. Hakeem (Eds.), *Plant micronutrients: Deficiency and toxicity management* (pp. 101–111). Switzerland AG: Springer Nature.

Banerjee, A., & Roychoudhury, A. (2020b). Regulation of inducible promoters during salinity stress in plants. In S. H. Wani (Ed.), *Transcription factors for abiotic stress tolerance in plants* (pp. 111–122). United Kingdom: Academic Press, Elsevier.

Banerjee, A., & Roychoudhury, A. (2020c). Oxylipins and strobilurins as protective chemical agents to generate abiotic stress tolerance in plants. In A. Roychoudhury, & D. K. Tripathi (Eds.), *Protective chemical agents in the amelioration of plant abiotic stress* (pp. 483–490). USA: John Wiley & Sons, Ltd.

Banerjee, A., & Roychoudhury, A. (2020d). Seed priming as a method to generate heat-stress tolerance in plants: A minireview. In S. H. Wani, & V. Kumar (Eds.), *Heat stress tolerance in plants: Physiological, molecular and genetic perspectives* (pp. 23–28). United Kingdom: John Wiley & Sons.

Banerjee, A., & Roychoudhury, A. (2020e). Gibberellic acid-priming promotes fluoride tolerance in a susceptible indica rice cultivar by regulating the antioxidant and phytohormone homeostasis. *Journal of Plant Growth Regulation*, *39*, 1476–1487.

Basu, M., Bhadoria, P. B. S., & Mahapatra, S. C. (2006). Influence of microbial culture in combination with micronutrient in improving the groundnut productivity under alluvial soil of India. *Acta Agriculturae Slovenica*, *87*, 435–444.

Baxter, I. R., Vitek, O., Lahner, B., Muthukumar, B., Borghi, M., Morrissey, J., ... Salt, D. E. (2008). The leaf ionome as a multivariable system to detect a plant's physiological status.

Proceedings of the National Academy of Sciences of the United States of America, 105, 12081–12086.

Bhattacharya, P. T., Misra, S. R., & Hussain, M. (2016). Nutritional aspects of essential trace elements in oral health and disease: An extensive review. *Scientifica (Cairo), 2016*, 5464373.

Chandra, S., Banerjee, A., & Roychoudhury, A. (2020). Quantitative trait loci for rice grain quality improvement. In A. Roychoudhury (Ed.), *Rice research for quality improvement: Genomics and genetic engineering* (vol. 2, pp. 687–697). Singapore: Springer Nature.

Chen, Z., Watanabe, T., Shinano, T., Okazaki, K., Osaki, M., & Shinano, T. (2009). Rapid characterization of plant mutants with an altered ionprofile: A case study using *Lotus japonicus*. *The New Phytologist, 181*, 795–801.

Das, P., Banerjee, A., & Roychoudhury, A. (2019). Polyamines ameliorate oxidative stress by regulating antioxidant systems and interacting with plant growth regulators. In A. Roychoudhury, & D. K. Tripathi (Eds.), *Molecular plant abiotic stress: Biology and Biotechnology* (pp. 135–143). USA: John Wiley & Sons.

Delwiche, C. C., Johnson, C. M., & Reisenauer, H. M. (1961). Influence of cobalt on nitrogen fixation by *Medicago*. *Plant Physiology, 36*, 73–78.

Frieden, E. (1974). The evolution of metals as essential elements [with special reference to iron and copper]. In M. Friedman (Ed.), *Protein-metal interactions* (vol. 48, pp. 1–31). New York: Springer.

Gad, N. (2012). Role and importance of cobalt nutrition on groundnut (*Arachis hypogaea*) production. *World Applied Sciences Journal, 20*, 359–367.

Gad, N., El-Moez, Abd, Bekbayeva, L. K., Karabayeva, A. A., & Surif, M. (2013). Effect of cobalt supplement on the growth and productivity of soybean (*Glycine max* L. Merril). *World Applied Sciences Journal, 26*, 926–933.

Jaleel, C. A., Jayakumar, K., Chang-Xing, Z., & Azooz, M. M. (2008). Effect of soil applied cobalt on activities of anti-oxidant enzymes in *Arachis hypogaea*. *Global Journal of Molecular Sciences, 3*, 42–45.

Kandil, H. (2007). Effect of cobalt fertilizer on growth, yield and nutrient status of faba bean (*Vicia faba* L.) plants. *Journal of Applied Sciences Research, 3*, 867–872.

Kandil, H., Ihab, M. F., & El-Maghraby, A. (2013). Effect of cobalt level and nitrogen source on quantity and quality of soybean plant. *Journal of Basic and Applied Scientific Research, 3*, 185–192.

Khan, M. R., & Khan, M. M. (2010). Effect of varying concentration of nickel and cobalt on the plan growth and yield of chickpea. *Australian Journal of Basic and Applied Sciences, 4*, 1036–1046.

Lopez-Moreno, M. L., Aviles, L. L., Perez, N. G., Irizarry, B. A., Perales, O., et al. (2016). Effect of cobalt ferrite ($CoFe_2O_4$) nanoparticles on the growth and development of *Lycopersicon lycopersicum* (tomato plants). *Science of the Total Environment, 550*, 45–52.

Manal, A. A., Manal, A. A., & Allam, S. M. M. (2016). Effect of zinc and cobalt applied with different methods and rates on the yield components of *Vicia faba* L. *World Wide Journal of Multidisciplinary Research and Development, 2*, 52–58.

Minz, A., Sinha, A. K., Kumar, R., Kumar, B., Deep, K. P., & Kumar, S. B. (2018). A review on importance of cobalt in crop growth and production. *International Journal of Current Microbiology and Applied Sciences, 7*, 2978–2984.

Nair, S., Joshi-Saha, A., Singh, S., Ramachandran, V., Singh, S., Thorat, V., … D'Souza, S. F. (2012). Evaluation of transgenic tobacco plants expressing a bacterial Co-Ni transporter for acquisition of cobalt. *Journal of Biotechnology, 161*, 422–428.

Palit, S., Sharma, A., & Talukder, G. (1994). Effect of cobalt on plants. *The Botanical Review; Interpreting Botanical Progress*, *60*, 149–181.

Piton-Smits, E. A. H., Quinn, C. F., Tapken, W., Malagoli, M., & Schiavon, M. (2009). Physiological functions of beneficial elements. *Current Opinion in Plant Biology*, *12*, 267–274.

Sahay, N., & Singh, S. P. (2012). Effect of cobalt application on growth, yield attributes and yield of lentil cultivars. *Annals of Plant and Soil Research*, *14*, 39–41.

Vijyarengan, P. (2012). Changes in growth, biochemical constitution and anti-oxidant potential in cowpea (*Vigna unguiculata* L.) under cobalt stress. *International Journal of Environmental Science and Technology*, *2*, 74–82.

Watanabe, T., Broadley, M. R., Jansen, S., White, P. J., Takada, J., Satake, K., ... Osaki, M. (2007). Evolutionary control of leaf element composition in plants. *The New Phytologist*, *174*, 516–523.

Yadav, D. V., & Khanna, S. S. (1983). Effect of cobalt and phosphorous on their availability and nitrogen fixation by legumes. *Haryana Agricultural University Journal Of Research*, *13*, 396–401.

Yadav, D. V., & Khanna, S. S. (1988). Role of cobalt in nitrogen fixation: A review. *Agricultural Reviews*, *9*, 180–182.

Yadav, R. P., Yadav, D. V., & Khanna, S. S. (1984). Improved nodulation and nitrogen fixation in berseem with the application of cobalt and phosphorous. *Indian Journal of Agricultural Science*, *54*, 506–508.

Yamagata, N., Murata, S., & Torii, T. (1962). The cobalt content of human body. *Journal of Radiation Research*, *3*, 4–8.

Chapter 19

Iron in the soil–plant–human continuum

Abhik Patra[1], Vinod Kumar Sharma[2], Hanuman Singh Jatav[1,3], Asik Dutta[4], Ravindra Kumar Rekwar[2], Arghya Chattopadhyay[1], Ankita Trivedi[2], Kiran Kumar Mohapatra[5] and Ajin S. Anil[2]

[1]Department of Soil Science and Agricultural Chemistry, Institute of Agricultural Sciences, Banaras Hindu University, Varanasi, India, [2]Division of Soil Science and Agricultural Chemistry, ICAR-Indian Agricultural Research Institute, New Delhi, India, [3]Sri Karan Narendra Agriculture University, Jobner, Jaipur, India, [4]Crop Production Division, ICAR-Indian Institute of Pulse Research, Kanpur, India, [5]Department of Soil Science and Agricultural Chemistry, College of Agriculture, Odisha University of Agriculture and Technology, Bhubaneswar, India

19.1 Introduction

It is expected that after 30 years, the world's human population will rise to 9.7 billion, and India's population will probably reach 1.6 billion, from its present level of 1.2 billion (Kumar, Patel, Ramani, Shukla, & Meena, 2020a). Worldwide 3 billion people are afflicted with iron (Fe) undernourishment and 2 billion people (30% of the world's population) suffer from anemia (WHO, 1999). Fe is a vital element in every living organism and a key component of hemoglobin. The recommended dietary allowances for Fe are reported as 10 and 15 mg day^{-1} for an adult male and female, respectively (FAO, WHO, 2000). Anemia is mainly the reduction in hemoglobin in the blood, which adversely affects oxygen flow in the human body. According to the Indian Council of Medical Research (Gavaravarapu & Hemalatha, 2018; ICMR, 2010), Fe deficit in Indian children less than 5 years old is a common dietary problem. Soil resource provides humans with more than 90% of all the food we consume and its health through interactions with plant inputs creates a vigorous environment. Only 14.4% of Indian soils are deficient of available Fe, causing "iron deficiency anemia." There is a strong relationship between available Fe level in the soil and occurrence of iron deficiency anemia. For example, more than 80% pregnant women were reported to be suffering from iron deficiency anemia in Bikaner, Mainpuri, Srinagar, Patna, Dibrugarh, Nagaon, Mehboobnagar Lakhimpur Kheri,

Frontiers in Plant–Soil Interaction. DOI: https://doi.org/10.1016/B978-0-323-90943-3.00009-2

Badaun, Bishnupur, Mandi, Dehradun, and Kohima districts, which have diverse climatic conditions (Laxmaiah et al., 2013; NNMB National Nutrition Monitoring Bureau, National Institute of Nutrition ICMR, 2009). Meeting the food, fruit, fodder, fiber, and fuel (5F) requirements of humans is possible through further intensification of agriculture. However, it might lead to more and more pressure on the finite soil nutrient reserves that would lead to the appearances of the deficiencies of other micronutrients besides aggravating the existing ones. Before the "green revolution," the threat of widespread household food insecurity and chronic undernutrition was very common due to the extensive growth in population and insufficient food availability (Shukla & Pakhare, 2015).

Iron deficiencies in humans, especially vegetarians, are linked to the poor Fe concentrations in food grains and ultimately to the low concentrations in the soils supporting the crop growth. Essential nutrients in the human diet are supplied through food from plants that take up the elements from the soil during growth (Shukla, 2014). Thus the occurrence of soil Fe deficiencies is a growing concern in India, resulting in low productivity and poor nutritional quality of products, which in turn cause diverse health problems in human beings across the world (Takkar, 2011). In recent years, it has been observed that the rampant deficiencies of these nutrients in soils have been adversely affecting animal and human health. The survey report shows that our soil has been deficient in micronutrients (especially Fe) from the middle of the last century (Shukla, Tiwari, & Prakash, 2014). With the passage of time intensive agriculture, unbalanced fertilization, poor management practices, and a lack of knowledge increases its severity. For that reason, the agriculture sector has to play a significant role in enhancing dietary safety and improving human health.

19.2 Iron geochemistry

Iron is the second most abundant metal in the Earth's crust and the fourth most abundant element; its valence state range from $+2$ to $+3$ depending upon the redox potential. Iron has four stable isotopes (percent natural abundance): ^{58}Fe (0.28%), ^{57}Fe (2.12%), ^{54}Fe (5.84%), and ^{56}Fe (91.8%) (Wiederhold, Teutsch, Kraemer, Halliday, & Kretzschmar, 2007). Its abundance is mainly noticed in igneous rocks, limestone, and calcareous soil. Coarse textured, alkaline, and upland soils generally show Fe deficiency. In cultivated soils, Fe is mostly present as Fe^{3+} and Fe^{2+} forms under oxic and anoxic environments, respectively (Marschner, 2011). According to Lindsay and Schwab (1982), Fe should have present in the range of $>7.7-10$ M in soil solution with a redox potential of soil-root environments under <12 to avoid its deficiency. It is not only an essential nutrient for almost all organisms, but also plays an important role in the behavior and fate of other elements in the environment. The conversion of iron into soil driven by existing geochemical and pedogenic conditions leads to the basic processes of weathering and soil formation (Wiederhold et al., 2007). Being an essential nutrient for most of the organisms iron plays a vital role in the behavior and fate of other elements in the environment. The conversion of iron in the soil, encouraged by the current geochemical and pedogenic conditions, leads to soil formation processes.

19.2.1 Iron status of Indian soils

Indian soils are high in total and available Fe content ranging from 13,000–18,000 mg kg^{-1} and that of available Fe is 3.40–68.1 mg kg^{-1} soil. Iron is less readily available in alkaline soils, which cover about 30% of the Earth's land surface. Many alkaline soils have high bicarbonate concentrations, which can hinder Fe earnings. Usually, high Fe content is found in acid and lateritic soils, this level sometimes is toxic to many crops. Distribution of iron in various states of India is depicted in Table 19.1 (Shukla, Tiwari, Pakhare, & Prakash, 2016).

19.2.2 Methods for assessing iron status in soils

The general methods of Fe estimation are based on the principle of Fe mobilization from soil particles through suitable extractant under defined conditions (Chao & Zhou, 1983). The extractant includes chelating agents such as diethylene triamine penta-acetic acid (DTPA) (Lindsay & Norvell, 1978; Soltanpour & Schwab, 1977) citrate, ascorbate (Reyes & Torrent, 1997), ammonium oxalate (Schwertmann, 1964), or reducing agents such as dydroxylamine (de Santiago, Díaz, del Campillo, Torrent, & Delgado, 2008). These extractants target only the available Fe pool, there are several other methods which are used for sequential extractions (Borggaard, 1988; Heron, Crouzet, Bourg, & Christensen, 1994). The sequential extraction steps are usually associated with crystalline Fe oxides, poorly crystalline Fe oxides, silicate-bound Fe, etc. The differences in a various isotopes of Fe of such pools have been observed and understood with respect to the Fe cycling in soils (Wiederhold et al., 2007) and plant uptake (Kiczka, Wiederhold, Kraemer, Bourdon, & Kretzschmar, 2007).

19.3 Iron in soil

19.3.1 Sources of iron in the soil

Iron is a highly reactive substance in soil and it form complexes with many ions. Iron is present in nature in oxide form, hydrooxide form, carbonate form, sulfate form, etc. Among different Fe-bearing minerals the highest amount of Fe is present in magnetite followed by hematite, goethite, and

TABLE 19.1 Status of iron deficiency in different states of India.

State	No. of soil samples analyzed	% Fe deficiency
Andhra Pradesh	7292	17.9
Assam	7208	0.0
Bihar	14,487	12.4
Chhattisgarh	4731	6.8
Goa	680	3.5
Gujarat	7587	25.9
Haryana	5637	21.7
Himachal Pradesh	10,586	0.5
Jammu and Kashmir	597	0.2
Jharkhand	995	0.0
Karnataka	7987	10.9
Kerala	894	1.2
Madhya Pradesh	14,236	8.8
Maharashtra	13,663	23.7
Manipur	1860	0.0
Odisha	4591	8.0
Punjab	6166	11.1
Rajasthan	6320	34.4
Tamil Nadu	34,315	12.6
Telangana	4939	16.7
Uttar Pradesh	8072	10.2
Uttarakhand	2575	1.4
West Bengal	3872	0.1
India (average)	169,290	14.4

pyrite (Table 19.2). Lepidocrocite and maghamite also contain a good amount of Fe but they are not abundant in nature.

19.3.2 Factors affecting Fe availability in soil

19.3.2.1 Soil pH

At low pH value, Fe ions are highly soluble; as pH increases, Fe ions are first converted to hydroxyl ions and finally to insoluble iron hydroxides. So,

TABLE 19.2 Different iron-bearing minerals in nature.

Sl. no.	Fe minerals	Formula	Type of minerals	Fe (%)
1.	Wüstite	FeO		79.0
2.	Magnetite	Fe_3O_4	Oxide mineral	72.4
3.	Maghamite	γ-Fe_2O_3	Oxide mineral	69.9
4.	Hematite	α-Fe_2O_3	Oxide mineral	69.9
5.	Goethite	α-FeOOH	Hydro-oxide mineral	62.9
6.	Lepidocrocite	γ-FeOOH	Hydro-oxide mineral	62.9
7.	Limonite	$FeO(OH)\cdot n(H_2O)$		52.0
8.	Sedarite	$FeCO_3$	Carbonate mineral	48.2
9.	Pyrite	FeS_2	Sulfur bearing mineral	46.5
10.	Ilmenite	$FeTiO_3$	Titanium bearing mineral	36.8
11.	Chalcopyrite	$CuFeS_2$	Sulfur bearing mineral	30.4
12.	Chamosite	$(Mg, Fe, Al)_6(Si, Al)_{414}(OH)_8$	Magnesium–aluminum-bearing mineral	29.6
13.	Bornite	$CuFeS_4$	Sulfur-bearing mineral	11.1

Fe insufficiency is mostly detected in high pH, calcareous soils of arid regions, but it may also occur in acid soils where total Fe is inherently low. Fe^{3+} species become dominant in the soil at low pH (pH <2), but $FeOH^{2+}$ becomes dominant as the pH increases to about 3. At pH 4, $FeOH^{+2}$ and $Fe(OH)_2{}^{+}$ remain in balance, and with a further rise in pH, the $Fe(OH)_3$ form predominates.

Iron becomes more available to the plant in the pH range of pH 4.5 to 5.5, and decreases sharply above the pH 5.5. The dissolution and precipitation of ferric oxides are the key factors that control the solubility of Fe in soils. Thus the activity of Fe^{3+} is very much dependent on pH and the iron oxides content of the medium. For each unit increment in pH, the solubility of Fe^{3+} decreases by 1000-fold, and it decreases to the levels of below 10–20 M when the pH goes as high as 7.5. The solubility of various Fe oxides decreases in the order of $Fe(OH)_3$ (amorphous) > $Fe(OH)_3$ (soil) > γ-Fe_2O_3 (maghemite) > γ-FeOOH (lepidocrocite) > α-Fe_2O_3 (hematite) > α-FeOOH (goethite).

19.3.2.2 Calcium carbonate and bicarbonate

Calcareous soil contain a high amount of $CaCO_3$ at pH 8.5. At that high pH Fe is precipitated as the hydroxide form, $Fe(OH)_3$, and it leads to Fe deficiency. Bicarbonate ions neutralize the H^+ in soil and maintain high pH of the soil.

19.3.2.3 Carbon dioxide

Carbon dioxide (CO_2) accumulates in poorly drained compact fine textured soils and reacts with free $CaCO_3$ in calcareous soil to yield HCO^{3-}. These bicarbonate ions are associated with high pH at which Fe is precipitated as iron hydroxide. Thus iron becomes unavailable.

19.3.2.4 Microorganisms

Iron is ubiquitous in the biosphere, hydrosphere, lithosphere, and atmosphere of our planet, either as ferrous (Fe(1l)), particulate ferric (Fe(1ll)), Fe-bearing minerals, or as dissolved ions. The redox transformations of Fe, mobilization, dissolution, and precipitation are caused by chemical and microbial processes (Fig. 19.1). Microorganisms catalyze the oxidation of Fe(II) and reduction of Fe(III) under oxic and anoxic conditions, respectively (Table 19.3). Microbial-mediated transformations of Fe are often much faster than the chemical process, and happen mainly in soils, sediments, marine, and freshwater environments. Microbial-mediated Fe transformations also play an important role in biogeochemical cycles, particular in the carbon cycle.

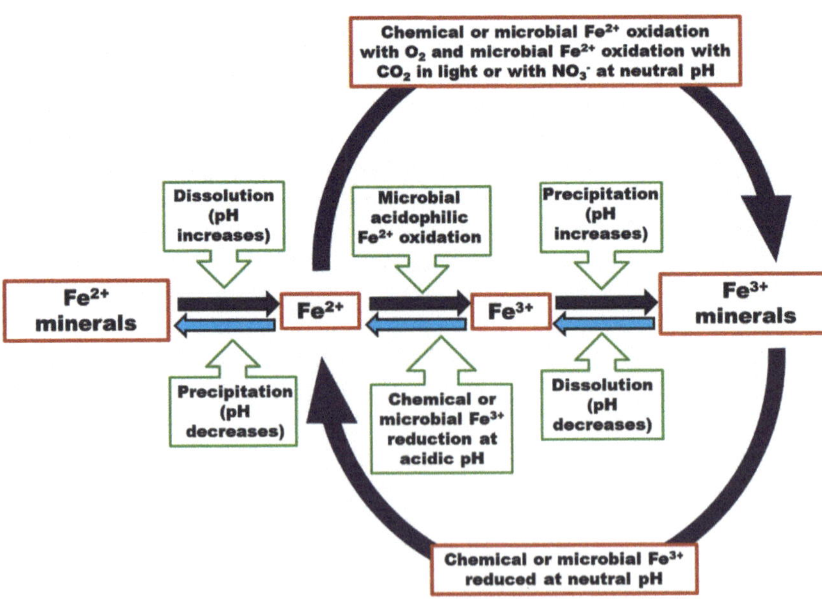

FIGURE 19.1 Interrelation among microbial and chemical iron cycle.

TABLE 19.3 Physiological groups of prokaryotes that catalyze iron redox transformations.

Habitat	Electron donor	Electron acceptor	pH	Microbial metabolism	Representative strains
Oxic	Fe(II)	O_2	Acidic	-	*Thiobacillus ferreoxidase* *Sulfobacillus acidophilus*
	Fe(II)	O_2	Neutral	Fe(II) oxidase	*Gallionella ferruginea* *Leptothrix ochracea*
Anoxic	Fe(II)	NO_3^-	Neutral	NO_3^- dependent Fe(II) oxidation	*Acidovorax sp.* *Azospira oryzae*
	Fe(II)	CO_2	Neutral	Phototrophic Fe(II) oxidation	*Rhodobacter ferrooxidance*
	Organic or inorganic compounds	Fe(III)	Acidic	Fe(II) reduction	*Acidiphilium cryptum* *Thiobacillus thiooxidance*
	Organic or inorganic compounds	Fe(III)	Neutral	Fe(II) reduction	*Geobacter metallireducer* *Shewanella onedensis*

19.4 Iron in plant

19.4.1 Iron solubilization in the rhizosphere

The following mechanisms for Fe availability:

1. Minimizing the pH of the rhizosphere via the releae of different acidifying agents like enzymes, organic acids (H^+- extrusion).
2. Secretion of different chelating molecules like phytosiderophores with a varying affinity for Fe.
3. Release of multiple reducing agents or promoting activity of reductases hence developing a potential reducing environment for Fe solubility.

19.4.2 Iron uptake by plants

Iron is the fourth most abundant element on the Earth's crust (5.1%), although it is highly insoluble and poorly available, especially in high pH and aerobic soils. Takagi (1976) first reported the secretion of Fe(III) solubilizing molecules, such as mugineic acid (MA) family phytosiderophores, largely from Fe-unfertilized graminaceous plants and later, based on experimental reports. Römheld and Marschner (1986) classified the Fe-acquisition mechanisms into two categories (Fig. 19.2):

1. Strategy I: Nongraminaceous plants
2. Strategy II: Graminaceous plants

The main mechanism can be explained very simply by reducing Fe^{3+} into Fe^{2+} ions in the rhizosphere and channelizing it across the plasma membrane. The major genes responsible for strategy I are ferric-chelate reductase oxidase gene (FRO2) and iron regulated transporter gene (IRT1), which were isolated from *Arabidopsis* in the early 1990s and later were reported from other plants too (Eide, Broderius, Fett, & Guerinot, 1996; Robinson, Procter, Connolly, & Guerinot, 1999). Other mechanisms, such as rhizospheric acidification either through acids, phenolic (−OH) compounds, or by creating reducing environments are reported, like CsHA1 in cucumber and AHA2 in *Arabidopsis* (Santi & Schmidt, 2009; Santi, Cesco, Varanini, & Pinton, 2005). According to, Ishimaru et al. (2011) the identification of PHENOLICS EFFLUX ZERO 1 (PEZ1) genes involved in secretion of protocatechuic acid in rice raises the possibility of phenolics secretion in strategy I nongraminaceous plants but it requires further validation and concrete results. Apart from genetic regulations root morphological changes do exist under Fe deficiency like the development of transfer cells and root hair formations (Marschner, 2011; Santi & Schmidt, 2008).

Strategy II is linked with cereal crops (graminaceous family) and based on the biogenesis and release of MAs from 2'-deoxymugineic acid (DMA), which is common to all the nine MAs identified to date (Bashir et al., 2011).

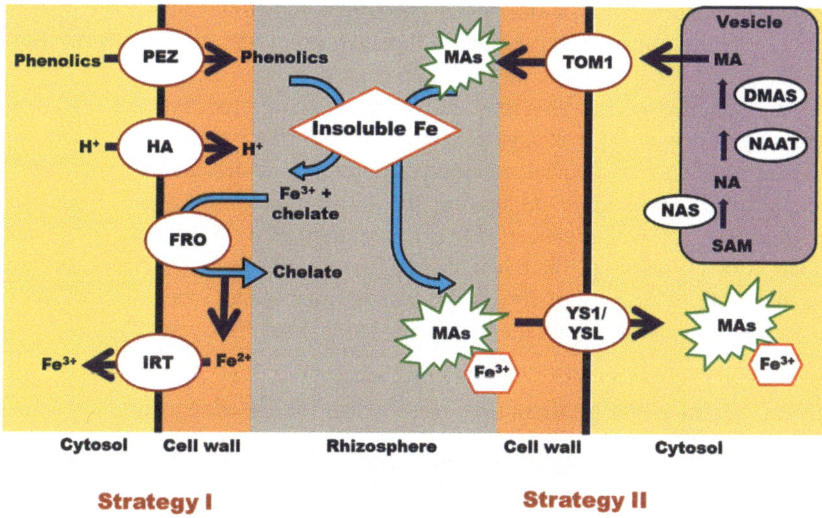

FIGURE 19.2 Iron acquisition strategies in higher plants: Strategy I in nongraminaceous plants (*left*) and Strategy II in graminaceous plants (*right*). Ovals represent the transporters and enzymes that play central roles in these strategies, all of which are induced in response to Fe deficiency. *DMAS*, Deoxymugineic acid synthase; *FRO*, ferric-chelate reductase oxidase; *HA*, H^+-ATPase; *IRT*, iron-regulated transporter; *MAs*, mugineic acid family phytosiderophores; *NA*, nicotianamine; *NAAT*, nicotianamine aminotransferase; *NAS*, nicotianamine synthase; *PEZ*, phenolics efflux zero; *SAM*, *S*-adenosyl-L-methionine; *TOM1*, transporter of mugineic acid family phytosiderophores 1; *YS1/YSL*, yellow stripe 1/yellow stripe 1-like.

In some species like barley and rye, DMA was further hydroxylated to other MAs via hydroxylases (Bashir et al., 2011). There must be an unhindered supply of methionine which via a series of cycles, commonly known as the methionine cycle or Yang cycle, sustains the production of MAs. Recently, the responsible genes for inducing MAs production upon Fe deficiency have been characterized by using the same genetic information from bacteria and yeast. The activity of MAs follows a circadian pattern with a maximum in the early hours and a low in the evening (Takagi, Nomoto, & Takemoto, 1984). The MAs production pathway includes three sequential enzymatic reactions mediated by nicotinamide synthase (NAS), nicotianamine aminotransferase (NAAT), and deoxymugineic acid synthase (DMAS) which generating 2′-deoxymugineic acid (DMA), the precursor of all other MAs (Nakanishi, Yamaguchi, Sasakuma, Nishizawa, & Mori, 2000). Identifying and characterizing the genes is a major drawback for understanding strategy II efficiently, although recent scientific findings are unfolding many mysteries (Nozoye et al., 2011). In rice, the mechanism follows like this, after releasing MAs from the roots, solubilized Fe(III) forms a metal–MA complex which is progressively taken up by the plant root using transporters like YELLOW STRIPE 1 (YS1) and YELLOW STRIPE 1–like (YSL) (Inoue et al., 2009).

Although predominantly a strategy II plant, rice sometimes also follows strategy I by using transporters like OsIRT1 which allow the crop to uptake Fe^{2+} which is the major form in the submerged soils (Inoue et al., 2009). Nonetheless, researchers also reported various type of exporters like ZIP (zinc-regulated transporter, iron-regulated transporter–like protein) and NRAMP (natural resistance-associated macrophage protein), which are homologous to the IRT1 of *Arabidopsis*, regulating the movement of divalent metals like Fe^{2+}, zinc (Zn^{2+}), manganese (Mn^{2+}), cadmium (Cd^{2+}), nickel (Ni^{2+}), and cobalt (Co^{2+}), although their location in the plant body is different, such as, the main site of activity of IRT is the plasma membrane, whereas NRAMP is located in the vesicle (Cailliatte, Schikora, Briat, Mari, & Curie, 2010).

19.4.3 Iron translocation and regulation inside plant body

Due to its insolubility and reactivity, proper translocation of iron inside the plant body must have two important components: (1) chelators like citrate, nicotianamine, and MAs; and (2) actual redox state for interconversion between the two forms of Fe. Inside the plant, Fe moves via symplastic movement, that is, from Casparian strip to xylem followed by phloem and in the end to the site of actual demand and if needed retranslation from senescence to new tissue (Kim & Guerinot, 2007). Recently, Rellań-Álvarez, Giner-Martinez-Sierra, Orduna, Orera, and Rodriguez-Castrillon (2010) reported the original structure of a Fe-citrate chelating structure in tomato which is a three Fe metal core linked with oxo-bonding. Similarly, in *Arabidopsis* ferric reductase defective 3 (FRD3) facilitates metal loading in the xylem and is highly specific to a particular nutrient (Durrett, Gassmann, & Rogers, 2007). In rice also, OsFRDL1, a citrate effluxer outlined in the root pericycle is responsible for Fe translocation, although information regarding OsFRDL1 for alleviating Fe deficiency is restricted up to a certain limit and indicates the importance of other Fe-absorption pathways (Koike et al., 2004). But before the formation of the metal–chelate complex, it is important to mention the transporters responsible for xylem loading, and to date the most promising loader is Arabidopsis ferro-portin 1/iron regulated 1 (AtFPN1/AtIREG1), which works similarly to mammalian Fe efflux transporters (Morrissey et al., 2009). However, the transport role of AtFPN1 is yet to be confirmed. Upon deficiency of Fe, YS1 in maize uptakes Fe via a chelating complex (Fe^{3+}-DMA) by roots and translocates it to the above-ground portions. Nongraminaceous plants belonging to the YSL group use NAs to chelate the metal cations (Curie et al., 2009). In graminaceous plants OsYSL18, a regulator gene has been reported to take part in Fe translocation but also to promote activity in reproductive plant parts like pollen germination (Inoue et al., 2009). The translocation pattern of Fe(III)-DMA is different according to age but it plays an important role to sustain Fe uptake and nutrition in plants (Tsukamoto et al., 2009). Gene regulation is a very important and modern practice for improving plant functional traits with regard to the changing environment. Targeting specific genes under Fe-deficiency is very important irrespective of any crop species and details of these genes are presented in Fig. 19.3. Ling, Bauer, Bereczky, Keller, and Ganal (2002) first reported a distorted strategy I in tomato mutant T3238FER. Yuan et al. (2008) reported coexpression of Fe-deficiency-induced genes (FIT) along with others like bHLH, AtbHLH38, or AtbHLH39 which are efficient Fe-translocators. Wang et al. (2007) reported four major important genes—AtbHLH38, AtbHLH39, AtbHLH100, and AtbHLH101—expressed in the above = and belowground plant portion due to Fe deficiency regulating Fe movement. Alternately, Barberon et al. (2011)

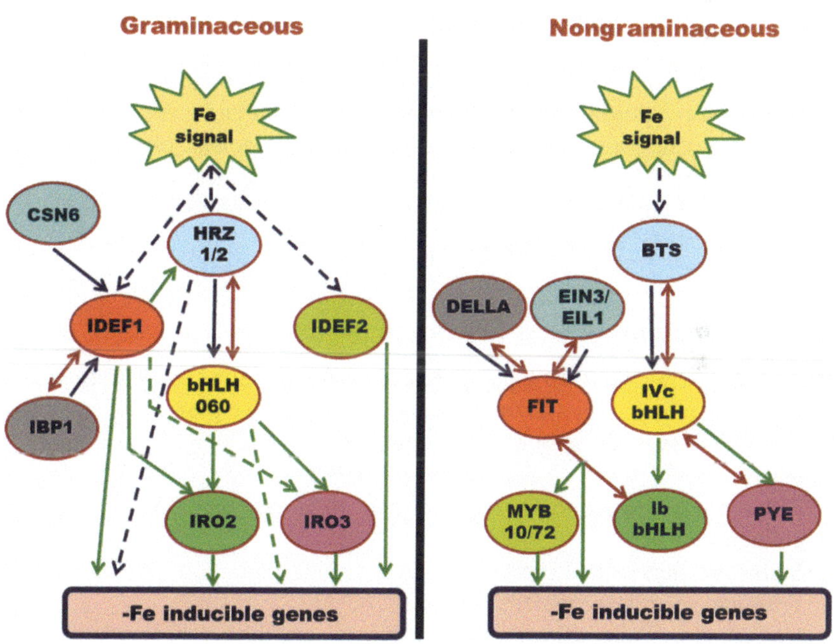

FIGURE 19.3 Regulation of iron deficiency responses in graminaceous and nongraminaceous plants. Ovals indicate regulatory proteins, and colors indicate their types. Double-lined ovals indicate genes clearly induced under Fe deficiency at the transcriptional level. Broken lines indicate putative pathways. Line colors, type of regulation: green lines, transcriptional regulation; red lines, protein–protein interaction; blue lines, protein-level or unknown-type regulation. *HRZ1/2*, Ubiquitin ligase; *bHLH060*, basic helix-loop-helix; *IRO2*, iron-related transcription factor 2; *IRO3*, iron-related transcription factor 3; *IDEF1*, iron deficiency–responsive element-binding factor 1; *IDEF2*, iron deficiency–responsive element-binding factor 2; *CSN6*, COP9 signalosome subunit 6; *IBP1*, IDEF1-binding protein 1; *BTS*, ubiquitin ligase; *IVcbHLH*, basic helix-loop-helix sub group IVc; *IbbHLH*, basic helix-loop-helix subgroup Ib; *PYE*, popeye; *FIT*, T3238FER/FER-like iron deficiency–induced transcription factor; *EIN3/EIL1*, ethylene insensitive 3/ethylene insensitive 3–like.

opposed the protein level regulation of IRT1 and it is also reported that irrespective of Fe-content IRT1 protein binds Fe present in the media and this can be justified by the expression of N-terminus-truncated IRT1 protein. The research work of Dinneny et al. (2008) reported that huge genetic variation between cellular layers and genes is mostly generated in the epidermal regions, whereas genes responsible for stress and signaling have their origin in the stele regions of the stem. Long et al. (2010) found that transcriptional factors like POPEYE (PYE) and BRUTUS (BTS) play a vital role in Fe translocation, particularly in deficient condition. The exact role of these mentioned transcriptional factors and their negative correlation is still a mystery, but understanding the numerous feedback pathways might be an alternate way to advance scientific understanding.

19.4.4 Functions of iron in plants

Iron (Fe) is a ubiquitous component of two groups of proteins: (1) heme protein and (2) Fe-S protein. It helps in plant metabolism and passes on energy-rich molecules in plant parts. Iron is a structural component of enzymes, for example, aminolevolinic acid coproporphyrinogen oxidase and nitrogenase. It is an integral component of symbiotic nitrogen fixation and is essential for chlorophyll formation (synthesis).

19.4.5 Symptoms of Fe deficiency in plants

1. A plant having <50 mg kg^{-1} of Fe is usually classified as Fe-deficient.
2. Fe-deficiency mainly shows in younger leaves of the plant.
3. Younger leaves show interveinal chlorosis or yellowing or whitening of leaves.
4. Chlorophyll formation is reduced and it is popularly known as Fe-chlorosis.
5. Stalks are slender and short.
6. Fruit and seed development is poor.
7. Under severe conditions, growth ceases and the whole plant turns necrotic.

19.4.6 Sources of Fe for the plants

To remedy Fe deficiency in plants, it is advisable to use Fe-containing fertilizers which can supply a sufficient amount of Fe to plants to overcome the Fe deficiency (Table 19.4). Among the Fe-containing fertilizers, ferrous sulfate is mainly used which contains 20% Fe. Application in conjugation with chelators is an improved practice but not an economical option for the farmers.

TABLE 19.4 Different iron fertilizers for plant.

Sources	Formula/product name	Iron (%)	Method of application
Chelates	Fe DTPA	10	Soil, Foliar
	Fe EDTA	9–12	Soil, Foliar
	Fe EDDHA	6	Soil, Foliar
	Fe HEDTA	5–9	Soil, Foliar
Ferrous ammonium sulfate	$(NH_4)_2(Fe\ SO_4)_2 \cdot 6H_2O$	14	Foliar
Ferrous sulfate	$FeSO_4 \cdot 7H_2O$	20	Foliar
Ferrous ammonium phosphate	$Fe(NH_4)PO_4 \cdot H_2O$	29	Foliar
Nano Fe fertilizer	Nano Micro Nutrient (Eco Star)	6	Foliar
	Nanomol (F) Micronutrient	-	Foliar
	Plant Nutrition Powder (Green Nano)	1	Foliar
	Nano Capsule (The Best)	2	Foliar

19.5 Conclusion

Iron is mostly present in the Fe^{3+} and Fe^{2+} forms under oxic and anoxic environments, respectively. As the pH increases the Fe deficiency in soil increases. Plants develop active iron acquisition strategies based on the acidification and reduction of Fe(III) (strategy I) and on the release of phytosiderophores (strategy II). Foliar application of $FeSO_4$ gives a high value of vegetative and reproductive growth of plants. Biofortification offers a rapid solution to improve Fe concentration in the edible portion of crops. Iron is an essential micronutrient for almost all living organisms because it plays a critical role in metabolic processes such as DNA synthesis, respiration, and photosynthesis. Iron deficiency causes chlorosis (yellowing) in the newest plant growth because Fe is not remobilized in the plant from old to new leaves. The dissolution and availability of soil Fe minerals is usually very low; however, the interaction with microorganisms, plant, and organic matter can improve the formation of soluble Fe(III) complexes and increase the availability of Fe for plant growth. Microorganisms release siderophores and plant exudates like phytosiderophores, organic acids, and flavonoids, which can bind with Fe present in minerals and solubilize it. Because of the very low concentrations of Fe^{3+} and Fe^{2+} in soils, chelated forms of Fe fertilizers generally performed better for plant response and Fe uptake than iron oxides or sulfates. Also foliar applications of Fe have resulted in more uniform success for correcting plant Fe deficiencies.

References

Barberon, M., Zelazny, E., Robert, S., Conéjéro, G., Curie, C., Friml, J., & Vert, G. (2011). Monoubiquitin-dependent endocytosis of the iron-regulated transporter 1 (IRT1) transporter controls iron uptake in plants. *Proceedings of the National Academy of Sciences of the United States of America, 108*, E450–E458.

Bashir, K., Ishimaru, Y., Shimo, H., Nagasaka, S., Fujimoto, M., Takanashi, H., ... Nishizawa, N. K. (2011). The rice mitochondrial iron transporter is essential for plant growth. *Nature Communications, 2*, 1–7.

Borggaard, O. K. (1988). Phase identification by selective dissolution techniques. In *Iron in soils and clay minerals* (83–98). Springer, Dordrecht.

Cailliatte, R., Schikora, A., Briat, J. F., Mari, S., & Curie, C. (2010). High-affinity manganese uptake by the metal transporter NRAMP1 is essential for Arabidopsis growth in low manganese conditions. *The Plant Cell, 22*, 904–917.

Chao, T. T., & Zhou, L. (1983). Extraction techniques for selective dissolution of amorphous iron oxides from soils and sediments. *Soil Science Society of America Journal, 47*, 225–232.

Curie, C., Cassin, G., Couch, D., Divol, F., Higuchi, K., Le Jean, M., ... Mari, S. (2009). Metal movement within the plant: Contribution of nicotianamine and yellow stripe 1-like transporters. *Annals of Botany, 103*, 1–11.

de Santiago, A., Díaz, I., del Campillo, M. D. C., Torrent, J., & Delgado, A. (2008). Predicting the incidence of iron deficiency chlorosis from hydroxylamine-extractable iron in soil. *Soil Science Society of America Journal, 72*, 1493–1499.

Dinneny, J. R., Long, T. A., Wang, J. Y., Jung, J. W., Mace, D., Pointer, S., ... Benfey, P. N. (2008). Cell identity mediates the response of *Arabidopsis* roots to abiotic stress. *Science, 320*, 942–945.

Durrett, T. P., Gassmann, W., & Rogers, E. E. (2007). The FRD3-mediated efflux of citrate into the root vasculature is necessary for efficient iron translocation. *Plant Physiology, 144*, 197–205.

Eide, D., Broderius, M., Fett, J., & Guerinot, M. L. (1996). A novel iron-regulated metal transporter from plants identified by functional expression in yeast. *Proceedings of the National Academy of Sciences of the United States of America, 93*, 5624–5628.

FAO, WHO. (2000). *Evaluation of certain food additives and contaminants: Fifty-third report of the Joint FAO/WHO Expert Committee on Food Additives*. World Health Organization. Available from https://www.who.int/ipcs/publications/jecfa/reports/trs940.pdf.

Gavaravarapu, S. M., & Hemalatha, R. (2018). National Institute of Nutrition: 100 years of empowering the nation through nutrition. *The Indian Journal of Medical Research, 148*, 477.

Heron, G., Crouzet, C., Bourg, A. C., & Christensen, T. H. (1994). Speciation of Fe (II) and Fe (III) in contaminated aquifer sediments using chemical extraction techniques. *Environmental Science & Technology, 28*, 1698–1705.

ICMR. (2010). *Dietary guidelines for Indians—A manual*. New Delhi: National Institute of Nutrition, Indian Council of Medical Research. Available from https://www.nin.res.in/downloads/DietaryGuidelinesforNINwebsite.pdf.

Inoue, H., Kobayashi, T., Nozoye, T., Takahashi, M., Kakei, Y., Suzuki, K., ... Nishizawa, N. K. (2009). Rice OsYSL15 is an iron-regulated iron (III)-deoxymugineic acid transporter expressed in the roots and is essential for iron uptake in early growth of the seedlings. *The Journal of Biological Chemistry, 284*, 3470–3479.

Ishimaru, Y., Kakei, Y., Shimo, H., Bashir, K., Sato, Y., Sato, Y., ... Nishizawa, N. K. (2011). A rice phenolic efflux transporter is essential for solubilizing precipitated apoplasmic iron in the plant stele. *The Journal of Biological Chemistry*, *286*, 24649–24655.

Kiczka, M., Wiederhold, J. G., Kraemer, S. M., Bourdon, B., & Kretzschmar, R. (2007). The impact of Fe isotope fractionation by plants on the isotopic signature of soils. *Geochimica et Cosmochimica Acta*, *71*, A482.

Kim, S. A., & Guerinot, M. L. (2007). Mining iron: Iron uptake and transport in plants. *FEBS Letters*, *581*, 2273–2280.

Koike, S., Inoue, H., Mizuno, D., Takahashi, M., Nakanishi, H., Mori, S., & Nishizawa, N. K. (2004). OsYSL2 is a rice metal-nicotianamine transporter that is regulated by iron and expressed in the phloem. *The Plant Journal: for Cell and Molecular Biology*, *39*, 415–424.

Kumar, D., Patel, K. P., Ramani, V. P., Shukla, A. K., & Meena, R. S. (2020a). Management of micronutrients in soil for the nutritional security. In *Nutrient dynamics for sustainable crop production* (103–134). Springer, Singapore.

Laxmaiah, A., Arlappa, N., Balakrishna, N., Rao, K. M., Galreddy, C., Kumar, S., ... Brahmam, G. N. (2013). Prevalence and determinants of micronutrient deficiencies among rural children of eight states in India. *Annals of Nutrition & Metabolism*, *62*, 231–241.

Lindsay, W. L., & Norvell, W. A. (1978). Development of a DTPA soil test for zinc, iron, manganese, and copper. *Soil Science Society of America Journal*, *42*, 421–428.

Lindsay, W. L., & Schwab, A. P. (1982). The chemistry of iron in soils and its availability to plants. *Journal of Plant Nutrition*, *5*, 821–840.

Ling, H. Q., Bauer, P., Bereczky, Z., Keller, B., & Ganal, M. (2002). The tomato fer gene encoding a bHLH protein controls iron-uptake responses in roots. *Proceedings of the National Academy of Sciences of the United States of America*, *99*, 13938–13943.

Long, T. A., Tsukagoshi, H., Busch, W., Lahner, B., Salt, D. E., & Benfey, P. N. (2010). The bHLH transcription factor POPEYE regulates response to iron deficiency in *Arabidopsis* roots. *The Plant Cell*, *22*, 2219–2236.

Marschner, H. (2011). *Marschner's mineral nutrition of higher plants*. U.S.A: Academic Press.

Morrissey, J., Baxter, I. R., Lee, J., Li, L., Lahner, B., Grotz, N., ... Guerinot, M. L. (2009). The ferroportin metal efflux proteins function in iron and cobalt homeostasis in Arabidopsis. *The Plant Cell*, *21*, 3326–3338.

Nakanishi, H., Yamaguchi, H., Sasakuma, T., Nishizawa, N. K., & Mori, S. (2000). Two dioxygenase genes, Ids3 and Ids2, from *Hordeumvulgare* are involved in the biosynthesis of mugineic acid family phytosiderophores. *Plant Molecular Biology*, *44*, 199–207.

NNMB (National Nutrition Monitoring Bureau, National Institute of Nutrition ICMR) (2009). *Diet and nutritional status of tribal population and prevalence of hypertension among adults*. Report on Second Repeat Survey. Hyderabad. pp. 265.

Nozoye, T., Nagasaka, S., Kobayashi, T., Takahashi, M., Sato, Y., Sato, Y., ... Nishizawa, N. K. (2011). Phytosiderophore efflux transporters are crucial for iron acquisition in graminaceous plants. *The Journal of Biological Chemistry*, *286*, 5446–5454.

Rellań-Álvarez, R., Giner-Martinez-Sierra, J., Orduna, J., Orera, I., & Rodriguez-Castrillon, J. A. (2010). Identification of a tri-iron(III), tri-citrate complex in the xylem sap of iron-deficient tomato resupplied with iron: New insights into plant iron long-distance transport. *Plant & Cell Physiology*, *51*, 91–102.

Reyes, I., & Torrent, J. (1997). Citrate-ascorbate as a highly selective extractant for poorly crystalline iron oxides. *Soil Science Society of America Journal*, *61*, 1647–1654.

Robinson, N. J., Procter, C. M., Connolly, E. L., & Guerinot, M. L. (1999). A ferric-chelate reductase for iron uptake from soils. *Nature*, *397*, 694–697.

Römheld, V., & Marschner, H. (1986). Evidence for a specific uptake system for iron phytosiderophores in roots of grasses. *Plant Physiology, 80*, 175–180.

Santi, S., Cesco, S., Varanini, Z., & Pinton, R. (2005). Two plasma membrane H^{+}-ATPase genes are differentially expressed in iron-deficient cucumber plants. *Plant Physiology and Biochemistry, 43*, 287–292.

Santi, S., & Schmidt, W. (2008). Laser microdissection-assisted analysis of the functional fate of iron deficiency-induced root hairs in cucumber. *Journal of Experimental Botany, 59*, 697–704.

Santi, S., & Schmidt, W. (2009). Dissecting iron deficiency-induced proton extrusion in *Arabidopsis* roots. *The New Phytologist, 183*, 1072–1084.

Schwertmann, U. (1964). Differenzierung der Eisenoxide des Bodens durch Extraktion mit Ammoniumoxalat-Lösung. *Zeitschrift für Pflanzenernährung, Düngung, Bodenkunde, 105*, 194–202.

Shukla, A. K. (2014). Understanding the mechanism of variation in status of a few nutritionally important micronutrients in some important food crops and the mechanism of micronutrient enrichment in plant parts, NAIP Funded Research Project. AICRP on Micronutrients, IISS, Nabibagh, Berasia Road, Bhopal. 1–97.

Shukla, A. K. & Pakhare, A. (2015). Trace elements in soil-plant-human continuum. In *Soil science: An introduction*. ISSS, New Delhi.

Shukla, A. K., Tiwari, P. K., Pakhare, A., & Prakash, C. (2016). Zinc and iron in soil, plant, animal and human health. *Indian Journal of Fertilisers, 12*, 133–149.

Shukla, A. K., Tiwari, P. K., & Prakash, C. (2014). Micronutrients deficiencies vis-a-vis food and nutritional security of India. *Indian Journal of Fertilisers, 10*, 94–112.

Soltanpour, P. N., & Schwab, A. P. (1977). A new soil test for simultaneous extraction of macro-and micro-nutrients in alkaline soils. *Communications in Soil Science and Plant Analysis, 8*, 195–207.

Takagi, S. I. (1976). Naturally occurring iron-chelating compounds in oat-and rice-root washings: I. Activity measurement and preliminary characterization. *Soil Science and Plant Nutrition, 22*, 423–433.

Takagi, S. I., Nomoto, K., & Takemoto, T. (1984). Physiological aspect of mugineic acid, a possible phytosiderophore of graminaceous plants. *Journal of Plant Nutrition, 7*, 469–477.

Takkar, P. N. (2011). Zinc in human and animal health, zinc special: Improving crop production and human health. *Indian Journal of Fertilisers, 7*, 46–69.

Tsukamoto, T., Nakanishi, H., Uchida, H., Watanabe, S., Matsuhashi, S., Mori, S., & Nishizawa, N. K. (2009). ^{52}Fe translocation in barley as monitored by a positron-emitting tracer imaging system (PETIS): Evidence for the direct translocation of Fe from roots to young leaves via phloem. *Plant and Cell Physiology, 50*, 48–57.

Wang, H. Y., Klatte, M., Jakoby, M., Bäumlein, H., Weisshaar, B., & Bauer, P. (2007). Iron deficiency-mediated stress regulation of four subgroup Ib BHLH genes in *Arabidopsis thaliana. Planta, 226*, 897–908.

WHO. (1999). *Malnutrition worldwide*. Geneva: World Health Organization.

Wiederhold, J. G., Teutsch, N., Kraemer, S. M., Halliday, A. N., & Kretzschmar, R. (2007). Iron isotope fractionation in oxic soils by mineral weathering and podzolization. *Geochimica et Cosmochimica Acta, 71*, 5821–5833.

Yuan, Y., Wu, H., Wang, N., Li, J., Zhao, W., Du, J., … Ling, H. Q. (2008). FIT interacts with AtbHLH38 and AtbHLH39 in regulating iron uptake gene expression for iron homeostasis in Arabidopsis. *Cell Research, 18*, 385–397.

Chapter 20

The molecular mechanism of brassinosteroids in mediating the abiotic stress responses of plants

Saima Liaqat, Peer Saffeullah, Shahid Umar and Tariq Omar Siddiqi
Department of Botany, School of Chemical and Life Sciences, Jamia Hamdard, New Delhi, India

20.1 Introduction

Brassinosteroids (BRs) area group of plant hormones that were first discovered in pollen of *Brassica napus* around 40 years ago (Peres et al., 2019). The key components of the canonical BR signaling pathway are identified via numerous genetic and biochemical screens since they were discovered (Planas-Riverola et al., 2019). Mutations in genes that encode the key components of the BR synthesis and signaling pathways result in extreme dwarfism, impaired organ growth and development, and reduced plant fertility and yield (Gruszka, 2020). Despite the aforesaid understanding about the BR pathway, it still remains unclear how BRs function in a cell-specific manner and also how they interact with the other hormonal pathways in normal and stressed environmental conditions (Planas-Riverola et al., 2019; Vukašinović & Russinova, 2018). BR are analogous to animal steroid hormones and like their animal counter parts, BRs regulate thousands of genes such as cell division and elongation, male sterility, vascular differentiation, photomorphogenesis, leaf expansion, stomatal development, and seed germination. BRs also regulate certain physiological and developmental processes that lead to the overall developmental processes. They confer resistance to different abiotic (salinity, drought, heat, chilling, and heavy metal) and biotic stresses (Anwar et al., 2018). The detailed BR signaling pathway and related signaling components were elucidated by genetic and molecular studies in *Arabidopsis*. In outline, BRs directly bind to the cell surface receptor-like kinase (RLK) BRASSINOSTEROID-INSENSITIVE 1 (BRI1) and trigger a signal transduction cascade leading to the activation of two main transcription factors,

Frontiers in Plant–Soil Interaction. DOI: https://doi.org/10.1016/B978-0-323-90943-3.00003-1

BRASSINAZOLE-RESISTANT1 (BZR1) and BRI1-EMS-SUPPRESSOR 1 (BES1), also known as BZR2 (Zhang et al., 2018). Several genetic and proteomic studies in *Arabidopsis* and rice lead to the unraveling of events of BR signal transduction and BR perception from receptor kinase activation to transcriptional networks. Further characterization of BR signals in rice and other crops, particularly cereals, would likely reveal new mechanisms that could be used to improve crops or provide insight into the BR signaling evolution. In this chapter we characterize the BR signaling network and its association with other phytohormones in managing the plant growth and development. In addition, we are also addressing the control of stress responses by BRs in interaction with the other hormones.

20.2 Brassinosteroids perception and receptor kinases

BR signal is perceived extracellularly by leucine-rich repeat (LRR) RLK, BRI 1 (BRASSINOSTEROID INSENSITIVE 1), which is localized in the plasma membrane. It comprises an extracellular ligand-binding domain of 25 LRRs, an island domain of 70 amino acids between LRR21 and LRR22, a single transmembrane domain, and a cytoplasmic domain that has kinase activity (Chen et al., 2020). BRI 1 is widely accepted as the plasma membrane bound receptor for BRs and acts as an inactive homodimer in the absence of BRs (Yusuf, Khan, & Fariduddin, 2017). It remains bound through its cytoplasmic domain with its negative regulator, BRI1 KINASE INHIBITOR 1 (BKI1) (Martins, Vert, & Jaillais, 2017). As the BR binds at the 93 amino acid long region in the extracellular domain of BRI 1, it activates its kinase activity leading to the phosphorylation of BRI 1 at Tyr211 converting it to a heterodimer by associating with its coreceptor kinase, BRI1-ASSOCIATED RECEPTOR KINASE 1 (BAK1)/SOMATIC EMBRYOGENESIS RECEPTOR KINASE3 (SERK3) (Sun et al., 2013), This phosphorylation in turn activates the intracellular phosphorylation cascade. Recently it was discovered that Ser270 and Ser274 in the C-terminal region of BKI1 is responsible for the subsequent phosphorylation of Tyr211 in BRI1 and also its disassociation from the plasma membrane (Wang et al., 2014). The BR signaling pathway gets inactivated when BIN2 (BRASSINOSTEROID-INSENSITIVE 2), GSK3-like kinase, phosphorylates BZR1/BES1 proteins in the absence of BRs, leading to pathway inactivation by facilitating their binding to 14-3-3 proteins, which as a result are therefore retained in cytoplasm. In the presence of BRs, the phosphorylation of BKI1 then relieves BES1(BRI1-EMSSUPPRESSOR 1) and BZR1(BRASSINAZOLE RESISTANT 1) from its inhibition by phosphopeptide-binding proteins, 14-3-3 (Li et al., 2020). These plant-specific transcription factors BZR1 and BES1 control the transcription of thousands of BR-responsive genes and thus regulate a multitude of developmental events in the plant (Nolan, Vukašinović, Liu, Russinova, & Yin, 2020). The phosphorylation site mapping and functional studies have helped in identifying the phosphorylation sites at Ser/Thr and Tyr of both BRI1 and BAKI (Hao, Yin, & Fei, 2013). There are three homologs of

BRI1 based on the presence of the extracellular BR-binding domain, namely BRL1, BRL2, and BRL3 (BRI1-LIKE 1, 2, and 3). Like BRI1, BRL1 and BRL3 function as high affinity functional receptors binding to steroid molecules while BRL2 does not appear to be a functional receptor. BRI1 is ubiquitously expressed in the root, whereas BRL1 and BRL3 are limited in vascular stem cells governing the cell-specific BR-response pathways (Planas-Riverola et al., 2019; Salazar-Henao, Lehner, Betegón-Putze, Vilarrasa-Blasi, & Caño-Delgado, 2016). Both BRL1 and BRL3 forms complex by heterodimerization with the BAK1 coreceptor and not with BRI1 suggesting that BRI1 and BRLs form different complexes in different cell types and hence perform different signaling roles (Anwar et al., 2018). It is suggested that BRs cannot be transported to longer distances as revealed from a study carried out in pea, and therefore it is necessary to maintain the internal active BRs by carrying out BR biosynthesis and inactivation in various tissues at various developmental stages (Hao et al., 2013). BRs are widely known to have interactions with other plant hormones, such as abscisic acid (ABA), auxin, gibberellins (GA), cytokinin, salicylic acid (SA), jasmonic acid (JA), and ethylene, for the regulation of many biological processes (Hao et al., 2013) (Fig. 20.1).

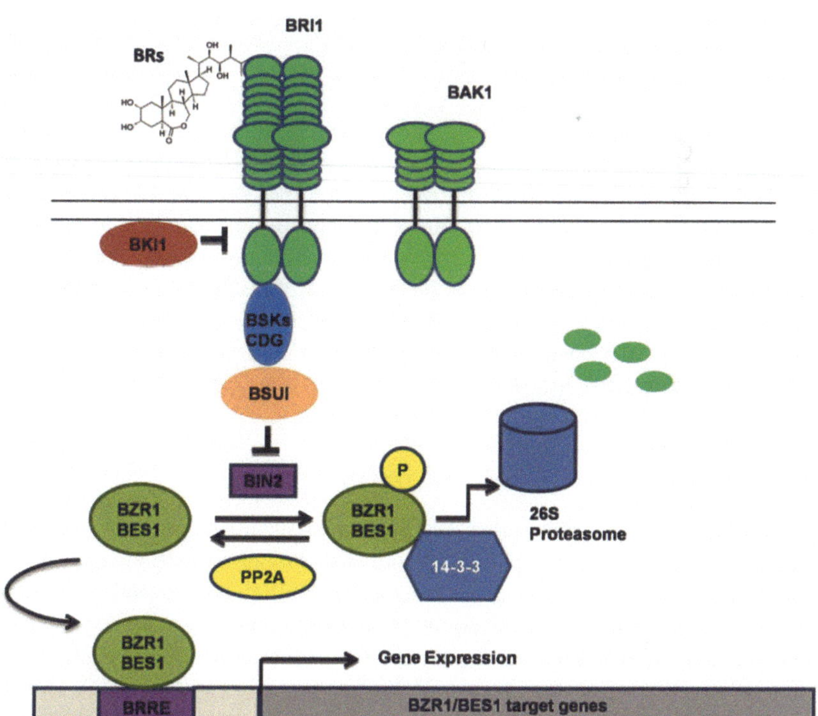

FIGURE 20.1 Brassinosteroids signaling model in *Arabidopsis*, as reported by *Li Q.F. and He J.X., Mechanisms of signaling crosstalk between brassinosteroids and gibberellins, Plant Signaling & Behavior* ***8*** *(7), 2013, e24686.*

20.3 Interaction of abscisic acid and brassinosteroids

ABA regulates several biological processes, including stomatal closure, seed germination, and responses to environmental stresses. Previous studies suggest an antagonist interaction between BRs and ABA in the case of seed germination, however ABA and BRs do not have opposing effects in the case of root elongation. The constitutively active mutants impeding ABA (ABA-deficient *aba1-1*) after treatment with BRs conferred tolerance for salinity and heat. Steber and McCourt (2001) demonstrated BR-related mutants (*det2-1* and *bri1-1*) showed increased sensitivity during seed germination to the inhibitory effects of ABA when compared to WT. Zhang, Cai, & Wang (2009) proposed that after BRs perception, signaling cascades of ABA and BRs crosstalk occurs before their transcriptional activation. The subcellular localization of BRI1 and *bri1 116* mutant for BR signaling revealed that ABA does not require the BR receptor complex to act on BR signaling outputs. The function of ABA was however dependent on BIN2 because after the application of BIN2 inhibitor, LiCl, ABA was incapable to restrain BR signaling outputs. This study was further supported by ABA insensitive *abi1* and *abi2* mutants (without ABA) being unable to act on BR signaling outputs, which suggested the participation of early ABA signaling components (ABI1 and ABI2) in ABA regulation of the BR signaling (Choudhary, Yu, Yamaguchi-Shinozaki, Shinozaki, & Tran, 2012; Rodriguez, Lozano-Juste, & Albert, 2019).

The interaction between ABA and BRs is varied, for example, in *Chlorella vulgaris* and canola (*B. napus*) plants BRs induced the ABA accumulation when exposed to heat stress and also the BR-induced stress tolerance was reduced due to the inhibition of ABA biosynthesis (Ahammed, Li, Zhou, Zhou, & Yu, 2016), which suggests a conflicting nature of interaction of ABA and BRs under stress. To answer the discrepancies on the BRs and ABA H_2O_2 accumulation, the authors argue that ABA biosynthesis is required for sustained tolerance to stress, whereas the BRs lead to the Respiratory Burst Oxidase Homolog 1 (*RBOH1*) induction and production of H_2O_2 without the ABA assistance (Tian et al., 2018) (Fig. 20.2).

20.4 Interaction of auxin and brassinosteroids

In *Arabidopsis* comprehensive genome wide microarray analyses have revealed a number of genes that are induced by both BRs and auxins. BRs and auxin are synergistic in improving plant growth responses and the regulation of transcription (Ohri et al., 2019). An example of photomorphogenesis represents the synergistic effects of BRs and Aux by direct interaction of BIN2 and ARF2 (Hao et al., 2013). BIN2 phosphorylates ARF2 resulting in a loss of its DNA-binding repression activities and thus BIN2 through direct inactivation of ARF repressors regulates the expression of Aux-induced genes, and hence causes the

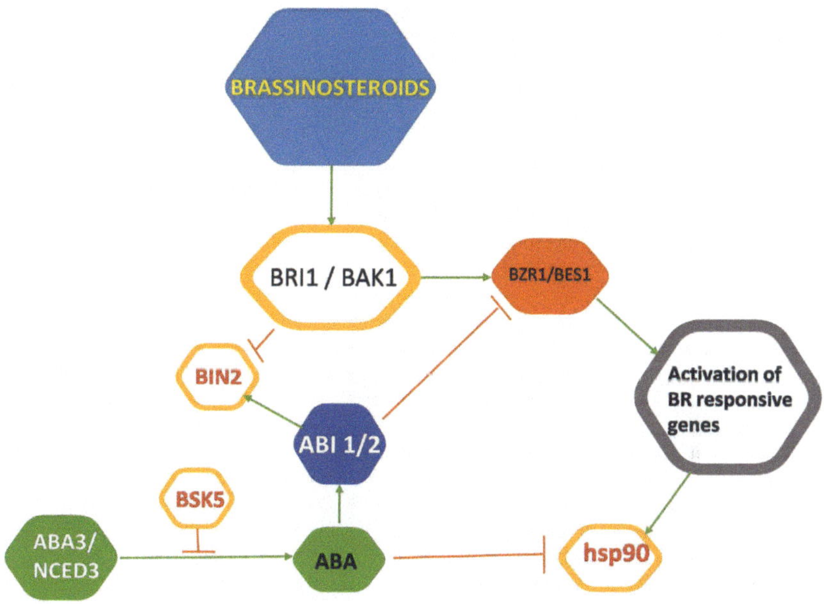

FIGURE 20.2 A schematic model showing brassinosteroid–abscisic acid crosstalk. Involvement of different genes and transcription factors in the transcriptional regulation of plant development has been represented. Positive effects are indicated by *green arrows*, *red bars* indicate repression, and unknown interactions are represented by *dashed arrows*.

synergistic effect (Vert, Walcher, Chory, & Nemhauser, 2008). The BRs are also responsible for the endogenous Aux distribution and their polar transport by regulating the expression of *PIN* and *ROP* genes (Sun, Feraru, Feraru, & Kleine-Vehn, 2019). The PIN Aux efflux carriers controlling the mitotic and cellular differentiation are influenced by BRs suggesting a possible role of BRs in mediating root growth through regulation of Aux distribution. In the recent study in *Arabidopsis* Aux was shown to function as a biosynthetic signal for BRs (Choudhary, Yu, Yamaguchi-Shinozaki, Shinozaki, & Tran, 2012). Also, BES1 and BZR1transcription factors have many auxin responsive genes as target genes. Identification of BES1 and BZR1 target genes also showed that many auxin-responsive genes are regulated by these BR-regulated transcription factors (Galstyan & Nemhauser, 2019). The BES1 directly binds to the promoter of *PIN7* (Pinformed 7) and *SHY2* (*IAA3/Short Hypocotyl 2*) gene responsible for the growth of root meristem by upregulating PIN7 and downregulating SHY2 (Li et al., 2020). Another study reported the role of PILS proteins in the interaction of BRs and auxins. It was revealed that imperial pils (*imp*) mutants, defective in the BR-receptor BRASSINOSTEROID INSENSITIVE 1 (BRI1), enhance PILS5 overexpression phenotypes and lead to nuclear abundance and auxin signaling (Sun et al., 2020) (Fig. 20.3).

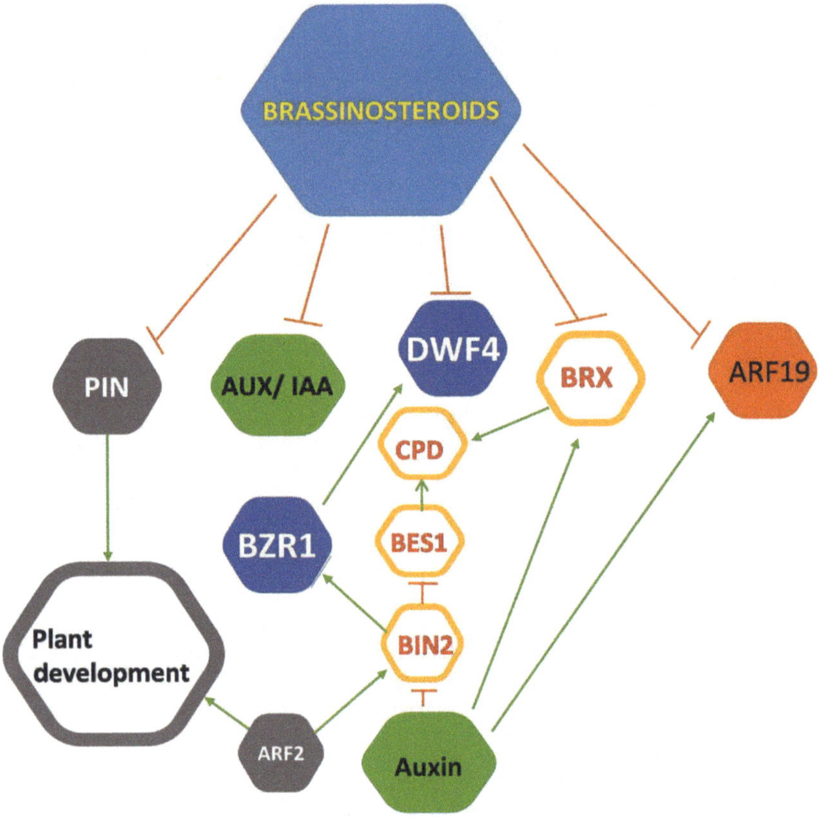

FIGURE 20.3 A schematic model showing brassinosteroid–auxin crosstalk. Involvement of different genes and transcription factors in the transcriptional regulation of plant development has been represented. Positive effects are indicated by *green arrows*, *red bars* indicate repression, and unknown interactions are represented by *dashed arrows*.

20.5 Interaction of ethylene and brassinosteroids

Ethylene is ubiquitously found in all plant parts and plays an important role in plant growth, development, and also in defense mechanisms to a wide range of abiotic and biotic stresses. Another study suggested that BR significantly increased the transcription level of ethylene signaling biosynthesis genes including those that were ripening related (Anwar et al., 2018). The Ethylene Insensitive2 (*ein2*) mutant was found to be hypersensitive to salt stress and improved resistance after treatment with BRs. BRs induce the ethylene biosynthesis to confer tolerance to oxidative stress by inducing the formation of H_2O_2 to enhance the level of AOX for efficient ROS scavenging. Therefore ethylene acts synergistically with the BRs for improved stress tolerance. BRs positively regulate ACS and ACC oxidase activities, and thus suggest a role in the ethylene biosynthesis (Park et al., 2020). Another study in cucumber revealed that BRs significantly increased the transcription levels of 1-aminocyclopropane-1-carboxylic acid (*ACC*), 1-aminocyclopropane-1-carboxylate synthesis1 (*ACS1*), 1-aminocyclopropane-1-carboxylate synthesis2 (*ACS2*), 1-aminocyclopropane-1-carboxylate synthesis3 (*ACS3*), (1-aminocyclopropane-1-carboxylate oxidase1) (*ACO1*), (1-aminocyclopropane-1-carboxylate oxidase1) (*ACO2*), and alternative oxidase (*AOX*) in stressed conditions. Also, recently it was revealed that BRs are also involved in control of FER to promote the ET-induced growth (Anwar et al., 2018). The function of FER is to encode the RLK used in cell elongation and the perception of pollen tube for normal ET responses in plant. BRs and ET have strong interactions in plant defense responses. It is reported that Eix proteins in several cultivars are strong elicitors of plant defense responses. An example of this was provided in the tomato (*Solanum lycopersicum*), where two Eix (LeEix1 and LeEix2) RLK receptors come into play for mediating defense responses. The major role in plant defense is provided by LeEix2, whereas LeEix1 performs the attenuating function by heterodimerizing with LeEix2. Moreover, LeEix1 requires BAK1 for attenuating with LeEix2 and thus provides a strong evidence for the interaction of BRs and ET (Choudhary, Yu, Yamaguchi-Shinozaki, Shinozaki, & Tran, 2012) (Fig. 20.4).

20.6 Interaction of gibberellins and brassinosteroids

An interaction between BRs and GA has been well established in the regulation of a broad spectrum of light responses. A model has been proposed for BR and GA interaction through the DELLA BZR PIF module (Hao et al., 2013). BZR1 and PIF4 are both inactivated by GA-mediated DELLA degradation and both can regulate GA responsive gene expression independently and interdependently established through RNA sequencing genome-wide gene expression analysis (Thomas, Blázquez, & Alabadí, 2018). Another study by Wang et al. (2009) suggesting the crosstalk between BRs and GA demonstrated that *OsGSR1* (member of GAST family in rice) is induced by GA, whereas plants having suppressed *OsGSR1* expression exhibit characteristics similar to the genotypes with BR-deficient plants which can only be rescued by application of BRs. OsGSR1 directly interacted with DWF1 leading to the activation of BR biosynthesis further suggesting the role of OsGSR1 in crosstalk between GA and BR signaling pathways. The enhanced stabilization of OsSLR1 by exogenous BR application is another piece of evidence showing the negative crosstalk of BRs with GAs. OsSLR1 is the only negative key regulator of DELLA GA-signaling rice repressor that acts in resistance to *Pythium graminicola* and also its expression increases in response to pathogen infection and BR application (Choudhary, Yu, Yamaguchi-Shinozaki, Shinozaki, & Tran, 2012) (Fig. 20.5).

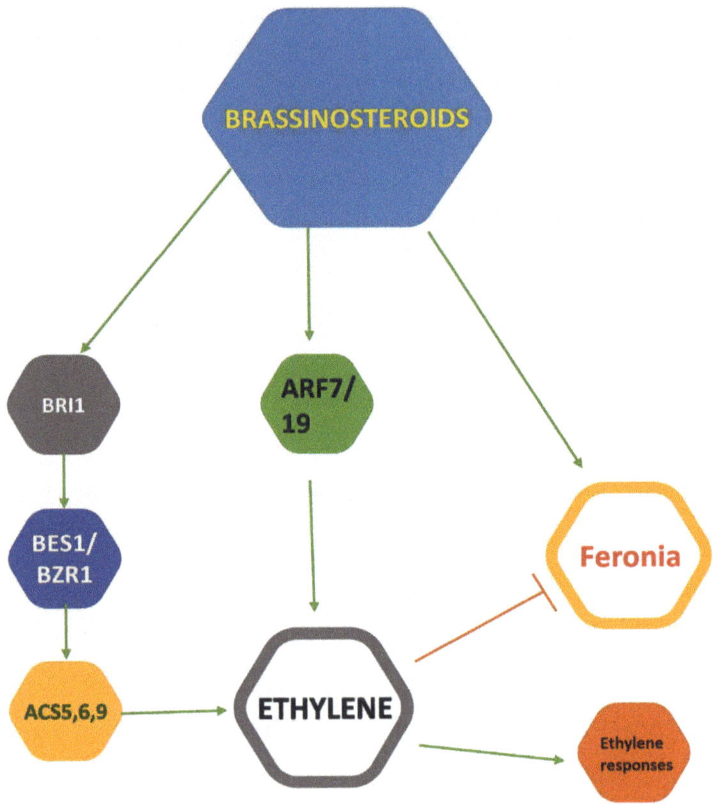

FIGURE 20.4 A schematic model showing brassinosteroid–ethylene (ET) crosstalk. Involvement of different genes and transcription factors in the transcriptional regulation of plant development has been represented. Positive effects are indicated by *green arrows*, *red bars* indicate repression, and unknown interactions are represented by *dashed arrows*.

20.7 Interaction of salicylic acid and brassinosteroids

It has been shown that the crosstalk between BR and SA is mainly involved in the regulation of environmental stress responses of plant. Nonexpressor of Pathogenesis-Related Genes1-1 (*npr1-1*) genotypes with signaling defective for SA-mediated systemic acquired resistance (SAR) showed minimal tolerance to heat when exposed to BRs. In *Arabidopsis*, BR induced tolerance to salt and high temperature is mediated by NONEXPRESSOR OF PATHOGENESIS-RELATED GENES1 (NPR1) which is SA regulatory protein (Banerjee & Roychoudhury, 2018), for example, in *Brassica*, the BR and SA application together conferred resistance to plant salt tolerance (Hao et al., 2013). Thus NPR1, which plays a key role in SA-mediated SAR, was found to be critical to BR-mediated increase in hyperthermal and salt tolerance. The use of BR biosynthesis inhibitor, Bzr, in rice seedlings revealed

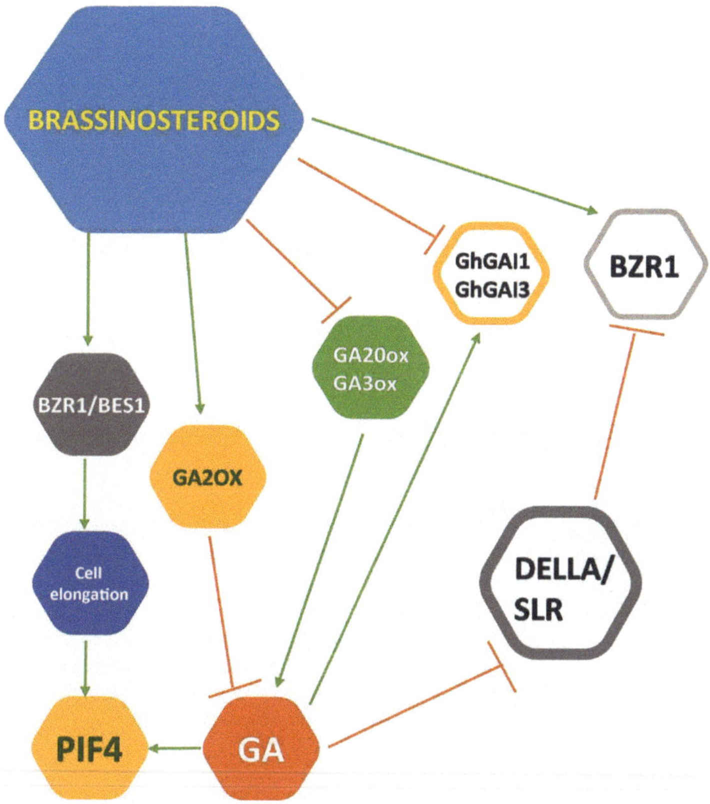

FIGURE 20.5 A schematic model showing brassinosteroid–gibberellin crosstalk. Involvement of different genes and transcription factors in transcriptional regulation of plant development has been represented. Positive effects are indicated by *green arrows*, *red bars* indicate repression, and unknown interactions are represented by *dashed arrows*.

reduced susceptibility against the root oomycete *P. graminicola*. This was reported to be as a result of downstream SA defense response repression of SA biosynthesis but upstream of key defense regulators (*OsNPR1* and *OsWRKY45*). This further supports the hypothesis of *P. graminicola* exploiting the BR pathway as a decoy to counteract the SA-mediated defense responses in rice (De Vleesschauwer, Gheysen, & Höfte, 2013). There has been an increasing amount of evidence regarding the BR crosstalk with SA in response to abiotic stress. SA-insensitive *npr1-1* mutant on exogenous application of BRs was unable to confer salt stress tolerance compared with WT (Divi, Rahman, & Krishna, 2010), indicating that in *Arabidopsis* BR-induced salt tolerance in partly depends on NPR1, the main regulator of the SA-mediated defense signaling pathway (Choudhary, Yu, Yamaguchi-Shinozaki, Shinozaki, & Tran, 2012) (Fig. 20.6).

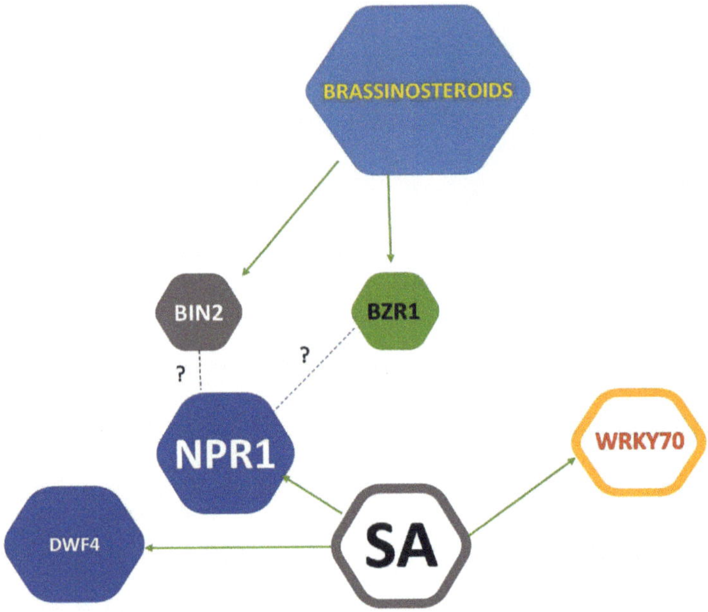

FIGURE 20.6 A schematic model showing brassinosteroid–salicylic acid crosstalk. Involvement of different genes and transcription factors in transcriptional regulation of plant development has been represented. Positive effects are indicated by *green arrows*, *red bars* indicate repression, and unknown interactions are represented by *dashed arrows*.

20.8 Interaction of jasmonic acid and brassinosteroids

As stated by Ren et al. (2009), there is an antagonistic relationship between BRs and JA in controlling plant production. They used the *psc1* mutant with coi1 insensitivity to root inhibition induced by JA. Several genetic and sequence analysis techniques revealed that *psc1* is the leaky mutant of DWF4, encoding the key enzymes in BR biosynthesis, whereas in *Arabidopsis* JA inhibition of root growth was negatively regulated by BRs. The two important genes Allene Oxide Synthase (*OsAOS2*) and JA-Inducible Rice MYB (*OsJAMYB*) in the rice JA-signaling pathway were suppressed by the application of BRs. On the contrary, BR-related gene expression was suppressed by exogenous JA application (Hao et al., 2013). BRs and JA interaction play a central role in plant stress responses. The F-box protein CORONATINE INSENSITIVE1 (COI1) is crucial for the JA responses in *Arabidopsis* and the execution of JA signaling (Pérez-Salamó, Krasauskas, Gates, Díaz-Sánchez, & Devoto, 2019). BRs application restored JA sensitivity of the *psc1* mutant in *coi12* and JA hypersensitivity of *psc1* in WT. Furthermore, the repression of *DWF4* expression by JA in a *COI1*-dependent manner demonstrated the negative regulation of BRs to JA induced

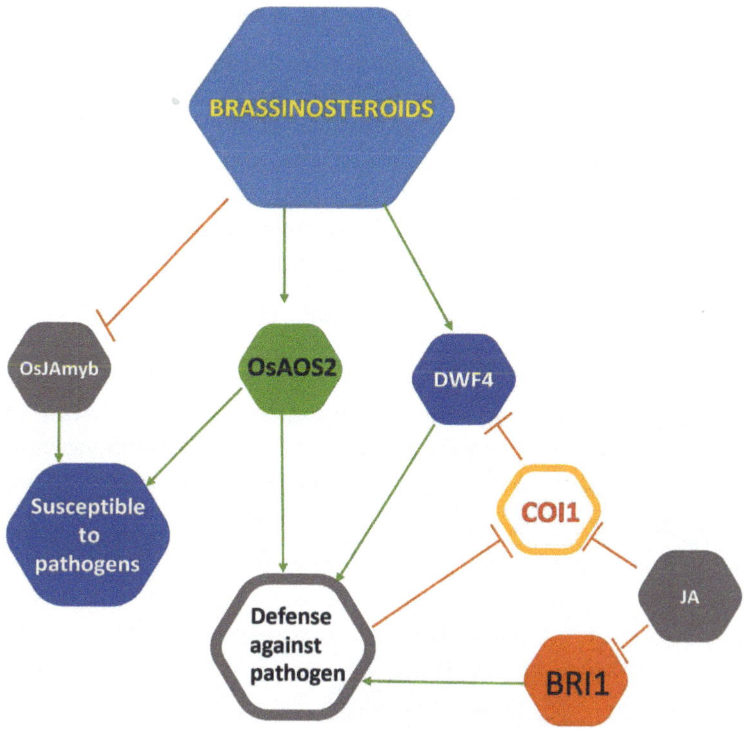

FIGURE 20.7 A schematic model showing brassinosteroid–jasmonic acid crosstalk. Involvement of different genes and transcription factors in transcriptional regulation of plant development has been represented. Positive effects are indicated by *green arrows*, *red bars* indicate repression, and unknown interactions are represented by *dashed arrows*.

root inhibition. Also, the downregulation of *DWF4* by JA occurs downstream of *COI1* in JA signaling (Ahanger, Ashraf, Bajguz, & Ahmad, 2018). In another study on *Nicotiana attenuata* altered expression of *NaBAK1* was observed in plants exposed to oral secretions of *Manduca sexta* (specialist herbivore) larvae (Kohli et al., 2019). When *NaBAK1*-silenced plants were elicitated with oral secretions, the attenuated JA and JA isoleucine bursts were observed but this observation was not as a result of compromised MAPK activity or SA levels. After the stimulated herbivory, *NaBAK1*-silenced plants had higher levels of secondary metabolites, for example, defensive trypsin proteinase inhibitors. Also, the *NaBAK1*-silenced plants were observed to have reduced *NaTD* transcription, whereas *NaJAR4* and *NaJAR6* expression was increased. Therefore BAK1 plays an essential role in modulating herbivory-induced JA accumulation essential for the resistance of *N. attenuata* to *M. sexta* (Choudhary, Yu, Yamaguchi-Shinozaki, Shinozaki, & Tran, 2012; Kohli et al., 2019) (Fig. 20.7).

20.9 Interaction of cytokinin and brassinosteroids

There have been several lines of evidence regarding many biological processes involving the interaction and crosstalk of BRs and CKs. The study regarding the involvement of BRs in CK regulation was performed in wheat seedlings where the application of BRs led to the increase in the gene expression of cytokinin oxidase, which ultimately contributed to the increased levels of cytokinins, and hence suggested the BR involvement in regulation of cytokinin activity (Anwar et al., 2018; Hao et al., 2013). Werner et al. (2010) reported that the overexpression of *CKX3* gene with *PYK10* promoter in *Arabidopsis* roots decreased the CK levels and hence enhanced the root growth. BRI1 overexpression in *PYK10*: *CKX3* increased root and leaf growth relative to wild type and comparable to tobacco Cytokinin Dehydrogenase Oxidase 3 (*CKX3*) overexpression. This suggests a strong correlation with BRs and CKs in the regulation of plant growth. Also, under the water stress in rice plants, *P(SARK):IPT* plants increased the expression of BR genes, such as *BAK1*, *BSK1*, *SERK1*, and *DWF5*, signifying crosstalk between BRs and CKs leading to increased drought tolerance (Choudhary, Yu, Yamaguchi-Shinozaki, Shinozaki, & Tran, 2012) (Fig. 20.8).

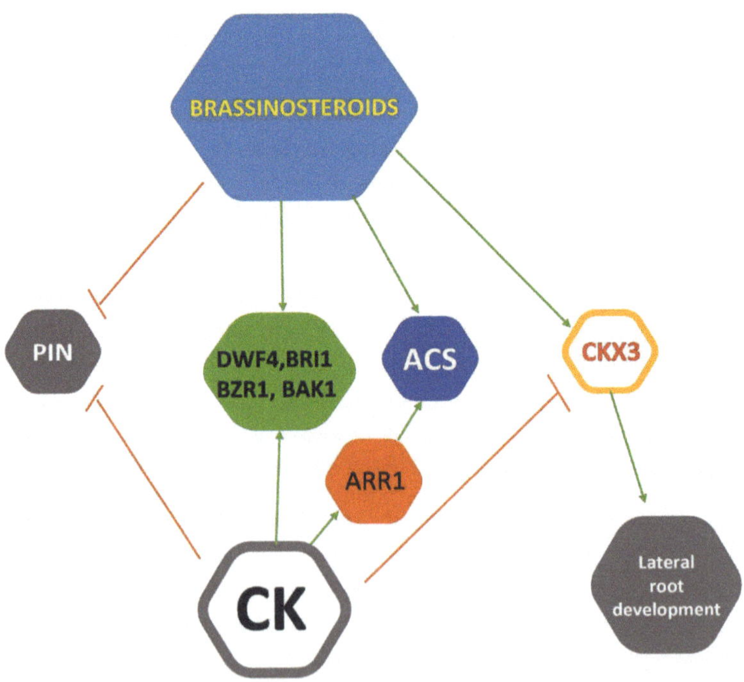

FIGURE 20.8 A schematic model showing brassinosteroid–cytokinin crosstalk. The involvement of different genes and transcription factors in the transcriptional regulation of plant development has been represented. Positive effects are indicated by *green arrows*, *red bars* indicate repression, and unknown interactions are represented by *dashed arrows*.

In various plant processes, phytohormones play a central role in adapting to environmental changes by regulating growth, nutrient allocation, and source/sink interactions (Fernie et al., 2020). Phytohormone levels are differentially modulated by abiotic stresses altering their biosynthesis and metabolism. BRs regulate the abiotic stresses by genetically deactivating the negative regulators of the BR signaling pathway (Kaur & Pati, 2019), as well as by signal transduction and transcriptional regulation, and hence the stress ameliorative effects of BRs with the crosstalk with other stress hormones (Ahanger et al., 2018). There is evidence where the BRs along with other phytophormones like ABA, polyamines, and SA act synergistically and give greater protection in combined doses of hormones from the effect of stress than using the individual hormone treatment (Ahanger et al., 2018). BRs regulate the complex network of genes that overlaps with other hormones like auxin, ABA, and JA. Different hormonal signaling pathways undergo crosstalks among themselves and between different stress signaling pathways under natural conditions to generate a customized stress response that is most suitable for plant survival under given environmental conditions.

BR are plant steroid hormones that play an important role in adaptation to abiotic and biotic stresses by targeting certain genes responsible for defense mechanism in plants. Numerous studies indicate that BR controls defense enzymes and hormones during the response of plant stress induction.

20.10 Response of brassinosteroids in abiotic stress using different transcription factors

Several biotechnological strategies have led to the study of the influence of BRs and tolerance of plants to abiotic and biotic stress. In *Arabidopsis* the overexpression of seed-specific *AtDWF4* led to the upregulation of *COR15A* (Cold Responsive Gene) and enhanced the cold resistance (Sahni et al., 2016). Another study in rice where *OsGSK1* (*BIN2* homolog) loss of function led to the improved tolerance to salt, drought, heat, and cold stresses when compared to WT (Gao et al., 2019). Likewise, pepper seedlings under cold stress upregulated thousands of genes when treated with BRs, as revealed by transcriptome analysis, suggesting BR signaling components may serve as potential targets to improve abiotic stress tolerance through genetic engineering techniques (Choudhary, Yu, Yamaguchi-Shinozaki, Shinozaki, & Tran, 2012). Further, to reduce the effect of chilling stress CAT2 (catalase isozyme 2), GSTX1 (glutathione S-transferase), and PER72 (peroxidase 72) are activated by BRs (Anwar et al., 2018).

BES1 interacts with transcription factors like EARLY FLOWERING 6 (ELF6) and RELATIVE OF EARLY FLOWERING 6 (REF6), (homolog of ELF6), the MYB factor MYB30, INTERACTING-WITH-SPT6−1 (IWS1), and bHLH factor BES1-INTERACTING MYC-LIKE 1 (BIM1) to promote the gene expression of several genes involved in RNA polymerase II functioning. However, in a recent study BES1 is shown to inhibit the Myeloblastosis Family Factor Like-2 (*MYBL2*) expression (Hao et al., 2013).

Various genes involved in important cellular processes, such as cytoskeleton rearrangement, vesicular trafficking, signaling, and biosynthesis of various hormones have been induced in most plants due to abiotic stress. BR induces the regulation of most of these gene products, that is, heat shock proteins, dehydrins, glutathione-S-transferase tubulin, actin, ACC oxidase, and allene-oxidase cyclase (Divi, Rahman, & Krishna, 2016). Certain genes induced by abiotic stress regulated by BRs are *WRKY33*, Acid Phosphatase5 (*ACP5*), Jacalin-Related Lectin1−3 (*JAC-LEC1−3*), and a BR-Responsive-Receptor-Like Kinase (*BRRLK*) (Divi et al., 2016).

In rice DWARF AND LOW-TILLERING (DLT), LEAF AND TILLER ANGLE INCREASED CONTROLLER (LIC), DWARF1 (D1), and TAIHU DWARF1 (TUD1), CYP90D2/D2 are known to be induced by BRs in stress (Liu et al., 2018). Also, *OsmiR397* microRNA downregulates *OsLAC*, encoding a laccase-like protein regulating the BR sensitivity (Tang & Chu, 2017). Heat shock proteins (OsHSP17.0 and OsHSP23.7) are known to be upregulated and have recently been found to be direct targets of BZR1/BES1 (Sharma, Kaur, & Pati, 2017). Another heat shock protein, HSP90.3 is found to provide affinity with the amino and central parts of BES1 in vitro. Furthermore, BZR1 had been recognized as a novel HSP90 and thus is known to form the complex with two main transcription factors BES1 and BIN2 in BR signaling (Shigeta et al., 2015). LATE EMBRYOGENESIS PROTEINS are enhanced upon BR treatment by enhanced transcript level of Responsive To Desiccation 29A (*RD29A*) and Early Response To Dehydration10 (*ERD10*). LATE EMBRYOGENESIS PROTEINS (LEA) preventing protein aggregation and stabilization of membranes during abiotic stress, which are accumulated in rice, are also found to be targeted by activated BES1 in BR-signaling (Kagale, Divi, Krochko, Keller, & Krishna, 2007). In the chilling stress, the signal transduction of ABSCISIC ACID STRESS RIPENING (ASR)-like protein and LIPOCALINS. REMORIN, skeleton protein associated drought-stress responsive-, and dehydrin proteins are some of the key proteins involved in BRs-induced stress tolerance. FERRITIN protein in *Arabidopsis* helps to sequester excess iron and the effect of salt when given brassinolide treatment (Sharma et al., 2017). It could very well be a mechanism not only for upholding iron homeostasis but for prevention of the production of hydroxyl radicals through the Fenton action as well and thus it could also prove to be vital in maintaining ion balance and general enzyme activity (Lobréaux, Thoiron, & Briat, 1995) under stress conditions. A very fascinating study performed on the *SALT* gene located on the *SalTol* QTL in rice was highly expressed in salinity stress (Ali et al., 2013). While the expression of the *SALT* gene was seen to rise under salt treatment, its expression was found to be severely reduced in rice seedlings cotreated with salt and BRs. The motifs like *ABRE* (Abscisic Acid Responsive Element), *MBS* (a MYB Binding Site involved in drought-inducibility), and *LTR* (involved in Low-Temperature Responsiveness) present in the promoter region of the *SALT* gene are involved in different types of stresses and are regulated by BRs; they could very well be examined as one of the stress protection mechanisms of BRs. Furthermore, administration of BR biosynthetic genes has also brought evidence for the role of endogenous BRs in stress tolerance. The study in *Arabidopsis* showed that the inhibition of *DWARF4* (*DWF4*) by brazzinozole led to the susceptibility of plants to abiotic stress and overexpression of *AtDWF4* and *HYDROXYSTEROID DEHYDROGENASE*, genes involved in BR biosynthesis increased the resistance to the stress. In *B. napus* it was revealed that AtDWF4 is likely to be mediated by BES1/BZR1, therefore it was established that BRs can regulate abiotic stress tolerance mechanisms in plants (Sahni et al., 2016). However, the study in rice using RNA interference-mediated showed otherwise, in which SQUALENE SYNTHASE (SQS), an enzyme catalyzing the initial reactions in the BR biosynthesis, decreased stomatal conductance and enhanced drought tolerance (Xue, Lung, & Chye, 2016), thus signifying that reduced BR levels might augment stress tolerance, leading to a doubt in accepting the endogenous BRs role in stress tolerance.

In addition, rice BR biosynthetic mutant *d2-1* showed increased tolerance to Fe-deficiency and this response could be partially reversed by exogenous BR application. The authors argue that the lower endogenous BRs promote the transmission and translocation of Fe from roots to shoots, while complementing the seedlings with BR suppresses the translocation of Fe from roots to shoots, showing a severe Fe shoot deficiency phenotype (Wang, Li, & Zhang, 2015). However, the stress response may not be manifested in the form of changes in BR levels, but could be reflected in changes in BR signaling components' activity (sensitivity) as explained by Xia et al. (2009).

There are many other essential stress-regulated genes playing a pivotal role in stress tolerance, such as CHLORIDE CHANNEL (CLC), transporters and symporters (OsAKT, OsHKT, OsNHX), NON-SELECTIVE CATION CHANNEL (NCC), MYO-INOSITOL SYNTHESIS (PINO1), etc. (Kumar, Kumar, Kim, Ryu, & Cho, 2013). Further characterization of these genes under the BR effect will enhance our understanding of their role in the tolerance of BR-mediated stress (Sharma et al., 2017).

It has been shown that RD26 is also a direct target of BES1 and is repressed by BR under drought conditions; correspondingly it also modulates the transcription of BES1-regulated genes for inhibition of BRs (Ye et al., 2017). The transcription factors WRKY46, 54, and 70 also interact with BES1 directly repressing drought-inducible global transcripts to inhibit drought tolerance and also lead to the promotion of BR-regulated plant growth (Chen and Yin, 2017). BIN2 negatively regulates its effect on the BES1-mediated BR response by phosphorylating and destabilizing WRKY54 (Chen et al., 2017). it was also revealed recently that BR signaling coordinate plant growth and survival under drought stress and starvation by interacting with autophagy pathways through BIN2 (Nolan, Chen, & Yin, 2017). In this context, BIN2 phosphorylates and activates DSK2, a ubiquitin receptor protein, which further interacts with BES1 and targets it for degradation via autophagy (Nolan, Chen, & Yin, 2017), thus highlighting the complexity of BR-mediated responses to drought.

The BR-regulated basic helix-loophelix (bHLH) transcription factor CESTA controls Core Cold Responsive (*COR*) genes transcription by activating the expression of C-Repeat/Dehydration-Responsive Element Binding Factor (*CBF*) (Eremina et al., 2016). BR-ENHANCED EXPRESSION 1 (BEE1), a BR-regulated transcription factor, influences the transcription of *MYBbHLH-WD40* complex components, indirectly promoting cold acclimation (Petridis, Döll, Nichelmann, Bilger, & Mock, 2016). In line with these findings, BIN2 overexpression has been shown to cause hypersensitivity to freezing stress under both nonacclimated and acclimated conditions, whereas *bin2–3 bil1 bil2* triple mutants and also *bzr1-1D* and *bes1-D* mutants gain-of-function have led to the freezing tolerance (Li et al., 2017). BZR1 dephosphorylation is also induced upon cold treatment and can regulate *COR* genes affecting their downstream targets by binding directly or indirectly to CBF1 and CBF2. BR signaling also regulates the plant growth under high temperature stress. Upon elevated temperature, BZR1 accumulates in the nucleus and induces the expression of growth-promoting genes, to regulate thermomorphogenesis by binding directly or via a promoter of PHYTOCHROME INTERACTING FACTOR 4 (PIF4) (Ibañez et al., 2018; Oh, Zhu, & Wang, 2012). Elevated temperature has been shown to increase the accumulation of active PIF4, thereby shifting the balance of nuclear protein complexes from BES1 homodimers to BES1-PIF heterodimers (Martinez et al., 2018). The de-repression of BR biosynthesis is caused by the reduced availability of active BES1 homodimers and it also leads to the feedback inhibition of BR signaling output. In contrast, the genes involved in thermomorphogenesis are activated by abundant levels of BES1-PIF4 complexes (Martinez et al., 2018). Elevated ambient temperatures can affect primary root elongation growth by reducing the BRI1 levels (Martins et al., 2017). Recently, the kinase defective BRI1 protein from *bri1–301* mutants was found to show less stability and biochemical activity under elevated temperature (29°C) and also the mutated version of this protein undergoes protein misfolding and degradation by elevated temperatures through an as-yet unknown mechanism (Zhang et al., 2018).

The role for BR in salt stress tolerance may be conciliated by BRI1; inhibiting the endoplasmic reticulum-associated protein degradation system is able to partially rescue the *bri1–9* mutants salt hypersensitivity, confirming the role of BRI1 signaling complex in the salinity response (Cui et al., 2012). Suppression of nuclear accumulation of BZR1in high salinity causes growth quiescence in roots and also subsequent BR signaling functions (Planas-Riverola et al., 2019).

References

Ahammed, G. J., Li, X., Zhou, J., Zhou, Y. H., & Yu, J. Q. (2016). *Role of hormones in plant adaptation to heat stress. Plant hormones under challenging environmental factors* (pp. 1–21). Dordrecht: Springer.

Ahanger, M. A., Ashraf, M., Bajguz, A., & Ahmad, P. (2018). Brassinosteroids regulate growth in plants under stressful environments and crosstalk with other potential phytohormones. *Journal of Plant Growth Regulation*, *37*(4), 1007–1024.

Ali, S., Gautam, R. K., Mahajan, R., Krishnamurthy, S. L., Sharma, S. K., Singh, R. K., & Ismail, A. M. (2013). Stress indices and selectable traits in SALTOL QTL introgressed rice genotypes for reproductive stage tolerance to sodicity and salinity stresses. *Field Crops Research*, *154*, 65–73.

Anwar, A., Liu, Y., Dong, R., Bai, L., Yu, X., & Li, Y. (2018). The physiological and molecular mechanism of brassinosteroid in response to stress: A review. *Biological Research. BioMed Central*, *51*, 1–15. Available from https://doi.org/10.1186/s40659-018-0195-2.

Banerjee, A., & Roychoudhury, A. (2018). Interactions of brassinosteroids with major phytohormones: Antagonistic effects. *Journal of Plant Growth Regulation*, *37*(4), 1025–1032.

Chen, J., & Yin, Y. (2017). WRKY transcription factors are involved in brassinosteroid signaling and mediate the crosstalk between plant growth and drought tolerance. *Plant Signaling & Behavior*, *12*(11), 1365212.

Chen, J., Nolan, T. M., Ye, H., Zhang, M., Tong, H., Xin, P., ... Yin, Y. (2017). Arabidopsis WRKY46, WRKY54, and WRKY70 transcription factors are involved in brassinosteroid-regulated plant growth and drought responses. *The Plant Cell*, *29*(6), 1425–1439.

Chen, T., Wang, B., Wang, F., Niu, G., Zhang, S., Li, J., & Hong, Z. (2020). The evolutionarily conserved serine residues in BRI1 LRR motifs are critical for protein secretion. *Frontiers in Plant Science*, *11*, 32.

Choudhary, S. P., Yu, Y. Q., Yamaguchi-Shinozaki, K., Shinozaki, K., & Tran, L. S. P. (2012). Benefits of brassinosteroid crosstalk. *Trends in Plant Science*, *17*(10), 594–605. Available from https://doi.org/10.1016/j.tplants.2012.05.012, Elsevier Ltd.

Cui, F., Liu, L., Zhao, Q., Zhang, Z., Li, Q., Lin, B., ... Xie, Q. (2012). Arabidopsis ubiquitin conjugase UBC32 is an ERAD component that functions in brassinosteroid-mediated salt stress tolerance. *The Plant Cell*, *24*(1), 233–244.

De Vleesschauwer, D., Gheysen, G., & Höfte, M. (2013). Hormone defense networking in rice: Tales from a different world. *Trends in Plant Science*, *18*(10), 555–565.

Divi, U. K., Rahman, T., & Krishna, P. (2010). Brassinosteroid-mediated stress tolerance in *Arabidopsis* shows interactions with abscisic acid, ethylene and salicylic acid pathways. *BMC Plant Biology*, *10*(1), 151.

Divi, U. K., Rahman, T., & Krishna, P. (2016). Gene expression and functional analyses in brassinosteroid-mediated stress tolerance. *Plant Biotechnology Journal*, *14*(1), 419–432.

Eremina, M., Unterholzner, S. J., Rathnayake, A. I., Castellanos, M., Khan, M., Kugler, K. G., ... Poppenberger, B. (2016). Brassinosteroids participate in the control of basal and acquired freezing tolerance of plants. *Proceedings of the National Academy of Sciences*, *113*(40), E5982–E5991.

Fernie, A. R., Bachem, C. W., Helariutta, Y., Neuhaus, H. E., Prat, S., Ruan, Y. L., ... Sonnewald, S. (2020). Synchronization of developmental, molecular and metabolic aspects of source–sink interactions. *Nature Plants*, 1–12.

Galstyan, A., & Nemhauser, J. L. (2019). Auxin promotion of seedling growth via ARF5 is dependent on the brassinosteroid-regulated transcription factors BES1 and BEH4. *Plant Direct*, *3*(9), e00166.

Gao, X., Zhang, J. Q., Zhang, X., Zhou, J., Jiang, Z., Huang, P., … Zhang, W. (2019). Rice qGL3/OsPPKL1 functions with the GSK3/SHAGGY-like kinase OsGSK3 to modulate brassinosteroid signaling. *The Plant Cell*, *31*(5), 1077–1093.

Gruszka, D. (2020). Exploring the brassinosteroid signaling in monocots reveals novel components of the pathway and implications for plant breeding. *International Journal of Molecular Sciences*, *21*(1), 354.

Hao, J., Yin, Y., & Fei, S. Z. (2013). Brassinosteroid signaling network: Implications on yield and stress tolerance. *Plant Cell Reports*, *32*(7), 1017–1030. Available from https://doi.org/10.1007/s00299-013-1438-x.

Ibañez, C., Delker, C., Martinez, C., Bürstenbinder, K., Janitza, P., Lippmann, R., … Grossjohann, A. (2018). Brassinosteroids dominate hormonal regulation of plant thermomorphogenesis via BZR1. *Current Biology*, *28*(2), 303–310.

Kagale, S., Divi, U. K., Krochko, J. E., Keller, W. A., & Krishna, P. (2007). Brassinosteroid confers tolerance in *Arabidopsis thaliana* and *Brassica napus* to a range of abiotic stresses. *Planta*, *225*(2), 353–364.

Kaur, N., & Pati, P. K. (2019). *Harnessing the potential of brassinosteroids in abiotic stress tolerance in plants. Brassinosteroids: Plant growth and development* (pp. 407–423). Singapore: Springer.

Kohli, S. K., Bali, S., Khanna, K., Bakshi, P., Sharma, P., Sharma, A., … Bhardwaj, R. (2019). *A current scenario on role of brassinosteroids in plant defense triggered in response to biotic challenges. Brassinosteroids: Plant growth and development* (pp. 367–388). Singapore: Springer.

Kumar, K., Kumar, M., Kim, S. R., Ryu, H., & Cho, Y. G. (2013). Insights into genomics of salt stress response in rice. *Rice*, *6*(1), 1–15.

Li, H., Ye, K., Shi, Y., Cheng, J., Zhang, X., & Yang, S. (2017). BZR1 positively regulates freezing tolerance via CBF-dependent and CBF-independent pathways in *Arabidopsis*. *Molecular Plant*, *10*(4), 545–559.

Li, Q. F., Zhou, Y., Xiong, M., Ren, X. Y., Han, L., Wang, J. D., … Liu, Q. Q. (2020). Gibberellin recovers seed germination in rice with impaired brassinosteroid signalling. *Plant Science*, *293*110435.

Li, T., Kang, X., Lei, W., Yao, X., Zou, L., Zhang, D., & Lin, H. (2020). SHY2 as a node in the regulation of root meristem development by auxin, brassinosteroids, and cytokinin. *Journal of Integrative Plant Biology*, *62*, 1500–1517.

Liu, F., Wang, P., Zhang, X., Li, X., Yan, X., Fu, D., & Wu, G. (2018). The genetic and molecular basis of crop height based on a rice model. *Planta*, *247*(1), 1–26.

Lobréaux, S., Thoiron, S., & Briat, J. F. (1995). Induction of ferritin synthesis in maize leaves by an iron-mediated oxidative stress. *The Plant Journal*, *8*(3), 443–449.

Martínez, C., Espinosa-Ruíz, A., de Lucas, M., Bernardo-García, S., Franco-Zorrilla, J. M., & Prat, S. (2018). PIF 4-induced BR synthesis is critical to diurnal and thermomorphogenic growth. *The EMBO Journal*, *37*(23), 99552.

Martins, S., Vert, G., & Jaillais, Y. (2017). *Probing activation and deactivation of the BRASSINOSTEROID INSENSITIVE1 receptor kinase by immunoprecipitation. Brassinosteroids* (pp. 169–180). New York, NY: Humana Press.

Nolan, T., Chen, J., & Yin, Y. (2017). Cross-talk of Brassinosteroid signaling in controlling growth and stress responses. *Biochemical Journal*, *474*(16), 2641–2661.

Nolan, T. M., Vukašinović, N., Liu, D., Russinova, E., & Yin, Y. (2020). Brassinosteroids: Multidimensional regulators of plant growth, development, and stress responses. *The Plant Cell, 32*(2), 295–318.

Oh, E., Zhu, J. Y., & Wang, Z. Y. (2012). Interaction between BZR1 and PIF4 integrates brassinosteroid and environmental responses. *Nature Cell Biology, 14*(8), 802–809.

Ohri, P., Bhardwaj, R., Kaur, R., Jasrotia, S., Parihar, R. D., Khajuria, A., & Sharma, N. (2019). *Emerging trends on crosstalk of BRS with other phytohormones. Brassinosteroids: Plant growth and development* (pp. 425–441). Singapore: Springer.

Park, C. H., Seo, C., Park, Y. J., Youn, J. H., Roh, J., Moon, J., & Kim, S. K. (2020). BES1 directly binds to the promoter of the ACC oxidase 1 gene to regulate gravitropic response in the roots of *Arabidopsis thaliana. Plant Signaling & Behavior, 15*(1)1690724.

Peres, A. L. G., Soares, J. S., Tavares, R. G., Righetto, G., Zullo, M. A., Mandava, N. B., & Menossi, M. (2019). Brassinosteroids, the sixth class of phytohormones: A molecular view from the discovery to hormonal interactions in plant development and stress adaptation. *International Journal of Molecular Sciences, 20*(2), 331.

Pérez-Salamó, I., Krasauskas, J., Gates, S., Díaz-Sánchez, E. K., & Devoto, A. (2019). An update on core jasmonate signalling networks, physiological scenarios, and health applications. *Annual Plant Reviews Online, 2*, 387–452.

Petridis, A., Döll, S., Nichelmann, L., Bilger, W., & Mock, H. P. (2016). *Arabidopsis thaliana* G2-LIKE Flavonoid Regulator and Brassinosteroid Enhanced Expression1 are low-temperature regulators of flavonoid accumulation. *New Phytologist, 211*(3), 912–925.

Planas-Riverola, A., Gupta, A., Betegón-Putze, I., Bosch, N., Ibañes, M., & Caño-Delgado, A. I. (2019). Brassinosteroid signaling in plant development and adaptation to stress. *Development, 146*(5).

Ren, C., Han, C., Peng, W., Huang, Y., Peng, Z., Xiong, X., ... Xie, D. (2009). A leaky mutation in DWARF4 reveals an antagonistic role of brassinosteroid in the inhibition of root growth by jasmonate in *Arabidopsis. Plant Physiology, 151*(3), 1412–1420.

Rodriguez, P. L., Lozano-Juste, J., & Albert, A. (2019). *PYR/PYL/RCAR ABA receptors. Advances in Botanical Research* (92, pp. 51–82). Academic Press.

Sahni, S., Prasad, B. D., Liu, Q., Grbic, V., Sharpe, A., Singh, S. P., & Krishna, P. (2016). Overexpression of the brassinosteroid biosynthetic gene DWF4 in *Brassica napus* simultaneously increases seed yield and stress tolerance. *Scientific Reports, 6*, 28298.

Salazar-Henao, J. E., Lehner, R., Betegón-Putze, I., Vilarrasa-Blasi, J., & Caño-Delgado, A. I. (2016). BES1 regulates the localization of the brassinosteroid receptor BRL3 within the provascular tissue of the *Arabidopsis* primary root. *Journal of Experimental Botany, 67*(17), 4951–4961.

Sharma, I., Kaur, N., & Pati, P. K. (2017). Brassinosteroids: A promising option in deciphering remedial strategies for abiotic stress tolerance in rice. *Frontiers in Plant Science, 8*, 2151.

Shigeta, T., Zaizen, Y., Sugimoto, Y., Nakamura, Y., Matsuo, T., & Okamoto, S. (2015). Heat shock protein 90 acts in brassinosteroid signaling through interaction with BES1/BZR1 transcription factor. *Journal of Plant Physiology, 178*, 69–73.

Steber, C. M., & McCourt, P. (2001). A role for brassinosteroids in germination in *Arabidopsis. Plant Physiology, 125*, 763–769.

Sun, L., Feraru, E., Feraru, M. I., & Kleine-Vehn, J. (2019). Brassinosteroid signaling controls PILS-dependent nuclear auxin input in *Arabidopsis thaliana. bioRxiv*. Available from https://doi.org/10.1101/646489.

Sun, L., Feraru, E., Feraru, M. I., Waidmann, S., Wang, W., Passaia, G., ... Kleine-Vehn, J. (2020). PIN-LIKES coordinate brassinosteroid signaling with nuclear auxin input in *Arabidopsis thaliana. Current Biology, 30*, 1579–1588.

Sun, Y., Han, Z., Tang, J., Hu, Z., Chai, C., Zhou, B., & Chai, J. (2013). Structure reveals that BAK1 as a co-receptor recognizes the BRI1-bound brassinolide. *Cell Research*, *23*(11), 1326–1329.

Tang, J., & Chu, C. (2017). MicroRNAs in crop improvement: Fine-tuners for complex traits. *Nature Plants*, *3*(7), 1–11.

Thomas, S. G., Blázquez, M. A., & Alabadí, D. (2018). Della proteins: Master regulators of gibberellin-responsive growth and development. *Annual Plant Reviews Online*, *49*, 189–227.

Tian, Y., Fan, M., Qin, Z., Lv, H., Wang, M., Zhang, Z., ... Ding, Z. (2018). Hydrogen peroxide positively regulates brassinosteroid signaling through oxidation of the BRASSINAZOLE-RESISTANT1 transcription factor. *Nature Communications*, *9*(1), 1–13.

Vert, G., Walcher, C. L., Chory, J., & Nemhauser, J. L. (2008). Integration of auxin and brassinosteroid pathways by Auxin Response Factor 2. *Proceedings of the National Academy of Sciences*, *105*(28), 9829–9834.

Vukašinović, N., & Russinova, E. (2018). BRexit: Possible brassinosteroid export and transport routes. *Trends in Plant Science*, *23*(4), 285–292.

Wang, B., Li, G., & Zhang, W. H. (2015). Brassinosteroids are involved in Fe homeostasis in rice (*Oryza sativa* L.). *Journal of Experimental Botany*, *66*(9), 2749–2761.

Wang, J., Jiang, J., Wang, J., Chen, L., Fan, S. L., Wu, J. W., ... Wang, Z. X. (2014). Structural insights into the negative regulation of BRI1 signaling by BRI1-interacting protein BKI1. *Cell Research*, *24*(11), 1328–1341.

Wang, L., Wang, Z., Xu, Y., Joo, S. H., Kim, S. K., Xue, Z., ... Chong, K. (2009). OsGSR1 is involved in crosstalk between gibberellins and brassinosteroids in rice. *The Plant Journal*, *57*(3), 498–510.

Werner, T., Nehnevajova, E., Köllmer, I., Novák, O., Strnad, M., Krämer, U., & Schmülling, T. (2010). Root-specific reduction of cytokinin causes enhanced root growth, drought tolerance, and leaf mineral enrichment in *Arabidopsis* and tobacco. *The Plant Cell*, *22*(12), 3905–3920.

Xia, X. J., Huang, L. F., Zhou, Y. H., Mao, W. H., Shi, K., Wu, J. X., ... Yu, J. Q. (2009). Brassinosteroids promote photosynthesis and growth by enhancing activation of Rubisco and expression of photosynthetic genes in *Cucumis sativus*. *Planta*, *230*(6), 1185–1196.

Xue, Y., Lung, S. C., & Chye, M. L. (2016). *Present status and future prospects of transgenic approaches for drought tolerance. Drought stress tolerance in plants* (2, pp. 549–569). Cham: Springer.

Ye, H., Liu, S., Tang, B., Chen, J., Xie, Z., Nolan, T. M., ... Wang, Y. (2017). RD26 mediates crosstalk between drought and brassinosteroid signalling pathways. *Nature Communications*, *8*(1), 1–13.

Yusuf, M., Khan, T. A., & Fariduddin, Q. (2017). *Brassinosteroids: Physiological roles and its signalling in plants. Stress signaling in plants: Genomics and proteomics perspective* (Volume 2, pp. 241–260). Cham: Springer.

Zhang, S., Cai, Z., & Wang, X. (2009). The primary signaling outputs of brassinosteroids are regulated by abscisic acid signaling. *Proceedings of the National Academy of Sciences of the United States of America*, *106*(11), 4543–4548.

Zhang, T., Xu, P., Wang, W., Wang, S., Caruana, J. C., Yang, H. Q., & Lian, H. (2018). *Arabidopsis* G-protein β subunit AGB1 interacts with BES1 to regulate brassinosteroid signaling and cell elongation. *Frontiers in Plant Science*, *8*, 2225.

Chapter 21

Emerging roles of osmoprotectant glycine betaine against salt-induced oxidative stress in plants: a major outlook of maize (*Zea mays* L.)

Akbar Hossain[1], Biswajit Pramanick[2], Karma L. Bhutia[3], Zahoor Ahmad[4], Debojyoti Moulick[5], Sagar Maitra[6], Adeel Ahmad[7] and Tariq Aftab[8]

[1]Bangladesh Wheat and Maize Research Institute, Dinajpur, Bangladesh, [2]Department of Agronomy, Dr. Rajendra Prasad Central Agricultural University, Pusa, India, [3]Department of Agricultural Biotechnology & Molecular Breeding, College of Basic Science and Humanities, Dr. Rajendra Prasad Central Agricultural University, Pusa, India, [4]Department of Botany, University of Central Punjab, Punjab Group of College Bahawalpur, Bahawalpur, Pakistan, [5]Plant Stress and Metabolomics Laboratory, Assam University, Silchar, India, [6]Department of Agronomy, Centurion University of Technology and Management, Paralakhemundi, India, [7]Institute of Soil and Environmental Science, University of Agriculture Faisalabad, Faisalabad, Pakistan, [8]Department of Botany, Aligarh Muslim University, Aligarh, India

Abbreviations

APX Ascorbate peroxidase
CAT Catalase
CDH Choline dehydrogenase
GB Glycine betaine
NO Nitric oxide
NOS Nitric oxide synthase
POD Peroxidase
QACs Quaternary ammonium compounds
ROS Reactive oxygen species
SOD Superoxide dismutase

Frontiers in Plant–Soil Interaction. DOI: https://doi.org/10.1016/B978-0-323-90943-3.00015-8

21.1 Introduction

Salinity is one of the greatest abiotic stresses that restrict agricultural productivity globally (Isayenkov & Maathuis, 2019). Salt stress interrupts morphophysiological processes in plants through osmotic and ionic toxicity (Flowers & Colmer, 2015; Flowers, Munns, & Colmer, 2014). Salinity stress diminishes the action of numerous enzymes that are linked to nitrogen metabolism, thus reducing nitrogen status in plant tissues (Ashraf, Shahzad, Imtiaz, & Rizwan, 2018). Besides these consequences, the excessive Na^+ ions in plant cells also limit photosynthesis by reducing green pigments (Demidchik et al., 2014; Gupta & Huang, 2014). During salt-induced oxidative stress, the unwarranted accumulation of reactive oxygen species (ROS) hastens toxicity in plant cells, such as protein collapse, lipid peroxidation, and DNA alteration (Hasanuzzaman et al., 2017) (Fig. 21.1). To affluence these salt induced-oxidative effects, plants synthesize diverse types of antioxidants like catalase (CAT), ascorbate peroxidase (APX), peroxidase (POD), and superoxide dismutase (SOD) (Noreen & Ashraf, 2009).

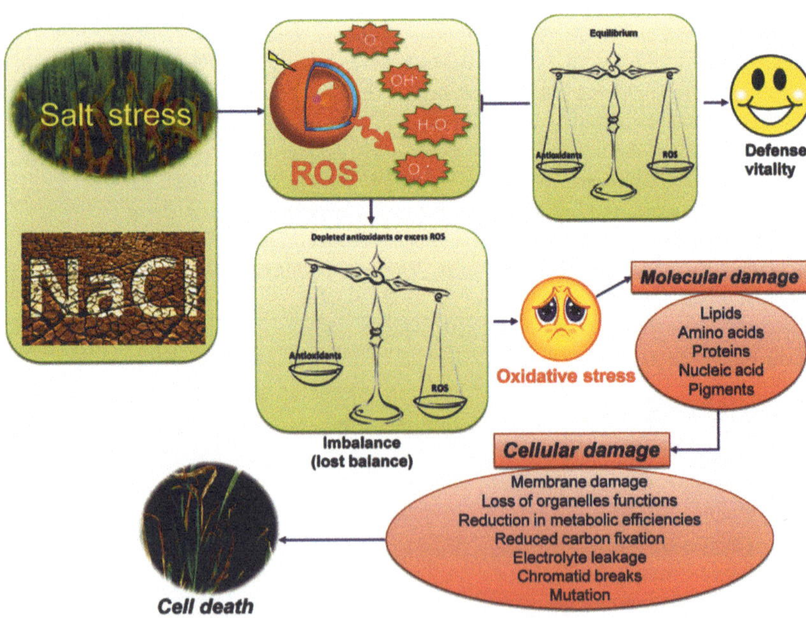

FIGURE 21.1 Excessive production of reactive oxygen species as a consequence of salt-induced oxidative stress alters the physiobiochemical process in plants. *Modified and adapted from Hasanuzzaman M., Nahar K., Rahman A., Anee T.I., Alam M.U., Bhuiyan T.F., ... Fujita M. (2017). Approaches to enhance salt stress tolerance in wheat. In Wanyera, R., Owuoche, J. (Eds.) Wheat improvement, management and utilizationy. IntechOpen, 24, 151-187. http://doi.org/10.5772/67247. with permission.*

Maize is a cross-pollinated polymorphic C_4 plant. Due to its capability to grow under a wide range of agroclimatic conditions, it is cultivated in numerous countries as a major source of food and feed (Langner et al., 2019). Although it can grow in diverse environmental conditions, but the initial stage of maize is highly sensitive to salinity stress. Soil salinity is a serious threat to its production worldwide (Gill & Tuteja, 2010). Similar to other plants soil salinity also hamper the biological processes including transpiration, respiration, stomatal conductance, photosynthesis, hormones, antioxidants, osmoprotectants, and other enzymatic activity, eventually reducing the growth and development of maize leading to a decline in the yield of affected plants (Halo et al., 2015). Under salinity stress, the increased production of ROS alters redox state, shrinks flexibility of membrane, denatures and alters the formation of membrane proteins, damages DNA, abolishes enzymatic activity, and destroys cell homeostasis, leading to cell death of affected plants (Hasanuzzaman et al., 2017; Fig. 21.2). As a C_4 plant and also moderately sensitive to soil salinity, maize plants experience diverse

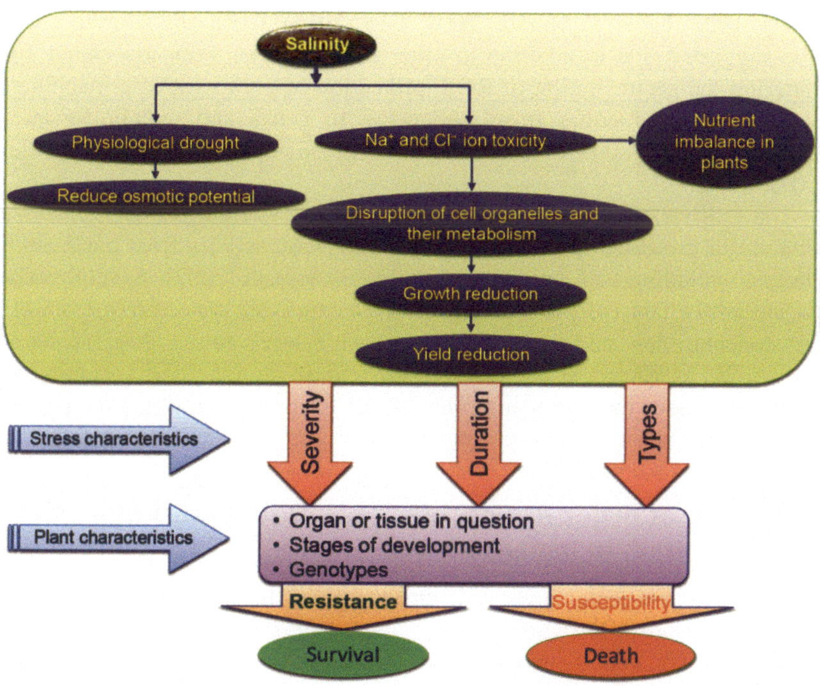

FIGURE 21.2 Consequences and adaptive mechanisms of plants against salt-induced oxidative stress. *Modified and adapted from Hasanuzzaman M., Nahar K., Rahman A., Anee T.I., Alam M. U., Bhuiyan T.F., ... Fujita M. (2017). Approaches to enhance salt stress tolerance in wheat. In Wanyera, R., Owuoche, J. (Eds.) Wheat improvement, management and utilizationy. IntechOpen, 24, 151-187. http://doi.org/10.5772/67247. with permission.*

adaptations at cellular, subcellular, and organ levels (Halo et al., 2015), through controlling stomatal conductance, hormonal equilibrium in cellular and organ levels, stimulation of the antioxidant protection system, adjustment of osmotic balance in plant cells, maintaining the tissue water status, and various mechanisms through the prohibition of poisonous ions (Iqbal, Hussain, Qayyaum, & Ashraf, 2020).

Osmoregulation or osmoprotection is mainly encountered with the deposit of organic and inorganic solutes under salinity to lower water potential without decreasing actual water status in plant cells (Serraj & Sinclair, 2002). Various osmolytes including soluble sugars, sugar alcohols, proline, glycine betaine (GB), organic acids, and trehalose have mainly maintained the osmoregulation or osmoprotection. Among these osmolytes, GB is the most important one, which is responsible for osmoregulation in maize under salt stress (Lv, Yang, Zhang, Wang, & Zhang, 2007; Quan, Shang, Zhang, Zhao, & Zhang, 2004a,b; Rajendrakumar, Suryanarayana, & Reddy, 1997). Plants known to accrue GB have been described to grow well under saline conditions (Chen & Murata, 2002; 2008). The accumulation of GB in the plant cell is known to be an osmoprotectant against abiotic stresses (Bohnert & Jensen, 1996; Chen & Murata, 2002; Giri, 2011). Chen and Murata (2008) have suggested that in addition to other roles, GB has been associated with hindering the accumulation of ROS, protection of photosynthetic machinery, the inspiration of some stress connected to genes and membrane safety (Diamant et al., 2003; Murata, Takahashi, Nishiyama, & Allakhverdiev, 2007).

The GB has also been concerned with the safeguarding of the quaternary structure of proteins, thus protecting the enzymatic activity from harsh environments including soil salinity (Sakamoto & Murata, 2002). Several earlier studies also established the ability of GB against abiotic stresses even at a low concentration under abiotic stress (Al-Taweel et al., 2007; Chen & Murata, 2011; Murata et al., 2007; Takahashi & Murata, 2008). The current chapter discusses the promising roles of GB for protection of plants, particularly maize, against soil salinity stress.

21.2 Role of major osmoprotectants during salt-induced oxidative stress in maize

Different environmental stresses disturb the plants throughout their life cycle, such stresses are UV radiation, toxic metal ion concentration, temperature extremes, water deficit, and salinity. The severity of these stresses defines the degree of limitation in productivity and growth of the plants. Salt-affected soils comprise excessive exchangeable sodium or soluble salts in the root zone. In semiarid and arid regions, salt stress, besides water management practices, poor soil health, high evapotranspiration, and limited rainfall, is a serious threat to the production of crops (Flowers & Yeo, 1995; Munns,

2002). None of the climatic zones is free from the problem of salinity, though commonly this problem is linked to the arid and semiarid areas (Rengasamy, 2006). In the world, about 434 million hectares are under sodicity and 397 million hectares are under salinity, thus in total sodicity and salinity are affecting more than 800 million hectares (FAO, 2005; Munns, 2005). Different salts originating from the weathering of parent material and various anthropogenic practices like power generation, combustion of fuels, organic wastes, industrial discharges, improper irrigation practices, and geogenic processes account for the problem of salinity (Galvani, 2007; Le & McQueen-Mason, 2006; Munns & Tester, 2008; Szabolcs, 1989).

The ROS production (like $OH^{\cdot}$, $O^{2\cdot}$, H_2O_2) is one of the responses of plants to stresses. Due to the interaction with macromolecules and the peroxidation of lipid components of the membrane, ROS causes considerable damage to plants. However, in signal transduction, low quantities of ROS are required (Foyer & Noctor, 2005). Low-molecular-weight antioxidants like antioxidative enzymes (like SOD, APX, and catalase), α-tocopherol, ascorbic acid, and glutathione scavenge ROS. Drought and salinity cause oxidative stress, to counter the stress effects plants also induce antioxidant enzymes including APX, CAT, POD, and SOD (Gill & Tuteja, 2010; Hossain & Fujita, 2010; Kadioglu, Saruhan, Saglam, Terzi, & Acet, 2011; Kaya, Sonmez, Aydemir, Ashraf, & Dikilitas, 2013; Kaya Sönmez Aydemir & Dikilitaş, 2013; Kubis', Floryszak-Wieczorek, & Arasimowicz-Jelonek, 2014; Vardharajula, Ali, Grover, Reddy, & Bandi, 2011). However, oxidative stress results due to the generation of excessive free radicals compared to the cellular antioxidative potential under stressed conditions, which adversely affects the growth of plants. Thus plants accumulate various compatible solutes in large quantities under different stresses (Serraj & Sinclair, 2002).

Compatible solutes are usually nontoxic at higher concentrations; they are highly soluble and low-molecular-weight organic compounds. By contributing to protein/enzymes stabilization, protection of membrane integrity, ROS detoxification, and cellular osmotic adjustment, these solutes protect plants (Ashraf & Foolad 2007; Bohnert & Jensen, 1996; Yancey, 1994). Such solutes include quaternary ammonium compounds (like pipecolate betaine, proline betaine, alanine betaine, and GB), trehalose, polyols, sucrose, and proline (Ashraf & Harris 2004; Rhodes & Hanson 1993). Generally, proline accumulation occurs under heavy metal exposure (Bassi & Sharma, 1993a,b; Schat, Sharma, & Vooijs, 1997; Sharma & Dietz, 2006), low temperature (Naidu, Paleg, Aspinall, Jennings, & Jones, 1991), UV radiations, salinity (Munns, 2005; Rhodes, Nadolska-Orczyk, & Rich, 2002), and water deficit (Hare, Cress, & Van Staden, 1998) conditions.

Despite the osmotic adjustment, proline is also important in maintaining appropriate $NADP^+$/NADPH ratios, alleviating cytoplasmic acidosis (Hare & Cress, 1997), buffering cellular redox potential (Ashraf & Foolad, 2007), as protein compatible hydrotrope (Strizhov et al., 1997), scavenging free radicals and stabilizing subcellular structures (e.g., membranes and proteins) under stressed conditions (Foyer & Noctor, 2005). In salt-tolerant plant species, the concentration of proline is generally higher, as the accumulation of proline is linked to salt tolerance. In sensitive plants, the concentration of proline was much lower (Fougère, Le Rudulier, & Streeter, 1991; Gangopadhyay, Basu, Mukherjee, & Gupta, 1997; Madan, Nainawatee, Jain, & Chowdhury, 1995; Petrusa & Winicov, 1997). The accumulation of proline is a stress signal, it accumulates in the cytoplasm where it maintains cell redox status, buffers cytosolic pH, and stabilizes the structure of proteins. Nawaz and Ashraf (2007) described that growth of maize cultivars was improved due to the effective mitigation of adverse effects of salinity by the foliar application of GB. In comparison to other stages, foliar GB applied at the vegetative stage was more effective in maize. They also explained that under salt stress, exogenous application of GB improved the water status of maize plants in addition to slight changes in leaf osmotic potential, which led to improved growth.

21.3 Physiological mechanism of glycine betaine during salt-induced oxidative stress in maize

Salt-induced oxidative stress, as well as salinity stress, is categorized as abiotic stress to the plant that adversely affects the growth and development process, and ultimately the productivity of many crops (Ashraf & Harris 2004). ROS are produced under oxidative stress when there is a higher salt concentration in the plant (Cramer, Alberieo, & Schmidt, 1994). Namjooyan, Khavari-Nejad, Bernard, Namdjoyan, and Piri (2012) stated that the production of ROS under oxidative stress can be highly detrimental to the physiological activities of the plant as ROS can damage the protein molecules, fats, etc. present in the plant body. However, the defense mechanism of producing antioxidant enzymes inside the plant under oxidative stress can minimize these detrimental effects of ROS (Tavallali, Rahemi, Esghi, Kholdebarin, & Ramezanian, 2010). Salt-tolerant plants, in particular, can sustain the movements of water and ions (Rout & Shaw, 2001), and thus they may produce antioxidants to minimize the negative impact of ROS. This mechanism minimizes the ROS level in the stress-tolerant plants so that they can grow under salinity stress (Masood, Shah, Zeeshan, & Abraham, 2006). ROS such as H_2O_2 can be scavenged by the enzyme POD. Thus the adverse impact of salinity stress in the plant is reduced (Dionisio-Sese & Tobita, 1998). Development of suitable agronomic management strategies to combat such oxidative stress, particularly under arid and semiarid climates, is a global agrarian challenge now. Exogenous application of GB is a promising option to manage the oxidative stresses in the higher plants. It can adjust ion fluxes in the plants. Moreover, some recent studies also show that GB can protect the PS-II of photosynthesis against salt-induced stresses.

21.3.1 Glycine betaine during salt-induced oxidative stress in maize

GB, a solute, is involved in several pathways that are responsible for the defense mechanism of many plants, including maize, against salt-induced oxidative stress (Fig. 21.3). However, Chen and Murata (2011) reported that many of the pathways regulated by GB in maize under salt stress are still unknown. The major function of GB under salt-induced oxidative stress in maize that have been identified is osmoregulation (osmotic adjustment and cellular compatibility) (Giri, 2011). Salt stress significantly induces the expression of the CMO and BADH genes which consequently favor the accumulation of GB (Annunziata, Ciarmiello, Woodrow, Dell'Aversana, & Carillo, 2019; Rhodes, Nadolska-Orczyk, & Rich, 2002). Salt stress causes the leakages of electrons from mitochondria and the chloroplast electron transport chain system and this phenomenon is responsible for the formation of ROS like superoxide (O^-), hydrogen peroxide (H_2O_2), and hydroxyl radicals (OH^-) by interacting with electron acceptors, that is, O_2 (Hasanuzzaman et al., 2019). The oxidative stress to maize cells by ROS induces peroxidation of lipids and other damage such as distortion of protein and nucleic acid structures (Waszczak, Carmody, & Kangasjarvi, 2018).

The salt-induced stress causes ROS to build up in the cells and salt-tolerant maize plants accumulate GB at a significantly higher level (Menezes-Benavente, Kernodle, Margis-Pinheiro, & Scandalios, 2004; Rhodes et al., 1989). The GB prevents chlorosis and reduces the lipid peroxidation of the cell membranes in maize under salt stress (Chen, Li, & Chen,

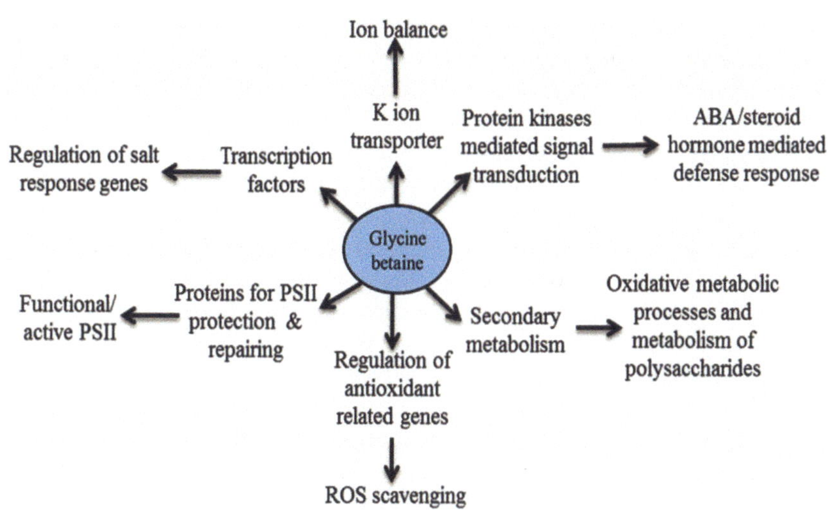

FIGURE 21.3 Most probable pathways regulated by the accumulation of glycine betaine in maize (*Zea mays* L.) plants imparting tolerance against salt stress.

2000). However, GB itself is not involved in scavenging of ROS directly but helps in reducing the oxidative damage by employing the ROS-scavenging enzymes and through other mechanisms (Chen & Murata, 2008). Transcriptome analysis based on RNA sequencing of salt-treated maize genotypes revealed the possible mechanisms by which GB improves the tolerance capacity of maize plants under salt stress (Chen et al., 2020). The accumulation of GB has been associated with the upregulation of the potassium (K) transporter gene that is essential for the better uptake of K and reducing the biased uptake of Na^+ under salt stress. This ultimately creates a balance in ion concentration, thereby reducing the problem of ion imbalance in the cell. Enzymes that regulate the secondary metabolites pathways are also reported to be upregulated by the presence of an increased level of GB in maize cells. The upregulation of genes encoding enzymes like Ent-cassadiene C_2-murhydroxylase-like (a member of Cytochrome P450) engaged in xenobiotic metabolism and steroid biosynthesis (Maves & Sligar, 2001) and cyanidin 3-O-rutinoside 5-O-glucosyltransferase involved in flavonoid biosynthesis by the accumulation of GB suggests that these genes could be attributed to salt tolerance in maize plants by regulating oxidative metabolic processes and metabolism of polysaccharides, respectively, in the presence of GB (Chen et al., 2020). GB also seems to impart salt tolerance in maize plants by inducing signal transduction that is mediated by the genes encoding protein kinases such as Ca^{2+}-dependent protein kinase 3 (CPK3), CBL-interacting protein kinase, mitogen-activated protein kinases (MAPK), and protein phosphatase 2C. In the presence of GB, these proteins are imparting tolerance against salt stress through ABA and steroid hormone-mediated signaling pathways and signal transduction (Chen et al., 2020). Transcription factors (TFs) like the members of AP2-EREBP, NAC, WRKY, MYB, and MYB-related families showed significant upregulation in GB-accumulating maize plants under salt stress. The members of these TFs families seem to be regulating the expression of genes with defensive roles in imparting salt tolerance in maize. Similarly, GB buildup upregulated the expression of antioxidant-related genes like POD, glutathione S-transferases, and SOD, suggesting that GB protects the maize cell from oxidative damage by scavenging the ROS with antioxidant-related proteins (Chen et al., 2020; Qi et al., 2010; Rohman et al., 2016). The maize genotypes that lack GB can also induce the GB-mediated salt tolerance responses by the foliar spray of GB (Kaya, Sonmez, Aydemir, Ashraf, & Dikilitas, 2013; Kaya Sönmez Aydemir & Dikilitaş, 2013; Nawaz & Ashraf, 2007).

From all the reports it can be concluded that biosynthesis of GB in tolerant maize and other plant genotypes is induced by salt stress. GB mediates different pathways that lead to the expression of different ROS scavenging genes, secondary metabolites, TF associated with stress-responsive genes, ion transporters, proteins involved in PSII protection or repair, etc. The products of these pathways protect maize plants by osmotic adjustment of the cell, preventing lipid peroxidation, repairing the PSII system, taking up Na^+ competitors like K^+, providing membrane stability, and preventing chlorosis and apoptosis of the maize cell, which ultimately leads to a healthy maize plant even under oxidative stress conditions.

21.4 Molecular mechanism of glycine betaine during salt-induced oxidative stress in maize

Maize, also known as corn, is an important cereal crop that is grown in large parts of the world. It is a domesticated variant of Teosinte (Benz, 2001) with extensive genetic diversity, and is cultivated around the world (Prasanna, 2012). Southern Mexico is considered as its primary center of origin/diversity (Keleman, Hellin, & Bellon, 2009) and Sikkim (India) is considered to be a secondary center of diversity for maize (Ramaiyan & Avasthe, 2016) due to the rich genetic resources available in these regions. The area under maize cultivation in the world is around 193.73 million ha with a global production of 1147.62 million tons (FAOSTAT, 2020). The United States leads the world in terms of production followed by China and Brazil (FAOSTAT, 2020). Maize is one of the oldest and is still a staple food in many parts of the world due to its high nutritional values. Maize grains contain good amounts of starch, dietary fibers, proteins, and other elements like beta-carotene, vitamins, magnesium, zinc, phosphorus, iron, and so on (Mukhala, De Jager, Van Rensburg, & Walker, 1999). Due to its high nutritional values, maize grains are consumed by humans as well as used as animal feed. For human consumption, there are several ways of making different dishes using either whole grains, such as popcorn, or corn-flour for making dishes like masa, bread, or corn syrup, and corn starch. It is also used for the production of beverages like whiskey and biofuel, that is, corn ethanol (Foley, 2017).

The production and productivity of maize are affected by several biotic and abiotic factors. Among the abiotic factors the primary determinant of crop productivity is climate. Climatic changes like the variation in rainfall pattern, extremes of temperature, deterioration of soil conditions like salinity, acidity, nutrient deficiency, elements toxicity, and the climatic stresses like drought, floods, etc. obstruct crop growth and productivity (Sarada, Ratnam, Naidu, Ramana, & Rajani, 2015). One of the abiotic stresses that severely affects the growth and development of maize is soil salinity or salt stress. Due to poor irrigation and soil management practices coupled with changing climates, such as higher temperature and limited rainfall, which is causing an increased level of evapotranspiration in dryland areas, the world is facing a threat of increasing soil salinity leading to poor crop yields (Farooq, Hussain, Wakeel, & Siddique, 2015; Munns, 2002). However, the salinity stress is not confined to the arid and semiarid regions of the world, no continent in the world is free from the problem of soil salinity (Hossain, 2019; Shahid, Zaman, & Heng, 2018). The data on the total area of the global salt-affected soil varies according to different data sources. According to Ghassemi, Jakeman, and Nix (1995) and Pitman and Läuchli (2002) about 20% of the highly irrigated-soil of the world is facing a soil salinity problem and it is expected to increase with every passing year due to the change in climate.

21.4.1 Salt induced oxidative stress in maize

Maize is moderately susceptible to soil salinity; salt stress causes poor yield performance. However, due to the availability of high genetic resources in maize, many of the genotypes are tolerant to salt stress and these are being used as sources to develop tolerant genotypes and to understand the salt tolerance mechanism of maize. The soil salinity affects several physiological processes, such as the process of photosynthesis, respiration, the functioning of stomata, regulation of plant hormones, and germination of seeds, of crop plants including maize (Iqbal, Hussain, Qayyaum, & Ashraf, 2020). The different genotypes of plants show different responses to salt stress and undergo various physiological, biochemical, and molecular changes (Greenway & Munns, 1980). The vast genetic diversity of maize has been screened to identify the maize genotypes tolerant against salt stress (Akram et al., 2010; Ali, Khan, Shah, Naveedullah, & Jalal, 2019; Cha-um & Kirdmanee, 2010; Collado, Aulicino, Arturi, & Molina, 2015; Devi & Chitdeshwari, 2019; Galvani, 2007; Hoque, Jun, & Guoying, 2015; Khodarahmpour, 2011; Niu, Xu, Rodriguez, & Sun, 2012; Qayyum et al., 2016; Shanko, Degefa, Kere, & Hayeto, 2016). All the morphological and physiological screening of maize genotypes under salt stress has highlighted that the traits like potassium ion (K^+)/sodium ion (Na^+) ratio, potassium ion (K^+) content, calcium ion content (Ca^{2+}), chlorophyll content, root and shoot fresh weight, as well as dry weight, root–shoot length, germination percentage, gas exchange parameters, etc., are changed when a plant is exposed to oxidative stress; tolerant maize plants show better performance for all these physiological traits under salt stress (Devi & Chitdeshwari, 2019; Farooq et al., 2015). Salt stress induces a reduction in root biomass with high accumulation of Na^+ in the roots. Accumulation of Na^+ induces the ROS-mediated oxidative stress in the roots. The effect is moderate in older leaves and less in younger leaves, suggesting that ROS accumulated in roots could be acting as a systemic signaling response from roots to leaves and inducing the expression/accumulation of salt stress-responsive gene/s or other biochemical (AbdElgawad et al., 2016).

At the molecular level, the production of ROS is enhanced at salt-induced oxidative stress owing to changes in the activities of certain enzymes and metabolic processes. The tolerant genotypes of maize would express some antioxidant-related genes to defend the cell damage by ROS (Menezes-Benavente et al., 2004). Several reports have highlighted the molecular mechanism that could be playing a pivotal role in imparting tolerance against salt stress. *Zea mays* high-affinity potassium transporter (ZmHKT) genes were found to be highly expressed in tolerant genotypes with the role to maintain optimal Na^+/K^+ balance and reduce the accumulation of ROS by increasing the antioxidant activity in tolerant genotypes (Jiang et al., 2018). In another experiment, the transgenic *Arabidopsis* with a constitutive expression of maize ABA (abscisic acid)-responsive element (ABRE) binding protein 2 (ABP2) showed enhanced tolerance against salt stress (Na et al., 2018). ABP2 is a member of a large transcription factor family bZIP (basic leucine zipper) whose members in maize are reported to regulate the stress-responsive genes like LEA2, RD20, RD21, Rab18, NHX3, GEA6, and ERD (Ying et al., 2012). Overexpression of a member of maize WRKY transcription factor (ZmWRKY114) was found to be associated with salt stress sensitivity and it is assumed that the ZmWRKY114 could be functioning as a negative regulator that makes plants susceptible to salt stress by regulating the ABA-mediated pathways (Bo et al., 2020). Similarly, a transcription factor ZmRAV1 (related to ABI3/VP1) showed association with the salt-tolerant mechanism. Overexpression of ZmRAV1 transcription factor enhanced the expression of genes related to ROS scavenging which could be one of the reasons for salt tolerance in maize genotypes (Min, Zheng, & Wang, 2014). Transgenic plant expressing maize transcription factor, that is, ZmMYB showed tolerance against salt stress by transcriptionally upregulating the expression of several stress-responsive genes (Chen et al., 2018; Wu et al., 2019). Salt stress also influences the expression of several microRNAs (miRNAs) in maize, leading to either sensitivity or tolerance of maize genotype against the salt-induced oxidative stress. Maize miRNAs' expression, that is, mir-250, mir-205, mir-330, and mir-17, was downregulated, which resulted in upregulation of their target genes and the activation of salt stress-tolerant pathways (Fu et al., 2017). Transcriptome analysis using the microarray technique revealed the altered expression of many miRNAs of maize under salt stress, among which many of the miRNAs were found to be expressed only in the tolerant genotypes, suggesting that genotype-specific expression of miRNAs in maize could be the reason for the differential response of maize genotypes to salt stress (Ding et al., 2009). Several proteins were identified by proteomic analysis of maize genotypes grown under salt stress. The proteins that are upregulated under salt stress were mostly related to antioxidation and lignin synthesis (Cheng et al., 2014). Upregulation or downregulation of stress-responsive genes in maize genotypes tolerant to salt stress ultimately leads to the osmotic adjustment or osmoregulation at the cellular level which makes maize plants to withstand the stress (Serraj & Sinclair, 2002). Many osmolytes like organic acids, proline, soluble sugars, alcohol, GB, etc. have been reported for osmotic adjustment under salt-induced stress (Rohman et al., 2019), among which GB is one of the major soluble solutes reported in maize for osmotic adjustment in the tolerant plant under salt-induced stress (Iqbal, Hussain, Qayyaum, & Ashraf, 2020).

21.4.2 Biosynthesis of glycine betaine in maize

Maize is a betaine-accumulating species. However, many genotypes of maize completely lack betaine. A study reported that betaine deficiency in diverse maize genotypes resulted from the absence of single genetic locus. The deficient locus was responsible for the formation of betaine aldehyde from chlorine (Lerma et al., 1991; Saneoka et al., 1995). Under normal conditions, a lower amount of GB is presented in maize leaves (Ashraf & Foolad, 2007; Rhodes & Hanson, 1993). Choline, a derivative of proline and a precursor of betaine, is first converted to betaine aldehyde by the enzyme choline monooxygenase (Khan, Yu, Kikuchi, Asahina, & Watanabe, 2009; Quan et al., 2004a). The enzyme betaine aldehyde dehydrogenase (BADH) which is a member of the superfamily aldehyde dehydrogenase then catalyzes the betaine aldehyde oxidation to betaine (the extensively studied type is GB, also known as dimethyl-glycine). The conversion of betaine aldehyde to GB by BADH requires oxidizing cofactor either NAD^+ (nicotinamide adenine dinucleotide) or $NADP^+$ (nicotinamide adenine dinucleotide phosphate). In plants, NAD^+ as an oxidizing cofactor showed higher activity of BADHs over $NADP^+$ (Nakamura et al., 1997) (Fig. 21.4).

Nitric oxide synthase (NOS) is involved in the synthesis of nitric oxide (NO) which induces salt stress tolerance in maize by increasing the activity of proton pumps and Na^+/H^+ antiports in tonoplast (Zhang et al., 2006). The

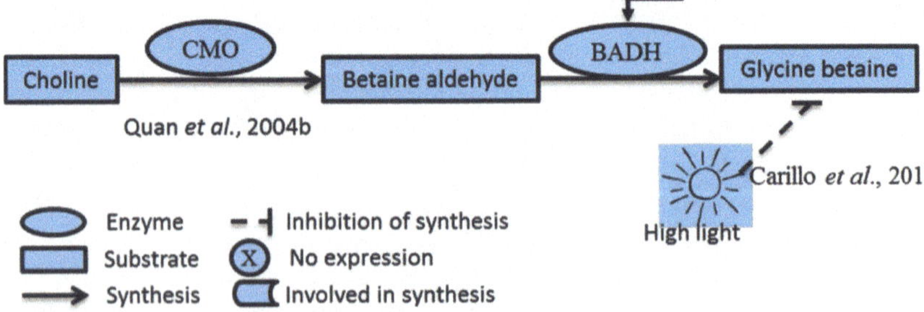

FIGURE 21.4 Suitable pathway showing the biosynthesis of glycine betaine from its precursor by different enzymes in maize (*Zea mays* L.).

NO influences the accumulation of GB during salt stress in maize by enhancing the synthesis of BADH (Ullah, Kolo, Egbichi, Keyster, & Ludidi, 2016). In maize, the synthesis of BADH is suppressed due to the inhibition of NOS synthesis by the NOS inhibitor (Phillips et al., 2018). Transgenic maize lines expressing transgenes like betA gene encoding choline dehydrogenase resulted in a higher concentration of GB in maize leaves under oxidative stress. However, the accumulation of GB was significant in the cytosol of maize leaves cell, which is generally in the chloroplast (Quan et al., 2004a, b). Similarly, transgenic maize expressing BADH of *Suaeda liaotungensis* enhanced the synthesis and accumulation of GB in maize leaf tissue under salt stress, suggesting that BADH is an essential enzyme that is involved in the biosynthesis of GB that is required for the protection of plants under salt stress (Wu et al., 2019). Similarly, an experiment identified the differential posttranscriptional processing in BADH genes which could be potentially contributing to the variation in the synthesis of GB in different genotypes of maize and other crops (Niu et al., 2007).

21.5 Conclusions

Maize is a cross-pollinated polymorphic C_4 plant and has the capability to grow in a wide range of agroclimatic conditions. It is cultivated in numerous countries as a major source of food and feed. Although it can grow in diverse environmental conditions, maize is moderately susceptible to salinity stress. Similar to other plants, soil salinity also hampers the biological processes of maize including transpiration, respiration, stomatal conductance, photosynthesis, hormones, antioxidants, osmoprotectants, and other enzymatic activity, eventually condensing the growth and development of the plant leading to a decline in the yield of maize. Besides these consequences, the excessive Na^+ ions in plant cells also limit photosynthesis by reducing green pigments. During salt-induced oxidative stress, the unwarranted accumulation of ROS hastens toxicity in plant cells, leading to protein collapse, lipid peroxidation, and DNA alteration. To ameliorate these salt-induced oxidative effects, plants synthesize diverse types of antioxidants, like CAT, APX, POD, and SOD. However, the accumulation of various osmolytes in plant cells including soluble sugars, sugar alcohols, proline, GB, organic acids, and trehalose could maintain the osmoregulation. Among these osmolytes, GB is the most important one, which is responsible for osmoregulation in maize under salt stress. GB has also been concerned with the safeguarding of the quaternary structure of proteins, thus upholding the enzymatic activity from harsh environments including soil salinity. Several earlier studies also established the ability of GB against abiotic stresses even at a low concentration under abiotic stress. Therefore an understanding of GB with salt-induced oxidative stress could mitigate the adverse effects of maize under soil salinity stress.

References

AbdElgawad, H., Zinta, G., Hegab, M. M., Pandey, R., Asard, H., & Abuelsoud, W. (2016). High salinity induces different oxidative stress and antioxidant responses in maize seedlings organs. *Frontiers in Plant Science*, *7*, 276.

Akram, M., Ashraf, M. Y., Ahmad, R., Waraich, E. A., Iqbal, J., & Mohsan, M. (2010). Screening for salt tolerance in maize (*Zea mays* L.) hybrids at an early seedling stage. *Pakistan Journal of Botany*, *42*(1), 141–154.

Ali, S., Khan, M. J., Shah, Z., Naveedullah, N., & Jalal, A. (2019). Genotypic screening of maize (*Zea mays* L.) for salt tolerance at early growth stage under different salinity levels. *Sarhad Journal of Agriculture*, *35*(1), 208–215.

Al-Taweel, K., Iwaki, T., Yabuta, Y., Shigeoka, S., Murata, N., & Wadano, A. (2007). A bacterial transgene for catalase protects translation of D1 protein during exposure of salt-stressed tobacco leaves to strong light. *Plant Physiology*, *145*, 258–265. Available from https://doi.org/10.1104/pp.107.101733.

Annunziata, M. G., Ciarmiello, L. F., Woodrow, P., Dell'Aversana, E., & Carillo, P. (2019). Spatial and temporal profile of glycine betaine accumulation in plants under abiotic stresses. *Frontiers in Plant Science*, *10*, 230.

Ashraf, M., & Foolad, M. R. (2007). Roles of glycine betaine and proline in improving plant abiotic stress resistance. *Environmental and Experimental Botany*, *59*, 206–216. Available from http://doi.org/10.1016/j.envexpbot.2005.12.006.

Ashraf, M., & Harris, P. J. C. (2004). Potential biochemical indicators of salinity tolerance in plants. *Plant Science (Shannon, Ireland)*, *166*, 3–16. Available from http://doi.org/10.1016/j.plantsci.2003.10.024.

Ashraf, M., Shahzad, S. M., Imtiaz, M., & Rizwan, M. (2018). Salinity effects on nitrogen metabolism in plants – Focusing on the activities of nitrogen metabolizing enzymes: A review. *Journal of Plant Nutrition*, *41*, 1–17. Available from https://doi.org/10.1080/01904167.2018.1431670.

Bassi, R., & Sharma, S. S. (1993a). Changes in proline content accompanying the uptake of zinc and copper by *Lemna minor*. *Annals of Botany (London)*, *72*, 151–154. Available from http://doi.org/10.1006/anbo.1993.1093.

Bassi, R., & Sharma, S. S. (1993b). Proline accumulation in wheat seedlings exposed to zinc and copper. *Phytochemistry*, *33*, 1339–1342. Available from http://doi.org/10.1016/0031-9422(93)85086-7.

Benz, B. F. (2001). Archaeological evidence of teosinte domestication from Guilá Naquitz, Oaxaca. *Proceedings of the National Academy of Sciences of the United States of America*, *98*(4), 2104–2106.

Bo, C., Chen, H., Luo, G., Li, W., Zhang, X., Ma, Q., ... Cai, R. (2020). Maize WRKY114 gene negatively regulates salt-stress tolerance in transgenic rice. *Plant Cell Reports*, *39*(1), 135–148.

Bohnert, H. J., & Jensen, R. G. (1996). Strategies for engineering water-stress tolerance in plants. *Trends in Biotechnology*, *14*, 89–97. Available from http://doi.org/10.1016/0167-7799(96)80929-2.

Cha-um, S., & Kirdmanee, C. (2010). Salt tolerance screening in six maize (*Zea mays* L.) genotypes using multivariate cluster analysis. *The Philippine Agricultural Scientist*, *93*, 156–164.

Chen, F., Fang, P., Zeng, W., Ding, Y., Zhuang, Z., & Peng, Y. (2020). Comparing transcriptome expression profiles to reveal the mechanisms of salt tolerance and exogenous glycine betaine mitigation in maize seedlings. *PLOS One*, *15*(5), e0233616.

Chen, T. H., & Murata, N. (2002). Enhancement of tolerance of abiotic stress by metabolic engineering of betaines and other compatible solutes. *Current Opinion in Plant Biology, 2002* (5), 250–257. Available from https://doi.org/10.1016/S1369-5266(02)00255-8.

Chen, T. H., & Murata, N. (2008). Glycinebetaine: An effective protectant against abiotic stress in plants. *Trends in Plant Science, 13*, 499–505. Available from https://doi.org/10.1016/j.tplants.2008.06.007.

Chen, T. H., & Murata, N. (2011). Glycinebetaine protects plants against abiotic stress: Mechanisms and biotechnological applications. *Plant, Cell & Environment, 34*, 1–20. Available from https://doi.org/10.1111/j.1365-3040.2010.02232.x.

Chen, W. P., Li, P. H., & Chen, T. H. H. (2000). Glycinebetaine increases chilling tolerance and reduces chilling-induced lipid peroxidation in *Zea mays* L. *Plant, Cell & Environment, 23* (6), 609–618.

Chen, Y. H., Cao, Y. Y., Wang, L. J., Li, L. M., Yang, J., & Zou, M. X. (2018). Identification of MYB transcription factor genes and their expression during abiotic stresses in maize. *Plant Biology, 62*(2), 222–230.

Cheng, Y., Chen, G., Hao, D., Lu, H., Shi, M., Mao, Y., ... Xue, L. (2014). Salt-induced root protein profile changes in seedlings of maize inbred lines with differing salt tolerances. *Chilean Journal of Agricultural Research, 74*(4), 468–476.

Collado, M.B., Aulicino, M.B., Arturi, M.J., & Molina, M.D.C. (2015). Evaluation of salinity tolerance indices in seedling of maize (*Zea mays* L.). <http://revista.agro.unlp.edu.ar/index.php/revagro/article/view/412> Accessed 18.10.20.

Cramer, G. R., Alberieo, G. J., & Schmidt, C. (1994). Salt tolerance is not associated with the sodium accumulation of two maize hybrids. *Australian Journal of Plant Physiology, 21*, 675–692.

Demidchik, V., Straltsova, D., Medvedev, S. S., Pozhvanov, G. A., Sokolik, A., & Yurin, V. (2014). Stress-induced electrolyte leakage: The role of K+-permeable channels and involvement in programmed cell death and metabolic adjustment. *Journal of Experimental Botany, 65*, 1259–1270. Available from https://doi.org/10.1093/jxb/eru004.

Devi, B. N., & Chitdeshwari, T. (2019). Physiological and morphological traits of Maize hybrids under saline water irrigation. *Indian Journal of Experimental Biology, 57*, 188–194.

Diamant, S., Rosenthal, D., Azem, A., Eliahu, N., Ben-Zvi, A. P., & Goloubinoff, P. (2003). Dicarboxylic amino acids and glycine-betaine regulate chaperone-mediated protein-disaggregation under stress. *Molecular Microbiology, 49*, 401–410. Available from https://doi.org/10.1046/j.1365-2958.2003.03553.x.

Ding, D., Zhang, L., Wang, H., Liu, Z., Zhang, Z., & Zheng, Y. (2009). Differential expression of miRNAs in response to salt stress in maize roots. *Annals of Botany, 103*(1), 29–38.

Dionisio-Sese, M. L., & Tobita, S. (1998). Antioxidant responses of rice seedlings to salinity stress. *Plant Science (Shannon, Ireland), 135*, 1–9.

FAO. (2005). Global network on integrated soil management for sustainable use of salt-affected soils. FAO Land and Plant Nutrition Management Service, Rome, Italy, <http://www.fao.org/ag/agl/agll/spush>.

FAOSTAT. (2020). Production quantities of maize by country. <http://www.fao.org/faostat/en/#data/QC/visualize> Accessed 27.10.20.

Farooq, M., Hussain, M., Wakeel, A., & Siddique, K. H. (2015). Salt stress in maize: Effects, resistance mechanisms, and management. A review. *Agronomy for Sustainable Development, 35*(2), 461–481.

Flowers, T. J., & Colmer, T. D. (2015). Plant salt tolerance: Adaptations in halophytes. *Annals of Botany, 115*, 327–331. Available from https://doi.org/10.1093/aob/mcu267.

Flowers, T. J., Munns, R., & Colmer, T. D. (2014). Sodium chloride toxicity and the cellular basis of salt tolerance in halophytes. *Annals of Botany*, *115*, 419–431. Available from https://doi.org/10.1093/aob/mcu217.

Flowers, T. J., & Yeo, A. (1995). Breeding for salinity resistance in crops. Where next? *Australian Journal of Plant Physiology*, *22*, 875–884. Available from https://doi.org/10.1071/PP9950875.

Foley, J. (2017). It's time to rethink America's corn system. Scientific American.

Fougère, F., Le Rudulier, D., & Streeter, J. G. (1991). Effects of salt stress on amino acid, organic acid, and carbohydrate composition of roots, bacteroids, and cytosol of alfalfa (*Medicago sativa* L.). *Plant Physiology*, *96*, 1228–1236. Available from http://doi.org/10.1104/pp.96.4.1228.

Foyer, C. H., & Noctor, G. (2005). Redox homeostasis and antioxidant signaling: A metabolic interface between stress perception and physiological responses. *The Plant Cell*, *17*, 1866–1875. Available from http://doi.org/10.1105/tpc.105.033589.

Fu, R., Zhang, M., Zhao, Y., He, X., Ding, C., Wang, S., ... Wang, B. (2017). Identification of salt tolerance-related microRNAs and their targets in maize (*Zea mays* L.) using high-throughput sequencing and degradome analysis. *Frontiers in Plant Science*, *8*, 864.

Galvani, A. (2007). The challenge of the food sufficiency through salt tolerant crops. *Reviews in Environmental Science and Biotechnology*, *6*, 3–16.

Gangopadhyay, G., Basu, S., Mukherjee, B., & Gupta, S. (1997). Effects of salt and osmotic shocks on unadapted and adapted callus lines of tobacco. *Plant Cell, Tissue and Organ Culture*, *49*, 45–52. Available from http://doi.org/10.1023/A:1005860718585.

Ghassemi, F., Jakeman, A. J., & Nix, H. A. (1995). *Salinisation of land and water resources: Human causes, extent, management and case studies*. CAB international.

Gill, S. S., & Tuteja, N. (2010). Reactive oxygen species and antioxidant machinery in abiotic stress tolerance in crop plants. *Plant Physiology and Biochemistry*, *48*, 909–930.

Giri, J. (2011). Glycinebetaine and abiotic stress tolerance in plants. *Plant Signaling & Behavior*, *6*(11), 1746–1751. Available from https://doi.org/10.4161/psb.6.11.17801.

Greenway, H., & Munns, R. (1980). Mechanisms of salt tolerance in nonhalophytes. *Annual Review of Plant Physiology*, *31*, 149–190.

Gupta, B., & Huang, B. (2014). Mechanism of salinity tolerance in plants: Physiological, biochemical, and molecular characterization. *International Journal of Genomics*, Article ID 701596. Available from https://doi.org/10.1155/2014/701596.

Halo, B. A., Khan, A. L., Waqas, M., Al-Harras, A., Hussain, J., Ali, L., et al. (2015). Endophytic bacteria (*Sphingomonas* sp. LK11) and gibberellin can improve *Solanum lycopersicum* growth and oxidative stress under salinity. *Journal of Plant Interactions*, *10*(1), 117–125.

Hare, P. D., & Cress, W. A. (1997). Metabolic implications of stress-induced proline accumulation in plants. *Plant Growth Regulation*, *21*, 79–102. Available from http://doi.org/10.1023/A:1005703923347.

Hare, P. D., Cress, W. A., & Van Staden, J. (1998). Dissecting the roles of osmolyte accumulation during stress. *Plant, Cell & Environment*, *21*, 535–553. Available from http://doi.org/10.1046/j.1365-3040.1998.00309.x.

Hasanuzzaman, M., Banerjee, A., Bhuyan, M. B., Roychoudhury, A., Al Mahmud, J., & Fujita, M. (2019). Targeting glycinebetaine for abiotic stress tolerance in crop plants: Physiological mechanism, molecular interaction and signaling. *Phyton – Annales Rei Botanicae*, *88*(3), 185.

Hasanuzzaman, M., Nahar, K., Rahman, A., Anee, T. I., Alam, M. U., Bhuiyan, T. F., ... Fujita, M. (2017). Approaches to enhance salt stress tolerance in wheat. In R. Wanyera, &

J. Owuoche (Eds.), *Wheat improvement, management and utilization* (24, pp. 151–187). IntechOpen. Available from http://doi.org/10.5772/67247.

Hoque, M. M. I., Jun, Z., & Guoying, W. (2015). Evaluation of salinity tolerance in maize (*Zea mays* L.) genotypes at seedling stage. *Journal of BioScience and Biotechnology*, *4*(1), 405–411.

Hossain, M. A., & Fujita, M. (2010). Evidence for a role of exogenous glycine betaine and proline in antioxidant defense and methyl-glyoxal detoxification systems in mung bean seedlings under salt stress. *Physiology and Molecular Biology of Plants*, *16*, 19–29.

Hossain, M. S. (2019). Present scenario of global salt affected soils, its management and importance of salinity research. *International Research Journal of Biological Sciences*, *1*, 1–3.

Iqbal, S., Hussain, S., Qayyaum, M. A., & Ashraf, M. (2020). *The response of maize physiology under salinity stress and its coping strategies. Plant stress physiology*. IntechOpen.

Isayenkov, S. V., & Maathuis, F. J. M. (2019). Plant salinity stress: Many unanswered questions remain. *Frontiers in Plant Science*, *10*, 80. Available from https://doi.org/10.3389/fpls.2019.00080.

Jiang, Z., Song, G., Shan, X., Wei, Z., Liu, Y., Jiang, C., ... Li, Y. (2018). Association analysis and identification of ZmHKT1; 5 variation with salt-stress tolerance. *Frontiers in Plant Science*, *9*, 1485.

Kadioglu, A., Saruhan, N., Saglam, A., Terzi, R., & Acet, T. (2011). Exogenous salicylic acid alleviates effects of long term drought stress and delays leaf rolling by inducing antioxidant system. *Plant Growth Regulation*, *64*, 27–37.

Kaya, C., Sonmez, O., Aydemir, S., Ashraf, M., & Dikilitas, M. (2013). Exogenous application of mannitol and thiourea regulates plant growth and oxidative stress responses in salt-stressed maize (*Zea mays* L.). *Journal of Plant Interactions*, *3*, 234–241.

Kaya, C., Sönmez, O., Aydemir, S., & Dikilitaş, M. (2013). Mitigation effects of glycinebetaine on oxidative stress and some key growth parameters of maize exposed to salt stress. *Turkish Journal of Agriculture and Forestry*, *37*(2), 188–194.

Keleman, A., Hellin, J., & Bellon, M. R. (2009). Maize diversity, rural development policy, and farmers' practices: Lessons from Chiapas, Mexico. *The Geographical Journal*, *175*(1), 52–70.

Khan, M. S., Yu, X., Kikuchi, A., Asahina, M., & Watanabe, K. N. (2009). Genetic engineering of glycine betaine biosynthesis to enhance abiotic stress tolerance in plants. *Plant Biotechnology*, *26*(1), 125–134.

Khodarahmpour, Z. (2011). Screening maize (*Zea mays* L.) hybrids for salt stress tolerance at germination stage. *African Journal of Biotechnology*, *10*(71), 15959–15965.

Kubiś, J., Floryszak-Wieczorek, J., & Arasimowicz-Jelonek, M. (2014). Polyamines induce adaptive responses in water deficit stressed cucumber roots. *Journal of Plant Research*, *127*, 151–158.

Langner, J. A., Zanon, A. J., Streck, N. A., Reiniger, L. R. S., Kaufmann, M. P., & Alves, A. F. (2019). Maize: Key agricultural crop in food security and sovereignty in a future with water scarcity. *Revista Brasileira de Engenharia Agrícola e Ambiental*, *23*(9), 648–654, Epub August 12, 2019. Available from https://doi.org/10.1590/1807-1929/agriambi.v23n9p648-654.

Le, T. N., & McQueen-Mason, S. J. (2006). Desiccation-tolerant plants in dry environments. *Reviews in Environmental Science and Biotechnology*, *5*, 269–279.

Lerma, C., Rich, P. J., Ju, G. C., Yang, W. J., Hanson, A. D., & Rhodes, D. (1991). Betaine deficiency in maize: Complementation tests and metabolic basis. *Plant Physiology*, *95*(4), 1113–1119.

Lv, S., Yang, A., Zhang, K., Wang, L., & Zhang, J. (2007). Increase of glycinebetaine synthesis improves drought tolerance in cotton. *Molecular Breeding*, *20*, 233–248. Available from https://doi.org/10.1007/s11032-007-9086-x.

Madan, S., Nainawatee, H. S., Jain, R. K., & Chowdhury, J. B. (1995). Proline and proline metabolising enzymes in in-vitro selected NaCl-tolerant *Brassica juncea* L. under salt stress. *Annals of Botany (London)*, *76*, 5–17. Available from https://doi.org/10.1006/anbo.1995.1077.

Masood, A., Shah, N. A., Zeeshan, M., & Abraham, G. (2006). Differential response of antioxidant enzymes to salinity stress in two varieties of Azolla (*Azolla pinnata* and *Azolla filiculoides*). *Environmental and Experimental Botany*, *58*, 216–222.

Maves, S. A., & Sligar, S. G. (2001). Understanding thermostability in cytochrome P450 by combinatorial mutagenesis. *Protein Science: A Publication of the Protein Society*, *10*(1), 161–168.

Menezes-Benavente, L., Kernodle, S. P., Margis-Pinheiro, M., & Scandalios, J. G. (2004). Salt-induced antioxidant metabolism defenses in maize (*Zea mays* L.) seedlings. *Redox Report: Communications in Free Radical Research*, *9*(1), 29–36.

Min, H., Zheng, J., & Wang, J. (2014). Maize ZmRAV1 contributes to salt and osmotic stress tolerance in transgenic arabidopsis. *Journal of Plant Biology*, *57*(1), 28–42.

Mukhala, E., De Jager, J. M., Van Rensburg, L. D., & Walker, S. (1999). Dietary nutrient deficiency in small-scale farming communities in South Africa: Benefits of intercropping maize (*Zea mays*) and beans (*Phaseolus vulgaris*). *Nutrition Research (New York, N.Y.)*, *19*(4), 629–641.

Munns, R. (2002). Comparative physiology of salt and water stress. *Plant, Cell & Environment*, *25*, 239–250. Available from https://doi.org/10.1046/j.0016-8025.2001.00808.x.

Munns, R. (2005). Genes and salt tolerance: Bringing them together. *The New Phytologist*, *167*, 645–663. Available from http://doi.org/10.1111/j.1469-8137.2005.01487.x.

Munns, R., & Tester, M. (2008). Mechanisms of salinity tolerance. *Annual Review of Plant Biology*, *59*, 651–681.

Murata, N., Takahashi, S., Nishiyama, Y., & Allakhverdiev, S. I. (2007). Photoinhibition of photosystem II under environmental stress. *Biochimica et Biophysica Acta*, *1767*, 414–421.

Na, Z. O. N. G., Li, X. J., Lei, W. A. N. G., Ying, W. A. N. G., Wen, H. T., Ling, L. I., … Jun, Z. H. A. O. (2018). Maize ABP2 enhances tolerance to drought and salt stress in transgenic *Arabidopsis*. *Journal of Integrative Agriculture*, *17*(11), 2379–2393.

Naidu, B. P., Paleg, L. G., Aspinall, D., Jennings, A. C., & Jones, G. P. (1991). Amino acid and glycine betaine accumulation in coldstressed wheat seedlings. *Phytochemistry*, *30*, 407–409. Available from http://doi.org/10.1016/0031-9422(91)83693-F.

Nakamura, T., Yokota, S., Muramoto, Y., Tsutsui, K., Oguri, Y., Fukui, K., & Takabe, T. (1997). Expression of a betaine aldehyde dehydrogenase gene in rice, a glycine betaine nonaccumulator, and possible localization of its protein in peroxisomes. *The Plant Journal: For Cell and Molecular Biology*, *11*, 1115–1120.

Namjooyan, S., Khavari-Nejad, R., Bernard, F., Namdjoyan, S., & Piri, H. (2012). The effect of cadmium on growth and antioxidant responses in the safflower (*Carthamus tinctorius* L.) callus. *Turkish Journal of Agriculture and Forestry*, *36*, 145–152.

Nawaz, K., & Ashraf, M. (2007). Improvement in salt tolerance of maize by exogenous application of Glycinebetaine: Growth and water relations. *Pakistan Journal of Botany*, *39*(5), 1647–1653.

Niu, G., Xu, W., Rodriguez, D., & Sun, Y. (2012). Growth and physiological responses of maize and sorghum genotypes to salt stress. *International Scholarly Research Notices*, *2012*, 145072.

Niu, X., Zheng, W., Lu, B. R., Ren, G., Huang, W., Wang, S., ... Liu, Y. (2007). An unusual post-transcriptional processing in two betaine aldehyde dehydrogenase loci of cereal crops directed by short, direct repeats in response to stress conditions. *Plant Physiology*, *143*(4), 1929–1942.

Noreen, Z., & Ashraf, M. (2009). Assessment of variation in antioxidative defense system in salt-treated pea (*Pisum sativum*) cultivars and its putative use as salinity tolerance markers. *Journal of Plant Physiology*, *166*(16), 1764–1774. Available from https://doi.org/10.1016/j.jplph.2009.05.005.

Petrusa, L. M., & Winicov, I. (1997). Proline status in salt tolerant and salt sensitive alfalfa cell lines and plants in response to NaCl. *Plant Physiology and Biochemistry: PPB / Societe Francaise de Physiologie Vegetale*, *35*, 303–310.

Phillips, K., Majola, A., Gokul, A., Keyster, M., Ludidi, N., & Egbichi, I. (2018). Inhibition of NOS-like activity in maize alters the expression of genes involved in H_2O_2 scavenging and glycine betaine biosynthesis. *Scientific Reports*, *8*(1), 1–9.

Pitman, M. G., & Läuchli, A. (2002). *Global impact of salinity and agricultural ecosystems. Salinity: Environment-plants-molecules* (pp. 3–20). Dordrecht: Springer.

Prasanna, B. M. (2012). Diversity in global maize germplasm: Characterization and utilization. *Journal of Biosciences*, *37*(5), 843–855.

Qayyum, A., Saeed, H. M., Hanif, M., Noor, E., Malik, W., Liaqat, S., ... Murtaza, N. (2016). Exploitation of variability for salinity tolerance in maize hybrids (*Zea mays* L.) at early growth stage. *African Journal of Agricultural Research*, *11*(42), 4206–4213.

Qi, Y. C., Liu, W. Q., Qiu, L. Y., Zhang, S. M., Ma, L., & Zhang, H. (2010). Overexpression of glutathione S-transferase gene increases salt tolerance of *Arabidopsis*. *Russian Journal of Plant Physiology*, *57*(2), 233–240.

Quan, R., Shang, M., Zhang, H., Zhao, Y., & Zhang, J. (2004a). Improved chilling tolerance by transformation with betA gene for the enhancement of glycinebetaine synthesis in maize. *Plant Science (Shannon, Ireland)*, *166*(1), 141–149.

Quan, R., Shang, M., Zhang, H., Zhao, Y., & Zhang, J. (2004b). Engineering of enhanced glycine betaine synthesis improves drought tolerance in maize. *Plant Biotechnology Journal*, *2* (6), 477–486. Available from https://doi.org/10.1111/j.1467-7652.2004.00093.x.

Rajendrakumar, C. S., Suryanarayana, T., & Reddy, A. R. (1997). DNA helix destabilization by proline and betaine: Possible role in the salinity tolerance process. *FEBS Letters*, *410*, 201–205. Available from https://doi.org/10.1016/S0014-5793(97)00588-7.

Ramaiyan, K. & Avasthe, R. (2016). Maize genetic resources in Sikkim, their adaptability and usefulness for hill agriculture. In *Proceedings of the national seminar on physiological and molecular approaches for the improvement of agricultural, horticultural and forestry crops*, Kerala Agricultural University, Vellanikkara, Thrissur.

Rengasamy, P. (2006). World salinization with emphasis on Australia. *Journal of Experimental Botany*, *57*, 1017–1023. Available from https://doi.org/10.1093/jxb/erj108.

Rhodes, D., & Hanson, A. D. (1993). Quaternary ammonium and tertiary sulfonium compounds in higher plants. *Annual Review of Plant Physiology and Plant Molecular Biology*, *44*, 357–384. Available from http://doi.org/10.1146/annurev.pp.44.060193.002041.

Rhodes, D., Nadolska-Orczyk, A., & Rich, P. J. (2002). *Salinity, osmolytes and compatible solutes. Salinity: Environment-plants-molecules* (pp. 181–204). Dordrecht: Springer.

Rhodes, D., Rich, P. J., Brunk, D. G., Ju, G. C., Rhodes, J. C., Pauly, M. H., & Hansen, L. A. (1989). Development of two isogenic sweet corn hybrids differing for glycinebetaine content. *Plant Physiology*, *91*, 1112–1121.

Rohman, M., Islam, M., Monsur, M. B., Amiruzzaman, M., Fujita, M., & Hasanuzzaman, M. (2019). Trehalose protects maize plants from salt stress and phosphorus deficiency. *Plants*, *8* (12), 568.

Rohman, M. M., Talukder, M. Z. A., Hossain, M. G., Uddin, M. S., Amiruzzaman, M., Biswas, A., . . . Chowdhury, M. A. Z. (2016). Saline sensitivity leads to oxidative stress and increases the antioxidants in presence of proline and betaine in maize (*Zea mays* L.) inbred. *Plant Omics*, *9*(1), 35.

Rout, N. P., & Shaw, B. P. (2001). Salt tolerance in aquatic macrophytes: Possible involvement of the antioxidative enzymes. *Plant Science (Shannon, Ireland)*, *160*, 415–423.

Sakamoto, A., & Murata, N. (2002). The role of glycine betaine in the protection of plants from stress: Clues from transgenic plants. *Plant, Cell & Environment*, *25*, 163–171. Available from https://doi.org/10.1046/j.0016-8025.2001.00790.x.

Saneoka, H., Nagasaka, C., Hahn, D. T., Yang, W. J., Premachandra, G. S., Joly, R. J., & Rhodes, D. (1995). Salt tolerance of glycinebetaine-deficient and-containing maize lines. *Plant Physiology*, *107*(2), 631–638.

Sarada, C., Ratnam, M., Naidu, L. N., Ramana, C. V., Rajani, A., et al. (2015). Chilli production and productivity in relation to seasonal weather conditions in Guntur district of Andhra Pradesh. *Indian Journal of Pure & Applied Biosciences*, *3*, 207–213.

Schat, H., Sharma, S. S., & Vooijs, R. (1997). Heavy metalinduced accumulation of free proline in a metaltolerant and a nontolerant ecotype of *Silene vulgaris*. *Physiologia Plantarum*, *101*, 477–482. Available from http://doi.org/10.1111/j.1399-3054.1997.tb01026.x.

Serraj, R., & Sinclair, T. R. (2002). Osmolyte accumulation: Can it really help increase crop yield under drought conditions? *Plant, Cell & Environment*, *25*(2), 333–341. Available from http://doi.org/10.1046/j.1365-3040.2002.00754.x.

Shahid, S. A., Zaman, M., & Heng, L. (2018). *Soil salinity: Historical perspectives and a world overview of the problem. Guideline for salinity assessment, mitigation and adaptation using nuclear and related techniques* (pp. 43–53). Cham: Springer.

Shanko, D., Degefa, I., kere, B., & Hayeto, G. (2016). Effect of different salinity levels on maize (*Zea mays* L.) genotypes during germination. *International Journal of Current Science*, *19* (4), E52–E61.

Sharma, S. S., & Dietz, K. J. (2006). The significance of amino acids and amino acid-derived molecules in plant responses and adaptation to heavy metal stress. *Journal of Experimental Botany*, *57*, 711–726. Available from http://doi.org/10.1093/jxb/erj073.

Strizhov, N., Abrahám, E., Okrész, L., Blickling, S., Zilberstein, A., Schell, J., et al. (1997). Differential expression of two P5CS genes controlling proline accumulation during salt-stress requires ABA and is regulated by ABA1, ABI1 and AXR2 in *Arabidopsis*. *The Plant Journal: For Cell and Molecular Biology*, *12*, 557–569. Available from http://doi.org/10.1046/j.1365-313X.1997.00557.x.

Szabolcs, I. (1989). *Salt affected soils*. Boca Raton: CRC Press.

Takahashi, S., & Murata, N. (2008). How do environmental stresses accelerate photoinhibition? *Trends in Plant Science*, *13*, 178–182. Available from https://doi.org/10.1016/j.tplants.2008.01.005.

Tavallali, V., Rahemi, M., Esghi, S., Kholdebarin, B., & Ramezanian, A. (2010). Zinc alleviates salt stress and increases antioxidant enzyme activity in the leaves of pistachio (*Pistacia vera* L. '*Badami*') seedlings. *Turkish Journal of Agriculture and Forestry*, *34*, 349–359.

Ullah, S., Kolo, Z., Egbichi, I., Keyster, M., & Ludidi, N. (2016). Nitric oxide influences glycine betaine content and ascorbate peroxidase activity in maize. *South African Journal of Botany*, *105*, 218–225.

Vardharajula, S., Ali, S. Z., Grover, M., Reddy, G., & Bandi, V. (2011). Drought-tolerant plant growth promoting Bacillus spp.: Effect on growth, osmolytes, and antioxidant status of maize under drought stress. *Journal of Plant Interactions*, *1*, 1–14.

Waszczak, C., Carmody, M., & Kangasjarvi, J. (2018). Reactive oxygen species in plant signaling. *Annual Review of Plant Biology*, *69*, 209–236.

Wu, J., Jiang, Y., Liang, Y., Chen, L., Chen, W., & Cheng, B. (2019). Expression of the maize MYB transcription factor ZmMYB3R enhances drought and salt stress tolerance in transgenic plants. *Plant Physiology and Biochemistry: PPB / Societe Francaise de Physiologie Vegetale*, *137*, 179–188.

Yancey, P. H. (1994). Compatible and counteracting solutes. In K. Strange (Ed.), *Cellular and molecular physiology of cell volume regulation* (1994, pp. 81–109). Boca Raton, FL: CRC Press.

Ying, S., Zhang, D. F., Fu, J., Shi, Y. S., Song, Y. C., Wang, T. Y., & Li, Y. (2012). Cloning and characterization of a maize bZIP transcription factor, $ZmbZIP_{72}$, confers drought and salt tolerance in transgenic *Arabidopsis*. *Planta*, *235*(2), 253–266.

Zhang, Y., Wang, L., Liu, Y., Zhang, Q., Wei, Q., & Zhang, W. (2006). Nitric oxide enhances salt tolerance in maize seedlings through increasing activities of proton-pump and Na + / H + antiport in the tonoplast. *Planta*, *224*(3), 545–555.

Chapter 22

Ecodesigning for ecological sustainability

Nazish Huma Khan[1], Mohammad Nafees[2], Amjad ur Rahman[3] and Tooba Saeed[4]
[1]Department of Environmental Sciences, University of Swabi, Swabi, Pakistan, [2]Department of Environmental Sciences, University of Peshawar, Peshawar, Pakistan, [3]Department of Botany, University of Swabi, Swabi, Pakistan, [4]National Centre of Excellence in Physical Chemistry, University of Peshawar, Peshawar, Pakistan

List of abbreviations

LiDs wheel	The life cycle design wheel
UN	United Nations
UNEP	United Nations Environmental Program
WCED	World Commission on Environment and Development
IWRM	Integrated water resource management
INRM	Integrated natural resource management
ICM	Integrated catchment management
IRBM	Integrated river basin management
RS	Remote sensing
GIS	Geographic information system
GPS	Global positioning system
OECD	Organization for economic cooperation and development
HIV/AIDS	Human immunodeficiency virus/acquired immunodeficiency syndrome

22.1 Introduction

History revealed that in the mid of the 20th century, our planet came into being for the first time. It was a small fragile ball subjected not by humans or any organization, but by a configuration of clouds, oceans, soils, and greenery. While with the changing pattern of the planet, humans were unable to fit their activities into that specific pattern. With time, many such changes are occupied by life-threatening exposures. The truth from which there is no escape must be managed and recognized as our Earth (Morelli, 2011). This earth gave life to the living beings by providing the life-supporting services necessary for

Frontiers in Plant–Soil Interaction. DOI: https://doi.org/10.1016/B978-0-323-90943-3.00019-5

survival. But the human involvement rapid increase of population reduced the availability of the resources. To manage these resources and maintain their availability for the future, the concept of "Sustainable Development" was proposed. The idea of sustainable development was developed in 1970 and was explained in the book "Limits to Growth." Later on the concept gained importance with the publication of the Bruntland report (1987), an initiative of the UN World Commission on Environment and Development.

The term sustainable development has two components, sustainability and development. Sustainability is the capacity for continuance into the future (Robinson, 2004). Sustainable development is the interaction between three systems as social, economic, and environment. This concept arose from increasing the perception that human activities have disturbed the global equilibrium and cannot be sustained forever. The term environmental sustainability has been regarded as the upkeep of "natural capital" and connects both economic and social sustainability in the environment. Environmental sustainability is closely related to ecological or agricultural sustainability. The concept of ecological sustainability was developed by Callicott and Mumford (1997) as a useful concept for conservation that connects ecosystem services and human needs. Humans have intensively changed and modified the ecosystems in the form of agricultural lands and urban areas. The term sustainability maintains the equilibrium between human needs and ecosystem services due to human activities. Economic sustainability comprises analysis to reduce the social expenditures of meeting standards for the protection of environmental resources. It involves the flow of material, energy, and resources in a sustainable way (Rothenberg, 2007). While social sustainability states the positive condition of human communities to meet their needs without affecting their livelihood.

The concept of environmental sustainability is a broad term which includes "the condition of resilience, balance, and connection to allow society (humans) to meet its requirements without surpassing the capacity of its supporting ecosystem to keep continue the regeneration of the services without harming the biological diversity" (Vos, 2007). Many countries of the world are adopting the concept of sustainable development to manage ecological resources, biodiversity, and environmental preservation. Incompetent land use, water, soil, energy, and improper crop pattern and domestic animals are the major obstacles to sustainable development in improving the environment and living standards. The human population, energy, and environment are the factors that can affect sustainability. The majority of the global population lives in poverty and are lacking the basic needs of food, water, safety, shelter, education, and health (Xue, 1997). Sustainability and biodiversity evoke beneficial associations. But both are at greater risk due to human activities. Sustainable production and ecological product innovations are important for long-term financial success. The scientists suggest the ecological definition of sustainability is "to meet the human necessities without conceding the health of the ecosystem." Collectively sustainability is an approach toward the conservation of ecological services, such as air, water, soil, biodiversity, and wildlife (Callicott & Mumford, 1997).

22.2 Ecological design

The term ecodesign or ecological design is referred to as any form of design that minimizes the adverse environmental impacts by integrating with natural ecosystems. Ecodesign tries to find a background for environmental systems of appropriate design and management by keeping both ecological and anthropogenic values (Shu-Yang, Freedman, & Cote, 2004). The concept of ecodesign has an ancient history and can be articulated by different means. For example, the old traditional agricultural patterns where crops are grown with the cultivation of various species in a diverse mixture allow a stable agroecosystem with a probable yield. In China the traditional farming practices of small plot farming have a long practice of diverse patterns of crop rotation while using a good quality compost. This pattern of farming is called as "organic agriculture." The American farmers prefer to grow a staple crop (beans, maize, *Zea mays, Phaseolus vulgaris, Cucurbita pepo*) known as "three sisters" in an integrated system of organic agriculture. Besides agriculture, the concept of ecodesign is likewise adopted in the construction of numerous traditional buildings by using the local material and these are designed to obtain the optimality of cooling and heating. Such buildings are constructed with adobe and living spaces are excavated from soft rocks (Van der Ryn & Cowan, 1995). There are seven key principles of ecodesign as given below:

1. To sustain the consistency of the structure and function of ecosystems.
2. The need to meet the basic necessities of humans with economic development.
3. The appropriate designs of natural systems in anthropogenic management systems.
4. A progress toward sustainable economy through renewable resources while keeping focus on reuse, recycling, and efficient use of energy and materials.
5. To address the environmental issues, that is, depletion of natural resources by adopting the practice of ecological economics.
6. Focus on the conservation of biodiversity and natural ecosystems.
7. To build social support for the sustainability of resources by promoting environmental literacy.

By implementing the ecodesign attributes to the management and formation communities of humans, architecture, industrialized networks, and products contribute noticeable progress toward sustainable human economy. The field of ecological designing was promoted by McHarg (1967) and is also known as ecological planning. These designs integrate components intended to give space for native species, their habitations, and ecosystems.

22.2.1 Applications of ecological design

The abovementioned doctrines of ecodesign can be useful within a diversity of spatial scales ranging from individual households to neighborhood, industrialized parks, and particular products.

22.2.1.1 *Urban planning*

The idea of ecodesign can be used by improving the present urban communities and to plan for new areas. While to improve the existing urban areas, it is important to identify the environmental issues, problems with indigenous biodiversity, and use of energy and material. Such designs focus to mitigate these problems and to neutralize the natural and urban ecosystems.

22.2.1.2 *Industrial networks*

Industrial networks are also called industrial parks. These are complex structures of industries located in an area away from human settlements. The development of environmental synergies among industrial enterprises is one of the key aspects of ecodesign. This planning involves the interconnection of web of industries for sharing of resources to form a competent and interdependent system. In these systems, the waste of one firm is utilized by another one as a raw material. In this way, wastes have value to other kinds of production permitting them to be recycled and reused as resources, and getting the improvement of the zero-emission idea, leading to minimizing of the environmental pollution problems (Cote & Cohen-Rosenthal, 1998).

22.2.1.3 *Architecture*

It is important to pay more attention to the resource and energy efficiency during the operation and construction of buildings. This can be possible by adopting land use planning or minimize the land use, save heat, energy, and minimize the pollutant emission. Currently, the practice of "Green Building" is adopting by some countries using environment-friendly materials. Energy can be conserved by using solar technologies and pollution can be controlled by planting trees (Cote & Crawford, 2003).

22.2.1.4 *Products*

Several products can be manufactured by applying the ecodesign practice. The practice is extensively implemented in Western countries. The progressive increase in human population and industrialization has raised various environmental concerns with the associated burden on the consumption of resources for product manufacturing. In this regard, ecodesign has been recognized as a tool to minimize the effects associated with the production and consumption of products. For example, the life cycle design Wheel compares the existing product to an environmentally improved one in terms of reducing material use, choosing low impact materials, improving durability, recovery, production, and distribution system. Similarly, Xerox Corporation has set its product goals as "zero waste" and established an Assets Recovery program as it leases photocopiers for use with replacement with new. The Interface Corporation is working under the concept of a "green lease" program in which it leases modular carpets to customers and substituting worn modules when necessary. Recovered carpets are reimbursed to the company for recycling and converted into new carpets (Shu-Yang et al, 2004).

22.3 Soil conservation and restoration

Soil is one of the significant natural resources playing a vital role in sustaining the climate systems and multifaceted terrestrial ecosystems of the earth. Humans are also dependent on soil for food production, fiber, feed, raw materials, and renewable energy. But a progressive increase of the human population poses negative impacts on soil resources. Approximately 11% of the world's land is used for agriculture and livestock (Fig. 22.1).

Data show that about 8.7 billion ha of land is available to feed 5.9 billion people globally (Jie, Jing-Zhang, Man-ZhL, & Zi-Tong, 2002). Soil is facing the degradation of various types due to intensive agricultural practices and land overuse. Nowadays, soil degradation has become a major agricultural and environmental concern. The problem of soil degradation is resulting in the considerably increasing challenge to feed an increasing population from a declining land area of poorer quality, thus causing food insecurity and low

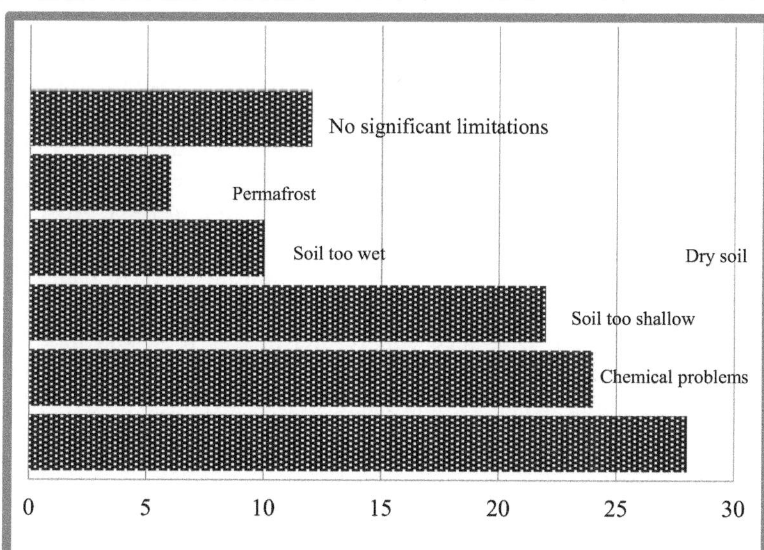

FIGURE 22.1 Percentage of Soil cover (only 11% of worlds land is farmed).

agricultural incomes. Soil degradation also leads to several other problems such as climate change, sedimentation, poor watershed system, and loss of biodiversity and genetic stock (due to changes in natural habitats) (Oldeman, 2000).

22.3.1 Soil degradation, a global issue

In the past 50 years, the problem of soil degradation has accelerated considerably. Around the world, about 2 billion hectares of soil resources have been degraded, including 22% of forests, pastures, croplands, and woodland. The rate of soil loss has continued progressively with a further loss of 5–6 billion ha since 1990. This rate is still in progress and has been estimated to be up to 200 million ha for the upcoming 20 years.

Soil loses its important nutrients due to the erosion process, making the soil infertile and unfit to obtain a good yield (Roose, 2008). Erosion and active mineralization of organic matter have been considered as important factors of soil degradation. About two thirds of the human population lives in developing countries where food production does not meet human requirements. This ratio can be increased by extending the cultivable areas, introducing good yielding varieties, and advancing technology in the field of agriculture. Moreover, improper cropping patterns, nonscientific agricultural techniques, unsuitable use of machines, improper irrigation and drainage systems, and misuse of natural resources have been considered as the most serious problems in the way of agricultural development. Another problem of low agricultural food is soil erosion which leads to soil degradation and the loss of soil fertility. Soil degradation is caused by a variety of factors including erosion as the most dominant type. About 50% of the world's land is degraded by erosion. About 70% of total erosion is accelerated by water erosion. Wind erosion and desertification are next to water erosion in participating soil degradation. About 10 million ha of productive land vanishes every year due to desertification. The intensity of soil loss due to desertification is high in Asia, Africa, and South America. The trend of continent-wide soil degradation is recorded as Asia > Africa > America > Oceania. These continents have had 73% of rangelands, 30% irrigated, and 47% of rain-fed drylands affected, respectively (Poulsen, 2001). Degraded soil results in less food and more cost. To increase the total global food production, an increase in use of fertilizers, higher cropping densities, and expansion of irrigated lands will lead to soil degradation on crop yields (WRI, 2001). The soil degradation has reduced the crop production by 17% globally. Literature has revealed that in sub-Saharan Africa, Ethiopia has the highest rate of soil loss and nutrient-deficient soil through erosion. Erosion such as sheet, rill, and gully erosions are common in high rainfall areas, where rain carries away the fine soil particles and organic content from fertile lands. Other causes of soil degradation are the progressive increase in population growth, deforestation, livestock production, and unbalanced crops. A great deal of attention is required to promote sustainable soil quality and productivity.

Earth was created first while mountains have been shaped by the erosion of soil. Man has tried to prevent this land erosion for 7000 years. Therefore the rural population has adopted various techniques to stop the soil and water resources and to improve soil fertility. In the past, the soil conservation was associated with the restoration of lands or the prevention of erosion. In the modern world, the focus is on the preservation of natural resources. The developing countries should be educated and trained about the conservation methods and expansion of safe and productive agricultural areas (Thampapillai & Anderson, 1994). The management practices of soil such as crop rotation can improve soil health. In this technique, different sequences of crops are cultivated on the same plot of land. This can improve the soil fertility by reducing soil erosion and control weeds, pests, and diseases (Degu, Melese, & Tena, 2019). Physical soil and water conservation structures are designed to intercept and minimize runoff intensity, to store runoff water, trap sediments and nutrients, to form natural terraces, stop flooding of nearby lands, and lower the sedimentation of water channels. Regular monitoring of soil is important for its physical, chemical, and biological characteristics (Degu et al., 2019).

The terrestrial land surface has soil that provides essential services to humans. The total land use of the Earth's surface as arable farmland is about 15 million km^2. Furthermore, 30 million km^2 of land is used for grazing. Humans use the lands to get food and support necessary livelihoods (Ramankutty, Evan, Monfreda, & Foley, 2008). Soil is a supporting system of humanity's global life as it is host to a diversity of fauna, water, and organic contents. Due to human activities and modern agricultural practices, soils are under intensive pressure with the loss of organic carbon, compaction, salinization, and erosion (Montgomery, 2007). Soils, after the removal of natural vegetation, are often disturbed by tillage, thus accelerating the erosion rates by wind and water. Soil erosion threatens agricultural land. The literature has revealed that France and Iceland faced environmental degradation by intensive soil erosion where the hillslopes have ruined the livelihoods of entire communities due to excessive water and wind erosion. The key contributing practices behind soil degradation were identified as deforestation and grazing. In response to this situation, France started a broad program in 1860, and Iceland started the first soil conservation service of the world in 1907 to recover the losses (Govers, Merckx, Wesemael, & Oost, 2017).

22.3.2 Soil conservation technology

Over the last decades, soil conservation technology tools have been effectively working to reduce soil erosion rates. Soil conservation tillage is a tool to use crop residues for soil conservation to enhance soil quality. This is a sustainable use of a soil technique for mitigating the greenhouse effect and improving soil quality. Another definition of conservation tillage is "any tillage and planting system that maintains at least 30% of soil covered with residues after planting to reduce water or wind erosion," according to Conservation Tillage Information Center (Lal, 1997). Conservation tillage tool is in practice in many areas especially in the United States, Europe, and the Pacific. The practice of conservation tillage has shown effective results for the reduction of water erosion by up to 1–2 orders of magnitude. This technique also helped in controlling the rate of wind erosion as residue cover keeps the soil wet and thus improves its shear strength (Montgomery, 2007). But conservation is not too very effective in controlling gulley erosion. In such cases, infrastructure measures (terrace building, stone bunds) and vegetation measures in combination with crop rotation and modification in tillage practices can be helpful to reduce the soil erosion rate (Nyssen et al., 2009).

To lower the grazing pressure, the strategies of land restoration with effective and sustainable vegetation cover and reforestation can be adopted. The process of erosion is not limited to agricultural lands, but also infrastructure related to these activities like roads and field boundaries. For this, the necessary technological solution is to improve the water drainage system, check dams, implement field buffer zones, and effective landscape organization to minimize the sediment production on roads and in built up areas (Rijsdijk, Bruijnzeel, & Sutoto, 2007).

22.3.3 Why soil conservation is not generally more adopted?

Soil conservation is important and is needed today to protect the loss of this natural resource and improve agricultural production. But there is a lack of implementation of conservation of technology as the soil restoration and conservation measures are not in practice on most agricultural lands. Several factors contribute to the failure of conservation techniques, such as scientific knowledge, poverty, technical institutions, human attitude, untrained farmers, farm size, and work organization. Changing farming practices must be inherently difficult due to such factors. This problem is not common to all countries and the practice of conservation tillage is slow due to a multitude of factors (Lahmar, 2010). Farming systems rely on well-tried technology, labor, and crop types and are therefore difficult to change. But on the other side of the picture, there are some countries where farming systems are changing with the strong adoption of conservation technology. In Brazil and Argentina the tillage practice has been adopted rapidly and is successful in controlling soil erosion problems. Therefore the adoption of conservation technology is economically beneficial to farmers (Peiretti & Dumanski, 2014); the soil quality shows its potential for productivity and environmental quality. Soil quality is affected by tillage methods and crop residue management. Therefore adoption of a sustainable management system is important

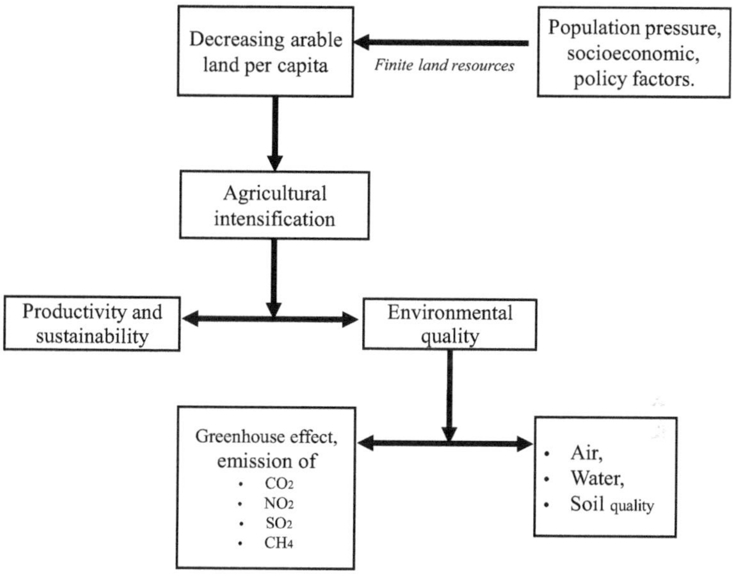

FIGURE 22.2 Land degradation, it effects on agriculture, environment and soil quality.

to reduce the adverse impacts of soil degradation and improve soil quality and resilience, as shown in Fig. 22.2 (Lal, 1997).

22.4 Watershed management toward sustainability

Water is an essential input in agrarian livelihood and agricultural growth. It is one of the basic needs of life, therefore its importance cannot be ignored. But some countries are facing the problem of water scarcity and have inadequate water resources. This is because of poor governmental policies and ineffective institutions, especially in developing and poor countries (Falkenmark & Rockström, 2006). In the situation of growing water stress, it is important to improve the management of water resources to secure and maximize the benefits from water to users. For this purpose, innovative methods are needed to manage the land and water with social consideration and ecological functions to sustain such developments (Grey & Sadoff, 2004).

The watershed is a topographically defined zone that is drained by a stream or river. Or it is a drainage basin or a catchment area which has a ridgeline on three sides and a drainage point. Watershed management is a tool of managing watershed resources to get maximum benefits without affecting the ecological sustainability. It is the process of organizing water, land, and other natural resources to provide goods and services to alleviate the impacts on soil and watershed resources (Sudha, 2007). This approach is known as a holistic approach as it integrates various components such as water and soil conservation, agriculture, livestock, and forestry development. It tries to form a balance between natural resources and the environment to humans and other living organisms. Watershed management constitutes human, institutional, socioeconomic, and biophysical linkages among water, soil, and land use which connect uplands and downstream parts (Wang et al., 2016). Therefore it is also called the ever-evolving practice that keeps comprising the land, water, and biota management in an area for economic, social, and ecological purposes. Most of the researchers and experts consider watershed management as an important tool toward conservation, preservation, and sustainability. It is effective for the mitigation of erosion, provides safe water, a stabilizing landscape, and improves the agrarian production system. The success of watershed management depends upon the will of the people and the level of activities involved in it. Therefore people participation is an important core component of watershed management (Tennyson & Achouri, 2004).

22.4.1 History of watershed management

The approach of watershed management dates back to 2000 BCE and it has regularly improved with time. The Atharva Veda text from 800 BCE shows watershed management by stating the appropriate measures about water usage and its conservation. Benjamin Franklin recognized the importance of watershed management in 1970 in the West (Kumar et al., 2014). Watershed management is aimed at sustainable dissemination of its resources and to enhance the effectiveness of watershed function. Through the development of watershed management, the concept of integrated watershed management (IWM) has become more pronounced as it integrates several technical, social, institutional, and economic dimensions (German et al., 2007). The approach of watershed management was properly defined by the late 20th century when rapid population growth resulted in a shortage of water, land, and other natural resources. The progressive shrinkage of the Aral Sea basin illustrated the importance of watershed management furthermore. Many countries of the world are adopting this management approach within the framework of natural and human systems. The UN Food and Agriculture Organization and World Bank also use the watershed management approach to measure the ecological benefits of developmental projects and to maximize the availability of goods and services in a sustainable way (Sha et al., 2014). Watershed management states that societies are affected by water interaction with additional resources and thus people can affect the nature and extent of those linkages (Manuelli, Hofer, & Vita, 2014).

22.4.2 Land water and livelihood challenges

Sub-Saharan Africa is facing the developmental challenges of water scarcity, poverty, low agricultural production, poor health facilities, and poor water management. This is the poorest region in the world where the situation is getting worse day by day. Literature has revealed that 70% of its population is rural and 33% of people are undernourished. Due to lacking safe drinking water and proper sanitation, health problems such as malaria, diarrhea, HIV/AIDS, and child malnutrition are of major concern. About 77% of the population lives in 63 transboundary river basins which imply high water interdependence. Therefore integrated planning and management of the river basins are important (Bahri et al., 2011). Agriculture is the backbone of the African economy, providing 60% of employment. Unfortunately, agricultural growth and per capita food production has fallen over the past 40 years. This is associated with agricultural vulnerable to variability in rainfall and mismanagement of water resources. To cope with all these problems, some ideal ideas were adopted. In Africa, the most complex and extensive transformation of river flows at the international level were the Nile (Conway, 2000). Furthermore, other countries (Ghana, Nigeria, Egypt) established national and international river basin authorities in 1960 to cover the equitable balance between urban and rural economic development. The main focus was the construction of irrigation schemes and dams. In the mid-1990s water management was considered important to keep water management in all river basins or watersheds to study the associated hydrologic, socioeconomic, and environmental aspects of water management at multiple scales. This management also helped to keep a balance between resource use and its conservation (Ashton, 1999). The concepts of watershed management and river basin evolved parallel. Watershed management is a holistic approach aiming to stop land degradation, to improve the management of agriculture and natural resources, and to secure downstream water-related services and ecosystems (Falkenmark & Rockstrom, 2008). In the 1990s the practice of participatory IWM and concepts of Integrated Water Resource Management (IWRM), Integrated Natural Resource Management (INRM), IWM, Integrated Catchment Management (ICM), and Integrated River Basin Management (IRBM) were introduced for sustainable development of water and land resources (German et al., 2007). All these concepts have some common features as they are integrated, holistic, and participatory and based on hydrological, biogeophysical units with a principal goal of improving and managing water resources. Unfortunately, the implementation of these practices is still a challenge due to institutional arrangements and the need for cooperation across institutions (Bahri et al., 2011).

22.4.3 Tools for watershed management

Physical changes in a watershed can lead to various responses including both short-term and long-term events. The responses are land-sliding, flood, point and nonpoint source pollution, soil and water degradation, etc. To know about all these responses, a massive array of spatial and temporal data is required by adopting the following techniques.

22.4.3.1 Remote sensing

This technique derives the required information and images about the land and water surface by employing electromagnetic radiation emitted or reflected from the Earth's surface (Brooks, Ffolliott, & Magner, 2013). Such data provides a resource of identifying, surveying, and monitoring different characteristics inside a watershed.

22.4.3.2 Geographic information system

This technique is used to assess watershed conditions through modeling and visualizing impacts of human activities and therefore is considered a critical tool for watershed management (Singh & Woolhiser, 2002). This approach enables watersheds to be divided into more discrete units in vertical and horizontal patterns to enhance the model outputs. Geographic information system can be used in combination with remote sensing to provide spatial input data for watershed management models. While in combination with the internet, it increases data accessibility (Choi, Engel, & Farnsworth, 2005).

22.4.3.3 Global positioning system

Global positioning system (GPS) is an easily accessible space-based global navigation system that provides information about location and time for almost anywhere on Earth (Brooks et al., 2013). GPS in combination with radiotracking remains valuable to track animal movements within a watershed.

22.4.3.4 Internet

Information, tools, and data regarding watersheds are easily available through the Internet. The practice of Internet use is common everywhere. Several online tools are available on watershed management and planning.

22.4.3.5 Big data

This tool is used to define the exponential growth and data availability, both unstructured and structured. It is used to find velocity, variability, volume, variety, and complexity. This data helps to reduce cost and time and thus large sizes of data can deliver accurate analysis that will lead to firm decision making (Wang et al, 2016).

22.4.3.6 Metaanalysis technique

This is s statistical tool for assimilating quantitative findings from various studies and then distilling data into comprehensive conclusions. This will give statistical final data and the facts can be helpful for policy and decision makers (Locatelli & Vignola, 2009).

22.4.3.7 Multilevel social–ecological system analysis

This analysis is required for operative watershed management and involves the accountable government agencies, partnerships, and coordination among institutions and stakeholders. In analytical models, there have been numerous advances for watershed management. These advances are watershed simulation modeling, watershed decision-making systems, system dynamic modeling, stakeholder analysis modeling, etc. (Mutekanga, Kessler, Leber, & Visser, 2013).

22.4.3.8 Sustainable practices in watershed management

Sustainable development is possible when the needs of people and the capacity of natural resources meet their needs are balanced with time. The concept of watershed management provides an appropriate base toward this balance. A set of complex developmental problems involving the use of natural resources can be managed by adopting the practice of watershed management. It also offers a universal solution to the conflicts arise from environmental safety and economic development. Watershed management concepts can be introduced effectively by conducting trainings by policy makers, managers, experts, farmers, and land users. In this regard, international development organizations have to take the initiative in developing and supporting the state-of-the-art training programs to provide a background to all those who will integrate watershed management practices in developmental work (Conable & Repetto, 1988). There are certain sustainable models of watershed management adopted successfully in many parts of the world. Five models are presented here include the following:

1. *Ecological Economic Model*: the effective utilization of environmental services is closely associated with common pool resources by local stakeholders to get maximum benefits without harming the natural balance of the ecosystem. In this regard, the Ecological Economic Model was practiced in the Tonameca Catchment area of Mexico with active public participation and the use of environmental services. It helped to get better employment through ecotourism, fishing, and development of forestry. So, they utilized the environmental goods and services to improve the livelihood of the coastal inhabitants of Mexico.
2. *The United Nations Environmental Program Funded Watershed Project*: this project was implemented in the Dominican Republic with the active participation of the local community. This modern technique helped the conservation of biodiversity and eradication of industrial pollutants of watersheds. It helped to mitigate the environmental and biological diversity of the zone with the welfare of the locality.
3. *Magat Watershed Project*: the project was successfully implemented in the Philippines combined with social, financial, political, and technical

capital to enhance water management, soil conservation, forest conservation, and land management of the area.

4. *IWRM*: the IWRM approach was practiced in Burkino Faso for the first time. The approach was participatory and the results were effective to resolve the water-related issues and conflicts of the area.
5. *Wuhan Watershed Project*: this participatory approach was implemented by the Chinese government in the city of Wuhan, China. The government and public policy intervention toward the development of integrated watershed showed significant improvements in the economy, agriculture, and land use pattern of the region.

The above project studies revealed that public participation is important and is required by watershed management approaches. The success of such practices depends upon the local knowledge about indigenous soil taxonomy, soil and water conservation, biodiversity, and traditional experimental ethics of the locality. Watershed management is based on the concept that various environmental problems are best addressed at the watershed level because the specific natures of problems need complex policy solutions that come from experts and authorities of various sectors at government level (Sudha, 2007).

22.5 Ecological modeling and sustainability

Ecological modeling has become a necessary tool for environmental systems. Model is an illustration of any specific thing, condition or idea. A model can be simple or complex about a particular field composed of graphical and mathematical descriptions such as nitrogen transformation in the ecological system. Ecological modeling is a series of stages taken to convert an idea into a conceptual model and formerly into a quantitative model. Because the ecologists first get the hypothesis, collect fresh data, and then based on the findings they undergo many variations from the initial concept to what is supposed to be the end product (Jackson, Trebitz, & Cottingham, 2000). The rapid increase of the global population and urbanization has become a challenging issue of this era. Overpopulation is putting significant pressure on environmental resources and is threatening sustainability. In return, sustainability has become one of the challenges to the global population due to the overconsumption of resources. To address these changes, there is a dire need for new developmental approaches for resource saving and synergy (Kharrazi & Masaru, 2012). The ecological planning approach has long been recognized as an expedient technique. The ecological approach to environmental information was discussed by Rapport and Regier (1980). This concept includes classifications of ecosystems, stresses on the ecosystems, identification of ecological indicators, and impact of the stresses. They also suggested the geocoding for this comprehensive database. The technique of the ecosystem was further defined by Vallentyne and Beeton (1988) as an integrated set of policies and management practices that must be developed to integrate people into the ecosystems. The concept of the ecosystem approach and its application to planning and managing human impact on the environment was further developed by a report "The Royal Commission on the future of the Toronto Waterfront" by Crombie (1990). However, the human participation, activities, and planning for modification of an ecosystem can affect the ecological functions and services. The scientist's predictions regarding the impacts of human disturbances in ecological systems have been steadily increasing over the years. But there is a lot to be learned about the complex networking of ecological interactions and dynamics. The potential of policy makers and planners to accommodate the ecological dynamics is less developed than the ability of scientists to predict and understand such interactions. Hence, more work can be done with the present ecological modeling and new approaches toward sustainability and ecological responses to human activities (David & Tufgar, 1991).

22.5.1 The conceptual model

In research, the development of a conceptual model is an important part. Such models are generally designed in the form of diagrams represented with arrows and boxes to give a clear picture of the research statement. The boxes show the state variables for the condition ecosystem components, while the arrows represent the ecological interactions among state variables for material flow (Cale, 1995). The conceptual model can be developed after the formulation of a general research question. The model is based on secondary data or consultation with ecologists who have studied the relationships among species or systems. Fig. 22.3 shows the relationship between various strategies for stocking exotic salmon in the Great Lakes and the toxic contaminants in the salmon and their alewife prey. The boxes in the model show alewife and Chinook salmon and the arrows represent stocking. Fig. 22.3A shows sample arrows for the flow of biomass and contaminants between two species. Fig. 22.3B refers to the interactions among four species based on available data. To reflect the ecological features more accurately, size structure might be added to the model when the predators need a specific size of prey as shown in Fig. 22.3C (Jackson et al, 2000).

22.5.2 Quantitative ecological model

This model is a set of mathematical expressions in which data and coefficients are attached to the boxes and arrows of the conceptual model. Quantitative models are used by ecologists for different purposes, such as formulation predictions, explaining the existing data and guiding research. Ecologists usually used the quantitative models to formulate predictions

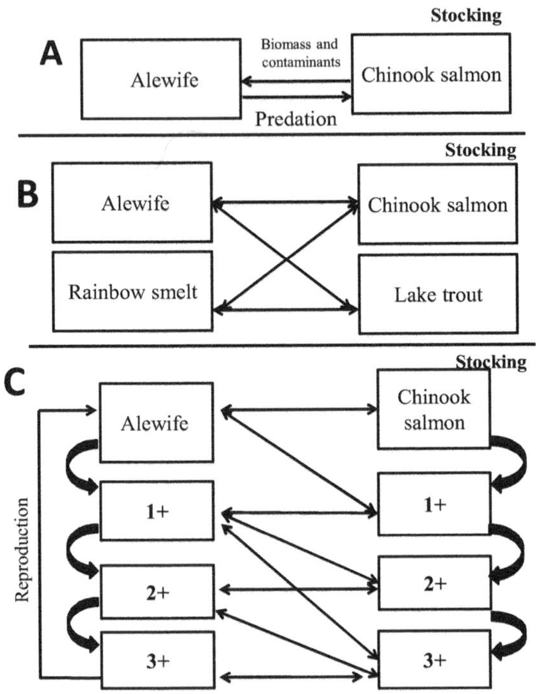

FIGURE 22.3 Conceptual model of interactive nature of building (an example).

about the systems they study (Weiner, 1995). The simple forms of quantitative models can be easily solved with pencil and paper but most of the ecological models are computer based. Some predictive models are empirical which state the interactions determined by data. Empirical models are not most valuable within the bounds of data. An empirical model of the aquatic ecology will show the level of summer chlorophyll from spring total phosphorus. Quantitative models elaborate the research in several ways which can help in designing the experiments (Hilborn & Mangel, 1997). These models can even help to develop surrogate data to assess the potential sampling schemes and to check potential environmental indicators. The most important quantitative ecological models translate an ecological hypothesis into predictions based on new existing data. Suppose, an ecologist will use three equations when he wants to evaluate how predator consumption is different as a function of prey availability, such as:

$$C = aP \tag{22.1}$$

$$C = bP \tag{22.2}$$

$$\begin{gathered} C = 1 + cP \\ dPe - fP \end{gathered} \tag{22.3}$$

where C is the consumption rate of the predator, P is the measure of prey availability, a, b, c, f are the coefficients, and e is the base of natural logarithms. Equation (22.1) shows that consumption increases as a linear function of the prey availability. Equation (22.2) shows that the consumption rate increases at low prey availability and gets saturated at high prey availability. While Eq. (22.3) states that consumption increases at low prey availability and decreases at high prey availability (Jackson et al, 2000).

In ecology, there are some general models, such as energy and mass balance models (e.g., Hewett, 1989), individual-based population models (e.g., DeAngelis & Gross, 1992), optimization and game theory models (e.g., Mangel & Clark, 1988), population genetics models (e.g., Roughgarden, 1979), community and ecosystem models (e.g., Scavia & Robertson, 1980), landscape models (e.g., Baker, 1989), and size/age-structured population models (e.g., Caswell, 1976).

22.5.3 Urban ecological modeling

Both marine and terrestrial biodiversity provide the basis for ecosystem services that support human well-being. But biodiversity and ecosystem services have been receiving alarming threats for a long time. Therefore the sustainability of biodiversity needs to maintain a balance in ecological conservation. To maintain the biological system, energy and matter input and output can be quantified. In the urban ecosystem, most of the organisms are heterotrophic. Such systems require a highly diverse source of energy. In urban ecological modeling, the heterotrophic nature of the system and the flows of matter and energy pose a difficult challenge (Jiang et al., 2009). For this purpose, the modeling technique of Ecological Network Analysis focused on the transfer of energy flow, material flow, and sharing information between different nodes in a system. This technique allows the modeling of complex ecosystems from a system perspective to direct resources efficiently and maintain a reserve capacity to address the disruptions. Climate change and improper use of land planning have induced significant impacts on ecological biodiversity at the global level. These changes also put consequences on human societies with significant threats and risks to the economy and economic services (Butchart, 2010). It needs to be managed sustainably. In this regard, the concept of ecological economics can be helpful. The discipline of ecological economics sustains the control of environmental–economic systems by designing and analyzing decision plans, public policies, and scenarios for biodiversity and ecosystem services (Thebaud et al., 2013). It focuses on the main features, strengths, and challenges of ecological–economic approaches. The benefit of the ecological–economic model is to contribute to clear enactments of dynamic interactions between socioeconomic and ecological systems. It also helps in adopting the concept of sustainable development for conflicting targets through constrains. Scientists place emphasize on the balance between efficiency and resiliency in a network system. By doing so, the sustainability of different urban infrastructure networks can be easily assessed (Zorach & Ulanowicz, 2003).

22.5.4 Waste to energy network model

Overpopulation across the world has led to a shortage of land and problems with waste generation. People generally oppose the construction of landfills and countries are increasingly taking into consideration waste incineration technologies. Such techniques have hazardous environmental impacts by affecting the air quality. To eradicate the waste and its environmental impacts, the incineration technique generates heat and electricity to be used as a resource flow (Liu, Yang, & Chen, 2010). On the other hand, incineration requires a certain amount of waste as input and produces greenhouse gases into the atmosphere. To maintain the ecological waste system and energy system, the concept of sustainable development can be employed as an evaluatory approach toward waste–energy systems. The integration of energy and waste disposal systems are helpful in incineration technology to generate electricity (Ulanowicz, Goerner, Lietaer, & Gomez, 2009). A conceptual ecological model of two integrated network is presented to evaluate the waste–energy systems in an area. The model is composed of eight different sectors. These sectors include household, industrial, agricultural, landfills, incinerators, electricity generation plants, recycling centers, and the environment. The last sector (the environment) represents the area with material energy for processing and also gets unconvertable waste and emissions from the rest of the sectors inside the network. All these sectors are interconnected to each other through multiple pathways in which waste flow and energy exchanges occur, representing the waste into the energy system. In such a system, the waste of one sector becomes the input of another and the process is continuing inside the network by analyzing a balance between network efficiency and resilience. In this model, there are 28 different pathways for waste sharing among eight sectors (Fig. 22.4). Such types of ecological models can be applied inside a city to maintain an advanced waste–energy network (Kharrazi & Masaru, 2012).

22.5.5 Industrial symbiosis an ecological model of sustainability

Industry interacts with the environment in different ways. By considering the industry as a black box with inputs and outputs, the most understandable interactions will have marked themselves as raw materials extracted from and waste that is inserted into the environment (Hemmer, 2011). The interaction of industry with the environment can be studied in various ways, including industry ecology. Industrial ecology maintains sustainability by viewing the industrial system and its interaction with the environment

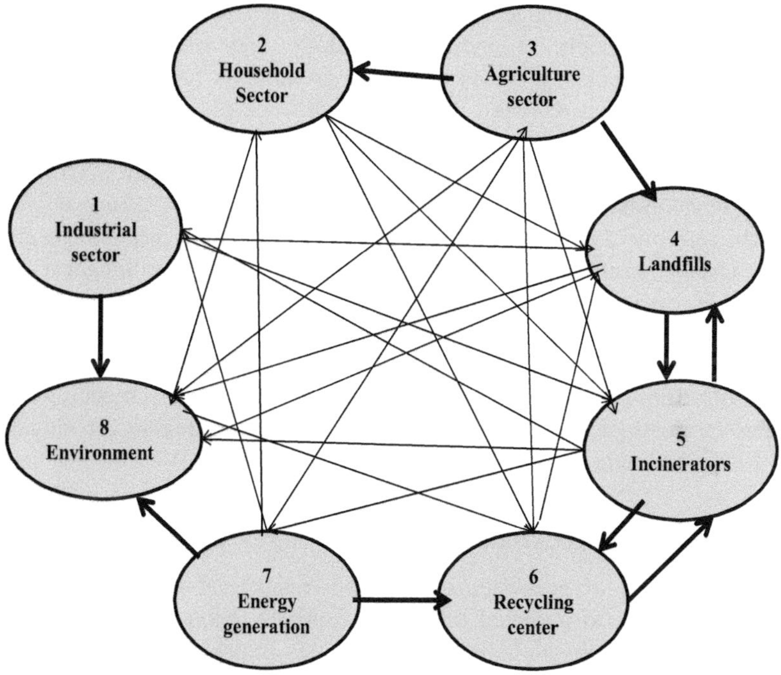

FIGURE 22.4 Ecological model of waste to energy system.

(Watkins, Husgafvel, Pajunen, Dahl, & Heindustrial, 2013). The concept of industrial symbiosis was proposed by Frosh and Gallopoulos in 1989 in an article entitled "Strategies for Manufacturing." They presented the industrial symbiosis as an important industrial ecological component, where waste of an industry can be utilized in another as a resource (Liu et al, 2011). Therefore industrial symbiosis has been considered as a concrete tool of industrial ecology which relates the reuse of waste and by-products systematically and minimizes the demand for fresh material to save the environmental services. The practice of industrial symbiosis remained effective for resource conservation and recycling. (Wu, Li, & Yang, 2015). Energy recovering from waste is an important and useful policy for waste integration and energy management in an ecoindustrial park. It helps to reduce waste, energy use, and business cost. These strategies help in achieving the complex and effective industrial systems (Dong, Satoshi, Fujita, Geng, & Dong, 2014).

Many countries of the world are adopting the practice of industrial symbiosis to mitigate pollution and save resources. For example, waste generated by a sugar mill is used as an input in the manufacturing of alcohol, cement, fertilizer, and paper pulp. Moreover, the effluents of the paper industry consisted of cellulose and were used as a resource in the paper molding industry. Environmental sustainability is responsible for long-term consideration of reducing the negative impacts of industries and it can be made possible by applying the concepts of industrial ecology and industrial symbiosis. Reducing the impacts on the environment promotes a better environmental performance (Persson & Ivner, 2011). This practice plays an important role in association amongst organizations as it cannot be adopted individually. It is driven by the community initiatives by facilitating the multiple mechanisms to develop the beneficial connections among partners. Industrial symbiosis can improve the advantage of industry by reducing costs (Leigh & Li, 2008). Under the concept of industrial symbiosis, waste is no longer a waste when it is marketable as a useful environment-friendly product (Puente, Arozamena, & Evans, 2015). Industrial symbiosis defines the shearing of services and waste among industries to minimize cost and improve the environmental conditions. Industrial symbiosis is a subset of industrial ecology with a focus on material and energy exchange, which are valuable for industries and the environment (Simboli, Taddeo, & Morgante, 2014).

22.5.6 Bioeconomic challenges

The management of ecosystems and biodiversity and ecological services remain inadequate at a global level. For example, about two thirds of fish stocks need to rebuild globally. Therefore it is important to emphasize the existing ecological–economic systems and to develop new methods to meet new challenges (Gutierrez, Hilborn, & Defeo, 2011). To activate the ecosystem-related approaches, the major complexities of systems should be considered to elucidate the use of ecosystem services, and biodiversity indicators. The decision process, risk management, and governance issues for public policies and developmental methods for integration are other challenges in ecological–economics that need to be addressed (Larkin, Alvarez, Sylvia, & Harte, 2011; Worm, 2009). The researchers believe that the present limitations of public policies for ecological and biodiversity management can be tackled by an inadequate account for complexity. Such as for fisheries management, an ecosystem-based tool is applied. Although, it is a tough job to start from monospecies toward multispecies, communities, trophic web management, and explicit models. In this regard, graph and network theory might play an important part in the study of the food webs and ecosystem management (Doyen et al, 2013).

22.6 Strategies and approaches toward environmental sustainability

Environmental sustainability is the recognition of linkages between humans to the ecosystems for life-supporting services. In return, it expresses the human's impacts on ecosystems. The concept of sustainability is increasingly disgraced as a useful practice for the conservation of natural resources, biodiversity, and other life-supporting systems. Environmental sustainability is a system of interconnections of living organisms to the ecosystems that they inhabit and depend on air, water, soil, and other environmental resources. It is also closely linked to the socioeconomic condition of living beings. Because it is difficult to have a sustainable society without a sustainable productive environment that provides the base for resources. Similarly, a sustainable economic system is also dependent on the sustainable flow of energy, material, and environmental resources. Hence, a sustainable environment can stand alone as a sustainable system and is not dependent on social or economic systems. While humans are completely dependent on the flow of ecosystem services as:

1. *Provisioning services*: include food, fiber, water, biochemical, pharmaceutical, medicines, and energy resources.
2. *Supporting services*: include soil formation, photosynthesis, nutrient cycling, seed dispersal, hydrological cycle.
3. *Regulating services*: include water purification, waste treatment, regulation of air quality, disease, pest, erosion, water, pollination, climate, and natural hazards.
4. *Cultural services*: include all nonmaterial benefits obtained from the environment such as cognitive development, recreation, reflection, and esthetic values.

22.6.1 Principles of environmental sustainability

Environmental sustainability is a conservation concept which is the meeting of services and resources of present and future generations without affecting the health of the ecosystems that provide them. The principles for Strategy of environmental sustainability are given below as:

1. *Biodiversity conservation*: this is the preservation of biodiversity and energy resources.
2. *Social needs*: the availability of basic needs, products, and services for present and future generations. Support local employment, fair trade and environmental attributes of raw material.
3. *Regenerative capacity*: protect the depletion of natural resources and keep the harvest rate of renewable resources within the capacity of regeneration.
4. *Reuse, recycling*: support the reuse, recycling practices to reduce waste, emissions, and cost and improve product efficiency.
5. *Limitations of nonrenewable resources and waste generation*: the human economic system should be within the carrying capacity, the emissions should be within the assimilative capacity of the ecosystem, prioritize low-impact transportation, and effective decisions with consideration of environmental quality (Morelli, 2011).

22.6.2 Strategic sustainable development

The concepts of environmental sustainability and social sustainability combine to build a concrete, flexible, and durable development strategy. Strategic sustainability provides an important template to know how an approach aimed to develop by getting a sustainable society where the people are fully facilitated with the availability of resources. Under the Brundtland definition, the concept of sustainability has been widely adopted by various developmental assistance donors and agencies. Organization for economic cooperation and development sustainable development keeps the economy, society, and environment in a single band where the distribution of resources (resources in any form such as social, economic, and environmental) across generations should be fair (Strange & Bayley, 2008). Strategic sustainable development introduced a scientific-based definition of socioecological sustainability which is composed of four principles including one social and three ecological. These principles comprise a holistic view of sustainability that finds a way toward development and unforeseen impacts and describes the natural functional limits of a sustainable society. These principles are:

1. Concentration of materials/substances extracted from earth's crust including oil, gas, and metals (*ecological*).
2. Concentrations of substances generated by society including chemicals that are hard and do not break down quickly (*ecological*).
3. Degradation of resources by physically such as deforestation, loss of wetlands, and damage from mining (*ecological*).
4. People are not subject to conditions that undermine their capacity to meet their needs systematically (*social*) (Ny, MacDonald, Broman, Yamamoto, & Robert, 2006; Robert, 2000).

22.6.3 Global sustainability challenge

To maintain environmental, social, and economic sustainability has become a global challenge. The environment is facing issues due to overconsumption and misuse of environmental resources. Currently, climate change has become the most pressing issue placing environmental concerns to both society and the environment. The effects of this hazard will be felt by the industrial countries of the world (Karl-Henrik, 2000). It will affect the economies of countries as well as refugee migration, resource-driven conflict specifically for water, food crisis, increasing commodity prices, loss of agricultural resources, and poor health. The weakening of the biosphere is resulting from the industrialization of rich countries but the results of the climate crisis will be faced by the poor nations. It is in the interest of all countries of the world to tackle this issue, to safeguard their interests in isolation, and to support each other in a sustainable way. Development should no longer be described as assistance but rather as a mechanism to maximize the resilience of society against the most serious threats facing the global stability (Balaskas, Lima, & Seed, 2009).

22.7 Conclusion

Our planet Earth is facing a so-called environmental crisis. The crisis is characterized by overpopulation, depletion of natural resources, and intensive damage to biodiversity and ecosystems. These environmental crises are due to improper planning and ecological design. The practice of ecodesigning is projected to the conservation of ecological integrity, thus allowing an ecosystem with sustainable flow of life-supporting services. Technically, the concept of ecodesign contributes to the wise use of natural resources, conservation ecosystems, and providing an equitable lifestyle. Ecodesign provides a structure by considering the ecological concerns by uniting the conventional and management of the environment. An ecologically sustainable economic development may be possible by implementing the principles of ecodesign. By doing so, a small number of species will be enough to build an active functional ecosystem. In the future, the social and environmental benefits can be possible with the practice of ecodesigning. Therefore we need to think ecologically and with an integrated ecologicalism that defines the services of all life systems. This approach would acknowledge the need for ecological sustainability where the interest of human and nature come together. It is therefore intended to implement the practice of sustainability by maintaining the value of ecological services.

References

Ashton, P. (1999). *Integrated catchment management: Balancing resource utilization and conservation*. African Water Issues Research Unit (AWIRU).

Bahri, A., Sally, H., McCartney, M., Namara, R., Awulachew, S.B., Koppen, B.V., & Rooijen, D.V. (2011). Integrated watershed management towards sustainable solutions in Africa. Water for Food. (pp. 50–72).

Baker, W. L. (1989). A review of models of landscape change. *Landscape Ecology*, *2*, 111–133.

Balaskas, A., Lima, E., & Seed, T. (2009). A strategic approach to sustainable development through official development assistance. (pp. 1–9).

Brooks, K. N., Ffolliott, P. F., & Magner, J. A. (2013). *Hydrology and the management of watersheds* (4th ed., p. 552)Ames: Wily-Blackwell.

Butchart, S. H. (2010). Global biodiversity: Indicators of recent declines. *Science (New York, N. Y.)*, *328*(5982), 1164–1168.

Cale, W. G. (1995). Model aggregation: Ecological perspectives. pp:. In B. C. Patten, & S. E. Jorgensen (Eds.), *Complex ecology: The part–whole relation in ecosystems* (pp. 230–241). Englewood Cliffs, NJ: Prentice Hall.

Callicott, J. B., & Mumford, K. (1997). Ecological sustainability as a conservation concept. *Conservation Biology*, *11*(1), 32–40.

Caswell, H. (1976). The validation problem. In B. C. Patten (Ed.), *Systems analysis and simulation in ecology* (IV, pp. 313–325). New York: Academic Press.

Choi, J. Y., Engel, B. A., & Farnsworth, R. L. (2005). Web-based GIS and spatial decision support system for watershed management. *Journal of Hydroinformatics*, *7*(3), 165–174.

Conable, B. & Repetto, R. (1988). The role of watershed management in sustainable development by the working group on watershed management and development. Working paper. (pp. 1–22).

Conway, D. (2000). The climate and hydrology of the Upper Blue Nile River. *The Geographical Journal*, *166*(1), 49–62.

Cote, R., & Cohen-Rosenthal, E. (1998). Designing eco-industrial parks: A synthesis of some experiences. *Journal of Cleaner Production*, *6*, 181–188.

Cote, R., & Crawford, P. (2003). A case study in eco-industrial development: The transformation of Burnside Industrial Park. In E. Cohen-Rosenthal, & J. Muskinow (Eds.), *Eco-industrial strategies: Synergy between economic development and the environment* (pp. 308–315). Sheffield: Greenleaf Publishing.

Crombie D., Watershed Interim report. 1990, Royal Commission on the Future of the Toronto Waterfront, Toronto, 201.

David, L. C., & Tufgar, R. (1991). Integrated watershed management approach for small southern Ontario rural/urban watersheds. *Canadian Water Resources Journal*, *16*(4), 421–432. Available from https://doi.org/10.4296/cwrj1604421.

DeAngelis, D. L., & Gross, L. J. (1992). *Individual-based models and approaches in ecology: Populations, communities, and ecosystems*. NewYork: Chapman & Hall.

Degu, M., Melese, A., & Tena, W. (2019). Effects of soil conservation practice and crop rotation on selected soil physicochemical properties: The case of Dembecha district, northwestern Ethiopia. *Applied and Environmental Soil Science*. Available from https://doi.org/10.1155/2019/6910879.

Dong, H., Satoshi, T., Fujita, Y., Geng, M. F., & Dong, L. (2014). Achieving carbon emission reduction through industrial & urban industrial symbiosis: A case of Kawasaki. *Energy*, *64*(1), 277–286.

Doyen, L., Cisse, A., Gourguet, S., Mouysset, L., Hardy, P. Y., Bene, C., ... Thebaud, O. (2013). Ecological-economic modelling for the sustainable management of biodiversity. *Computational Management Science*, *10*, 353–364. Available from https://doi.org/10.1007/s10287-013-0194-2.

Falkenmark, M., & Rockström, J. (2006). The new Green and Blue paradigm: Breaking new ground for water resource planning and management. *Journal of Water Resource Planning and Management*, *123*(3), 129–132.

Falkenmark, M. & Rockstrom, J. (2008). Integrating agricultural water use with the global water budget. In *Proceedings of the Rosenberg water policy forum*, Zaragoza, Spain.

German, L., Mansoor, H., Alemu, G., Mazengia, W., Amede, T., & Stroud, A. (2007). Participatory integrated watershed management: Evolution of concepts and methods in an eco-regional program of the eastern African highlands. *Agricultural Systems*, *94*, 189–204.

Govers, G., Merckx, R., Wesemael, B. V., & Oost, K. V. (2017). Soil conservation in the 21st century: Why we need smart agricultural intensification. *Soil*, *3*, 45–59.

Grey, D., & Sadoff, C. (2004). Sink or swim? Water security for growth and development. *Water Policy*, *9*(6), 545–571.

Gutierrez, N. L., Hilborn, R., & Defeo, O. (2011). Leadership, social capital and incentives promote successful fisheries. *Nature*, *470*, 386–389.

Hemmer F., An adapted approach to industrial symbiosis with a case study on the northern Stockholm region, 2011, Master of Science Thesis, 1–59.

Hewett, S. W. (1989). Ecological applications of bioenergetics model. *American Fisheries Society Symposium*, *6*, 113–120.

Hilborn, R., & Mangel, M. (1997). *The ecological detective: Confronting models with data*. Princeton, NJ: Princeton University Press.

Jackson, L. J., Trebitz, A. S., & Cottingham, K. (2000). An introduction to the practice of ecological modeling. *Bioscience*, *50*(8), 694–706.

Jiang, M., Zhou, J., Chen, B., Yang, Z., Ji, X., Zhang, L., & Chen, G. (2009). Ecological evaluation of Beijing economy based on energy indices. *Communications in Nonlinear Science and Numerical Simulatio*, *14*, 2482–2494.

Jie, C., Jing-Zhang, C., Man-ZhL, T., & Zi-Tong, G. (2002). Soil degradation: A global problem endangering sustainable development. *Journal of Geographical Sciences*, *12*(2), 243–252.

Karl-Henrik, R. (2000). Tools and concepts for sustainable development, how do they relate to a general framework for sustainable development, and to each other? *Journal of Cleaner Production*, *8*, 243–254.

Kharrazi, A., & Masaru, Y. (2012). Quantifying the sustainability of integrated urban waste and energy networks: Seeking an optimal balance between network efficiency and resilience. *Procedia Environmental Sciences*, *13*(2012), 1663–1667.

Kumar, G., Sena, D. R., Kurothe, R. S., Pande, V. C., Rao, B. K., Vishwakarma, A. K., ... Mishra, P. K. (2014). Watershed impact evaluation using remote sensing. *Current Science*, *106*(10), 1369–1379.

Lahmar, R. (2010). Adoption of conservation agriculture in Europe: Lessons of the KASSA project. *Land Use Policy*, *27*, 4–10.

Lal, R. (1997). Residue management, conservation tillage and soil restoration for mitigating greenhouse effect by CO-enrichment. *Soil & Tillage Research*, *43*, 81–107.

Larkin, S.L., Alvarez, S., Sylvia, G., & Harte, M. (2011). Practical considerations in using bioeconomic modelling for rebuilding fisheries. OECD food, agriculture and fisheries working papers.

Leigh, M. & Li, X. (2008). Industrial ecology, industrial symbiosis and supply chain environmental sustainability: A case study of a large UK distributor.

Liu, G., Yang, Z., & Chen, B. (2010). Extended exergy-based urban ecosystem network analysis: A case study of Beijing, China. *Procedia Environmental Sciences*, *2*, 243–251.

Liu, Q., Jiang, P., Zhao, J., Zhang, B., Bian, H., & Qian, G. (2011). Life cycle assessment of an industrial symbiosis based on energy recovery from dried sludge and used oil. *Journal of Cleaner Production*, *19*(1), 1700–1708.

Locatelli, B., & Vignola, R. (2009). Managing watershed services of tropical forests and plantations: Can meta-analyses help? *Forest Ecology and Management*, *258*, 1864–1870.

Mangel, M., & Clark, C. W. (1988). *Dynamic modelling in behavioral ecology*. Princeton, NJ: Princeton University Press.

Manuelli, S., Hofer, T., & Vita, A. (2014). FAO's work on sustainable mountain development and watershed management. *Mountain Research and Development*, *34*(1), 66–70.

McHarg, I. L. (1967). *An ecological method for landscape architecture. The ecological design and planning reader*, (pp. 341–347).

Montgomery, D. R. (2007). Soil erosion and agricultural sustainability. *Proceedings of the National Academy of Sciences of the United States of America*, *104*, 13268–13272.

Morelli, J. (2011). Environmental sustainability: A definition for environmental professionals. *Journal of Environmental Sustainability*, *1*(1), 1–12. Available from https://doi.org/10.14448/jes.01.0002.

Mutekanga, F. P., Kessler, A., Leber, K., & Visser, S. (2013). The use of stakeholder analysis in integrated watershed management. *Mountain Research and Development, 33*(2), 122–131.

Ny, H., MacDonald, J. P., Broman, G., Yamamoto, R., & Robert, K.-H. (2006). Sustainability constraints as system boundaries – An approach to make life cycle management strategic. *Journal of Industrial Ecology, 10*(1–2), 61–77.

Nyssen, J., Clymans, W., Poesen, J., Vandecasteele, I., De-Raets, S., Haregeweyn, N., ... Deckers, J. (2009). How soil conservation affects the catchment sediment budget – A comprehensive study in the north Ethiopian highlands. *Earth Surface Processes and Landforms, 34*, 1216–1233.

Oldeman, L.R. (2000). GLASOD classification of soil degradation, ESCAP environment statistics course (draft).

Peiretti, R., & Dumanski, J. (2014). The transformation of agriculture in Argentina through soil conservation. *International Soil and Water Conservation Research, 2*, 14–20. Available from https://doi.org/10.1016/S2095-6339(15)30010-1.

Persson, S. & Ivner, J. (2011). Uncovering industrial symbiosis in Sweden – Exploring a possible approach. In *Proceedings of the world renewable energy congress on industrial energy efficiency* (pp. 8–11). Linköping, Sweden.

Poulsen L., Soil degradation, poverty and migration in Africa: Example of the International Community's Response: The UN convention to combat desertification introductory note to the session on 11 January 2001, 2001.

Puente, M. C. R., Arozamena, E. R., & Evans, S. (2015). Industrial symbiosis opportunities for small and medium sized enterprises: Preliminary study in the Besaya region Cantabria, Northern Spain. *Journal of Cleaner Production, 87*(1), 357–374.

Ramankutty, N., Evan, A. T., Monfreda, C., & Foley, J. A. (2008). Farming the planet: 1. Geographic distribution of global agricultural lands in the year 2000. *Global Biogeochemical Cycles, 22*, GB1003.

Rapport, D. J., & Regier, H. A. (1980). An ecological approach to environmental information. *Ambio, 9*(1), 22–27.

Rijsdijk, A., Bruijnzeel, L. A. S., & Sutoto, C. K. (2007). Runoff and sediment yield from rural roads, trails and settlements in the upper Konto catchment. *East Java, Indonesia, Geomorphology, 87*, 28–37.

Robert, K.-H. (2000). Tools and concepts for sustainable development, how do they relate to a general framework for sustainable development, and to each other? *Journal of Cleaner Production, 8*, 243–254.

Robinson, J. (2004). Squaring the circle? Some thoughts on the idea of sustainable development. *Ecological Economics, 48*, 369–384.

Roose, E. (2008). Soil erosion, conservation and restoration: A few lessons from 50 years of research in Africa. (pp. 159–180).

Rothenberg, S. (2007). Environmental managers as institutional entrepreneurs: The influence of institutional and technical pressures on waste management. *Journal of Business Research, 60*, 751, ABI/ Inform & ProQuest.

Roughgarden, J. (1979). *Theory of population genetics and evolutionary ecology: An introduction*. NewYork: Macmillan.

Scavia, D. & Robertson, A. (1980). Perspectives on lake ecosystem modeling. Ann-Arbor (MI): Ann Arbor Science.

Sha, J., Li, Z., Swaney, D. P., Hong, B., Wang, W., & Wang, Y. (2014). Application of a Bayesian watershed model linking multivariate statistical analysis to support watershed-scale nitrogen management in China. *Water Resources Management, 28*, 3681–3695.

Shu-Yang, F., Freedman, B., & Cote, R. (2004). Principles and practice of ecological design. *Environmental Review, 12*, 97–112.

Simboli, A., Taddeo, R., & Morgante, A. (2014). Analyzing the development of industrial symbiosis in a motorcycle local industrial network: The role of contextual factors. *Journal of Cleaner Production, 66*(1), 372–383.

Singh, V. P., & Woolhiser, D. A. (2002). Mathematical modeling of watershed hydrology. *Journal of Hydrologic Engineering, 7*, 270–292.

Strange, T. & Bayley, A. (2008). Sustainable development: Linking economy, society, environment. Organization for Economic Co-operation and Development.

Sudha, M. (2007). Sustainable practices in watershed management: Global experiences. MPRA Paper No. 5854. https://mpra.ub.uni-muenchen.de/5854/

Tennyson, L.C. & Achouri, M. (2004). The next generation of watershed programmes: An overview of FAO review. Watershed Development Coordination Unit.

Thampapillai, D. J., & Anderson, J. R. (1994). Soil conservation in developing countries: A review of causes and remedies. *Review of Marketing and Agricultural Economics, 62*(3), 291–315.

Thebaud, O., Smith, T., Doyen, L., Planque, B., Lample, M., Mahevas, S., ... Innes, J. (2013). Building ecological-economic models and scenarios of marine resource systems: Workshop report. *Marine Policy*. Available from https://doi.org/10.1016/j.marpol.2013.05.010.

Ulanowicz, R. E., Goerner, S., Lietaer, B., & Gomez, R. (2009). Quantifying sustainability: Resilience, efficiency and the return of information theory. *Ecological Complexity, 6*, 27–36.

Vallentyne, J., & Beeton, A. (1988). The 'Ecosystem' approach to managing human uses and abuses of natural resources in the Great Lakes Basin. *Environmental Conservation, 15*(1), 58–62.

Van der Ryn, S., & Cowan, S. (1995). *Ecological design*. Washington, D.C: Island Press.

Vos, R. O. (2007). Defining sustainability: A conceptual orientation. *Journal of Chemical Technology and Biotechnology, 82*, 334–339.

Wang, G., Mang, S., Cail, H., Liu, S., Zhang, Z., Wang, L., & Innes, J. L. (2016). Integrated watershed management: Evolution, development and emerging trends. *Journal of Forestry Research, 27*(5), 967–994. Available from https://doi.org/10.1007/s11676-016-0293-3. (2016).

Watkins, G., Husgafvel, R., Pajunen, N., Dahl, O., & Heindustrial, K. (2013). Overcoming institutional barriers in the development of novel process industry residue based industrial symbiosis products – Case study at the EU level. *Minerals Engineering, 41*(2), 31–40.

Weiner, J. (1995). On the practice of ecology. *Journal of Ecology, 83*, 153–158.

Worm, B. (2009). Rebuilding global fisheries. *Science (New York, N.Y.), 325*(5940), 578–585.

WRI (World Resources Institute), Disappearing land: Soil degradation, sustainable development service, 2001, Global Trends, Washington.

Wu, J., Li, C., & Yang, F. (2015). The decomposition of chromite ore processing residue (COPR) incorporating industrial symbiosis. Journal of Cleaner Production. Available online. In Press, Corrected Proof.

Xue, B. (1997). Environmental, economic, and ecological sustainable development. *Environment and Development, I*(1), 1–4.

Zorach, A. C., & Ulanowicz, R. E. (2003). Quantifying the complexity of flow networks: How many roles are there? *Complexity, 8*, 68–76.

Further reading

Cisse, A., Gourguet, S., Blanchard, F., Doyen, L., & Pereau, J.-C. (2013). A bio-economic model for the viable management of the coastal fishery in French Guyana. *Environment and Development Economics*, *18*, 245–269.

David, C. (1990). Royal commission on the future of the Toronto waterfront.

Hardy, P.-Y., Doyen, L., Bene, C., & Schwartz, A.M. (2013). Food security-environment conservation nexus: A case study of Solomon Islands' small-scale fisheries. Environmental Development. http://www.sciencedirect.com/science/article/pii/S2211464513000584

John, H., & Robert, K.-H. (2000). Backcasting from non-overlapping sustainability principles – A framework for strategic planning. *International Journal of Sustainable Development and World Ecology*, *7*(1), 291–308.

Zhang, Y., Zheng, H., & Fath, B. D. (2014). Ecological network analysis of an industrial symbiosis system: A case study of the Shandong Lubei Eco-Industrial Park. *Ecological Modelling*, *306*, 174–184.

Index

Note: Page numbers followed by "*f*" and "*t*" refer to figures and tables, respectively.

B

C

D

E

F

G

H

I

J

K

L

M

N

T

U

V

W

FSC
www.fsc.org
MIX
Papier | Fördert
gute Waldnutzung
FSC® C083411

Kolibri 360 GmbH
Ferdinand-Jühlke-Straße 7
99095 Erfurt, Deutschland
produktsicherheit@kolibri360.de